SELECTED PHYSICAL QUANTITIES AND MEASUREMENT UNITS

Physical quantity	Quantity symbol	Measurement unit	Unit symbol	Unit dimensions
		Fundamental (base) units		
length	l	meter	m	m
mass	m	kilogram	kg	kg
time	t	second	s	s
electric charge*	Q	coulomb	C	C
temperature	T	kelvin	K	K
amount of substance	n	mole	mol	mol
luminous intensity	I	candela	cd	cd
		Derived units		
acceleration	a	meter per second per second	m/s^2	m/s^2
area	A	square meter	m^2	m^2
capacitance	C	farad	F	$C^2 \cdot s^2/kg \cdot m^2$
density	D	kilogram per cubic meter	kg/m^3	kg/m^3
electric current	I	ampere	A	C/s
electric field intensity	E	newton per coulomb	N/C	$kg \cdot m/C \cdot s^2$
electric resistance	R	ohm	Ω	$kg \cdot m^2/C^2 \cdot s$
emf	$\mathcal{E}$	volt	V	$kg \cdot m^2/C \cdot s^2$
energy	E	joule	J	$kg \cdot m^2/s^2$
force	F	newton	N	$kg \cdot m/s^2$
frequency	f	hertz	Hz	s^{-1}
heat	Q	joule	J	$kg \cdot m^2/s^2$
illuminance	E	lumen per square meter (lux)	lm/m^2 (lx)	cd/m^2
inductance	L	henry	H	$kg \cdot m^2/C^2$
luminous flux	F	lumen	lm	cd · sr
magnetic flux	Φ	weber	Wb	$kg \cdot m^2/C \cdot s$
magnetic flux density	B	weber per square meter (tesla)	Wb/m^2 (T)	$kg/C \cdot s$
potential difference	V	volt	V	$kg \cdot m^2/C \cdot s^2$
power	P	watt	W	$kg \cdot m^2/s^3$
pressure	p	newton per square meter (pascal)	N/m^2 (Pa)	$kg/m \cdot s^2$
velocity	v	meter per second	m/s	m/s
volume	V	cubic meter	m^3	m^3
work	W	joule	J	$kg \cdot m^2/s^2$

*The SI fundamental (base) electric unit is the ampere. For pedagogical reasons the coulomb of electric charge is introduced in this book as the fundamental unit and the ampere as a derived unit having the dimensions, coulomb per second.

FREDERICK E. TRINKLEIN

MODERN PHYSICS

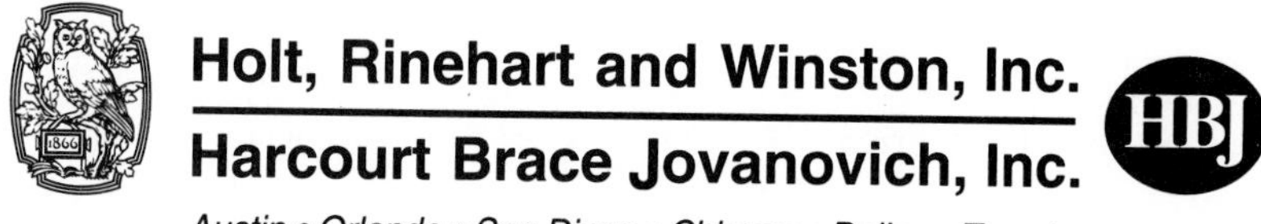
Holt, Rinehart and Winston, Inc.
Harcourt Brace Jovanovich, Inc.
Austin • Orlando • San Diego • Chicago • Dallas • Toronto

FREDERICK E. TRINKLEIN, Adjunct Professor, Department of Physical Sciences, Nassau Community College, Garden City, New York; former Dean of Faculty and physics teacher, Long Island Lutheran High School, Brookville, New York.

Cover: A computer microchip, highly magnified, superimposed on a bolt of lighting. The chip is an example of integrated circuits, which are the fundamental components of today's computers. The same basic laws of electricity apply to the power of a bolt of lighting to knock down trees and to the ability of a microchip to do millions of calculations per second. Computer chip: Roland Birke/G & J Image/ The Image Bank. Electrical storm in San Francisco, California: Garry Gay/The Image Bank.

Photo credits appear on pages 751–752.

Printed in the United States of America

ISBN 0-03-074317-6

234567890 041 98765432

TO THE STUDENT

In studying physics, you are embarking on an exploration of the very nature of matter and energy. You are exposing yourself to a wealth of knowledge contributed by scientists since the beginning of civilization. Scientists such as Newton, Curie, and Einstein may come to mind when you think about physics. These and many other scientists have advanced our understanding of the physical world, and this understanding has affected nearly every aspect of our daily lives.

In many respects, however, our knowledge of physics is still in its infancy. New knowledge continues to spawn new questions. Old questions continue to be asked anew, as further light is shed on our understanding. Physicists today are trying to answer such questions as "What is the smallest unit of matter?" and "How did the universe begin?" These and other questions continue to capture the imagination of the scientists who study physics.

In your study of physics, you will engage in a quest for knowledge in much the same way that scientists do. *Modern Physics* will help you develop science skills that will allow you to observe natural phenomena with new insight. You will be encouraged to think beyond the obvious and to question information and data. You will learn how to use mathematics to explore and explain phenomena and to predict outcomes. You will form hypotheses and devise methods of testing their validity. And you will work in a laboratory to discover for yourself the principles that govern matter and energy.

Think of *Modern Physics* as one of the many scientific tools you will need to be successful in your search for knowledge about physics. To get the best use of any tool, it is important for you become familiar with its parts and purpose. The following information will help you to get better acquainted with the textbook.

About the Text

As you thumb through *Modern Physics,* you will see that it is organized by chapters. You will also note that each chapter begins with a list of objectives. When you start your study of a chapter, be sure to read these objectives. They will help you identify and focus on what is important in the chapter. You should also notice that each chapter is divided into major subdivisions that contain several numbered sections. Before you begin your reading of a chapter, go through the chapter and read the titles of the sections. This will help you mentally prepare for what is coming up as you study the chapter.

As you read, you will notice the use of *italic* and **boldface** type. These typefaces are used to identify important terms and definitions. Special notes in the margins are printed in italic type. These notes provide you with additional information related to the text material.

You will also notice that many chapters contain mathematical concepts. In these chapters, example problems are provided that include step-by-step solutions. You can expect to see an example problem after each explanation of a math-related concept. Be sure to take time to work these problems and any practice problems before going on. The answers to the practice problems are provided so that you can check your work.

At the end of each major subdivision are sets of questions and problems. Group A questions and problems are designed to check your reading and basic understanding of the section content. Group B questions and problems are more challenging, requiring more critical thinking and higher-order math skills. The end of each chapter consists of a summary and a vocabulary list, both of which are ideal for reviewing the concepts emphasized in the chapter. You may also want to use the summary and vocabulary list as a preview of the chapter.

About the Features

Several features are included in *Modern Physics* to help spark interest. A full-color insert is included in the middle of the textbook (following page 336). The first eight pages of this insert are devoted to color plates illustrating the behavior of light. The remaining sixteen pages comprise a unique feature, called *Intra-Science: How the Sciences Work Together,* that links physics to other scientific disciplines. Here you can discover vital connections between the sciences by exploring such technologies as fiber optics, lasers, and ultrasound.

Four other features are located at various points in the text to help you explore career opportunities in physics. Each of these features, called *Careers in Physics,* examines the tools, techniques, and knowledge that physicists use to develop new products and technologies related to four broad topics of physics: lasers, sound, superconductors, and magnetic resonance imaging. These features show you professionals at work and provide information about working conditions and academic requirements of various careers in physics.

In-Text Investigations

To encourage firsthand exploration of physics, laboratory investigations have been included in the textbook. A total of 25 detailed investigations are provided—one for each chapter—starting on page 653. These investigations will help guide you in your hands-on discovery of the characteristics of matter and energy. Detailed information is also provided about process skills and safety in the laboratory. Be sure to review the safety information and the safety symbols that identify all procedures for which extra precautions must be taken.

Reference Information

The back of the book contains a variety of reference materials that you will use often in your study of physics. A mathematics refresher, starting on page 685, is provided to give you additional help in working on various types of physics problems. Appendix A, starting on page 691, contains a list of the major equations used in the textbook. The section number is given for each equation so that you can quickly refer to its explanation in the body of the text. Appendix B, starting on page 699, consists of the tables you will need in the study of physics. A glossary and index conclude *Modern Physics.* Page numbers are given in the glossary so that you can refer to where the term is first presented.

CONTENTS

6 CONSERVATION OF ENERGY AND MOMENTUM

7 PHASES OF MATTER

8 HEAT MEASUREMENTS

9 HEAT ENGINES

10 WAVES

11 SOUND WAVES

12 THE NATURE OF LIGHT

13 REFLECTION

18 HEATING AND CHEMICAL EFFECTS

19 MAGNETIC EFFECTS

20 ELECTROMAGNETIC INDUCTION

25 HIGH-ENERGY PHYSICS

1

Physics: The Science of Energy

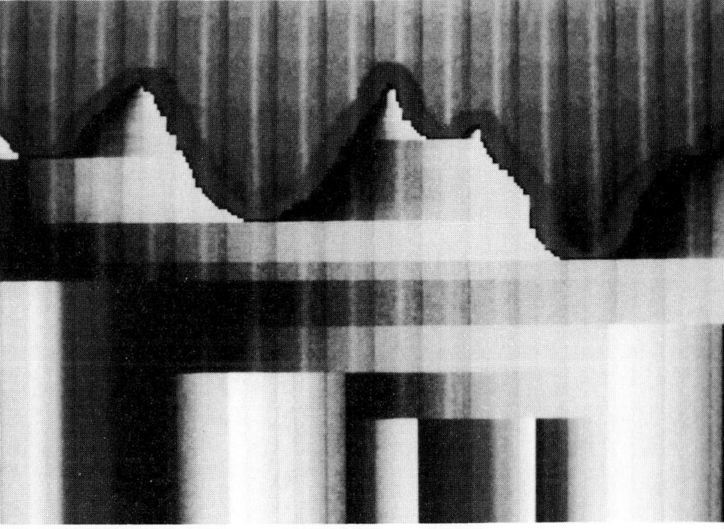

physics (FIZ-iks) n.: the branch of science concerned with matter and energy, especially as they relate to mechanics, heat, electricity, magnetism, and atomic phenomena.

THE METHODS OF SCIENCE

OBJECTIVES

- Describe the nature of science.
- Describe the basic properties of matter.
- Discuss the relationship between matter and energy.
- Describe and discuss the subdivisions of physics.

1.1 Science and Daily Life We live in a scientific age. The discoveries of science and their application to everyday life have direct and profound effects on our standard of living.

The development of air transportation is a good example of this cause-and-effect relationship. When the Wright brothers made the first flight of a heavier-than-air device in 1903, they applied scientific principles that were discovered by the eighteenth-century Swiss physicist Daniel Bernoulli (1700–1782), and other scientists. Today, routine flights are essential to industry and make worldwide traveling commonplace.

The extensive use of air travel also illustrates the connection between our standard of living and the energy required to sustain it. Airplanes consume enormous quantities of fuel, and the price of fuel is a large factor in the cost of an airline ticket. The search for new energy sources, therefore, is an important part of science, particularly of modern physics.

Figure 1-1 shows the rapid increase in energy usage in the United States since 1950. Note that the relative amount of energy derived from various sources is also changing, especially the reliance on oil and gas. (In the graph, energy is measured in joules, a unit that will be more fully explained in Chapter 6.)

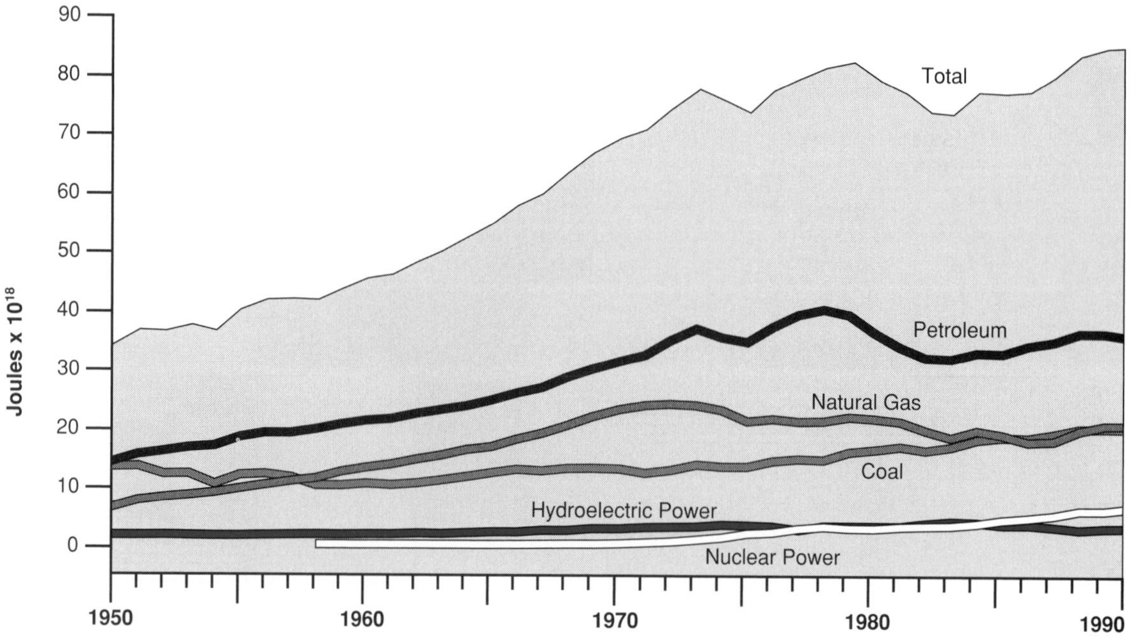

Figure 1-1. Energy consumption in the United States. The graph shows, in joules x 10^{18}, the amounts of energy obtained from various fuels since 1950.

1.2 Science and Engineering Briefly stated, *science* is the search for relationships that explain and predict natural phenomena. *Engineering* is the application of these relationships to our needs and goals. Finding new energy sources falls into the realm of science. Developing and utilizing these discoveries are matters of engineering.

Government and private industry allocate large sums of money to support the research and development activities of scientists and engineers. Without such support, there would be little improvement in the processes, products, and services that are derived from these activities. The field of engineering is also referred to as *technology*.

The development of the miniaturized computer is a good example of the cooperative effort between science and engineering. At the heart of this amazing device is a tiny chip, such as the one shown in Figure 1-2, that contains thousands of transistors. This tiny chip has the calculating capacity of a vacuum tube computer that, thirty years ago, needed an entire room for space.

Figure 1-2. Magnified view of a modern computer chip containing thousands of transistors.

One application of the miniaturized computer is the supermarket electronic checkout. This system reads and records the names and prices of items labeled with the Universal Product Code (UPC). This code, a series of black lines underscored by numbers, is read by a laser beam scanner, and the coded information is relayed to the store's computer. This coded information is processed by the computer, which then displays the product's name and price and provides an itemized receipt. While it is recording each purchase, the computer is also checking the store's inventory for reordering purposes.

Computer-assisted instruction demonstrates the application of computers in the field of education. Individualized instruction in many subjects is available to the student. The student retrieves the stored information and then converses with the computer by means of a keyboard and monitoring screen.

In banks, computers already handle most of the billions of checks Americans write each year. In hospitals, computers are used to analyze the medical histories and symptoms of patients as well as to detect and monitor their ailments. A computer program can order prescriptions and alert nurses to administer them at the right time and in the right dosage. A major advance is the use of the computer to assemble thousands of X rays of any part of the body into a single, two-dimensional view. See Figure 1-3.

In 1979, Allan M. Cormack of the United States and Godfrey N. Hounsfield of Great Britain received the Nobel Prize for developing the CAT scan.

In the home, computer devices may someday be as commonplace as the kitchen sink. When mass production and new technology bring the cost within reach of most budgets, routine chores of all kinds can be handled quickly and accurately by the computer chip. In cars, computers can automatically apply the brakes, determine gas mileage, and check the oil and tire pressure; a miniradar can warn the driver of impending hazards.

Many applications of computers still await discovery. Today's high school students, who have grown up in our highly technological society, can meet this challenge as they become the engineers and scientists of the future.

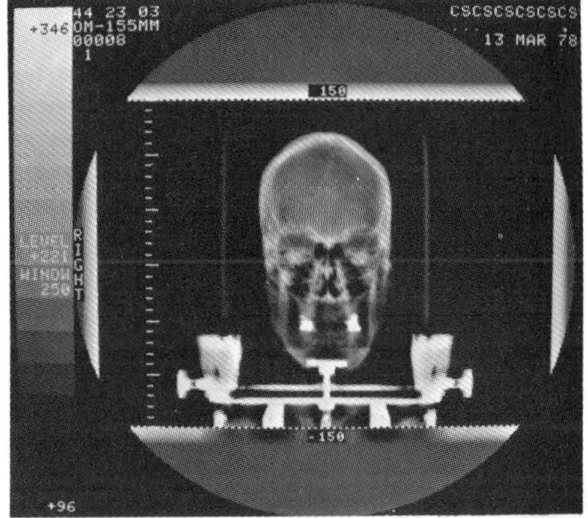

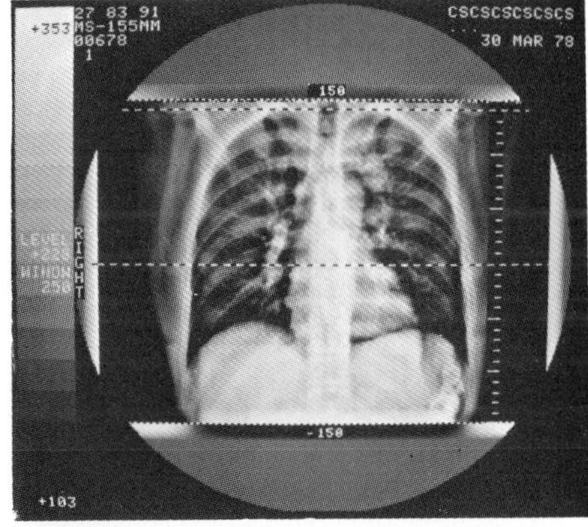

Figure 1-3. Computerized axial tomography (CAT) scan. Cross-sectional views of the head and chest cavity are obtained from the computer assembly of tens of thousands of X-ray readings.

1.3 Scientific Laws and Theories

In science, *a* ***law*** *(or* ***principle****) is a statement that describes the relationship between various phenomena.* Unlike laws that *restrict* behavior (i.e., traffic laws), scientific laws *describe* behavior. For example, Jacques Charles (1746–1823) observed, under certain controlled conditions, a relationship between the temperature and the volume of most gases. This regular behavior of gases, which you will study in greater detail in Chapter 8, is now known as Charles' law. A scientific law may be stated in words. However, in physics a law is usually expressed by a mathematical equation.

A ***theory*** *is a reasonable explanation of observed events that are related.* A theory often involves an imaginary *model* that helps scientists picture the way an observed event could be produced. A good example of this is our modern atomic theory, which you will study in Chapters 7, 23, 24, and 25. Another example is the kinetic molecular theory. In this theory, gases are pictured as being made up of many small particles called molecules that are in constant motion. You will study the kinetic molecular theory in Chapter 7.

A useful theory, in addition to explaining past observations, helps to predict events that have not as yet been observed. After a theory has been publicized, scientists design experiments to test the theory. For example, if matter is composed of small particles called atoms, it should behave in a certain way. If observations confirm the scientists' predictions, the theory is supported. If observations do not confirm the predictions, the scientists must search further. There may be a fault in the experiment, or the theory may have to be revised or rejected.

Figure 1-4. Dr. Walter E. Massey (born 1938), theoretical physicist and educator. Dr. Massey (right) served as director of the Argonne National Laboratory (1979–1984) and is currently Vice President for Research at the University of Chicago and for Argonne National Laboratory. He became professor of physics and dean of the College at Brown University at age 36. He is committed to creating opportunities in science for minority students.

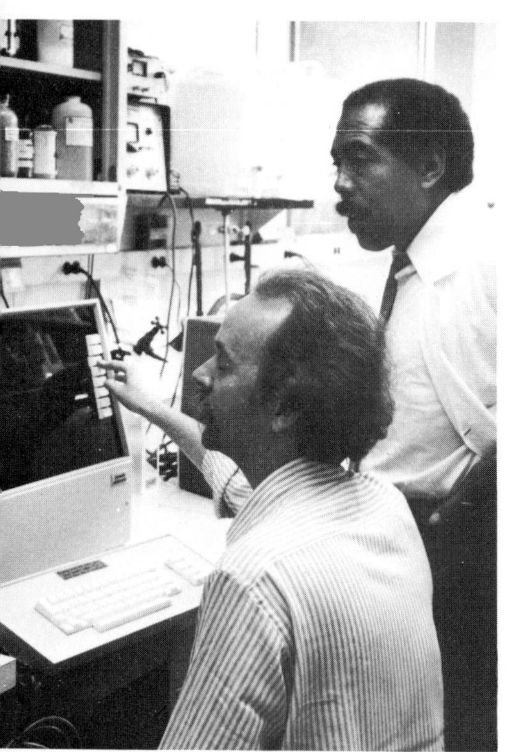

1.4 Scientific Hypotheses Albert Einstein (1879–1955) was one of the greatest theoretical scientists of all time. Theoretical scientists work mostly with ideas. They use the results of other scientists' observations to develop their theories. Einstein's theories and predictions changed the course of modern science. Once he was asked to explain how a scientist works. "If you want to know the essence of scientific method," he said, "don't listen to what a scientist may tell you. Watch what he does."

Science is a way of doing things, a way that involves imagination and creative thinking as well as collecting information and performing experiments. Facts (observations, principles, laws, and theories) by themselves are not science, but science deals with facts. The mathematician Jules Henri Poincaré (1854–1912) said: "Science is built with facts just as a house is built with bricks, but a collection of facts cannot be called a science any more than a pile of bricks can be called a house."

Most scientists start an investigation by finding out what other scientists have learned about the problem. After gathering the known facts, the scientist comes to the part of the investigation that requires considerable imagination. He or she must formulate possible solutions to the problem. These possible solutions are called ***hypotheses.***

For example, at the time of Johannes Kepler (1571–1630), it was generally believed that the planets move in circular orbits. But the observed movements of the planets could not be satisfactorily explained by circular orbits. So Kepler formulated other hypotheses to explain planetary motion. The one that best fit the observations was: All the planets have elliptic orbits.

In a way, any hypothesis is a leap into the unknown. It

extends the scientist's thinking beyond the known facts. The scientist plans experiments, calculations, and observations to test hypotheses. For without hypotheses, further investigation lacks purpose and direction. In the case of Kepler, he continued his observations and calculations over a period of ten years after the announcement of his hypothesis about elliptic orbits. Eventually he was able to relate, in the form of an equation, the time required for the orbit of a planet with its distance from the sun.

When hypotheses are confirmed, they are incorporated into theories or laws. Thus Kepler's hypotheses have become parts of Kepler's laws of planetary motion.

A hypothesis is a possible answer to a question.

The scientist's laboratory is any place where an investigation can be conducted. To the astronomer the sky is a laboratory. The biologist may do experimental work in a swamp or an ocean. The physicist and chemist are often surrounded by a maze of apparatus housed in specially designed facilities.

1.5 Certainty in Science

There is no such thing as absolute certainty in science. The validity of a scientific conclusion is always limited by the method of observation and by the skill of the person who made it. If a nurse takes a patient's temperature with a thermometer that reads 1° too high, the physicians might reach an invalid conclusion about the condition of the patient. The certainty of every measurement is limited by the precision of the apparatus used. Furthermore, these limits are not always apparent to the experimenter. For centuries, the speed of light was thought to be instantaneous, since its speed could not be measured. Later experiments showed this conclusion to be wrong.

Hence, it is important to keep an open mind about the validity of scientific laws, theories, or hypotheses. No amount of experimentation can ever prove any one of them absolutely, whereas a single crucial experiment can disprove any of them. When someone is unwilling to question a statement in science, no matter how well established it is, that person has lost a very important part of the scientific method. For this reason, the famous physicist Niels Bohr (1885–1962) told his associates: "Every sentence I utter must be understood not as an affirmation but as a question."

Do you think it is frustrating to deal with scientific uncertainties?

The methods of science are not restricted to the sciences. They are used in many other subjects as well. This means that the attitudes necessary for successful work in science are universally productive. Among these attitudes are imagination, a thirst for knowledge, a regard for data and their meaning, a respect for logical thinking, patience in reaching conclusions, and the willingness to work with new ideas.

Not all questions can be answered with the methods of science, however. For example, science does not deal with spiritual or religious topics. This does not mean that there is an inherent conflict between science and other subjects. It simply means that science is concerned with the "how" rather than the "why."

QUESTIONS: GROUP A

1. Use Figure 1-1 to answer the following questions. (a) From which fuel has the greatest amount of our energy been obtained in the last thirty years? (b) Which of the fuels listed would be considered "renewable" resources? (c) List some energy sources not shown.
2. Where and how are computers used in your school? List as many situations as you can.
3. How does Charles' law differ from the law requiring you to stop your bike at a stop sign?

GROUP B

4. What are two functions of a useful scientific theory?
5. List the steps a scientist might use to solve a problem.
6. What is the difference between experimental and theoretical scientists?
7. You have decided to select a new car by using the scientific method. How might you proceed?
8. Interview someone in a science or engineering career. What do they do? What preparation did they need? Present your findings to the class.

PHYSICS ACTIVITY

At the supermarket, compare the UPC markings on two items that differ in only one way, such as two different size cans of the same brand of peas. Compare your information with classmates to see if you can determine what the lines mean.

WHAT IS PHYSICS?

1.6 Matter A good way to describe something is to list its various forms. A description is not a definition in the real sense of the word, but it helps to bring an abstract idea down to familiar terms.

Matter can be described in terms of its measurable *properties.* In describing a person, you might refer to height, weight, eye color, hair color, etc. Similarly, all forms of matter possess properties. And just as a person can be identified by listing various characteristics, so a specimen of matter can be singled out by listing its properties.

The number of properties that can be measured for specimens of matter is very large. Entire handbooks of chemistry and physics are devoted to listings of various properties of matter.

In the study of physics it is important to recognize that *unless a property can be measured and compared with some kind of standard, it is of limited use to the scientist. Without measurement there can be no science.*

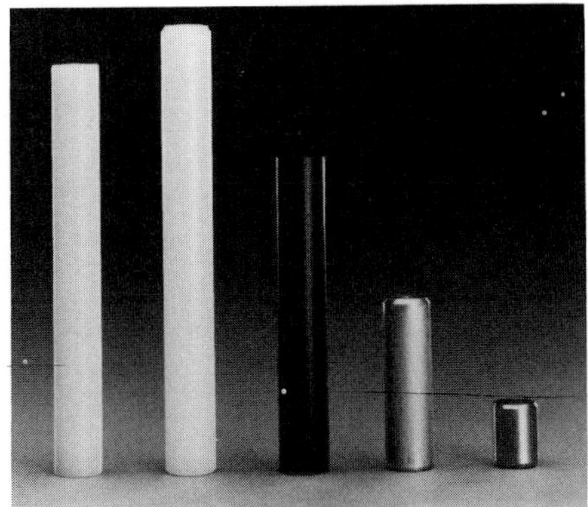

Figure 1-5. Distinction between mass and size. The cylinders have the same mass but not the same volume.

1.7 Mass

An important property of matter is its *mass*. ***Mass** is a measure of the amount of material in an object.* This is another way of saying that matter takes up space. But the mass of an object cannot be determined merely by measuring its size. Two objects may have the same size and be composed of the same material, yet one may contain more hollow spaces and thus have less mass. Or, one material may have more mass in the same amount of space than does another material. For example, a piece of brass and a piece of gold may have the same size but not the same mass. Or, a piece of lead and a piece of aluminum may have the same mass but not the same size. See Figure 1-5.

The mass of an object is determined by comparing it with known masses. Mass measurements will be more fully described in Chapter 2.

Figure 1-6. An inertia balance, as seen from above. When the instrument is set in horizontal motion (bottom), the period of vibration can be used to compute the magnitudes of masses attached to it.

1.8 Inertia

The mass of an object can also be measured by using a property of matter called *inertia*. ***Inertia** is the property of matter that opposes any change in its state of motion.* Inertia shows itself when objects are standing still as well as when they are moving. A baseball resting on the ground will not start moving by itself. A baseball in flight will keep moving unless something stops it. This does not mean that there are two kinds of inertia—a stationary kind and a moving kind; the same property of matter is merely showing itself in different circumstances.

The inertia of matter can be used to measure mass with a device called an inertia balance, shown in Figure 1-6. One end of the balance is clamped to a table. The pan on the other end can be made to vibrate horizontally. The number of times this device vibrates in a second of time depends upon the length and stiffness of the two supporting metal blades and upon the total mass of the pan and any objects fastened to it. Since the blades vibrate horizontally, the action of the inertia balance is entirely independent of gravity, which acts vertically. The more massive the pan and its contents, the slower will be the changes in their motion and the longer will be the time they will take to go through a complete vibration. The time required for a single vibration is called the *period*.

Unknown masses can be measured by comparing their periods with the periods of objects having known masses. First, several of the known objects are used and their periods are plotted as functions of their masses, as shown in Figure 1-7. Then the period of an unknown object is measured and its mass is read from the graph.

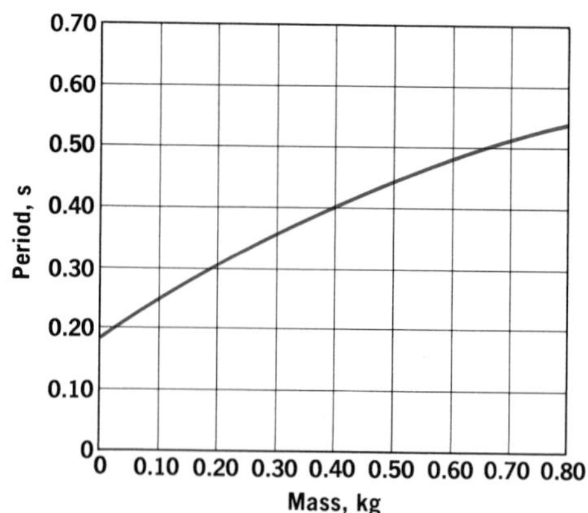

Figure 1-7. Graph of inertia balance data. If an object is attached to the pan of the balance and the period is measured, this graph can be used to find the mass of the object.

The graph in Figure 1-7 is a curved line. By analyzing the data, the relationship can be expressed by the following equation:

$$\frac{m_1}{m_2} = \frac{T_1^2}{T_2^2}$$

in which m_1 and m_2 are two masses (*including* the mass of the pan in each case) and T_1 and T_2 are their respective periods of vibration.

Masses can also be compared using either a spring balance or a platform balance, as shown in Figure 1-8. For comparing masses, these devices depend upon the pull of the earth's gravity. The stretch of a spring depends upon the amount of pull on the spring. The greater the pull, the greater is the amount of stretch. Since the pull of gravity is greater on a large mass than it is on a small mass, the spring in a spring balance stretches farther for a large mass than it does for a small mass. The amount of stretch for standard masses can be marked on a scale. The stretch for an unknown mass can then be compared with the stretch produced by the standard mass.

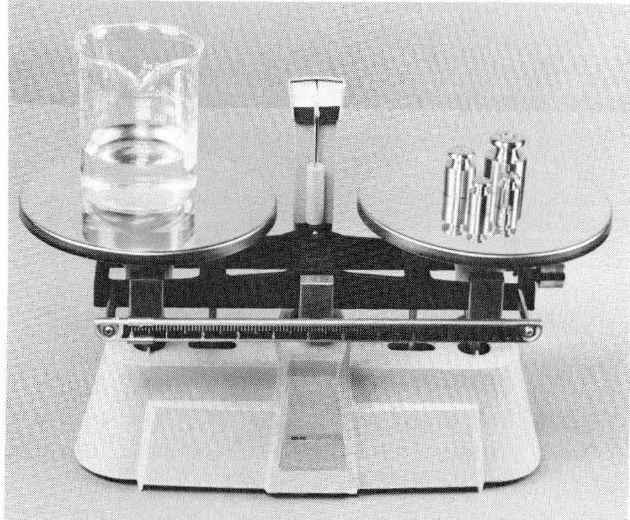

Figure 1-8. A spring balance (left) and a platform balance (right) in use.

In the case of a platform balance, a uniform beam is balanced at its center on a sharp blade. The blade acts as a pivot. If a mass is added to either side of the beam, the beam tips. Platforms of equal masses are placed on the beam at equal distances from the pivot. If an unknown mass is placed on one platform (usually the left one), known masses can be added to the other platform until the beam again balances. At that point, the known and the unknown masses are equal.

When the mass of an object is measured by its resistance to a change in its motion, as in the inertia balance, the mass is called *inertial mass*. Mass that is measured with a spring balance or a platform balance, on the other hand, is called *gravitational mass*. Many experiments have shown that the inertial mass and gravitational mass of a substance are always equal, even though they may seem to be unrelated. The single term "mass" can therefore be used to describe both forms.

Now that we have discussed two important properties of matter, we can use these properties to define matter. ***Matter** is anything that has mass (both gravitational and inertial).*

1.9 Mass Density A property of matter that is closely related to mass is *mass density*. ***Mass density*** refers to the amount of matter in a given amount of space and is defined as the *mass per volume of a substance*. Thus if a substance has a mass of 45 g and occupies a space of 15 cm^3, its mass density is 3.0 g/cm^3. The mathematical equation is

Volume: quantity of space occupied by a substance, measured in cubic units of length.

$$\textbf{mass density} = \frac{\textbf{mass}}{\textbf{volume}}$$

In giving the mass density of a substance, it is important to include the units (kilograms per cubic meter, grams per cubic centimeter, or some other unit of mass per unit of volume) in order that it may be compared with other values of mass densities. For example, the mass densities of solids and liquids may be compared to the mass density of water, which is equal to 1 g/cm^3. The ratio of such a comparison is called the *specific gravity* of the substance.

Some stars have mass densities of thousands of kilograms per cubic centimeter!

Mass density is a property that varies with the environment. The mass density of a gas increases when the gas is placed under pressure and decreases when the pressure is reduced. Such quantities as pressure and temperature, which describe the environment of an object, are called *conditions*. Very often a measurable property of an object will change if the conditions are changed.

1.10 Energy It is said that, as a youngster, James Watt (1736–1819) became interested in steam power by watching water boil in a teakettle over a fire. When he plugged the spout of the kettle, the lid rose. Perhaps this observation eventually led Watt to the improvements of steam engines for which he is famous. In any case, this observation is a good example of how energy is related to the concept of work.

Energy is latent work.

Work is done when an object is lifted against the pull of the earth's gravity. Later we will define work more completely; for now this common use of the word will be adequate. You do work when you lift the lid of a teakettle. When you plug the spout, the steam lifts the lid. The work is done by the steam. So we reason that the steam in the kettle had the *capacity* for doing work before it lifted the lid. The steam had *energy*. Thus we say that ***energy*** *is the capacity to do work,* if the conditions are right. Since energy can often be transformed into mechanical work, the same units describe both energy and work.

Energy is sometimes quite noticeable because we have sense organs that are able to detect its presence in various forms. Our eyes respond to visible light energy. Our ears detect sound energy. Special nerves are sensitive to heat energy. Other nerves respond to electric and mechanical energy.

Scientists have discovered forms of energy in addition to those that can be detected by our sense organs. Special detecting and measuring instruments had to be developed to record these forms of energy in a form that we can sense. For example, we cannot directly sense X rays, a form of energy similar to visible light. But X rays can affect special photographic film. Under visible light, we can look at a piece of exposed and developed X-ray film and tell where the X rays have affected the film. As another example, we cannot directly sense small amounts of infrared radiation, an invisible form of energy similar to visible light. (We can feel large amounts of infrared radiation as heat.) However, a special thermometer exposed to infrared radiation registers an increase in temperature. We can study the behavior of the thermometer and say that invisible radiation is present. By means of devices such as these, scientists have extended the range and sensitivity of the human senses. Among the forms of energy that fall into the category that humans cannot directly detect are chemical energy, gravitational energy, nuclear energy, and forms of energy similar to visible light, such as X rays and infrared rays.

The concept of energy is very important in the study of

physics. Physicists study and measure the transformation of energy, for the forms of energy are mutually interchangeable. For example, consider the energy you expend when you walk. Traced backward, it came from the food you ate. The food got its energy from nutrients in the ground and radiation from the sun. The sun got its energy from nuclear reactions in its interior, etc. Tracing your walking energy forward, the friction and motion of your feet on the ground heat the ground slightly. This heat radiates into space and helps to evaporate water from the earth, making rain possible, etc.

1.11 Potential Energy Two important forms of energy that an object may have are its energy of position and its energy of motion. Energy of position is *potential energy*. Energy of motion is *kinetic energy*. A book resting on a desk top, as in Figure 1-9, acquired potential energy when someone did the work of lifting it to the desk top. The book has potential energy, or energy of position, while it is on the desk top. If the book is pushed off the desk top, it falls to the floor. While falling, it may strike some other object to which it gives some of its energy. This is kinetic energy, or energy of motion. When the book is stopped by the floor, the rest of its kinetic energy is transformed into other forms of energy, such as sound and heat.

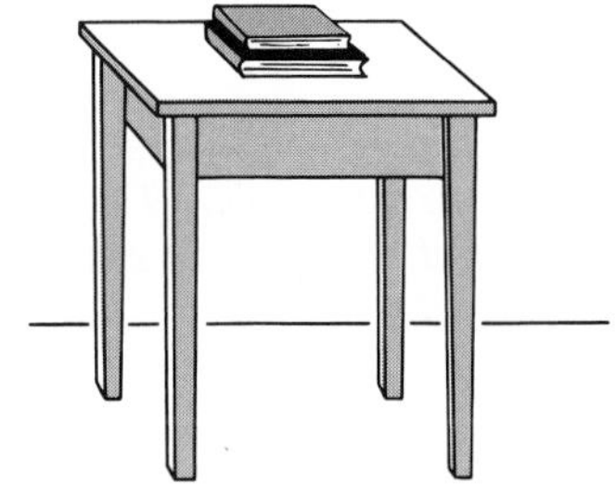

Figure 1-9. An example of gravitational potential energy. The amount of energy of the books depends upon the reference point chosen for the calculation.

The floor is an arbitrary reference point for our example. If a hole were cut in the floor under the book, the book would continue to fall and do more work when its motion is stopped. For each problem in *gravitational* potential energy, a reference point must be specified.

An object may also have potential energy that is not connected with gravitation in any way. For example, a compressed spring gets its energy from the push that is exerted on it to get it into a compressed position. This energy is called *elastic* potential energy. Elastic potential energy is not as easy to compute as the gravitational variety, since one must be familiar with the transformations of various forms of energy. Some of these forms of energy, especially those concerning the interior of the atom, can take the problem to the very frontiers of modern physics.

1.12 Kinetic Energy Every moving object has kinetic energy. This is the same as saying that everything has kinetic energy, because everything in the universe is in relative motion. Since the description of the motion of an object depends on another object that is designated as being "motionless," it is necessary to choose a reference point or, in this case, an arbitrary *stationary* point. In ordinary, earthbound physics the surface of the earth is considered stationary. An object resting on the earth's surface is said to have zero kinetic energy even though the object rotates and revolves with the earth.

Potential energy is energy due to the position of an object. Kinetic energy is energy due to the motion of an object.

Potential energy and kinetic energy are interchangeable. The total amount of energy in a situation remains constant.

1.13 Conservation of Energy As Figure 1-10 shows, there is a constant interplay between potential energy and kinetic energy in physics. Consider the swinging of a ball at the end of a string. When the ball is at the top of its swing, it is momentarily stationary. At that point the energy of the ball is gravitational potential (except for internal energy). As the ball begins its downward swing, some of the gravitational potential energy changes into kinetic energy. At the bottom of the swing, which we will consider as the reference point for gravitational potential energy, the kinetic energy of the ball is at a maximum because it is moving at its maximum speed. As the ball swings up the other side of its arc, the energy interchange is reversed. The total amount of energy is always the same—it is merely changing form. This discussion demonstrates the *law of conservation of energy*—the total amount of energy in the universe is constant.

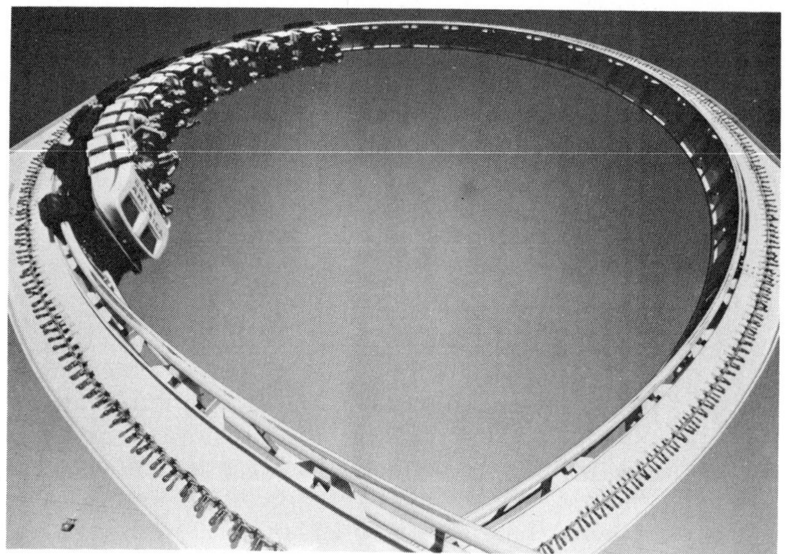

Figure 1-10. Energy transformation. As the roller coaster and its occupants descend, some of their gravitational potential energy changes into kinetic energy.

The conservation of energy was demonstrated in a striking way by the classic experiments with heat energy conducted by Count Rumford (1753–1814) and James Prescott Joule (1818–1889) in the early part of the nineteenth century. At that time most scientists thought of heat as a special kind of "fluid" called *caloric*, which was able to flow in

Figure 1-11. A model of Count Rumford's cannon-boring experiment. A team of horses furnished the energy to rotate the cannon barrel against a stationary boring tool. Water poured into the barrel during the boring process quickly came to a boil, especially when the boring tool became dull.

and out of objects without affecting their weights. Count Rumford was in charge of the boring of brass cylinders for the construction of cannon barrels in Bavaria. Figure 1-11 shows a model of Rumford's apparatus. During the boring process, the barrels became so hot that water poured into them could be brought to a boil. According to the caloric theory, the brass contained a specific amount of caloric. This caloric was being released from the brass as the metal was broken down into chips by the boring tool. Count Rumford questioned this explanation since the amount of heat produced seemed limitless. As long as the horses provided the work to turn the boring tool, more heat developed. Rumford demonstrated that if the boring tools were dull and no metal were ground to chips that supposedly contained caloric, heat continued to pour out. In fact, the metal heated up even more. Rumford concluded that the heat produced was a result of the work done on the cannon barrel by the horses; it was not the result of a substance present in the brass. In spite of Rumford's experiments, the caloric theory persisted for fifty more years.

Joule continued the investigation of heat by establishing the quantitative relationship between heat and mechanical energy. This relationship will be discussed more fully in Chapter 9. The equivalence between heat and mechanical energy provided strong evidence that heat is a form of energy. It also supported the theory that the total amount of energy remains constant when it is transformed from one kind to another.

1.14 Relationship Between Matter and Energy So far, we have discussed matter and energy as though they were two entirely different entities. Actually, the two are inseparably related. All matter contains some form of energy. Ordinarily, the idea of energy is associated with a substance such as a hot gas or an electrified object. In

Figure 1-12. Einstein did not spend all of his time thinking about the theory of relativity.

some cases it is necessary to associate energy with the empty space surrounding an electrified object.

In 1905, Einstein expressed the equivalence of matter and energy with the following equation:

$$E = mc^2$$

in which E stands for units of energy (work units), m stands for mass, and c for the speed of light. Einstein (Figure 1-12) developed this equation entirely from theoretical considerations, and there was no way of verifying it in the laboratory at the time. Subsequent experiments, however, have shown that the equation is correct.

Einstein's equation states that the units of mass and energy are directly proportional to each other. When one increases, the other increases; when one decreases, so does the other. The equation is also interpreted to mean that a given amount of mass is equivalent to a specific amount of energy.

A simple example will serve to illustrate the relationship between mass and energy. When you throw a ball, you impart energy to it. Energy is transferred from you to the ball. The mass-energy equation says that since c is constant, the increase in energy E will be accompanied by an increase in mass m. That is, while the ball is moving, its mass will be greater than it was while the ball was at rest. Both the energy and the mass of the ball have increased. The extra energy and mass of the ball came from you.

When an object is at rest with respect to the observer and the observer's measuring instruments, the mass of the object is said to be the *rest mass*. When the object is moving, its mass increases. This new mass, which increases rapidly as the object approaches the speed of light, is called the *relativistic mass* because it is in keeping with Einstein's *theory of relativity*. In the mass-energy equation, m is the relativistic mass. Table 1-1 shows the relationship between the speed of an object and its relativistic mass.

Table 1-1
RELATIVISTIC MASS

Speed of object (% of speed of light)	*Relativistic mass (compared to rest mass)*
0	1.00
25	1.03
50	1.15
75	1.51
80	1.67
85	1.90
90	2.29
95	3.20
99	7.09
99.9	22.4

Actually, what Einstein's equation means is that the terms "matter" and "energy" describe different aspects of the same quantity. This is one of the most important concepts of contemporary physics.

It should not be assumed, however, that Einstein's mass-energy equation is a statement of the conservation of matter and energy. The equation has value even if the matter and energy of the universe were constantly changing. The conservation laws rest on repeated laboratory measurements that show mass and energy are not lost in chemical and physical changes. Some scientists think the law of conservation may not hold for large energies and masses in outer space, but this hypothesis has not been verified. The deviations predicted by scientists are too small for measurement with present instruments.

The equivalence of matter and energy will become evident in the study of physics in still another way. We usually think of light energy in terms of waves and of matter in terms of particles. Actually, there are times when light acts as though it has granular properties. In other words, there are aspects of light that can be explained only by assuming that it is made up of discrete particles. Similarly, particles of matter exhibit wave properties. This wave-particle duality of nature is central to the study of physics, although some aspects of this concept are still not fully understood. Perhaps you will play a role in explaining this fundamental secret of nature.

Figure 1-13. Dr. Shirley Ann Jackson (born 1946), physicist. The first American black woman to earn a doctorate from the Massachusetts Institute of Technology, Dr. Jackson currently serves on that school's board of trustees. Originally involved in high energy physics, she now specializes in solid or condensed state physics at Bell Laboratories, the research and development arm of AT&T, where she works to expand the frontiers of knowledge in physics.

1.15 Subdivisions of Physics The study of the physical universe is called *physical science. Chemistry* is a physical science. It deals with the composition of matter and reactions among various forms of matter. *Physics* is also a physical science. It is concerned with the relationship between matter and energy. The ultimate goal of physics is to explain the physical universe in terms of basic interactions and simple particles.

Since both chemistry and physics deal with matter, these two sciences overlap to some extent. But a distinguishing consideration in the study of physics is always the idea of energy: what it is, how it affects matter and how matter affects it, and how it can be changed from one form to another.

The study of physics can be subdivided in a number of ways. The sequence of topics followed in this book is one that has been used for many years because of its logical progression and because it leads from concepts developed many years ago to those at the very frontiers of physics.

The chapters in your text fall into four major categories: mechanics and heat, waves, electricity and magnetism, and nuclear physics. The first category includes the study of motion, forces, work, power, and certain aspects of heat. These topics, and the ways in which they involve the transfer of energy, are considered in Chapters 3 through 9. (Chapter 2 describes measurement and problem solving, which are essential to the entire subject of physics.)

Waves are considered in Chapters 10 through 15. Sound energy and light energy are transmitted as waves. The nature of waves as a mechanism for energy transfer and the characteristics of sound and light comprise the subject matter of this section of your text.

Electricity and magnetism involve forms of energy that are increasingly important to our way of life. In Chapters 16 through 22, the ways of generating and transmitting this vital commodity are explained. This section of the text also contains an introduction to the study of electronics, which is the basis for radio, television, and computers.

In the last three chapters of the book, the principles of nuclear and particle physics are introduced. It is here that the physicist tries to get to the very heart of the composition of matter and of the energy forms that bind the building blocks of the universe together. With topics ranging from "atom smashers" to the most distant star systems, this part of the course is usually called modern physics.

Before a scientist can specialize in any one of the many branches of physics, he or she must have a good background in all areas of physics. Furthermore, interests in any one area should not stifle important exploration into other areas. Progress in science requires diversity as well as specialization.

The author hopes that as you study the many topics in this course, you will catch some of the excitement that motivates the scientists who work in this important field. The possibilities are truly unlimited.

QUESTIONS: GROUP A

1. List some properties you would use to identify your classmates. Describe several of your classmates according to this list. Can others in the class identify the person you described?
2. Which properties on your list for Question 1 would be useful to scientists?
3. (a) What are the three types of balances? (b) What property does each measure?
4. (a) Using Figure 1-7, find the mass on the balance that has a period of 0.40 s. (b) Find the period when the mass is 0.70 kg.
5. (a) Predict the shape of the graph you would obtain by plotting the square of the period as a function of mass for the data in Figure 1-7. (b) Construct the graph. Does it agree with your prediction?
6. Describe the shape of a graph of inertial mass versus gravitational mass.
7. What is the difference between *rest mass* and *relativistic mass?*
8. List some of the forms of energy you encounter during the day.

GROUP B

9. How did Count Rumford use the steps in the scientific method to show that heat is a form of energy?
10. Make a graph of relativistic mass as a function of speed using the data provided in Table 1-1. Describe the shape of the graph.
11. Use Figure 1-10 to answer the following questions relating to energy. (a) At what point is the gravitational potential energy of the riders at its highest value? (b) Where is the kinetic energy at its maximum?

(c) Where is the potential energy at its least? (d) At what point is the total energy at its least?

12. (a) How is specific gravity related to density? (b) What is the mass density of water? What is its specific gravity? (c) The density of mercury is 13.6 g/cm^3. What is the specific gravity of mercury?

PHYSICS ACTIVITY

Use a chemistry and physics reference manual to make a list of some of the properties used to describe matter. Then select a common element and describe it using its chemical and/or physical properties.

SUMMARY

Modern civilization requires ever-larger amounts of energy. By discovering and developing new energy sources, science and engineering can help to meet this need. Engineering is also called technology. The miniaturized computer and its many applications are good examples of the impact of science and technology on our way of life. Even so, many applications of computers still await discovery.

Scientific laws, theories, and hypotheses are steps in the search for truth in the physical universe. A hypothesis is a possible solution to a scientific problem. Confirmed hypotheses become laws or theories, but even these are never absolutely certain.

Science deals with the "how" but not the "why" of the universe. Among the properties of matter are mass (the measure of the quantity of material), inertia (the resistance of matter to change in its state of motion), and mass density (mass per volume). The environment of a sample of matter is called a condition. Physics further deals with the various types of energy and their transformations. Energy is the capacity for doing work. Potential energy is energy of position. Gravitational energy and elastic energy are forms of potential energy. Kinetic energy is energy of motion. Potential and kinetic energy can be readily interchanged. However, the law of conservation of energy states that energy may be changed from one form to another without loss. The experiments of Count Rumford and James Prescott Joule verified this law.

Einstein's equation $E = mc^2$ states that the units of energy and mass are directly proportional to each other and that they are interchangeable. Mass can thus be considered to be equivalent to energy.

Physics is the physical science that deals with the various forms of energy. Its four main subdivisions are mechanics and heat, wave motion, electricity and magnetism, and nuclear and particle physics. The last of these is also referred to as modern physics.

VOCABULARY

caloric
conditions
energy
engineering
gravitational potential energy
hypothesis
inertia
kinetic energy
law
law of conservation of energy
mass
mass density
matter
period
physics
potential energy
relativistic mass
rest mass
science
specific gravity
technology
theory

Lasers

It may be hard to believe, but the same instrument used to measure the distance between the earth and moon is also used to detect individual molecules in a car's exhaust, repair a torn retina, and weld steel.

The laser is an extremely accurate device for measuring distance because the intense light that it emits can be directed in a straight line over great distances.

Using a laser, scientists have been able to measure the constantly changing distance between the earth and moon with an uncertainty of only 1.7 cm. This is accomplished by shooting a pulse of laser light from a ground station on earth to reflectors left on the moon by the Apollo astronauts. The pulse bounces off a reflector and returns to the ground station. Scientists then use the round-trip travel time and the speed of light to compute the precise distance between the earth and moon.

On a microscopic scale, laser technology is being used to analyze the composition of gases in car exhaust. Molecules of a particular compound or element

Laser cutting through a metal plate.

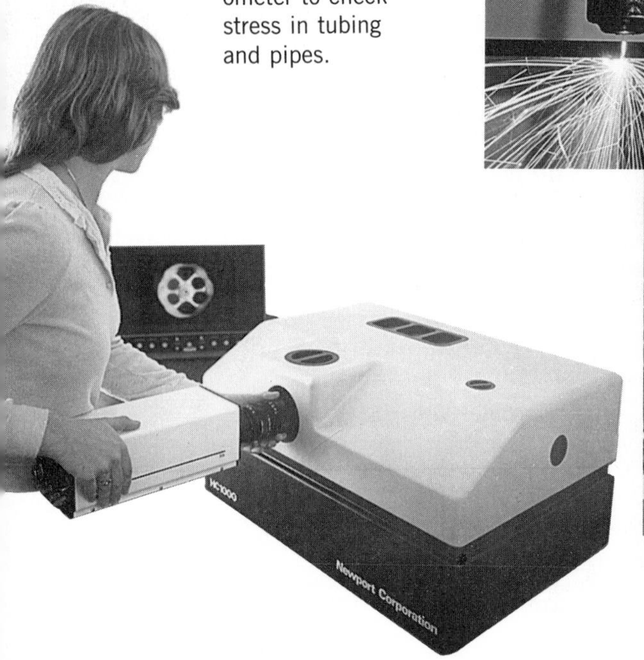

Use of a holographic interferometer to check stress in tubing and pipes.

Automobile exhaust gas analysis using an infrared diode laser, the beam of which is not visible. Here, carbon monoxide is monitored with very rapid response.

in the exhaust will absorb only certain energies of light. Since lasers produce light with a precise energy, the presence of specific substances in the gas can be inferred by the amount of laser light they absorb.

Laser Technicians

In the new and growing field of laser technology, laser technicians are in great demand. A laser technician must understand complicated wiring charts and blueprints of laser devices.

Some laser technicians work in the manufacturing industry. These technicians operate laser equipment that cuts, solders and drills through metal. Other technicians operate lasers that cut diamonds or fabric for clothing. In the field of holography, laser technicians maintain the cameras and equipment that produce holograms and assist in devising new techniques and applications.

A laser technician usually has a technical or community-college associate's degree in electronics. A solid understanding of applied electronics, math, and laser technology is necessary. Hands-on experience is a definite advantage. Because of the growing need for laser technicians, many companies provide on-the-job training.

Laser Physicists

Physicists involved in the laser field are mainly responsible for the research and development of lasers for scientific, medical, industrial, and military purposes. Laser physicists develop the specific requirements necessary for a laser to perform a certain task. They then outline the theoretical framework from which the design engineers develop the laser device desired. Some laser physicists with engineering backgrounds also help design the laser equipment.

Laser physicists are employed in independent and government laboratories and in university research centers. Most laser physicists have Ph.D. degrees in nuclear or optical physics. An understanding of mechanical and electrical engineering is also necessary. ■

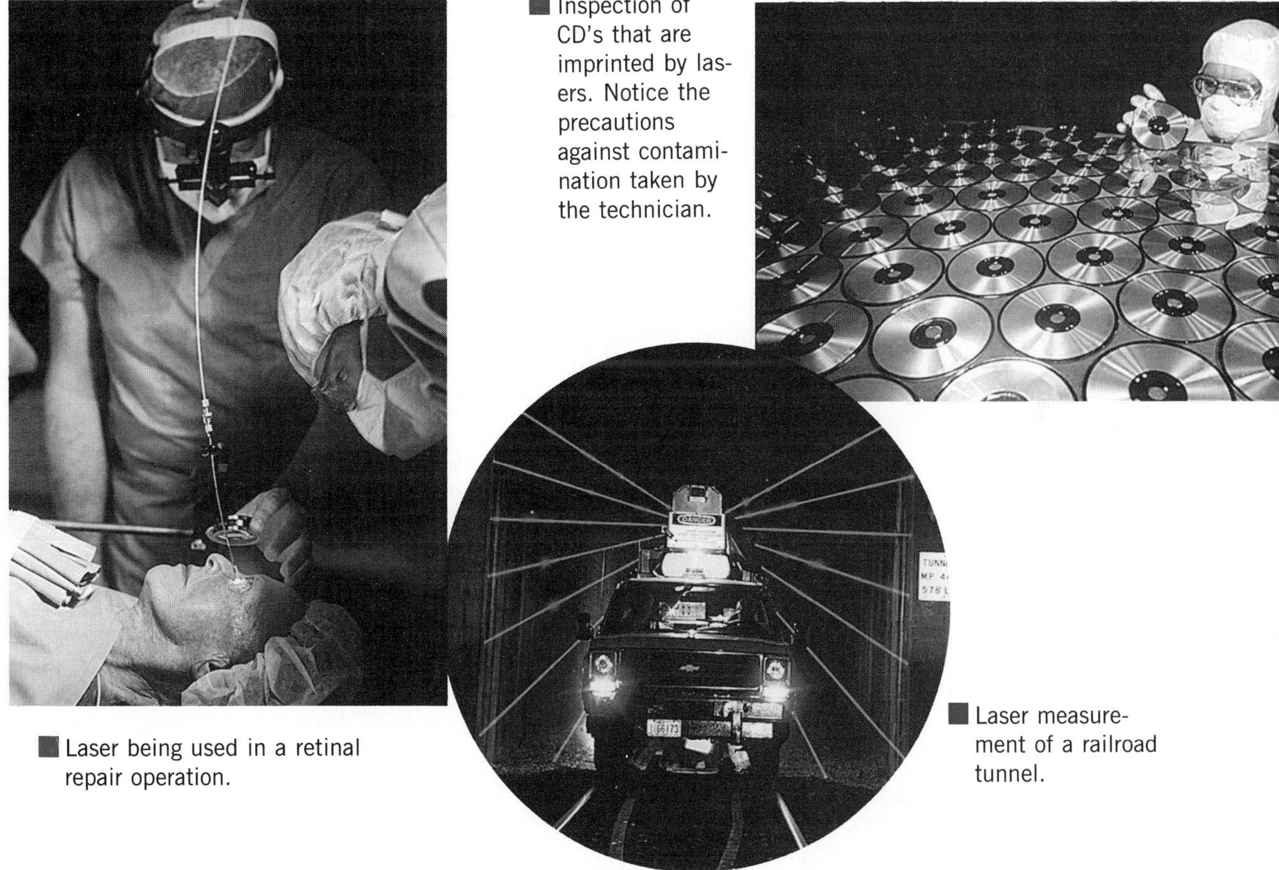

■ Inspection of CD's that are imprinted by lasers. Notice the precautions against contamination taken by the technician.

■ Laser being used in a retinal repair operation.

■ Laser measurement of a railroad tunnel.

2 Measurement and Problem Solving

accuracy (AK-yoo-ra-see) n.: the closeness of a measurement to the accepted value for a specific physical quantity.

OBJECTIVES

- Describe the metric units of measure.
- Define and calculate the accuracy and precision of a group of measurements.
- Identify and apply appropriate mathematical operations to solve physics problems.

Figure 2-1. A highway sign that shows distances in both miles and kilometers.

UNITS OF MEASURE

2.1 The Metric System The system of measurement that is best suited to scientific purposes is the *metric system*. Developed in France near the end of the eighteenth century, the metric system uses a decimal basis for multiples and subdivisions of the basic units of measure, which greatly simplifies calculations. In 1866, Congress legalized the metric system for use in the United States, and since 1893 the yard and the pound have been defined in terms of metric measures. In 1975, a law was passed to promote more extensive use of the metric system in the United States, but no date for complete conversion was specified. (See Figure 2-1.)

The modern metric system is officially called the International System of Units (abbreviated SI for *Système International d'Unités*). The International Bureau of Weights and Measures, which oversees the system, is located in Sèvres, France, a suburb of Paris. The property of the Bureau is internationally owned, much like the buildings and grounds of the United Nations in New York.

In a system of measurement, it is important to distinguish between *physical quantities* and *units of measure*. A physical quantity is a measurable aspect of the universe, such as *length*. The metric unit of measure for length is the *meter*.

There are seven *fundamental units* of measure. These

Table 2-1
THE FUNDAMENTAL UNITS OF MEASURE

Physical quantity	*Quantity symbol*	*Unit of measure*	*Unit symbol*
length	l	meter	m
mass	m	kilogram	kg
time	t	second	s
electric current	I	ampere	A
temperature	T	kelvin	K
amount of substance	n	mole	mol
luminous intensity	I	candela	cd

seven units, the physical quantities they measure, and the symbol for each are shown in Table 2-1. (Because this information is so important, it is also shown inside the front cover of the book and in Appendix B, Table 4.) The meter, kilogram, and second are basic to the study of mechanics. The kelvin and the mole are used in the study of heat. The candela is used in measurements of light. The ampere will be used in the section of the course on electricity. Since the *m*eter, *k*ilogram, and *s*econd are three of the fundamental units, the modern metric system is sometimes also called the MKS system.

Physical quantities other than those in Table 2-1 are measured by using combinations of the fundamental units of measure. For example, velocity is measured in meters per second and mass density in kilograms per cubic meter (or grams per cubic centimeter). Units that consist of combinations of the fundamental units are called *derived units.*

The prefixes used in the metric system, with their corresponding symbols and values, are shown in Table 2-2. The most commonly used of these prefixes are kilo (k), centi (c), and milli (m).

The use of units and symbols in this book is based on the latest specifications of the International Bureau of Weights and Measures. Though Table 2-1 lists the ampere as the fundamental unit of electric current, in this text the concepts of electricity will be based on the unit of electric *charge*, the coulomb (C). According to SI, the coulomb is a *derived* unit.

In SI usage, the abbreviations for units that are named after scientists are capitalized.

Table 2-2
METRIC SYSTEM PREFIXES

Prefix	*Symbol*	*Factor*
exa	E	10^{18}
peta (*pet*-ah)	P	10^{15}
tera (*tair*-ah)	T	10^{12}
giga (*jig*-ah)	G	10^{9}
mega	M	10^{6}
kilo	k	10^{3}
hecto	h	10^{2}
deka	da	10^{1}
deci	d	10^{-1}
centi	c	10^{-2}
milli	m	10^{-3}
micro	μ	10^{-6}
nano (*nan*-oh)	n	10^{-9}
pico (*pea*-koh)	p	10^{-12}
femto (*fem*-toh)	f	10^{-15}
atto (*at*-oh)	a	10^{-18}

2.2 The Meter The length of the meter was originally calculated to be one ten-millionth of the distance from the north pole to the equator along a line running through

Figure 2-2. The best approximation to the ideal meter is provided by the light from a helium-neon laser that is stabilized by relating it to the spectrum of methane or iodine. Here an iodine laser (long object, center) is being used to calibrate the wavelength of another laser.

France and Spain. Until 1960, the standard for the meter was a metal bar kept in a vault at the International Bureau of Weights and Measures.

In 1960, the definition of the standard meter was changed to a laboratory procedure that did not require comparison with a physical object. Instead, the meter was defined as a specified number of wavelengths of the orange-red light given off by a glowing tube of krypton gas. These wavelengths can be measured much more precisely than the standard meter bar in France.

In 1983, the standard meter was defined still more precisely, this time in terms of the speed of light. The new definition is

1 meter (m) = the distance that light travels in a vacuum in 1/299 792 458th of a second

In this definition of the meter, the speed of light is fixed arbitrarily. The standard meter is, therefore, not an independent fundamental unit but is related to the second. The new definition of the meter was adopted because it is possible to measure it more accurately than the previous standard. The new definition also means that the speed of light is no longer subject to revision. Any change in the measurement of the speed of light will, instead, change the definition of the standard meter.

In most track and field events, such as the Olympic Games, distances are now given in meters and kilometers rather than yards and miles.

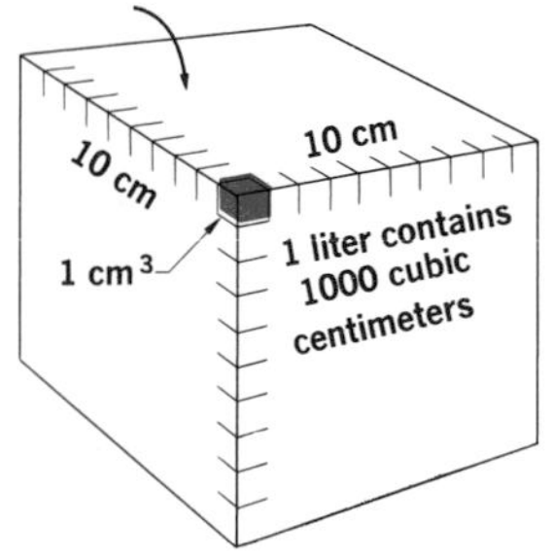

Figure 2-3. Relationship among the metric units of length, area, and volume.

The metric units for the measurement of area and volume are derived from the meter, as Figure 2-3 indicates. A rectangular floor 3 meters long and 2 meters wide has an area of 6 square meters. Similarly, a rectangular tank 4 meters long, 3 meters wide, and 2 meters deep has a volume of 24 cubic meters.

The liter (L) *is a special name for a cubic decimeter* (1000 cm^3). Since one milliliter (1 mL) is 1/1000th of a liter, 1 mL is equivalent to one cubic centimeter (1 cm^3).

Although SI does not include the liter as a unit, it has authorized its use as a special name for the cubic decimeter and has adopted both l *and* L *as symbols to be used for the unit liter.*

2.3 The Kilogram

The SI standard of mass is the *kilogram.* It is defined as the mass of the standard kilogram kept by the International Bureau of Weights and Measures, as shown in Figure 2-4. Duplicates of the standard kilogram are kept in various countries around the world. These duplicates are occasionally checked against the standard kilogram.

The standard for the kilogram is the only one of the seven standards that still consists of a natural object representing a unit of measure. One reason why the standard

Figure 2-4. The international standard of mass. Three evacuated bell jars cover the standard kilogram at the bottom center. Six comparison standards are under double bell jars on either side. The standards are cylinders made of platinum-iridium alloy, which does not oxidize and has a mass density almost twice that of lead. The cylinders are 3.9 cm high and 3.9 cm in diameter. (Equal height and diameter gives a cylinder the smallest possible area-to-volume ratio.) The standard meter bar used before 1960 is in a protective case on the shelf above the masses.

kilogram is still a natural object is because masses can presently be compared with the standard with greater precision than is possible by means of other laboratory processes. All other units of measure are presently determined through reproducible processes in the laboratory, such as the technique used in determining the standard meter.

It is useful to know that one milliliter of water has a mass of approximately one gram and that a five-cent United States coin has a mass of about five grams. A metric ton contains 1000 kg.

2.4 Force and Weight The SI unit of force is the *newton* (N). The newton is a derived unit that will be more fully described in Chapter 4. But it is important at this point to know the relationship between a newton and a kilogram, namely, that a force of about 9.8 newtons is required to lift a mass of one kilogram in most places on the earth. (The exact number of newtons required at a specific location depends on the pull of gravity at that location, but 9.80 is a useful approximation.)

In measurements with absolute quantities greater than 9999, spaces, not commas, are used to separate groups of three digits to the left of the decimal (indicated or understood). Spaces are also used to separate groups of three digits to the right of the decimal.

In most countries that use the metric system, the weights of people and objects are commonly described in kilograms. Since weight is a force, it should be described in newtons, and the use of kilograms for weights is not consistent with the rules of the International System of Units. Perhaps a new definition of the kilogram will eventually be adopted to accommodate its widespread usage with weights. But in scientific literature at the present time, weights are correctly expressed in newtons.

2.5 The Second The unit of time is the *second* (s). Originally, the standard second was defined in terms of the earth's motion. This standard has been replaced by one that is based on atomic vibrations, which can be measured with an accuracy of one part in ten trillion.

Figure 2-5. A decimal clock. This watch was made in anticipation of the adoption of decimal days, which never took place.

The standard second is based on the rate of vibration of the cesium atom. The vibration is produced by radio waves, and it is not affected by such external influences as temperature and gravity. The rate of vibration can be tuned precisely to the radio waves that are producing it. The definition adopted in 1967 is

1 second (s) = 9 192 631 770 vibrations of cesium-133 atoms

When the metric system was first proposed, an effort was made to redefine other time intervals as well. One proposal called for a month of three ten-day weeks. In this system, the day was divided into ten hours, the hour into 100 minutes, and each minute into 100 seconds. A decimal watch of this kind is shown in Figure 2-5. However, decimal days were never officially adopted in any country. Thus, the numbers on our clocks and calendars remain conspicuous exceptions to the metric system.

QUESTIONS: GROUP A

1. (a) What is the mathematical basis for the metric system? (b) What is its official name? (c) Why is it the system used in science?
2. What is the most recent definition of the standard meter?
3. What is the only fundamental quantity measured by comparison with a physical object? Why?
4. A box of crackers at the grocery store is labeled "1 pound (454 g)." What is wrong with this label?

GROUP B

5. (a) What was the original standard for a second? (b) What is it now? (c) Why was it changed?
6. Identify the following units as fundamental or derived. In which branch of physics is each used? (a) newton, (b) liter/mole, (c) candela, (d) kilogram, (e) ampere.
7. A box of cookies is labeled, "0.5 kg (4.9 N)." Would the label be correct if you took the cookies to the moon, where the force of gravity is less?

PROBLEMS: GROUP A

1. Convert each of the following to the units shown:
   (a) 2 dm^3 = ____ mL
   (b) 350 cm^3 = ____ L
   (c) 16 g = ____ μg
   (d) 0.75 km = ____ mm
   (e) 675 mg = ____ g
   (f) 0.596 m^3 = ____ dm^3
   = ____ L
   (g) 75 m^2 = ____ cm^2
   (h) 2 h 10 min = ____ s
   (i) 462 μm = ____ cm
   (j) 35 km/h = ____ m/s
2. From the data in Figure 2-1, determine how many kilometers there are in a mile by making a ratio of a distance measured in miles to the corresponding one in kilometers.
3. (a) Calculate the cross-sectional area of one of the standard kilogram cylinders shown in Figure 2-4. (b) What is the volume? (The data for the height and diameter of the cylinders are in the caption.)
4. How many two-liter bottles of carbonated water would it take to fill a jug that holds 5 dm^3?

GROUP B

5. A crane must lift a crate with a mass of 3.5×10^3 kg. (a) How much force will be required? (b) On the moon, do you think it would take more or less force to lift the crate? Explain.

6. (a) When you buy a two-liter bottle of soda pop, how many dm^3 have you purchased? How many cm^3? How many mL? (b) What is the mass in grams of the soda, if it is mostly water? (c) What is its weight in newtons?

7. How much force will it take in order for you to lift this book, which has a mass of 1350 g?

8. How many four-liter buckets of water do you need to fill your bath tub, which is 123 cm long, 57.2 cm deep, and 33.0 cm wide?

MAKING AND RECORDING MEASUREMENTS

2.6 Accuracy The measurement of a physical quantity is always subject to some degree of uncertainty. There are several reasons for this: the limitations inherent in the construction of the measuring instrument or device, the conditions under which the measurement is made, and the different ways in which the person uses or reads the instrument, as you can see in Figure 2-6. Consequently, in reporting the measurements made during a scientific experiment, it is necessary to indicate the degree of uncertainty so far as it is known.

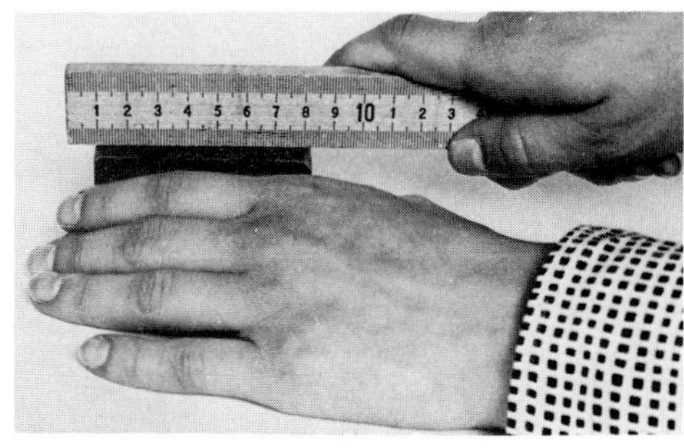

Figure 2-6. The accuracy of measurement depends upon the instrument used and the care with which the reading is made.

One way to express the uncertainty of a measurement is in terms of *accuracy*. ***Accuracy*** *refers to the closeness of a measurement to the accepted value for a specific physical quantity.* It is expressed as either an *absolute* or a *relative error*. *Absolute error* is the actual difference between the measured value and the accepted value. The equation for absolute error is

$$\mathbf{E_a} = |\mathbf{O} - \mathbf{A}|$$

where $\mathbf{E_a}$ is the absolute error, **O** is the observed (measured) value, and **A** is the accepted value. (The vertical lines mean the absolute value is used, regardless of sign.)

Relative error is expressed as a percentage and is often called the percentage error. It is calculated as follows:

Accuracy is expressed as absolute error or relative error.

$$\mathbf{E_r} = \frac{\mathbf{E_a}}{\mathbf{A}} \times 100\%$$

where $\mathbf{E_r}$ is the relative error, $\mathbf{E_a}$ the absolute error, and **A** the accepted value.

2.7 Precision In common usage, accuracy and *precision* are often used synonymously. But in science it is important to make a distinction between them. You should learn

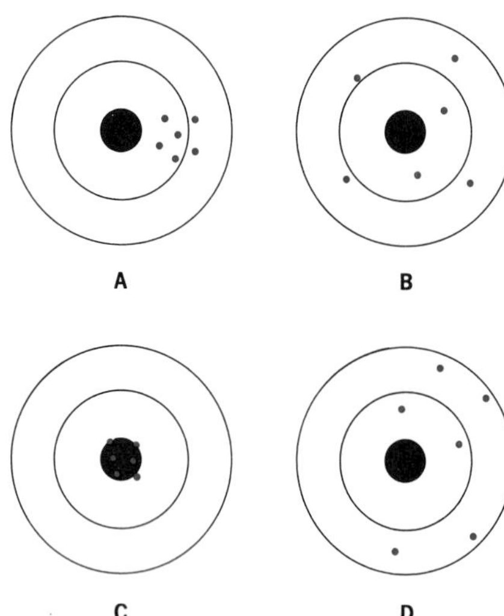

Figure 2-7. Distinction between accuracy and precision. Six shots were fired at each target. In A, the precision is good, but the accuracy is poor. In B, the average accuracy is good, but the precision is poor. In C, both the accuracy and precision are good. In D, both the average accuracy and the precision are poor.

to use the two terms correctly and consistently. ***Precision*** *is the agreement among several measurements that have been made in the same way.* It tells how reproducible the measurements are and is expressed in terms of *deviation.* Figure 2-7 should help you to distinguish between accuracy and precision as used in science. As in the case of accuracy, deviations in precision can be given as absolute or relative (percentage). *Absolute deviation* is the difference between a single measured value and the average of several measurements made in the same way, or

$$\mathbf{D_a = |O - M|}$$

where $\mathbf{D_a}$ is the absolute deviation, **O** is the observed or measured value, and **M** is the mean, or average, of several readings.

Relative deviation is the percentage average deviation of a set of measurements and is calculated as follows:

$$\mathbf{D_r = \frac{D_a\ (average)}{M} \times 100\%}$$

where $\mathbf{D_r}$ is the relative deviation, $\mathbf{D_a}$ (average) is the average absolute deviation of a set of measurements, and **M** is the mean, or average, of the set of readings.

2.8 Significant Figures Another way of indicating the precision of a measurement is by means of *significant figures.* ***Significant figures*** *are those digits in a number that are known with certainty plus the first digit that is uncertain.* For example, if the length of a laboratory table was measured with a meter stick graduated in millimeters and found to be 1623.5 millimeters, the measurement has five significant figures. The 5 is the uncertain digit. It was an estimation of the value between two millimeter marks on the meter stick.

In working with significant figures, you can usually assume to know the last figure of a number, the uncertain figure, to within 1. For instance, when the value for π is given as 3.14, you can assume that the true value lies between 3.13 and 3.15.

The following rules will help you to understand how significant figures are used in physics problems. Even these rules do not cover all cases. However, common sense usually shows how a specific case should be handled.

1. All nonzero figures are significant: 112.6°C has four significant figures.

2. All zeros between nonzero figures are significant: 108.005 m has six significant figures.

3. Zeros to the right of a nonzero figure, but to the left of an understood decimal point, are not significant unless specifically indicated to be significant. In this book the rightmost such zero which is significant is indicated by a *bar* placed above it: 109 000 km contains *three* significant figures; 109 0$\bar{0}$0 km contains *five* significant figures.

4. All zeros to the right of a decimal point but to the left of a nonzero figure are not significant: 0.000 647 kg has three significant figures. The single zero placed to the left of the decimal point in such an expression serves to call attention to the decimal point and is never significant.

5. All zeros to the right of a decimal point and following a nonzero figure are significant: both 0.070 80 cm and 20.00 cm have four significant figures.

Further examples of these rules are given in Table 2-3.

Table 2-3
SIGNIFICANT FIGURES

Two significant figures	*Three significant figures*	*Four significant figures*
2300	5420	1521
4$\bar{0}$00	60$\bar{0}$0	504$\bar{0}$
5.2	73.0	4.050
0.16	0.915	0.378 0
0.007 8	0.046 7	0.015 20

The rules for significant figures should be observed throughout the course.

6. Rule for addition and subtraction. Remember that the rightmost significant figure in a measurement is uncertain. The rightmost significant figure in a sum or difference will be determined by the leftmost place at which an uncertain figure occurs in any of the measurements being added or subtracted. The following example of addition illustrates this rule:

13.05 cm
309.2 cm ← **leftmost place of uncertain figure in measurements being added**
3.785 cm
326.035 cm ← **rightmost significant figure in the answer**

Since the answer should have only one uncertain figure (the rightmost one), it should be recorded as 326.0 cm.

7. Rule for multiplication and division. Remember that when an uncertain figure is multiplied or divided by a number, the answer is likewise uncertain. Therefore the product or quotient should not have more significant figures than the least precise factor. The following example illustrates the rule for multiplication:

$$3.54 \text{ cm} \times 4.8 \text{ cm} \times 0.5421 \text{ cm} = 9.211\,363\,2 \text{ cm}^2$$

least precise factor (4.8) — **rightmost significant figure in the answer** (the 2 in 9.2)

Consequently, the answer should be recorded as 9.2 cm^3.

8. Rule for rounding. If the first figure to be dropped in rounding off is 4 or less, the preceding figure is not

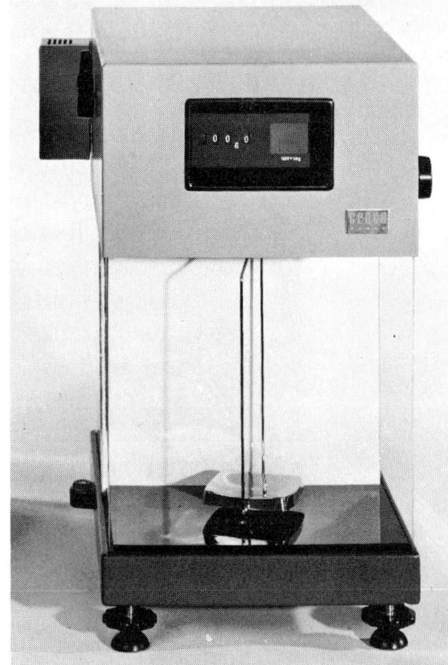

Figure 2-8. The tolerance of this analytical balance makes it possible to measure differences as small as 10^{-5} g.

changed; if it is 6 or more, the preceding figure is raised by 1. If the figures to be dropped in rounding off are a 5 followed by figures other than zeros, the preceding figure is raised by 1. If the figure to be dropped in rounding off is a 5 followed by zeros (or if the figure is exactly 5), the preceding figure is not changed if it is even; but if it is odd, it is raised by 1.

The number of significant figures in your laboratory measurements will be governed by the instruments you use. The number of significant digits obtainable from a measuring instrument is determined largely by the *tolerance* of the device. Thus the figure listed for the tolerance of an instrument indicates its limitations. The instrument manufacturer assumes that the instrument is used properly and that human errors are held to a minimum. The typical platform balance, for example, has a tolerance of 0.1 gram. However, research instruments like the one shown in Figure 2–8 have tolerances of 0.000 01 g or better!

PRACTICE PROBLEMS **1.** Add 165 g, 0.23 g, 9.6 g, and 36 g. *Ans.* 211 g

2. Subtract 3.8 s from 27.54 s. *Ans.* 23.7 s

3. Multiply 760 cm by 93.2 cm. *Ans.* 71 000 cm^2

4. Divide 23 000 m^2 by 75.62 m. *Ans.* $3\bar{0}0$ m

If you have not worked with significant figures before, you may find it hard to believe that an answer with, let us say, two figures is more correct than an answer with five or more figures. This may especially be true when you work with an eight-place calculator, in which all the displayed figures look equally correct. But remember that calculators "crunch" pure numbers and not measured quantities. The last figure of a pure number is not uncertain and is, therefore, not subject to the rules of significant figures.

The significant figures in a measurement reveal the precision with which the measurement was made. This precision cannot be changed by mathematical operations. An answer with too many figures is, therefore, misleading.

2.9 Scientific Notation Scientists frequently deal with very large and very small numbers. The velocity of light in air is about 300 000 000 meters per second. The mass of the earth is about 6 000 000 000 000 000 000 000 000 kilograms. The rest mass of an electron is 0.000 000 000 000 000 000 000 000 000 000 910 953 4 kilogram. These numbers are inconvenient to write and difficult to read. Through the use of the mathematical rules of exponents, however, it is possible to express these numbers in a much simpler way. *Using powers of ten in writing a number is called* ***scientific notation.*** Such numbers have the form

$$\mathbf{M} \times 10^{\mathbf{n}}$$

where **M** is a number having a single nonzero figure to the left of the decimal point and **n** is a positive or negative exponent. To write a number in scientific notation:

1. Determine **M** by shifting the decimal point in the original number to the left or to the right until only one nonzero digit is to the left of it. **M** should contain only the significant figures of the original number.
2. Determine **n** by counting the number of places the decimal point has been shifted; if it has been moved to the left, **n** is positive; if to the right, **n** is negative.

For example, we see that the velocity of light can be written as 3×10^8 m/s, the mass of the earth as 6×10^{24} kg, and the mass of the electron as $9.109\,534 \times 10^{-31}$ kg.

It is often helpful to know the magnitude of a number in terms of its powers of ten. In this form it can be compared quickly with other numbers. A numerical approximation to the nearest power of ten of a number is called its *order of magnitude* (Figure 2-9). The order of magnitude of the velocity of light in m/s is 10^8. The order of magnitude of the mass of the earth in kg is 10^{25}, since its mass is nearer to 10^{25} kg than it is to 10^{24} kg. Similarly, the order of magnitude of the mass of the electron in kg is 10^{-30}. Orders of magnitude of length for a number of objects in the universe are shown in Figure 2-10 on the next page.

Figure 2-9. The great spiral galaxy in the constellation Andromeda. This huge formation of stars, which is similar in shape to our own Milky Way galaxy, is estimated to be more than 10^{18} km in diameter and 2×10^{19} km from the earth. The mass of the Andromeda galaxy has been calculated to be 8×10^{41} kg, which is 4×10^{11} times the mass of the sun.

Scientific notation is sometimes called exponential notation. Can you explain why?

SOLVING PROBLEMS

2.10 Data, Equations, and Graphs Scientists usually use the term "data" to refer to the numbers, units, and other information recorded when measurements are made. Much of your homework in physics will involve solving problems in which the necessary data are already provided for you. The problems at the ends of the chapters in this book are like that. Your objective is to use the data properly in order to find the quantity required.

The first step in solving a problem of this kind is to list carefully all the data provided. Be sure to include the correct symbol and unit for each physical quantity. If the symbol is not given in the problem, look it up inside the front cover of this book or in Appendix B, Table 4. Record these symbols correctly at the start. As you continue working the problem, be sure you use only these symbols. Also,

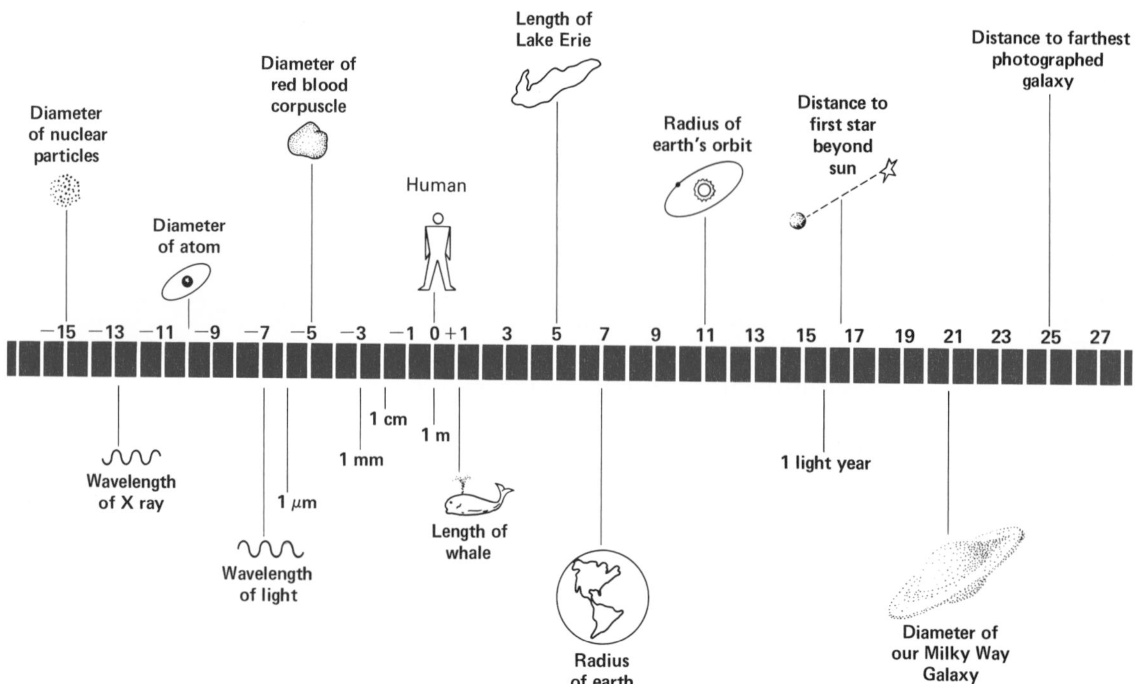

Figure 2-10. Orders of magnitude of length in meters of various objects in the universe. The numbers above the scale are powers of ten.

check to make sure that you have copied the right number of significant figures in each case.

In the laboratory, most of the data needed must be obtained by direct measurement. In some experiments you will be told what to measure, what instruments to use, and even how many significant figures to use. But in other experiments you must make these decisions for yourself.

The Mathematics Refresher is a handy reference. Consult it frequently.

In recording and interpreting data in physics, you will use mathematical skills that you have acquired in previous courses. The Mathematics Refresher, following Chapter 25, contains a summary of information and skills from arithmetic, elementary algebra, plane geometry, and trigonometry that is useful in the study of physics. At this time, you should review the information contained there and try some of the examples. Study carefully any procedure with which you are not familiar.

The purpose of a scientific experiment is to discover relationships that may exist among the quantities measured. The experimenter controls the conditions of the experiment in order to obtain the most accurate measurements. Uncontrolled conditions might make the measurements so confusing that relationships would not be discovered. As an example, suppose you wish to discover any relationships that might exist between the mass of a homogeneous substance (the same throughout) and its volume. You measure the volumes of several samples of different substances and determine the mass of each. Since temperature affects the volume of most substances, you control this condition by taking all of your measurements at the same temperature, such as 20°C. Your data for three substances, with the volume measured in cm^3 at 20°C and the mass measured in grams, might look like Table 2-4.

Now you look for mathematical relationships among these numbers. As the volume of each substance increases, the mass also increases. But this statement is very general. You are looking for a specific equation. Sometimes, plotting the data on a graph reveals the equation that expresses the relationship among the measurements.

Table 2-4
VOLUME-MASS RELATIONSHIPS AT 20°C

	Mass				Mass		
Volume (cm^3)	***aluminum*** (g)	***water*** (g)	***maple wood*** (g)	***Volume*** (cm^3)	***aluminum*** (g)	***water*** (g)	***maple wood*** (g)
$1\bar{0}$	27	$1\bar{0}$	6	$6\bar{0}$	162	$6\bar{0}$	36
$2\bar{0}$	54	$2\bar{0}$	12	$7\bar{0}$	189	$7\bar{0}$	42
$3\bar{0}$	81	$3\bar{0}$	18	$8\bar{0}$	216	$8\bar{0}$	48
$4\bar{0}$	108	$4\bar{0}$	24	$9\bar{0}$	243	$9\bar{0}$	54
$5\bar{0}$	135	$5\bar{0}$	$3\bar{0}$	$10\bar{0}$	$27\bar{0}$	$10\bar{0}$	$6\bar{0}$

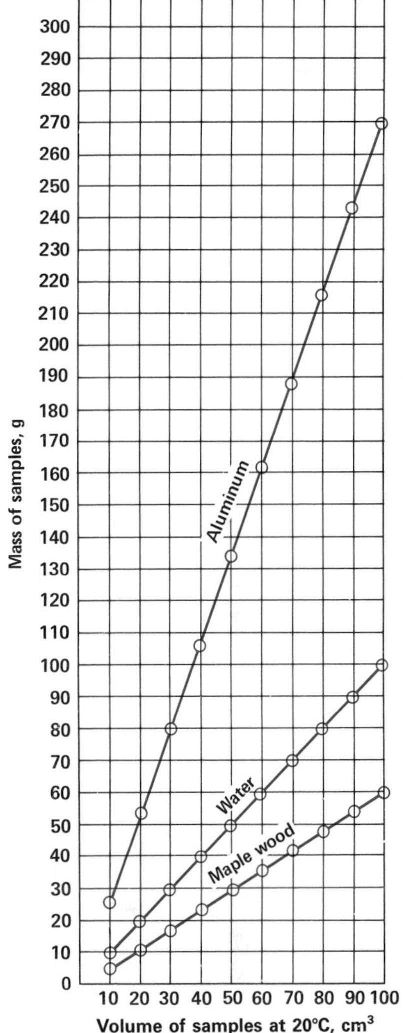

Figure 2-11. Graph of three direct proportions. Straight lines are typical of these relationships.

As you see, Figure 2-11 is a graph of the data in Table 2-4. The zero point is the origin of the graph. The horizontal line through the origin is called the x axis. On it are measured distances from the origin, or *abscissas*, equivalent to the various volumes. For convenience, use one small interval to represent $1\bar{0}$ cm^3. The vertical line through the origin is called the y axis. On it are measured distances from the origin, or *ordinates*, equivalent to the various masses. Use one small interval to represent $1\bar{0}$ g.

When the data for aluminum, water, and maple wood are plotted on the graph, the data for each substance can be connected by a straight line. The mathematical relationship between the mass and the volume is called a *direct proportion*. The symbol that means "is proportional to" is $\propto$. Hence the relationship between mass and volume is expressed mathematically as

$$\textbf{mass } (m) \propto \textbf{ volume } (V)$$

Table 2-5
SPEED-TIME RELATIONSHIPS

Speedometer reading (km/h)	*Time for 100-km trip* (h)
20.0	5.00
30.0	3.33
40.0	2.50
50.0	2.00
60.0	1.67
70.0	1.43
80.0	1.25
90.0	1.11
100.0	1.00

Plotting data that are in ***direct proportion*** *always results in the graph of a straight line.*

To illustrate another kind of graph, let us plot the data given in Table 2-5, which shows the time required to make a trip of 10$\bar{0}$ kilometers at various speeds. A line drawn through the plotted points, as shown in Figure 2-12, is a *hyperbola*. Furthermore, the product of the speed and time is constant. *Two quantities whose product is a constant are in* ***inverse proportion.***

A proportion can be changed into an equation by introducing a constant. In the volume-mass relationship the constant for each substance can be calculated from the data in Table 2-4. (For example, if each measurement in the volume column is multiplied by 2.7, the measured mass of aluminum results.) The equation for this relationship is $m = DV$ where D is a constant. Solving the equation for D,

$$D = \frac{m}{V}$$

This is the equation for mass density that was given in Section 1.9. With this equation you can find any one of the three physical quantities (D, m, or V) provided the measurements for the other two are given. The mass densities of most substances range from 0.1 g/cm^3 to 10 g/cm^3. Mass densities are usually given in g/cm^3 or in kg/m^3. The relationship between these two combinations of units is calculated as follows:

$$\left(\frac{1\ \text{g}}{1\ \text{cm}^3}\right)\left(\frac{1\ \text{kg}}{10^3\ \text{g}}\right)\left(\frac{10^6\ \text{cm}^3}{1\ \text{m}^3}\right) = \frac{10^3\ \text{kg}}{\text{m}^3} = 1000\ \text{kg/m}^3$$

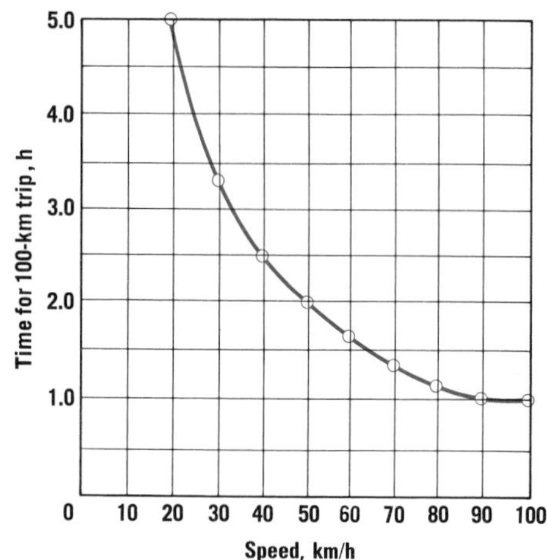

Figure 2-12. Graph of an inverse proportion. The curve is a hyperbola.

A carefully drawn graph provides you with another helpful tool. You can use it to estimate values between the plotted points (interpolation) and beyond the measurements made in the experiment (extrapolation).

Estimation of answers is not limited to interpolation and extrapolation. When you solve a problem in physics, you should make an intelligent guess about the reasonableness of your answer. Such a "check for reasonableness" will frequently disclose major errors in arithmetic or algebra.

2.11 Vectors Physical quantities such as length, area, volume, mass, density, and time can be expressed in terms of magnitude alone as single numbers with suitable units. For instance, the length of a table can be completely described as 1.5 m. The mass of a steel block is completely described as 54 kg. *Quantities,* such as these, *that can be expressed completely by single numbers with appropriate units are called* ***scalar quantities,*** *or simply,* ***scalars.*** ("Scalar" is derived from the Latin word *scala,* which means "ladder" or "steps" and implies magnitude.) Additional examples of scalars in physics are energy and temperature.

Other physical quantities, such as force, velocity, acceleration, electric field strength, and magnetic induction, cannot be completely described in terms of magnitude alone. In addition to magnitude these quantities always have a specific direction. *Quantities that require magnitude and direction for their complete description, and whose behavior can be described by certain mathematical rules, are called* ***vector quantities.*** *They can be represented by* ***vectors.*** ("Vector" is derived from a Latin word meaning "carrier," which implies displacement.) The ability to work with vectors is an important mathematical skill that is necessary in solving problems that deal with vector quantities.

A vector is usually shown as an arrow. The length of the arrow represents the magnitude of the vector, while the orientation of the arrow shows direction. North is usually toward the top of the page, east is toward the right, south is toward the bottom, and west is toward the left.

When two or more vectors act at the same point, it is possible to find a single vector that produces the same effect as the combination of separate vectors. In such a calculation, each separate vector is called a *component* and the single vector that produces the same result as the combined components is called the *resultant.* The following examples show basic rules in solving vector problems.

In Figure 2-13, a vector of 8.0 units and a vector of 3.0 units are shown, both directed northward. (These vectors might represent velocities, or forces, or any other vector quantities. But in this introductory treatment of vectors, we shall simply designate them as "units.") When two or more vectors act at the same point and in the same direction, the magnitude of the resultant is the algebraic sum of the magnitudes of the components. In Figure 2-13, the resultant has a magnitude of 11.0 units. (The vectors in the figure should be considered as acting at the same point, even though they are shown side by side for the sake of clarity.) The resultant has the same direction as the components, as shown in the figure.

A different situation is shown in Figure 2-14. This time the 8.0-unit vector acts northward, as before, but the 3.0-unit vector acts southward. When these two vectors act at the same point, the magnitude of the resultant is the algebraic difference of the component magnitudes, or 5.0 units. Another way of saying this is that the magnitude of

Figure 2-13. The magnitude of the resultant of two component vectors acting in the same direction is the sum of the magnitudes of the components. The resultant has the same direction as the components.

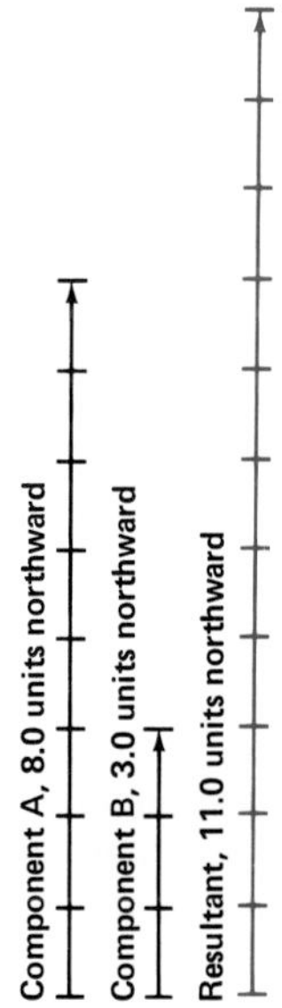

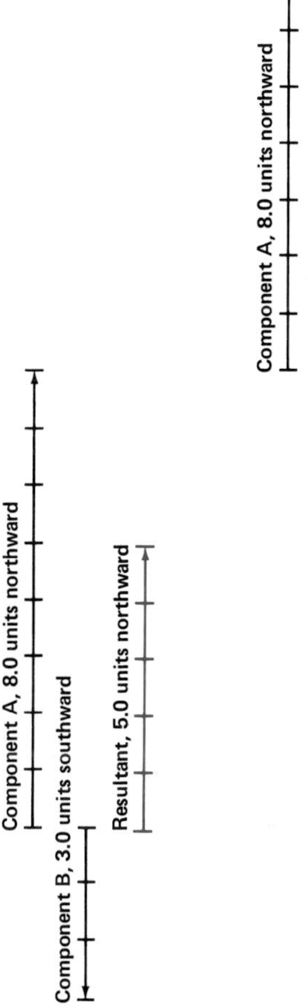

Figure 2-14. The magnitude of the resultant of two component vectors acting in opposite directions is the difference between the magnitudes of the components. The resultant has the same direction as the larger component.

the resultant is the algebraic sum of the component magnitudes, with the 3.0-unit vector considered as negative, since it is acting in the opposite direction. The resultant has the direction of the larger component, or northward.

A third case is shown in Figure 2-15. The 8.0-unit vector is again acting northward, but the 3.0-unit vector is now directed eastward. In this case, the components do not act along a straight line, and the resultant cannot be found by simple addition or subtraction. Instead, the *parallelogram method* is used. (Experiments confirm that the parallelogram method accurately represents the behavior of vectors that do not act along a straight line.) To find the resultant, we construct the parallelogram **ONRE.** The diagonal of this parallelogram, **OR,** is the resultant.

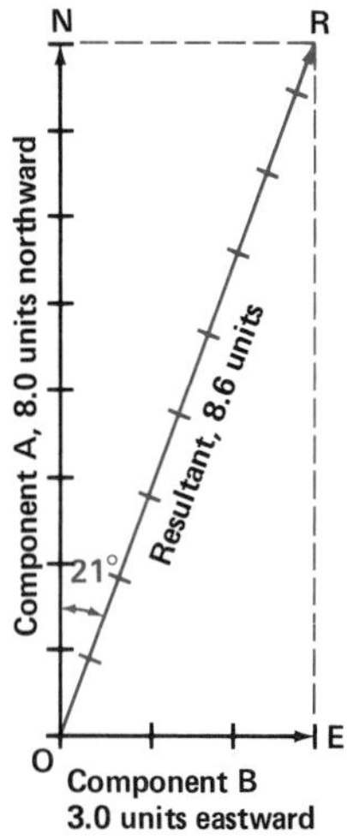

Figure 2-15. The resultant of two component vectors acting at right angles to each other is the diagonal of the parallelogram of which the components are the sides. The magnitude and direction of the resultant can be found either graphically or trigonometrically.

The magnitude and direction of **OR** can be found in either of two ways. If the vector diagram is drawn accurately to scale, **OR** can be measured directly with a ruler and a protractor. This is called the *graphic solution*. Of course, in using a ruler it is important to use the same scale in measuring each vector. The ruler and protractor should be read to the limits of their precision and the rules of significant figures should be followed in rounding the answer.

A second way of finding **OR** is to use trigonometry. To understand this method, you should review the Mathematics Refresher, 12. Triangles (in the back of the book). The use of the Pythagorean theorem is not recommended for finding the resultant, since it gives only the magnitude and not the direction. The direction of a vector can be calculated only by the use of trigonometric ratios. This method of finding resultants is called the *trigonometric solution*. Be sure to follow the rules of significant figures.

By either the graphic or trigonometric method, **OR** in Figure 2-15 has a magnitude of 8.6 units and a direction of 21° east of north. The direction could also be designated as 69° north of east, but in the answers to the vector problems in this book, the smaller of the two complementary angles will always be given.

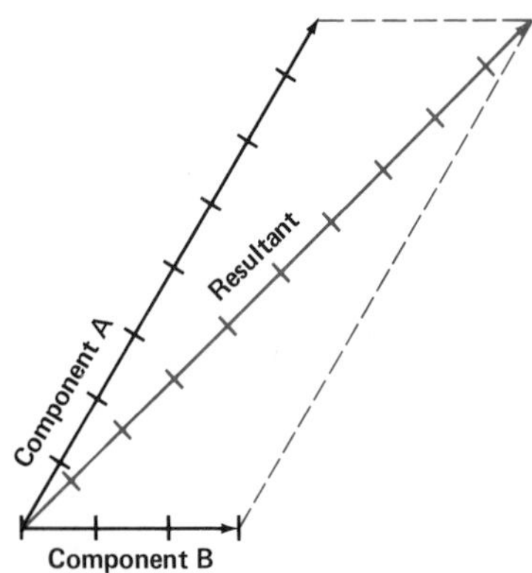

Figure 2-16. The parallelogram method can be used to find the resultant of two component vectors even though the components do not act perpendicularly to each other.

The parallelogram method can also be used when vectors act at angles other than right angles, as in Figure 2-16. The graphic solution is similar to the one in the previous example, but the trigonometric solution is lengthier, since it involves the law of cosines and law of sines. Examples and solutions of this kind are given in Chapter 3.

Vectors may also be combined by placing them head to tail, as shown in Figure 2-17. This method is especially helpful in finding the resultant of three or more vectors that act at the same point. If the parallelogram method were used in such a case, it would first be necessary to find the resultant of any two of the vectors, then use this resultant with a third vector to find a second resultant, and continue in this way until all the component vectors have been applied. The final resultant would then be the required answer. The procedure shown in Figure 2-17 is really a shortcut for the parallelogram method. In using this method, it does not matter in what sequence the arrows are drawn, as long as the direction of each arrow is maintained when it is transposed. After all the components have been transposed, the resultant is the arrow that is made by starting at the tail of the first component and drawing a straight line to the head of the last component.

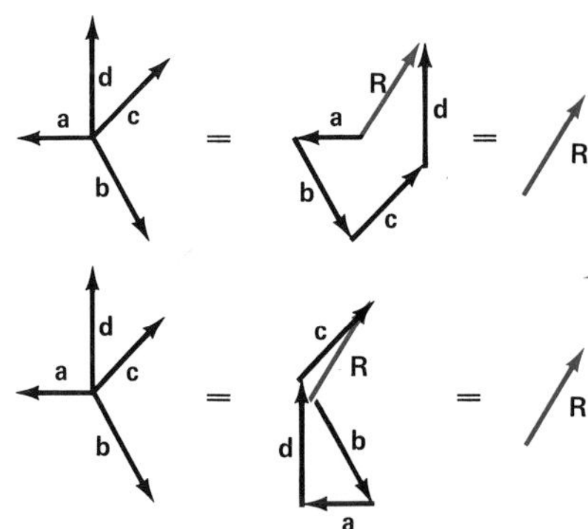

Figure 2-17. The resultant of component vectors acting at the same point can also be found by placing the components head to tail. The resultant is the vector drawn from the tail of the first component to the head of the last one. The sequence in which the components are transposed does not matter, but their directions must be maintained.

2.12 Rules of Problem Solving The problems in this text range from those that are quite simple and involve only a single computation to those that require several derivations and a number of separate mathematical steps. In general, problems of the first type will be found in Group A at the ends of the chapters. Examples of this type are given within the chapters themselves. Problems of greater difficulty involving several steps are included in Group B, and examples of these are also in the chapters.

The method used to solve physics problems consists of a number of logical steps. These steps are described below and they are illustrated in examples throughout the book.

1. Read the problem carefully and make sure that you know what is being asked and that you understand all the terms and symbols that are used in the problem. Write down all given data.

2. Write down the symbols for the physical quantity or quantities called for in the problem, together with the appropriate unit symbols.

3. Write down the equation relating the known and unknown quantities of the problem. This is called the *Basic equation.* In this step you will have to draw upon your understanding of the physical principles involved in the problem. It is helpful to draw a sketch of the problem and to label it with the given data.

In this book the symbols for quantities are always italicized. The symbols for units are not.

4. Solve the basic equation for the unknown quantity in the problem, expressing this quantity in terms of those given in the problem. This is called the *Working equation.*

5. Substitute the given data into the working equation. Be sure to use the proper units and carefully check the significant figures in this step.

6. Perform the indicated mathematical operations with the units alone to make sure that the answer will be in the

Even though your calculator or computer printout may give you an answer to many decimal places, do not forget to consider only the significant figures according to the measured quantities. Rounding off correctly is more important than using all available digits.

units called for in the problem. This process is called *unit analysis.* (The same procedure can also be used with the symbols for the physical quantities in a problem. In that case, the process is called *dimensional analysis.)*

7. Perform the indicated numerical operations. Be sure to observe the rules of significant figures.

8. Check the answer for reasonableness. For example, if your answer shows a car moving at a speed of 1000 km/h, you made a mistake in one of the previous steps.

9. Review the entire solution.

Some problems do not involve all of these steps and may require other skills instead.

EXAMPLE The mass of a rectangular metal block is 63.2 g. The block is 3.42 cm long, 1.09 cm wide, and 2.56 cm high. Calculate the mass density of the metal block.

Given	Unknown	Basic equations
$m = 63.2$ g $l = 3.42$ cm $w = 1.09$ cm $h = 2.56$ cm	D	$D = \frac{m}{V}$ $V = lwh$

Solution

To get a working equation for this problem, substitute the expression for V in the second Basic equation for V in the Basic equation for mass density.

$$\textit{Working equation: } D = \frac{m}{lwh}$$

Now substitute the given data into the Working equation:

$$D = \frac{63.2 \text{ g}}{(3.42 \text{ cm})(1.09 \text{ cm})(2.56 \text{ cm})}$$

$$\textit{Unit analysis: } D = \frac{\text{g}}{(\text{cm})(\text{cm})(\text{cm})} = \frac{\text{g}}{\text{cm}^3}$$

$$\textit{Numerical operations: } \frac{63.2}{(3.42)(1.09)(2.56)} = 6.62 \text{ (rounded to three figures)}$$

Finally, combine the numerical answer with the units obtained in the unit analysis, or $D = 6.62$ g/cm^3. This answer is reasonable for the mass density of a metal, when compared with the values given for aluminum in Section 2.10.

PRACTICE PROBLEMS 1. A metal cube has a mass of 452 g. The mass density of the metal is 11.3 g/cm^3. Find the length of one side of the metal cube. *Ans.* 3.42 cm

2. A liquid has a volume of 2.47 L and a mass of 3.78 kg. Determine the mass density of the liquid. *Ans.* 1.53 g/cm^3

QUESTIONS: GROUP A

1. (a) What is accuracy? (b) How is accuracy expressed? (c) What is the person in Figure 2-6 doing to improve the accuracy of the measurement?
2. Can a set of measurements be precise but not accurate? Explain.
3. What are significant figures?
4. (a) What is the purpose of a scientific experiment? (b) What does an experimenter do to ensure the most accurate measurements?
5. What is the primary purpose for drawing a graph of experimental data?
6. (a) What is the difference between extrapolation and interpolation? (b) Which one is more likely to give you an accurate estimate? Why?

GROUP B

7. Identify the following quantities as vectors or scalars. (a) 5 m/s east, (b) 6 kg, (c) 27°C, (d) 735 N down, (e) 17.8 J of energy.
8. A small airplane is flying at 50 m/s toward the east. A wind of 20 m/s toward the east suddenly begins to blow giving the plane a velocity of 70 m/s east. (a) Which of these are component vectors? (b) Which is the resultant? (c) What is the "magnitude" of the wind velocity? (d) What would be the resultant velocity if the wind had been 20 m/s toward the west?
9. What are the two methods for finding the resultant to two vectors that don't act in the same direction?
10. A physics student is dropping an object from the top of the lab table to the floor. She makes a calculation and writes in her data chart that the object fell a distance of 97 m. Which of her problem-solving step(s) does she need to check?
11. Give an example of the importance of using units in describing physical quantities.
12. You are told to make a measurement using an ordinary meterstick. How many significant figures past the decimal point could you be expected to have in a measurement made with this meterstick?

PROBLEMS: GROUP A

1. Each of the following measurements listed below has been made and recorded as accurately as possible. How many significant figures are there in each measurement?
   (a) 300 000 000 m/s
   (b) 25.030°C
   (c) 0.006 070°C
   (d) 1.004 g/cm^3
   (e) 9.109 534 × 10^{-31} kg
2. For each of the measurements given in Problem 1, express it in scientific notation. In each case, be sure to retain the correct number of significant digits.
3. What is the order of magnitude for each of the measurements in Problem 2?
4. Using the graph in Figure 2-11, answer the following questions.

(a) What part of the graph represents the density of the material? (b) Which material shown would you expect to be able to float on water? How does the graph show this? (c) What is the mass of 55 cm^3 of Al? (d) Find the mass density of maple wood from the graph.

5. Using Figure 2-12, answer the following questions: (a) How long will the 100-km trip take traveling at 55 km/h? (b) If you wish to make the trip in 2.5 h, how fast must you travel? (c) Extrapolate to find the time the trip will take at a speed of 15 km/h.

6. Graphically add a vector of 6.00 units north to one 5.00 units east. (b) Add the same vectors trigonometrically.

7. Calculate the mass density of the standard kilogram cylinders shown in Figure 2-4.

8. (a) What is the mass of the air in a physics classroom that is 8.0 m wide, 10.0 m long, and 2.8 m high? (The mass density of air is 1.29 g/dm^3.) (b) If the mass density of lead is 11.4 g/cm^3, what volume would be occupied by the same mass of lead?

9. Set up working equations for each of the following basic equations:
(a) $F = ma$, solve for a
(b) $E = mc^2$, solve for m
(c) $V_{min} = \sqrt{rg}$, solve for g
(d) $F_c = mv^2/r$, solve for v
(e) $\frac{pV}{T_K} = \frac{p'V'}{T'_K}$, solve for T_K

10. Add the following vectors graphically by the head-to-tail method: 5.0 units north, 4.0 units northeast, and 3.0 units 30.0° south of east.

11. You move 26 m at an angle of 40.0° west of south. (a) How far south of your starting point are you? (b) How far west are you?

GROUP B

12. A student measures the mass of a rubber stopper on a balance for three trials as 14.25 g, 14.51 g, and 13.95 g. The mass of the stopper is stamped as 14.03 g. (a) What is the absolute deviation for trial one? (b) What is the average absolute deviation for all three trials? (c) What is the average absolute error for all three trials? (d) Calculate the average percent deviation for this experiment. (e) Which trial was the most accurate?

13.

Temperature (°C)	Vapor pressure of water (mm Hg)
0.00	4.60
10.0	9.20
20.0	17.50
30.0	31.80
40.0	55.30
60.0	149.00

(a) Graph the data using the temperature values as abscissas and the pressures as ordinates. (b) Describe the shape of the graph. (c) Does the graph show an inverse or direct relationship? (d) From your graph, predict the vapor pressure at 35.0°C.

SUMMARY

The metric system of measurement is a decimal system well suited to scientific purposes. The United States is in the process of converting to the metric system.

Physical quantities are aspects of the universe. They are measured in units. All units of measure are derived from seven fundamental units that are used to measure length, mass, time, electric charge, temperature, amount of substance, and luminous intensity. Many other units are derived from the fundamental units. The names and symbols for physical quantities and units have been standardized in the International System of Units (abbreviated SI).

The SI standard unit of length is the meter. It is based on the distance that light travels in a specified period of time. The standard unit of mass is the kilogram, which is based on the mass of a cylinder kept in Sèvres, France. The unit of force is the newton, which is a derived unit. The unit of time is the second, which is based on the vibrations of cesium-133 atoms.

All measurements are uncertain to some degree. The closeness of a measurement to an accepted value is called accuracy. It is expressed as absolute error or relative (percentage) error. Precision is the agreement among measurements that have been made in the same way. Precision is expressed as absolute deviation, relative deviation, or tolerance. It may also be shown by significant figures, which are all the figures of a measurement that are known with certainty plus the first figure that is uncertain. Specific rules govern operations with significant figures.

Scientific notation is a convenient way of writing very large and very small numbers. Such numbers have the form $M \times 10^n$, in which M is a number with one nonzero figure to the left of the decimal point and n is a positive or negative exponent. The order of magnitude of a number is the approximation of the number to the nearest power of ten.

Data is the term used for the numbers, units, and other information obtained in experiments. Tables, graphs, and equations are used to express the relationships among data. A straight-line graph illustrates a direct proportion. A hyperbola is indicative of an inverse proportion, in which the product of two quantities is a constant.

A scalar is a quantity that needs only a number with appropriate units to describe it. A quantity that requires magnitude and direction for its description is a vector quantity. It is usually represented by an arrow in which the length represents magnitude and the orientation of the arrow specifies direction. Vectors in the same or opposite direction can be combined by simple addition or subtraction. The resultant of vectors at other angles is found by the parallelogram, head-to-tail, or trigonometric methods.

The solving of a physics problem involves a number of logical steps. Among these are the use of a basic equation, which relates the known and unknown quantities of the problem, and a working equation, which expresses the unknown quantity in terms of the known ones. Another important step is unit analysis, in which the mathematical operations are performed with only the units.

VOCABULARY

absolute deviation
absolute error
accuracy
basic equation
component
derived unit
dimensional analysis
direct proportion
fundamental unit
inverse proportion
kilogram
length
liter
meter
metric system
newton
order of magnitude
parallelogram method
physical quantity
precision
relative deviation
relative error
scalar quantity
scientific notation
second
significant figure
tolerance
unit analysis
vector quantity
working equation

3

Velocity and Acceleration

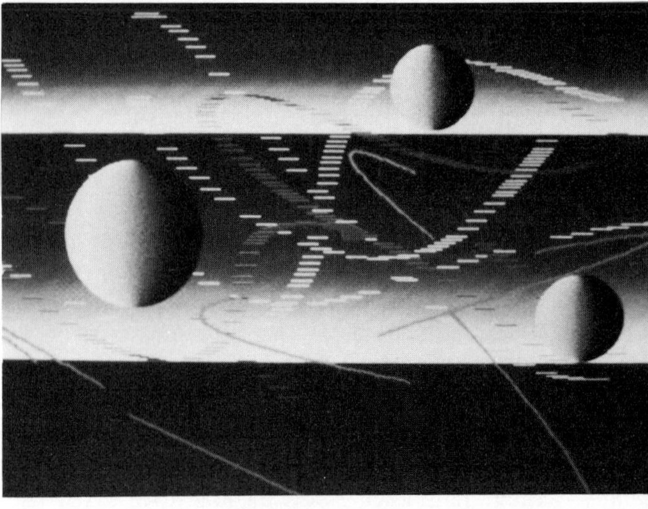

motion (MO-shun) n.: the displacement of an object in relation to objects that are considered to be stationary.

OBJECTIVES

- Define displacement, speed, velocity, and acceleration.
- Solve problems dealing with accelerated motion.
- Discuss and apply Newton's three laws of motion.
- Discuss and apply Newton's law of universal gravitation.

VELOCITY

3.1 The Nature of Motion If you see a car in front of your house and later see it farther along the street, you are correct in assuming that the car has moved. To reach this conclusion, you observed two positions of the car and you also noted the passage of time. You might not know how the car got from one position to the other. It might have moved at a steady rate or it might have speeded up and then slowed down before it got to its second position. It might also have been moving when you first noticed it and might still have been moving at its second location. The car may have moved from one location to the other in a straight line, or it may not have. But none of these possibilities changes the truth of the statement that the car has moved.

Usually, when we see something move, we do not just make two observations. Unless we are only concerned with the beginning and end of an object's motion (for example, in determining average speed), we usually observe the object moving continuously. However, even a case of continuous observations can be broken down into a series of shorter observations for any two points along an object's trajectory.

Figure 3-1 shows several important things about the nature of motion. First, it shows the changing positions of a moving object. Such *a change of position in a particular direction is called a* ***displacement.*** A displacement has magni-

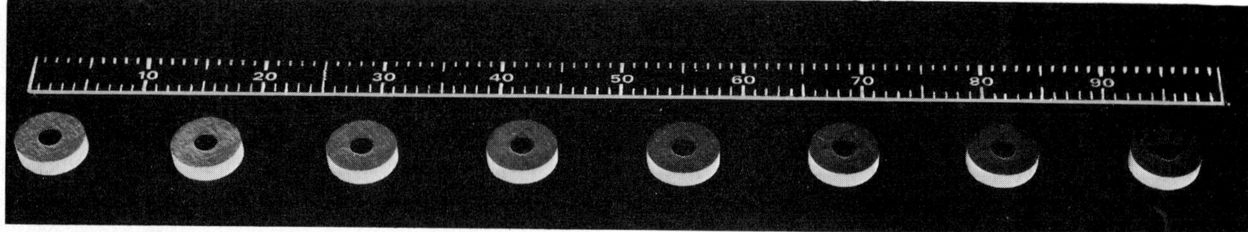

Figure 3-1. A study of motion. The disk was photographed every 0.1 s as it moved from left to right along the meterstick. Note that the displacement of the disk is equal in equal time periods.

tude and direction and is, therefore, a vector quantity. A displacement is always a straight line segment from one point to another point, even though the actual path of the moving object between the two points is curved. Displacement vectors can be combined like other vector quantities.

For example, the statement that an airplane flew 500 kilometers south describes the airplane's displacement, since the statement specifies the magnitude and direction of the airplane's motion. Whether the airplane flew in a straight line or not, the airplane's displacement vector is a straight arrow directed south and representing a magnitude of 500 kilometers.

Displacement is a change of position.

Motion is relative displacement.

Speed is the time rate of motion.

Figure 3-1 also shows that relative motion can be detected by comparing the displacement of one object with respect to another. The reference object is usually considered to be stationary. Thus in the case of the moving car, the street and houses are the stationary references. So ***motion*** *may be defined as the displacement of an object in relation to objects that are considered to be stationary.*

This chapter deals with the simplest type of motion—the motion of an object along a straight line. Motion in a curved line will be discussed in Chapter 5.

3.2 Speed *The time rate of motion is called* ***speed.*** For example, if you are walking at a speed of 3.0 m/s, in 1.0 second you travel 3.0 meters. In 2.0 seconds, you travel 6.0 meters, and in 3.0 seconds you travel 9.0 meters. A graph of this motion, and of several other motions with constant speeds, is shown in Figure 3-2. Notice that the lines that represent greater speeds are more steeply inclined. The *slope* of a line indicates how steep the line is. The slope is the ratio of the vertical component of a line to the corresponding horizontal component. For example, the blue segment in line **A** has a vertical component of 3.0 units and a horizontal component of 1.0 unit. The slope of the line is $\frac{3.0}{1.0} = 3.0$.

This definition of the slope of a line on a graph may be used to represent a physical quantity such as speed. If the

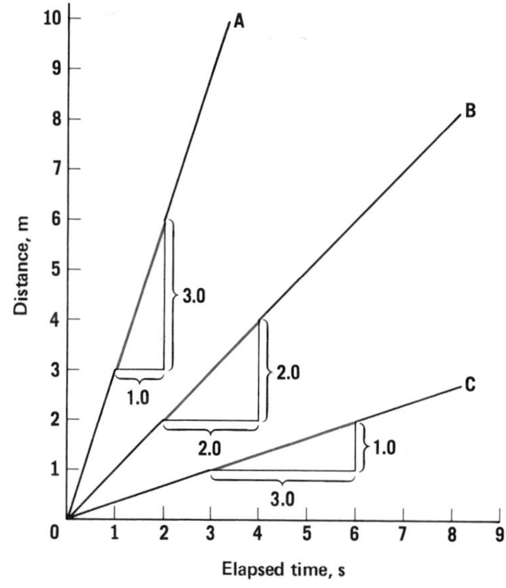

Figure 3-2. Three graphs of motion with different constant speeds. A is for 3.0 m/s, B is for 1.0 m/s, and C is for 0.33 m/s. In each case, speed is represented by the slope of the line.

vertical axis represents distance and the horizontal axis represents elapsed time, then the slope of the line represents speed. For each of the speeds graphed in Figure 3-2 the data for computing the slope are given. A graph such as this and a single computation of the slope will determine the speed over the entire motion only if the speed is constant. *If the determination of speed over any randomly chosen interval gives the same value, the* ***speed*** *is* ***constant.*** The graph of constant speed is a straight line.

Average speed *is found by dividing the total distance by the elapsed time.* For example, if you run 112 m in 20.0 s, you might not run at constant speed. You might speed up and slow down. But as long as you move a total distance of 112 m in 20.0 s, your average speed is $\frac{112 \text{ m}}{20.0 \text{ s}} = 5.60$ m/s.

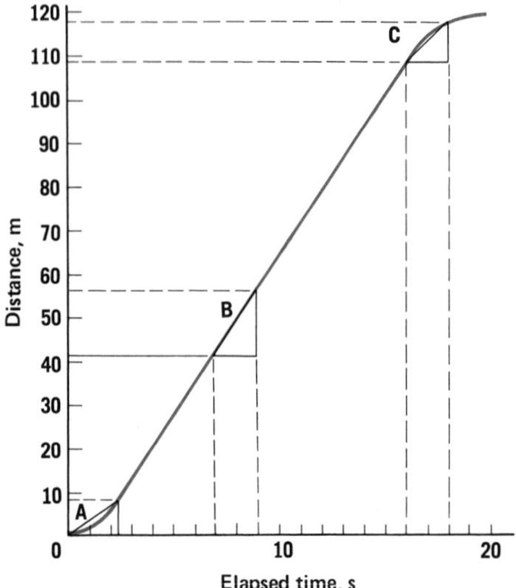

Figure 3-3. A graph of motion with variable speed.

Suppose you ride a bicycle 120 m in $2\bar{0}$ s. Your average speed is $\frac{120 \text{ m}}{2\bar{0} \text{ s}} = 6.0$ m/s. A graph of your speed might look like Figure 3-3. You start at a speed of 0.0 m/s and increase your speed. Then from about 3 s to 15 s you travel at constant speed. Then at about 16 s you begin to tire and slow down until at $2\bar{0}$ s you stop. ***Instantaneous speed*** *is the slope of the line that is tangent to the curve at a given point.* Without a speedometer it is difficult to determine your instantaneous speed. It is the speed that you would have if your speed did not change from that point on.

Instantaneous speed can be measured approximately by timing the motion over a short distance. In Figure 3-3 instantaneous speed is found approximately by dividing a small amount of time, measured on the graph, into the corresponding distance for that time. The region of the curve marked **A** represents your change in speed as you move faster. An interval of time of 2.3 s is indicated on the horizontal axis. The corresponding distance for this time interval, approximated to the nearest 0.1 m, is 9.0 m. The slope of the blue line changes over the interval **A,** but the average is $\frac{9.0 \text{ m}}{2.3 \text{ s}} = 3.9$ m/s. This method gives only an approximation of the instantaneous speed because we found the *average speed* for the time interval indicated. This value for the average speed is the same as the instantaneous speed (the slope of the tangent line) at some point in region **A,** but we cannot precisely determine that point.

Since the graph of your motion is a straight line in the region **B,** it represents constant speed. The elapsed time in region **B** is 9.0 s − 7.0 s = 2.0 s. The corresponding distance is 56 m − 41 m = 15 m. The instantaneous speed in region **B** can be determined precisely because it is the same as the average speed. It is $\frac{15 \text{ m}}{2.0 \text{ s}} = 7.5 \text{ m/s}$.

Region **C** on the graph is curved. The elapsed time is 18.0 s − 16.0 s = 2.0 s. The corresponding distance is 118.0 m − 109.0 m = 9.0 m. The average speed over the interval **C** (as an approximation of the instantaneous speed) is $\frac{9.0 \text{ m}}{2.0 \text{ s}} = 4.5 \text{ m/s}$.

Both instantaneous speed and average speed are completely described in terms of magnitude alone. Hence speed is a scalar quantity.

3.3 Velocity When both speed and direction are specified for the motion, the term *velocity* is used. In other words, ***velocity** is speed in a particular direction.* A person walking eastward does not have the same velocity as a person walking northward, even though their speeds are the same. Two persons also have different velocities if they walk in the same direction at different speeds. As stated in Section 2.11, velocities are vector quantities. Magnitude and direction are necessary to describe velocities.

Velocity is speed in a given direction.

For the problems in this chapter, in which the motion is along a straight line, the word "velocity" and the symbol for velocity will be used without stating the direction angle. The values given for velocity will indicate only the magnitude and not the direction. In these cases the velocity may be either a positive or a negative number to indicate in which direction the object is moving. A negative sign as part of velocity, therefore, does not mean that the speed is negative; speed always represents the magnitude of the velocity. Hence, when the direction angle of velocity is not given, the equations for velocity and speed appear the same.

***Average velocity** is defined as the total displacement divided by the total elapsed time.*

$$v_{\mathrm{av}} = \frac{d_{\mathrm{f}} - d_{\mathrm{i}}}{t_{\mathrm{f}} - t_{\mathrm{i}}}$$

where v_{av} is the average velocity, d_{i} is the initial position at time t_{i}, and d_{f} is the final position at time t_{f}. This relationship may also be written

$$v_{\mathrm{av}} = \frac{\Delta d}{\Delta t}$$

The Greek letter Δ in an equation means "change of."

where Δd represents the *change* of position (displacement) during the time interval Δt. The Greek letter Δ (delta) designates a change in the quantity of variables d and t.

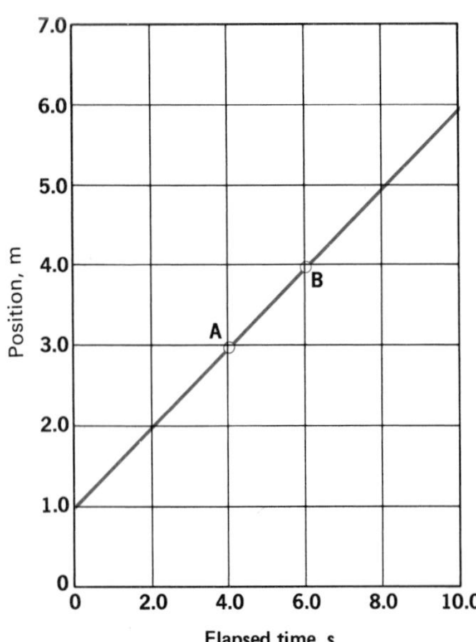

Figure 3-4. A graph of motion with constant velocity.

Graphic representations of the velocity of motion along a straight line are shown in Figure 3-4 and Figure 3-5. These graphs indicate that the direction of motion does not change in either case because the magnitude of displacement is continuously increasing. Furthermore, since the graphs do not show the direction of motion, they have the same form as the speed graphs in Figure 3-2 and Figure 3-3. In Figure 3-4, Δt for the entire range of the graph is $1\bar{0}$ s and Δd is 5.0 m. Consequently, $v_{av} = \frac{5.0 \text{ m}}{1\bar{0} \text{ s}}$, or 0.50 m/s. The same value for v_{av} is obtained by choosing any other two points on the graph, such as **A** and **B**. In this case, Δt is 2.0 s and Δd is 1.0 m and v_{av} again equals 0.50 m/s. The velocity, represented by the slope of the straight line, is *constant*. The motion of the object is uniform.

In Figure 3-5, the velocity is *variable*. The average velocity for the entire range of the graph $v_{av} = \frac{5.0 \text{ m}}{1\bar{0} \text{ s}}$, or 0.50 m/s, as before. However, the average velocity between points **C** and **D** is $\frac{2.0 \text{ m}}{2.0 \text{ s}}$, or 1.0 m/s.

To find the instantaneous velocity at any point on the graph in Figure 3-5, we use the same method that was used to find instantaneous speed in Section 3.2. For example, the instantaneous velocity at point **C** is found by drawing a tangent to the curve at **C** and measuring the slope of the tangent. The slope is 0.5 m/s, the instantaneous velocity at **C**.

The reason that the slope method can be used to measure instantaneous velocity at a specific point on a curved line is that a very small segment of a curve is almost identical to a straight line segment. This becomes apparent when you look closely at the area around point **C** in Figure 3-5 and imagine it to be magnified several times. The blue and black line segments near **C** would then virtually coincide.

Instantaneous velocity *is the velocity that a moving body would have if it did not speed up, slow down, or change direction from that point on.* The speedometer of a car shows instantaneous speed. But since a speedometer does not indicate direction, it is not a velocity meter. To determine velocity, a speedometer must be used in conjunction with a compass.

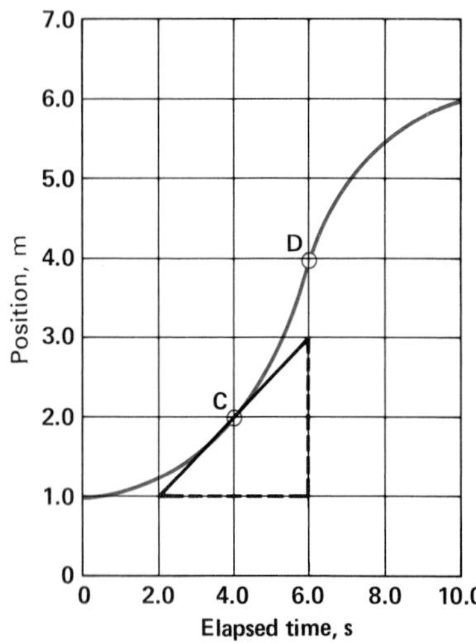

Figure 3-5. A graph of motion with variable velocity.

3.4 Solving Velocity Vector Problems As vector quantities, velocities can be combined by the parallelogram method discussed in Section 2.11. The following examples show how to solve velocity problems of this kind.

EXAMPLE A pitcher can throw a ball with a velocity of 125 km/h relative to the ground in still air. If he throws the ball with that velocity when a crosswind of 28 km/h is blowing from the left, what is the velocity of the ball with respect to the ground?

Given	Unknown	Basic equations
$v_B = 125$ km/h	θ	$\tan\theta = \dfrac{v_W}{v_B}$
$v_W = 28$ km/h	v_R	$\sin\theta = \dfrac{v_W}{v_R}$

Solution

Since this is a vector problem, the parallelogram method described in Section 2.11 can be used to find the answer. Both the graphic and trigonometric solutions are shown below.

Graphic solution. Refer to Figure 3-6. Using an appropriate scale (such as 10 mm = 1.0 km/h), draw the velocity vector v_B to represent the velocity of the ball in still air. Then draw the velocity vector v_W to the same scale at a right angle to v_B to represent the velocity of the crosswind. Complete the parallelogram formed by these two vectors, and draw the vector v_R. This represents the resultant velocity.

Measure the length of v_R and convert it to km/h. The answer should be 130 km/h. Measure θ with a protractor. The answer should be 12° clockwise from v_B.

Trigonometric solution. Use the Basic equations above as follows:

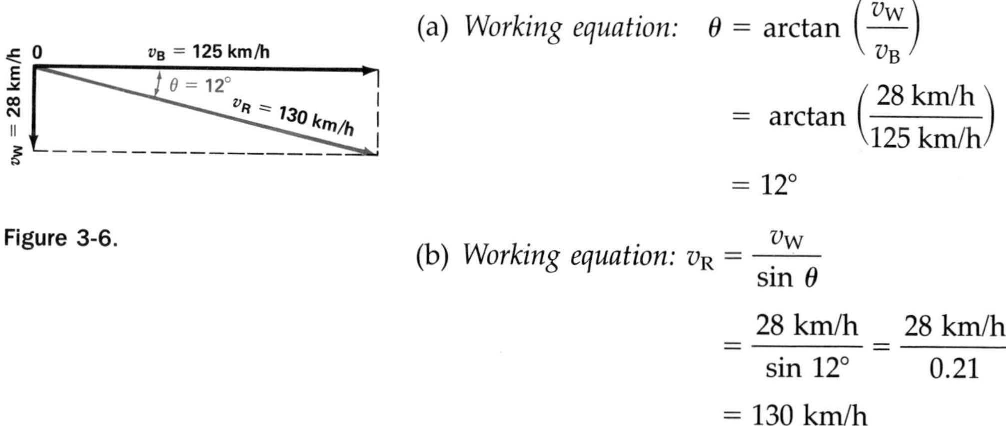

Figure 3-6.

(a) *Working equation:* $\theta = \arctan\left(\dfrac{v_W}{v_B}\right)$

$$= \arctan\left(\frac{28\text{ km/h}}{125\text{ km/h}}\right)$$

$$= 12°$$

(b) *Working equation:* $v_R = \dfrac{v_W}{\sin\theta}$

$$= \frac{28\text{ km/h}}{\sin 12°} = \frac{28\text{ km/h}}{0.21}$$

$$= 130\text{ km/h}$$

EXAMPLE A pilot wants to fly a plane at a velocity of 500.0 km/h eastward with respect to the ground. A wind is blowing southward at a velocity of 75.0 km/h. What velocity must the pilot maintain with respect to the air to achieve the desired ground velocity?

Given	Unknown	Basic equations
$v_R = 500.0$ km/h	θ	$\tan\theta = \dfrac{v_W}{v_R}$
$v_W = 75.0$ km/h	v_P	$\sin\theta = \dfrac{v_W}{v_P}$

Solution

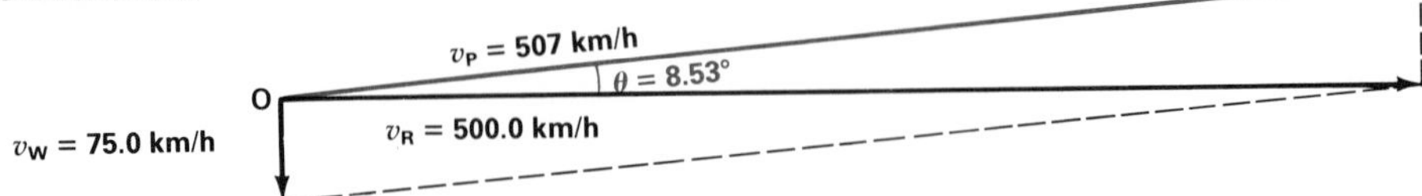

Figure 3-7.

The vector diagram for this problem is shown in Figure 3-7. The desired velocity, v_R, is the diagonal of the parallelogram formed by the vectors v_W (the velocity of the wind) and v_P (the velocity needed to keep the plane at 500.0 km/h). The trigonometric solution is shown below:

(a) *Working equation:* $\theta = \arctan\left(\dfrac{v_W}{v_R}\right) = \arctan\left(\dfrac{75.0\text{ km/h}}{500.0\text{ km/h}}\right)$

$$= 8.53^\circ \text{ north of east}$$

(b) *Working equation:* $v_P = \dfrac{v_W}{\sin\theta} = \dfrac{75.0\text{ km/h}}{\sin 8.53^\circ}$

$$= 507\text{ km/h}$$

EXAMPLE The velocity of an airplane with respect to the air is 425 km/h in a direction 45.0° north of east. A wind is blowing northward at 75.0 km/h. What is the resultant velocity of the airplane?

Given	Unknown	Basic equations
$v_P = 425$ km/h	v_R	$v_R = \sqrt{v_W^2 + v_P^2 - 2v_Wv_P\cos\theta'}$
$v_W = 75.0$ km/h	θ	$\dfrac{v_W}{\sin\theta} = \dfrac{v_R}{\sin\theta'}$

Solution

The vector diagram for this problem is shown in Figure 3-8. Since there are no right angles in the problem, the laws of sines and cosines must be used for the trigonometric solution, as follows:

(a) $\theta' = \dfrac{[(360.0° - 2(45.0°)]}{2} = 135.0°$

Working equation:

$$v_R = \sqrt{v_W^2 + v_P^2 - 2v_Wv_P \cos \theta'}$$
$$= \sqrt{(75.0 \text{ km/h})^2 + (425 \text{ km/h})^2 - 2(75.0 \text{ km/h})(425 \text{ km/h})(\cos 135.0°)}$$
$$= 481 \text{ km/h}$$

(b) *Working equation:*

$$\sin \theta = \frac{v_W \sin \theta'}{v_R}$$

$$\theta = \arcsin \left[\frac{(75.0 \text{ km/h})(\sin 135.0°)}{(481 \text{ km/h})} \right]$$
$$= 6.33°$$

Hence, the resultant velocity is 481 km/h in a direction 45.0° + 6.33°, or 51.3°, north of east.

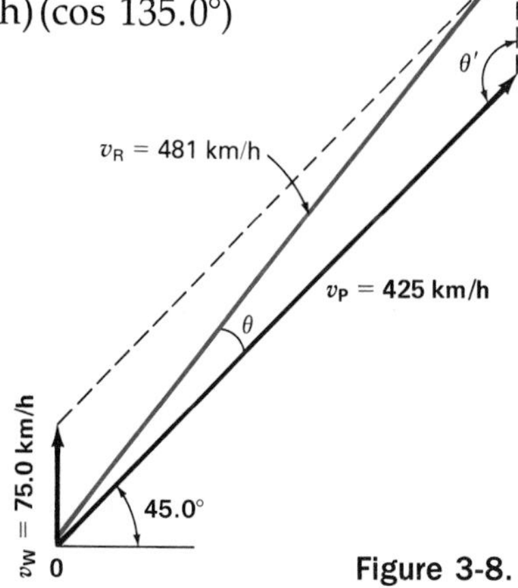

Figure 3-8.

QUESTIONS: GROUP A

1. Explain the statement "All motion is relative."
2. (a) What is a reference object? (b) When we describe motion, what is a common reference object?
3. On a distance-versus-time graph, what represents the speed?
4. What is the shape of a distance-versus-time graph for an object moving at constant speed?
5. What is the difference between average speed and instantaneous speed?
6. How are instantaneous speed and average speed related for an object with constant speed?

GROUP B

7. What is the relationship between speed and velocity?
8. Answer the following by referring to Figure 3-5. (a) Over which interval did the object have its greatest velocity? (b) Over which interval was the object's velocity decreasing? (c) Over which interval was its speed increasing?
9. Figures 3-4 and 3-5 show two objects in motion. Compare the distance each traveled. Compare their displacements.
10. Sketch a graph of motion (distance as a function of time) that would correspond to the following bicycle trip. You start your bicycle from rest and increase your speed until you reach a comfortable traveling speed. You continue at this speed for a few minutes and complete your trip by slowing down uniformly to a stop.

PROBLEMS: GROUP A

1. If you live 10.0 km from your school and it takes the school bus 0.53 h to reach the school, what is the average speed of the bus?
2. Determine the average speed of the disk in Figure 3-1.

3. A small plane is traveling toward the east at 155 km/h. It encounters a wind blowing toward the east (a tail wind) of 35 km/h. What is the velocity of the plane relative to an observer on the ground?
4. The pilot of a plane measures an air speed of 165 km/h north. An observer on the ground sees the plane pass overhead at a velocity of 145 km/h toward the north. What is the velocity of the wind that is affecting the plane?

GROUP B

5. A motorboat heads due east at 12 m/s across a river that flows toward the south with a current of 3.5 m/s. (a) What is the resultant velocity relative to an observer on the shore? (b) If the river is 1360 m wide, how long does it take the boat to cross? (c) How far downstream is the boat when it reaches the other side?
6. Find the air velocity of a plane that must have a relative ground velocity of 250.0 km/h north if it encounters a wind pushing it toward the northeast at 75.0 km/h.
7. A swimmer can swim in still water at a speed of 9.50 m/s. He intends to swim directly across a river that has a downstream current of 3.75 m/s. (a) What must his heading be? (b) What is his velocity relative to the bank?

ACCELERATION

Acceleration is a change in velocity over time. It can be positive or negative.

3.5 The Nature of Acceleration The gas pedal of a car is sometimes called the accelerator. When you drive a car along a level highway and increase the pressure on the accelerator, the speed of the car increases. We say that the car is *accelerating*. When you apply the brakes of the car that you are driving, the speed of the car decreases. The speedometer indicates that the car's speed is decreasing, and we say that the car is *decelerating*.

***Acceleration** is defined as the time rate of change of velocity.* An automobile that goes from 0.0 km/h to $6\bar{0}$ km/h in $1\bar{0}$ seconds has a greater acceleration than a car that goes from 0.0 km/h to $6\bar{0}$ km/h in 15 seconds. The first car has an average acceleration of $\dfrac{6\bar{0}\text{ km/h}}{1\bar{0}\text{ s}} = 6.0\text{ (km/h)/s}$. The second car has an average acceleration of 4.0 (km/h)/s.

Acceleration, like velocity, is a vector quantity. In the case of a car moving along a straight road, acceleration can be represented by a vector in the direction the car is moving. Deceleration occurs in the opposite direction. For the cases in which the motion is known to be along a straight line, the direction of the acceleration is described as positive or negative. In Section 3.3 we saw that the equations for speed and velocity appear the same, since the equations do not specify the direction of displacement. For the same reason, the equations for rate of change of speed and acceleration appear the same. These equations involve only the magnitude and not the direction of the motion.

Average acceleration is found by the relationship

$$a_{av} = \frac{v_f - v_i}{t_f - t_i}$$

where a_{av} is the average acceleration, v_i is the initial velocity at time t_i, and v_f is the final velocity at time t_f. This relationship may also be stated

$$a_{av} = \frac{\Delta v}{\Delta t}$$

where Δv is the change in velocity for time interval Δt.

Figures 3-9 and 3-10 are velocity-time graphs for accelerated motion. In Figure 3-9 the velocity curve is a straight line. Hence the acceleration does not change. This is an illustration of *constant acceleration.*

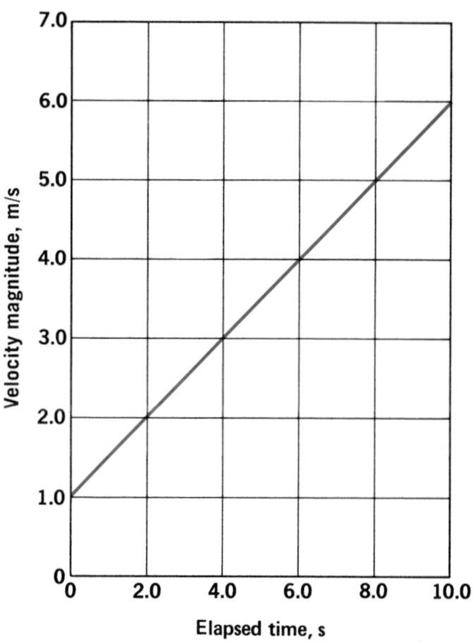

Figure 3-9. A graph of velocity versus time for constant acceleration.

The motion illustrated in Figure 3-10 is an example of *variable acceleration.* As we mentioned in the case of finding instantaneous speed or instantaneous velocity, finding the average acceleration between two points on the curve only approximates the instantaneous acceleration. Instantaneous acceleration is equal to the slope of the line that is tangent to the curve at the given point.

Notice that the slope of the curve in Figure 3-10 is not always positive. That is, the velocity does not increase during all time intervals. For the interval between points **A** and **B**, Δt is 3.0 s and Δv is -2.0 m/s. Consequently,

$$a_{av} = \frac{-2.0 \text{ m/s}}{3.0 \text{ s}} = -0.67\frac{\text{m/s}}{\text{s}}$$

This reduction in speed is called deceleration, or negative acceleration.

The dimensions of acceleration should be carefully examined. Acceleration is the rate of change of the rate of displacement, or $(\Delta d/\Delta t)/\Delta t$. If Δd is in meters and Δt is in seconds, acceleration is given in m/s/s, or m/s^2.

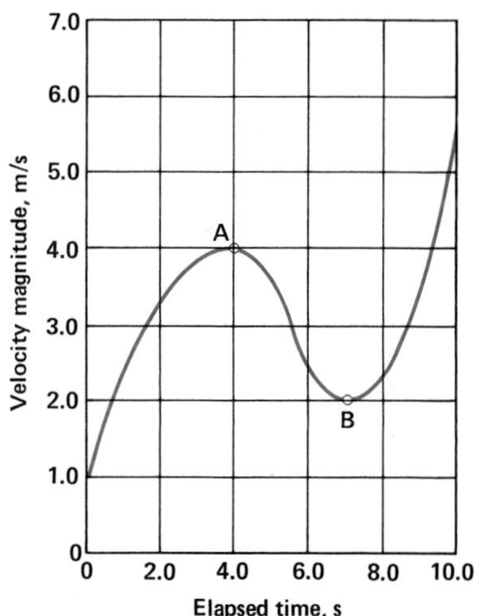

Figure 3-10. A graph of velocity versus time for motion with variable acceleration.

3.6 Solving Acceleration Problems Galileo Galilei (1564–1642) was the first scientist to understand clearly the concept of acceleration. He found that the acceleration of a ball rolling down an inclined plane and the acceleration of a falling object were both examples of the same natural phenomenon and were described by the same mathematical rules.

Let us consider one of Galileo's experiments. One end of a board is raised so that it is just steep enough to allow a ball to roll from rest to the 0.5-m position in the first second. The apparatus is shown in Figure 3-11. As the ball rolls down the board, its position is shown at one-second

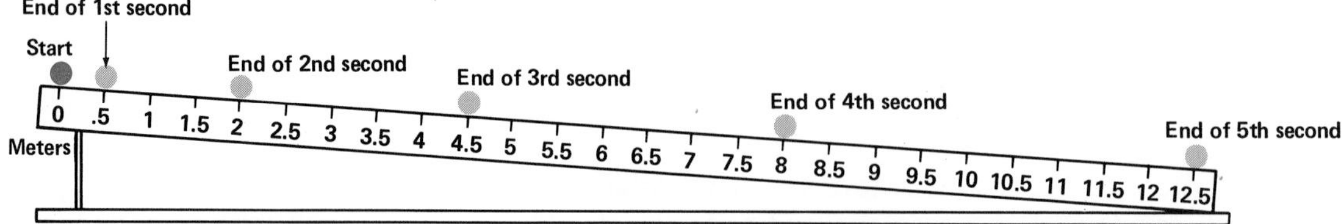

Figure 3-11. The ball rolls down the plane with constant acceleration.

intervals. The data obtained are shown in the first six columns of Table 3-1.

The data in the last four columns of Table 3-1 show that the velocity at the end of each second is 1.0 m/s greater than at the end of the previous second. Thus the acceleration is 1.0 (m/s)/s, or 1.0 m/s^2. The motion of a ball rolling down an inclined plane is an example of constant acceleration.

Table 3-1
CONSTANT ACCELERATION

Interval number	*Time at start of interval* (s)	*Time, t, at end of interval* (s)	*Position at start of interval* (m)	*Position at end of interval* (m)	*Distance, Δd, traveled in interval* (m)	*Average velocity, v_{av}, during interval* (m/s)	*Instantaneous velocity, v_i, at start of interval* (m/s)	*Instantaneous velocity, v_f, at end of interval* (m/s)	*Average acceleration, a_{av}, during interval* (m/s^2)
1	0.0	1.0	0.0	0.5	0.5	0.5	0.0	1.0	1.0
2	1.0	2.0	0.5	2.0	1.5	1.5	1.0	2.0	1.0
3	2.0	3.0	2.0	4.5	2.5	2.5	2.0	3.0	1.0
4	3.0	4.0	4.5	8.0	3.5	3.5	3.0	4.0	1.0
5	4.0	5.0	8.0	12.5	4.5	4.5	4.0	5.0	1.0

Galileo's experiment verified the property of inertia.

The initial velocity of the ball is zero. Its acceleration, or rate of gain in velocity, is 1.0 m/s^2. Therefore its velocity at the end of the *first second* is 1.0 m/s. (If the ball at that instant were no longer accelerated, it would continue to move with a constant velocity of 1.0 m/s.) At the end of the next second, its velocity is 2.0 m/s. (2.0 s × 1.0 m/s^2 = 2.0 m/s.) Similarly the ball's velocity at the end of the fifth second is 5.0 m/s. These data show that *the final velocity of an object starting from rest and accelerating at a constant rate equals the product of the acceleration and the elapsed time. If an object does not start from rest, its final velocity will equal the sum of its initial velocity and the increase in velocity produced by the acceleration.*

Galileo was among the first scientists to use experiments in testing hypotheses.

We may arrive at this same conclusion by mathematical reasoning. The equation defining acceleration is

$$a_{av} = \frac{v_f - v_i}{t_f - t_i}$$

For constant acceleration a_{av} becomes a. Then assuming that v_i is the velocity in the direction of the acceleration when $t_i = 0$ and t_f is elapsed time Δt,

$$a = \frac{v_f - v_i}{\Delta t}$$

Multiplying both sides of the equation by Δt,

$$a\,\Delta t = v_f - v_i$$

Isolating the term v_f yields

$$v_f = v_i + a\,\Delta t \qquad \text{(Equation 1)}$$

If the object starts from rest, $v_i = 0$

and $$v_f = a\,\Delta t$$

In Section 3.3 we saw that

$$v_{av} = \frac{\Delta d}{\Delta t}$$

or $$\Delta d = v_{av}\,\Delta t \qquad \text{(Equation 2)}$$

For constant acceleration, the average velocity for any given interval of time can be found by taking one-half the sum of the initial and final velocities:

$$v_{av} = \frac{v_i + v_f}{2} \qquad \text{(Equation 3)}$$

By substituting the value of v_f (obtained in Equation 1) in Equation 3,

$$v_{av} = \frac{v_i + (v_i + a\,\Delta t)}{2} \qquad \text{(Equation 4)}$$

Now, substituting the value of v_{av} (obtained in Equation 4) in Equation 2 gives

$$\Delta d = v_i\,\Delta t + \tfrac{1}{2}a\,\Delta t^2 \qquad \text{(Equation 5)}$$

If the object starts from rest,

$$v_i \Delta t = 0$$

and $$\Delta d = \tfrac{1}{2}a\,\Delta t^2$$

From this equation we see that if an object starts from rest and travels with constant acceleration, the displacement of the object during the first second is numerically equal to one-half the acceleration.

Now let us solve Equation 1 for Δt, and substitute this value of Δt in Equation 5. From Equation 1,

$$\Delta t = \frac{v_f - v_i}{a}$$

The Mathematics Refresher may come in handy here.

Substituting in Equation 5,

$$\Delta d = v_i\left(\frac{v_f - v_i}{a}\right) + \tfrac{1}{2}a\left(\frac{v_f - v_i}{a}\right)^2$$

Multiplying by $2a$ and expanding terms,

$$2a\,\Delta d = \cancel{2v_iv_f} - 2v_i^2 + v_f^2 - \cancel{2v_fv_i} + v_i^2$$

Combining terms,

$$2a\,\Delta d = v_f^2 - v_i^2$$

Solving for v_f,

$$v_f = \sqrt{v_i^2 + 2a\,\Delta d} \qquad \text{(Equation 6)}$$

With constant acceleration, the final velocity equals the square root of the sum of the square of the initial velocity and twice the acceleration times the displacement. If the object starts from rest,

$$v_i^2 = 0$$

and

$$v_f = \sqrt{2a\,\Delta d}$$

Table 3-2
ACCELERATION EQUATIONS

With initial velocity	*Starting from rest*
$v_f = v_i + a\,\Delta t$	$v_f = a\,\Delta t$
$\Delta d = v_i\,\Delta t + \frac{1}{2}a\,\Delta t^2$	$\Delta d = \frac{1}{2}a\,\Delta t^2$
$v_f = \sqrt{v_i^2 + 2a\,\Delta d}$	$v_f = \sqrt{2a\,\Delta d}$

The equations just given are summarized in Table 3-2. They apply whether the constant acceleration is positive or negative; hence it is important to note the sign of the displacement. When it is negative, the equations may be used to calculate the distance an object will travel when its motion is constantly decelerated and the time that elapses in bringing the object to a stop. The following example problems show how the equations are used in problems dealing with both positively and negatively accelerated motion.

EXAMPLE Starting from rest, a ball rolls down an incline at a constant acceleration of 2.00 m/s^2. (a) What is the velocity of the ball after 8.5 s? (b) How far does the ball roll in 10.0 s?

Given	Unknown	Basic equations
$v_i = 0.0$ m/s	v_f	$v_f = v_i + a\Delta t$
$a = 2.00$ m/s^2		
$\Delta t_1 = 8.5$ s		
$\Delta t_2 = 10.0$ s	Δd	$\Delta d = v_i\Delta t + \frac{1}{2}a\Delta t^2$

Solution

(a) *Working equation:* $v_f = v_i + a\Delta t_1 = 0.0 \text{ m/s} + (2.00 \text{ m/s}^2)(8.5 \text{ s})$
$= 17$ m/s

(b) *Working equation:* $\Delta d = v_i\Delta t_2 + \frac{1}{2}a\Delta t_2^{\,2} = 0.0 + \frac{1}{2}(2.00 \text{ m/s}^2)(10.0 \text{ s})^2$
$= 1\bar{0}0$ m

EXAMPLE A car traveling at 88 km/h undergoes a constant deceleration of 8.0 m/s^2. (a) How long does it take the car to come to a stop? (b) How far does the car move after the brakes are applied?

Given	Unknown	Basic equations
$v_i = 88$ km/h	Δt	$a = \dfrac{v_f - v_i}{\Delta t}$
$v_f = 0.0$ km/h		
$a = -8.0$ m/s^2	Δd	$\Delta d = v_i\Delta t + \frac{1}{2}a\Delta t^2$

Solution

(a) *Working equation:* $\Delta t = \dfrac{v_f - v_i}{a}$

$$= \frac{(0.0 \text{ km/h} - 88 \text{ km/h})(1000 \text{ m/km})}{(-8.0 \text{ m/s}^2)(3600 \text{ s/h})}$$

$= 3.1$ s

(b) *Working equation:* $\Delta d = v_i\Delta t + \frac{1}{2}a\Delta t^2$

$$= \frac{(88 \text{ km/h})(3.1 \text{ s})(1000 \text{ m/km})}{3600 \text{ s/h}} + \tfrac{1}{2}(-8.0 \text{ m/s}^2)(3.1 \text{ s})^2$$

$= 37$ m

PRACTICE PROBLEM An object starts from rest and moves with constant acceleration for a distance of 150 m in 5.0 s. What is the acceleration of the object? *Ans.* 12 m/s^2

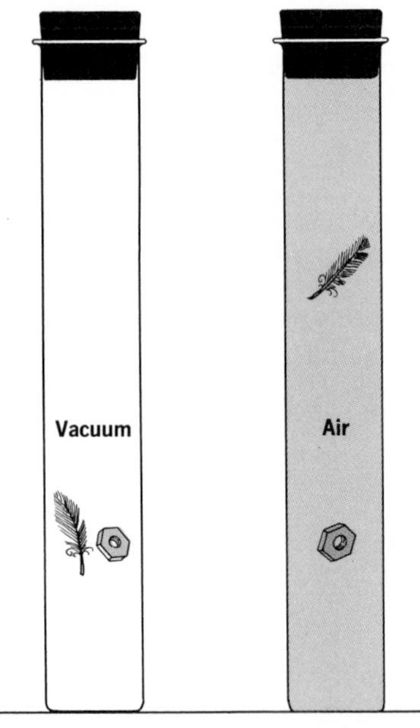

Figure 3-12. Objects fall with the same acceleration in a vacuum (left) but with different accelerations through air (right).

Figure 3-13. Skydivers in free fall. Air resistance allows skilled divers to control position and direction of their bodies even before the parachutes are opened.

3.7 Freely Falling Bodies In New York City if a body falls freely from rest, it reaches a velocity of 9.803 m/s in one second of time. Since $a = v_f/\Delta t$, the acceleration equals 9.803 m/s^2 at this location. We must consider the location because the value of the acceleration depends on the earth's gravity at that location. Since the gravity on the earth may vary from place to place, the value of the acceleration may also vary. (The nature of gravity will be discussed more fully in Sections 3.11–3.13.) A rounded value of 9.80 m/s^2 is approximately valid for most locations. (Air friction is not considered in this example.)

The equations for accelerated motion apply to freely falling bodies. Since the acceleration due to gravity, g, is the same for all objects at a given location, we may substitute g for a in these equations. Thus

$$v_f = v_i + g\,\Delta t$$
$$\Delta d = v_i\,\Delta t + \tfrac{1}{2}g\,\Delta t^2$$
$$v_f = \sqrt{v_i^2 + 2g\,\Delta d}$$

When freely falling bodies start from rest, $v_i = 0$, and the motion is described by the following equations:

$$\mathbf{v_f = g\,\Delta t}$$
$$\mathbf{\Delta d = \tfrac{1}{2}g\,\Delta t^2}$$
$$\mathbf{v_f = \sqrt{2g\,\Delta d}}$$

In the case of free-falling bodies, Δd is always a vertical distance. The vector quantities v and Δd in these equations are customarily assigned plus signs if they are directed downward and minus signs if they are directed upward. (The description of the motion is simplified by considering the downward direction of gravity as positive.) These equations apply only to objects that are falling freely in a vacuum because they do not take into consideration the resistance of air, as illustrated in Figure 3-12. These equations do not apply to objects whose air resistance is great compared to their weight, such as leaves falling in air.

An object thrown upward is constantly decelerated by the force of gravity until it finally stops rising. Then as the object falls, it is constantly accelerated by the force of gravity. If the effect of the atmosphere is neglected, then the time required for the object to fall is the same as the time required for it to rise. Study the following example.

EXAMPLE A ball is thrown vertically upward with a velocity of 115 m/s. (a) To what height will it rise? (b) How long will it take for the ball to fall back to the earth?

Given	Unknown	Basic equations
$v_i = -115$ m/s $v_f = 0.0$ m/s	Δd	$v_f = \sqrt{v_i^2 + 2g\Delta d}$
$g = 9.80$ m/s^2	Δt	$v_f = v_i + g\Delta t$

Solution

(a) *Working equation:* $\Delta d = \dfrac{v_f^2 - v_i^2}{2g}$

$= \dfrac{0 - (-115 \text{ m/s})^2}{(2)(9.80 \text{ m/s}^2)}$ (The value for v_i is negative because the velocity is upward.)

$= -675$ m (The negative value obtained for Δd indicates an upward displacement.)

(b) *Working equation:* $\Delta t = \dfrac{v_f - v_i}{g} = \dfrac{0 - (-115 \text{ m/s})}{9.80 \text{ m/s}^2}$

$= 11.7$ s

Neglecting the effect of the atmosphere, the total time will be twice the amount shown above, or 23.4 s.

PRACTICE PROBLEMS **1.** A ball drops from rest and attains a velocity of 53 m/s. How much time has elapsed? *Ans.* 5.4 s

2. How far did the ball in the previous problem fall during the third second? *Ans.* 25 m

QUESTIONS: GROUP A

1. (a) Define acceleration verbally and mathematically. (b) What type of quantity is acceleration? (c) What is deceleration?
2. A certain car is accelerating at 5.0 km/h/s. Explain what is happening to the motion of the car.
3. Give an example of a moving object that has a velocity vector and an acceleration vector in the same direction and one that has velocity and acceleration vectors in opposite directions.

GROUP B

4. Can an object be moving at a constant speed and accelerating?
5. What would be the shape of a speed-versus-time graph for an object in free fall?
6. What is the algebraic sign (positive or negative) for designating the displacement and velocity vectors of objects in free fall?

PROBLEMS: GROUP A

Note: Use $g = 9.80$ m/s^2. Disregard the effect of the atmosphere.

1. A car moving along a level road increases its speed uniformly from 16 m/s to 32 m/s in 10.0 s. (a) What is the car's acceleration? (b) What is its

average speed? (c) How far did it move while accelerating?

2. A ball initially at rest rolls down a hill with an acceleration of 3.3 m/s^2. (a) If it accelerates for 7.5 s, how far will it move? (b) How far would it have moved if it had started at 4.0 m/s rather than from rest?
3. A penny dropped from the roof of a building takes 6.3 s to strike the ground. How high is the roof?
4. A flower pot falls from a windowsill 25.0 m above the sidewalk. (a) How fast is the flower pot moving when it strikes the ground? (b) How much time does a passerby on the sidewalk below have to move out of the way before the flower pot hits the ground?

GROUP B

5. A boy sliding down a hill accelerates at 1.40 m/s^2. If he started from rest, in what distance would he reach a speed of 7.00 m/s?
6. A plane, starting from rest at one end of a runway, undergoes a constant acceleration of 1.6 m/s^2 for a distance of 1600 m before takeoff. (a) What is its speed upon takeoff? (b) What is the time required for takeoff?
7. A physics student throws a softball straight up into the air. Her friend has a stopwatch and determines that the ball was in the air for a total of 3.56 s (a) With what speed was it thrown? (b) How high did it rise?
8. A water rocket starts from rest and rises straight up with an acceleration of 5.00 m/s^2 until it runs out of water 2.50 s later. If it continues to rise until its speed is zero, what will be its maximum height?

PHYSICS ACTIVITY

Determine your reaction time by having a friend unexpectedly drop a vertically held meterstick between your hands. Catch the meterstick and calculate the time from the distance the meterstick has fallen through your grasp.

NEWTON'S LAWS OF MOTION

Figure 3-14. Sir Isaac Newton formulated the law of gravitation and three laws of motion that describe how forces act on matter.

3.8 Law of Inertia Thus far in Chapter 3, we have discussed motion apart from its causes. Now we shall study the effects that forces have on motion. *A* **force** *is a physical quantity that can affect the state of motion of an object.* A push or a pull is a force. Since a force has magnitude and direction, and can be combined with other forces the way displacements are combined, it is a vector quantity.

The relationships among force, mass, and motion were described clearly for the first time by Sir Isaac Newton (1642–1727) in three laws of motion that bear his name. While some aspects of Newton's laws of motion can be tested only under carefully controlled conditions, repeated experiments and observations show that the laws are universally true. That is, the laws apply to objects throughout the universe.

Newton's first law of motion deals with the motion of a body on which no net force is acting. That is, either there is no force at all acting on the body or the *vector sum* of all forces acting on the body is zero. The word "net" refers to the second situation. Even though a body may have many forces acting on it, these forces may act against each other. They may balance each other in such a way that the body does not change its state of motion. If such a body is at rest, it will remain at rest. If it is in motion, it will continue moving *in a straight line with constant speed.* ***Newton's first law of motion*** may therefore be stated as follows: *If there is no net force acting on a body, it will continue in its state of rest or will continue moving along a straight line with constant speed.* As we saw in Section 1.8, this property of a body that opposes any change in its state of motion is called inertia. Hence Newton's first law of motion is known as the law of inertia.

The law of inertia states that unless a net force acts on an object, the motion of an object (or lack of it) does not change.

At first glance, this law seems to contradict our everyday experiences. If a car is to be kept moving with a constant velocity, the car's engine must apply a constant force to it. If the engine stops applying this force, the car comes to a stop. Only then does the car seem to obey the part of Newton's law that states that objects at rest will remain at rest unless acted upon by an unbalanced force.

Close study of a moving car shows, however, that it is the force of friction that brings the car to a stop and not the absence of the force provided by the engine. If it were possible to remove this friction, the car would keep rolling without applying a constant force. That is what happens to a space vehicle beyond the earth's atmosphere. The vehicle keeps moving even when its rocket engine is shut off, and the ship's velocity is changed only by the force of gravity of other objects in space.

The study of the motion of a car in the absence of friction is an example of a *thought experiment* since the study cannot be performed under actual conditions. It was a thought experiment of this kind that led Galileo to an understanding of inertia even before Newton described it. Galileo noticed that if a ball rolls down one incline and up a second one, as shown in Figure 3-15, the ball will reach

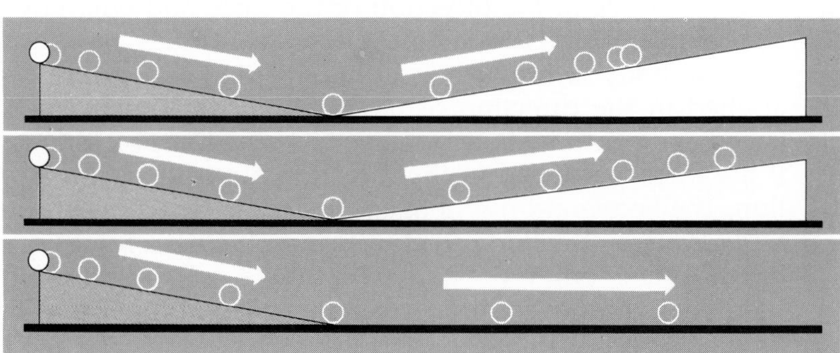

Figure 3-15. A diagram of Galileo's thought experiment. In the upper and middle drawings, the ball reaches almost the height at which it started. In the lower drawing, the ball continues to move indefinitely at constant velocity. In each drawing, the successive images show the positions of the ball after equal time intervals.

almost the same height on the second incline as the height from which it started on the first incline. Galileo concluded that the difference in height is caused by friction and that if friction could be eliminated the heights would be exactly alike. Then he reasoned that the ball would reach the same height no matter how shallow the slope of the second incline. Finally, if the second slope were eliminated altogether, the ball would keep rolling indefinitely with constant velocity. This is the same idea as the one expressed in Newton's first law of motion. Inertia keeps a stationary object stationary and a moving object moving.

The greater the mass (and inertia) of an object, the greater is the force required to produce a given acceleration. A pencil lying on the floor has relatively little mass and therefore little inertia. You can produce acceleration easily by kicking it with your foot. A brick has much more mass and more inertia. You can easily tell its difference from the pencil if you try to kick it.

Newton's first law specifies which forms of motion have a *cause* and which forms do not. Constant motion in a straight line is the only motion possible for an object far removed from other objects. Variable motion is always *caused* by the presence of some other object.

Figure 3-16. A demonstration of inertia. In the bottom photo, the table has been quickly pulled away, and the chicken dinner is momentarily suspended in midair.

3.9 Law of Acceleration

Newton's first law of motion tells us how a body acts when there is no net applied force. Let us consider what happens when there *is* either a single applied force or two or more applied forces whose *vector sum* is not zero. (Remember that forces are vector quantities.) In the following discussion, we shall use "applied force" or just "force" to mean the vector sum of all forces applied to the body.

Newton's second law of motion states: *The effect of an applied force is to cause the body to accelerate in the direction of the force. The acceleration is in direct proportion to the force and in inverse proportion to the mass of the body.*

If the body is at rest when the force is applied, it will begin to move in the direction of the force and will move faster and faster as long as the force continues.

If the body is moving in a straight line and a force is applied in the direction of its motion, it will increase in speed and continue to do so as long as the force continues. If the force is applied in the direction opposite to the motion, the acceleration will again be in the direction of the force, causing the body to slow down. If such a force continues long enough, the body will slow down to a stop and then begin to move with increasing speed in the opposite direction.

If the force applied to a moving object is not along the line of its motion, the acceleration will still be in the direction of the force. The effect in general will be to change both the direction and the speed of the motion. We shall postpone detailed consideration of this more complex case until Chapter 5. For the present we shall consider only the speed of an object initially at rest or one that is moving in a straight line that is changed only by a force acting along that line.

It should be emphasized again that a body to which no force is applied has zero acceleration. If such a body is in motion, its velocity does not change. If it is at rest, its velocity remains unchanged; its velocity stays at zero. And when there is no change in velocity, the acceleration is zero. When a force is applied to a body, the acceleration is not zero, the velocity changes, and the acceleration is in the direction of the applied force.

Because of friction and gravity, stepping on the accelerator of a car does not necessarily accelerate the car.

If different forces are applied to the same body, the magnitude of the acceleration is directly proportional to the amount of the force. That is, doubling the force causes the acceleration to double, and so on.

We can express such a direct proportion as an equation by saying that one of the quantities equals the other one times a constant, or

$$\mathbf{F} = k\mathbf{a}$$

in which F is the applied force, a is the resulting acceleration, and k is the constant of proportionality. Since F and a are both vectors, this equation also says that the applied force and the resulting acceleration are in the same direction (which is a restatement of the first part of Newton's second law of motion).

What happens when forces are applied to *different* objects? This question brings us to the third part of Newton's second law. It turns out that the equation $F = ka$ is always true but that the value of k is different for different objects. In fact, we find that if we properly define the units of force and mass, the value of k for any object is identical with the mass, m, of that object. Thus the equation is written

$$F = ma$$

In the metric system, the unit of mass is the kilogram and acceleration has the units of meters per second squared. (See Section 3.5.) The force required to accelerate 1 kilogram of mass at 1 meter per second squared is 1 kg · m/s^2. This relationship defines the newton in terms of fundamental units. That is,

$$\mathbf{1\ N = 1\ kg \cdot m/s^2}$$

Since the acceleration acquired by a particular object is directly proportional to the amount of force applied, this law of acceleration can be expressed as the proportion

$$\frac{F}{F'} = \frac{a}{a'}$$

where F and F' are two different forces and a and a' are the corresponding accelerations.

In the case of a freely falling object, one of the forces is known. It equals the weight, F_W, of the object. The acceleration, g, is also known. It is the acceleration due to gravitation, 9.80 m/s^2. Making these substitutions, the proportion becomes

$$\frac{F}{F_W} = \frac{a}{g}$$

and

$$F = \frac{F_W a}{g}$$

Therefore, we can calculate the force needed to give any desired acceleration to an object of known weight.

Since $F = ma$, we can substitute this value for F in the above equation and get

Notice the similarity between $F_W = mg$ and $F = ma$. What is the reason for this similarity?

$$ma = \frac{F_W a}{g}$$

Dividing both sides of the equation by a, we have

$$m = \frac{F_W}{g}$$

and

$$F_W = mg$$

Provided the value of g is known, these two equations are used to convert from newtons to kilograms, and vice versa. The following examples show how Newton's second law is used in solving acceleration problems.

EXAMPLE A car weighs 1.50×10^4 N. What force is required to accelerate the car at 4 m/s^2? (Neglect the effect of friction and air resistance in this problem.)

Given	Unknown	Basic equations
$F_W = 1.50 \times 10^4$ N $a = 4$ m/s^2	F	$F_W = mg$ $F = ma$

Solution

Substituting the first Basic equation into the second, we get the

Working equation: $F = \dfrac{(F_W)(a)}{g}$

$$= \frac{(1.50 \times 10^4 \text{ N})(4 \text{ m/s}^2)}{9.80 \text{ m/s}^2}$$

$$= 6 \times 10^3 \text{ N}$$

PRACTICE PROBLEMS **1.** What is the weight of a person whose mass at sea level is 72 kg? *Ans.* 710 N

2. A car weighs 1.15×10^4 N. A force of 895 N is used to accelerate the car from rest to 25 m/s. How much time is required for the acceleration? *Ans.* 33 s

3. What force is required to accelerate the car in Question 2 from 25 m/s to 50 m/s in 5 s? *Ans.* 6×10^3 N

3.10 Law of Interaction ***Newton's third law of motion*** may be stated as follows: *When one body exerts a force on another, the second body exerts on the first a force of equal magnitude in the opposite direction.* To illustrate this law, consider some of the forces that are exerted when a book is resting on the top of a level table. The book's weight acts downward against the table. The table top exerts an upward force on the book. These forces are equal in magnitude and opposite in direction. When you walk forward on a level floor your feet exert a horizontal force against the floor, and the floor pushes against your feet with a force of equal magnitude, but in the opposite direction.

For every action there is an equal and opposite reaction.

In each of these situations we have two objects. In the first instance, the objects are the book and the table. In the second instance, they are the foot and the floor. Two forces are involved in each situation. In the first, they are the weight of the book on the table and the force of the table against the book. In the second, they are the force of the foot against the floor and the force of the floor against the foot. In Newton's own words: "To every action there is always opposed an equal reaction." Unaccompanied forces do not exist in nature.

Newton's third law holds true for all objects at all times, whether they are stationary or moving. Every force is accompanied by an equal and opposite force, independent

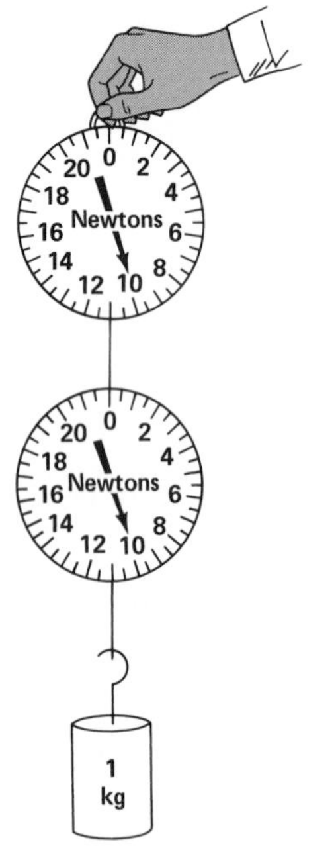

Figure 3-17. A demonstration of action-reaction. The kilogram mass exerts a downward force of 9.8 newtons, while the persons hand exerts an equal upward force of 9.8 newtons. (The weights of the balances are not considered in this illustration.)

of the motion of the objects involved. Let us consider another example. A boy rows a boat toward the shore of a lake and, when he is a meter or so from shore, he attempts to leap ashore. He exerts a force against the boat and the boat exerts an equal but opposite force against him. The force exerted by the boat against the boy accelerates him. The direction of his acceleration is opposite to the direction in which his force caused the boat to be accelerated. If we assume the resistance to the motion of the boat by the water to be negligible, the amount of acceleration of each object is inversely proportional to its mass. The boy might judge the force he must exert to reach shore on the basis of his experience in jumping the same distance from an object fixed on the earth. If so, when he jumps from the boat, he might not reach shore, but instead fall into the water.

Suppose a kilogram mass is suspended from two spring balances that are connected, as shown in Figure 3-17. As we learned in Section 2.4, a kilogram weighs about 9.8 newtons near the earth's surface. Then why do *both* balances have a reading of 9.8 newtons? The answer is Newton's third law. The downward force exerted by the mass is accompanied by an equal upward force exerted by the person's hand. Neither force can exist without the other. If additional balances are added in this experiment, each balance will still read 9.8 newtons (provided the weights of the balances are neglected). Can you explain why?

You may think that motion cannot occur if opposing forces are equal. If two people pull on a lightweight wagon with equal force in opposite directions, the wagon will not move. But this is not an example of action and reaction. There are two forces, it is true, and they have the same magnitude and opposite directions, *but they are both exerted on the same object.* Action and reaction apply when forces are exerted on *different* objects. For example, action and reaction are involved in the force that each person's feet exert against the ground and the equal but opposite force the ground exerts against the feet.

QUESTIONS: GROUP A

1. (a) What is a force? (b) What type of quantity is a force?
2. Explain the term *net* force.
3. State the law of inertia.
4. Draw a sketch of your physics book lying at rest on the desk. Draw arrows representing all the forces acting on the book. (This sketch is called a *free-body diagram.*) What is the net force acting on the book?
5. Push your physics book across the desk at a constant speed. Draw a free-body diagram and indicate the net force acting on the book.
6. In your diagram for Question 5, is there a force of which people were unaware before Galileo's time?

7. Give your physics book a sudden push so that it moves across the desk. What happens to its motion? Draw a free-body diagram. Is there an unbalanced force acting?
8. What physical quantity is a measure of the amount of inertia an object has?
9. You are having trouble starting your car. A friend offers to give a push. A second friend comes along and also offers to help. How does the acceleration change when two people push compared to when only one person pushes?

GROUP B

10. A 70-kg defensive back and a 110-kg nose tackle are chasing you down the football field. If you do some broken-field running (change direction continually), which one will have a more difficult time changing his direction to follow you? Explain.
11. An astronaut performed an experiment on the moon in which he dropped a feather and a hammer from the same height. He found they had the same acceleration and landed on the surface at the same time. Why did both objects have the same acceleration?
12. Two children are pulling on a lightweight wagon to move it across the floor. If the force of the children on the wagon equals the force of the wagon on the children, why does it move? Draw a free-body diagram to show all the forces on the children and wagon.
13. An astronaut on the moon has a 110-kg crate and a 220-kg crate. (a) Which crate weighs more on the earth? On the moon? (b) How do the forces needed to lift the crates straight up on the moon compare to the forces needed to lift them on the earth? (c) Compare the forces that the astronaut would have to exert to push the 110-kg crate across a level, frictionless surface on the moon and on the earth.

PROBLEMS: GROUP A

Note: Use $g = 9.80 \text{ m/s}^2$. All forces are net forces.

1. What force is needed to give a mass of 25 kg an acceleration of 20.0 m/s^2?
2. A force of 30.0 N acting on an object gives it an acceleration of 5.00 m/s^2. (a) What is the mass of the object? (b) What is its weight?
3. What acceleration will you give to a 24.3-kg box if you push it with a force of 90.0 N?
4. A car weighs 19 600 N. (a) What is the mass of the car? (b) If a braking force of 1250 N is needed to stop the car, what is its acceleration?

GROUP B

Note: Use $g = 9.80 \text{ m/s}^2$. All forces are net forces.

5. An 1800-kg car starting from rest accelerates constantly during the first 9.50 s of its motion. If the force acting on the car is 2150 N, what is the car's speed at the end of this time?
6. What force must be exerted on an electron to move it from a state of rest to a speed of 3.5×10^7 m/s in a distance of 0.75 m?
7. During a throw, a baseball pitcher exerts a force on the ball through a distance of 3.20 m before releasing it. If the 0.12-kg ball leaves the pitcher's hand with a speed of 35.0 m/s, how much of a force did the pitcher exert on the ball?
8. A catcher stops the ball in Problem 7 by pulling back 25.0 cm on the mitt during the catch. How much force did the catcher exert on the ball?

GRAVITATION

Newton developed his laws of motion and gravitation when he was only twenty-three years old and while he was home from college on an extended vacation during the Great Plague in 1665.

3.11 Newton's Law of Universal Gravitation In addition to formulating his three laws of motion, Newton described the force that makes falling bodies accelerate toward the earth. In doing so, he made use of the laws of planetary motion that were developed by Johannes Kepler almost a century before Newton's time. See Section 1.4. In Newton's account of his study of falling bodies, he states that he wondered whether the force that makes an apple fall to the ground was related to the force that keeps the planets in their orbits. If so, a single law could be used to describe the attraction between objects in the entire universe.

Newton explained Kepler's laws by stating that the *force of attraction between two objects is directly proportional to the product of the masses of the objects and inversely proportional to the square of the distance between their centers of mass.* (The center of mass of an object is that point at which all its mass can be considered to be concentrated. This concept is similar to that of "center of gravity," which will be discussed in Chapter 4.) Newton called this attractive force the *force of gravitation.* Hence the statement is known as Newton's ***law of universal gravitation.*** In equation form the law is

$$F = G\frac{m_1 m_2}{d^2}$$

where F is the force of gravitation, m_1 and m_2 are the masses of the attracting objects, d is the distance between their centers of mass, and G is a proportionality constant called the *universal gravitational constant.*

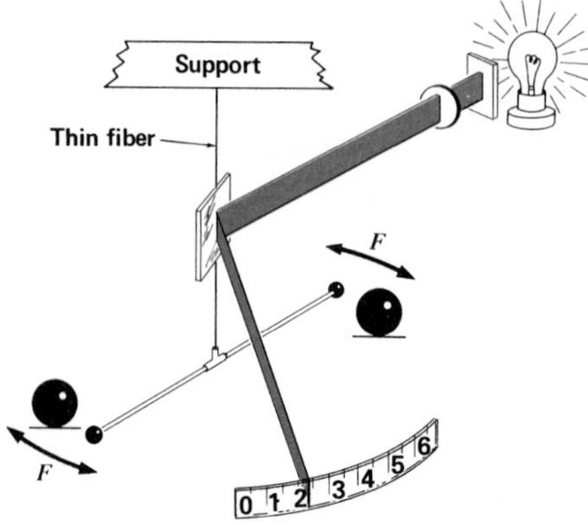

Figure 3-18. The Cavendish experiment. The force of gravitation, F, between the movable and stationary spheres can be measured by observing, with the help of a mirror and light beam, the amount of twist in the suspending fiber.

Newton was only able to confirm his theory with astronomical observations. He knew the approximate value of G, but the first precise measurement was made in 1797 by the English scientist Henry Cavendish (1731–1810). A schematic diagram of this experiment is shown in Figure 3-18. A small lead sphere is attached to each end of a lightweight rod. The rod is suspended by a thin quartz fiber that has a mirror attached to it. Two large lead spheres are placed in fixed positions near the small spheres. The gravitational force between the fixed and movable spheres draws the movable spheres toward the fixed ones. This motion causes the suspending fiber to twist. The fiber offers a slight resistance to twisting. The resistance increases as the twisting increases, and the angle of twist is proportional to the force of gravitation between the fixed and the movable spheres. The angle of twist could be measured directly from the rod's movement, but the sensitivity of the instrument can be increased by shining a light into the mirror. The light is reflected onto a distant scale, and small movements of the mirror result in large movements of the reflected light across the scale. Such an arrangement is called an *optical lever*. Since G has a very small value, extreme care must be used to isolate the Cavendish apparatus from outside forces such as those produced by air currents and electric charges. The present value of G is $6.67 \times 10^{-11}\ \text{N} \cdot \text{m}^2/\text{kg}^2$. (Some scientists think that this value will decrease as the universe keeps expanding.)

3.12 The Mass of the Earth Once G has been measured, the equation for Newton's law of universal gravitation can be used to find the force of gravitation between any two objects of known mass with a known distance between their centers of mass. Or if the force of gravitation, the distance, and one of the masses are known, the equation can be used to find the value of the other mass. For example, the equation can be used to find the mass of the earth. In the equation, use m_e, the mass of the earth, for m_1. For m_2, use m_p, the mass of a particle on the earth's surface. Newton showed that, for a spherical mass, we can consider all of the mass to be located at the center. So in this case d will be the radius of the earth. The force of gravitation, F, will be the weight of the particle, so we shall replace F by F_w.

$$F_w = G\frac{m_e m_p}{d^2} \qquad \text{(Equation 1)}$$

Dividing both sides of the equation by m_p,

$$\frac{F_w}{m_p} = \frac{Gm_e}{d^2} \qquad \text{(Equation 2)}$$

But

$$F_w = m_p g \qquad \text{(Equation 3)}$$

Or

$$\frac{F_w}{m_p} = g \qquad \text{(Equation 4)}$$

Substituting Equation 4 in Equation 2,

$$g = \frac{Gm_e}{d^2} \qquad \text{(Equation 5)}$$

Since both G and m_e are constants, Equation 5 shows that the acceleration due to gravity in a given location depends only on the square of the distance from the center of the earth.

How does this value compare with the one given in Section 2.9?

Equation 5 may be used to determine the mass of the earth. Solving this equation for m_e,

$$m_e = \frac{gd^2}{G}$$

Substituting 9.80 m/s^2 for g, 6.37×10^6 m for d, and 6.67×10^{-11} N·m^2/kg^2 for G, $m_e = 5.96 \times 10^{24}$ kg, the mass of the earth.

Figure 3-19. Saturn's rings. Each of the millions of particles circling the planet is held in a specific orbit by the force of gravitation between the planet and the particle.

3.13 Relation Between Gravity and Weight

*The term **gravity** is used to describe the force of gravitation on an object on or near the surface of a celestial body, such as the earth.* Thus we say that the moon has less gravity than the earth because the force of gravitation near the moon's surface is less than the force of gravitation near the earth's surface. A calculation of the gravity of a celestial body near the surface must take into consideration the size of the body since the force of gravitation varies inversely with the square of the distance to the center of mass of the body. See Table 3-3 for the surface gravities of the moon and planets.

The value of the mass of the earth can be used to determine the value of the gravity at any given location on or above the earth's surface. Such additional factors as local variations in the composition of the earth's crust and the effect of the earth's rotation must also be taken into consideration, however.

Equation 1, Section 3.12 shows that the *weight* of an object is the measure of the force of attraction between the object and the earth. When we say that a person weighs 900 N, we mean that the force exerted on that person by the earth is 900 N. The person also exerts a force of 900 N on the earth. Equation 3 tells us that the weight of an object equals the product of its mass and the acceleration due to gravity. The mass of an object is constant. But Equation 5 indicates that the acceleration due to gravity is inversely proportional to the square of the distance from the center of the body to the center of the earth. All parts of the earth's surface are not the same distance from its center. The variations range from 393 m below sea level at the shore of the Dead Sea to 8848 m above sea level at the top of Mount Everest. An object at the top of a mountain, where gravity is less, will weight less there than it does at sea level. Also, since the earth is slightly flattened at the poles and since the earth's rotation counteracts the force of gravitation, an object weighs a little more at the North Pole than it does at the equator.

Table 3-3
SURFACE GRAVITIES OF THE MOON AND PLANETS

Body	*Relative gravity (Earth = 1.00)*
Jupiter	2.53
Neptune	1.18
Saturn	1.07
Uranus	0.92
Earth	1.00
Venus	0.91
Mercury	0.38
Mars	0.38
Pluto	0.09
Moon	0.16

3.14 Gravitational Fields

The value of g, the acceleration due to gravity, at a particular point is sometimes called the *gravitational field strength* at that point. The effect of rotation, if any, is usually included in determining g. *A region of space in which each point is associated with the value of g at that point is called a* ***gravitational field.*** Since g is an acceleration, it is a vector. Hence, a gravitational field is a region of space in which each point has associated with it a vector equal to the value of g at that point and which is called the gravitational field strength.

If an object of mass m is located at any point in a gravitational field, the force of gravitation on the object can be calculated by the equation in Section 3.9

$$F_W = mg$$

If the mass of the object is 1 kg, then F_W is numerically equal to g. Another way to think of the gravitational field strength at a point is that it equals the force of gravitation that would be exerted on a mass of one kilogram if the mass were located at that point, or

$$g = \frac{F_W}{m}$$

The value of g for the earth varies with the distance from the center of the earth and with certain other aspects of the earth's motion and composition. In Figure 3-20, vectors are used to represent the gravitational field strength at various altitudes. Table 3-4 lists the values of gravitational field strength for various locations on the earth's surface, as determined by careful measurements.

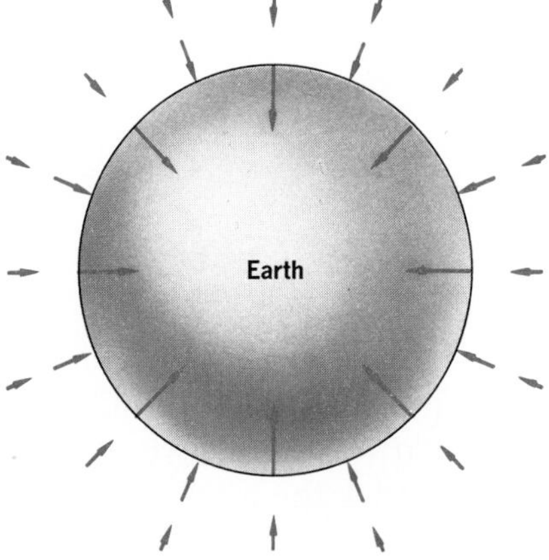

Figure 3-20. Simplified diagram of the earth's gravitational field. The inner circle of vectors represents the magnitude and direction of the field at various places on the earth's surface. The outer vectors represent the field at different altitudes.

Table 3-4
GRAVITATIONAL FIELD STRENGTH OF THE EARTH

Location	*Latitude*	*Altitude* (m)	*Field strength* (N/kg)
North Pole	90° North	0	9.832
Greenland	70° North	20	9.825
Stockholm	59° North	45	9.818
Brussels	51° North	102	9.811
Banff	51° North	1376	9.808
Chicago	42° North	182	9.803
New York	41° North	38	9.803
Denver	40° North	1638	9.796
San Francisco	38° North	114	9.800
Canal Zone	9° North	6	9.782
Java	6° South	7	9.782
New Zealand	37° South	3	9.800

The concept of force fields is useful in describing various natural phenomena.

The concept of a field is more complex than the idea of forces pushing on objects. Yet the field concept can explain more physical phenomena (such as electromagnetic waves) than can the simpler concept of forces. In the field concept, objects are surrounded by regions that exert forces on masses in those regions. Thus a ball falls to the earth because of interaction between the earth's gravitational field and that of the ball.

Another way to describe a gravitational field is to say that an object changes the properties of the space in its vicinity. However, the field concept does not explain the cause of gravitation any more than the idea of acceleration does. Even though scientists are able to express certain aspects of gravitation in terms of acceleration and force fields, the real nature and origin of the force of gravitation are still unknown.

QUESTIONS: GROUP A

1. (a) What force pulled the apple to earth in the story of Sir Isaac Newton under the apple tree? (b) Was there another force acting here? Explain.
2. (a) If you constructed a graph of the force of gravity as a function of distance from the earth, what would it look like? (b) Would the curve ever reach the abscissa?
3. Comment on this statement, "There is no gravity in outer space."
4. Where would you live if you wanted to have the smallest possible weight and still remain on the surface of the earth? (Refer to Table 3-4.)
5. If all objects have a gravitational field, why do you not find yourself attracted to the lockers in the school's hallway as you walk to class?
6. Your textbook weighs about 13 N. If you carry it from the kitchen table (1 m from the earth's surface) to the second floor (3 m from the earth's surface), is its weight now one ninth of 13 N? Explain.
7. The legend is that Galileo dropped two objects from the top of the Tower of Pisa to show that objects in free fall have the same acceleration. (a) What are the action/reaction forces involved here? (b) What type of forces are they?
8. (a) What is a gravitational field? (b) How does field theory explain why a falling object drops to earth?

PROBLEMS: GROUP A

1. Approximate the gravitational force of attraction between a 50.0-kg girl and a 60.0-kg boy if they are sitting 2.50 m apart in a physics class?
2. What is the weight of a 50.0-kg girl if the acceleration due to gravity is 9.80 m/s^2?
3. What would a 70.0-kg boy weigh on the moon, where $g = 1.62$ m/s^2?

GROUP B

4. What is the force of gravitation between the proton (mass = 1.67×10^{-27} kg) and the electron in a hydrogen atom if they are 1.00×10^{-10} m apart?
5. Calculate g at a point 6.37×10^6 m above the surface of the earth.
6. Two satellites of equal mass are orbit-

ing the earth 50.0 m apart. If the gravitational force between them is 1.50×10^{-7} N, find the mass of each.

7. The moon orbits the earth at a distance of 3.90×10^{8} m. If the gravitational force of attraction between them is 1.90×10^{20} N, what is the mass of the moon?

SUMMARY

Displacement is a change of position of an object in a particular direction. Motion is the displacement of an object in relation to stationary objects. Speed is the time rate of change of position (displacement). Velocity is speed in a particular direction. Speed is a scalar quantity, but velocity is a vector quantity. Graphs can be used to determine average and instantaneous speed and velocity. Velocity problems can be worked either graphically or trigonometrically by use of the parallelogram method of vector addition.

Acceleration is the rate of change of velocity. The relationships among displacement, time, initial and final velocity, and acceleration can be expressed as equations. In the case of freely falling bodies, the acceleration caused by the earth's gravity is used.

A force is a quantity that can affect the state of motion of an object. Newton's laws of motion are (1) a body will continue in its state of rest of uniform motion in a straight line unless a net force acts on it; (2) the acceleration of a body is directly proportional to the net force exerted on it, inversely proportional to its mass, and in the same direction as the force; (3) when one body exerts a force on another body, the second body exerts an equal force on the first body but in the opposite direction.

The force of gravitation is the mutual force of attraction between bodies. Newton's law of universal gravitation states that the force of gravitation between two objects is directly proportional to the product of their masses and inversely proportional to the square of the distance between their centers of mass. The mass of the earth can be calculated by means of this law. The force of gravitation near the surface of a celestial body is called gravity. The values of gravity at various distances from a celestial body make up the gravitational field of the body.

VOCABULARY

acceleration
average acceleration
average speed
average velocity
constant speed
constant velocity
displacement
gravitational field
gravity
instantaneous speed
instantaneous velocity
law of acceleration
law of inertia
law of interaction
law of universal gravitation
motion
slope
speed
thought experiment
variable acceleration
variable velocity
velocity

Concurrent and Parallel Forces

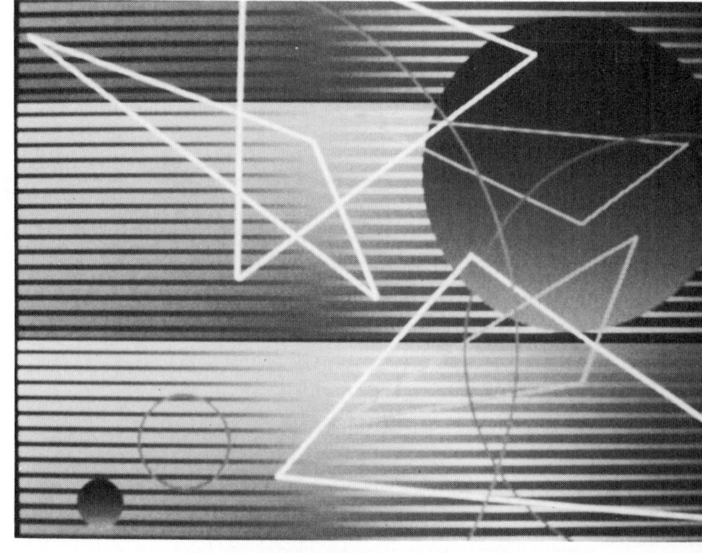

force (FŌRS) *n.: a physical quantity that can affect the motion of an object.*

OBJECTIVES

- Identify forces as vectors.
- Define and calculate resultant and equilibrant forces.
- Resolve forces into components.
- Define and identify frictional forces.
- Solve problems involving frictional forces.
- Define and calculate torque.
- Solve motion problems by applying the two conditions of equilibrium.

COMPOSITION OF FORCES

4.1 Describing Forces In Chapter 3, we studied the relationship between forces and motion. Now we shall take a look at some other characteristics of forces, and see how these characteristics are used in solving force problems.

When you push a door shut with your hand, your hand exerts a force on the door. The door also exerts a force on your hand. When you sit in a chair, you push on the chair and the chair pushes on you. In both of these cases, there is physical contact between the objects that are exerting forces on each other.

Forces can also be exerted without such physical contact. While an object is falling toward the earth, the earth exerts a gravitational force on the object and the object exerts a gravitational force on the earth. Yet there is no physical contact between the earth and the falling object.

These examples illustrate several important characteristics of forces:

1. *A net force will change the state of motion of an object.* The door moves because the force exerted by your hand is sufficient to overcome friction and other forces acting on the door. An object falls because a force is pulling it toward the earth. As we saw in Chapter 3, the application of a net force to an object always produces an acceleration.

2. *Forces can be exerted through long distances.* Gravitational and magnetic forces have this characteristic.

A force never exists by itself.

3. *Forces always occur in pairs.* When one object pushes

or pulls on another object, there is a force on *each* of the two objects. In the given examples, the two objects were your hand and the door, you and the chair, and the falling object and the earth.

4. *In each pair of forces, the two forces act in exactly opposite directions.* You pushed on the door, and the door pushed back. You pushed down, and the chair pushed up. The earth pulled the falling object toward the earth's center, and the object pulled the earth toward the object's center.

Now let us see how the magnitudes of forces are measured. (You will remember from Chapter 3 that forces are vector quantities and have both magnitude and direction.) When an object is suspended from a spring, it is pulled toward the earth by the force of gravitation. The spring stretches until the restoring force of the spring is equal to the force of gravitation on the object. Another object having the same weight stretches the spring by the same amount. Both objects together stretch the spring twice as far, and an object with three times the weight stretches it three times as far, etc. This characteristic of coiled springs, that the amount of stretch is proportional to the force pulling on the spring, means that we can use the amount of stretch to measure the size of a force. A device that measures forces in this way is called a *spring balance*. The results of such measurements can always be expressed in terms of newtons.

Figure 4-1. This telescoping boom crane can lift weights of more than 17 000 newtons to a height of sixteen stories.

Refer to Section 1.8 for a discussion of spring and platform balances.

4.2 Combining Force Vectors

Since forces have magnitude and direction, and can combine like displacements, they are vector quantities. When a vector is used to represent a force, the magnitude of the force is represented by the length of the arrow. The direction of the force can be deduced from the physical situation. For example, suppose a barge is being towed through still water by a tugboat. The tugboat applies a force of 10 $\bar{0}$00 N to the barge through the towline. The length of the arrow representing the force is proportional to the magnitude of the force, 10 $\bar{0}$00 N. Figure 4-2 is a diagram of this example. The point of application of the force is the point at which the rope is attached to the barge. A long rope can transmit only a pull in a direction along its length. It cannot transmit a push or a sideways force. The rope is in the direction of the force. The arrow shows this direction.

Force vectors are treated like velocity vectors. For example, suppose two tugboats are attached to the same barge. Tugboat **A** is pulling with a force of 10 $\bar{0}$00 N and tugboat **B** is pulling with a force of 7500 N in the same direction.

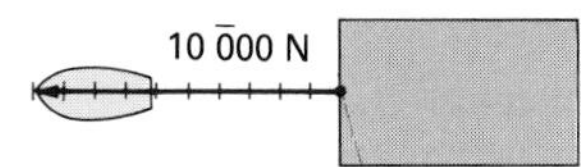

Figure 4-2. A vector diagram of a force of 10 $\bar{0}$00 N applied by a tugboat to a barge through a towline. The direction and point of application of the vector represent the line of action and point of application of the force.

Figure 4-3 represents this situation with vectors. The resultant force vector is 17 500 N in the direction of the towline. Figure 4-4 shows a vector diagram representing two tugboats pulling in opposite directions on a barge. Even though the forces act on extended objects such as opposite ends of the barge, the vectors may be considered as acting at the same point. The resultant force vector is 2500 N in the direction of tugboat **A.**

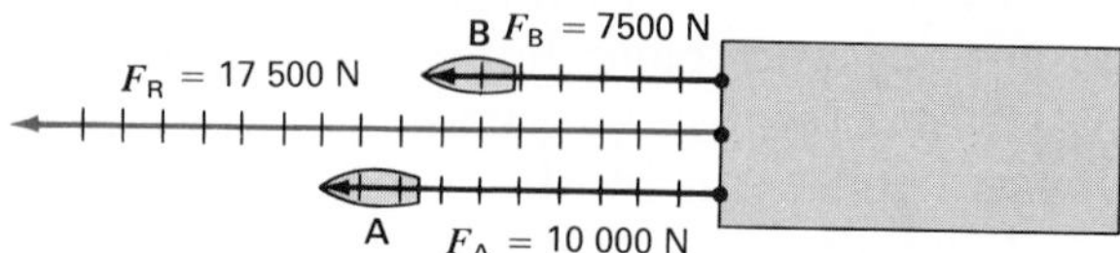

Figure 4-3. A vector diagram of two tugboats applying forces to a barge in the same direction. Tugboat **A** exerts a force of 10 000 N, and tugboat **B** exerts a force of 7500 N. The resultant force vector is 17 500 N in the direction both tugs are pulling.

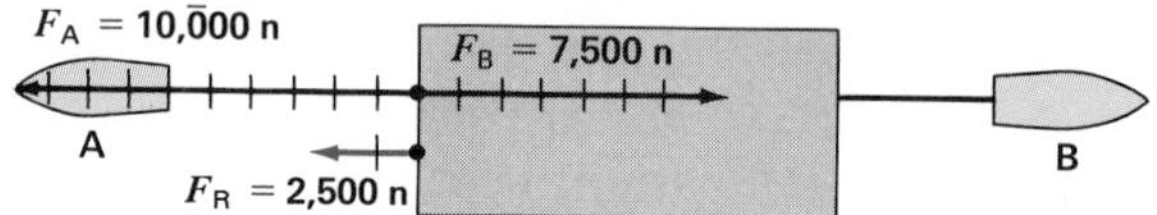

Figure 4-4. A vector diagram of two tugboats applying forces to a barge in opposite directions. Tugboat **A** exerts a force of 10 000 N. Tugboat **B** exerts a force of 7500 N. The resultant force vector is 2500 N in the direction of tugboat **A.**

The sum of two or more vectors is called the resultant.

Concurrent forces act through the same point at the same time.

Several concurrent forces can be combined into a single resultant that has the same effect.

When two or more forces act on the same point at the same time, they are called ***concurrent forces.*** *A* ***resultant force*** *is a single force that has the same effect as two or more concurrent forces.* When two forces act concurrently in the same or in opposite directions, the resultant has a magnitude equal to the algebraic sum of the forces and acts in the direction of the greater force.

When two forces act concurrently at an angle other than 0° or 180°, the resultant can be found by the parallelogram method, as in the velocity vector problems in Chapter 3. Suppose one force of 10.0 N, F_E, acts eastward upon an object at a point **O.** Another force of 15.0 N, F_S, acts southward upon the same point. Since these forces act concurrently upon point **O,** the vector diagram is constructed with the tails of both vectors at **O.** See Figure 4-5. F_E tends to move the object eastward. F_S tends to move the object southward. When the forces act simultaneously, the object tends to move along the diagonal of the parallelogram of which the two forces are sides. This is the vector F_R. The resultant force vector of two forces acting at an angle upon a given point is equal to the diagonal of a parallelogram of which the two force vectors are sides.

The graphic solution relies on a scale diagram. The trigonometric solution makes use of the facts that the opposite sides of a parallelogram are equal and that the diagonal of a parallelogram divides it into two congruent triangles. Vectors at right angles are a special case, as shown in the following example.

EXAMPLE Calculate the (a) direction and (b) magnitude of the resultant of the two forces acting at right angles on point **O** in Figure 4-5.

Given	Unknown	Basic equations
$F_E = 10.0\text{ N}$	θ	$\tan\theta = \dfrac{F_E}{F_S}$
$F_S = 15.0\text{ N}$	F_R	$\cos\theta = \dfrac{F_S}{F_R}$

Solution

(a) *Working equation:* $\theta = \arctan\left(\dfrac{F_E}{F_S}\right) = \arctan\left(\dfrac{10.0\text{ N}}{15.0\text{ N}}\right)$

$= 33.7°$ east of south

(b) *Working equation:* $F_R = \dfrac{F_S}{\cos\theta} = \dfrac{15.0\text{ N}}{0.833}$

$= 18.0\text{ N}$

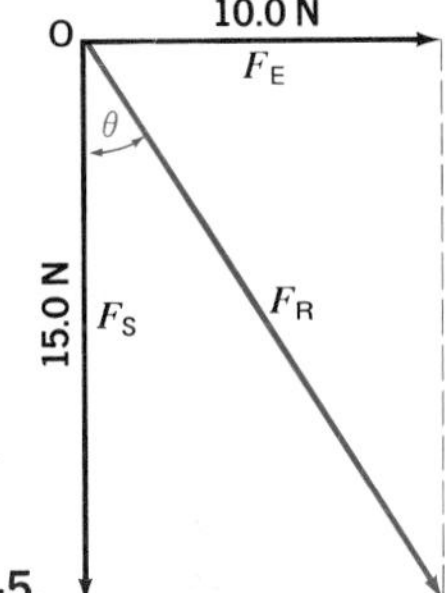

Figure 4-5.

The angle between two forces acting on the same point is often not a right angle (see Figure 4-6). The parallelogram is completed as shown. Observe that the resultant *is drawn from the point on which the two original forces are acting since the resultant will also act on this point*. The magnitude of $\boldsymbol{F_R}$ is found graphically to be 23 N. The direction is $3\bar{0}°$ south of east. (Compare these values to the corresponding ones in the example above.)

The resultant is very different if the angle between the same two forces is 140.0°. The parallelogram is constructed to scale in the same manner. The force vectors are the sides and the angle between them 140.0°. This parallelogram is shown in Figure 4-7. The diagonal must be drawn from **O,** the point at which the two forces act. The graphic solution for $\boldsymbol{F_R}$ yields $1\bar{0}$ N 8° west of south.

The resultant of two forces acting at an acute or obtuse angle can be found trigonometrically by the laws of sines and cosines. The use of these equations is shown in the following example.

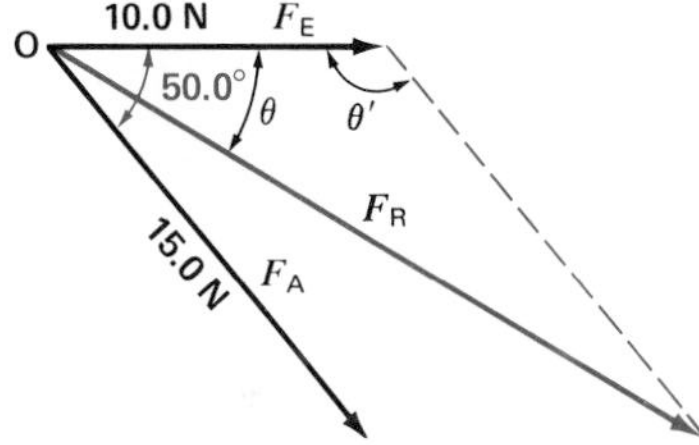

Figure 4-6. The resultant when two vectors are at acute angles.

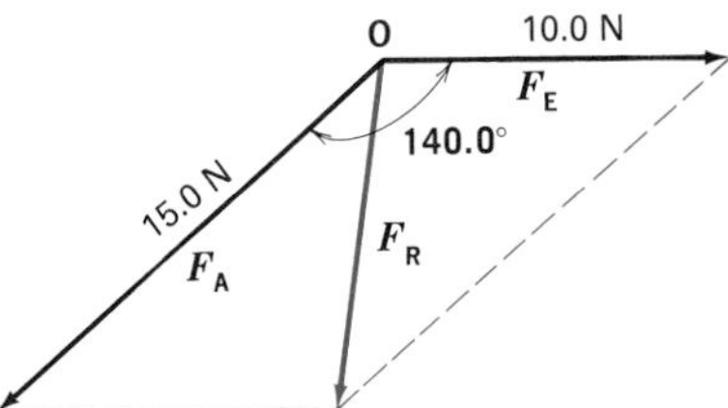

Figure 4-7. The resultant when two vectors are at obtuse angles.

EXAMPLE Calculate the (a) magnitude and (b) direction of the resultant of the two forces acting at an angle of 50.0° on point **O,** as shown in Figure 4-6.

Given	Unknown	Basic equations
$F_E = 10.0$ N	F_R	$F_R = \sqrt{F_E^2 + F_A^2 - 2F_EF_A \cos \theta'}$
$F_A = 15.0$ N	θ	$\dfrac{F_A}{\sin \theta} = \dfrac{F_R}{\sin \theta'}$
$\theta' = 130.0°$		

Solution

(a) *Working equation:*

$$\begin{aligned} F_R &= \sqrt{F_E^2 + F_A^2 - 2F_EF_A \cos \theta'} \\ &= \sqrt{(10.0 \text{ N})^2 + (15.0 \text{ N})^2 - 2(10.0 \text{ N})(15.0 \text{ N})(\cos 130.0°)} \\ &= 22.8 \text{ N} \end{aligned}$$

(b) *Working equation:* $\sin \theta = \dfrac{F_A \sin \theta'}{F_R}$

$$\begin{aligned} \theta &= \arcsin \left[\frac{(15.0 \text{ N})(\sin 130.0°)}{(22.8 \text{ N})}\right] \\ &= 30.3° \text{ south of east} \end{aligned}$$

PRACTICE PROBLEMS 1. Two forces act concurrently at right angles on point **O.** One force of 30.0 N acts south. The other force acts west with a magnitude of 40.0 N. Calculate the magnitude and direction of the resultant. *Ans.* 50.0 N 36.9° south of west

2. A force of 3.50 N acts north on point **O.** A second force of 8.75 N acts concurrently on point **O,** but at an angle of 30.0° west of north. Calculate the magnitude and direction of the resultant force.
Ans. 11.9 N 21.6° west of north

An equilibrant force is equal in magnitude to the resultant of two or more concurrent forces and acts in the opposite direction.

4.3 The Equilibrant Force

Equilibrium *is the state of a body in which there is no change in its motion.* A body in equilibrium is either at rest with respect to other bodies or moving at constant speed in a straight line. In this section we shall discuss the conditions for equilibrium of bodies at rest. The same conditions hold for the equilibrium of bodies that are in motion.

A body at rest must be in both translational and rotational equilibrium. ***Translational equilibrium** is the state in which there are no unbalanced (net) forces acting on a body.* The second condition of equilibrium deals with rotation. This will be discussed in Section 4.12.

When there are no unbalanced forces acting on a body, the vector sum of all the forces acting on the body is zero. For example, if a person pulls on a rope with a force of $8\bar{0}$ N and another person pulls on the same rope in the opposite direction with a force of $8\bar{0}$ N, the vector sum of the two forces is zero and the system is in equilibrium. We can also say that each force is the *equilibrant* of the other.

To find the equilibrant of two concurrent forces, we first find their resultant. Then, since the equilibrant must balance the effect of this resultant, it must have the same magnitude but act in the opposite direction. See Figure 4-8.

The equilibrant of three or more concurrent forces can be found in a similar way. First find the resultant by vector addition. Then the equilibrant is graphically drawn from the origin of the forces so that it is equal in magnitude to the resultant but extends in the opposite direction. *When two or more forces act concurrently at a point, the **equilibrant force** is that single force that if applied at the same point produces equilibrium.*

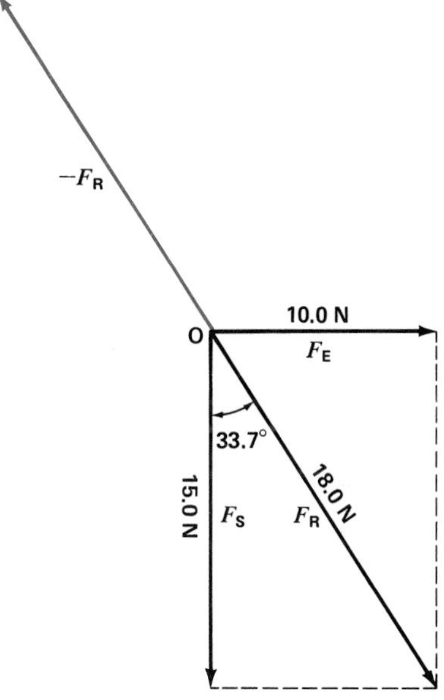

Figure 4-8. Equilibrium results when the resultant (F_R) is counterbalanced by its equilibrant ($-F_R$). Both forces must act on the same point (**O**).

QUESTIONS: GROUP A

1. What are the two ways in which objects can exert forces on one another? Give examples.
2. List four characteristics of forces.
3. What characteristic of springs allows us to use them to measure forces?
4. (a) How is the magnitude of a force shown? (b) How is the other part of this vector quantity shown?
5. (a) What are concurrent forces? (b) How are they related to the resultant force?
6. From what point is the resultant vector drawn?

GROUP B

7. (a) What is meant by *equilibrium*? (b) Is it possible for a moving object to be in equilibrium?
8. (a) If two forces acting on an object cause equilibrium, what is the sum of the two vectors? (b) What is the angle between the two vectors? (c) Is it possible for two concurrent forces, one of 5.0 N and another of 10.0 N, to cause equilibrium?
9. (a) What is an equilibrant force? (b) What is its relation to the resultant?

PROBLEMS: GROUP A

Note: Where appropriate, solve the problems both graphically and trigonometrically.

1. A classmate grabs your arm and pulls to the left with a force of 30 N. Another pulls your other arm to the right with a force of 50 N. Find the resultant force.

2. Two soccer players kick a ball at the same instant. One strikes with a force of 65 N north and the other 88 N east. Find the resultant force on the ball.
3. Find the magnitude of the resultant of the two force vectors shown in Figure 4-7 by the trigonometric method.
4. Two children pull a wagon by exerting forces of 15 N and 18 N at the same point. If the angle between them is 35.0°, what is the magnitude of the resultant force on the wagon?
5. A boy and a girl carry a 12.0-kg bucket of water by holding the ends of a rope with the bucket attached at the middle. If there is an angle of 100.0° between the two segments of the rope, what is the tension in each part?
6. Three men are pulling on ropes attached to a tree. The first man exerts a force of 6.0 N north, the second a force of 35 N east, and the third 40 N 30.0° east of south. (a) Find the resultant force on the tree by using the graphic solution. (b) What is the equilibrant force?

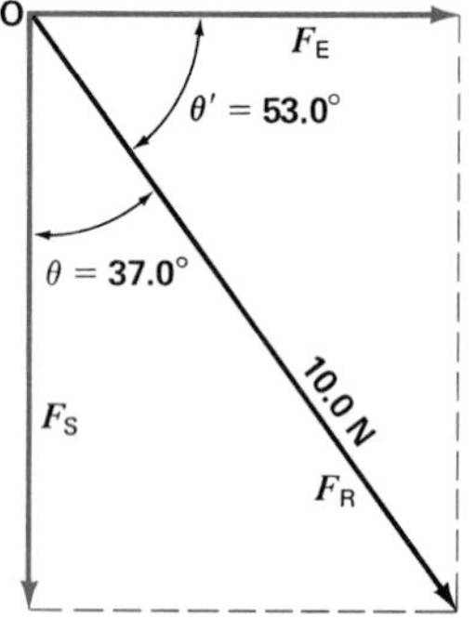

Figure 4-9. A diagram for the determination of the perpendicular components of a force.

A single force can be resolved into two or more components that have the same effect.

RESOLUTION OF FORCES

4.4 Components of Force Vectors Frequently a force acts on a body in a direction in which the body cannot move. For example, gravitational force pulls vertically downward on a wagon on an incline, but the wagon can move only along the incline. Finding the magnitude of the force that is pulling the wagon along the incline is an example of *resolution of forces.* Instead of a single force, two forces acting together can have the same effect as the original single force. One of the forces can be parallel to the surface of the incline and can pull the wagon along the incline. The other can be perpendicular to the surface of the incline. This force does not contribute to the force along the incline. These forces are at right angles to each other. As we saw in Section 2.11, two vectors that have the same effect as a single vector are called the *components* of the original vector. *This procedure of finding component forces is called **resolution of forces.*** Most of the examples we shall consider involve resolving a force into components that are at right angles to each other.

Before working problems with objects on an incline, study the following example, in which a force is resolved into two perpendicular components.

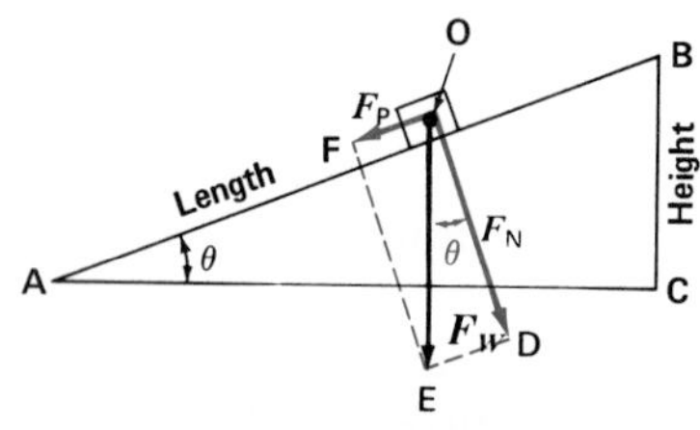

Figure 4-10. Resolution of gravitational force. One component acts parallel to the plane while the other component acts normal (perpendicular) to the plane.

4.5 Resolving Gravitational Forces An object placed on an inclined plane is attracted by the earth. The force of attraction is the weight of the object. See Figure 4-10. The plane prevents the motion of the object in the direction of F_W, the direction in which the earth's attraction acts. The vector representing the force of attraction can, however, be resolved into two components. One component acts in a direction perpendicular to the surface of the plane. In physics, the term *normal* is often used to mean perpendicular. Hence we label the normal component F_N. The other component, F_P, acts parallel to the plane. We choose these two components because they have physical significance. The vector F_N represents the force exerted by the object perpendicular to the incline or the amount of the object's weight supported by the incline. The vector F_P represents the component that tends to move the object down the incline. (The plane is assumed to be frictionless.)

Using F_W as the diagonal, we can construct the parallelogram **ODEF** and find the relative values of the sides F_P and F_N by plotting to scale. We can also express these values trigonometrically. Since right triangles **ABC** and **OED** have mutually perpendicular, or parallel, sides, the triangles are similar and $\angle \mathbf{EOD} = \theta$. Hence sin θ can be expressed either as $\frac{\mathbf{BC}}{\mathbf{AB}}$ or $\frac{F_P}{F_W}$. This equation means that the force vector parallel to the plane, F_P, is smaller in magnitude than the weight vector, F_W, in the same ratio as the height of the plane, **BC**, is smaller than its length, **AB**.

By using cos θ, it may be similarly shown that the magnitude of the force vector perpendicular to the plane, F_N is related to the weight vector of the object in the same way that the base of the plane is related to its length.

If θ and F_W are known,

$$F_P = F_W \sin \theta$$

and

$$F_N = F_W \cos \theta$$

Making the plane steeper increases the component F_P and decreases the component F_N. This steeper inclined plane is shown in Figure 4-11. The vector F_A, which is equal and opposite to F_P, represents the applied force needed to keep the object from sliding down the plane. The steeper the plane, the greater this force becomes. It should be noted that in both Figure 4-10 and Figure 4-11, the normal force that the plane exerts on the object in reaction to force F_N is not shown, since it is not involved in the calculations for finding F_A.

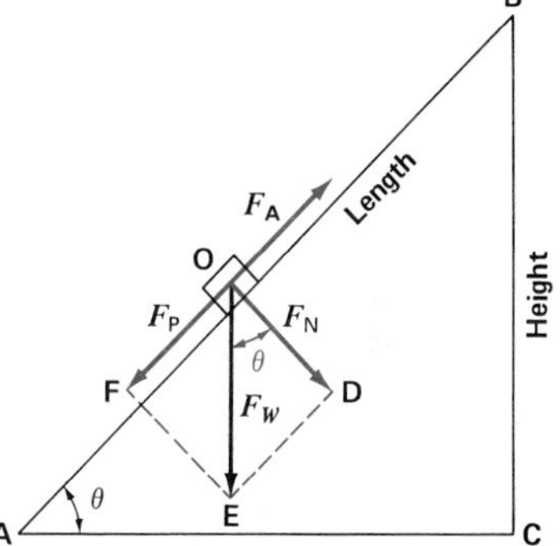

Figure 4-11. When the angle of the incline increases, the component of the weight acting parallel to the plane increases, while the component that acts normal (perpendicular) to the plane decreases.

EXAMPLE A force of 10.0 N acts on point **O** at an angle of 37.0° east of south, as shown in Figure 4-9. Find the magnitudes of the (a) eastward and (b) southward components of its force vectors.

Given	Unknown	Basic equations
$F_R = 10.0$ N	F_E	$\sin \theta = F_E/F_R$
$\theta = 37.0°$	F_S	$\cos \theta = F_S/F_R$

Solution

(a) *Working equation:*
$F_E = F_R \sin \theta$
$= (10.0\ \text{N})(\sin 37.0°)$
$= 6.02$ N due east

(b) *Working equation:*
$F_S = F_R \cos \theta$
$= (10.0\ \text{N})(\cos 37.0°)$
$= 7.99$ N due south

PRACTICE PROBLEM A lawn mower has a mass of 130 kg. A person tries to push the mower up a hill that is inclined 15° to the horizontal. How much force does the person have to exert along the handle just to keep the mower from rolling down the hill, assuming that the handle is parallel to the ground? *Ans.* 330 N

QUESTIONS: GROUP A

1. (a) A 5.0-N force and an 8.0-N force act at a point. What is the maximum resultant of these forces? (b) What is the minimum resultant? (c) Is it possible for these forces to cause equilibrium?
2. What is meant by "resolving" a force?
3. (a) Into what two components is the weight of an object on an incline resolved? (b) What is the physical significance of each?
4. (a) The handle of a lawn mower you are pushing makes an angle of 60.0° with the ground. How could you increase the horizontal forward force you are applying without increasing the total force? (b) What are some disadvantages of doing this?

GROUP B

5. What effect does making the plane in Figure 4-11 steeper have on F_N and F_P?
6. (a) What is the relationship between F_W and F_N when the angle of inclination is 0°? (b) At what angle will the components F_P and F_N have equal magnitude?
7. A skier is coming down a steep hill. (a) Which component of her weight pushes her against the hill? (b) Which component causes her to slide down the hill?
8. Which would require more force—pushing an object up to the top of the incline shown in Figure 4-11, or lifting it straight up? Why?

PROBLEMS: GROUP A

1. A child pulls a toy by exerting a force of 15 N on a string making an angle of 55° with the floor. Find the vertical and horizontal components of the force.
2. A person pushes a grocery cart by exerting a 76-N force on the handle inclined at 40.0° above the horizontal. (a) What component of the force pushes the cart against the floor? (b) What component of the force moves it forward?
3. A 20 000-N car is parked on an incline that makes an angle of 30.0° with the horizontal. If the maximum force the brakes can withstand is 12 000 N, will the car remain at rest?
4. Find the x and y components (i.e. the horizontal and vertical components) of an 88-N force making an angle of 22.0° with the x axis.

GROUP B

5. Two paramedics are carrying a person on a stretcher. One of the paramedics exerts a force of 350 N at 58° above the horizontal and the other exerts a force of 410 N at 43° above the horizontal. What is the total upward force exerted by the paramedics?
6. Ms. Jones has attached a sign that has a weight of 495 N to a wall outside her office, as shown in Figure 4-12. Determine (a) the magnitude of the tension in the chain and (b) the thrust force exerted by the rod.
7. A traffic light is supported by two wires, as shown in Figure 4-13. If the maximum tension in each wire is 750 N, what is the maximum weight of the light they can support?
8. A 2.00×10^3-kg car is to be held on a 20.0° incline by a rope in which the maximum tension is 8.00×10^3 N.
(a) Will the rope support the car?
(b) If the rope is released, how far will the car have moved down the incline by the time its speed reaches 35.0 m/s?

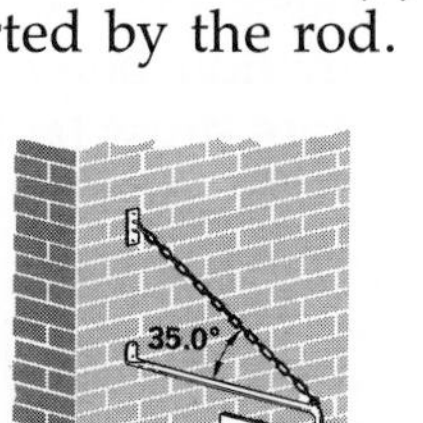
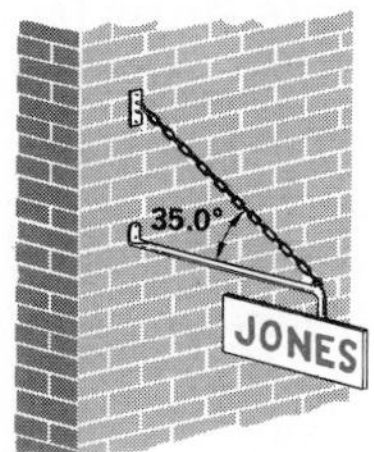

Figure 4-12.

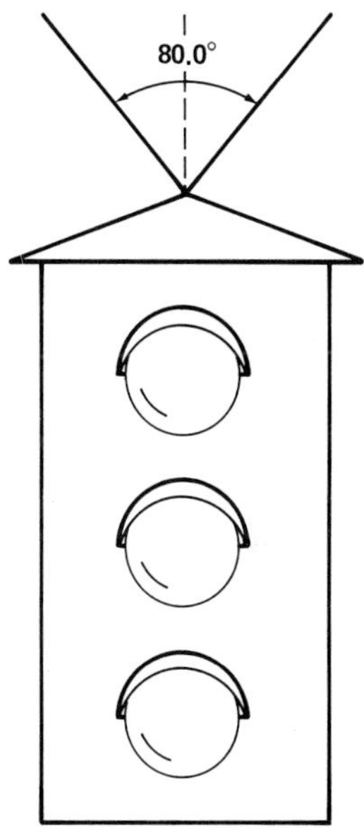

Figure 4-13.

FRICTION

4.6 The Nature of Friction In Section 4.5, we discussed the resolution of the weight of an object resting on an incline. One of the components of the weight tends to pull the object down the incline. As the angle of the incline increases, this component also increases. The slightest angle of incline will produce the component that pulls the object down the incline *if* there is no restraining force on the object. However, in performing experiments of this type, we find that the object does not begin to slide until the component parallel to the incline reaches a certain value. This means that forces must exist between the object and the incline that prevent the object from sliding. These forces are called *forces of friction*, or simply *friction*. ***Friction** is a force that resists motion. It involves objects that are in contact with each other.*

The causes of friction are sometimes complicated. Take, for example, a book lying on a table: the book's weight slightly deforms the surface of the table, along with that of the book. A "plowing" force is required to move the book over these deformations. The irregularities on both surfaces tend to interlock and offer resistance to the sliding of the book. In the process, tiny particles are torn from one surface and become imbedded in the other.

From this example one would expect that if the two surfaces are carefully polished, sliding friction between them would be lessened. Experiments have shown, however, that there is a limit to the amount by which friction may be reduced by polishing the surfaces. If they are made very smooth, the friction between them actually increases. This observation shows that some cases of sliding friction are caused by the forces of attraction between the molecules of substances.

Without friction you couldn't write your homework.

In many instances, friction is very desirable. We would be unable to walk if there were no friction between the soles of our shoes and the ground. There must be friction between the tires of an automobile and the road before the automobile can move. When we apply the brakes on the automobile, the friction between the brake linings and the brake drums, or disks, slows down the wheels. Friction between the tires and the road brings the car to a stop. In a less obvious way, friction holds screws and nails in place and it keeps dishes from sliding off a table if the table is not perfectly level. On the other hand, friction can also be a disadvantage, as it is when we try to move a heavy piece of furniture by sliding it across the floor.

4.7 Measuring Friction

Friction experiments are not difficult to perform, but the results are not always easy to express as equations or laws. The following statements, therefore, should be understood as approximate descriptions only. Furthermore, they deal exclusively with solid objects. Frictional forces involving liquids and gases are beyond the scope of this book. Also, our discussion is restricted to starting and sliding friction. *Starting friction* is the maximum frictional force between stationary objects. *Sliding friction* is the frictional force between objects that are sliding with respect to one another. Static friction (which varies from 0 to the value of starting friction) and rolling friction are not considered in this text.

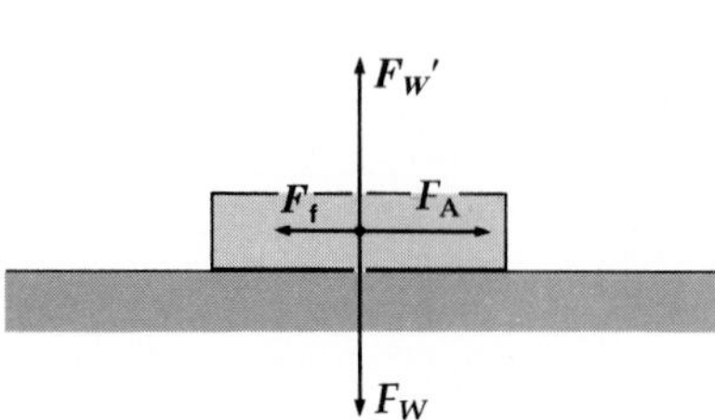

Figure 4-14. The force on a block being pulled along a surface $\mathbf{F_f}$ is the force of sliding friction. The block does not move until $\mathbf{F_A}$ exceeds the force of starting friction, which is usually greater than $\mathbf{F_f}$.

1. *Friction acts parallel to the surfaces that are in contact and in the direction opposite to the motion of the object or to the net force tending to produce such motion.* Figure 4-14 illustrates this principle. The weight of the block, F_W, is balanced by the upward force of the table, F_W'. The force F_A is sliding the block along the table top. In this case, F_A is parallel to the table top. The sliding frictional force, F_f, also parallel

to the table top, resists the motion and is exerted in a direction opposite to that of F_A.

2. *Friction depends on the nature of the materials in contact and the smoothness of their surfaces.* The friction between two pieces of wood is different from the friction between wood and metal.

3. *Sliding friction is less than or equal to starting friction.* Starting friction prevents motion until the surfaces begin to slide. When the object begins to slide, less force is required to keep it sliding than was needed to start it sliding.

4. *Friction is practically independent of the area of contact.* The force needed to slide a block along a table is almost the same whether the block lies on its side or on its end. Figure 4-15 illustrates this principle. When the block is on its end, the increased pressure causes the actual area of contact to be the same as when the block is on its side.

5. *Starting or sliding friction is directly proportional to the force pressing the two surfaces together.* It does not require as much force to slide an empty chair across the floor as it does to slide the same chair when a person is sitting on it. The reason for this is that the extra force actually deforms the surfaces to some extent and thus increases the friction.

A simple way to measure starting and sliding friction is with a spring balance, as shown in Figure 4-15. Blocks of the same substance, but with different sizes and shapes, are pulled along a smooth surface. If the surfaces in contact are consistently smooth, the ratio between the force of sliding friction and the weight of the block is the same in each trial. The ratio depends only on the substances used and not on the area of contact or the weight of the block. This ratio is called the ***coefficient of sliding friction.*** It may be defined as *the ratio of the force of sliding friction to the normal* (perpendicular) *force pressing the surfaces together.* As an equation, it can be written as

$$\mu = \frac{F_f}{F_N}$$

where F_f is the force of sliding friction, μ (the Greek letter mu) is the coefficient of sliding friction, and F_N is the nor-

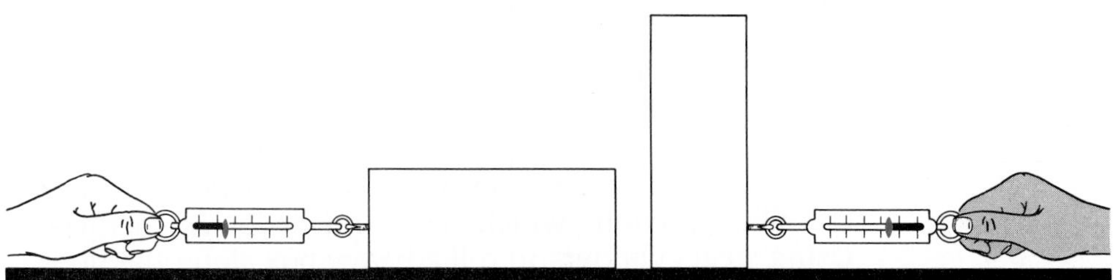

Figure 4-15. The force of friction does not vary significantly with the area of contact if the two blocks have the same weight and all surfaces are equally smooth.

Table 4-1
COEFFICIENTS OF FRICTION

Surfaces	*Starting friction*	*Sliding friction*
steel on steel	0.74	0.57
glass on glass	0.94	0.40
wood on wood	0.50	0.30
rubber tire on dry road		0.70
rubber tire on wet road		0.50
Teflon on Teflon	0.04	0.04

Figure 4-16. Friction between skis and snow is appreciably reduced with the application of a wax layer on the wood or metal surface of the skis.

mal (perpendicular) force between the surfaces. (Within certain limits, this equation is an approximate summary of friction measurements.) The coefficient of starting friction is determined in a similar fashion, except that F_f is the force of starting friction. The approximate values for the coefficients of friction of various surfaces in contact with each other are given in Table 4-1.

4.8 Changing Friction In winter, we sand icy sidewalks and streets in order to increase friction. Tire chains and snow tires are used for the same reason. In baseball, pitchers often use rosin to get more friction between their fingers and the ball. Many more examples could be given in which friction is purposely increased by changing the nature of the surfaces that are in contact.

The most common method of reducing sliding friction is by lubrication. The skier pictured in Figure 4-16 applied a layer of wax to the skis to reduce friction. A thin film of oil between rubbing surfaces reduces friction. The lesser friction between a liquid and a solid has replaced the greater friction between two solids. Alloys have also been developed that are in effect self-lubricating. For example, when steel slides over an alloy of lead and antimony, the coefficient of friction is less than when steel slides over steel. Bearings lined with such an alloy reduce friction. From Table 4-1 it is also obvious that if a bearing is coated with a plastic such as Teflon, there is very little friction. Such bearings are used in electric motors where the use of a liquid lubricant is undesirable.

Friction may also be greatly reduced through the use of ball bearings or roller bearings. Sliding friction is changed to rolling friction, which has a much lower coefficient. Using steel cylinders to roll a heavy box along the floor is another example of changing sliding to rolling friction.

The squeaking wheel gets the grease.

4.9 Solving Friction Problems The force required to slide an object along a level surface can be computed easily from the weight of the object and the coefficient of sliding friction between the two surfaces. However, when the force applied to the object is not applied in the direction of the motion, it is necessary to resolve forces in the calculation. This resolution of forces is illustrated in the following example.

In the case of an object resting on an incline, the forces of starting and sliding friction will determine whether the object remains at rest, slides down the incline with constant speed, or accelerates as it descends. How the angle of the incline and coefficient of sliding friction are used in such a problem is illustrated in another example.

EXAMPLE A box weighing 45$\bar{0}$ N is pulled along a level floor at constant speed by a rope that makes an angle of 30.0° with the floor, as shown in Figure 4-17. If the force on the rope is 26$\bar{0}$ N, (a) what is the horizontal component (F_h) of this force? (b) What is the normal force (F_N)? (c) What is the coefficient of sliding friction (μ)?

Given	Unknown	Basic equations
$F_W = 45\bar{0}$ N	F_h	$\cos\theta = \dfrac{F_h}{F_A}$
$\theta = 30.0°$	F_N	$F_N = F_W - F_A \sin\theta$
$F_A = 26\bar{0}$ N	μ	$\mu = \dfrac{F_f}{F_N}$
$F_f = F_h$		

Solution

(a) *Working equation:*
$$F_h = F_A \cos\theta = (26\bar{0}\text{ N})(0.866) = 225\text{ N}$$

(b) The normal force (F_N) is the difference between the downward force of the block's weight (F_W) and the vertical component of the force of the rope (F_v):

Working equation:
$$F_N = F_W - F_A \sin\theta = 45\bar{0}\text{ N} - (26\bar{0}\text{ N})(0.500) = 32\bar{0}\text{ N}$$

(c) *Working equation:*
$$\mu = \frac{F_f}{F_N} = \frac{F_h}{F_N} = \frac{225\text{ N}}{32\bar{0}\text{ N}} = 0.703$$

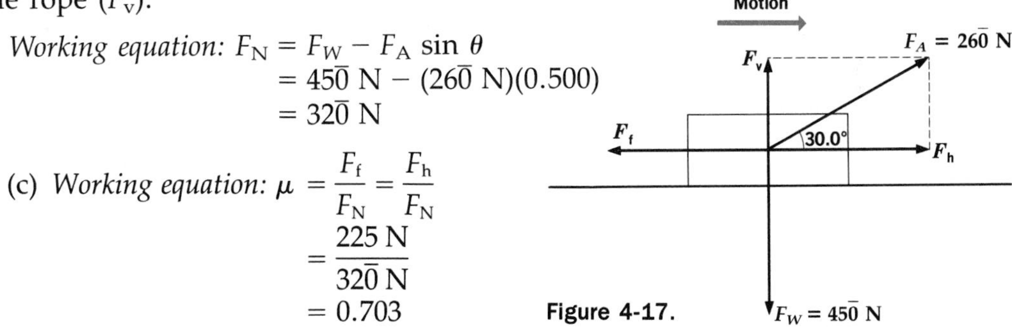

Figure 4-17.

EXAMPLE A wooden block weighing 13$\bar{0}$ N rests on an inclined plane, as shown in Figure 4-18. The coefficient of sliding friction between the block and the plane is 0.620. Find the angle of the inclined plane at which the block will slide down the plane at constant speed once it has started moving.

Given	Unknown	Basic equations
$F_W = 13\bar{0}$ N	θ	$\tan\theta = \dfrac{F_P}{F_N}$
$\mu = 0.620$		$\mu = \dfrac{F_f}{F_N}$

Solution

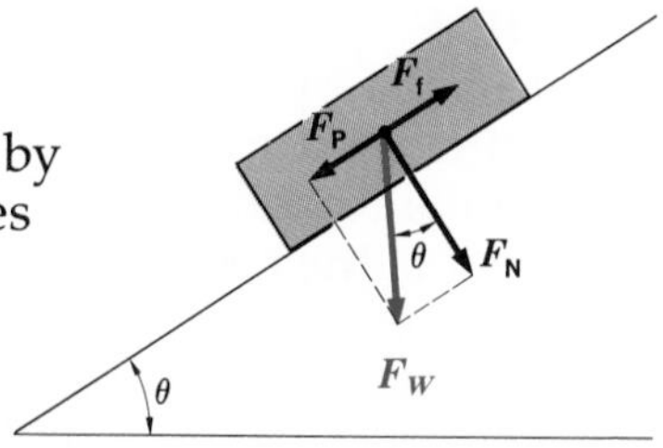

The force along the incline (F_P) is opposed by the force of friction (F_f). When the block slides down at constant speed, $F_P = F_f$.

Substituting this value into the first Basic equation: $\tan\theta = \dfrac{F_f}{F_N}$

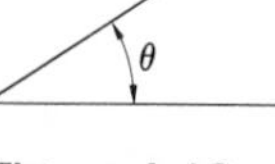

Figure 4-18.

Notice that the term on the right is also the definition for the coefficient of friction. From this we get the

Working equation: $\tan\theta = \mu$

$$\theta = \arctan \mu = \arctan (0.620)$$
$$= 31.8°$$

In other words, the block will slide at constant speed when the angle of the incline is 31.8°, no matter what the block weighs. (The weight of the block cancels out.) If the angle is greater than 31.8°, the block will accelerate as it slides down the plane. The angle at which the block slides depends only on the coefficient of friction.

PRACTICE PROBLEMS **1.** A box with a mass of 175 kg is pulled along a level floor with constant velocity. If the coefficient of friction between the box and the floor is 0.34, what horizontal force is exerted in pulling the box? *Ans.* 583 N

2. A crate is pulled with constant velocity up an inclined floor that makes an angle of 12° with the horizontal. The crate weighs 950 N and the pulling force parallel to the floor is 460 N. Find the coefficient of friction between the crate and the floor. *Ans.* 0.28

QUESTIONS: GROUP A

1. What is friction?
2. State two theories that scientists use to explain the causes of friction.
3. Make a list of the places you encounter frictional effects as you get ready to go to school. Indicate if the effects are useful.
4. When an object slides across a surface, what is the direction of the frictional force on the object?
5. What are the factors affecting solid, sliding friction?
6. Why is the amount of friction seemingly independent of the surface areas in contact?

GROUP B

7. (a) What happens to the coefficient of friction between tires and the road on a rainy day? (b) How should a driver compensate for this effect when approaching a stop sign?
8. Compare friction between solids sliding and solids rolling across a surface.
9. How does static friction change as you push downward on a stationary object?
10. Indicate if the force of friction will increase or decrease when (a) sand is thrown on icy streets, (b) bearings in machinery are lubricated, and (c) car tires are worn down.

PROBLEMS: GROUP A

1. A horizontal force of 400.0 N is required to pull a 1760-N trunk across the floor at constant speed. Find the coefficient of sliding friction.

2. How much force must be applied to push a 1.35-kg book across the desk at constant speed if the coefficient of sliding friction is 0.30?

3. A force of 105 N is applied horizontally to a 20.0-kg box to move it across a horizontal floor. If the box has an acceleration of 3.00 m/s^2, find the coefficient of friction.

4. A 1500.0-N force is exerted on a 200.0-kg crate to move it across the floor. If the coefficient of friction is 0.250, what is the crate's acceleration?

5. A 100.0-kg commuter is standing on a train accelerating at 3.70 m/s^2. What coefficient of friction must exist between the commuter's feet and the floor to avoid sliding?

GROUP B

6. A 146-N force is used to pull a 350-N wood block at constant speed by a rope making an angle of 50.0° with the floor. Find the coefficient of sliding friction.

7. A 75.0-kg baby carriage is pushed along a level sidewalk by exerting a force of 50.0 N on the handle, which makes an angle of 60.0° with the horizontal. What is the coefficient of friction between the carriage and the sidewalk?

8. A 3.00-kg wood box slides from rest down a 35.0° inclined plane. How long does it take the box to reach the bottom of the 4.75-m wood incline? (See Table 4-1 for the coefficient of friction.)

9. A 65.0-kg crate is to be accelerated at 7.00 m/s^2 up an incline making a 25.0° angle with the horizontal. If the coefficient of sliding friction between the crate and the incline is 0.200, how much force is required?

10. A 60.0-kg crate is attached to a weight by a cord that passes over a frictionless pulley, as shown in Figure 4-19. (a) If the coefficient of friction is 0.500, what weight will keep the crate moving up the 40.0° incline at a constant speed? (b) If the cord is cut when the crate is at rest at the top of the incline, how far would the crate have slid by the time its speed reached 7.50 m/s?

11. If the coefficient of friction between a set of waxed skis and the snow is 0.10, at what angle will a 90.0-kg skier move at a constant speed down the slope?

PARALLEL FORCES

4.10 Center of Gravity Thus far in our study of forces, we have been treating all the forces acting on a body as if they were acting at a single point. However, there can be many forces, each acting at a different point on the object. For example, Figure 4-20 represents a stone lying on the ground. Since every part of the stone has mass, every part is attracted to the center of the earth. Because of the large size of the earth, all the downward forces exerted on the stone are virtually parallel. The weight of the stone can be thought of as a force vector that is the vector sum, or resultant, of all these parallel force vectors. ***Parallel forces***

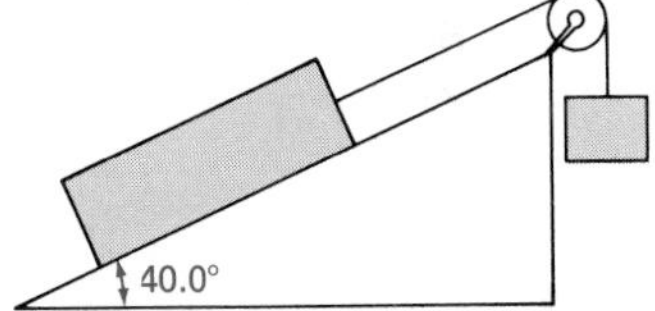

Figure 4-19.

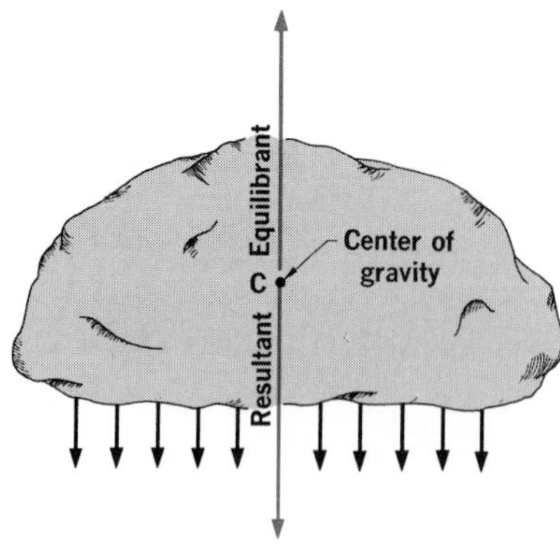

Figure 4-20. The stone's center of gravity is the point where all the weight seems to be concentrated.

act in the same or in opposite directions at different points on an object. The resultant of parallel forces has a magnitude equal to the algebraic sum of all the forces. The resultant acts in the direction of this net force.

But where in the stone is this resultant force acting? Experiments show that if the proper point of application is chosen, the stone can be lifted without producing rotation. As shown in Figure 4-20, the equilibrant lifting force vector is then in line with the resultant (weight) vector of the stone. In other words, the stone acts as if all its weight were located at one point, which is called the *center of gravity. The* ***center of gravity*** *of any object is that point at which all of its weight can be considered to be concentrated.*

In Figure 4-21, the center of gravity of the bar is at **C.** In a bar of uniform construction, **C** is at the geometric center. But if the density or shape of the bar is not uniform, **C** is not at the geometric center.

Since the weight of the bar, F_W, can be considered to be acting at **C,** the bar can be suspended without changing its rotation by an equilibrant force, F_E, applied at **C.** Since F_W and F_E are equal but opposite vectors, they counterbalance each other. In this condition, the bar is in both *translational and rotational equilibrium.* This means that the bar is not accelerating and its rotation (if any) is constant.

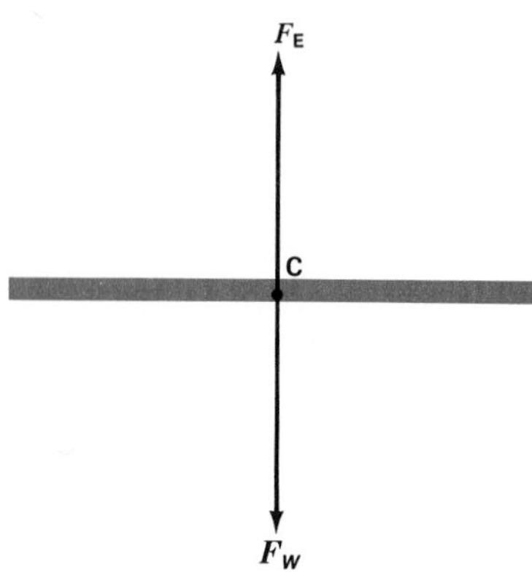

Figure 4-21. The weight of the bar, F_W, is apparently concentrated at the center of gravity, **C**, and can be balanced by an equal and opposite force, F_E, applied at **C**.

4.11 Torques The two forces represented in Figure 4-22 by the vectors F_A and F_B are parallel. They do not act on the same point as did the concurrent forces we studied earlier in this chapter. To measure the rotating effect, or *torque,* of such parallel forces in a given plane, it is first necessary to choose a stationary reference point for the measurements. We shall refer to this stationary reference point as the *pivot point.*

Sometimes, as in the case of a seesaw, there is a "natural" point about which the rotating effects can be measured. However, such a pivot point is "natural" only when the seesaw is in motion. When it is motionless there is no "natural" pivot point. Any point on the seesaw, or even beyond it, can be chosen.

Once a suitable pivot point is chosen, a perpendicular line is drawn on the vector diagram from it to each of the lines along which force vectors act on the object. Each such line is called a *torque arm.* In some cases, the force vectors must be extended in order to meet the perpendicular. ***Torque,*** *T, is the product of a force and the length of its torque arm.* The unit of torque is the meter-newton.

To illustrate the concept of torque, consider the bar in Figure 4-22 with the application of an additional force, F_C,

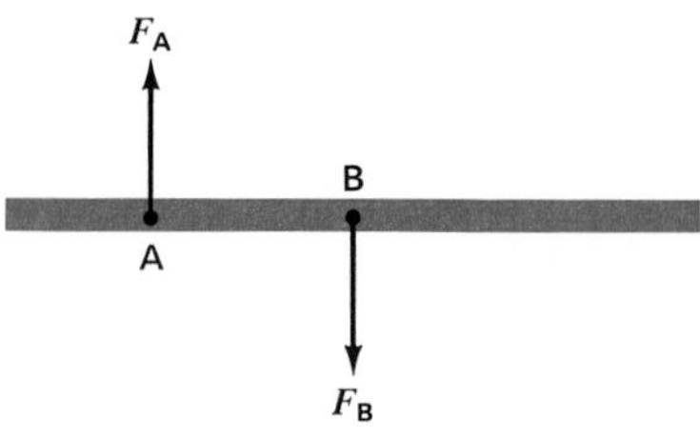

Figure 4-22. Two parallel forces, F_A and F_W, act at different points. A tendency to rotate results.

as shown in Figure 4-23. Choosing **A** as the pivot point, **BA** is the torque arm of F_B and **CA** is the torque arm of the additional upward force, F_C. The clockwise torque around **A** is the product of F_B and **BA**. The counterclockwise torque is the product of F_C and **CA.** Since F_A has a torque arm of zero, it produces no torque and does not enter into the calculations.

To identify a torque as clockwise or counterclockwise, imagine that the bar is free to rotate around a stationary pivot point. Further imagine that the force producing the torque is the only force acting on the bar. The direction in which the bar would rotate is the direction of the torque.

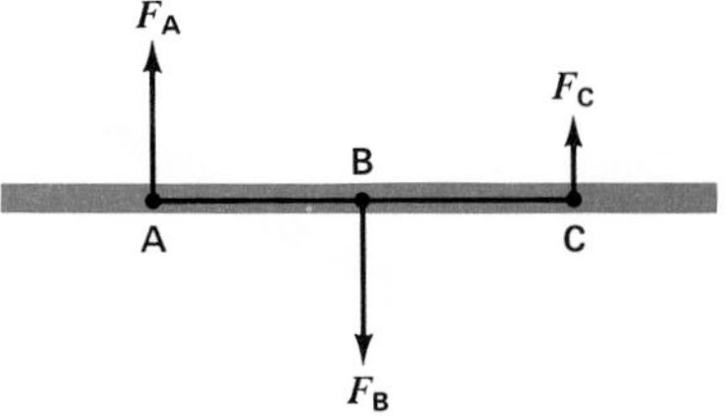

Figure 4-23. Rotational equilibrium results when the sum of the clockwise torques is equal to the sum of the counterclockwise torques. Any point may be chosen as the pivot point in making the computation provided the vector sum of the forces is zero.

4.12 Rotational Equilibrium In Section 4.3 we discussed the conditions necessary for the translational equilibrium of an object. These conditions, however, do not prevent the *rotary motion* of an object that is subjected to torques. To prevent rotation in a given plane a second condition of equilibrium must be met. ***Rotational equilibrium*** *in a given plane is the state in which the sum of all the clockwise torques equals the sum of all the counterclockwise torques about any pivot point.*

Both conditions of equilibrium are illustrated in Figure 4.24. The sum of the force vectors is zero (3$\bar{0}$ N upward

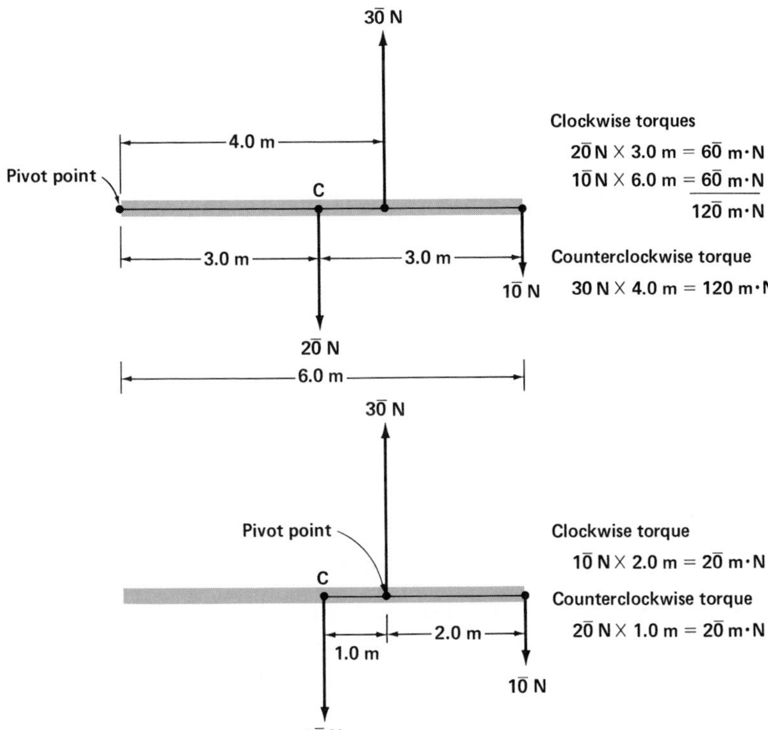

Figure 4-24. The calculation of torques can be simplified by setting one of the torque arms equal to zero. In the upper diagram, the left end of the bar is used as the pivot point. In the lower diagram, the point of application of the upward force is used, thereby reducing the number of torques.

against $2\bar{0}$ N + $1\bar{0}$ N downward). The sum of the clockwise torques is equal to the sum of the counterclockwise torques. Two methods of computing the torques are shown. In the upper drawing of Figure 4-24, the left end of the bar is chosen as a pivot point. In the lower drawing, the point of application of the upward force is used as the pivot point. This simplifies the calculation.

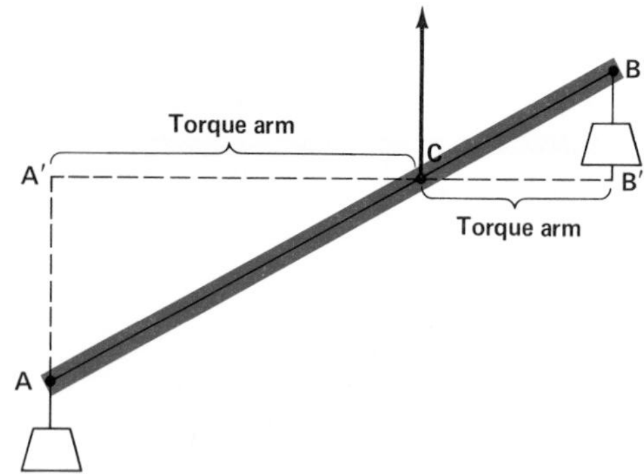

Figure 4-25. The torque arm is the perpendicular distance from the pivot point to the line indicating the direction of the applied force.

When forces are applied to a bar at an angle other than perpendicular, the distances along the bar measured from the points of application of the forces cannot be used to measure the torque arms. *The torque arms must always be measured perpendicular to the directions of the forces.* Figure 4-25 shows such a situation. A meter stick is under the influence of three parallel forces. To find the torque arms, it is necessary to draw a horizontal line through the pivot point. Since the weights hang vertically, a horizontal line is perpendicular to the force vectors of the weights. The problem is simplified by placing the pivot point at the center of gravity. If the bar is not homogeneous, this center of gravity may not be located at the geometric center of the bar. (Experimentally, the center of gravity can be approximately located by finding the point where the bar balances.) Placing the pivot point at the center of gravity eliminates two torques in the equation. It eliminates the torque produced by the weight of the bar located at some distance from the geometric center of the bar. And it also eliminates the torque produced by the force acting upward at **C**. The required torque arms, **CA′** and **CB′**, are then found by multiplying the distances **CA** and **CB** by the cosine of the angle **ACA′** or **BCB′**.

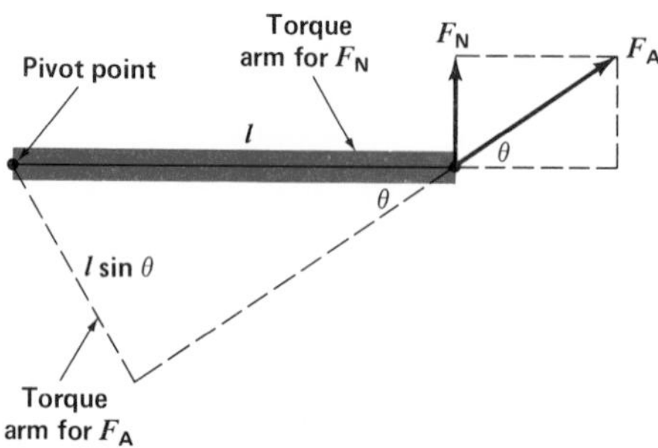

Figure 4-26. The resolution of forces is used to find the torque produced by a force acting on the bar at an angle other than perpendicular.

In Figure 4-26, F_A is applied to the right end of the bar at an angle other than perpendicular. In order to find the counterclockwise torque produced by F_A, we choose the left end of the bar as the pivot point. Then we find the vertical component, F_N, of F_A. By trigonometry,

$$F_N = F_A \sin \theta$$

Since F_N is perpendicular to the bar, the required torque is

$$T = F_N l$$

Substituting,

$$T = F_A l \sin \theta$$

Hence, $l \sin \theta$ is the torque arm of F_A. The result is further verified in Figure 4-26, where $l \sin \theta$ is the length of the perpendicular from the pivot point to the extended line of direction of F_A. This is in accord with the definition of torque arm in Section 4.11.

Figure 4-27. A pair of parallel forces of equal magnitude acting in opposite directions but not on the same point is called a couple. (The weight of the bar is not considered in this example.)

4.13 Coupled Forces The conditions of equilibrium hold true no matter how many forces are involved. An interesting example is one in which *two forces of equal magnitude act in opposite directions in the same plane, but not on the same point.* Such a pair of forces is called a ***couple.*** A diagram of a couple is shown in Figure 4-27. The torque is equal to the product of one of the forces and the perpendicular distance between them. (This can be proved by computing the sum of the torques produced by the action of the separate forces about any desired pivot point.) A good example of a couple is the pair of forces acting on the opposite poles of a compass needle when the needle is not pointing north and south.

A couple cannot be balanced by a single force since this single force would be unbalanced and would produce linear motion where it was applied. The only way to balance a couple is with another couple; the torques of the two couples must have equal magnitudes but opposite directions.

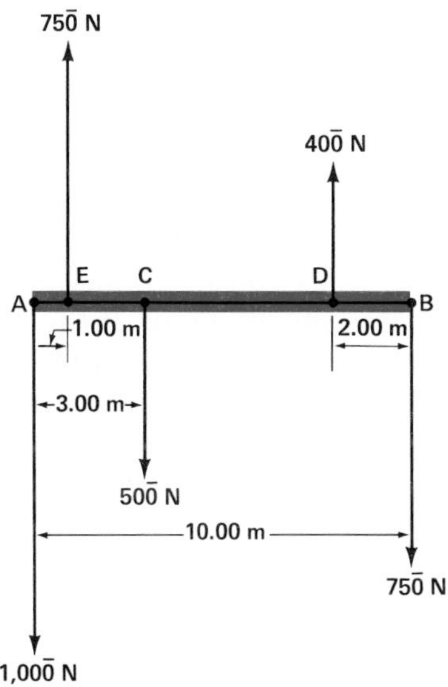

Figure 4-28.

EXAMPLE A horizontal rod, **AB,** is 10.00 m long. It weighs $50\bar{0}$ N and its center of gravity, **C,** is 3.00 m from **A.** At **A** a force of $100\bar{0}$ N acts downward. At **B** a force of $75\bar{0}$ N acts downward. At **D,** 2.00 m from **B,** a force of $40\bar{0}$ N acts upward. At **E,** 1.00 m from **A,** a force of $75\bar{0}$ N acts upward. (a) What is the magnitude and direction of the force that must be used to produce equilibrium? (b) Where must it be applied? (See Figure 4-28.)

Solution

(a) Consider the known upward force vectors as being positive and the known downward force vectors as being negative. The algebraic sum of these force vectors will then be the resultant force vector. This resultant force vector must be counterbalanced by a force vector in the opposite direction in order to establish translational equilibrium.

$$75\bar{0}\text{ N} + 40\bar{0}\text{ N} - 100\bar{0}\text{ N} - 50\bar{0}\text{ N} - 75\bar{0}\text{ N} = -110\bar{0}\text{ N}$$

Therefore, $110\bar{0}$ N must be applied upward to establish translational equilibrium.

(b) Use **A** as the pivot point, and let x be the distance from **A** to the point where the $110\bar{0}$-N force must be applied to prevent rotary motion.

$$\text{clockwise torque} = (50\bar{0}\text{ N})(3.00\text{ m}) + (75\bar{0}\text{ N})(10.00\text{ m})$$
$$\text{counterclockwise torque} = (75\bar{0}\text{ N})(1.00\text{ m}) + (40\bar{0}\text{ N})(8.00\text{ m}) + (110\bar{0}\text{ N})x$$
$$90\bar{0}0\text{ m}\cdot\text{N} = 3950\text{ m}\cdot\text{N} + (110\bar{0}\text{ N})x$$
$$x = 4.59\text{ m}$$

4.59 m is the distance from **A** to the point where the $110\bar{0}$-N upward force must be applied.

PRACTICE PROBLEMS **1.** A nonuniform bar is 3.8 m long and has a weight of 560 N. The bar is balanced in a horizontal position when it is supported at its geometric center and a 340-N weight is hung 0.70 m from the bar's light end. Find the bar's center of gravity.
Ans. 1.2 m from heavy end

2. A large wooden beam weighs 820 N and is 3.2 m long. The beam's center of gravity is 1.4 m from one end. Two workers begin carrying the beam away. If they lift the beam at its ends, what part of its weight does each worker lift? *Ans.* 460 N and 360 N

QUESTIONS: GROUP A

1. (a) What is meant by the term *center of gravity?* (b) Where is the center of gravity of a meterstick? A bowling ball? An ice cube? A doughnut? A banana?
2. (a) What are the conditions for equilibrium? (b) Explain how they apply to children attempting to balance a seesaw.
3. How is torque calculated?
4. Why is it easier to loosen the lid from the top of a can of paint with a long-handled screwdriver than with a short-handled screwdriver?
5. How would the force needed to open a door change if you put the handle in the middle of the door?

GROUP B

6. How does an orthodontist use torque in realigning teeth?
7. What factor determines the location of the pivot point in a torque problem?
8. How can an object on which a couple is acting be placed in equilibrium?
9. What must be true for a moving object to be in equilibrium?
10. A twirler throws a baton straight up into the air. (a) Describe the motion of the ends of the baton. (b) Describe the motion of the center of gravity of the baton.

PROBLEMS: GROUP A

Note: For each problem, draw and label an appropriate force diagram. Unless otherwise noted, the center of gravity is at the geometric center of the object.

1. A 400.0-N child and a 300.0-N child sit on either end of a 2.00-m-long seesaw. Where along the seesaw should the pivot support be placed to ensure rotational equilibrium?
2. Based on the information in Problem 1 and its solution, suppose a 225-N child sits 0.200 m from the 400.0-N child. Where must a 325-N child sit to maintain rotational equilibrium?
3. A uniform meterstick, supported at the 30.0-cm mark, is balanced when a 0.50-N weight is hung at the 0.0-cm mark. What is the weight of the meterstick?
4. A 650-N boy and a 490-N girl sit on a 150-N porch swing that is 1.70 m long. If the swing is supported by a chain at each end, what is the tension in each chain when the boy sits 0.750 m from one end and the girl 0.500 m from the other?
5. A uniform bridge, 20.0 m long and weighing 4.00×10^5 N, is supported by two pillars located 3.00 m from each end. If a 1.96×10^4-N car is parked 8.00 m from one end of the bridge, how much force does each pillar exert?

GROUP B

6. A 30.0-N fishing pole is 2.00 m long and has its center of gravity 0.350 m from the heavy end. A fisherman holds the end of the pole in his left hand as he lifts a 100.0-N fish. If his right hand is 0.800 m from the heavy end, how much force must he exert with his right hand to maintain equilibrium?
7. A uniform 2.50-N meterstick is hung from the ceiling by a single rope. A 500.0-g mass is hung at the 25.0-cm mark and a 650.0-g mass at the 70.0-cm mark. (a) What is the tension in the rope? (b) Where is the rope attached to the meterstick?
8. An 850-N painter stands 1.20 m from one end of a 3.00-m scaffold supported at each end by a stepladder. The scaffold weighs 250 N and there is a 40.0-N can of paint 0.50 m from the end opposite the painter. How much force is exerted by each stepladder?
9. A 10.0-N meterstick is suspended by two spring scales, one at the 8.00-cm mark and the other at the 90.0-cm mark. If a weight of 5.00 N is hung at the 20.0-cm mark and a weight of 17.0 N is hung at the 55.0-cm mark, what will be the reading on each scale?

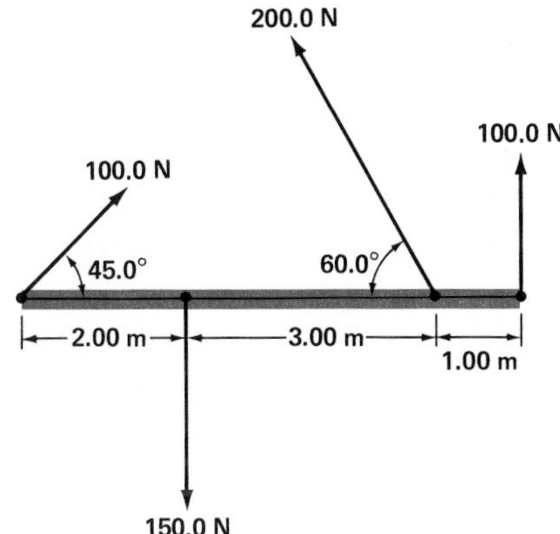

Figure 4-29.

10. (a) Find the torques exerted on the rod in Figure 4-29. (b) Find the magnitude and direction of the additional force that must be exerted at the right end, perpendicular to the rod, to maintain rotational equilibrium.

PHYSICS ACTIVITY

(a) Find the center of gravity of a broom by balancing it lengthwise on your hand. Is the center of gravity closer to the bristles or to the end of the handle?
(b) Balance the broom vertically on the palm of your hand, first with the top of the handle and then with the bristles. Which way is easier? Explain.

SUMMARY

Forces are vector quantities. A net force will change the state of motion of an object. Forces can be exerted over long distances. Every force is accompanied by an opposite force.

A resultant force is a single force that produces the same effect as several forces acting along lines that pass through the same point. The equilibrant force is the single force that produces equilibrium when applied at a point at which two or more concurrent forces are acting. A single force may be resolved into two components that usually act at right angles to each other. Resultant and component forces are found by the parallelogram method.

Friction is a force that resists the motion of objects that are in contact with each other. It acts parallel to the surfaces that are in contact, depends on the nature and smoothness of the surfaces, is virtually independent of position, and is directly proportional to the force pressing the surfaces together. Starting friction is usually greater than sliding friction. The coefficient of friction is the ratio of the force of friction to the perpendicular force pressing the surfaces together. The resolution of forces is used in solving friction problems.

Parallel forces act in the same or in opposite directions. The center of gravity of an object is that point at which all of the object's weight can be considered to be concentrated. A stationary object is in translational and rotational equilibrium. The torque produced by a force is the product of the force and the length of the torque arm on which it acts. To produce equilibrium in parallel forces, the sums of the forces in opposite directions must be equal and the sum of all the clockwise torques must equal the sum of all the counterclockwise torques about a pivot point. Two forces of equal magnitude that act in opposite directions but not along the same line are called a couple.

VOCABULARY

center of gravity
coefficient of sliding friction
concurrent forces
couple
equilibrant force
equilibrium
friction
parallel forces
resolution of forces
resultant force
rotational equilibrium
torque
torque arm
translational equilibrium

5 Two-Dimensional and Periodic Motion

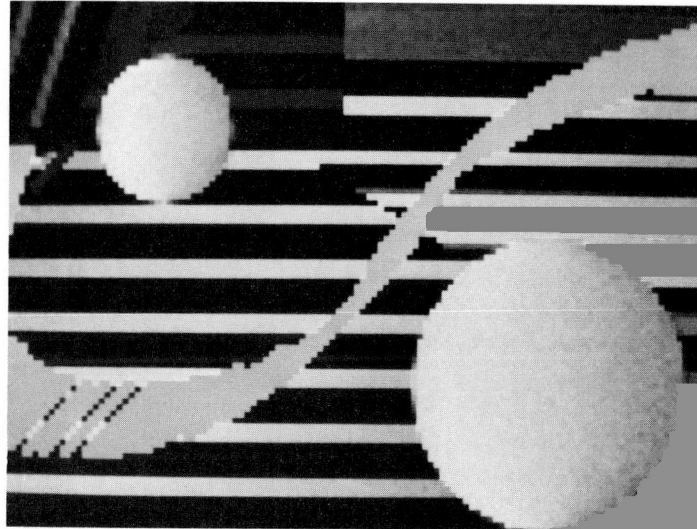

centripetal (sen-TRIP-et-al) adj.: motion that is directed toward a center or axis.

CIRCULAR MOTION

Objectives

- Describe and identify motion in two dimensions.
- Define and calculate centripetal acceleration.
- Distinguish between centripetal and centrifugal forces.
- Define and calculate angular displacement, angular velocity, and angular acceleration.
- Define and identify harmonic motion.

5.1 Motion in a Curved Path In Chapters 3 and 4, we studied the motion of objects along a straight line. Now let us examine what happens to the motion of an object that moves horizontally with constant velocity but is also accelerated vertically. That is, there is a right angle between the direction of the constant velocity and the direction of the acceleration. What will be the path of such an object? The ball in Figure 5-1 illustrates such motion. It was released at the same time as the ball on the left, but had a constant horizontal velocity. The horizontal distances between adjacent images of the right-hand ball are equal. This indicates that the ball is still traveling horizontally with constant velocity. At the same time, the ball on the right covered the same vertical distance between each image as did the ball on the left, which had no horizontal velocity. Each position of the ball on the right is the resultant of motion with constant velocity in one direction and with constant acceleration in the other. This combination produces motion along a curved path.

Two-dimensional motion has a curved path.

Suppose a rifle bullet is fired horizontally with a velocity of 1250 m/s. Neglecting air resistance, the bullet travels 1250 m horizontally by the end of the first second. Immediately after it leaves the muzzle of the gun, the force of gravity begins to accelerate the bullet toward the earth's center of gravity. This force is vertical. During the first second, a freely falling body drops 4.90 m. The bullet

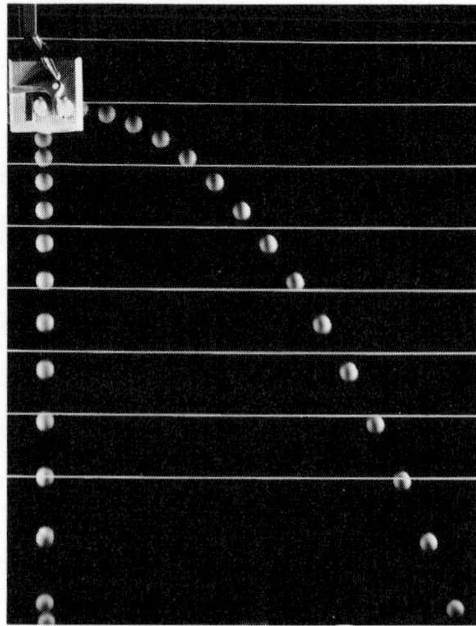

Figure 5-1. Two-dimensional motion. Both balls have the same vertical acceleration due to the force of gravity. But the ball at the right also has a constant horizontal motion. The resultant of motion in two dimensions is a curved path.

drops 4.90 m while traveling the first 1250 m horizontally. See Figure 5-2. In two seconds the bullet travels $25\bar{0}0$ m horizontally, at which time it also drops vertically through a distance of 19.6 m. Over short distances the path of a high-velocity projectile approximates a straight line, but over greater distances its path is noticeably curved.

If the line of sight to a target is horizontal, a projectile must be fired at a small upward angle in order to hit the target. This angle compensates for the downward acceleration due to the force of gravity on the projectile. Provided the same ammunition is used, the size of this angle depends upon the distance to the target. The front sight of most rifles is fixed at the end of the barrel while the rear sight is movable. Since the line of sight is a straight line to the target, the angle the barrel makes with this line is increased by raising the rear sight. When the rifle is aimed, its muzzle is directed upward at the predetermined angle. When the gun is fired, this angle gives the bullet the upward velocity component necessary to compensate for the bullet's drop on its way to the distant target.

The path a projectile takes if it is fired at an upward angle is shown in Figure 5-3. The muzzle velocity, v_m, is resolved into the horizontal component, v_h, and the vertical component, v_v. **AC** is called the range. The path of the projectile, **ABC**, is called the trajectory.

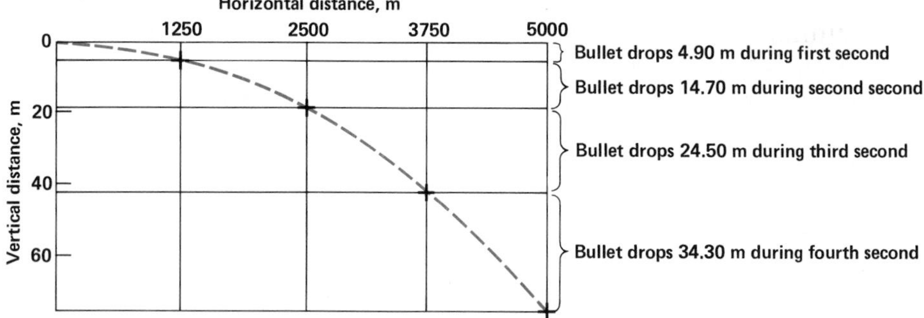

Figure 5-2. The path of a bullet is curved no matter how great the horizontal velocity is.

5.2 Motion in a Circular Path An important type of two-dimensional motion is motion in a circular path, as seen in Figure 5-4. Ball **A** is attached to the end of string **CA.** The string is fastened at **C.** If the speed of the ball in the circular path is constant, the ball is said to describe *constant circular motion.* If the speed of the ball in the circular path varies, its motion is *variable circular motion.*

Imagine that you are at a location where there is no force of gravity. You attach a ball to a string as shown in Figure 5-4 and hold the string at **C.** Then you give the ball an

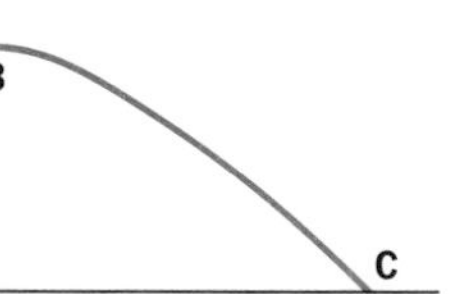

Figure 5-3. When air resistance is not considered, the path of a projectile is a parabola.

initial velocity in the direction of **B,** tangent to the circumference of the circle. The ball whirls in a circular path around **C.** You can feel yourself pulling continuously on the string to keep the ball in the circular path.

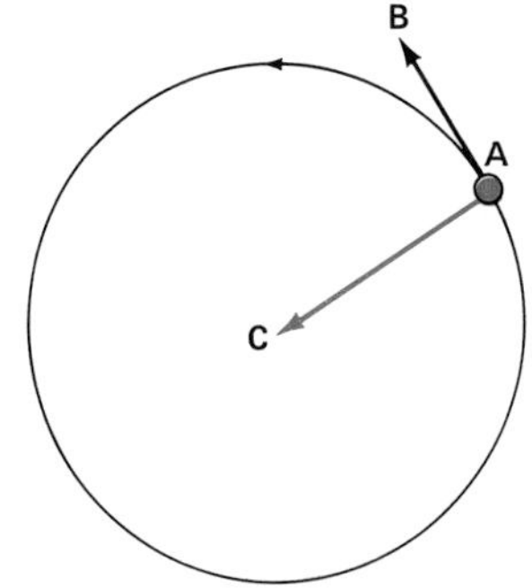

Figure 5-4. Constant circular motion. When the force along **AC** stops acting, the ball moves in a direction that is indicated by the tangent **AB.**

Let us analyze what you would observe in this situation. We know that the radius of a circle, **AC** for example, is perpendicular to the tangent drawn through the end of the radius. If the string is in the direction of the radius and the velocity is always directed along a tangent, your pull on the string is always directed perpendicularly to the velocity. Your pull accelerates the ball into a circular path, but the ball does not speed up or slow down. Your pull changes only the direction of the velocity but not the magnitude of the velocity. If your pull were not perpendicular to the velocity, a component of the acceleration in the direction of the ball's motion would exist and the speed of the ball would change.

In this example, the acceleration is directed toward the center of a circle. *Acceleration directed toward a central point is called* ***centripetal acceleration.*** (The word centripetal means "directed toward a center.")

Centripetal: toward a central point.

This example can also be analyzed by means of velocity vectors. Even though the *speed* of the ball in its circular path is uniform, the *velocity* is constantly changing. In Figure 5-5(A), **A** and **B** represent two successive positions of an object moving with constant circular motion about point **O.** The velocity vector v_i indicates the velocity of the object at point **A,** and the velocity vector v_f indicates its velocity at point **B.** The velocity vectors v_i and v_f are tangent to the circle at **A** and **B** respectively, and are thus perpendicular to the respective radii **OA** and **OB.**

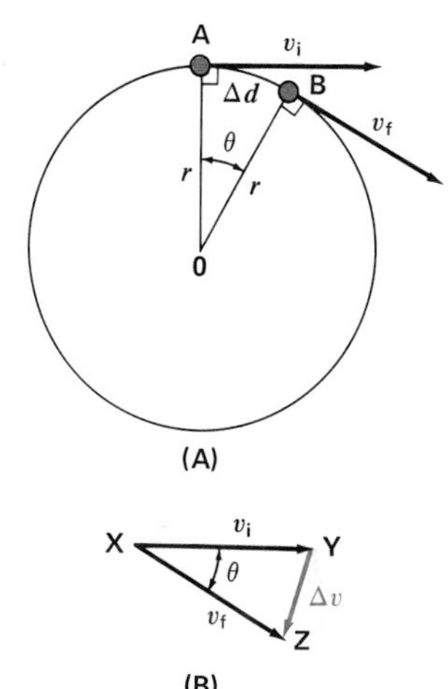

Figure 5-5. Centripetal acceleration. The change of velocity of an object in circular motion is directed toward the center of the circle.

In order to study the change of velocity between v_i and v_f, a separate vector diagram, Figure 5-5(B), is drawn in which v_i and v_f originate at point **X.** If v_f is the resultant and v_i one of its components, then Δv is the other component. This vector Δv represents the change in velocity between v_i and v_f. Because the vectors v_i and v_f are equal in magnitude and are perpendicular to their respective radii, θ in Figure 5-5(A) equals θ in Figure 5-5(B) and triangle **ABO** is similar to triangle **XYZ.** Then

$$\frac{\Delta v}{v_i} = \frac{\text{chord AB}}{r}$$

If angle θ is made smaller, arc Δd becomes more nearly equal to chord **AB** and can be substituted for it.

$$\frac{\Delta v}{v_i} = \frac{\Delta d}{r}$$

The equation $v = \frac{\Delta d}{\Delta t}$ is derived in Section 3.3.

But v_i and v_f are equal in magnitude and can be represented by v. If we let Δt represent the time between v_i and v_f and use the equation $\Delta d = v\,\Delta t$, we can write

$$\frac{\Delta v}{v} = \frac{v\,\Delta t}{r}$$

Transposing terms

$$\frac{\Delta v}{\Delta t} = \frac{v^2}{r}$$

But $\Delta v/\Delta t$ is the centripetal acceleration directed along a radius of the circle and can be represented by a. Then

$$a = \frac{v^2}{r}$$

in which a is the centripetal acceleration of the object, v is its speed along the circular path, and r is the radius of its circular path.

Refer to Section 3.9.

By Newton's second law of motion, $F = ma$. Thus the force producing centripetal acceleration can be written

$$F_c = \frac{mv^2}{r}$$

and is called the *centripetal force*. The centripetal force is the net force directed toward the center of the circle. The equation for centripetal force indicates that it is directly proportional to the mass of the object, directly proportional to the square of the object's speed along the circular path, and inversely proportional to the radius of the object's circular path.

Because matter has inertia, a centripetal force is required to produce the centripetal acceleration that changes the direction of a moving object. If the centripetal force applied on the ball by the string ceases to act (for example, if the string breaks) the ball will move in a straight line tangent to the curved path at the point where the string breaks. This behavior is in accordance with Newton's first law of motion. When the string breaks, the centripetal force no longer acts on the ball; the ball has no unbalanced forces acting on it and thus moves in a straight line with constant speed.

EXAMPLE A rubber stopper of mass 0.013 kg is swung at the end of a cord 0.85 m long with a period of 0.65 s. Find the centripetal force that is exerted on the stopper.

Given	Unknown	Basic equations
$m = 0.013$ kg	F_c	$F_c = \frac{mv^2}{r}$
$r = 0.85$ m		
$T = 0.65$ s		$v = \frac{\Delta d}{\Delta t} = \frac{2\pi r}{T}$

Solution

Substituting the second Basic equation into the first gives the

$$\textit{Working equation: } F_c = \frac{m4\pi^2 r}{T^2}$$

$$= \frac{(0.013 \text{ kg})(4)(\pi^2)(0.85 \text{ m})}{(0.65 \text{ s})^2}$$

$$= 1.0 \text{ N}$$

PRACTICE PROBLEMS **1.** A mass moves in a circular path at a velocity of 1.5 m/s and a centripetal acceleration of 3.6 m/s^2. What is the radius of the circular motion of the mass? *Ans.* 0.63 m

2. A cord 0.65 m long exerts a centripetal force of 11.6 N on a whirling 0.10-kg mass tied to the end of the cord. What is the velocity of the whirling mass? *Ans.* 8.7 m/s

---

5.3 Motion in a Vertical Circle Thus far, in describing the motion of an object in a circle, we have ignored the effect of the force of gravitation. This was done in order to simplify the analysis of the relationship between centripetal acceleration and the magnitude of the object's velocity. However, if the object moves in a vertical circle with gravity acting on it, the situation is more complex.

Suppose a ball on the end of a string moves along a circular path. The plane of the circle is vertical. Figure 5-6 is a diagram of the ball's motion. Because of the force of gravity, the speed of the ball in the circular path is not constant. The ball accelerates on the downward part of its path and decelerates on the upward part. The speed of the ball is a minimum at the top of the circle and a maximum at the bottom. Consequently, the centripetal force is at a minimum at the top of the circle and at a maximum at the bottom. Let us see what forces comprise the centripetal force to make this so.

At the top of the circle the ball has velocity v_{min}. The

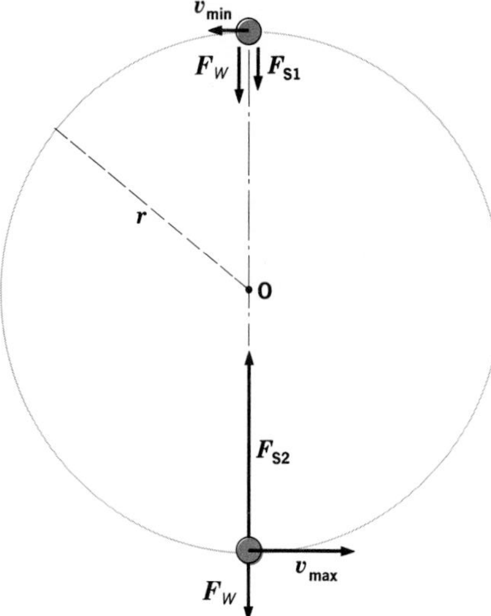

Figure 5-6. Motion in a vertical circle. When the ball is at its highest point, the centripetal force on the ball equals the tension of the string plus the weight of the ball. At its lowest point, the weight of the ball is subtracted from the tension to find the centripetal force.

centripetal force, F_{c1}, is the sum of the force the string exerts on the ball, F_{s1}, and the weight of the ball F_W, since these both act toward the center of the circle.

$$F_{c1}=\frac{mv_{min}^2}{r}=F_{s1}+F_W \qquad \text{(Equation 1)}$$

At the bottom of the circle, the ball has velocity v_{max}. The centripetal force, F_{c2}, is the difference between the magnitude of the force the string exerts on the ball, F_{s2}, and the weight of the ball, F_W, since these forces now act in opposite directions. The force the string exerts is inward, and the weight of the ball acts outward.

$$F_{c2} = \frac{mv_{max}^2}{r} = F_{s2} - F_W \qquad \text{(Equation 2)}$$

Observe that since v_{max} is greater than v_{min}, F_{c2} is greater than F_{c1}. Further, since the weight of the ball is added in Equation 1, but subtracted in Equation 2, F_{s2} is greater than F_{s1} by a ratio greater than the ratio F_{c2}/F_{c1}.

When a ball on a string moves in a vertical circle, there is a certain velocity below which the ball will not describe a circular path—the string slackens as the ball approaches its highest point. To find this minimum value of the velocity, we shall use Equation 1. When the string begins to slacken, $F_{s1} = 0$ and

$$\frac{mv_{min}^2}{r} = F_W$$

But

$$F_W = mg$$

So

$$\frac{mv_{min}^2}{r} = mg$$

Solving for v_{min}

$$v_{min} = \sqrt{rg}$$

A satellite in circular orbit around a planet moves with critical velocity.

This value of v_{min} is called the *critical velocity*. As the equation indicates, critical velocity depends only on the acceleration due to gravity and the radius of the vertical circle. The critical velocity does not depend on the mass of the object describing the motion. This is true because the force of gravity and the centripetal force are both proportional to the mass of the object.

5.4 Frames of Reference Figure 5-7 shows a chamber attached to the end of a long boom. The boom is designed to turn around a central point under the observation booth at the top of the picture. In operation, the chamber moves at constant speed in a large circle around the central point.

Suppose you are in the chamber while it is moving. You would still feel the pull of gravity. In addition, you would feel your body being pressed against the outside wall of the chamber. You would also observe that a ball placed on the floor would move across the floor toward the outside wall of the chamber. If you dropped an object in the chamber, you would see it fall to the floor in a curved path.

From these observations it would seem that Newton's law of inertia does not hold true in the moving chamber. These observations can be explained, however, by attributing them to a force that tends to move all particles toward the outside wall of the moving chamber. This force moves the ball across the floor and accelerates the falling object so that its path is curved. *A force that tends to move the particles of a spinning object away from the spin axis is called **centrifugal force.*** Centrifugal force exists only for an observer in an accelerating system that is considered to be stationary.

Centrifugal: away from a central point.

An observer in the booth above the moving chamber in Figure 5-7 interprets the situation quite differently, however. To this observer, the chamber and its occupant tend to continue their motion in a straight line because of their inertia. A force along the boom, a centripetal force, causes the chamber to follow a circular path. And a force exerted by the wall of the chamber on its occupant causes the occupant also to follow a circular path. The observer in the booth can explain the behavior of objects in the chamber in terms of inertia and centripetal force. To this observer, no centrifugal force is involved.

Figure 5-7. This centrifuge is used to study the effects of angular acceleration on the human body. At the left, a volunteer is entering the test chamber of the centrifuge.

To resolve this seeming contradiction, it is helpful to discuss this situation in terms of *frames of reference.* Each observer assumes that some surrounding objects are stationary because they do not move with respect to each other. To the observer in the moving chamber, the walls and floor of the chamber are stationary and are used as the basis of measurements. Other objects, such as the ball on the floor, move with respect to these stationary objects. The walls and floor of the chamber are the basis for this observer's *frame of reference.*

*A **frame of reference** is a system for describing the location of objects.* It is used to specify the positions and relative motions of objects. A frame of reference in which Newton's first law holds true is called an *inertial frame.* An accelerating frame of reference is *noninertial* because Newton's first law does not hold true.

Centripetal and centrifugal are never used to describe motion in the same frame of reference.

The observer in the booth in Figure 5-7 is stationary with respect to the earth. Strictly speaking, however, a frame of reference based on the earth is not an inertial frame. The earth spins and orbits around the sun and describes continuous acceleration in doing so. However, the earth is so close to being an inertial frame that we can consider a frame of reference based on the booth as an inertial frame. (For practical purposes, we often neglect the accelerations of frames of reference that are stationary with respect to the earth.)

The direction of motion of the moving chamber is continuously changing. With respect to the observer in the booth, the moving chamber is accelerating. It is, therefore, a noninertial frame. Using such a frame, one can observe acceleration that is not attributed to such forces as friction or gravitation. The concept of centrifugal force is necessary in such studies. Centrifugal force is sometimes called fictitious because it is not involved when we decide to use an inertial frame to describe motion.

QUESTIONS: GROUP A

1. (a) What is a projectile? (b) What is its path called?
2. Use Figure 5-2 to answer Questions 2 and 3. (a) What is the horizontal distance traveled by the bullet in the first second? In the third? (b) How far has the bullet traveled horizontally and vertically in the first two seconds? (c) How would the answers to (b) change if the bullet's initial speed had been 1700 m/s?
3. (a) Is there a force acting on the projectile? (b) If so, what is its direction?
4. What relationship between the direction of force and velocity is necessary to produce circular motion?
5. Is an object moving in a circle at constant speed accelerating?

GROUP B

6. Why does it take an unbalanced force to produce circular motion?
7. What provides the centripetal force (a) when a car goes around a curve? (b) when the moon orbits the earth? (c) when you are a passenger in a car going around a curve?
8. How does motion in a vertical circle differ from the motion of an object moving in a horizontal circle?
9. Explain how the concept of critical velocity could be used in the design of an amusement park ride.
10. Why do physicists call centrifugal force "fictitious"?
11. What is meant by a frame of reference?

PROBLEMS: GROUP A

1. A plane flying horizontally at 350 m/s releases a package at an altitude of 1.50×10^3 m. (a) How long will the package take to reach the ground? (b) How far will it move horizontally while falling?
2. In a TV tube, electrons are projected at a speed of 9.5×10^5 m/s toward the screen 0.45 m away. If gravity were the only force acting, how far would

the electron fall as it moves to the screen?

3. In setting up a chase scene for a movie, a stuntwoman builds a 45.0° ramp. If she drives a car off the top of the ramp at 50.0 m/s, what is the maximum distance a similar landing ramp should be placed?
4. A 25.0-kg child moves with a speed of 1.93 m/s when sitting 12.5 m from the center of a merry-go-round. Calculate (a) the centripetal acceleration and (b) the centripetal force.
5. The earth orbits the sun at a distance of 1.50×10^{11} m. (a) What is the centripetal acceleration of the earth in its orbit? (b) What is the centripetal force and what provides it?

GROUP B

6. A 13 500-N car traveling at 50.0 km/h rounds a curve of radius 2.00×10^2 m. Find (a) the centripetal acceleration of the car, (b) the centripetal force, and (c) the minimum coefficient of friction between the tires and the road so that the car can round the curve safely. (d) On a rainy day, the coefficient of friction is 0.050. What is the maximum safe speed of the car under these conditions?
7. A looping roller coaster ride at an amusement park has a radius of curvature of 7.50 m. At what minimum speed must the coaster be traveling at the top of the curve so the passengers will not fall out?
8. A physics student is twirling a 50.0-g rubber stopper attached to a 0.950-m length of cord at a uniform speed in a vertical circle. If its speed is 3.50 m/s, what is the tension in the cord at (a) the bottom of the circle and (b) the top of the circle?
9. A pilot pulls her jet out of a dive by swinging up in an arc of radius 3.80 km at a speed of 450.0 m/s. (a) What is the plane's centripetal acceleration? (b) How many *g*'s does the pilot experience?
10. A satellite is orbiting the earth at a distance of 1.00×10^3 km. What is its speed?

ROTARY MOTION

5.5 Motion Around an Axis ***Rotary motion*** *is the motion of a body about an internal axis.* Rotary motion occurs in a spinning bicycle wheel, the spinning crankshaft of an automobile engine, and a wheel attached to the spinning shaft of an electric motor. Note the difference between circular motion and rotary motion. In circular motion, the axis of the motion is not part of the object. In rotary motion, the axis of the motion is part of the moving object. A spinning wheel is in rotary motion; an object on the rim of the wheel describes circular motion.

Rotating and spinning are synonymous.

For rotary motion to be *constant,* the object must spin about a fixed axis at a steady rate. The movement of the hands of a clock is an example of constant rotary motion. If either the direction of the axis or the rate of spin varies, the rotary motion is *variable.* The movements of automobile wheels as a car is driven at different speeds and the move-

In radian measure,

$\theta = \Delta d / r$. *Hence,*

$$\omega = \frac{\Delta d / r}{\Delta t} = \frac{v}{r}$$

ments of a spinning top as it slows down are examples of variable rotary motion.

5.6 Angular Velocity For constant linear motion, velocity is defined as the time rate of displacement. Similarly, for constant rotary motion, ***angular velocity*** is defined as *the time rate of angular displacement.* Angular displacement is the angle about the axis of rotation through which the object turns. The symbol for angular velocity is the Greek letter ω (omega). The equation for angular velocity is

$$\omega = \frac{\Delta \theta}{\Delta t}$$

where $\Delta\theta$ is an angular displacement and Δt is the time interval in which the angular displacement occurs. The units of angular velocity are revolutions per second, degrees per second, or *radians per second.*

An angle of one radian is the angle that, when placed with its vertex at the center of a circle, subtends on the circumference an arc equal in length to the radius of the circle. Thus in Figure 5-8, if radius r' is rotated from radius r until arc Δd = radius r, the angle θ is 1 radian. Since the circumference of a circle is 2π times the radius

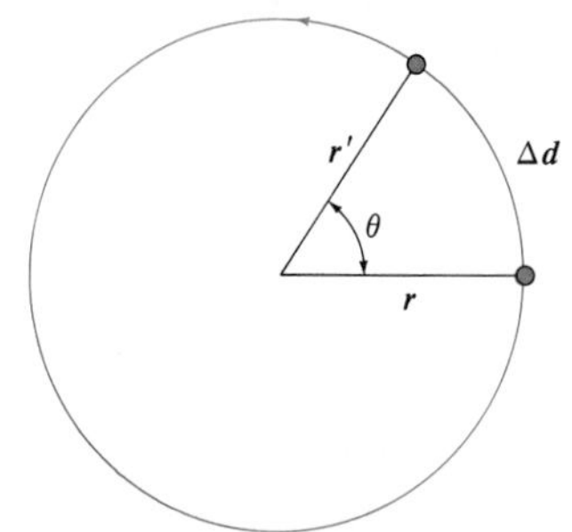

Figure 5-8. Radian measure. Since arc Δd equals radius r, angle θ is 1 radian. There are 2π radians in one revolution.

$$\textbf{1 revolution} = 360° = 2\pi \textbf{ radians}$$

and $$\textbf{1 radian} = \frac{360°}{2\pi} = \frac{180°}{\pi}$$

$$\textbf{1 radian} = 57.3°$$

Since an angle measured in radians equals the ratio of the length of the subtended arc to the length of the radius, the units cancel, and the angle in radians is a pure number.

The concept of angular velocity includes both the rate of rotation and the direction of the axis of rotation. Angular velocity is a vector quantity represented by a vector along the axis of rotation. The length of the vector indicates the magnitude of the angular velocity. The direction of the vector is the direction in which the thumb of the right hand points when the fingers of the right hand encircle the vector in the direction in which the body is rotating. See Figure 5-9.

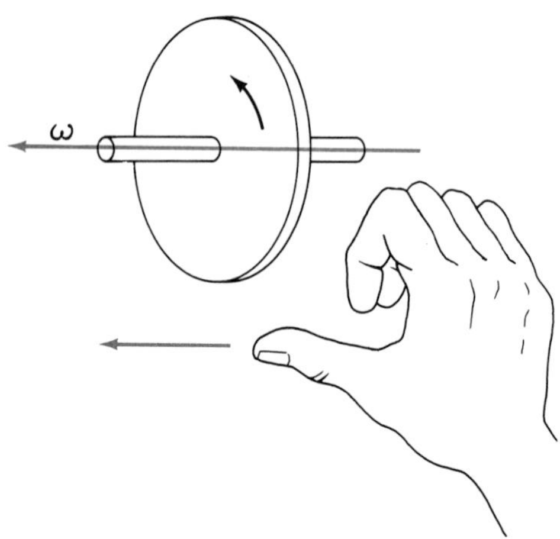

Figure 5-9. Right-hand rule of angular velocity. The vector representing angular velocity points in the direction of the thumb of the right hand when the fingers encircle the axis in the direction of rotation.

5.7 Angular Acceleration From the study of linear motion we know that a change of velocity defines acceleration. The same is true for rotary motion. Changing either the rate of rotation or the direction of the axis involves a change of angular velocity and thus defines angular acceleration. The constant rate of change of linear velocity is known as linear acceleration (Section 3.5). Similarly, *the constant rate of change of angular velocity* is known as ***angular acceleration*** and designated as α (alpha).

$$\alpha = \frac{\Delta\omega}{\Delta t}$$

The equations for constantly accelerated linear motion can be transformed into the corresponding equations for constantly accelerated rotary motion by substituting $\Delta\theta$ for Δd, ω for v, and α for a. See Table 5-1.

Table 5-1
EQUATIONS FOR UNIFORMLY ACCELERATED MOTION

Linear	*Rotary*
$v_f = v_i + a\Delta t$	$\omega_f = \omega_i + \alpha\Delta t$
$\Delta d = v_i\Delta t + \frac{1}{2}a\Delta t^2$	$\Delta\theta = \omega_i\Delta t + \frac{1}{2}\alpha\Delta t^2$
$v_f = \sqrt{v_i^2 + 2a\Delta d}$	$\omega_f = \sqrt{\omega_i^2 + 2\alpha\Delta\theta}$

EXAMPLE Find the angular displacement after 15.0 s (in radians) for a wheel that accelerates at a constant rate from rest to 725 rev/min in 10.0 min.

Given	Unknown	Basic equation
$\Delta t = 15.0$ s	$\Delta\theta$	$\Delta\theta = \frac{1}{2}\alpha\Delta t^2$
$\alpha = \dfrac{725 \text{ rev/min}}{10.0 \text{ min}}$		

Solution

Since α is expressed in minutes and t in seconds, make the time units consistent:

$$\Delta t = 15.0 \text{ s} = 15.0 \text{ s}(1 \text{ min}/60 \text{ s}) = 0.25 \text{ min}$$

We can now solve for $\Delta\theta$:

Working equation: $\Delta\theta = \frac{1}{2}\alpha\Delta t^2 = \frac{1}{2}$ (72.5 rev/min^2)(0.25 min)^2 (2π rad/rev)
$= 14.2$ rad

EXAMPLE A flywheel has a constant angular acceleration of 8.4 rad/s^2. How long will it take the wheel to attain an angular velocity of 6.0 rev/s, starting from rest?

Given	Unknown	Basic equation
$\alpha = 8.4\ \text{rad/s}^2$ $\omega = 6.0\ \text{rev/s}$	Δt	$\alpha = \dfrac{\Delta\omega}{\Delta t}$

Solution

$$\textit{Working equation: } \Delta t = \frac{\Delta\omega}{\alpha}$$

$$= \frac{(6.0\ \text{rev/s})(2\pi\ \text{rad/rev})}{8.4\ \text{rad/s}^2}$$

$$= 4.5\ \text{s}$$

PRACTICE PROBLEM A wheel starts from rest and attains an angular velocity of 750 rev/min in 1.5 min with constant angular acceleration. Find the angular acceleration in rad/s^2. *Ans.* 0.87 rad/s^2

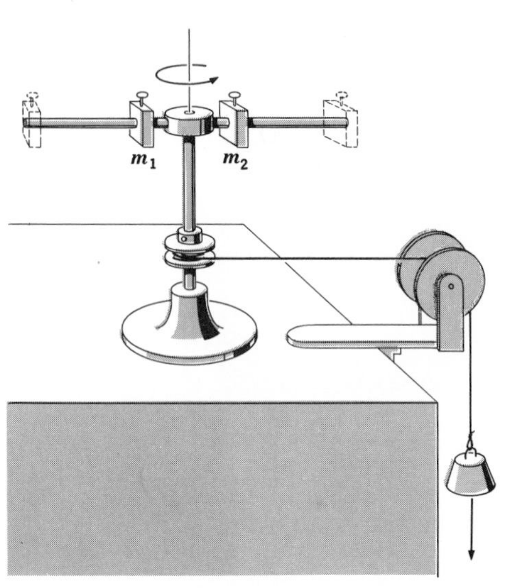

Figure 5-10. Rotational inertia. When the masses, m_1 and m_2, are moved outward on the rotating bar, the rotational inertia of the system increases.

5.8 Rotational Inertia

A wheel mounted on a shaft will not start to spin unless a torque is applied to the wheel. (See Section 4.11.) A wheel that is spinning will continue to spin at constant angular velocity unless a torque acts on it. Thus if we replace "force" with "torque," Newton's law of inertia also applies to rotary motion.

If we wish to change the rate of rotation of an object about an axis, we must apply a torque about the axis. The angular acceleration that this torque produces depends on the mass of the rotating object and upon the distribution of its mass with respect to the axis of rotation.

In Figure 5-10 two masses, m_1 and m_2, are mounted on a bar that is fastened to an axle. The masses can be placed on the bar at varying distances from the axle. If a cord is wound around the axle and attached to a weight that falls under the influence of gravity, a constant torque is applied to the axle. If different weights are attached to the cord, different torques are applied. If the masses mounted on the bar remain in a fixed position, the greater the torque, the greater will be the angular acceleration. If the masses are placed near the axis of rotation, the acceleration produced by a given torque is greater than if they are placed at the ends of the bar. Moving the masses farther apart does not change the amount of mass that is rotated. It does, however, change the distribution of mass. We say that it increases the *rotational inertia*. ***Rotational inertia*** *is the resistance of a rotating object to changes in its angular velocity.* The angular acceleration, α, is directly proportional to

Rotational inertia resists changes in rotation.

the torque, T, but inversely proportional to the rotational inertia, I.

$$\alpha = \frac{T}{I}$$

Transposing $$T = I\alpha$$

which for rotary motion is analogous to $F = ma$ for linear motion. The torque, T, is computed by

$$T = Fr$$

where F is the force applied tangentially at distance r from the pivot point. Like a force, a torque has magnitude and direction. As in the case of angular velocity, the direction of the torque vector is the direction in which the thumb of the right hand points when the fingers of the right hand encircle the vector in the direction in which the mass is accelerating. Torques resemble quantities that can be represented by vectors.

Rotational inertia takes into account both the shape and the mass of the rotating object. Equations for the rotational inertia of certain regularly shaped bodies are given in Figure 5-11. Rotational inertia has the unit designations kg · m². The following unit analysis verifies this.

$$I = \frac{T}{\alpha} = \frac{\mathbf{m \cdot N}}{\mathbf{1/s^2}} = \frac{\mathbf{kg \cdot m^2/s^2}}{\mathbf{1/s^2}} = \mathbf{kg \cdot m^2}$$

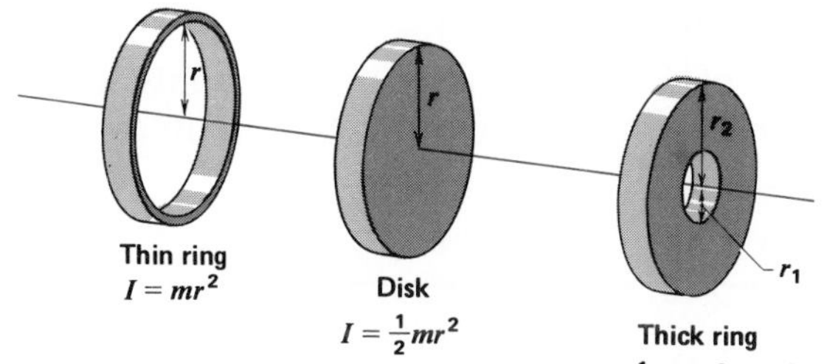

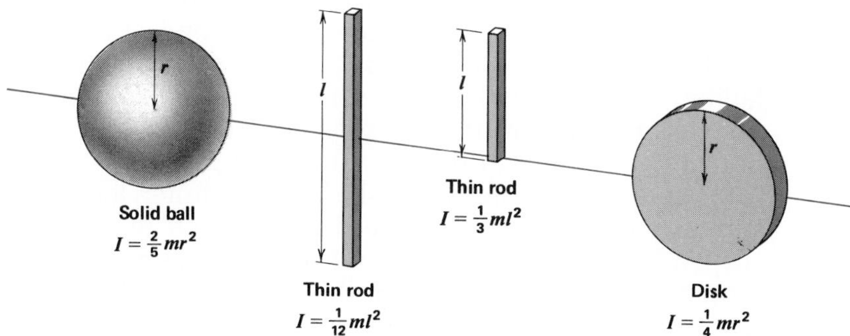

Figure 5-11. Rotational inertia equations for objects rotating about the indicated axes. (Be sure to note the distinction between the symbol **m** for meter and the symbol ***m*** for mass.)

EXAMPLE A force of 2.4 N is applied tangentially to the rim of a disk with a radius of 9.5 cm and a mass of 0.60 kg. Find the resulting angular acceleration (in rev/s^2) of the disk.

Given	Unknown	Basic equations
$F = 2.4$ N	α	$T = I\alpha$
$r = 9.5$ cm		
$m = 0.60$ kg		$T = Fr$

Solution

Since for a disk $I = \frac{1}{2}mr^2$, we get the

$$\textit{Working equation: } \alpha = \frac{Fr}{I} = \frac{2F}{mr} = \frac{2(2.4\text{ N})}{(0.60\text{ kg})(0.095\text{ m})}$$

$$= \frac{84\text{ rad/s}^2}{2\pi\text{ rad/rev}}$$

$$= 13\text{ rev/s}^2$$

PRACTICE PROBLEMS **1.** A ring has a mass of 0.80 kg, a diameter of 45 cm, and a hole 12 cm wide. Find the rotational inertia when it is rotating about an imaginary axis perpendicular to the plane of the ring and passing through its center. *Ans.* 0.022 kg · m^2

2. A flywheel is shaped like a thin ring and has a mass of 15 kg and a diameter of 0.50 m. A torque of 6.0 m·N is applied tangentially to the rim of the wheel. How long will it take the flywheel to attain an angular velocity of $2\bar{0}$ rad/s? *Ans.* 3.1 s

---

A rotating wheel is sometimes called a gyroscope.

A gyroscope does not precess in the absence of gravity or in a freely falling spaceship unless a torque is exerted on it.

5.9 Precession Figure 5-12 shows a spinning bicycle wheel that seems to be defying gravity. The only means of support of the wheel is the cord that is holding up the left end of the shaft. Yet if the wheel is spinning fast enough, its shaft will remain almost horizontal in space. Instead of falling, the shaft will slowly rotate counterclockwise (as seen from above) around the axis marked by the cord. This horizontal rotation of the shaft is called *precession.*

In Figure 5-12, **OA** is the shaft of a counterclockwise-spinning wheel. **OC** is the cord from which the shaft is suspended. The weight of the wheel produces a clockwise torque about **O** (as seen from the front). ω_s is the vector representing the angular velocity of the spinning wheel.

In Figure 5-13, α describes the angular acceleration due to the torque produced by Fw. (α is horizontal and directed into the paper.) Because of this angular acceleration there will be a change in the angular velocity during the short time Δt, or $\Delta\omega_\alpha = \alpha\Delta t$. This, added vectorially to ω_S, results in a new angular velocity ω'_S of the same magnitude but different direction. (Since the angular acceleration is small, it can be considered to be at right angles to both the initial and new angular velocities.) Because of the way the wheel is supported, if the new axis is now along ω'_S, the shaft of the wheel has to turn toward this new position. The analysis has to be repeated for successive time intervals and the result is the precessional motion of the wheel.

The motion of the earth is a good example of precession since it spins about an axis that is tilted with respect to its plane of revolution around the sun. The rotation of the earth produces an equatorial bulge. The gravitational pull of the sun on the part of the bulge closest to the sun is stronger than the pull on the part of the bulge on the other side of the earth. Because of the resulting torque, the earth's axis undergoes precessional motion and does not point constantly at the same place in space. See Figure 5-14. The earth's axis completes a single precessional cycle in 26 000 years.

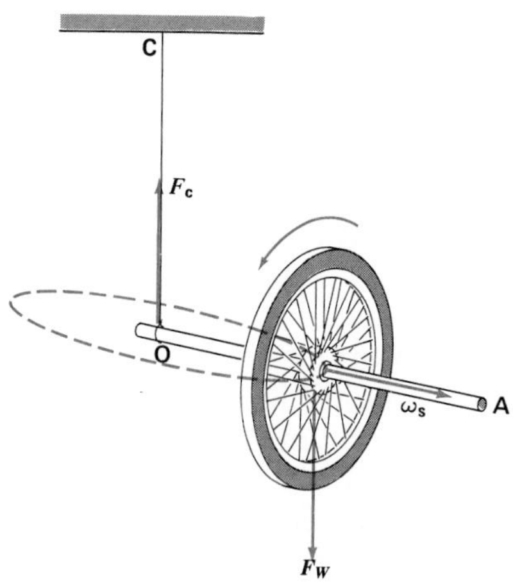

Figure 5-12. Precession. When the wheel rotates, the angular velocity combined with the torque produced by the weight of the wheel causes a precession.

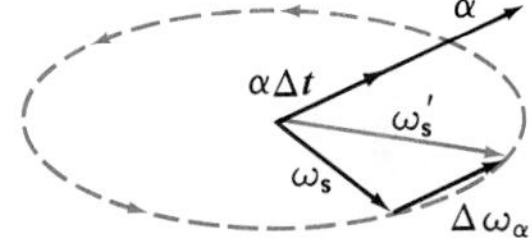

Figure 5-13. Vector analysis of precession. All vectors are in the same horizontal plane. When vector $\Delta\omega_\alpha$ is small, vectors α, $\alpha\,\Delta t$, and $\Delta\omega_\alpha$ are all perpendicular to vectors ω_S and ω_S'.

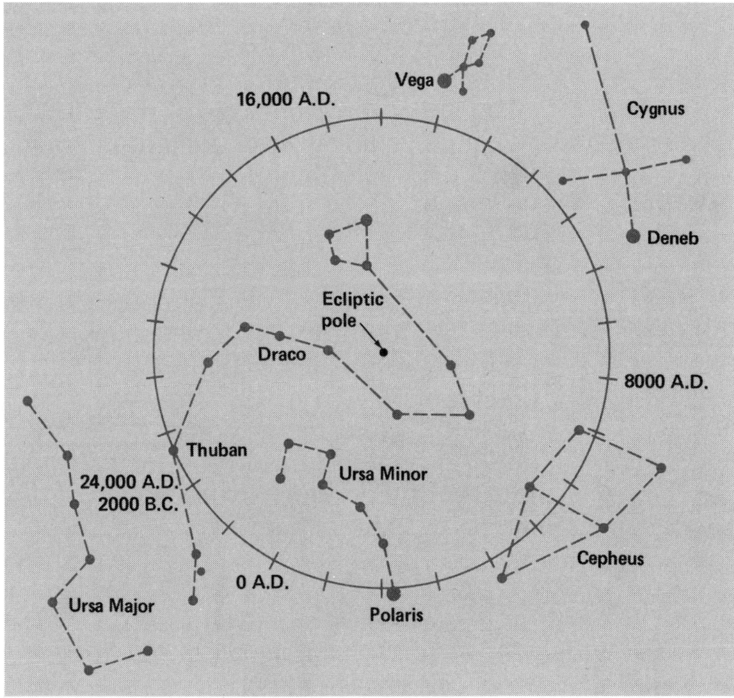

Figure 5-14. Precession in the northern sky. The present North Star is Polaris, because the northern end of the earth's axis points near that bright star. As the earth precesses, however, other stars will become "north stars."

QUESTIONS: GROUP A

1. (a) Distinguish between rotary and circular motion. (b) What are the earth's periods of rotation and revolution?
2. What is a radian?
3. How are angular and linear velocity related?
4. In an ice show, a line of ten skaters is moving in a large arc with Skater 1 as the pivot. (a) How does the angular speed of Skater 1 compare to that of Skater 5 and that of Skater 10? (b) How do their linear speeds compare?

GROUP B

5. (a) How does the rotational inertia of a high diver making a dive in a tuck (closed) position compare to that when the same diver is in a layout (open) position? (b) In which position would you expect the diver to be able to perform a 2.5-somersault dive in a shorter time? Why?
6. Under what circumstances are the equations for linear and rotary motion analogous to each other?
7. (a) What star was the "North Star" in 2000 B.C.? (b) Why is there a different North Star today? (c) What caused this change?

PROBLEMS: GROUP A

1. Convert the following angular displacements to radians. (a) 1.60 rev, (b) 235°, (c) 7.80 rev, (d) 75.0°, (e) 0.68 rev
2. (a) A diver leaps from a high tower. What is his angular displacement in both degrees and radians if he completes a dive with one and a half somersaults? (b) If his dive has two and a half twists?
3. If the diver in Problem 2 simultaneously completes his somersaults and twists in 1.50 s, what is his average angular velocity for each?
4. What is the rotational inertia of a spinning 0.20-kg baseball with a diameter of 15 cm? (Assume the baseball is a uniform sphere.)

GROUP B

5. (a) What is the angular acceleration of a 25.4-cm diameter record if it takes 1.25 s to reach 33.3 rev/min? (b) Through what angular displacement does it turn during this time?
6. A spinning bicycle wheel 65.0 cm in diameter is rotating at 2.00 rev/s and comes to rest in 15.0 s. (a) What is the angular acceleration of the wheel? (b) If the wheel has a mass of 1.75 kg, what frictional force in its bearing has caused it to stop? (Why can the mass of the hub be ignored?)
7. A merry-go-round has a mass of 500.0 kg and a radius of 2.00 m. What tangential force must be applied to change its speed from 0.00 to 2.5 rad/s in 5.0 s? The frictional force is 75.0 N.

PHYSICS ACTIVITY

Obtain four identical cans of soup. Securely tape two cans to a meterstick, centering one at the 40-cm mark and the other at the 60-cm mark. Similarly, tape cans to a second meterstick at the 5-cm and 95-cm marks. Hold each meterstick at the 50-cm mark and twirl it. Which meterstick is harder to twirl? Why?

HARMONIC MOTION

Periodic: repeated in the same way in equal time intervals.

5.10 Periodic Motion *When a body moves repeatedly over the same path in equal intervals of time, it is said to have*

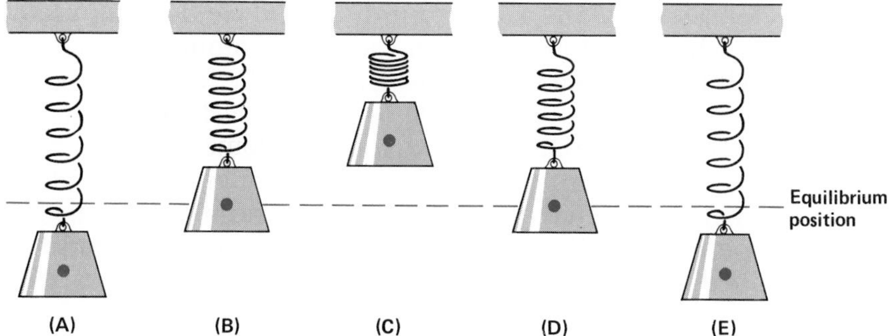

Figure 5-15. Simple harmonic motion. The frequency of vibration of the mass depends on the stiffness of the spring and the magnitude of the mass.

periodic motion. Figure 5-15 shows a mass attached to a spring that is supported from a horizontal beam. If we pull down on the mass and then let it go, it will vibrate up and down in periodic motion. Let us analyze the periodic motion of this mass.

If we exert a force, F, acting downward on the mass, we displace it from its equilibrium position by a distance d. We also find by experiment that if we exert a force $2F$ we displace the mass a distance $2d$. The downward displacement from the equilibrium position is directly proportional to the downward force we exert.

In exerting this force downward we pull against the spring, and the spring also pulls against our hand with a force $-F$ (law of interaction). When we release the mass (A), the upward force of the spring exceeds the downward pull of gravity on the mass. An upward acceleration that is directly proportional to the excess force is produced (law of acceleration).

As the mass approaches its equilibrium position (B), less and less force is exerted on it by the spring. Consequently, the acceleration decreases. The net force is zero when the mass is at its equilibrium position. The acceleration is also zero. The mass has acquired its maximum velocity at this point. It moves past its equilibrium position because of its inertia. As the overshoot occurs, the upward pull of the spring becomes less than the downward force on the mass. There is a net downward force on the mass, and this downward force decelerates it. The mass stops (C). (During the upward movement of the mass, the spring's pull may change to a downward force due to compression.) The downward force then accelerates it again (D and E). Because of the upward force exerted by the spring and the downward force due to gravity, the mass describes an up-and-down motion.

If we neglect the effect of friction in the spring, the mass moves above the equilibrium position the same distance it moves below the equilibrium position, and it completes each up-and-down cycle in the same amount of time. At each point in the up-and-down cycle, the force exerted on the mass, and therefore the resulting acceleration, are directly proportional to the displacement of the mass from the equilibrium position. Both the force and the acceleration are directed toward the equilibrium position. The type of periodic motion that has these characteristics is called *simple harmonic motion.*

Harmonic motion: acceleration directed toward an equilibrium position and proportional to the distance from that position.

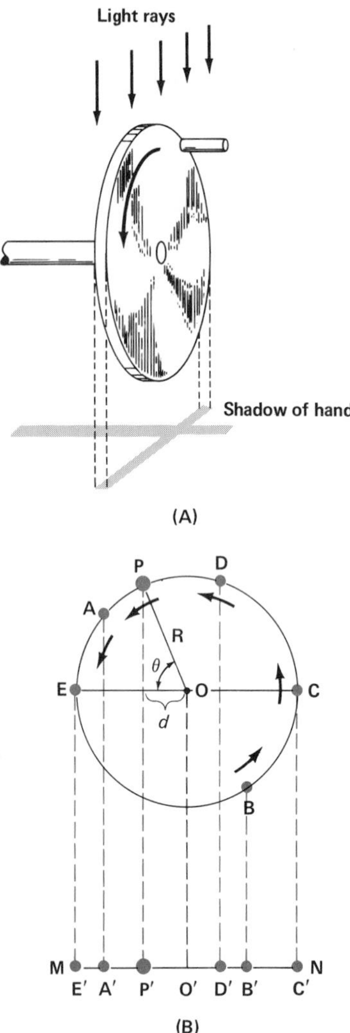

Figure 5-16. Simple harmonic motion can be analyzed in terms of circular motion.

5.11 Harmonic Motion

Simple harmonic motion *is linear motion in which the acceleration is proportional to the displacement from an equilibrium position and is directed toward that position.* Simple harmonic motion can be analyzed in terms of circular motion. In Figure 5.16(A) a light is shining directly down on a rotating wheel. As the wheel spins, the handle on the wheel describes circular motion. The shadow of the wheel and handle falls perpendicularly on the horizontal surface below. The shadow of the handle moves back and forth along the shadow of the wheel. The shadow of the handle slows down when it is going away from the center of the wheel's shadow and speeds up when it approaches the center point. In other words, the handle's shadow describes simple harmonic motion.

The diagram in Figure 5-16(B) shows more clearly the relationship between a point on the handle, **P,** and its shadow, **P′. P** is moving with uniform speed in a circular path around point **O. P′** is the shadow of **P** on line **MN.** When **P** is at **A,** its shadow is at **A′;** when it is at **B,** its shadow is at **B′;** when it is at **C,** its shadow is at **C′;** and so on. As **P** makes a single revolution, **P′** describes one *complete vibration.* **P′** moves with an acceleration proportional to **O′P′**, and the acceleration is always directed toward **O′.** Thus the motion of **P′** along **MN** is simple harmonic motion. When **P′** is at **B′**, the *displacement* of **P′** is the distance **B′O′**—its distance from the midpoint of its vibration at that particular instant. The *amplitude* of the vibration is the maximum displacement **O′M** or **O′N**—the radius of the reference circle. The *period* is the time of one complete vibration—the time required for the point to make one revolution on the reference circle. The *frequency* of a vibratory motion is the number of vibrations per second—the number of revolutions per second of a point on the reference circle. *The frequency is the reciprocal of the period.* (If the period is 25, the frequency is $\frac{1}{25}$.) The *equilibrium position* of an object that is describing simple harmonic motion is the midpoint of its path.

To express the displacement of **P′** along **MN** in Figure

5-16(B) as a function of time, we make use of the definition of angular velocity, $\omega = \theta/t$, or $\theta = \omega t$. If d is the displacement of **P′**, it can be seen in Figure 5-16(B) that

$$d = \mathbf{R} \cos \theta$$

Substituting the value of θ above yields

$$d = \mathbf{R} \cos \omega t$$

A graph of this relationship, which holds for all simple harmonic motion, is shown in Figure 5-17.

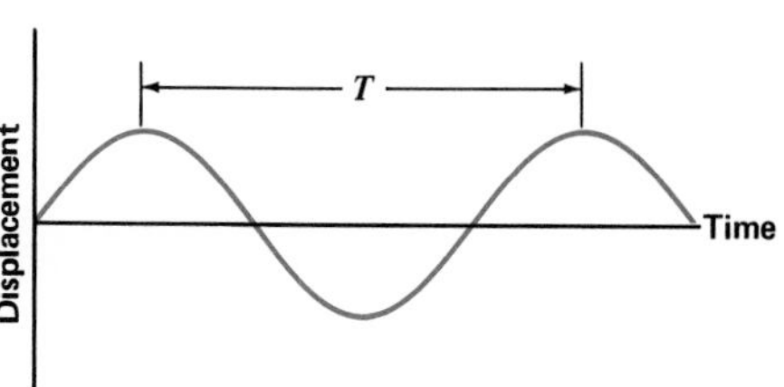

Figure 5-17. Graph of simple harmonic motion. T is the period of the motion.

5.12 The Pendulum *An object suspended so that it can swing back and forth about an axis is called a* ***pendulum.*** Figure 5-18 shows a simple pendulum in which a small dense mass, called the bob, is suspended by a cord. The mass of the cord is negligible in comparison to the mass of the bob. If the displacement of the bob is small in comparison to the length of the cord, the motion of a pendulum very closely approximates simple harmonic motion. As the pendulum bob moves from **A** to **B** and back again to **A,** it makes a complete vibration. **C** is the equilibrium position.

Galileo was probably the first scientist to make quantitative studies of the motion of a pendulum. It is said that he observed the gentle swaying of a chandelier in the cathedral at Pisa. Using his pulse as a timer, he found that successive vibrations of the chandelier were made in equal lengths of time, regardless of the amplitude of the vibrations. He later verified his observations experimentally and then suggested that a pendulum be used to time the pulse rates of medical patients.

In the simple pendulum shown in Figure 5-18, all the mass is considered to be concentrated in the bob. For an ideal pendulum of this type, the following statements hold true.

1. *The period of a pendulum is independent of the mass or material of the pendulum.* This statement is strictly true only if the pendulum vibrates in a vacuum. For example, air resistance has more effect on a pendulum bob made of cotton than it does on a bob made of lead.

2. *If the arc is small (10° or less), the period of a pendulum is independent of the amplitude.*

3. *The period of a pendulum is directly proportional to the square root of its length.* In the case of a pendulum 25 cm long and one 100 cm long, the period of the longer pendulum will be twice that of the shorter one. Since the square roots of 25 and 100 are 5 and 10 respectively, the periods are in the ratio of 5 to 10, or 1 to 2.

4. *The period of a pendulum is inversely proportional to the*

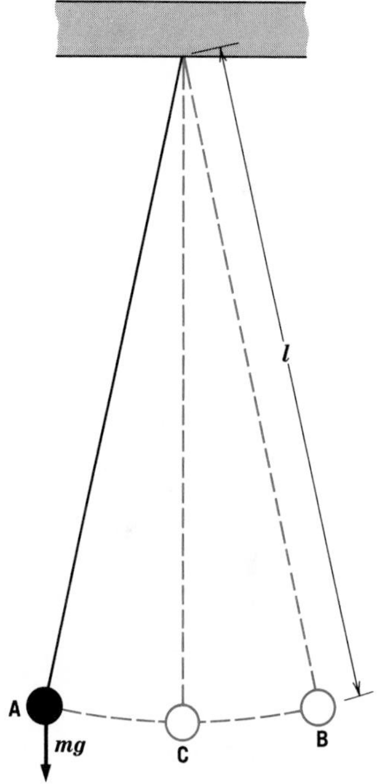

Figure 5-18. The simple pendulum. The frequency of vibration depends only on the values of l and g, provided that the displacement is small compared to the length of the cord and air resistance is negligible.

square root of the acceleration of gravity. Because the earth has an equatorial bulge, a pendulum vibrates slightly faster at the poles than at the equator. If we use the letter l to denote the length of a pendulum and g to denote the acceleration due to gravity, the period, T, is

$$T = 2\pi\sqrt{\frac{l}{g}}$$

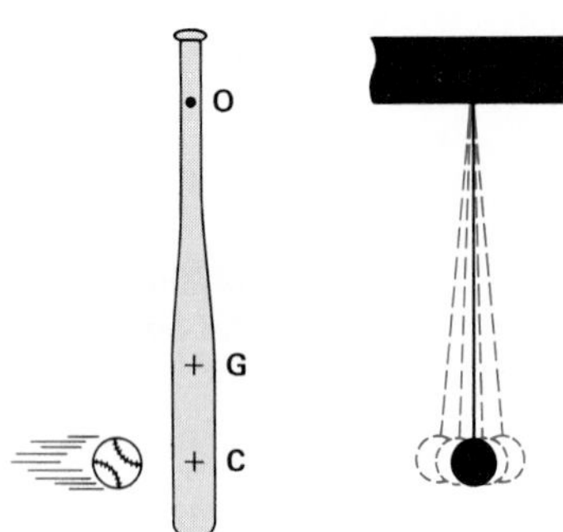

Figure 5-19. A physical pendulum and simple pendulum with the same period. **O** is the point of suspension, **G** is the center of gravity, and **C** is the center of oscillation and percussion.

A real object that can vibrate like a pendulum is called a *physical pendulum*, as distinguished from the theoretical *simple pendulum*. For example, a baseball bat is a physical pendulum as shown in Figure 5-19. When the bat is suspended from **O**, which is called the *center of suspension*, and set in vibration, its period will be the same as that of a simple pendulum with a length equal to the distance **OC**. Consequently, **C** is called the *center of oscillation*. **O** and **C** are interchangeable; that is, if the bat is suspended from **C**, the center of oscillation will be at **O**. (**G** is the center of gravity of the bat.)

An additional property of **C** is that it is the *center of percussion* of the bat. The batter's hands do not experience a "sting" if a ball strikes the bat at the center of percussion since the hands turn the bat about the center of suspension. If the ball hits the bat at a point other than the center of percussion, the bat rotates about some point other than the center of suspension. The part of the bat at the center of suspension then pushes against the hands, which accounts for the "sting."

EXAMPLE A simple pendulum has a period of 2.4 s at a location where the acceleration due to gravity is 9.7 m/s^2. What is the length of the pendulum?

Given	Unknown	Basic equation
$T = 2.4$ s $g = 9.7$ m/s^2	l	$T = 2\pi\sqrt{\frac{l}{g}}$

Solution

$$\textit{Working equation: } l = \frac{T^2 g}{4\pi^2}$$

$$= \frac{(2.4 \text{ s})^2(9.7 \text{ m/s}^2)}{(4)(3.14)^2}$$

$$= 1.4 \text{ m}$$

PRACTICE PROBLEM What is the period of a pendulum 0.30 m long at sea level? *Ans.* 1.1 s

QUESTIONS: GROUP A

1. What characterizes an object's motion as simple harmonic motion?
2. Is the earth's movement around the sun periodic motion or simple harmonic motion? Explain.
3. (a) When a mass moves at the end of a vibrating spring, where is the force acting on it maximum? (b) Where is the force minimum? (c) At the point where the force is minimum, what has happened to the object's velocity?
4. If it takes 2.0 s for the pendulum bob to move from point **A** to point **C** in Figure 5-18, what is its period?
5. (a) What happens to the period of a pendulum if its length is doubled? (b) If its amplitude decreases from 10° to 5°? (c) If its mass is doubled?
6. A simple pendulum can be used as an altimeter on a plane. How will the period of the pendulum vary as the plane moves from the runway to its cruising altitude of 1.00×10^4 m?

PROBLEMS: GROUP A

1. A pendulum makes 35 complete oscillations in 12 s. (a) What is its period? (b) What is its frequency?
2. (a) A pendulum is 3.500 m long. What is its period at the North Pole? (See Table 3-4.) (b) In Java?
3. A pendulum has a frequency of 5.50 Hz on earth at a point where $g = 9.80$ m/s^2. What would be its frequency on Jupiter? (See Table 3-3.)
4. A pendulum extends from the roof of a building almost to the floor. If the pendulum's period is 8.5 s, how tall is the building?
5. What is the period of a 1.00-m-long pendulum at a point 6.70×10^6 m above the earth's surface?
6. A pendulum is in a space craft to measure acceleration during lift off. Before the launch, its period is 6.7×10^{-3} s. At a point during lift off, its period is 3.1×10^{-3} s. (a) What is the acceleration? (b) How many g's is this?

SUMMARY

Uniform circular motion is motion at constant speed along a curved path of constant radius. An object moving in a curved path undergoes centripetal acceleration. Centripetal acceleration is produced by a force known as centripetal force which is directed toward a central point. When gravity is acting on an object, a minimum velocity is required to keep the object moving in a vertical circle.

A frame of reference is a system for specifying the location of objects. Centrifugal forces are required to explain the curved motions of objects in terms of a noninertial, or accelerating, frame of reference.

Rotary motion is the motion of a body about an internal axis. In such motion, the relationships among angular displacement, velocity, acceleration, and time are similar to the corresponding relationships for linear motion. The angular acceleration of a rotating body is directly proportional to the torque that produces motion and inversely proportional to the body's rotational inertia. The rotation of the shaft of a rotating body that is under the influence of gravity is called precession.

Periodic motion occurs when a body continually moves back and forth over a definite path in equal intervals of time. Simple harmonic motion is a special type of periodic motion. Simple harmonic motion can be analyzed in terms of circular motion. A swinging pendulum describes simple harmonic motion. The period of a pendulum is directly proportional to the square root of its length and inversely proportional to the square root of the acceleration of gravity.

VOCABULARY

angular acceleration
angular velocity
centrifugal force
centripetal acceleration
centripetal force
circular motion
critical velocity
frame of reference
pendulum
periodic motion
precession
radian
rotary motion
rotational inertia
simple harmonic motion

Conservation of Energy and Momentum

energy (EN-er-jee) n.: the capacity to do work.

WORK, MACHINES, AND POWER

6.1 Definition of Work The word *work* has a specific meaning in physics. ***Work** is done when a force is exerted on an object causing the object to move in the direction of a component of the applied force.* When you hold a heavy load on your shoulder, as long as you do not move you are not doing any work on the load. You are exerting an upward *force* that counteracts the downward force of gravity on the load. You do *work* when you raise the load to your shoulder, when you carry it up a flight of stairs, or when you pull it across the floor. In these cases, you exert a force that has a component in the direction in which the object moves.

Two factors must be considered in measuring work: the displacement of the object and the magnitude of the force in the direction of displacement. *The amount of work, W, equals the product of a displacement, Δd, and the force, F, in the direction of the displacement.*

$$W = F\,\Delta d$$

When the force is measured in newtons and the distance through which it acts is measured in meters, the work is expressed in *joules* (J). *A force of one newton acting through a distance of one meter does* ***one joule*** *of work.* This unit of work is named for the English physicist James Prescott Joule. Note that *a joule is a newton-meter.*

For example, let us compute the work required to lift a

OBJECTIVES

- Define and calculate work and power.
- Define and calculate kinetic energy and potential energy.
- Define and calculate impulse and momentum.
- Apply the laws of conservation of energy and momentum to solve motion problems.
- Differentiate between elastic and inelastic collisions.

1.0-kg mass to a height of 5.0 m. From the relationship between mass and weight discussed in previous chapters, we know that a force of about 9.8 N must be exerted to lift a mass of 1.0 kg at sea level. Thus, the work is

$$\begin{aligned} W &= F\Delta d \\ &= (9.8 \text{ N})(5.0 \text{ m}) \\ &= 49 \text{ J} \end{aligned}$$

If we slide the 1.0-kg mass at constant velocity along a horizontal surface having a coefficient of sliding friction 0.30 for 5.0 m, the work required is

The relationship $F_f = \mu F_N$ is explained in Section 4.7.

$$\begin{aligned} W &= F_f\Delta d \\ &= \mu F_N \Delta d \\ &= (0.30)(9.8 \text{ N})(5.0 \text{ m}) \\ &= 15 \text{ J} \end{aligned}$$

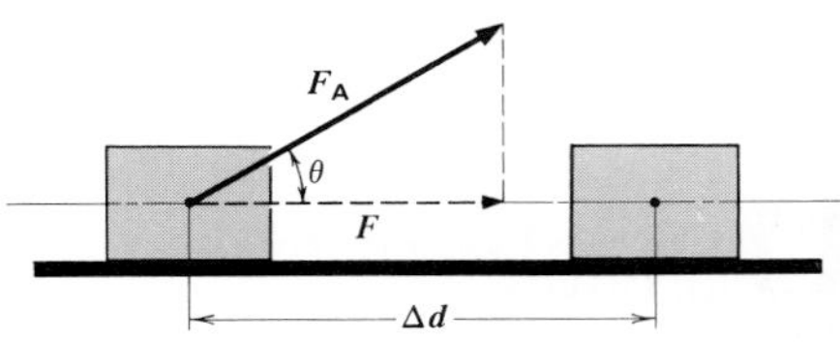

Figure 6-1. Definition of work. The work done by the applied force, F_A, is equal to the product of F, the component of F_A in the direction of the displacement, and Δd, the distance through which the mass moves.

In both instances, the force is applied in the direction in which the object moves.

Suppose, however, that the force is applied to the object in a direction other than that in which it moves. In that case, only the component of the applied force that acts in the direction the object moves is used to compute the work done on the object. Thus in Figure 6-1, the relationship between the applied force, F_A, and the component in the direction of motion, F, is

$$F = F_A \cos \theta$$

When the force is applied at an angle, θ, the normal force, F_N, is equal to the weight of the object, F_W, minus the vertical component of the applied force, or

$$F_N = F_W - F_A \sin \theta$$

This equation is used to calculate F_A when it is not given in a problem, as shown in the following example. Then

$$W = F_A \, \Delta d \cos \theta$$

EXAMPLE A 95.0-kg crate is pulled for 12.0 m on a horizontal surface at a constant velocity. The coefficient of friction between the crate and the ground is 0.260. Calculate the work done when the force is applied at an angle of 20.0°.

Given	Unknown	Basic equations
$m = 95.0$ kg $\Delta d = 12.0$ m $\mu = 0.260$ $\theta = 20.0°$	W	$W = \mu F_N \Delta d$ $F_f = \mu F_N$ $F_N = F_W - F_A \sin\theta$ $W = F_A \Delta d \cos\theta$

Solution

Step 1: $\mu = F_f/F_N = F_A \cos\theta/(mg - F_A \sin\theta)$

$$F_A = \mu mg/(\mu \sin\theta + \cos\theta)$$
$$= \frac{(0.260)(95.0 \text{ kg})(9.80 \text{ N/kg})}{(0.260)(0.342) + (0.940)} = 235 \text{ N}$$

Step 2: $W = F_A\,\Delta d \cos\theta$

$$= (235 \text{ N})(12.0 \text{ m})(0.940)$$
$$= 2.65 \times 10^3 \text{ J}$$

PRACTICE PROBLEMS **1.** A crate weighing 850 N is pushed up an inclined plane a distance of 10.0 m. The plane makes an angle of 15° with the horizontal and the crate moves with constant velocity. The coefficient of friction between the crate and the plane is 0.24. Calculate the work that is done in pushing the crate. *Ans.* 4.2×10^3 J

2. A girl pulls a wagon with constant velocity along a level path for a distance of 45 m. The handle of the wagon makes an angle of 20.0° with the horizontal, and she exerts a force of 85 N on the handle. Find the amount of work the girl does in pulling the wagon. *Ans.* 3.6×10^3 J

6.2 Work Done by Varying Forces In the examples of work in Section 6.1, the forces involved did not vary. In many problems involving work, however, the forces may vary in direction, in magnitude, or in both during the time that they are acting on an object. For example, when a force is used to stretch a spring, the magnitude of the force increases as the spring gets longer.

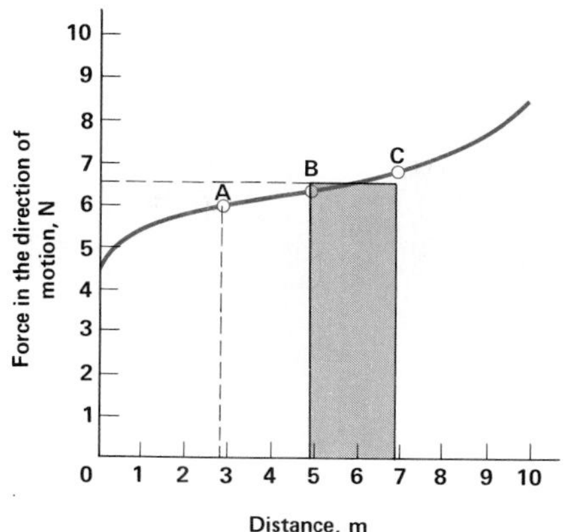

Figure 6-2. Work done by a variable force. The area under the curve represents the total work.

An easy way to determine the amount of work done by a varying force is to use a graph. In Figure 6-2, the area under the curved line represents the work done by a force that varies in magnitude. The horizontal axis is the distance, Δd, in meters and the vertical axis is the force that acts in the direction of motion, F, in newtons. The amount of work required to provide the displacement indicated by the curve up to point **A,** for example, is equal to the area bounded at the top by the curve, at the right by a vertical line from **A** to the horizontal axis, and at the left and bottom by the two coordinate axes.

To calculate the area of a geometric figure that is bounded by one or more curved lines requires a branch of mathematics called calculus. A good approximation is obtained, however, by the use of suitable rectangles. For example, the work indicated by the curve between points **B** and **C** is approximately equal to the area of the blue rectangle in the figure. Thus, the work is approximately equal to the product of 2.0 m and 6.5 N, or 13 J.

The total amount of work that is represented by the area under the curve can be found by adding the areas of many rectangles formed in the same way as the one in the figure. If the rectangles are made very narrow, the answer will be more accurate. An interesting way to measure the area under a curve is to cut it out of a piece of paper, weigh it, and compare it with the weight of a known area.

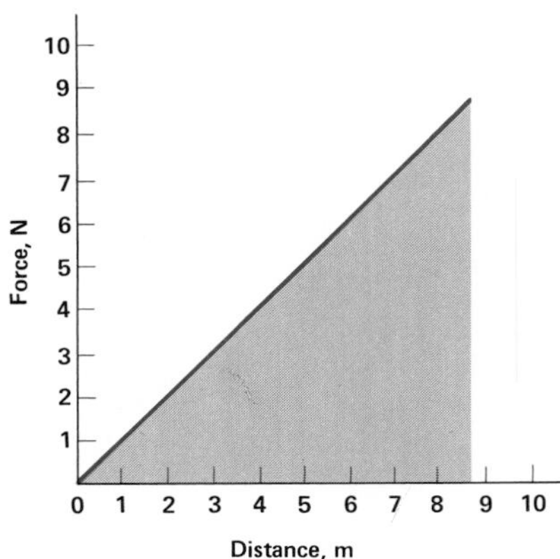

Figure 6-3. Work done by a constantly increasing force. The total work is represented by the blue triangle.

The problem of finding the total work done by a varying force is somewhat simpler in the case of a stretching spring, as long as the force is constantly applied in the direction of the stretching. The force required to stretch a spring depends on the stiffness of the spring, but the force is directly proportional to the amount of stretching. (The limits within which this relationship is true will be discussed in Chapter 7.) Consequently, a graph of the work done in stretching a spring is shown in Figure 6-3. No approximations are required in computing the total work because the area of the triangle under the curve is exactly equal to one-half the product of its base and height.

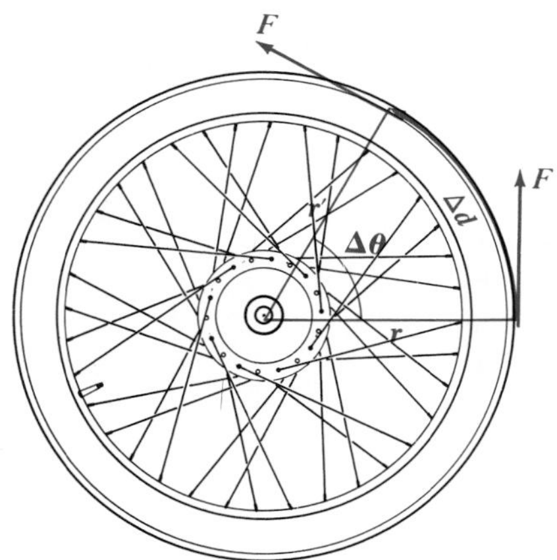

Figure 6-4. Work in rotary motion. The work done on the wheel is equal to the product of the applied force, F, the radius of the wheel, r, and the displacement of the rim, $\Delta\theta$.

6.3 Work in Rotary Motion To compute the work done in rotary motion, we make use of the principles of radian measure (Section 5.6). In Figure 6-4 the displacement of the rim of the wheel is designated by the arc Δd. If the angle $\boldsymbol{\Delta\theta}$ is expressed in radians, then $\Delta d = r\,\Delta\theta$. Substituting this expression for Δd in the work equation, we get

$$W = Fr\boldsymbol{\Delta\theta}$$

Furthermore, in Section 4.12 we saw that a torque, T, is equal to the product of a force and the length of its torque arm. In Figure 6-4, the torque arm is the radius of the circle, so $T = Fr$. The work equation can now be written

$$W = T\boldsymbol{\Delta\theta}$$

which means that the work done in rotary motion can be computed by finding the product of the torque producing the motion and the angular displacement in radians.

For example, if the radius of the wheel in Figure 6-4 is

2.0 m and a force of 12 N is applied tangentially to it, the work done in a single revolution is

$$\begin{aligned} W &= T\Delta\theta \\ &= Fr\Delta\theta \\ &= 2\pi Fr \\ &= 2(3.14)(12\ \text{N})(2.0\ \text{m}) \\ &= 150\ \text{J} \end{aligned}$$

6.4 Machines Six types of simple machines are shown in Figure 6.5. A *machine* can be used to multiply force. Other machines are either modifications of these simple machines or combinations of two or more of them. The six machines shown are actually variations of two basic types: the pulley and the wheel and axle are forms of the lever, and the wedge and screw are modified inclined planes.

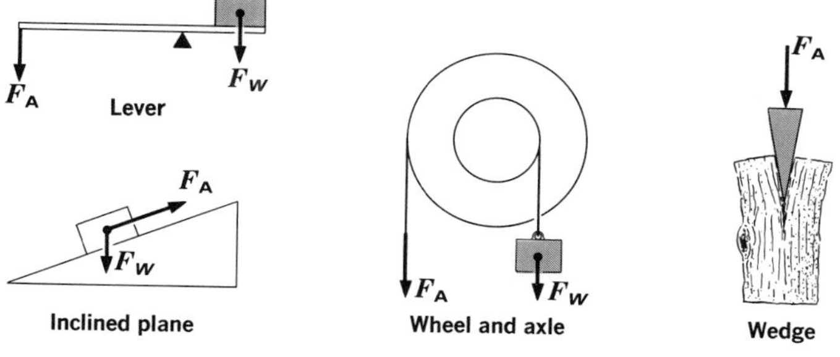

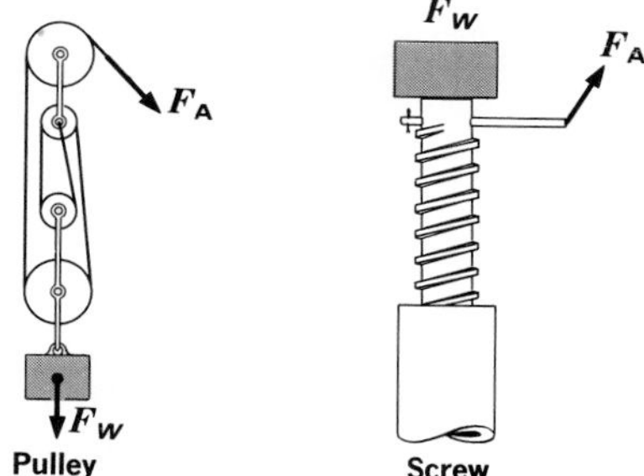

Figure 6-5. Simple machines. Each machine multiplies force at the expense of distance.

Although a machine can be used to multiply force, it cannot multiply work. The work output of a machine cannot exceed the work input. In a frictionless machine, work output and work input would be exactly equal. In a machine that multiplies force, this equality means that the distance over which the input force moves is always greater than the distance over which the load moves.

The ratio of the output force to the input force in a machine is called the mechanical advantage.

The ratio of the useful work output of a machine to total work input is called the ***efficiency.***

$$\text{Efficiency} = \frac{W_{\text{Output}}}{W_{\text{Input}}}$$

The efficiency of all machines is less than 100% because the work output is always less than the work input. This is due to the force of friction.

Thus in using a machine to lift an object, the efficiency equation becomes

$$\text{Efficiency} = \frac{F_W\ \Delta h}{F_A\ \Delta d}$$

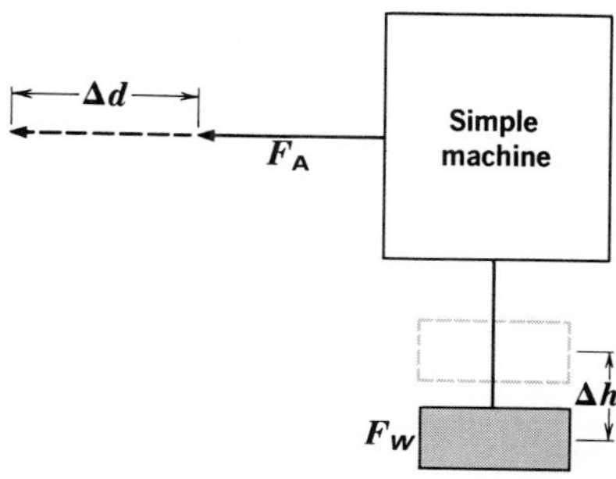

Figure 6-6. Principle of the simple machine. In the absence of friction, the work output, $F_W\ \Delta h$, is equal to the work input, $F_A\ \Delta d$.

where F_W is the weight of the object, Δh is the height through which it is lifted, F_A is the input force applied to the machine, and Δd is the distance through which F_A acts in the direction of the input motion.

The following example illustrates the use of the efficiency equation in solving problems dealing with simple machines.

EXAMPLE A crate is pulled 2 m with constant velocity along an incline that makes an angle of 15° with the horizontal. The coefficient of friction between the crate and the plane is 0.160. Calculate the efficiency that is achieved in this procedure. (Refer to Figure 4-18.)

Given	Unknown	Basic equations
$\mu = 0.160$	Efficiency	$\text{Eff.} = \dfrac{W_{\text{Output}}}{W_{\text{Input}}} = \dfrac{F_W\ \Delta h}{F_A\ \Delta d}$
$\theta = 15°$		$F_f = \mu F_W \cos\theta$
$\Delta d = 2\text{ m}$		$F_P = F_W \sin\theta$

Solution

(a) Finding W_{Output}:

The crate, with a weight of F_W, is lifted through a vertical distance $\Delta h = \Delta d \sin\theta$. Thus,

$$W_{\text{Output}} = F_W \Delta h = F_W \Delta d \sin\theta$$

(b) Finding W_{Input}:

The applied force is the sum of the force of friction, F_f, and the force along the incline, F_P:

$$W_{\text{Input}} = F_A \Delta d = (F_f + F_P)\Delta d$$

Substituting in the remaining Basic equations:

$$W_{\text{Input}} = (\mu F_W \cos\theta + F_W \sin\theta)\Delta d$$

(c) Combining both steps (a) and (b) gives the

Working equation: $\text{Efficiency} = \dfrac{F_W\ \Delta d \sin\theta}{F_W\ \Delta d\ (\mu\cos\theta + \sin\theta)}$

$$= \frac{\sin\theta}{(\mu\cos\theta + \sin\theta)}$$

$$= \frac{\sin 15°}{(0.160)(\cos 15°) + (\sin 15°)}$$

$$= 0.63$$

PRACTICE PROBLEM The efficiency of a pulley system is 73%. The pulleys are used to raise a mass of 58 kg to a height of 3.0 m. What force is exerted on the strand of the pulley if it is pulled for 18.0 m in order to raise the mass to the required height? *Ans.* 130 N

6.5 Definition of Power Like the term *work*, the term *power* has a scientific meaning that differs somewhat from its everyday meaning. When we say a person has great power, we usually mean that the person has great strength or wields great authority. In physics, the term ***power*** means *the time rate of doing work.*

You do the same amount of work whether you climb a flight of stairs in one minute or in five minutes, but your power output is not the same. See Figure 6-7. Power depends upon three factors: the displacement of the object, the force in the direction of the displacement, and the time required.

Since power is the time rate of doing work,

$$P = \frac{W}{\Delta t}$$

where P is power, W is work, and Δt is time. Or, because $W = F\Delta d$ and $v = \Delta d/\Delta t$,

$$P = \frac{F\Delta d}{\Delta t} = Fv$$

When work is measured in joules and time is measured in seconds, power is expressed in *watts* (W). A ***watt*** *is a joule per second.* This unit is named in honor of James Watt, who designed the first practical steam engine. Since the watt is a very small unit, power is more commonly measured in units of 1000 watts, or kilowatts (kW).

Figure 6-7. A power comparison. In both cases, the girl does the same amount of work in climbing the stairs. However, in the lower drawing her output of power is greater because she gets to the top in less time. (Kinetic energy is disregarded in these examples.)

Figure 6-8. James Watt in his workshop. How many pieces of apparatus can you identify?

The terms watt and kilowatt are used frequently in connection with electricity, but also apply to quantities of power other than electric power. The watt is the unit of power in the metric system and is used to express quantities of mechanical as well as electric power. Another frequently used unit is the "horsepower." It is equal to 746 watts.

EXAMPLE A woman drives her car up a parking ramp 12.0 m high at a constant velocity in 20.0 s. The mass of the car is 1.50 metric tons. Calculate the power output of the car's engine during this time, in horsepower (1 hp = 746 W).

Given	Unknown	Basic equations
$\Delta d = 12.0$ m	P	$P = \dfrac{F_W \, \Delta d}{\Delta t}$
$\Delta t = 20.0$ s		
$m = 1.50 \times 10^3$ kg		$F_W = mg$

Solution

Working equation: $P = \dfrac{mg \, \Delta d}{\Delta t}$

$$= \frac{(1.50 \times 10^3 \text{ kg})(9.80 \text{ m/s}^2)(12.0 \text{ m})}{(20.0 \text{ s})} \times \frac{(1 \text{ hp})}{(746 \text{ W})}$$

$$= 11.8 \text{ hp}$$

6.6 Power in Rotary Motion By using radian measure, we can compute the power involved in rotary motion the same way we calculated the work done in rotary motion in Section 6.3. Substituting the expression for work in rotary motion, $T\Delta\theta$, the power equation becomes

$$\boldsymbol{P = \frac{T\Delta\theta}{\Delta t}}$$

We saw in Section 5.6 that the time rate of angular displacement, $\Delta\theta/\Delta t$, is called the angular velocity, ω, so

$$P = T\omega$$

The power required to maintain rotary motion against an opposing torque is the product of the torque maintaining the rotary motion and the constant angular velocity.

QUESTIONS: GROUP A

1. Explain whether *work*, in the physics sense, is being done on a suitcase when you (a) pick it up from the floor, (b) carry it at a steady speed on a level street to the bus stop, (c) hold it above the ground while you wait for the bus, and (d) board the bus with the suitcase.
2. What represents the work done when a graph of force versus displacement is constructed?
3. What are three ways to determine the work that is done when a force-versus-displacement graph is not a straight line?
4. If you were to use a machine to increase the produced force, what factor would have to be sacrificed? Give an example.
5. (a) In calculating the work done in rotary motion, what is the expression that takes the place of Δd? (b) What is the equation used in finding the amount of rotary work?
6. For an object moving at a constant speed, list the two expressions for determining the object's power.

GROUP B

7. (a) Using Figure 6-3, determine the average force needed to stretch the spring a distance of 4.0 m. (b) How much work is done in stretching the spring 4.0 m?
8. Use the graph of force versus displacement shown in Figure 6-3 to derive an equation for the work done by the spring.
9. If a machine cannot multiply the amount of work, what is the advantage of using such a machine?
10. A heavy football player climbs a flight of stairs. Halfway up the stairs, a member of the girls' track team rushes past him. Is it possible for both to develop the same amount of power in climbing up the stairs?
11. Find the horsepower rating on a lawn mower or an electric tool. Convert this amount to units of kilowatts.

PROBLEMS: GROUP A

1. A weight lifter heaves a 200.0-kg barbell from the floor to a position directly over his head. If the distance from the floor to his extended arms is 2.50 m, how much work has the weight lifter done?
2. (a) How much work is done in lifting a 750-kg piano vertically 3.0 m to a large set of doors? (b) How much work would be done if the piano was pushed up a frictionless inclined plane to the same set of doors? (c) If the inclined plane was 5.0 m long, how much force would have been needed?
3. How much work is done in pushing a 45.5-kg wooden trunk a distance of 9.75 m across the floor if the coefficient of friction is 0.250?
4. Calculate the work done when a sled is pulled 20.0 m by a force of 105 N exerted on a rope that makes an angle of 50.0° with the horizontal.
5. A pulley system is used to lift the piano mentioned in Problem 2. If a force of 2.0×10^3 N is applied to the piano, and as a result the rope is pulled in 14 m, what is the efficiency of the machine?
6. How much power does a 63.0-kg athlete develop as he climbs a 5.20-m rope in 3.50 s?
7. A 45.0-kg cyclist exerts her full weight on the pedal with each stroke. How much work is done during 100.0 revolutions of the pedals as they turn in a 30.0-cm radius?
8. What is the power rating in kilowatts of a 1.20×10^3-kg elevator that moves

3.50 m from one floor to the one above it in 4.30 s?

9. A 23.0-cm screwdriver is to be used to pry open a can of paint. If the fulcrum is 2.00 cm from the end of the blade and a force of 84.3 N is exerted at the end of the handle, what force is applied to the lid?
10. A pulley system has an efficiency of 87.5%. How much of the rope must be pulled in if a force of 648 N is needed to lift a 105-kg desk 2.46 m?
11. A 175-N bucket of water is to be lifted from the bottom to the top of a 7.30-m well. If a force of 42.0 N is applied at the end of the 36.3-cm handle, how many times must the handle be turned to accomplish this?
12. A 0.50-kW motor moves a lawn tractor at a constant 1.2 m/s. What force is being applied to the tractor?

GROUP B

13. A force of 25.0 N is applied to a 4.50-kg object that is initially at rest. (a) How much work is done during the first 3.00 s of its motion? (b) How much power is developed during this same period of time?
14. A 65-kg crate is pushed at a constant speed up a 3.6-m plane inclined at 24° above the horizontal. If the coefficient of friction is 0.17, how much work is done?
15. A 175-kg flywheel is a uniform disk 1.80 m in diameter. (a) How much work is required to bring it from rest to 94.0 rev/min in 2.00 min? (b) What is the machine's power rating in kilowatts?
16. An elevator motor is rated at 25.0 kW. At what speed could the motor lift an 850.0-kg elevator with three passengers whose masses are 24.3 kg, 45.0 kg, and 64.0 kg?
17. What power must the engine of a 1680-kg car develop to move at a constant speed of 24.5 m/s up a 15° incline if the coefficient of friction between the tires and road is 0.090 0?
18. A 35.4-kg box falls off a truck moving at 40.0 km/h. The box slides to a stop after a distance of 17.5 m. Calculate (a) the force of friction on the box, (b) the work done in stopping it, and (c) the coefficient of friction between the box and pavement.
19. How much work is done in pushing an 85.4-kg grocery cart 2.05×10^2 m if a force is applied at a 40° angle to the horizontal and the coefficient of friction between the wheels and the floor is 0.025 0?

ENERGY

Potential and kinetic energy can each have a variety of forms.

6.7 Gravitational Potential Energy In Chapter 1 we saw that there are two kinds of energy, potential and kinetic. In the following sections, we will deal quantitatively with these concepts and see how the various forms of energy are expressed in terms of work units.

The potential energy acquired by an object equals the work done against gravity or other forces to place it in position, as shown in Figure 6-9. As we saw in Section 6.1, the equation for calculating work when the force acts in the same direction as the displacement is

$$W = F\Delta d$$

Therefore the equation for potential energy is

$$E_p = F\Delta d$$

In lifting an object, F is its weight, which from Newton's second law of motion equals mg, and Δd is the vertical distance Δh through which it is lifted. Hence the potential energy equation can be written

$$E_p = mg\Delta h$$

The *gravitational potential energy* defined by this equation is expressed in relation to an arbitrary reference level where $h = 0$. The reference level is usually determined by the nature of the problem. Sea level, street level, ground level, or floor level are all commonly used reference levels.

When the mass of an object is given in kilograms, the height in meters, and the acceleration due to gravity in m/s^2, the gravitational potential energy is expressed in joules. Thus if a 5$\bar{0}$-kg mass of steel is raised 5.0 m, its gravitational potential energy is

$$E_p = mg\Delta h$$
$$E_p = 5\bar{0}\text{ kg} \times 9.8\text{ m/s}^2 \times 5.0\text{ m}$$
$$E_p = 2.5 \times 10^3\text{ J}$$

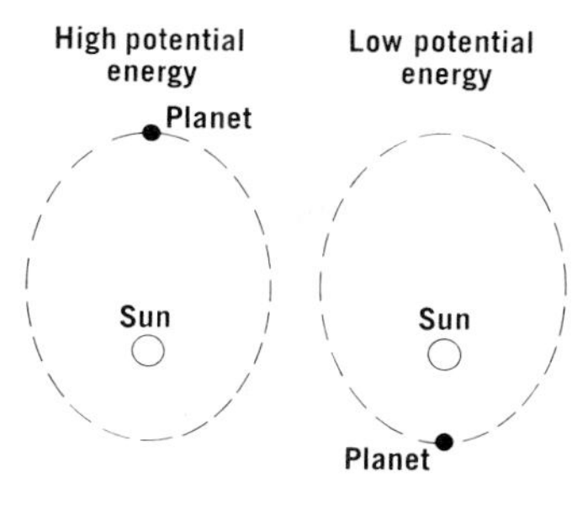

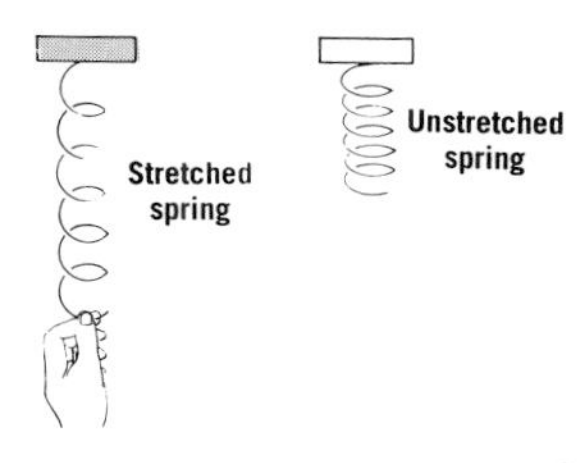

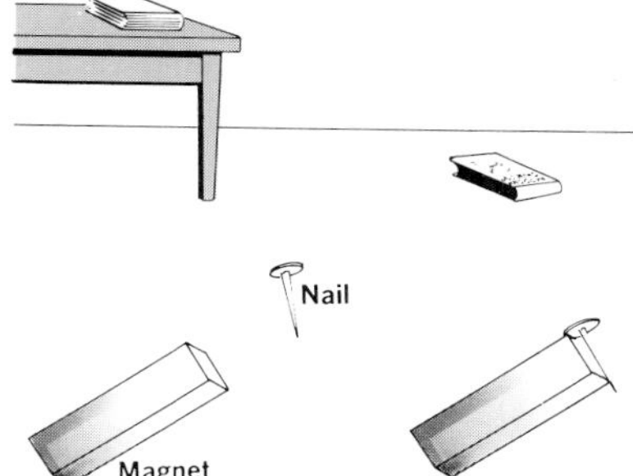

Figure 6-9. Examples of potential energy. Can you identify the reference level for each pair of situations?

6.8 Kinetic Energy in Linear Motion As we saw in Section 3.7, the velocity of a freely falling object that starts from rest, expressed in terms of the acceleration of gravity and the distance traveled, is given by the equation

$$v = \sqrt{2g\Delta d}$$

Solving for Δd, we obtain

$$\Delta d = \frac{v^2}{2g}$$

Since Δd in this equation corresponds to Δh in the equation $E_p = mg\Delta h$, let us substitute the expression we have just derived for Δd in place of Δh. The equation for the kinetic energy of a moving object then becomes

$$E_K = mg \times \frac{v^2}{2g}$$

$$E_K = \tfrac{1}{2}mv^2$$

Although this equation for kinetic energy was derived from the motion of a falling body, it applies to motion in any direction or from any cause.

As in the case of gravitational potential energy, kinetic

Figure 6-10. Examples of kinetic energy. What is the frame of reference for the velocity of the moving object in each case?

energy is expressed in joules if the mass is given in kilograms and the velocity in meters per second. Thus, if a baseball has a mass of 0.14 kg and it's thrown with a velocity of 26 m/s, its kinetic energy is

$$E_K = \tfrac{1}{2}mv^2$$

$$E_K = \frac{0.14\ \text{kg}\ (26\ \text{m/s})^2}{2}$$

$$E_K = 47\ \text{J}$$

6.9 Kinetic Energy in Rotary Motion As we saw in Section 5.6, the angular velocity, ω, of a rotating body that starts from rest is

$$\omega = \sqrt{2\alpha\Delta\theta}$$

where α is the angular acceleration and $\Delta\theta$ is the angular displacement. The equation given in Section 5.8 for the relationship among torque, rotational inertia, and angular acceleration is

$$T = I\alpha$$

Solving both of these expressions for angular acceleration,

$$\alpha = \frac{\omega^2}{2\Delta\theta} \text{ and } \alpha = \frac{T}{I}$$

Setting these two expressions equal,

$$\frac{T}{I} = \frac{\omega^2}{2\Delta\theta}$$

Solving for $T\Delta\theta$,

$$T\Delta\theta = \frac{I\omega^2}{2}$$

When only the net force and torque are considered in these equations, all of the work done to produce rotation appears as kinetic energy. Since for rotary motion $W = T\Delta\theta$, the equation for kinetic energy, E_K, is

$$E_K = \tfrac{1}{2}I\omega^2$$

The wheels of a moving car have both linear and rotational kinetic energy.

The wheel of a moving automobile has both linear motion and rotary motion. The wheel turns on its axle as the axle moves along parallel to the road. The kinetic energy of such an object is the sum of the kinetic energy due to linear motion and the kinetic energy due to rotary motion.

$$E_K = \tfrac{1}{2}mv^2 + \tfrac{1}{2}I\omega^2$$

EXAMPLE A metal disk with a mass of 1.30 kg rolls along a horizontal floor with a constant velocity of 1.68 m/s. Calculate the kinetic energy of the moving disk.

Given	Unknown	Basic equations
$m = 1.30$ kg $v = 1.68$ m/s	E_K	$E_K = \frac{1}{2}mv^2 + \frac{1}{2}I\omega^2$ $\omega = \Delta\theta/\Delta t$ $v = \Delta d/\Delta t$ $\theta = \Delta d/r$

Solution

From Fig. 5-11 we know that $I = \frac{1}{2}mr^2$.

$$\textit{Working equation: } E_K = \tfrac{1}{2}mv^2 + \tfrac{1}{2}(\tfrac{1}{2}mr^2)(v/r)^2$$
$$= \tfrac{3}{4}mv^2$$
$$= \tfrac{3}{4}(1.30 \text{ kg})(1.68 \text{ m/s})^2$$
$$= 2.75 \text{ J}$$

When energy is supplied to an object and simultaneously gives it both linear and rotary motion, the energy division depends on the rotational inertia of the object. If a ring and a solid disk of equal mass and diameter roll down the same incline, as in Figure 6-11, the disk will accelerate more. Its rotational inertia is less than that of the ring, thus its rotational kinetic energy is less. Its linear kinetic energy is therefore greater than that of the ring; the disk acquires a higher linear velocity and reaches the bottom of the incline first. However, at the bottom of the incline the total kinetic energy of the ring will be the same as that of the disk since both ring and disk had the same gravitational potential energy at the top.

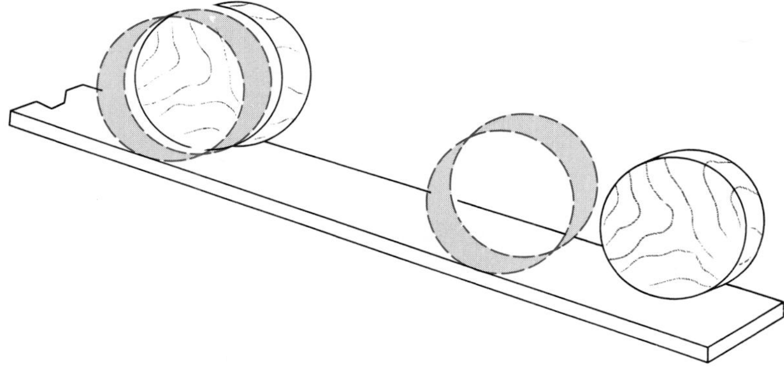

Figure 6-11. A kinetic energy race. If the ring and disk have equal masses and diameters, the rotational inertia of the disk is less than that of the ring. Hence the disk will accelerate more.

6.10 Elastic Potential Energy When a spring is compressed, energy is stored in the spring. At any instant during the compression, the elastic potential energy in the spring is equal to the work done on the spring. The potential energy in a stretched or compressed elastic object is called ***elastic potential energy.***

The work required to stretch or compress a spring does not depend on the weight of the spring. Consequently, gravity is not involved in the measurement of elastic potential energy. Instead, the work required to stretch or compress a spring is dependent upon a property of the spring known as the *force constant*. The force constant does not change for a specific spring so long as the spring is not permanently distorted.

Compare Equation 3 with the one for kinetic energy in Section 6.8. Why are they similar?

The force required to stretch a spring is written as

$$\boldsymbol{F = k\Delta d} \qquad \text{(Equation 1)}$$

where k is the force constant of the spring and Δd is the distance over which F is applied. We noted in Section 6.2 that the force is directly proportional to the amount of stretching (within limits). Consequently, the work done on the spring also varies directly with the amount of stretching. The total amount of work, therefore, is given by the equation.

$$\boldsymbol{W = \tfrac{1}{2}F\Delta d} \qquad \text{(Equation 2)}$$

where F is the force exerted on the spring at the end of the stretch through distance Δd. Because the force varies from zero to F, the equation gives the average of these two values, or $\frac{1}{2}F$.

Substituting the expression for F from Equation 1 in Equation 2, we get

$$\boldsymbol{W = \tfrac{1}{2}k\,(\Delta d)^2} \qquad \text{(Equation 3)}$$

This equation applies to the compression of a spring as well as the stretching of a spring. Since the potential energy of an object is equal to the work done on the object, Equation 3 can be rewritten as a potential energy equation

$$\boldsymbol{E_p = \tfrac{1}{2}k\,(\Delta d)^2}$$

where E_p is the elastic potential energy. This equation represents ideal conditions. In actual practice, a small fraction of the work of stretching or compression is converted into heat energy in the spring and does not show up as elastic potential energy.

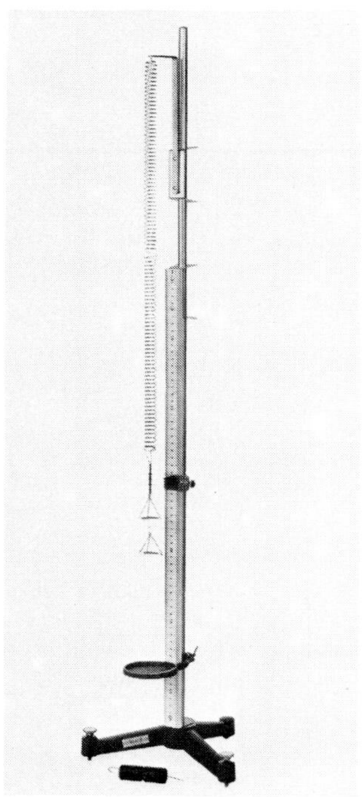

Figure 6-12. Apparatus for measuring the force constant of a spring. If the weight is made to move up and down, the interchange of potential and kinetic energy can also be measured.

6.11 Conservation of Mechanical Energy As we saw in Section 5.10, the vibration of a mass on a spring and the swinging of a pendulum are both examples of simple harmonic motion. Both can also be used to illustrate an important principle of physics called the ***law of conservation of mechanical energy.*** This law states that *the sum of the potential and kinetic energy of an energy system remains constant when no dissipative forces act on the system.*

This is a special case of the law of conservation of energy discussed in Section 1.13.

Gravitational forces and elastic forces are called *conservative forces* because they conform to the law of conservation of mechanical energy. There are forces, however, that produce deviations from the law of conservation of mechanical energy. The force of friction is an example. Forces of this type are called nonconservative, or *dissipative forces.*

The reason that friction is a dissipative force is that it produces a form of energy (heat) that is not mechanical. Energy is lost to the system. When considered in light of the more comprehensive law of conservation of *total* energy, there is no "lost" energy, of course. The calculation of heat energy and its role in energy transformations will be discussed in Chapter 8.

Another way to distinguish between conservative and dissipative forces is to observe the relationship between the force and the path over which it acts. In the case of a conservative force, the work done and energy involved are completely independent of the length of the path, provided the paths have the same end point. For example, in the illustration of gravitational potential energy in Figure 6-13, the gravitational potential energy of both men is the same, provided they have equal masses, even though one of them travels a greater distance than the other one. The mass of the men and their height above the reference level are the only factors required for the calculation of gravitational potential energy.

Figure 6-13. The work done by conservative forces is independent of the path. The potential energy of the man is, in each case, dependent only on his mass and his distance above the reference level.

The amount of heat energy lost through friction, a dissipative force, can be quite different for two objects moving through the same height, however. Where the path is longer, the amount of mechanical energy that is converted to heat energy will be greater. This is true even if the frictional forces remain constant. Thus the work done by a given dissipative force varies directly with the length of the object path.

QUESTIONS: GROUP A

1. What forms of energy are present in the following situations? (a) A diver who stands on the edge of a 10-m platform. (b) A bowstring that has been pulled back ready to launch an arrow. (c) A student who climbs the stairs to the school's library. (d) A penny that is dropped from the second floor. (e) The wheel of a wagon that is rotating as the wagon is pulled up a hill.

2. How does rotary kinetic energy differ from linear kinetic energy?
3. What would be the shape of a graph that shows an object's kinetic energy as a function of the object's speed? Be specific.

GROUP B

4. The kinetic and gravitational potential energy of a nail being driven into a piece of wood change very little. (a) What happens to the work done by the hammer? (b) Does this violate the law of conservation of energy?
5. Describe how the energy changes as a pole vaulter approaches the bar, clears the bar, and lands on the cushions below the bar.

PROBLEMS: GROUP A

1. A 65-kg diver is poised at the edge of a 10.0-m platform. Calculate the diver's gravitational potential energy relative to the pool.
2. What is the linear kinetic energy of a 1250-kg car moving at 45.0 km/h?
3. The force constant of a spring in a child's toy car is 550 N/m. How much elastic potential energy is in the spring if it is compressed a distance of 1.2 cm?
4. (a) What is the potential energy of a 1050-N rock on the edge of a cliff that is 20.4 m high? (b) If the rock falls, what is its kinetic energy when it strikes the ground? (c) How fast is it moving when it strikes the ground?
5. Calculate the rotary kinetic energy of a 32-cm diameter bicycle wheel with a mass of 5.5-kg as it spins at 65 rev/min.
6. A force of 22 N is exerted horizontally on an 18-kg box to move it 7.6 m across the floor. If the box was initially at rest and is now moving at 3.2 m/s, calculate (a) the work done, (b) the final kinetic energy of the box, and (c) the energy converted to thermal energy due to friction.
7. A 72-kg pole vaulter running at 8.4 m/s completes a vault. If all of his kinetic energy is transformed into gravitational potential energy, what is the maximum height of the bar?
8. To cut down on injuries, a highway guardrail is designed to be moved a maximum of 5.00 cm when struck by a car. What is the minimum force constant of the material in the guardrail if it is to withstand the impact of a 1250-kg car moving at 15.0 km/h?
9. (a) Calculate the rotary kinetic energy of the earth. (b) What is the earth's average linear kinetic energy as it orbits the sun?

GROUP B

10. A 3.00-kg ball rolls up a 45° incline. (a) If the ball is moving at 5.00 m/s at the bottom, what is its initial rotary and linear kinetic energy? (Ignore the effect of friction.) (b) How far does the ball roll before it stops?
11. A 100.0-g arrow is pulled back 30.0 cm against a bowstring. If the force constant of the bow and string is 1250 N/m, at what speed will the arrow leave the bow?
12. A 350.0-kg roller coaster car is poised at the top of a 42.0-m hill. (a) How fast will the car be moving at the bottom of the incline? (b) As it goes over the top of the next hill, 30.0 m high? (c) Could this problem be done without knowing the mass of the car?

PHYSICS ACTIVITY

Obtain two soup cans of equal mass and size, one having solid contents and the other loose. Hold the cans at the top of an inclined plank and release them simultaneously. Which one reaches the bottom first? Why?

MOMENTUM

6.12 *The Nature of Momentum* More force is needed to stop a train than to stop a car, even when both are moving with the same velocity. A bullet fired from a gun has more penetrating power than a bullet thrown by hand, even when both bullets have the same mass. The physical quantity that describes this aspect of the motion of an object is called *momentum*. ***Momentum** is the product of the mass of a moving body and its velocity.* The equation for momentum is

$$p = mv$$

where p is the momentum, m is the mass, and v is the velocity of an object.

In the example of the car and train, the greater mass of the train gives it more momentum than the car. Consequently, a greater change of momentum is involved in stopping the train than in stopping the car. In the case of the bullets, the greater momentum of the fired bullet is due to its greater velocity; a large change of momentum takes place when the speeding bullet is stopped.

From Newton's second law of motion, we can derive an important relationship involving momentum. In Section 3.5 we saw that the equation for average acceleration is $a_{av} = \Delta v / \Delta t$. When we substitute this value of a in the equation for Newton's second law, $F = ma$, we get

$$F = \frac{m\Delta v}{\Delta t} \quad \text{or} \quad F\Delta t = m\Delta v$$

The first of these equations states that force is the time rate of change of momentum. The second equation states that impulse equals the change in momentum.

The product of a force and the time interval during which it acts, $F\Delta t$, is called ***impulse.*** Hence from Newton's second law of motion we have established that impulse equals change in momentum. The equation $F = m\Delta v/\Delta t$ tells us that when a force is applied to a body, the body's rate of change of momentum is equal to the force. Since the equation is a vector equation, we also know that the body's rate of change of momentum is in the direction of the force.

A good example of the relationship between impulse and change in momentum is a bat hitting a baseball. See Figure 6-14. The impulse imparted to the ball depends on the force with which the ball is hit and the length of time during which the ball and bat are in contact. On leaving the bat, the ball has acquired a momentum equal to the product of its mass and its change of velocity. In a sense, the impulse produced the change of momentum; hence the two are equal.

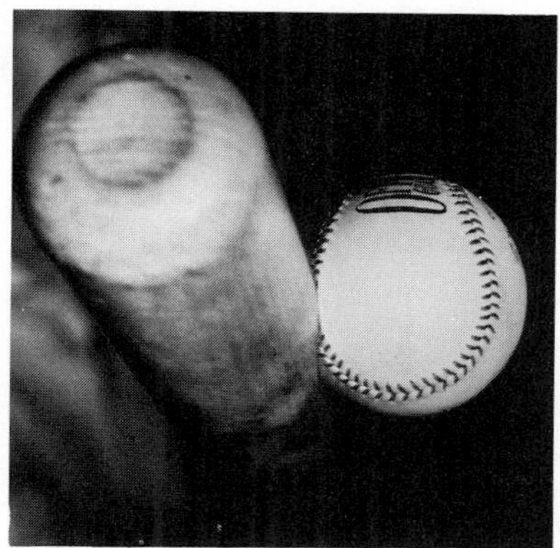

Figure 6-14. During the time of contact, much of the momentum of the bat is transferred to the baseball.

6.13 The Conservation of Momentum In Figure 6-15, a boy with a mass of $4\bar{0}$ kg and a man with a mass of $8\bar{0}$ kg are standing on a frictionless surface. When the man pushes on the boy from the back, the boy moves forward and the man moves backward.

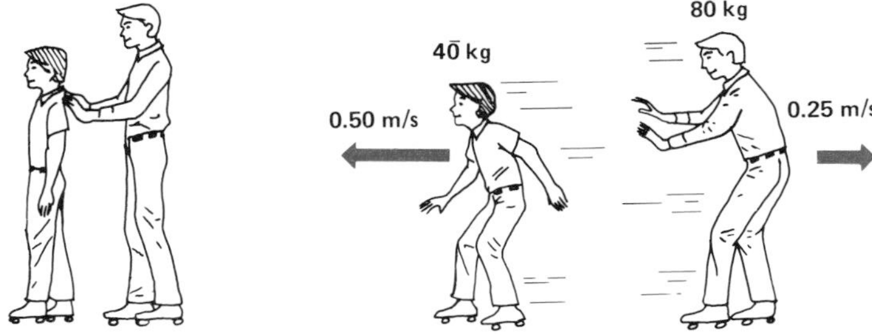

Figure 6-15. Conservation of momentum. On a frictionless surface, the momentum of the boy toward the left is equal to the momentum of the man toward the right.

Figure 6-16. Conservation of momentum in a rocket. The Space Shuttle is launched when momentum equal in magnitude to that of the exhaust gases is imparted to the space vehicle.

The velocities with which the boy and the man move are specified by one of the most important principles of physics, the ***law of conservation of momentum.*** This law states that *when no net external forces are acting on a system of objects, the total vector momentum of the system remains constant.*

Let us apply this law to the situation in Figure 6-15. Initially the man and the boy are at rest. The system, therefore, has zero momentum. When the man and the boy move apart, the law of conservation of momentum requires that the total vector momentum remains zero. Hence the momentum of the boy in one direction must equal that of the man in the other direction.

If the boy moves with a velocity of 0.50 m/s, the man will move with a velocity of 0.25 m/s since the mass of the man is twice as great as that of the boy. The momentum of the boy in one direction ($4\bar{0}$ kg $\times$ 0.50 m/s) must equal the momentum of the man in the opposite direction ($8\bar{0}$ kg $\times$ 0.25 m/s).

An important application of the law of conservation of momentum is the launching of a rocket. When a rocket fires, hot exhaust gases are expelled through the rocket nozzle. The gas particles have a momentum equal to the mass of the particles multiplied by their exhaust velocity. Momentum equal in magnitude is therefore imparted to the rocket in the opposite direction. See Figure 6-16. Newton's third law of motion (Section 3.10) is a special case of the law of conservation of momentum.

6.14 Inelastic Collisions The law of conservation of momentum is very helpful in studying the motions of colliding objects. Collisions can take place in various ways, and we shall see how the momentum conservation principles apply in several such cases.

In Figure 6-17(A), two carts of equal mass approach each other with velocities of equal magnitude. A lump of putty is attached to the front of each chart so that the two carts will stick together after the impact. This situation is an example of ***inelastic collision.*** Since the carts are traveling along the same straight line, it is also an example of a *collision in one dimension.*

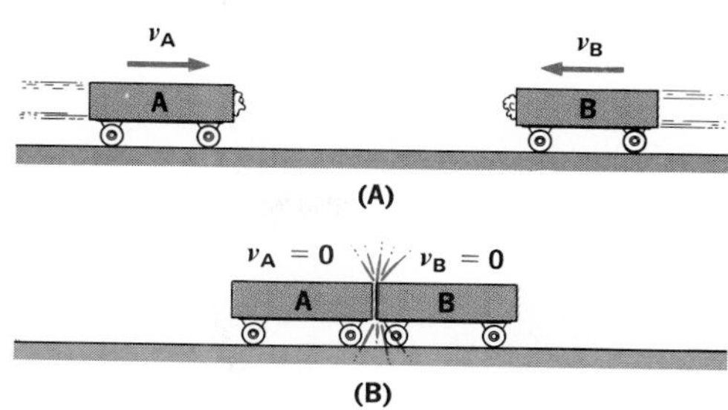

Figure 6-17. An inelastic collision. The carts have equal masses and approach each other with velocities of equal magnitude along the same straight line. The total momentum of the system is the same before and after the collision.

The momentum of cart **A** is $m_A v_A$. It is equal in magnitude to the momentum of cart **B**, $m_B v_B$. However, the direction of v_A is opposite to the direction of v_B, so

$$v_A = -v_B$$

Consequently,
$$m_A v_A = -m_B v_B$$

and
$$m_A v_A + m_B v_B = 0$$

This means that the total vector momentum of the system of two moving carts is zero. (We assume that the system is *isolated,* that is, there are no net external forces acting on it. In actual collision studies, the external force of friction is usually minimized by using rolling carts or air tracks, such as the one shown in Figure 6-18.)

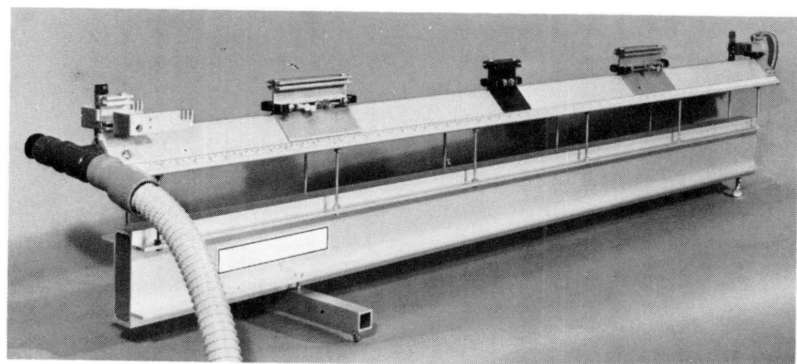

Figure 6-18. An air track designed for collision experiments. Air jets minimize the friction between the track and the masses placed on it.

After the carts in Figure 6-17(B) collide, they both come to rest. The cart velocities, v_A and v_B, are now both zero; the sum of the momenta of the two carts is zero, just as it was when the carts were in motion in opposite directions. Thus the total vector momentum of the system is unchanged by the collision.

If one of the carts has a greater mass than the other, although its velocity is still of equal magnitude but opposite sign, the outcome of the collision is different. After impact, the combined carts will move in the direction of the cart with the larger mass. The velocity of the combined carts will be such that the total momentum of the system remains unchanged.

For example, suppose

$$m_A = 2m_B$$

and $$v_A = -v_B, \text{ as before.}$$

Then, $$m_A v_A + m_B v_B \neq 0$$

Since $$m_B = \frac{m_A}{2}, \text{ by substitution,}$$

$$m_A v_A - \tfrac{1}{2} m_A v_A = \tfrac{1}{2} m_A v_A$$

This means that the total momentum of the system, before and after the collision, is $\frac{1}{2} m_A v_A$; the combined carts will move with this momentum in the direction of the original velocity of cart **A.** The velocity after the collision can be found by dividing the total momentum by the total mass

$$\frac{\frac{1}{2} m_A v_A}{\frac{3}{2} m_A} = \tfrac{1}{3} v_A$$

Kinetic energy is not conserved in an inelastic collision.

When carts of equal and opposite momenta collide inelastically, they come to rest. Before the collision they have kinetic energy; after the collision, they do not. This is typical of all inelastic and partially elastic collisions. Much or all of the kinetic energy that the moving objects have before collision is converted into heat or some other form of energy. If all these forms of energy are taken into account, the law of conservation of energy holds true for inelastic collisions. But kinetic energy alone is not conserved.

6.15 Elastic Collisions When colliding objects rebound from each other without a loss of kinetic energy, a perfectly ***elastic collision*** has just occurred. The only perfectly elastic collisions occur between atomic and subatomic particles. The situation can be approximated, however, by the use of hard steel balls or springs on an air track. The momentum conservation law holds for elastic collisions as well as inelastic ones and for collisions that are partly elastic and partly inelastic.

The law also holds for *collisions in two dimensions,* that is, when the colliding objects meet at an angle other than head-on. Figure 6-19(A) shows the elastic, two-dimensional collision of two balls of equal mass. Ball **B** collides at

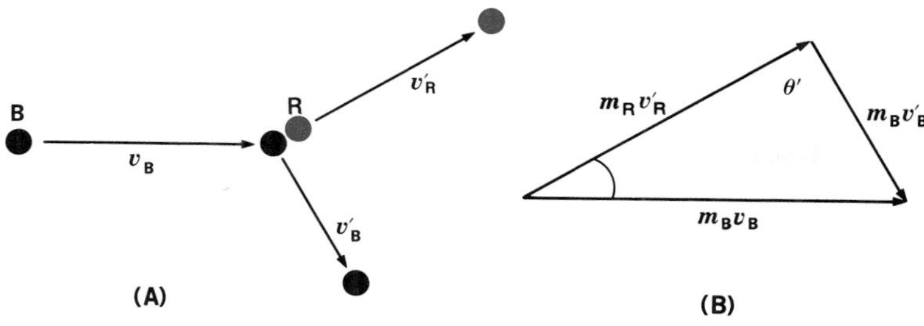

Figure 6-19. A vector diagram can be used to show that the total momentum and kinetic energy of a system both remain constant in a two-dimensional collision.

an angle with ball **R,** which is initially at rest. Figure 6-19(B) is the vector diagram representing the momenta of the balls before and after collision.

Review the parallelogram method of adding vectors explained in Section 2.11.

In Figure 6-19(B), $m_B v_B$ is the momentum of the black ball before the collision. Since the blue ball is stationary before the collision, its momentum is zero and is not represented by a vector. The total momentum of the system before collision is

$$m_B v_B + 0 = m_B v_B$$

After the collision, the measured velocities of the black and blue balls are v'_B and v'_R respectively, in the directions indicated. The total momentum now is the vector sum $m_B v'_B + m_R v'_R$. Since the momentum is conserved, the vector $m_B v_B$ must equal the vector sum of $m_B v'_B$ and $m_R v'_R$. The vector diagram appears as a closed triangle.

Since kinetic energy is also conserved in Figure 6-19,

$$\tfrac{1}{2} m_B v_B^2 = \tfrac{1}{2} m_B v'_B{}^2 + \tfrac{1}{2} m_R v'_R{}^2$$

Since the two balls have equal masses, this equation can be simplified to

$$v_B^2 = v'_B{}^2 + v'_R{}^2$$

This is the equation that relates the sides and hypotenuse of a right triangle, as in Figure 6-19(B). So it follows from the laws of conservation of energy and momentum that, when a moving ball strikes a stationary ball of equal mass other than head-on in an elastic collision, the two balls move away from each other at right angles.

The conservation of momentum also holds for partially elastic collisions, for collisions involving more than two bodies, and for three-dimensional situations. The following examlpes show how to solve collision problems.

EXAMPLE A 1.20-kg cart that has an eastward velocity of 0.50 m/s collides head-on inelastically with a 1.60-kg cart having a westward velocity of 0.70 m/s. Disregarding the slowing-down effects of friction, calculate the new velocity (speed and direction) of the two-cart system.

Given	Unknown	Basic equation
$m_E = 1.20$ kg $v_E = 0.50$ m/s $m_W = 1.60$ kg $v_W = -0.70$ m/s	v_{E+W}	$m_E v_E + m_W v_W = m_{E+W} v_{E+W}$

Solution

Working equation: $v_{E+W} = \dfrac{m_E v_E + m_W v_W}{m_{E+W}}$

$$= \frac{(1.20\text{ kg})(0.50\text{ m/s}) + (1.60\text{ kg})(-0.70\text{ m/s})}{(1.20\text{ kg} + 1.60\text{ kg})}$$

$= -0.19$ m/s (minus sign indicates westward motion)

EXAMPLE A ball moving eastward with a speed of 0.70 m/s hits a stationary ball of equal mass in an elastic collision. After the collision, the second ball moves away in a direction $3\bar{0}°$ north of east. Calculate the speed of this ball. (Refer to Figure 6-19(B).)

Given	Unknown	Basic equation
$m_B = m_R$ $v_B = 0.70$ m/s $\theta = 3\bar{0}°$ $\theta' = 9\bar{0}°$	v'_R	$m_R v'_R = m_B v_B \cos\theta$

Solution

Working equation: $v'_R = v_B \cos\theta$

$= (0.70\text{ m/s})(\cos 3\bar{0}°)$

$= 0.61$ m/s

6.16 Angular Momentum For rotary motion, the relationship between impulse and the change of angular momentum is similar to that for linear motion. Using the symbols for rotary motion, the equation becomes

$$\boldsymbol{T\Delta t = I\omega_f - I\omega_i}$$

where $T\Delta t$ is the ***angular impulse*** and $I\omega_f - I\omega_i$ is the change in ***angular momentum.*** The dimensions of both angular impulse and angular momentum are kg·m^2/s.

Just as the linear momentum of an object is unchanged unless a net external force acts on it, *the angular momentum of an object is unchanged unless a net external torque acts on it.* This is a statement of the *law of conservation of angular momentum.* A rotating flywheel, which helps maintain a constant angular velocity of the crankshaft of an automobile engine, is an illustration. The rotational inertia of a flywheel is large. Consequently torques acting on it do not produce rapid changes in its angular momentum. As the torque produced by the combustion in each cylinder tends to accelerate the crankshaft, the rotational inertia of the flywheel resists this action. Similarly, as the torques produced in the cylinders where compression is occurring tend to decelerate the crankshaft, the rotational inertia of the flywheel resists this action and the flywheel tends to maintain a uniform rate of crankshaft rotation.

If the distribution of mass of a rotating object is changed, its angular velocity changes so that the angular momentum remains constant. A skater spinning on the ice with arms folded, as in Figure 6-20, turns with relatively constant angular velocity. If she extends her arms, her rotational inertia increases. Since angular momentum is conserved, her angular velocity must decrease.

Figure 6-20. Conservation of angular momentum. When world-class figure skater Debi Thomas spins (left), her rotational inertia is small and her rotational velocity is high. When she extends her arms (right), the situation is reversed.

QUESTIONS: GROUP A

1. (a) Define impulse. (b) Define momentum. (c) Are they vectors or scalars? Explain.
2. Mathematically derive the relationship between impulse and momentum.
3. What is the difference between an elastic and an inelastic collision?
4. How does the law of conservation of momentum apply to the launch of a rocket ship?
5. If momentum is conserved, what else must happen when an object is dropped toward the earth?
6. Why does a fielder draw his hand back as he catches a baseball?

GROUP B

7. You are participating in a Physics Olympics event called the egg toss. How could you improve your chances of catching a tossed egg?
8. A diver leaps off a high platform in a layout position. If the diver pulls into a tuck position, what will happen to her rotary speed? Why?
9. What is the function of the long pole carried by a tightrope walker?
10. How do impulse and momentum explain why a 110-kg defensive lineman has more trouble changing direction than a 75-kg quarterback running at the same speed?
11. Can an object with a large mass and one with a small mass have the same momentum? Explain.
12. (a) Why does a child's toy top remain upright if it is spinning, but falls over as it slows down to a stop? (b) Why does it slow down?
13. When one billiard ball strikes another, there are two possible results. What are they and under what circumstances will each occur?
14. The safety net under a trapeze artist is loose. Use your knowledge of impulse and momentum to explain why.
15. Describe the situation in Figure 6-15 if (a) the boy instead of the man does the pushing and (b) the boy and the man push simultaneously.

PROBLEMS: GROUP A

1. An impulse of 20.0 N·s is applied to a 5.0-kg wagon initially at rest. What is its final speed?
2. (a) What impulse is required to stop a 0.250-kg baseball traveling 42.0 m/s? (b) If the ball is in the fielder's mitt for 0.100 s as it is being stopped, what is the average force acting on the ball?
3. A 60.0-g egg moving at 4.8 m/s is caught by a student. (a) If the time of interaction is 0.25 s, what is the average force on the egg? (b) If the maximum force the egg can withstand is 650 N, what minimum time is required to keep the egg intact?
4. A 50.0-kg person jumps from a window ledge 4.0 m above the pavement. (a) How fast is he moving as he hits the ground? (b) What impulse acts on the person's legs as he strikes the ground? (c) If the time of interaction is 0.060 seconds, how much force is acting?
5. The muzzle velocity of a 50.0-g shell leaving a 3.00-kg rifle is 400.0 m/s. What is the recoil velocity of the rifle?

GROUP B

6. A 1250-kg car is stopped at a traffic light. A 3550-kg truck moving at 8.33 m/s strikes the car from behind. What is the new velocity of the system if the bumpers lock during the collision?
7. A 65-kg person is skiing down a hill. The skier's speed at the bottom is 15 m/s. If the skier hits a snowdrift

and stops in 0.30 s, (a) how far does she go into the drift? (b) With what average force will she strike the drift?

8. Calculate the angular momentum of the rotating earth.
9. A 25-kg wagon moves eastward at 3.5 m/s. A force acting on the wagon for 4.0 s gives it a speed of 1.3 m/s to the west. Calculate (a) the impulse acting on the wagon and (b) the magnitude and direction of the force.
10. A 55.0-kg sailor jumps from a dock into a 100.0-kg rowboat at rest beside it. If the linear velocity of the sailor is 5.00 m/s as he leaves the dock, what is the resultant velocity of the sailor and the boat?
11. In the multiple-exposure photograph in Figure 6-21, a large ball approaches from the top and a smaller one from the bottom. The mass of the large ball is 150 g. The photo shows the balls at equal time interals. By means of a vector diagram, find the mass of the smaller ball.
12. An 85.0-g bullet is shot at a 3.00 kg piece of wood at rest at the edge of a counter 1.20 m high. If the bullet becomes embedded in the block and they land 5.0 m from the counter, what was the initial speed of the bullet?
13. A bowling ball of mass 8.00 kg, moving at 2.00 m/s collides with an identical ball at rest. If the first ball moves off at 30.0° from its original path, what are the speeds of the balls as they separate?

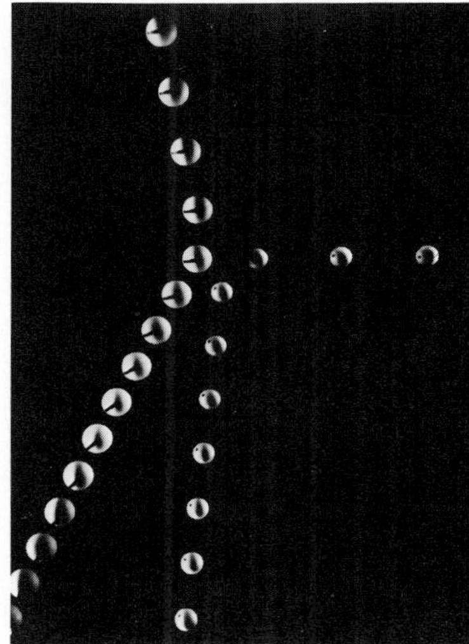

Figure 6-21.

SUMMARY

Work is the product of a displacement and the component of the force in the direction of the displacement. The unit of work is the joule. Work done by varying forces can be found by calculating the area under the curve of a graph in which the horizontal axis denotes the displacement and the vertical axis denotes the force. Radian measure is used to compute work done in rotary motion.

Machines can be used to multiply force at the expense of distance. The efficiency of a machine is the ratio of the useful work output to the total work input and is expressed as a percentage. Power is the time rate of doing work. It is measured in watts. Radian measure is used to compute power in rotary motion.

Gravitational potential energy is equal to the work done against gravity to place an object in position. Thus energy is measured in units of work. The kinetic energy of a body moving in a straight line is directly proportional to the mass of the body and the square of its velocity. In rotary motion, the kinetic energy is directly proportional to the rotational inertia and the square of the angular velocity.

Elastic potential energy is directly proportional to the force constant and the square of the amount of stretch. The law of conservation of mechanical energy states that the sum of the potential and kinetic energy of an ideal (friction-free) energy system remains constant.

Momentum is the product of the mass of a moving body and its velocity. The change of momentum of a moving body is equal to its impulse, which is the product of a force and the time interval during which the force acts. The momentum of a system of objects is conserved when no net external forces are acting on the system. This phenomenon is an example of the law of conservation of momentum. The reaction principle is an application of this law. The conservation of momentum also helps to describe the motion of objects colliding elastically or inelastically and in one, two, or three dimensions. Unless a net external torque acts on a rotating object, its angular momentum is also conserved.

VOCABULARY

angular impulse
angular momentum
efficiency
elastic collision
elastic potential energy
gravitational potential energy
impulse
inelastic collision
joule
law of conservation of mechanical energy
law of conservation of momentum
machine
momentum
power
watt
work

7

Phases of Matter

atom (A-tum) n.: the smallest unit of an element that can exist either alone or in combination with other atoms of the same or different elements.

THE STRUCTURE OF MATTER

OBJECTIVES

- Discuss the basic concepts of the kinetic theory of matter.
- Identify properties of materials in the solid, liquid, and gaseous phases.
- Apply Hooke's law to solve problems involving stress and strain.

7.1 Early Theories When a lump of sugar is crushed into smaller pieces, each of these pieces is still a particle of sugar. Only the size of the piece of sugar is changed. If the subdividing process is continued by grinding the material into a fine powder, the results are the same. Even when the sugar is dissolved in water and the pieces are too small to be seen with a microscope, the taste of the sugar is retained. Evaporation of the water returns the solid sugar's other identifying properties, such as color, crystalline shape, and density. Observations such as these, together with the quantitative evidence supporting the laws of definite and multiple proportions, which you may have already studied in chemistry, lead to these conclusions about the nature of matter:

1. *Matter is composed of particles.*
2. *The ultimate particles of matter are extremely small.*

As early as 400 B.C., Greek philosophers formulated ideas about the ultimate composition of matter. Democritus (460–370 B.C.) believed that matter is indestructible and that the subdividing process mentioned above would reach a limit beyond which no further separation is possible. He called these ultimate particles *atoms,* after the Greek word *atomos* meaning "indivisible." Democritus also believed that these particles differed in size and shape and that they were separated by empty space.

We still refer to wind and rain as the "elements of nature," especially during a storm.

Empedocles (490–430 B.C.) and Aristotle (384–322 B.C.) taught that matter consists of four basic "elements" (air, earth, fire, and water) that interact with each other through four "essences" (wet, dry, hot, and cold). Aristotle also believed that matter is infinitely divisible and that there is no need for "atoms" to explain its composition.

Figure 7-1. John Dalton postulated that each chemical element is composed of a different kind of atom. Dalton's ideas are the basis of modern atomic theory.

It was a long time before the idea of atoms and ultimate particles was actively studied again. Near the beginning of the 19th century, the English scientist John Dalton (1766–1844) conducted a series of experiments with gases that added greatly to our knowledge of the nature of atoms. Dalton's *atomic theory* explained the mechanisms of known reactions between substances and made it possible to extend the list of known elements.

The atomic theory has been modified and expanded many times since Dalton's day. Almost two hundred years of investigation have added much information to our knowledge of the makeup and properties of atoms, the way in which they interact, and the nature of the compounds they form. Modern theory includes data concerning the mass and size of atoms as well as the energy relationships among atoms. Experiments have also revealed the fact that atoms are not ultimate particles of matter after all. A whole array of subatomic particles is presently being studied by scientists. (See Chapters 23–25.)

7.2 Molecules *A **molecule** is the smallest chemical unit of a substance that is capable of stable, independent existence.* In the example in the previous section, the smallest particle of sugar that retains any identifying properties of the substance is a molecule of sugar. Not all substances, however, are composed of molecules. Some substances are composed of electrically charged particles known as ions.

To get an idea of the extremely small size of molecules, imagine a drop of water magnified until it is as large as the earth. With this tremendous increase in size, a single molecule of water would be about one meter in diameter. A simple molecule, like that of water, is about 3×10^{-10} m, or 3 angstroms (Å) in diameter.

1 Å = 10^{-10} m or 0.1 nanometers. The angstrom is not an SI unit, but it is widely used with units of that system. It is named after the Swedish physicist Anders Ångström (1814–1874).

Molecules of more complex substances may have sizes of more than 200 Å. The electron microscope, which is capable of magnifications of several million times, can be used to photograph some of these "giant" molecules.

7.3 Atoms If a molecule of sugar is analyzed, it is found to consist of particles of three simpler kinds of matter: carbon, hydrogen, and oxygen. These simpler forms of matter are called *elements*. *An **atom** is the smallest unit of an element that can exist either alone or in combination with other atoms of the same or different elements.* A molecule of sugar is made up of atoms of carbon, hydrogen, and oxygen.

Since atoms make up molecules, atoms are usually smaller than molecules. The smallest atom, an atom of hydrogen, has a diameter of about 0.6 Å. The largest atoms are a little more than 5 Å in size. The hydrogen atom is also the lightest atom. It has a mass of $1.673\ 559 \times 10^{-27}$ kg. The most common uranium atom, which is one of the heaviest atoms, has a mass of $3.952\ 989\ 1 \times 10^{-25}$ kg. There is about a tenfold range in the sizes of atoms and about a 250-fold range in their masses.

In 1970 the American scientist Albert Crewe (b. 1927) took the first photos of atoms. In 1976 Dr. Crewe obtained black and white movies of atoms. Two years later, Crewe used an electron microscope to take computer-enhanced color movies of atoms.

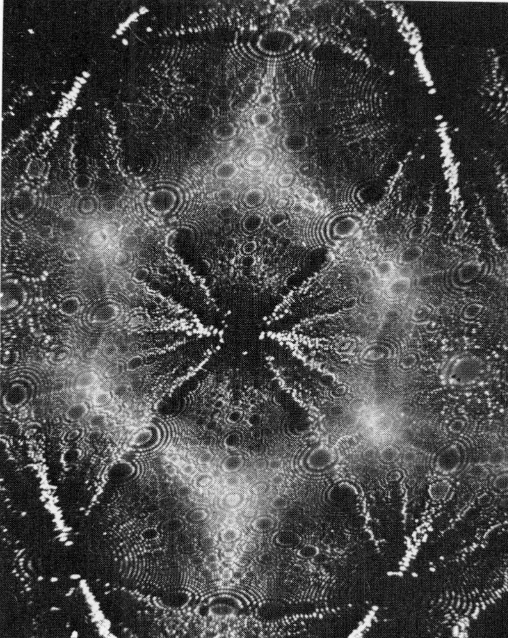

Figure 7-2. The point of a tungsten needle as seen through a field-ion microscope, which magnifies objects up to three million times. The photo shows the regular arrangement of tungsten atoms in the metal.

Even with the use of scientific notation, it is difficult to express conveniently the masses of individual atoms. Consequently, scientists express the *atomic mass* of an atom in ***atomic mass units*** *(u)*. One *u* is equal to $1.660\ 540\ 2 \times 10^{-27}$ kg. This is $\frac{1}{12}$ the mass of a carbon-12 atom. Carbon-12 is the most abundant form of carbon atoms. In other words, scientists use carbon-12 as the standard of mass for atoms.

In atomic mass units, the atomic mass of the most abundant form of hydrogen is 1.007 825 *u*, while that of the most abundant form of uranium is 238.050 784 *u*. You will notice that atomic masses are known very accurately. *The integer nearest to the atomic mass is called the **mass number** of an atom.* The mass number is represented by the symbol *A*. Thus for the common hydrogen atom, $A = 1$; for the common uranium atom, $A = 238$.

The concept of atoms is very useful in the study of chemical reactions. When substances interact chemically, atoms are regrouped, but the number of the various kinds of atoms does not change during the reaction.

7.4 Kinetic Theory of Matter

So far in our study of physics, we have been concerned almost entirely with the behavior of solid particles of matter. A solid is one of the three *phases* (or states) of matter—the solid phase, the liquid phase, and the gaseous phase. In the description of matter, ***phase*** indicates the way in which particles group together to form a substance. The structure of a substance can vary from compactly arranged particles to highly dispersed ones.

In a *solid,* the particles are close together in a fixed pattern. In a *liquid,* the particles are almost as close together as in a solid but are not held in any fixed pattern. In a *gas,* the average separation of the particles is relatively large, and as in a liquid, the particles are not held in any fixed pattern. (Frequently the term *vapor* is used to describe a gas that under ordinary conditions of temperature and pressure is in the liquid phase.) Both liquids and gases are also called *fluids* (from Latin meaning "to flow").

The kinetic theory was developed independently of the atomic theory.

To explain the motions of molecules and the energy molecules possess, particularly in the gaseous phase, scientists have developed the ***kinetic theory of matter***. Two basic concepts of this theory are:

1. *The molecules of a substance are in constant motion.* The amount of motion depends upon the average kinetic energy of the molecules; this energy depends upon the temperature.
2. *Collisions between molecules are perfectly elastic* (except when chemical changes or molecular excitations occur).

7.5 Forces Between Molecules

The forces required to pull a solid apart are generally much greater than the forces required to separate a similar amount of liquid. Liquids separate into drops. The forces of attraction between the molecules of liquids are not as great as the attractive forces of solids. Many solids have definite molecular structures, and the individual molecules are not as free to move as they are in a liquid. In some cases, the smaller attractive forces between liquid molecules are also due to greater intermolecular distances.

The forces that hold molecules together in crystals are known as van der Waals forces, after the Dutch scientist Johannes Diderik van der Waals (1837–1923).

In a gas, molecules separate from each other spontaneously; this accounts for the fact that a gas occupies a volume about 1000 times that of an equal mass of liquid. This spontaneous separation of gas molecules indicates that the kinetic energy of the molecules is great enough to keep them separated. As in the case of gravitation, forces between molecules decrease as the distance between them increases.

Solids and liquids are not easily compressed. Apparently, when molecules in solids and liquids are pushed closer together than their normal spacing, they repel each other. As the molecules are pushed still closer together, the repulsive forces become greater.

Intermolecular forces are mainly electric. They are about 10^{29} times as strong as the gravitational forces between molecules at the typical distances found in solids and liquids. Thus gravitational forces between molecules are negligible in comparison with intermolecular forces. Intermolecular forces are small compared to the weight of objects we can see and handle; however, the masses these intermolecular forces act on—the masses of molecules—are small, too. These forces can impart instantaneous accelerations 10^{14} times the acceleration of gravity. Such accelerations last only a very short time since one molecule, so accelerated, moves quickly out of the range of another.

Figure 7-3 shows how the force of interaction between molecules varies with the distance between their centers. If we imagine one molecule to be fixed at the intersection of the axes, the other molecule will be repelled until the distance of separation is such that their outer charges do not overlap. This condition occurs at r_0 where no net force acts between the molecules. The distance to r_0 is often called the equilibrium distance. This distance (about 2.5×10^{-10} meter) is, therefore, the distance between the centers of two "touching" molecules and is the diameter of a single molecule. As the separation between the molecules increases, the force of attraction between opposite charges first increases and then approaches zero. Different molecules have different sizes and charge configurations, but they always show the qualitative behavior indicated by Figure 7-3.

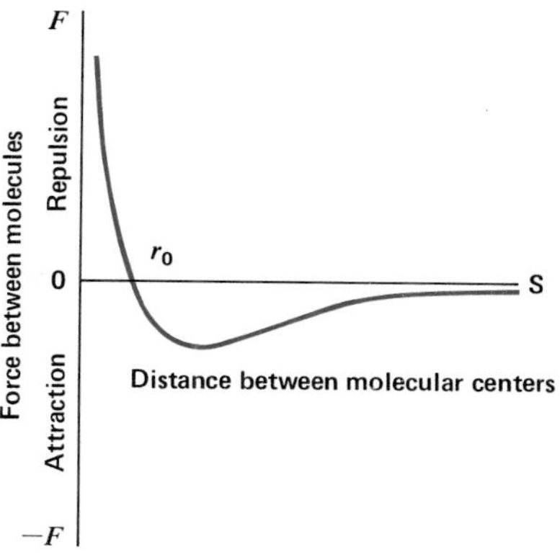

Figure 7-3. The force between molecules varies with the distance between molecular centers.

In the solid phase, molecules vibrate about the equilibrium position r_0. They do not have enough energy to overcome the attractive force. The equilibrium positions are fixed. In a liquid the molecules have greater vibrational energy about centers that are free to move, although the average distance between the centers of the molecules remains nearly the same. The average distance of separation between gas molecules is considerably greater than the range of intermolecular forces, and the molecules move in straight lines between collisions.

THE SOLID PHASE

7.6 The Nature of Solids Solids have definite shapes and definite volumes. Scientists usually describe solids as either *crystalline* or *amorphous.* Crystalline solids have a regular arrangement of particles; amorphous solids have a random particle arrangement.

In addition to the forces that bind particles of a solid together, the motion of particles of a solid is an important consideration. Particles of a solid are held in relatively fixed positions by the binding forces. However, they do have a vibratory motion about their fixed positions. The amplitude of their vibration and their resulting vibratory

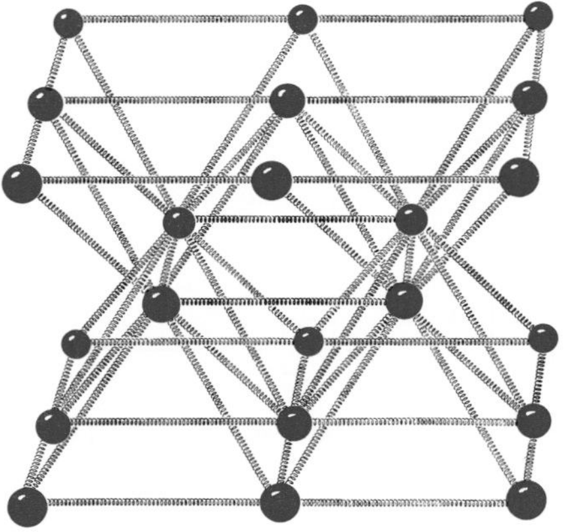

Figure 7-4. A crystalline solid. The particles of the solid are held in a specific pattern by intermolecular forces, which are shown here as springs. Each particle vibrates about its indicated position at a rate that depends on the temperature of the solid.

energy are related to the temperature of the solid. At low temperatures the kinetic energy is small; at higher temperatures it is larger.

Diffusion *is the penetration of one type of particle into a mass consisting of a second type of particle.* If a lead plate and a gold plate are in close contact for several months, particles of gold may be detected in the lead, and vice versa. This demonstrates that even solids can diffuse. Diffusion is slow in solids because of the limited motion of the particles and their close-packed, orderly arrangement.

7.7 Cohesion and Adhesion The general term for *the force of attraction between molecules of the same kind is* ***cohesion.*** Cohesion, the force that holds the close-packed molecules of a solid together, is a short-range force. If a solid is broken, layers of gas molecules from the air cling to the broken surfaces. These gas molecules prevent the rejoining of the solid surfaces. The molecules of the broken surfaces are not close enough to have sufficient attraction to hold. However, if the surfaces of two like solids are polished and then slid together, cohesion can cause the solids to stick together.

Molecules of different kinds sometimes attract each other strongly. Water wets clean glass and other materials. Glue sticks to wood. *The force of attraction between molecules of different kinds is called* ***adhesion.*** The forces of cohesion and adhesion have definite values for specific molecules.

Why is the tape used to hold a bandage in place called adhesive tape?

7.8 Tensile Strength Several properties of solids depend on cohesion; one of these is tensile strength. Suppose two wires of the same diameter, one copper and one steel, are put in a machine that pulls the wires until they break. When tested in this manner, steel wire proves stronger than copper wire of the same diameter. Therefore we say that steel has a higher *tensile strength* than copper. *The* ***tensile strength*** *of a material is the force per unit cross-sectional area applied perpendicularly to the cross section that is required to break a rod or wire of that material.* See Figure 7-5 and Appendix B, Table 7. Tensile strength is a measure of cohesion between adjacent molecules over the entire cross-sectional area.

Figure 7-5. Testing for tensile strength. The steel test rod elongates as large tensile forces act on it.

7.9 Ductility and Malleability If a metal rod can be drawn through a small opening, or die, to produce a wire, the metal is said to be *ductile* (*duk*-til) or to possess *ductility.* As the metal is pulled through the die, pressure decreases its diameter and increases its length and the rod becomes a wire. In one industrial application of ductility, more than 1000 meters of wire are formed per minute in a device in

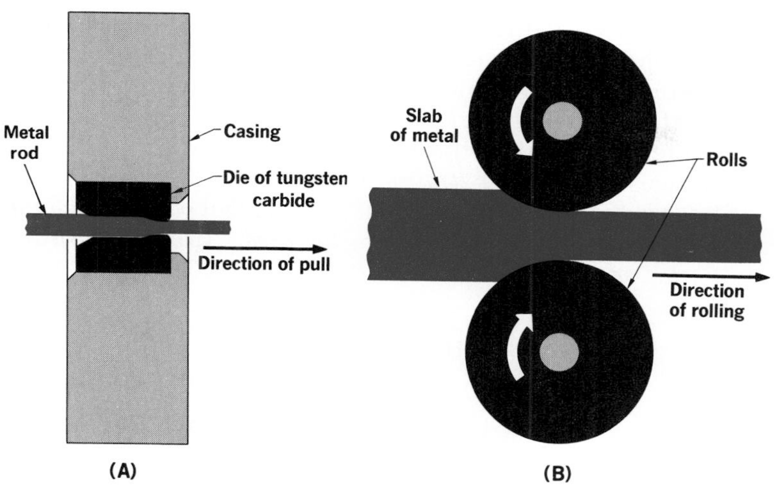

Figure 7-6. During the process of drawing (A) and rolling (B), the atoms of a metal are forced to move over each other from one position in the crystal pattern to another.

which the pressure is exerted by a fluid, and the wire does not touch a solid die. See Figure 7-6. Platinum is so ductile that a single gram of the metal can be drawn into a wire almost 600 kilometers long.

Metals that can be hammered or rolled into sheets are said to be *malleable* (*mal*-ee-uh-bul) or to have *malleability*. During the hammering or rolling, the shape of the metal is greatly changed.

In the process of hammering, rolling, or drawing, layers of atoms of a metal are forced to slide over one another, thus changing their positions in the crystal pattern. Since the cohesive forces are strong and the atoms do not become widely separated from each other during their rearrangement, the metal holds together while its shape is being changed. Silver, gold, platinum, copper, aluminum, and iron are all highly malleable and ductile.

7.10 Elasticity When opposing forces are applied to an object, the size and shape of the object are changed. *The ability of an object to return to its original size or shape when external forces are removed* is described as its ***elasticity.*** However, there is a limit beyond which the change produced by the applied forces does not disappear when the forces are removed. Ductility and malleability are properties of substances that can undergo such permanent changes without fracturing. *When a substance is on the verge of becoming permanently changed,* we say it has reached its ***elastic limit.***

Elastic potential energy was discussed in Section 6.10.

At the elastic limit, molecular forces are overcome to such an extent that particles slide past each other, and the shape of the material is permanently changed. Such a drastic change cannot be reversed by molecular forces. Every solid has a certain range through which it can be changed and yet return to its original condition before its elastic limit is reached.

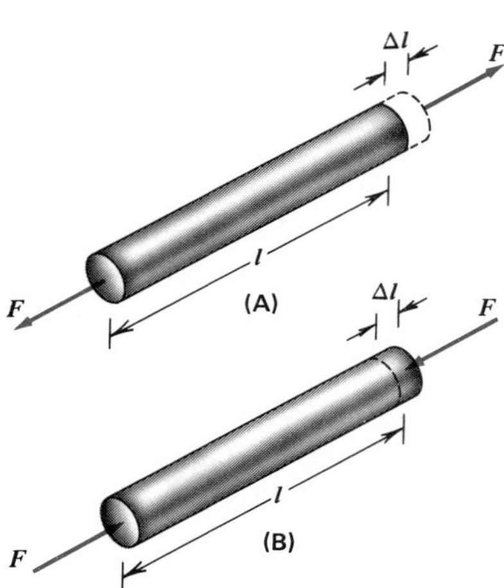

Figure 7-7. Two types of strain. Elongation strain (A) and compression strain (B) are both expressed by the ratio $\Delta l/l$.

Two terms are used to describe the elastic properties of substances: *stress* and *strain.*

Stress *is the ratio of the internal force,* ***F,*** *that occurs when the shape of a substance is changed to the area,* ***A,*** *over which the force acts, or*

$$\textbf{stress} = \frac{F}{A}$$

Stress represents the tendency of a substance to recover its original shape. The applied force changes the distance between molecules by either pulling them farther apart or pushing them closer together. When the force is removed, molecular forces restore the molecules to their normal spacing.

Strain *is the relative amount of deformation produced in a body under stress.* There are various kinds of strain, but its measure is always an absolute ratio—a number without units. When a rod is placed under tension, it stretches. *The ratio of the increase in length to the original length is called the* ***elongation strain*** *and is expressed as* $\Delta l/l$. Linear compression strain is the reverse of elongation, as shown in Figure 7-7. Elongation and compression are accompanied by small changes in the cross-sectional area, but their effect is small in situations where elastic limits are not exceeded.

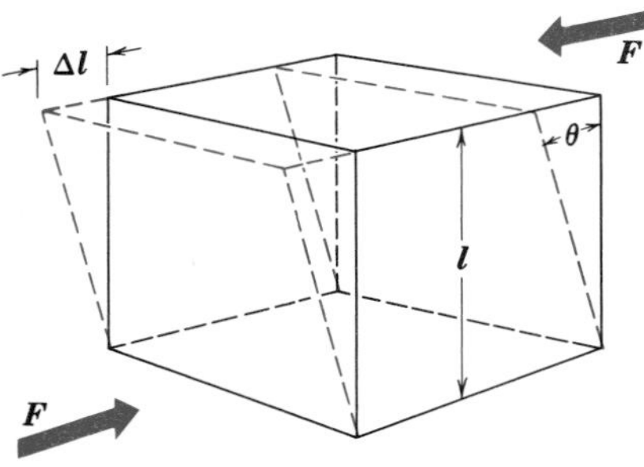

Figure 7-8. Shear strain. The amount of deformation produced by the forces, F, is equal to $\Delta l/l$, or $\tan \theta$.

In an *elongation strain,* the particles have been moved in a direction perpendicular to the area over which the forces act. In a ***shear strain,*** the particles move in a direction parallel to the area over which the forces act. A cube will take the shape of a rhombic prism when shear forces are applied to it. *The measure of shear strain is the ratio of the amount the top of the cube is moved to the side to the length of one side of the cube.* This ratio, shear strain, is expressed as the tangent of the angle through which the oblique edges of the imaginary cube have been rotated from their original direction, as shown in Figure 7-8.

Volume strain *is the ratio of the decrease in volume to the volume before the stress was applied.*

Flexure (bending) and torsion (twisting) are combinations of elongation and compression strains. A straight beam bent into a plane curve undergoes compression on one side and elongation on the other side. The layer of material down the center of the beam undergoes neither compression nor elongation.

7.11 Hooke's Law Beams of buildings and bridges are often subjected to varying forces or stresses. It is important for engineers to know what deformation, or strain, these forces will produce. This involves the measurement of the elasticity of materials.

If a coiled spring is stretched by a weight, as shown in Figure 7-9(A), and then returns to its original form after the stretching force is removed, the spring is said to be *perfectly elastic.* If the spring is stretched too far, it remains permanently deformed; its elastic limit has been exceeded.

Suppose one end of a steel wire is fastened to a beam, as in Figure 7-9(B), and weights are gradually added to the hanger attached to the lower end. As the weights are added one by one, the wire stretches gradually. The wire stretches by an amount that is exactly proportional to the weight pulling on it, and it returns to its previous length when the weight is removed. If more weights are added, the elastic limit is eventually reached. Then when the weights are removed, the wire remains deformed; it does not return to its original length.

By making such measurements, the English philosopher and scientist Robert Hooke (1635–1703) found that the amount of elongation in elastic solids is directly proportional to the deforming force provided the elastic limit is not exceeded. The elongation also depends on the length and cross-sectional area of the wire or rod. All these facts were combined by Hooke into one simple law. ***Hooke's law*** states that *within certain limits strain is directly proportional to stress.*

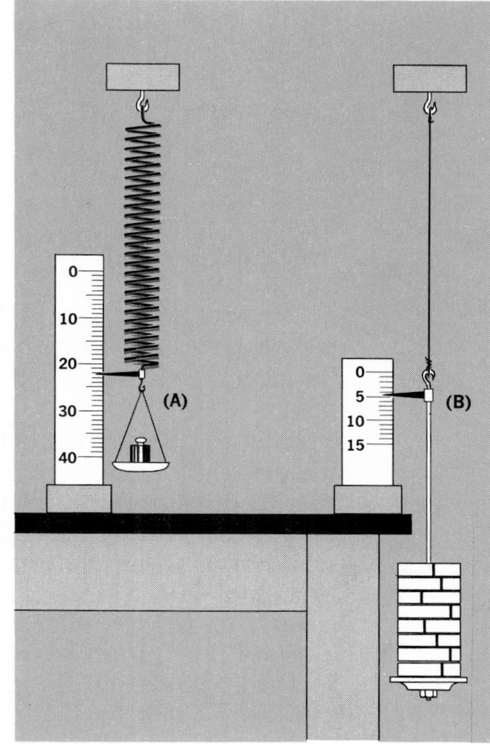

Figure 7-9. Hooke's law. In both cases, the pointers stood at zero before the weights were added, and the amount of stretch is proportional to the force exerted by the weights (provided that the elastic limit is not exceeded).

The value of the ratio stress/strain is different for different solids. However, the ratio is constant for a given substance, even when the substance has different shapes. This ratio gives us a means of comparing the elasticity of various solids. The numerical value of Hooke's law is called ***Young's modulus,*** *Y*, and is defined by the equation

Young's modulus is named after the English physicist Thomas Young, who also did experiments with light. Values of Young's modulus for various substances are given in Appendix B, Table 8.

$$Y = \frac{\textbf{stress}}{\textbf{strain}}$$

From the definitions of stress and strain given in the previous section, we get

$$Y = \frac{F/A}{\Delta l/l}$$

and

$$Y = \frac{Fl}{\Delta l A}$$

As an example, suppose that the steel "A" string of a piano is 43.0 cm long and has a cross-sectional area of

6.50×10^{-7} m^2. How much will the string stretch when a force of 710 N is applied to it by a piano tuner to bring it up to pitch?

$$\Delta l = \frac{Fl}{YA} = \frac{(7.10 \times 10^{2}\ \text{N})(4.30 \times 10^{-1}\ \text{m})}{(20.0 \times 10^{10}\ \text{N/m}^2)(6.50 \times 10^{-7}\ \text{m}^2)}$$

$$= 2.35 \times 10^{-3}\ \text{m}$$

QUESTIONS: GROUP A

1. (a) What is an atom? (b) What is the derivation of the word *atom?* (c) Does the original meaning still apply? Explain.
2. What is the difference between the process used by Aristotle and Democritus in developing their theories and the process used by Dalton?
3. (a) What is an atomic mass unit? (b) Upon what standard is this unit based?
4. (a) State the three phases of matter. (b) Which are considered fluids?
5. (a) How does a vapor differ from a gas? (b) Give an example of each.
6. List the two major concepts of the kinetic theory of matter.
7. (a) What are two types of solids? (b) How do they differ? (c) Give an example of each.
8. How is adhesion different from cohesion?
9. List and define the properties of a solid that depend on cohesion.
10. What is elasticity?
11. (a) Distinguish between stress and strain. (b) Give mathematical expressions for each. (c) What physical law relates stress and strain?
12. What is meant by the elastic limit?

GROUP B

13. (a) What force is most important in holding atoms and molecules together? (b) Is gravity an important force in intramolecular attraction? Explain.
14. List the common states of matter in order from lowest to highest average kinetic energy of an atom or molecule.
15. (a) Describe the shape of a graph of strain as a function of stress (within the elastic limits of a material). (b) What part of the graph would represent Young's modulus?
16. Which is more elastic: a rubber band or a steel wire of equal length and diameter? Explain.
17. Why are scissors called "shears"?
18. What happens to the crystal pattern of an elastic material when its elastic limit is exceeded?

PROBLEMS: GROUP A

1. (a) Calculate the mass in atomic mass units of a water molecule. (b) What is the mass of the molecule in grams? (c) How many molecules of water are there in 3.00 mL of water under standard conditions?
2. A 45.0-kg diver on the end of a diving board depresses it 3.67 cm below its equilibrium position. How much would a diver weighing 648 N depress the board?
3. The spring in a scale with no weight hanging from it is 12.0 cm long. A 7.50-kg mass stretches the spring to 17.0 cm. (a) How long would the spring be with a mass of 4.50-kg hanging from it? (b) If the scale is marked every 10.0 N, what is the distance between markings?
4. A platinum wire has a cross-sectional

area of 3.20×10^{-3} cm^2. If the tensile strength of platinum (from Table 7 in Appendix B) is 3.5×10^8 N/m^2, what force would snap the wire?

5. A weight of 2.50 N stretches a 4.30-m length of wire by 1.19×10^{-3} cm. If the cross-sectional area of the wire is 0.100 cm^2, calculate Young's modulus for this material.

GROUP B

6. Using Tables 7 and 21 in Appendix B, determine the perpendicular force necessary to break a 1.00-m long piece of No. 000 gauge copper wire.
7. A 2.00-m long piece of No. 7 gauge copper wire is suspended from the ceiling of a physics laboratory. If a 200.0-N weight is hung from the lower end of the wire, what will be the new length of the copper wire?
8. A brass wire that is 2.57 m long and 2.00 mm in diameter hangs from a fixed support attached to the ceiling. An 8.25-kg stool is hung from the lower end of the wire. What is the elongation of the wire?

THE LIQUID PHASE

7.12 The Nature of Liquids In 1827 the English botanist Robert Brown (1773–1858) put some pollen grains in water and placed a bit of this suspension on a small glass slide. When he examined the suspension through a microscope, he found that the pollen grains moved in a very random way. The path of a single particle resembled that shown in Figure 7-10. This so-called *Brownian movement* is caused by the continual bombardment of the suspended particles by molecules of the surrounding liquid. Brownian movement shows that molecules of a liquid are in constant, rapid, and random motion. This is in keeping with the kinetic theory of matter described in Section 7.4.

Brownian movement is also exhibited by solid particles, such as smoke, suspended in the atmosphere.

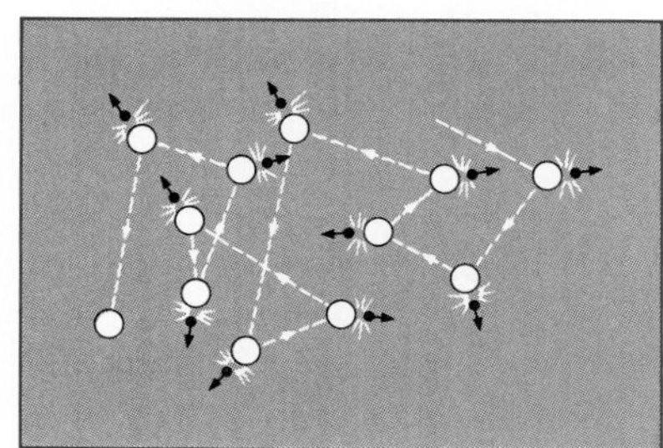

Figure 7-10. Brownian movement. The seemingly random motion of the white particle (starting at the upper right) is the result of collisions with molecules of the surrounding fluid. The directions of the fluid molecules after the collisions are pictured by the black arrows. (The distances between collisions are exaggerated for clarity.)

Liquids diffuse. This can be shown by the experiment in Figure 7-11. Enough concentrated copper(II) sulfate solution is poured into a tall cylinder to form a layer several centimeters deep. A flat cork is floated on the surface of the solution. Water is carefully poured through a funnel tube onto the top of the cork. The water flows around the edge of the cork and spreads out over the surface of the copper(II) sulfate solution, producing two distinct layers. The water "floats" on the copper(II) sulfate solution because the water has a lower density. After the cylinder stands for a few days, the interface between the layers is less distinct. Some of the copper(II) sulfate solution diffuses into the water above, and some of the water molecules diffuse into the copper(II) sulfate solution below. Even though weeks may pass before the diffusion is complete, we can see that diffusion does occur in liquids despite the force of gravity. Because of the slightly more

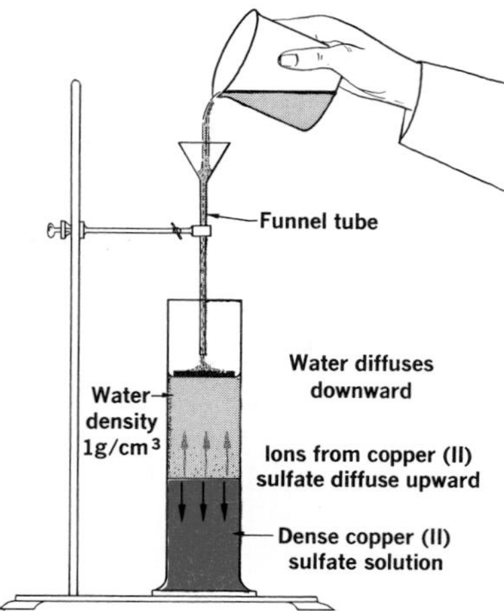

Figure 7-11. Diffusion in liquids. Water molecules diffuse downward into the heavier copper(II) sulfate solution, while copper(II) sulfate diffuses upward.

Figure 7-12. Surface tension keeps a razor blade afloat even though its mass density is much greater than that of water.

open molecular arrangement and greater molecular mobility of liquids, their rate of diffusion is considerably faster than that of solids.

7.13 Cohesion and Adhesion

If you push a spoon into a jar of honey and then pull it out, a certain amount of force is needed to separate the molecules of the honey. If you lick the honey from the spoon, you find that force is required to pull the molecules of the honey away from the molecules of the spoon. These are examples of cohesion within a liquid and adhesion between a liquid and a solid.

If a clean glass rod is dipped into water and then removed, some of the water clings to the glass rod. We say that the water *wets* the glass. The adhesion between water molecules and glass must therefore be greater than the cohesion between water molecules.

If the glass rod is dipped into mercury and then removed, the mercury does not cling to the glass because the cohesion between mercury molecules is greater than the adhesion between mercury and glass. Cohesion is usually smaller in liquids than in solids.

If you examine the surface of water in a glass container, you will find that it is not completely level. It is very *slightly concave* when viewed from above. The edge of the surface where water comes in contact with the glass is lifted a little above the general level. *The crescent-shaped surface of a liquid column is called the* ***meniscus***. The water rises at the edge because adhesion between water and glass is greater than cohesion between water molecules.

If the container contains mercury instead of water, the edges of the liquid are depressed and the surface is *slightly convex*. In this case, cohesion between mercury molecules is greater than adhesion between mercury and glass.

Cohesion is also responsible for a property of fluids called *viscosity*. ***Viscosity*** *is the ratio of shear stress to the rate of change of shear strain in a liquid or gas*. The viscosity of a fluid determines the rate of flow of the fluid. Viscosity is dependent on temperature. As the temperature increases, the viscosity of gases increases while the viscosity of liquids decreases. This explains the expression "as slow as molasses in January." Modern motor oils are specially formulated to minimize the effect of temperature on viscosity. Under normal conditions, gases are much less viscous than liquids.

7.14 Surface Tension

Have you ever seen a sewing needle or a razor blade floating on the surface of water? Even though they are about seven times as dense as water, if they are placed carefully on the surface, they remain there. A close look at the water surface shows that the needle or razor blade is supported in a hollow in the water surface, as seen in Figure 7-12. *The water acts as though it has a thin, flexible surface film.* The weight of the needle or razor blade is counterbalanced by the upward force that is exerted by the surface film. This property of liquids is due to *surface tension.*

All liquids show surface tension. Mercury has a very high surface tension. In many liquids the surface film is not as strong as that of water or mercury. Part of the cleaning action of detergents is due to their ability to lower the surface tension of water. This makes it possible for the water and detergent to penetrate more readily between the fibers of the articles being cleaned and the dirt particles.

Since particles in a liquid attract similar liquid particles that are nearby, they move as close together as possible. Hence the surface will tend to have a minimum area. The effect of this attraction is to make the liquid behave as if it were contained in a stretched elastic skin. The *tension* in this "skin" is the *surface tension.* When a force acts on a liquid surface film and distorts that film, the cohesion of the liquid molecules exerts an equal and opposite force that tends to restore the horizontal surface. Thus the weight of a supported needle produces a depression in the water surface film that increases the area of the film. In order to restore the surface of the liquid to its original horizontal condition, the cohesion of the water molecules exerts a counterbalancing upward force on the needle.

Surface tension produces contraction forces in liquid films. A liquid film has two free surfaces on which molecules are subject to an unbalanced force toward the inside of the film. Thus both free surfaces tend to assume a minimum area. The contraction of the film can be demonstrated by the device shown in Figure 7-13. A wire ring containing a loop of thread is dipped into a soap solution causing a film to form across the ring. If the film inside the loop of thread is broken with a hot wire, the unbroken film outside the loop contracts and pulls the thread equally on all sides to form a circle.

Surface tension causes a free liquid to assume a spherical shape. A free liquid is one that is not acted upon by any external force. This condition can be approximated by small drops of mercury on a table top, as shown in Figure 7-14. A sphere has the smallest surface area for a given volume. The unbalanced force acting on liquid surface molecules tends to pull them toward the center of the liquid, reducing the surface area and causing the liquid to

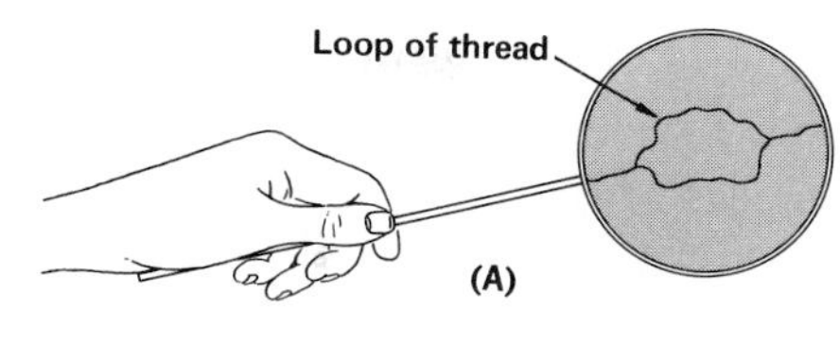

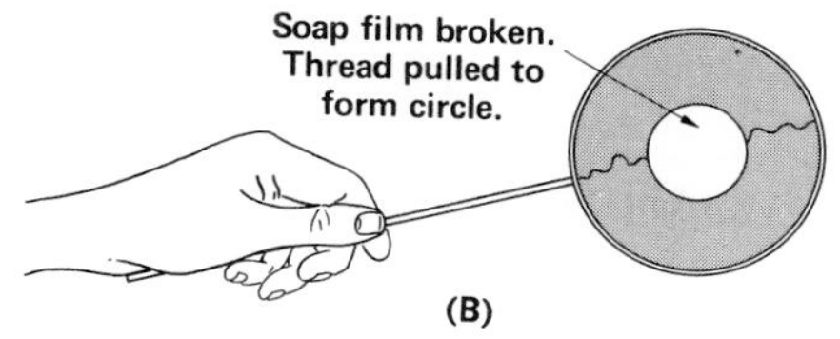

Figure 7-13. In (A), a soap film covers the entire area inside the ring. When the film inside the loop of thread is broken (B), the film outside the loop contracts and pulls the thread into a circle.

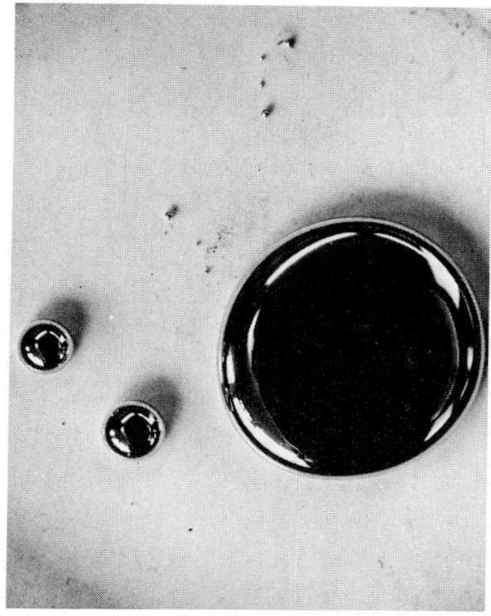

Figure 7-14. Small drops of mercury assume a spherical shape because of surface tension. Large drops are flattened because the downward force of gravitation is greater than the upward force produced by surface tension.

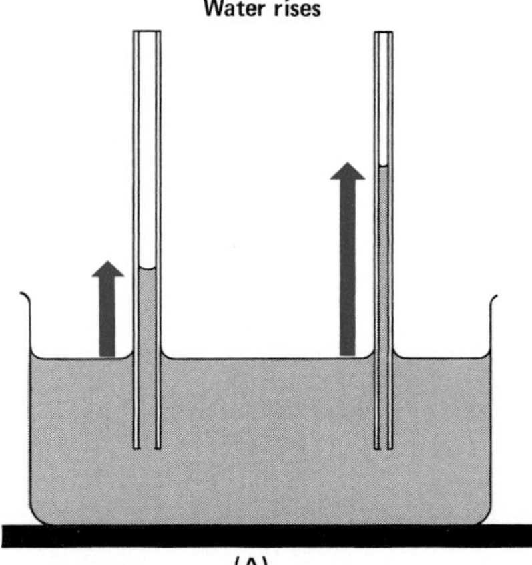

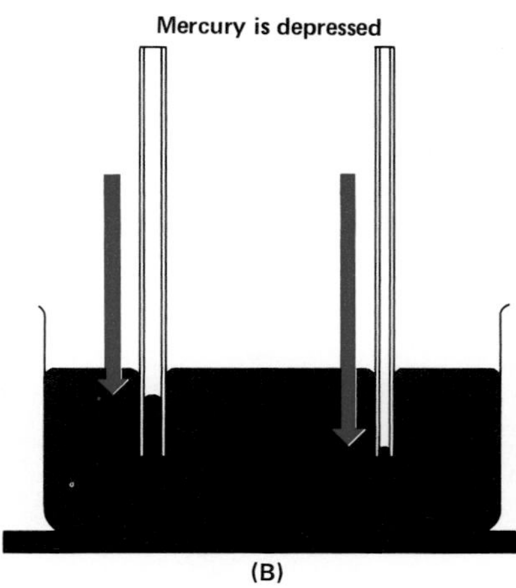

Figure 7-15. Capillarity. The elevation or depression is inversely proportional to the diameter of the tube. (The tube diameters and magnitudes of capillarity are exaggerated for clarity.)

assume a spherical shape. Since the cohesive force between mercury molecules is great and mercury has a very high surface tension, small drops of mercury are almost spherical. Larger drops of mercury, on which the effect of the force of gravity is greater, are noticeably flattened.

7.15 Capillarity When the lower ends of several tubes are immersed in water, the water will rise to the same height in each tube only if the tubes have large enough diameters so that the centers of the menisci are relatively flat. *Water does not rise equally in tubes of small diameters.* The height to which it rises increases as the diameter of the tube decreases. See Figure 7-15(A). When mercury is used, the depression of the surface is greater as the diameter of the tube is reduced. See Figure 7-15(B). *This elevation or depression of liquids in small-diameter tubes is called* ***capillarity.***

Capillarity depends on both adhesion and surface tension. Adhesion between water and glass causes water to creep up the glass walls and produce a concave surface; surface tension tends to flatten this surface by contraction. The combined action of these two forces raises the water above its surrounding level. The water level rises until the upward force is counterbalanced by the weight of the elevated liquid.

Experiments have verified the following: *(1) Liquids rise in capillary tubes they wet and are depressed in tubes they do not wet. (2) Elevation or depression is inversely proportional to the diameter of the tube. (3) The elevation or depression decreases as the temperature increases. (4) The elevation or depression depends on the surface tension of the liquid.*

7.16 Melting *The change of phase from a solid to a liquid is called* ***melting.*** Melting involves the breaking of bonds between the particles of a solid. The temperature at which this change occurs is called the ***melting point.*** Pure *crystalline* solids have definite melting points, and different solids have different melting points.

When a substance changes from a liquid to a solid, it is said to *freeze.* The temperature at which freezing occurs is known as the ***freezing point.*** For pure crystalline substances, the melting point and the freezing point are the same temperature at any given pressure.

Noncrystalline solids, like paraffin, have no definite melting point. When they are heated, they soften gradually. The temperature at which noncrystalline solids first soften and the temperature at which they flow freely are often greatly different.

In order for a solid to melt, energy must be supplied to it. This energy increases the energy of the particles of the solid and gives them the freedom of motion characteristic of the particles of a liquid. As the temperature of a solid increases, the vibrations of its particles increase in amplitude; thus more and more potential energy is stored in the average stretching of the bonds between the particles. Finally a point is reached at which the bonds between the particles cannot absorb any more energy without breaking. Thus crystalline solids have a definite melting point. When the liquid formed by melting a crystalline solid cools to a certain temperature, the energy of the liquid particles is reduced and the forces between the particles draw them into fixed positions in a crystal. Thus a liquid that forms a crystalline solid freezes at a definite temperature.

Figure 7-16. Noncrystalline solids, such as the glass in this building, do not have specific melting points.

All the energy supplied to a substance during melting is used to increase the potential energy of the particles in changing from a crystal structure to a liquid. The kinetic energy of the particles does not change. Since average kinetic energy depends on temperature, the temperature is unchanged during the melting process.

Usually, the particles of noncrystalline solids are held together by forces of attraction and by the physical entanglement of long-chain molecules. The bonding combination is not of such definite strength that the bonds are broken when the particles acquire a fixed amount of energy. The energy required to overcome these bonds varies with the extent of the bonding and entanglement of each molecule. As they are heated, noncrystalline substances soften at a lower temperature first. At some higher temperature, they flow freely. Similarly, as they are cooled, the molecules become bonded at various kinetic energies, and the liquid does not solidify at a definite temperature.

The degree of separation of particles of a substance is different in the solid and liquid phases because of the difference in the potential energy of the particles. If melted paraffin is poured into a vessel and allowed to harden, the center becomes indented, or depressed, as shown in Figure 7-17(A). The paraffin cools and contracts as it solidifies. Both kinetic energy and potential energy are lost in the process. The loss of kinetic energy is indicated by the decrease in temperature. Loss of potential energy permits the particles to move closer together and take up less space. Almost all substances behave in this manner; the particles of most substances are closer in the solid phase than in the liquid phase.

Water is the most important exception to the rule that a

substance contracts when it changes from a liquid to a solid. See Figure 7-17(B). When an ice cube tray is placed in the freezing compartment of a refrigerator, the level of the water in the sections of the tray is uniform. When the ice cubes are formed, however, each one has a slightly raised spot in the center. The volume occupied by ice is about 1.1 times that occupied by the water from which it was formed. The force of expansion when water freezes is enormous. Pressures close to 500 atmospheres have been measured in the laboratory! Antifreeze is used in water-cooled cars that are driven in areas where the atmospheric temperature drops below the freezing point of water. As its name suggests, a mixture of antifreeze and water freezes at a lower temperature and does not expand significantly in freezing if there is a sufficient amount of antifreeze in the mixture.

The force of expansion that accompanies freezing can break water pipes in winter.

Bismuth and antimony are two metals that expand rather than contract when they solidify. Substances such as ice, bismuth, and antimony have open crystal structures in which the particles are more widely separated than they are in the liquid phase. That is why the solid occupies a larger volume than the liquid.

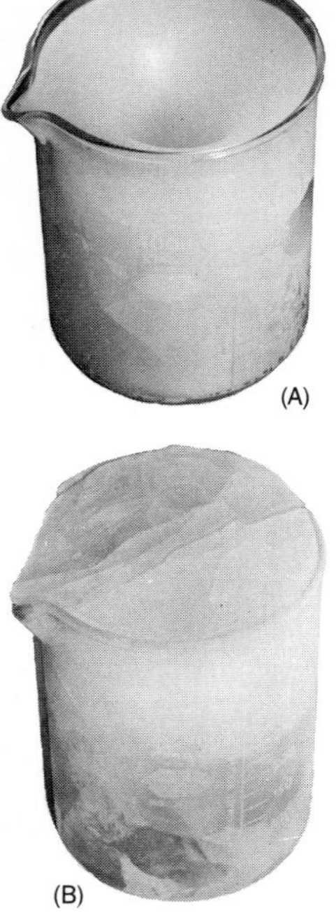

Figure 7-17. (A) Most substances, like paraffin, contract when they solidify. (B) Water is one of the few substances that expands on solidification.

7.17 Effect of Pressure on the Freezing Point In substances that contract when they solidify, like the paraffin in Figure 7-17(A), the molecules of the solid are closer together than the molecules of the liquid. If additional pressure is exerted, such a liquid can be made to solidify at a higher temperature than its normal freezing point under atmospheric pressure. An increase in pressure raises the freezing (melting) point of most substances.

Some of the rock in the interior of the earth is hot enough to melt at normal pressure, but most of it remains solid because of the tremendous pressure. If the pressure is released, as when a volcano erupts, more of the rock melts and forms lava.

An increase in pressure has the opposite effect on the freezing point of a substance like water that expands as it freezes. In such a substance the molecules are farther apart in the solid than they are in the liquid. Since an increase in pressure makes formation of the solid more difficult, the freezing point is lowered if the pressure is raised.

We can illustrate this effect by suspending two weights over the surface of a block of ice by means of a strong wire, preferably in a room where the temperature is below freezing. See Figure 7-18. The pressure of the wire on the ice lowers the melting point of the ice immediately below the wire. If the surrounding temperature is above the new melting point, this part of the ice melts, and the molecules of water are forced upward around the wire. When they reach a spot above the wire, the pressure returns to normal, the melting point rises, and the water freezes again. When the wire is embedded in the ice, the extra heat needed to melt the ice below the wire is supplied by the freezing of the water above the wire. In this way, the wire may cut its way through the block of ice, and yet leave the ice in one piece. This process is called *regelation* (ree-jeh-*lay*-shun).

When you ice-skate, regelation and friction melt the ice below the skates. You are, therefore, moving along on a thin layer of water. After the skates pass, the water freezes back to ice. Similarly, a moving ski melts the snow underneath, and a snowball becomes an "ice ball" when it is sufficiently compressed.

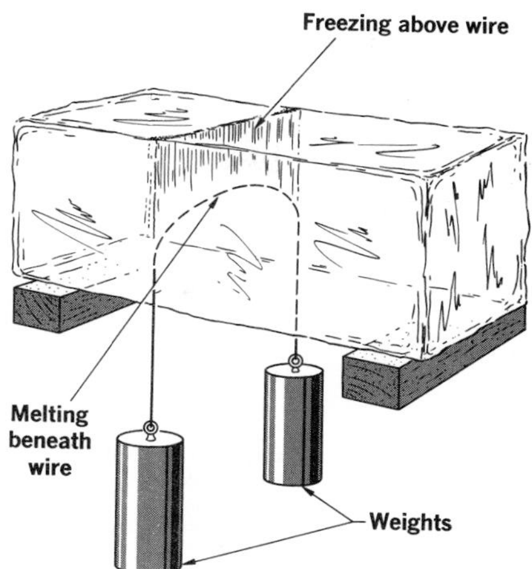

Figure 7-18. Regelation. The ice melts where pressure is applied by the wire and freezes again after the pressure is released.

7.18 Effect of Solutes on the Freezing Point The freezing point of a liquid is lowered whenever another substance is dissolved in it. The extent of the lowering depends on the nature of the liquid, the nature of the dissolved substance, and the relative amounts of each. The greater the amount of the solute in a fixed amount of liquid, the lower the freezing point of the liquid. We apply this principle when we use rock salt to prevent the formation of ice on roads and sidewalks, and to melt the ice that has formed. The dissolved substance interferes with crystal formation as the liquid cools. Before crystals form, the kinetic energy of the liquid particles must be reduced to a level below that at which they normally crystallize.

For best results, the experiment in Figure 7-18 should be conducted at a temperature slightly below the freezing point.

THE GASEOUS PHASE

7.19 The Nature of Gases The following properties of gases show that gas molecules move independently of each other at high speed:

1. *Expansion.* A gas has neither a definite shape nor a definite volume; it expands and completely fills any container. This characteristic indicates that gas molecules are independent particles, and that they have inertia.

2. *Pressure.* An inflated balloon may burst from the force that the air inside exerts on the balloon's inner surface. This force is a result of the continual bombardment of the inside surface by billions of moving molecules. If we increase the number of molecules within the balloon by blowing more air into it, the number of collisions against the inside surface increases, the pressure on the inside surface increases, and the balloon expands.

Figure 7-19. Mobile regelation. A thin film of water forms as the skates press on the ice. As the skates pass, the water film freezes again.

3. *Diffusion.* Hydrochloric acid is a water solution of the gas hydrogen chloride (HCl). Ammonia water is a solution of ammonia gas (NH_3). The molecular weight of HCl is 36.5; that of NH_3 is 17. When these gases react chemically, they form a cloud of fine, white particles of solid ammonium chloride. Suppose we put a few drops of hydrochloric acid in a warm bottle and an equal amount of ammonia water in a second warm bottle. We then cover the mouth of each bottle with a glass plate and invert the bottle containing NH_3 over the one containing HCl, as shown in Figure 7-20(A). After the bottles have stood this way for a minute or two, we remove the glass plates so that the bottles are now mouth-to-mouth. The formation of white smoke indicates that the less dense NH_3 descends and mixes with the more dense HCl. The HCl also rises and mixes with the NH_3 in the upper bottle. The movement of each of these gases is opposite to that which would be caused by the density of the gases. The movement must be due to molecular motion. Thus diffusion of gases is evidence of the rapid movement of gas molecules.

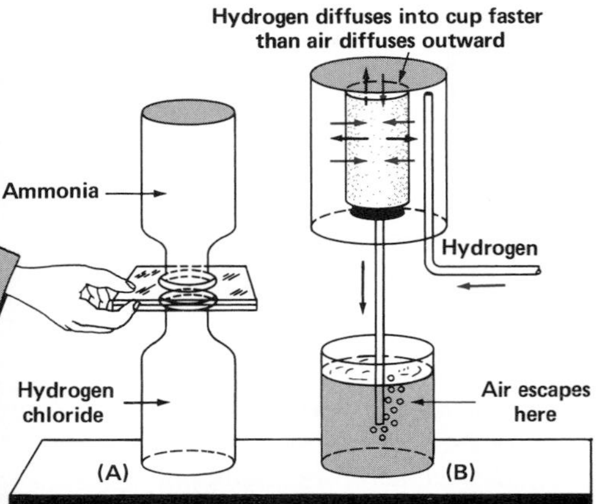

Gases also diffuse through porous solids. In Figure 7-20(B), an inverted, unglazed earthenware cup is closed with a rubber stopper through which passes a glass tube that dips into a blue liquid. When a low-density gas, such as hydrogen, is led into an inverted beaker placed over the cup, air immediately begins to bubble through the blue liquid. Evidently the lighter molecules of hydrogen move in through the porous walls of the cup faster than the heavier molecules in the air move out. Thus there is an accumulation of hydrogen molecules inside the cup. This accumulation increases the pressure inside the cup, thus pushing the liquid down the tube and forcing some of the air and hydrogen mixture in the tube to escape.

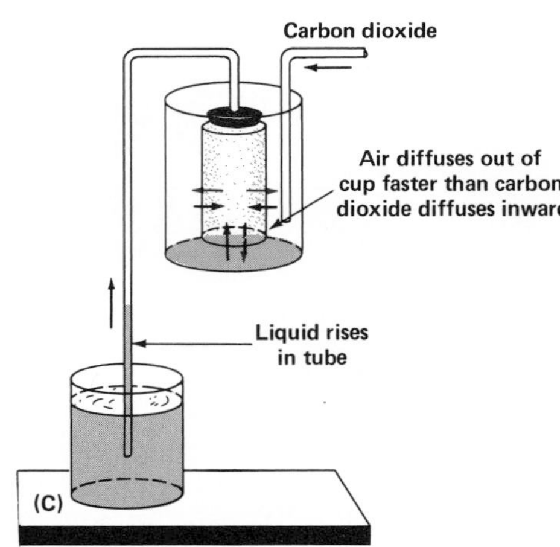

Figure 7-20. Diffusion in gases. (A) When the glass plates are removed, the gases mix by diffusion. (B) and (C) Differences in diffusion rates cause pressure differences between the upper and lower containers.

In Figure 7-20(C), a porous cup is surrounded with a dense gas, such as carbon dioxide. Since carbon dioxide is denser than air, the apparatus must be modified; the beaker surrounding the porous cup must be positioned with the open side up. When the beaker is filled with carbon dioxide, molecules from the air flow out through the porous cup faster than carbon dioxide molecules enter. This difference reduces the pressure inside the cup, and the blue liquid rises in the tube. Thus we see that there is an inverse relation between the rate of diffusion of a gas and its density.

Diffusion of gases through porous solids is an important process. Such diffusion occurs through the membranes of plants, animals, and humans, allowing oxygen to reach living cells and carbon dioxide to escape.

7.20 Vaporization When gasoline is placed in a shallow dish, within a short time the quantity of liquid decreases and the odor of gasoline becomes quite strong near the dish. Apparently molecules of gasoline liquid become molecules of gasoline vapor and mix with molecules of the gases in the surrounding air. In a similar manner, but at a much slower rate, moth balls left out in the air become smaller and smaller and eventually disappear while their characteristic odor is noticed in the air nearby. In this case, molecules of the solid moth balls turn into vapor and diffuse into the surrounding air. These are two examples of ***vaporization,*** *the production of a vapor or gas from matter in another phase.* See Figure 7-21.

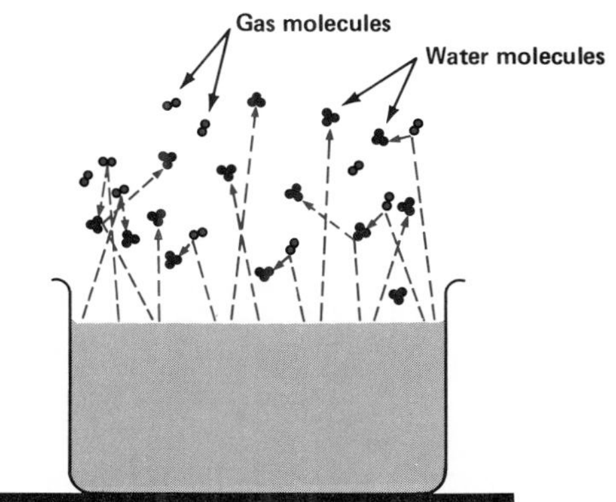

Figure 7-21. Vaporization. Water molecules escape into the air at the surface of the liquid. Some of them collide with molecules of gases in the air and rebound.

As in the case of melting (Section 7.16), vaporization is a constant-temperature process. All the energy supplied during vaporization goes to increase the potential energy of the particles. The average kinetic energy of the particles is unchanged.

The particles of all liquids and solids have an average kinetic energy that depends on the temperature of the liquid or solid. However, because of random collisions and vibratory motion, some particles have energies higher than average and others have energies lower than average. When the particle on the surface of a liquid or solid acquires enough energy to overcome the forces that hold it as part of the substance, the particle escapes and becomes a particle in the vapor phase. When vaporization occurs from liquids it is known as ***evaporation;*** when vaporization occurs from solids, it is known as ***sublimation.*** Sublimation is a direct change from solid to vapor without passing through the liquid phase.

The energy of the particles that evaporate from a liquid is obtained either from the surrounding atmosphere or from the remaining liquid. In either case, evaporation has a cooling effect on the environment. When perspiration evaporates from the skin, the process cools the skin. When rubbing alcohol is applied to the skin, the cooling effect is even greater because alcohol evaporates more rapidly than water.

Since the rate of evaporation or sublimation depends on the energy of the particles undergoing the change and in turn their energy depends upon their temperature, evaporation and sublimation occur more rapidly at higher temperatures and more slowly at lower temperatures.

7.21 Equilibrium Vapor Pressure A bell jar covering a container of water is shown in Figure 7-22. There are as many molecules of the gases of the air within the bell jar as there are in an equal volume of air outside it. The pressure (force per unit area) exerted by the gas molecules on the inside walls of the bell jar is the same as the pressure that such molecules exert on the outside walls.

When a molecule at the surface of the water inside the bell jar acquires sufficient kinetic energy, it escapes and becomes a water vapor molecule. As the water evaporates, water vapor molecules mix with the gas molecules in the bell jar. These water vapor molecules collide with gas molecules, with the walls of the bell jar, with the outside surface of the container of water, and with the surface on which the bell jar rests. They can also touch the water surface, be held by it, and become molecules of liquid again. The conversion of molecules of vapor to molecules of liquid is called ***condensation.*** Eventually the rate of evaporation equals the rate of condensation, and a condition of equilibrium prevails. At equilibrium, evaporation and condensation do not cease; they occur at the same rate. The number of water vapor molecules in the air in the bell jar remains constant. At equilibrium, the space above the water in the vessel is said to be *saturated* with water vapor.

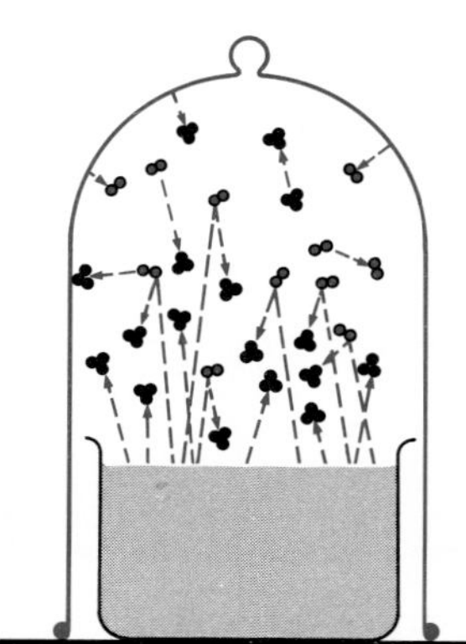

Figure 7-22. When the number of water molecules evaporating from the liquid equals the number of water vapor molecules returning to the liquid, the pressure exerted by the water vapor molecules against the walls of the bell jar is the equilibrium vapor pressure.

The collision of water vapor molecules against the walls of the bell jar increases the pressure on the bell jar so that it exceeds the pressure exerted by the gases of the air. *Added pressure exerted by vapor molecules in equilibrium with liquid* is called ***equilibrium vapor pressure.***

Since the kinetic energy of the particles of the liquid depends on both the temperature and mass of the particles, the equilibrium vapor pressure of a liquid depends on the composition as well as the temperature of the liquid. As the kinetic energy of the particles increases, so does the vapor pressure.

The ratio of the water vapor pressure in the atmosphere to the equilibrium vapor pressure at that temperature is called the *relative humidity.* Relative humidity is usually expressed as a percentage. For example, suppose that the air temperature on a given day is 30°C and that the water vapor pressure is 15.5 mm of mercury. The equilibrium vapor pressure at 30°C (as given in Appendix B, Table 12) is 31.8 mm of mercury. The relative humidity is

In summer, the humidity is often more oppressive than the temperature. Why?

$$\frac{\mathbf{15.5\ mm}}{\mathbf{31.8\ mm}} \times \mathbf{100\%} = \mathbf{48.7\%}$$

Relative humidity depends on both the temperature and the amount of water vapor in the atmosphere. The temperature at which a given amount of water vapor will exert equilibrium vapor pressure is called the *dew point.* In the above example, the dew point is 18°C because at that temperature the equilibrium vapor pressure of water is 15.5 mm of mercury.

7.22 Boiling When water is heated sufficiently, its vapor pressure will eventually equal the combined pressure of the atmosphere and the liquid pressure of the water. Vaporization then occurs at such a rapid rate throughout the water that the water becomes agitated. *Rapid vaporization that occurs when the vapor pressure of the liquid equals the pressure on its surface is called* ***boiling.*** If the pressure on the liquid surface is one atmosphere (1.01×10^5 Pa), the temperature at which boiling occurs is called the *normal boiling point.* If the pressure on the liquid surface is greater than one atmosphere, boiling occurs at a higher temperature than the normal boiling point; if the pressure on the liquid surface is less than one atmosphere, boiling occurs at a lower temperature than the normal boiling point. Greater pressure is largely the result of the greater number of collisions of the particles against the surface of the liquid. Consequently, the vibrational kinetic energy of the liquid molecules, and thus its temperature, must be raised to make boiling possible. The opposite is true when the pressure on a liquid is decreased.

Water or any other liquid that is boiling rapidly does not get hotter than when it is boiling slowly. While a liquid is boiling away, the boiling temperature remains constant until all the liquid has been vaporized.

Solids or gases dissolved in a liquid change the liquid's boiling temperature. For example, salt water boils at a higher temperature than pure water. In general, solids dissolved in liquids raise the boiling temperature and gases dissolved in liquids lower the boiling temperature.

7.23 Plasma A gas that is capable of conducting electricity is called a *plasma.* Gases do not ordinarily conduct electricity, but when they are heated to high temperatures or collide vigorously with each other, they form electrically charged particles called ions. These ions give the plasma the ability to conduct an electric current.

Plasma is sometimes called the fourth phase of matter. Under normal conditions, matter exists only as a solid, liquid, or gas. But at the high temperatures that prevail in the sun and other stars, their constituent matter exists almost entirely in the plasma phase. Because there are so many stars, most of the matter in the universe is in the form of plasma.

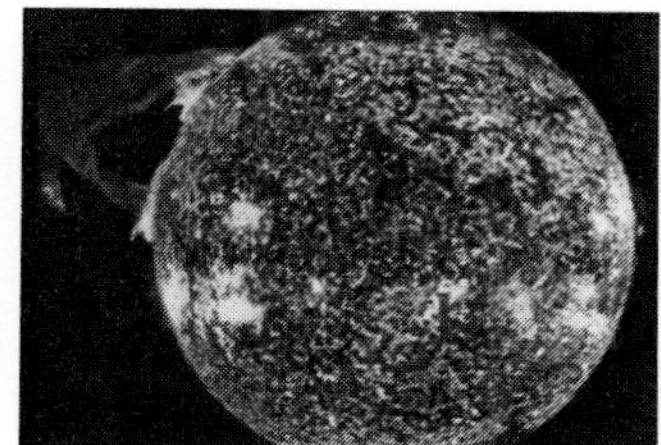

Figure 7-23. A solar eruption. A mass of plasma many times the size of the earth is ejected from the sun's surface at the upper left. The photo was taken through an ultraviolet filter.

FLUIDS IN MOTION

7.24 Common Properties of Fluids So far we have considered liquids and gases as separate states of matter. Now we shall describe some other properties that liquids and gases have in common. The term *fluid* is applied to both liquids and gases, and since the behavior of moving liquids is similar to the behavior of moving gases, we treat them together as fluids in motion.

7.25 The Buoyant Force of Fluids Cork and wood float on water. Balloons filled with helium float in air. These examples show that liquids and gases exert an upward force on objects placed in them; an object floats if the upward force the fluid exerts on it is greater than the weight of the object. This force is also observable when you place a dense object in water. This is why a person can lift a more massive stone under water than he or she can otherwise lift in air. *The upward force that any fluid exerts upon an object placed in it is called the* ***buoyant force.***

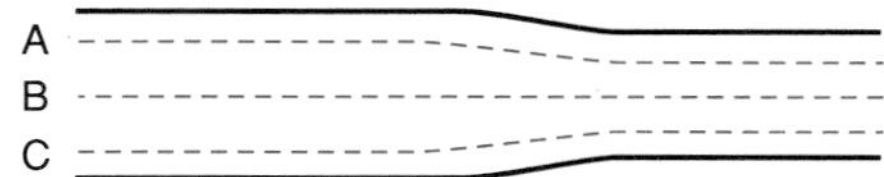

Figure 7-24. Streamline flow of water through a tube.

7.26 Streamline Flow The motion of fluids is very complex. We shall therefore limit our discussion of such motion to a few examples that can be explained in simplified terms.

Suppose we have a tube whose diameter varies, as shown in Figure 7-24. We find that the motion of water flowing through the tube is smooth and even. Water molecules that enter the tube at **A, B,** or **C** follow the paths shown by the dotted lines. Since the volume of water passing any cross-sectional area of the tube is the same in any given length of time, the water must flow more rapidly through the narrow portion of the tube than through the wide portion. This smooth flow of a fluid through a tube is called *streamline flow.* If the speed of the fluid becomes too great, or if changes in the diameter or direction of the tube are too abrupt, the fluid will not flow smoothly. Under these circumstances, the flow is said to be *turbulent*.

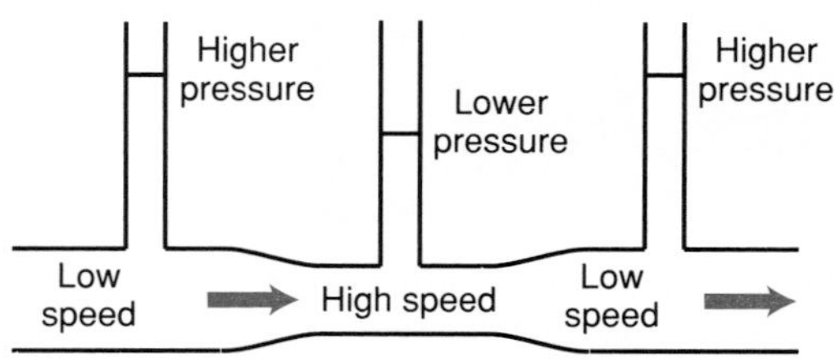

Figure 7-25. As the water flows more rapidly through the narrow portion of the tube, the pressure there is lowered.

7.27 Bernoulli's Principle Water is moving with streamline flow through the tube shown in Figure 7-25. Because the diameter of the center portion is less than that at either end, water flows faster through the narrow part of the tube. The vertical tubes act as pressure gauges, filling with liquid until the pressure due to the weight of the liquid in the tube equals that of the moving liquid.

The pressure gauges show that the pressure is higher where the speed of the fluid is lower; where the speed of the fluid is higher, the pressure it exerts is lower. This is the principle of the *Venturi meter*. This device enables us to calculate the speed of a fluid in the horizontal tube from the difference in pressure in the vertical tubes.

The explanation for the variation in pressure exerted by a moving fluid when its speed is changed was given by Daniel Bernoulli (1700–1782). He found that *for a flow of a fluid through a tube, the sum of the pressure and the kinetic energy per unit volume of the fluid is a constant.* This phenomenon is known as ***Bernoulli's principle.***

The kinetic energy of a moving fluid is directly proportional to the square of its speed. Bernoulli's principle states, however, that the sum of the pressure and the kinetic energy per unit volume is a constant. Therefore, as the speed of a moving fluid increases, its kinetic energy increases, and consequently, the pressure it exerts correspondingly decreases. This explains the results observed in the Venturi meter.

Bernoulli's principle has several important applications. For example, an automobile carburetor consists of a partially constricted area, called a venturi, where the gasoline is mixed with air. The venturi creates a low pressure area so that atmospheric pressure can force fuel from the float bowl into the air stream.

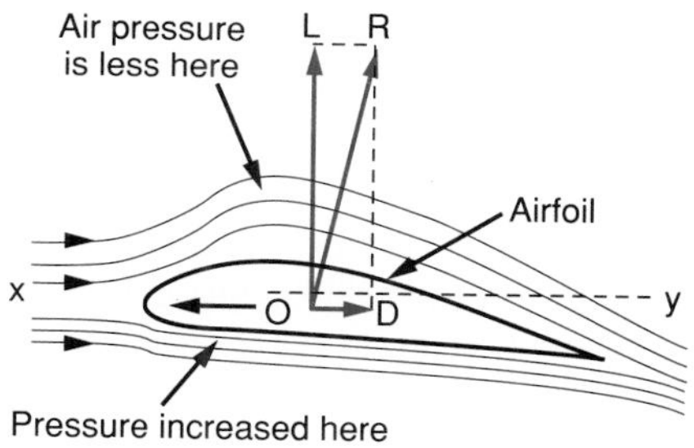

Figure 7-26. The movement of air over the wing produces a difference in pressure between the upper and lower surfaces. The force produced in this manner may be resolved into an upward lift force and a horizontal drag force.

An airplane wing also utilizes Bernoulli's principle. A moving airplane wing receives an upward force because of the motion of the air over it. Figure 7–26 shows a cross section of an airplane wing with air flowing around it from left to right. Though you might think that the air deflected downward by the bottom of the wing does the lifting, it actually accounts for only about 15 percent of the force required to lift an airplane. Most of the force needed to lift an airplane is produced by the movement of the air across the upper surface of the wings.

Because an airplane wing is shaped like one-half of a venturi, the air moving across the upper surface must travel faster than the air moving beneath. Consequently, the air moving over the top surface exerts less pressure than that moving beneath it. This difference in pressure provides most of the lifting force that causes an airplane to rise.

The vertical component of the forces on an airplane wing is called *lift*. The horizontal component of these forces, which tends to retard the movement of the airplane through the air, is called *drag*. As an airplane flies at constant velocity, the forces on its wings must be in equilibrium. The force of gravity is the equilibrant of the lift, while the thrust force produced by the airplane engines is the equilibrant of the drag.

7.28 Reducing Drag by Streamlining As was noted in Section 7.26, the flow of a fluid through or around an object

Figure 7-27. This skier is using the principles of streamlining. Note the similarity between the skier's position and that of the airfoil in Figure 7-26.

(or the flow of an object through a fluid) may be smooth or turbulent, depending on the shape of the object and on the relative speed. Designing an object to minimize the drag resulting from the turbulent motion of the surrounding fluid is called *streamlining*. The streamlining of automobiles and airplanes produces increased fuel efficiency as well as greater stability and maneuverability.

An interesting example of streamlining is shown in Figure 7-27. The skier is leaning as far forward as possible in order to minimize drag and thus obtain maximum jumping distance through the air. Some degree of lift can also be achieved in this manner.

QUESTIONS: GROUP A

1. (a) Compare the diffusion rates for solids and liquids. (b) Why do they differ?

2. (a) Define cohesion and adhesion. (b) Explain how these forces determine that water "wets" glass, whereas mercury does not.

3. (a) What is viscosity? (b) Which property is responsible for viscosity?

4. Why are free drops of mercury more spherical than drops of water of equal volume?

5. (a) What is capillarity? (b) Upon what factors does it depend?

6. Why does ice occupy a larger volume than an equal mass of water?

7. Define regelation.

8. How does the freezing point of a solution depend on the amount of solute present?

9. What property of gases shows they consist of independent molecules?

10. What makes it easier to lift a stone under water than in the air?

11. (a) What is plasma? (b) Why is plasma called the "fourth phase of matter"?

12. In terms of kinetic theory, explain why gases (a) expand when heated, (b) exert pressure, and (c) diffuse through porous substances.

GROUP B

13. If you lay a steel needle on its side, it will float on water. Yet if you place it in the water point first, it will sink. Explain.

14. What is the purpose of using detergents as cleaning agents?

15. If a thermometer is broken, small drops of mercury will roll across the table like little balls while larger drops are slightly flattened. Why?

16. Make a sketch of temperature as a function of time for ice being raised from $-20°C$ to $+20°C$.

17. What happens on the molecular level that causes a bottle of soda pop placed in the freezer to break?

18. What is the purpose of adding a handful of salt to the water in which you will be cooking spaghetti?

19. Why is it easier to ice skate when your skate blades are newly sharpened?

20. Why does a Mylar balloon inflated with helium stay inflated longer than an ordinary latex balloon?

21. If you increase the heat under a pan of water in which you are boiling corn, will the corn be cooked faster? Explain.

22. In terms of Bernoulli's principle, explain how the lift on an airplane increases as the plane's velocity increases.

Physics Activity

Place a pot of water on the stove and bring the water to a boil. Observe where the bubbles form and what happens as the water boils more rapidly.

SUMMARY

The smallest particle of any substance that is capable of stable, independent existence is a molecule. Molecules are composed of atoms. Atoms are the smallest particles of elements that can exist either alone or in combination with other atoms of the same or different elements. The masses of individual atoms are usually expressed in atomic mass units, rather than kilograms. The atomic mass unit is based on the mass of the carbon-12 atom.

The kinetic theory of matter states that molecules are in constant motion and undergo perfectly elastic collisions. The phase of matter is determined by the forces acting between molecules and the energy the molecules possess.

Properties of solids such as diffusion, cohesion, adhesion, tensile strength, ductility, malleability, and elasticity depend on molecular forces and/or molecular motion. Hooke's law states that, within limits, strain (the deformation of a solid) is directly proportional to the stress producing the strain. The ratio of stress to strain is called Young's modulus.

Molecular forces, molecular motion, and the weight of molecules are factors on which liquid properties such as diffusion, cohesion, adhesion, viscosity, surface tension, and capillarity depend. Melting is the change of phase from a solid to a liquid. Crystalline solids have characteristic melting points; noncrystalline solids do not have definite melting points. Most substances expand upon melting. Water is an important exception. Pressure and dissolved materials change the melting point and freezing point of substances.

Gases expand, exert pressure, and diffuse. Vaporization is the change of phase from a solid or liquid to a gas or vapor. The added pressure exerted by vapor molecules in equilibrium with liquid molecules is called equilibrium vapor pressure. Boiling is rapid vaporization that occurs when the vapor pressure of a liquid is equal to the pressure on its surface. At high temperatures, matter may exist in the form of plasma.

Fluids exert a buoyant force on objects. Bernoulli's principle states that the sum of the pressure and kinetic energy of a moving fluid in a tube is a constant. Streamlining can reduce fluid drag.

VOCABULARY

adhesion
atom
atomic mass unit
Bernoulli's principle
Brownian movement
buoyant force
capillarity
cohesion
condensation
diffusion
ductility
elastic limit
elasticity
elongation strain
equilibrium vapor pressure
evaporation
Hooke's law
kinetic theory
malleability
mass number
meniscus
molecule
phase
plasma
regelation
shear strain
strain
streamlining
stress
sublimation
surface tension
tensile strength
vaporization
viscosity
volume strain
Young's modulus

8

Heat Measurements

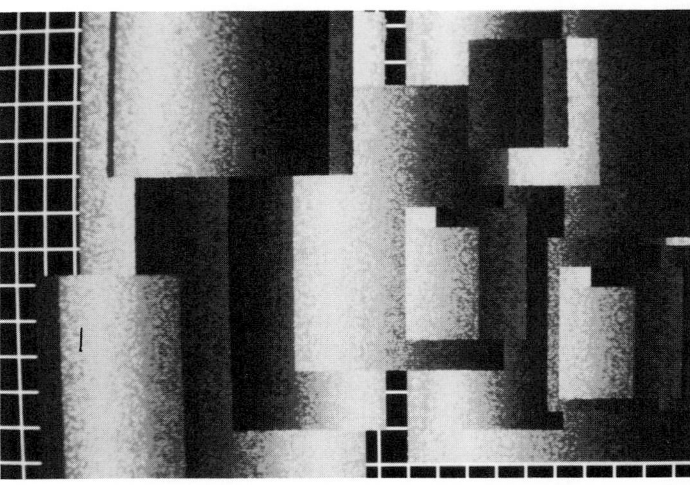

heat n.: the thermal energy that is absorbed given up, or transferred from one body to another.

OBJECTIVES

- Differentiate between heat and temperature.
- Solve problems involving thermal expansion of solids and liquids.
- State Charles' law and Boyle's law.
- Solve problems involving ideal gases.
- State and apply the law of heat exchange and the method of mixtures.
- Define and calculate specific heat, heat of fusion, and heat of vaporization.

UNITS OF TEMPERATURE AND HEAT

8.1 Relationship Between Heat and Temperature Suppose we fill a large beaker and a small beaker with boiling water, as shown in Figure 8-1. The water in the two beakers is at the same temperature, but the water in the large beaker can give off more heat. It could, for example, melt more ice than could the water in the small beaker. It is possible, therefore, for a body to be at a high temperature and give off little heat; to be at a high temperature and give off a large quantity of heat; to be at a low temperature and give off little heat; or to be at a low temperature and give off a large quantity of heat.

When a material is hot, it has more *thermal energy* than when it is cold. ***Thermal energy*** *is the total potential and kinetic energy associated with the random motion and arrangements of the particles of a material.*

Temperature is the "hotness" or "coldness" of a material. The quantity of thermal energy in a body affects its temperature. The same quantity of thermal energy present in different bodies, however, does not give each the same temperature. The ratio between temperature and thermal energy is different for different materials.

We saw in Chapter 7 that the temperature of a substance will rise if the average kinetic energy of its particles is increased. If the average kinetic energy is decreased, the temperature goes down. On the other hand, when the potential energy of the particles is increased or decreased without a change in the average kinetic energy, a change of phase takes place without a change in temperature.

Heat *is thermal energy that is absorbed, given up, or transferred from one body to another.* The temperature of a body is a measure of its ability to give up heat to or absorb heat from another body. Heat will flow from a body with a higher temperature to a body with a lower temperature, even if the cooler body contains more thermal energy.

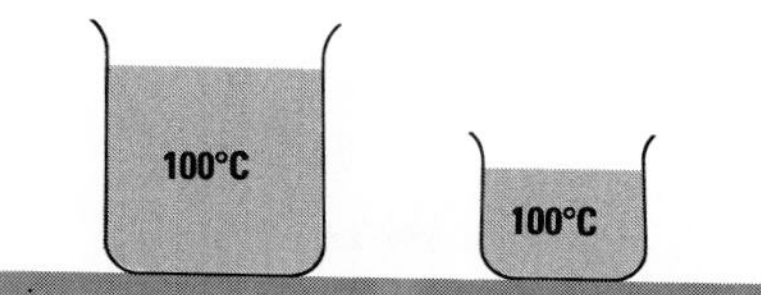

Figure 8-1. The water in the two beakers is at the same temperature, but the water in the large beaker can give off more heat.

The experiments of Count Rumford and James Prescott Joule (Section 1.13) show that mechanical energy and heat are equivalent and that *heat must be a form of energy.*

Since thermal energy is also defined as a form of energy, you may wonder why two different terms—thermal energy and heat—are used. An example will illustrate the difference. The temperature of the air in a bicycle tire will rise when the tire is being pumped up. It will also rise when the tire is out in the sun. In both cases the thermal energy and the temperature of the air are increased. In the first case the work done in pumping was converted to thermal energy. In the second case the rise in temperature was due to energy transferred from the sun to the tire. The term *heat* is used when there is a transfer of thermal energy from one body to another body at a different temperature.

Heat is thermal energy in motion.

Temperature is defined in terms of measurements made with thermometers that will be described later. But the following qualitative definition gives the relationship between temperature and energy. ***Temperature*** *is a physical property that determines the direction in which heat energy will flow between substances.*

8.2 Temperature Scales The Celsius temperature scale is often mentioned during weather reports. In Section 7.21, we used the °C in the calculation of relative humidity. Now we will see how the Celsius scale was developed.

To measure temperature, it is necessary to introduce a fourth fundamental unit. The unit of temperature difference, the degree, cannot be derived from length, mass, and time; a measurable physical property that changes with temperature must be used.

There are many physical properties that change with temperature. The length of a solid, the volume of a liquid, the pressure of a gas held at constant volume, the volume of a gas held at constant pressure, and the color of a solid heated to a high temperature are examples. Some of these properties of matter can be used in developing a temperature scale and constructing a thermometer.

To establish a temperature scale it is necessary to find a

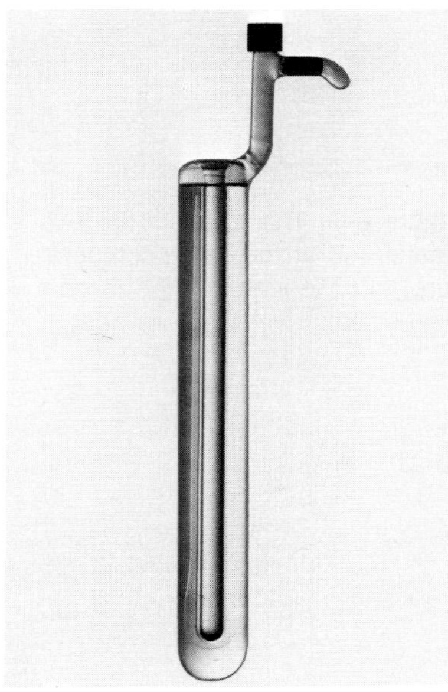

Figure 8-2. A triple-point cell. Pure water with air removed is sealed permanently in the cell, which is then immersed in a water-ice bath. The system is at the triple point when ice, water, and vapor are all present within the cell. A central well is provided for the insertion of a thermometer that is to be calibrated.

process that occurs without a change in temperature. The temperature at which such a process takes place can then be used as a fixed point on a temperature scale. A change of phase of a substance, such as melting or boiling, can be used. The temperature at which the solid phase of a substance is in thermal equilibrium with its liquid phase is a fixed value at a given pressure. This is the melting point. Similarly, at the boiling point the liquid is in thermal equilibrium with its vapor, and this point has a fixed value at a given pressure. The boiling temperature of the substance is always higher than the melting temperature.

There is only one pressure at which the solid, liquid, and vapor phases of a substance can be in contact and in thermal equilibrium. This equilibrium occurs at only one temperature, which is known as the *triple-point temperature*. For example, there is only one pressure and temperature condition at which ice, liquid water, and water vapor can exist in a vessel in thermal equilibrium. The ice, liquid water, and vapor are all at the same temperature and can continue to exist indefinitely in the constant volume of the sealed vessel. A device for determining the triple point of water is shown in Figure 8-2.

The triple point of water is the SI standard for defining temperature. Its assigned value is 273.16 K (kelvins).

The SI symbol for the unit of temperature is K. *The Kelvin scale was devised by Sir William Thomson (1824–1907), who is better known by his title, Lord Kelvin.*

Originally, two fixed points were used to define the standard temperature interval. They were the *steam point* (the boiling point of water at standard atmospheric pressure) and the *ice point* (the melting point of ice when in equilibrium with water saturated with air at standard atmospheric pressure). The *Celsius scale* (formerly called the centigrade scale), devised by the Swedish astronomer Anders Celsius (1701–1744), assigned the value 0°C to the ice point and 100°C to the steam point, as shown in Figure 8-3. Thus the interval between the two fixed points was 100 C°. (It is interesting that Celsius originally assigned 0° to the steam point and 100° to the ice point, but the scale was changed to its present sequence within a year.) Note that specific temperatures on the Celsius scale are expressed in the unit °C. Temperature *differences* on the Celsius scale are expressed in the unit C°. *Kelvin scale* temperatures have the unit K; Kelvin scale temperature *differences* are also expressed in K.

The magnitude of the kelvin (K) is the same as that of the Celsius degree (C°). These are arbitrarily established units for the measurement of temperature difference. 0 K is called the *absolute zero of temperature*. Absolute zero should not be thought of as a condition of matter with zero energy and no molecular motion. Molecular action

0 K (−273°C) *is the coldest possible temperature because at that temperature molecular energy is at a minimum.*

does not cease at absolute zero. The molecules of a substance at absolute zero have a minimum amount of kinetic energy, known as the zero-point energy. Molecular energy is a minimum, but not zero, at absolute zero.

Now let us consider the results obtained when a temperature is measured. To assign numbers on the temperature scale, it is assumed that a measurable physical property of a substance that changes with temperature, X, is proportional to the Kelvin temperature, T; thus $X \propto T$. If X_T is the measurement of the physical property at the standard fixed temperature, T_T, which is the triple point of water, and X is the measure of the same property of the same substance at the unknown temperature, T, then we can write the proportion

$$\frac{X}{X_T} = \frac{T}{T_T}$$

Since T_T is 273.16 K, we can rewrite the proportion as

$$T = 273.16\text{ K}\frac{X}{X_T}$$

Because X and X_T are measurable quantities, the value of the temperature, T, can be computed.

The relationship between the Celsius and Kelvin temperature scales, rounded to three significant figures, is given by the equation

$$\text{K} = {}^\circ\text{C} + 273^\circ$$

The degrees between the ice and steam points are numbered from zero degrees Celsius, 0°C, to one hundred degrees Celsius, 100°C. From one degree to the next is a temperature interval of one Celsius degree, 1 C°. Temperatures below 0°C and above 100°C are measured by extending the scale in 1 C° intervals. Temperatures below 0°C are represented by negative values.

The most commonly used thermometers contain either mercury or alcohol. In both cases, the liquid volume increases rather uniformly with temperature over the useful range of the instruments.

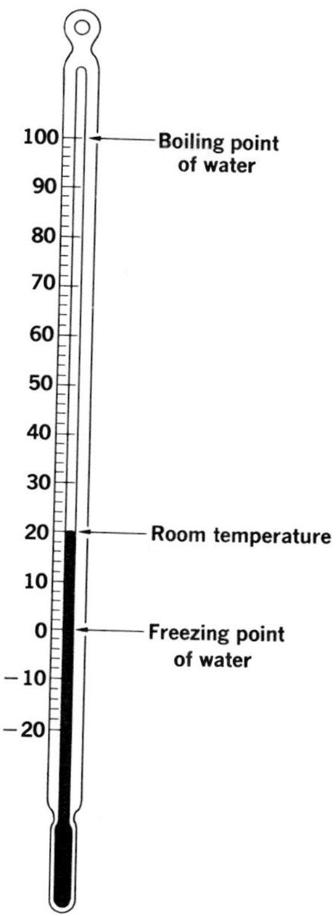

Figure 8-3. A Celsius thermometer of the type used in laboratory work. To convert to the Kelvin scale, add 273° to the Celsius reading.

Gabriel Fahrenheit (1686–1736) is credited with the first use of mercury in thermometers. He invented the temperature scale that bears his name.

8.3 Heat Units There is no instrument that directly measures the amount of thermal energy a body gives off or absorbs. Therefore, *quantities of heat must be measured by the effects they produce.* For example, the amount of heat given off when a fuel burns can be measured by measuring the temperature change in a known quantity of water that the burning produces. If one sample of coal warms 1.0 kg of water 1.0 C°, and another sample warms 1.0 kg of water 2.0 C°, then the second sample gives off twice as much heat.

In the past, water was the standard substance for defining heat units. In the metric system of units, the *kilocalorie* (kcal) was defined as the quantity of heat needed to raise the temperature of one kilogram of water one Celsius degree. The ***calorie*** (cal) was defined as the quantity of heat needed to raise the temperature of one gram of water one Celsius degree. Observe that a unit mass of water was used for defining each heat unit. Also note that the kilocalorie is one thousand times larger than the calorie. The kilocalorie is the "Calorie" used by biologists and dietitians to measure the energy value of foods.

The Calorie used in dietary tables is equal to one thousand of the calories used by physicists.

The foregoing definition of the calorie is still used in many laboratory measurements. However, as thermal measurements increased in precision, these older definitions became inadequate. Another reason is that the quantity of heat required to raise the temperature of 1 gram of water through 1 Celsius degree varies slightly for different water temperatures.

The calorie is also defined as a specific number of joules.

1 calorie = 4.186 joules

Defined in this way, the size of the calorie is nearly the same as the original calorie. When written in three significant figures, 4.19 joules, this slight difference disappears. Thus the relationships stated in the original definition are still useful in measuring thermal properties.

Though the calorie and kilocalorie were the first units of heat measurement used, the SI unit for heat is the joule. The conversion factor shown above is often used to compare specific joule measurements to their corresponding kilocalorie measure. Though the kilogram is the standard SI unit of mass, for practical purposes the gram is used instead. (For example, specific heat is measured in J/g · C°.)

QUESTIONS: GROUP A

1. (a) What is thermal energy? (b) How does it differ from heat?
2. What is the relationship between temperature and thermal energy?
3. What are some physical properties that could be used in developing a temperature scale?
4. Why are the steam and ice points of water better fixed points for a thermometer than the temperature of a human body?
5. What is the current standard for defining temperature?
6. (a) What is the SI scale for temperature? (b) How is it related to the Celsius scale?
7. What is the difference between 10°C and 10 C°?
8. Compare the amount of thermal energy of each of the following: (a) a soldering iron and a needle, both at 150°C; (b) a four-section and a ten-section radiator; (c) a kettle and a cup

of boiling water; (d) 20.0 kg of ice at a temperature of −10°C and 10.0 kg of ice at −10°C; (e) a liter and a milliliter of liquid air, both at a temperature of −189°C.

GROUP B

9. (a) Explain how a common mercury thermometer works. (b) What does a thermometer indicate?
10. How does a *calorie* differ from a *Calorie?*
11. Which contains more thermal energy: a pot with two liters of water at 90°C or a pot with one liter of water at 90°C?
12. What is the definition of the term *absolute zero*?
13. Use Table 15 in Appendix B to describe the difference in the spacing between degree marks on an alcohol thermometer and a mercury thermometer of similar size.

PROBLEMS: GROUP A

1. The triple point of water is 273.16 K. Express this as a Celsius temperature.
2. What are the ice point and steam point of water on the Kelvin scale?
3. If normal body temperature is 37.0°C, what is it in degrees Kelvin?
4. Helium liquefies at 4.10 K, while nitrogen becomes a liquid at 77.0 K. Express these temperatures in degrees Celsius.
5. How much heat is liberated when 750 g of water cools from 80.0°C to 50.0°C?
6. How many joules of heat will be needed to raise the temperature of 550 g of water from 20.0°C to 100.0°C?
7. A candy bar lists its "Calorie" content as 265 Calories. (a) How many joules of energy does this represent? (b) How many grams of water could be heated from 0°C to 100°C with the energy liberated?

THERMAL EXPANSION

8.4 Thermal Expansion of Solids With few exceptions, *solids expand when heated and contract when cooled.* They not only increase or decrease in length but also in width and thickness. When a solid is heated, the increase in thermal energy increases the average distance between the atoms and molecules of the solid, and it expands.

The expansion of solids can be measured experimentally. A metal rod is heated in an apparatus that has a precise measuring device. If the temperature of an aluminum rod 1.0 m long is raised 1.0 C°, the increase in length is 2.3×10^{-5} m. An iron rod of the same length expands only 1.1×10^{-5} m when its temperature is raised 1.0 C°. For the same increase in temperature, different materials of the same length expand by different amounts. *The change in length per unit length of a solid when its temperature is changed one degree is called its* ***coefficient of linear expansion.*** See Figure 8-4. While the coefficient of linear expansion of most solids varies with temperature, the change is slight and we shall neglect it in our discussion.

Why should ovenware be made of a substance with a low coefficient of linear expansion?

As mentioned, 1.0 m of aluminum expands 2.3 ×

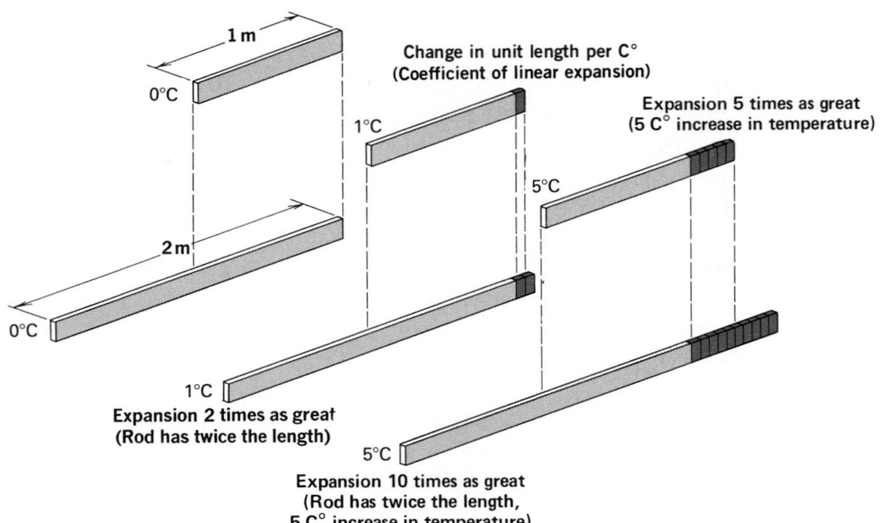

Figure 8-4. The change per unit length of a solid when its temperature is changed one degree is the coefficient of linear expansion. The change in length illustrated here is greatly exaggerated.

10^{-5} m when its temperature is raised 1.0 C°. The coefficient of linear expansion of aluminum is therefore 2.3×10^{-5}/C°. Likewise the coefficient of linear expansion of iron is 1.1×10^{-5}/C°. Since the coefficient of linear expansion is defined as *the change in length per unit length,* its value does not depend upon any particular length unit. Appendix B, Table 14 gives the value of the coefficient of linear expansion of several solids.

So far, we have been discussing 1.0-m lengths of aluminum and iron and a rise in temperature of 1.0 C°. If the temperature of 10.0 m of aluminum rod is raised 1.0 C°, the expansion is $1\bar{0}$ times as much as the expansion of the 1.0-m length: $1\bar{0} \times 2.3 \times 10^{-5}$ m $= 2.3 \times 10^{-4}$ m. If the temperature of this 10.0 m of aluminum is raised 10.0 C°, the increase is $1\bar{0}$ times as great as for 1.0 C°: $1\bar{0} \times 2.3 \times 10^{-4}$ m $= 2.3 \times 10^{-3}$ m. We can conclude from these observations that *the change in length of a solid equals the product of its original length, its change in temperature, and its coefficient of linear expansion.*

This can be given by the equation

$$\Delta l = \alpha l \Delta T$$

where Δl is the change in length, α is the coefficient of linear expansion, l is the original length, and ΔT is the difference between the final temperature, T_f, and the initial temperature, T_i.

How is the "clickety-clack" you usually hear when a train passes related to linear expansion?

In most practical situations, we are interested in the amount of *linear* expansion of solids. We must bear in mind, however, that when solids are heated they increase in all dimensions. The ***coefficient of area expansion,*** or *the* *change in area per unit area per degree change in temperature,* is *twice* the coefficient of linear expansion. The ***coefficient of cubic expansion,*** or *the change in volume per unit volume per degree change in temperature,* is *three times* the coefficient of linear expansion.

Expansion of solids is considered in the design and construction of any structure that will undergo temperature changes. When a concrete highway is built, provision is made for expansion by pouring the concrete in sections that are separated by small spaces. Steel rails for railroads can be laid with small spaces between the ends of the rails for the same reason. Bridges are also built so that the sections can expand and contract without distorting the entire structure, as shown in Figure 8-5.

Figure 8-5. Expansion and contraction of the roadway is allowed for by this special joint.

Suitable allowance must be made not only for changes in size due to expansion and contraction, but also for the different rates of expansion and contraction of different materials. For a tight seal, the wires that lead into the filament of an incandescent lamp must have the same coefficient of expansion as the glass of which the lamp is made. The principle of the expansion of solids is also applied in metallic thermometers, thermostats, and the balance wheels of some watches.

8.5 Thermal Expansion of Liquids If the gasoline tank of an automobile is filled on a cool morning and the car is then parked in the sun, some of the gasoline may overflow the tank. Heat causes the gasoline to expand. Here again the increased thermal energy of the molecules and their resultant increase in amplitude of vibration cause them to move away from each other slightly. Thus the principle of expansion of liquids has many useful applications. Thermometers contain either mercury or colored alcohol because these liquids expand and contract quite uniformly as the temperature changes.

Since liquids do not have a definite shape, but take the shape of their container, we are concerned only with their volume expansion. An apparatus as shown in Figure 8-6 can be used to measure the volume expansion of a liquid.

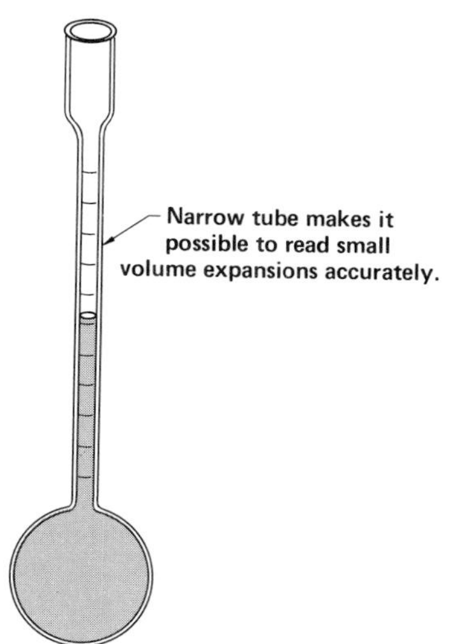

Figure 8-6. A tube for measuring the volume expansion of liquids. The scale takes into account the expansion of the container.

Liquids have greater coefficients of volume expansion than solids have. If this were not so, the liquid in a thermometer would not rise. The coefficients of volume expansion for some common liquids are given in Appendix B, Table 15.

The change in volume of a liquid can be given by the formula

$$\Delta V = \beta V \Delta T$$

where ΔV is the change in the volume, β is the coefficient of volume expansion, V is the original volume, and ΔT is the difference between the final and initial temperatures.

PRACTICE PROBLEMS **1.** An aluminum rod is initially 5.00 m long at a temperature of 20.0°C. How long will the rod be when it has been heated to a temperature of 100.0°C? *Ans.* 5.01 m

2. A 4.00-L sample of a liquid expands by 0.070 L when its temperature increases from 12.0°C to 27.0°C. Find the coefficient of volume expansion of the liquid. *Ans.* 1.17×10^{-3} L/L·C°

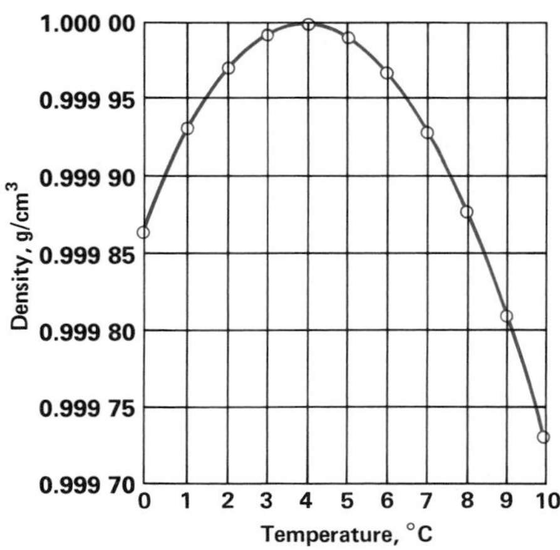

Figure 8-7. The density of water is greatest at 4°C.

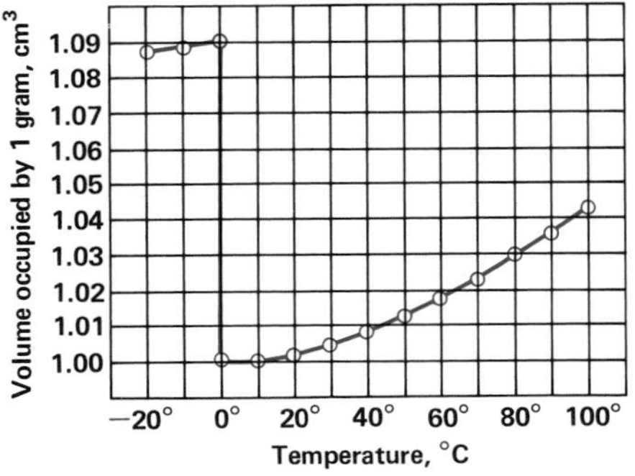

Figure 8-8. Graph of the volume of one gram of water between −20°C and 100°C. The abrupt change at 0°C is due to the change in phase.

8.6 Abnormal Expansion of Water Suppose an expansion bulb like that shown in Figure 8-6 is filled with pure water at 0°C. As the bulb and water are warmed, the water gradually *contracts* until a temperature of 4°C is reached. As the temperature of the water is raised above 4°C, the water *expands.* Because the volume of water decreases as the temperature is raised from 0°C to 4°C, the mass density of the water increases. (The mass of the water is constant.) Above 4°C, the volume of water increases as the temperature is raised, and the mass density decreases. Therefore, *water has its maximum mass density,* 1.000 00 g/cm^3, *at 4°C.* The variation of the density of water with the temperature is shown in Figure 8-7. The temperature range of the graph was chosen to include the temperature at which the density of water is a maximum.

This unusual variation of the density of water with the temperature can be explained as follows. When ice melts to water at 0°C, the water still contains groups of molecules bonded in the open crystal structure of ice. As the temperature of water is raised from 0°C to 4°C, these open crystal fragments begin to collapse and the molecules move closer together. The distances between the molecules tend to increase during the 0°C to 4°C interval, but the effect of the collapsing crystal structure predominates and the density increases. Above 4°C the effect of increasing intermolecular distances exceeds the effect of collapsing crystal structure and the volume increases, as shown in Figure 8-8.

If water did not expand slightly as it is cooled below 4°C and expand much more as it freezes, the ice that forms on the surface of a lake would sink to the bottom. During the cold winter months, ice would continue to form until the lake was frozen solid. In the summer months only a few feet of ice at the top of the lake would melt. However, because of the unusual properties of ice and water, no ice forms at the surface of a pond until all the water in it is cooled to 4°C. As the surface water cools below 4°C, it expands slightly and floats on the 4°C water. Upon freezing at 0°C, further expansion takes place and the ice floats on the 0°C water.

The graph in Figure 8-8 shows from right to left the gradual contraction of a given mass of water as it is cooled. Note the sharp expansion on freezing and the slight contraction as ice is cooled below 0°C. The slight expansion that occurs during cooling from 4°C to 0°C is too small to show in this graph.

8.7 Charles' Law Gases expand when heated because the increase in the thermal energy of the gas molecules results in an increase in their kinetic energy.

Different solids and liquids have different coefficients of expansion, but *all gases have approximately the same coefficient of expansion*. Also, *the coefficient of expansion of gases is nearly constant at all temperatures,* except for those near the liquefying temperature of the gas. This was demonstrated experimentally in 1787 by French scientist Jacques Charles (1746–1823) in a manner similar to that described below.

In Figure 8-9, a column of air is trapped by a globule of mercury in a capillary tube that is sealed at one end. The length of the air column is measured when the tube is immersed in a mixture of ice and water. Then the length of the air column is measured when the tube is immersed in boiling water. Comparing these two readings, we find that the air column increases by $\frac{100}{273}$ of its original length when heated from 0°C to 100°C. For each degree of temperature change the expansion is $\frac{1}{273}$ of the volume at 0°C. When gases other than air are used, similar results are obtained. The same fractional contraction occurs when gases are cooled below 0°C.

The coefficient of volume expansion for gases is $\frac{1}{273}$ of the volume at 0°C or 3.663×10^{-3}/C°. This is about 20 times the volume expansion of mercury and almost 60 times that of aluminum. All gases have approximately the same coefficient of expansion because at low densities they all consist of widely separated, exceedingly small molecules that are, in effect, independent particles. Gas molecules are separated by distances much greater than their molecular diameters. Consequently, the forces acting between them, and their volume in relation to the total gas volume, are negligible. For these reasons all gases have similar physical properties except at temperatures and pressures near those at which they liquefy.

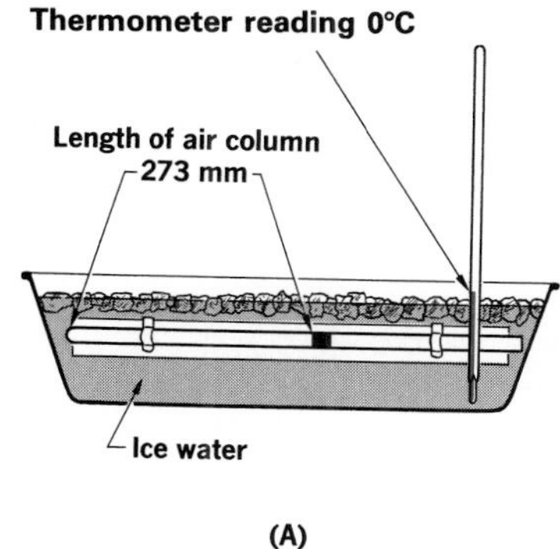

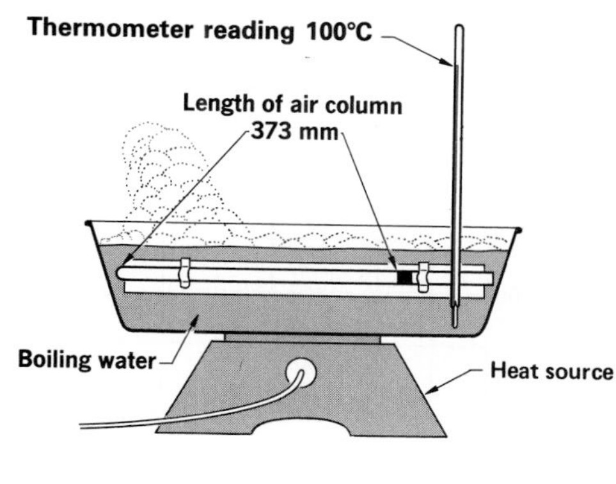

Figure 8-9. Verifying Charles' law. For each degree rise in temperature, a gas expands 1/273rd of its volume at 0°C.

Table 8-1
THE VOLUME-TEMPERATURE RELATIONSHIP OF A GAS

Volume (cm³)	Temperature (°C)	Temperature (K)
373	$10\bar{0}$	373
323	$5\bar{0}$	323
273	$\bar{0}$	273
223	$-5\bar{0}$	223
173	$-10\bar{0}$	173

Table 8-1 gives the volume occupied by a sample of gas at various Celsius and Kelvin temperatures. From the table we see that the volume of a gas varies directly with the Kelvin temperature. This relationship, which is known as ***Charles' law,*** can be stated thus: *The volume of a dry gas is directly proportional to its Kelvin temperature, provided the pressure remains constant.*

Unless a gas is under very high pressure or at very low temperature, Charles' law can be written as an equation in which the ratio of volume to Kelvin temperature is a constant,

$$\frac{V}{T} = k$$

or

$$\frac{V}{T_K} = \frac{V'}{T'_K}$$

and

$$V' = \frac{VT'_K}{T_K}$$

where V is the original volume, T_K is the original Kelvin temperature, V' is the new volume, and T_K' is the new Kelvin temperature.

Plotting the gas volumes given in Table 8-1 as a function of the Kelvin temperatures yields the curve shown in Figure 8-10. This linear relationship between the volume of a gas and its Kelvin temperature shows that these two quantities are directly proportional. Since 0°C (273 K) is frequently used as a reference point in calculations that involve temperature, this value is called *standard temperature.*

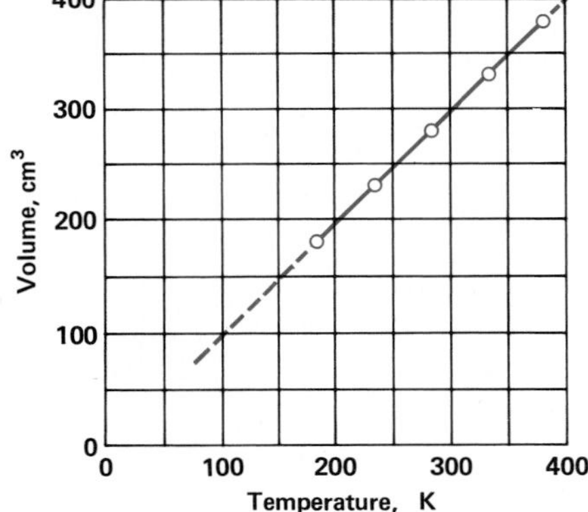

Figure 8-10. Graph of the data in Table 8-1. At constant pressure, the volume of a dry gas varies directly with the Kelvin temperature. (All gases liquefy before reaching 0 K.)

8.8 Boyle's Law

The English scientist Robert Boyle (1627–1691) was the first person to investigate what he called the "spring [elasticity] of the air." Other scientists at that time knew about compressed air, but none of them had performed experiments to learn how the volume of a gas is affected by the pressure exerted on it.

In his experiments, Boyle used a large J-shaped glass tube similar to that shown in Figure 8-11. He set up the apparatus in the stairwell of his laboratory at Oxford. The straight portion of the shorter arm was about one-third meter in length, while the longer arm had a length of about twenty-five meters. Enough mercury was poured into the tube to fill the bent portion. By tipping the tube to allow air to escape from the small arm, the mercury levels were then adjusted so that the mercury would stand at the same height in both arms. In this way Boyle trapped a column of air in the short arm of the tube. Next he measured the height of this air column. By assuming that the bore of the tube was uniform, Boyle used the height of the air column as a measure of the volume of air. The air in the short arm was at atmospheric pressure because the mercury levels were the same in both arms of the tube. See Figure 8-11(A).

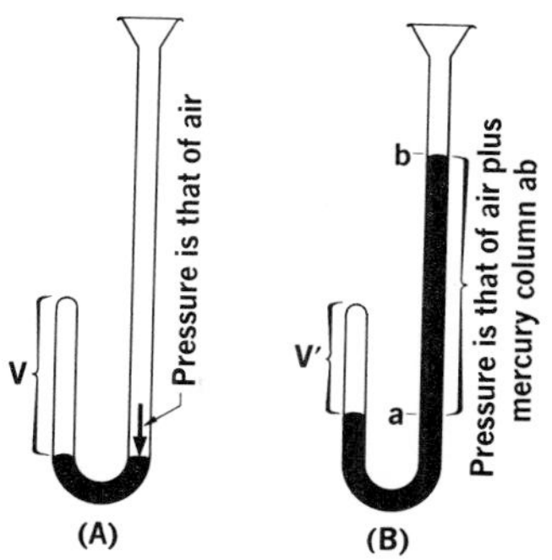

Figure 8-11. Verifying Boyle's law. The volume of the air enclosed in the short arm of the tube varies inversely with the pressure exerted on it.

In successive steps Boyle added more mercury to the long arm of the tube, as shown in Figure 8-11(B). By measuring the new height of the column of air in the short arm, he could determine each new volume. He found the corresponding pressure on this volume of air by measuring the height of the mercury column **ab** and adding that height to the atmospheric pressure as measured by a mercury barometer.

Boyle cooled the trapped, compressed air with a wet cloth, warmed it with a candle flame, and noted the resulting small changes in volume. These changes were so slight, however, that while it was obvious to Boyle that the temperature of the air during the experiment should be kept constant, small changes in temperature would not seriously affect the experimental results. Boyle's data, though not exceedingly accurate, convinced him and other scientists of his time that "the pressures and expansions . . . [are] in reciprocal proportion." Increasing the pressure on a column of confined air reduces its volume correspondingly. To reduce the volume to one-half, it is necessary to double the pressure; to reduce the volume to one-third, it is necessary to triple the pressure. Today we state ***Boyle's law*** as follows: *The volume of a dry gas varies inversely with the pressure exerted on it, provided the temperature remains constant.*

The mercury barometer was invented by Evangelista Torricelli (1608–1647), a student of Galileo.

Boyle's data are plotted in Figure 8-12. The total pressures are the abscissas, and the heights of the air columns (volumes) are the ordinates. You will recall from Chapter 2 that a graph of this shape, a hyperbola, suggests an inverse proportion.

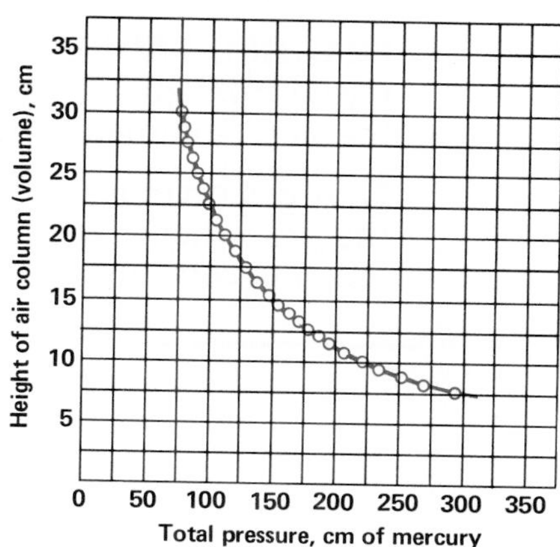

Figure 8-12. Graph of Boyle's original pressure-volume data. The shape of the curve is typical for an inverse proportion.

When two quantities are in inverse proportion, their product is a constant. At any given temperature, *the product of pressure and volume is always a constant.*

$$pV = k$$

Except during conditions of very high pressure, very low

temperature, or both, it is true in all cases that

$$pV = p'V'$$

and

$$V' = V\frac{p}{p'}$$

Here p is the original pressure and V is the original volume; p' represents the new pressure and V' the new volume.

Standard temperature and pressure (STP) is sometimes expressed as 0°C and 760 mm of mercury.

Even though Boyle used a column of mercury to increase pressure on a gas, *pressure* is really force per unit area. The SI unit of pressure is called the *pascal* (Pa). It is a force of one newton per square meter. The average atmospheric pressure at sea level is defined as one atmosphere and is 1.01×10^5 Pa. This is called *standard pressure*.

There is no change in the mass of a gas when its volume is changed by a difference in the exerted pressure. Since an increase in pressure produces a decrease in the volume of a gas, it must also increase the density of the gas. *The density of a gas varies directly with the pressure exerted on it*, or

$$\frac{D}{D'} = \frac{p}{p'}$$

A cubic meter of air has a mass of 1.29 kg at a pressure of one atmosphere. Under a pressure of four atmospheres, a container having a volume of one cubic meter can hold four times 1.29 kg, or 5.16 kg, of air because the air is four times as dense.

PRACTICE PROBLEMS **1.** The temperature of a 4.00-L sample of gas is raised from 20.0°C to 150.0°C. If the pressure remains constant, what is the new volume of the gas? *Ans.* 5.77 L

2. A gas occupies $50\bar{0}$ mL at a pressure of 1.00 atm. Find the volume of the gas when the pressure increases to 1.09 atm and the temperature remains constant. *Ans.* 459 mL

8.9 The Combined Gas Equation We can derive an equation that combines Boyle's and Charles' laws. At constant Kelvin temperature, T_{K1}, a certain mass of gas occupying volume V_1 is subject to a change in pressure from p_1 to p_2. The new volume, V_2, from Boyle's law is

$$V_2 = \frac{p_1 V_1}{p_2} \qquad \text{(Equation 1)}$$

Now if V_2 is subject to an increase in temperature from T_{K1} to T_{K2} at constant pressure, p_2, the new volume, V_3, from Charles' law is

$$V_3 = \frac{V_2 T_{K2}}{T_{K1}} \qquad \text{(Equation 2)}$$

Substituting the value of V_2 in Equation 1 into Equation 2, V_3 becomes

$$V_3 = \frac{p_1 V_1 T_{K2}}{p_2 T_{K1}}$$

and rearranging the terms, this expression becomes

$$\frac{p_1 V_1}{T_{K1}} = \frac{p_2 V_3}{T_{K2}}$$

But V_3 is the volume at pressure p_2 and temperature T_{K2}, and therefore we can write

The laws of Charles and Boyle can be combined into a single equation.

$$\frac{pV}{T_K} = \frac{p'V'}{T'_K}$$

where p, V, and T_K are original pressure, volume, and Kelvin temperature; and p', V', and T'_K are the new pressure, volume, and Kelvin temperature of a given mass of gas. You will note that when the pressure is constant in the above equation, it becomes Charles' law. When the temperature is constant, the equation becomes Boyle's law.

8.10 The Universal Gas Constant In 1811 the Italian physicist Amedeo Avogadro (1776–1856) recognized that at the same temperature and pressure, mass densities of different gases are proportional to their molecular weights. The molecular weight of a gas is the sum of the atomic weights of all the atoms comprising a molecule of that particular gas. Avogadro also postulated that the number of molecules in one gram-molecular weight (mole) of any substance was the same as that in a mole of any other substance. The number of molecules in one gram-molecular weight of a substance is called the *Avogadro number. A **mole** is the amount of a substance containing the Avogadro number of particles of that substance.*

The Avogadro number is 6.02×10^{23}.

The value of pV/T_K in the combined gas equation is designated by the symbol R and is the same for one mole of any gas. The value of R is independent of the chemical composition of the gas, except at very high pressure, and is known as the ***universal gas constant.*** If n moles are present in the sample we write

$$pV = nRT_K$$

This relationship defines the behavior of an ideal gas. An *ideal gas* is imagined to consist of infinitely small molecules that exert no forces on each other. The equation describes the behavior of real gases with reasonable accuracy except at low temperatures, extreme pressures, or both.

The numerical value of the universal gas constant R can be determined from the relationship $pV = nRT_K$. If the pressure is one atmosphere, one mole of an ideal gas will have a volume of 22.4 liters at the standard temperature of 273 K. R is then found as follows:

$$R = \frac{pV}{nT_K} = \frac{(1\ \text{atm})(22.4\ \text{L})}{(1\ \text{mol})(273\ \text{K})}$$

$$R = 8.21 \times 10^{-2}\ \text{L}\cdot\text{atm/mol}\cdot\text{K}$$

We may calculate the pressure, volume, mass, molecular weight, or Kelvin temperature of a gas provided four of these five quantities are known. To do this we must recognize that $n = m/M$, where m is the mass and M is the gram-molecular weight.

EXAMPLE A sample of ammonia gas (NH_3) has a mass of 15.0 g. What is the volume of the gas at standard temperature and pressure?

Given	Unknown	Basic equations
$m = 15.0$ g $M = 17$ g $T_K = 273$ K $p = 1$ atm $R = 8.21 \times 10^{-2}$ L·atm/mol·K	V	$pV = nRT_K$ $n = \frac{m}{M}$

Solution

$$\textit{Working equation: } V = \frac{nRT_K}{p} = \frac{mRT_K}{Mp}$$

$$= \frac{(15.0\ \text{g})(8.21 \times 10^{-2}\ \text{L}\cdot\text{atm/mol}\cdot\text{K})(273\ \text{K})}{(17.0\ \text{g})(1\ \text{atm})}$$

$$= 19.8\ \text{L}$$

PRATICE PROBLEM What is the volume of a mole of oxygen at 342 K and 8.46×10^4 Pa pressure (1 atm = 1.01×10^5 Pa)? *Ans.* 33.5 L

QUESTIONS: GROUP A

1. What causes most solids to expand when heated?
2. (a) What is the coefficient of linear expansion? (b) How are area and volume expansion related to linear expansion for a solid?
3. Why do liquids have coefficients of volume expansion only?
4. Why does the density of water increase as you heat it from 0°C to 4°C?
5. Is the coefficient of expansion a characteristic property of a gas? Explain.
6. How is the volume of a gas related to the temperature?
7. (a) At constant temperature, how is the volume of a gas related to its pressure? (b) What is the shape of a graph of density as a function of pressure for an ideal gas?
8. What were two of Avogadro's theories about gases?
9. What is the difference between a real and an ideal gas?
10. How is the number of moles of a gas present in a closed container calculated?

GROUP B

11. What does the fact that a metal pot full of water overflows when heated tell you about the relative expansion of solids and liquids?
12. A very deep body of water will not freeze even in the coldest weather. Why not?
13. In some cold areas, road signs warn of "frost heaves." Explain frost heaves.
14. (a) Use Figure 8-10 to determine the volume of gas when the temperature is 300 K. (b) Are expansion and contraction of gases related to how a Kelvin temperature scale is constructed? Explain.
15. (a) What happens to the size of a helium balloon as it rises? Why? (b) What would happen to it if it were held over a hot stove?
16. How would you use temperature to remove a tight ring from your finger?
17. How does an engineer accommodate thermal expansion or contraction when (a) laying a concrete sidewalk, (b) designing the roadway of a bridge, and (c) constructing a concrete highway?
18. A cooper makes wooden barrels. How does he use expansion properties to ensure leak-free barrels?

PROBLEMS: GROUP A

1. When the temperature is 29.40°C, 15.00-m iron rails are laid so that they just touch. How much space will there be between them at 1.00°C?
2. A steel tape measure is 10.000 m long at 20.0°C. What is the error in length when the temperature is 36.0°C?
3. A copper pipe is exactly 2.500 m long at 5.00°C. At what temperature would it be 2.501 m?
4. The diameter of a hole drilled through a piece of brass is 1.500 cm at 20.00°C. What is its diameter at 150.00°C?
5. A quantity of carbon tetrachloride occupies 500.00 mL at 20.0°C. What is its volume at 45.0°C?
6. (a) A helium-filled balloon has a volume of 2.00 L at STP. If it rises to a point where the temperature is −20.0°C, what is its volume (assuming pressure is constant)? (b) If the temperature is constant but the pressure decreases to 0.15 atm, what is the new volume of the balloon? (c) What is the density of the gas in the balloon in (b)?
7. An aluminum cookie pan is 25.0 cm long and 19.5 cm wide at 20.0°C. What is its area at 475.0°C?

8. A 2.00-L pot is filled with water at 10.5°C. If heated to 95.0°C, how much water will overflow? (Neglect the expansion of the pot.)
9. The density of air is 1.29 g/L at a pressure of 1 atm. What is its density at 0.40 atm?
10. A gas station attendant pumps 125 L of gasoline from an underground storage tank to your car, enough gasoline to fill the gas tank. If the temperature of the gasoline rises from 15.5°C to 35.5°C, how much gasoline would spill out of the tank?

GROUP B

11. A plumber is going to fit a copper ring with a diameter of 3.980 cm onto a pipe 4.000 cm in diameter. If the initial temperature of the ring and pipe is 20.00°C, to what temperature does the ring have to be heated to just slip over the pipe?
12. A 2.00-L tank of helium gas contains 1.785 g at a pressure of 202 kPa. What is the temperature of the gas in the tank?
13. A large calibrated flask contains 1000.0 cm^3 of water at 20.00°C. If the Pyrex glass flask and the water are both heated to a temperature of 80.00°C, what will be the new volume reading?
14. The density of hydrogen chloride gas is 0.820 g/L at 0.00°C. What pressure will this gas exert?
15. Mercury has a mass density of 13.60 g/cm^3 at 20.00°C. What is its mass density when the temperature is 250.0°C?
16. A steel tape measure is calibrated at room temperature, 20.0°C. When the temperature reaches 40.0°C, a workman uses it to measure the length of a room and obtains a reading of 33.50 m. What is the correct length of the room?
17. A certain gas has a volume of 3.00 L at a temperature of 10.0°C and at a pressure of 101 kPa. If the Kelvin temperature of the gas is doubled and the pressure on it is halved, what is the new volume of the gas?

PHYSICS ACTIVITY

Attach a balloon to the top of an empty soda bottle so that no air can escape from the bottle. Slowly place the bottle in a pan of boiling water. (Be cautious of steam coming off the surface of the water.) Observe the balloon. Using a hot pad or pair of tongs, carefully remove the bottle from the pan and allow it to cool on the counter. Once again observe the balloon. Explain your observations.

HEAT EXCHANGE

8.11 Heat Capacity Blocks of five different metals—aluminum, iron, copper, zinc, and lead—are shown in Figure 8-13. They all have the same mass and the same cross-sectional area, but the pieces have different heights because the metals have different densities. First the blocks are put in a pan of boiling water to heat them all to the same temperature. Then they are transferred to a block of paraffin. The diagram shows the relative depths to which the metals melt the paraffin. The aluminum block melts the most paraffin, iron follows as a poor second, copper and zinc are tied for third, and lead melts the least

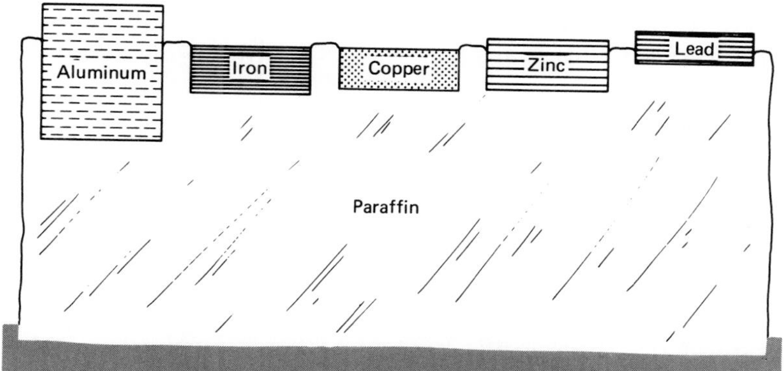

Figure 8-13. Because metals have different heat capacities, these blocks, all with equal masses and temperatures, melt the paraffin to different depths.

paraffin. This demonstration shows that different materials absorb or give off different amounts of heat, even though the materials have the same mass and undergo the same temperature change. Similarly, different amounts of heat are absorbed by blocks of the same material if their mass is different and their temperature change is the same, if their mass is the same and their temperature change is different, or if they have different masses and undergo different temperature changes. Such objects are then said to differ in *heat capacity.* Those with a high heat capacity warm more slowly because they absorb a greater quantity of heat; they also cool more slowly because they give off more heat. *The **heat capacity** of a body is the quantity of heat needed to raise its temperature 1°.*

$$\textbf{heat capacity} = \frac{Q}{\Delta T}$$

where Q is the quantity of heat needed to produce a change in the temperature of the body, ΔT. The units we shall use for heat capacity are J/C°.

8.12 Specific Heat The heat capacity of an object does not describe the thermal properties of the material of which it is made. For example, the heat capacity of 1.0 kg of copper differs from that of 1.0 kg of aluminum, but the heat capacity of 1.0 kg of aluminum also differs from that of 2.0 kg of aluminum. In order to obtain a quantity that is characteristic of copper, aluminum, or any material, the heat capacities of *equal masses* of the materials must be compared. This comparison yields a more useful quantity known as *specific heat.*

***Specific heat** is the heat capacity of a material per unit mass.* It is numerically equal to the quantity of heat that must be supplied to a unit mass of a material to raise its temperature one degree. If Q represents the quantity of heat needed to produce a temperature change, ΔT, in a quantity of material of mass m, the specific heat, c, is given by

ΔT *is always positive. Its value is found by subtracting the lower temperature from the higher temperature, regardless of which one is the initial or final temperature.*

$$c = \frac{Q/\Delta T}{m}$$

which when simplified yields

$$c = \frac{Q}{m\,\Delta T}$$

Since 1 calorie of heat raises the temperature of 1 g of water 1 C°, the specific heat of water is 1 cal/g·C°. In SI units the specific heat of water is 4.19 J/g·C°. Appendix B, Table 13, shows that the specific heat of most substances is less than that of water.

Let us solve the specific heat equation for Q.

$$Q = mc\Delta T$$

Thus the quantity of heat needed to produce a certain temperature change in a body equals the product of the mass of the material, its specific heat, and its temperature change. If m is in g, c in J/g·C°, and ΔT in C°, Q will be expressed in joules.

Heat lost = heat gained.

8.13 Law of Heat Exchange

Hot water can be cooled by the addition of cold water. As the two mix, the temperature of the hot water is lowered while the temperature of the added cold water is raised. The final temperature of the mixture lies between the original temperatures of the hot and cold water. Each time two substances of unequal temperature are mixed, the warmer one loses heat and the cooler one gains heat until both finally reach the same temperature. *A process that absorbs heat as it progresses is* ***endothermic.*** *An* ***exothermic*** *process gives off heat as it progresses.*

No thermal energy is lost when substances of unequal temperatures are mixed. *In any heat-transfer system, the heat lost by hot substances equals the heat gained by cold substances.* This is known as the ***law of heat exchange.*** The total number of heat units liberated by warmer substances equals the total number of heat units absorbed by cooler substances. This can be expressed as

$$Q_{Lost} = Q_{Gained}$$

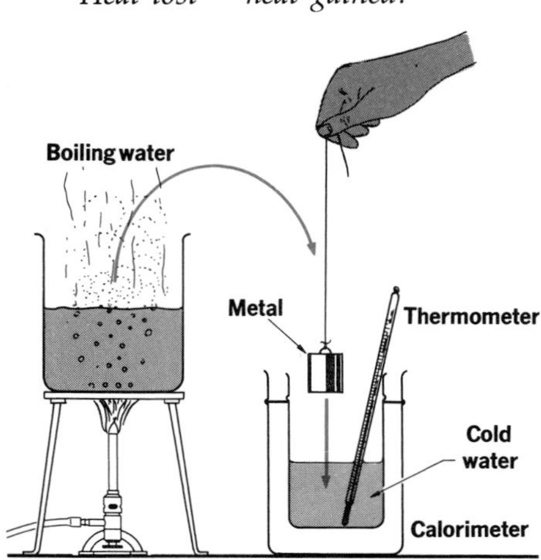

Figure 8-14. Apparatus for measuring the specific heat of a metal by the method of mixtures. The metal should be dried before it is transferred.

This equality is the basis of a simple technique known as the *method of mixtures* for measuring a quantity of heat in transit from one substance to another. The method of mixtures and the law of heat exchange can be used to determine the specific heat of a solid, as shown in Figure 8-14.

The hot solid of unknown specific heat, but of known mass and temperature, is "mixed" with water of known mass and temperature in a *calorimeter* (usually nested metal cups separated by an insulating air space) of known mass and temperature. The final temperature of the mixture is measured. All the data for the law of heat exchange equation are known except the specific heat of the solid; it can be calculated as in the example that follows.

In practical situations, some heat energy is usually transferred to the surroundings. As a result, this decrease in the amount of thermal energy that can be measured affects the accuracy of most heat experiments in the high school laboratory.

EXAMPLE A copper calorimeter with a mass of 150.0 g contains 350.0 g of water. The temperature of both the calorimeter and the water is 20.0°C. A metal cylinder, with a mass of 200.0 g and a temperature of 99.5°C, is placed in the calorimeter. The final temperature of the calorimeter, water, and metal cylinder is 26.7°C. Find the specific heat of the metal cylinder.

Given	Unknown	Basic equations
$m_C = 150.0$ g	c_M	$Q_{Lost} = Q_{Gained}$
$c_C = 0.387$ J/g·C°		$Q = mc\Delta T$
$m_M = 200.0$ g		
$T_M = 99.5$°C		
$m_W = 350.0$ g		
$c_W = 4.19$ J/g·C°		
$T_W = T_C = 20.0$°C		
$T_f = 26.7$°C		

Solution

Working equation:

$$\overbrace{m_M c_M \Delta T_M}^{\text{Heat lost}} = \overbrace{m_C c_C\, \Delta T_C + m_W c_W\, \Delta T_W}^{\text{Heat gained}}$$

$$m_M c_M (T_M - T_f) = m_C c_C (T_f - T_W) + m_W c_W (T_f - T_W)$$

$$c_M = \frac{m_C c_C (T_f - T_W) + m_W c_W (T_f - T_W)}{m_M (T_M - T_f)}$$

$$= \frac{(150.0\ \text{g})(0.387\ \text{J/g}\cdot\text{C}^\circ)(26.7^\circ\text{C} - 20.0^\circ\text{C}) + (350.0\ \text{g})(4.19\ \text{J/g}\cdot\text{C}^\circ)(26.7^\circ\text{C} - 20.0^\circ\text{C})}{(200.0\ \text{g})(99.5^\circ\text{C} - 26.7^\circ\text{C})}$$

$$= \frac{(389\ \text{J}) + (9830\ \text{J})}{(14\ 560\ \text{g}\cdot{}^\circ\text{C})}$$

$$= 0.702\ \text{J/g}\cdot\text{C}^\circ$$

EXAMPLE An aluminum calorimeter with a mass of 125 g contains 110.0 g of water at a temperature of 15.0°C. A 200.0-g mass of lead buckshot, at a temperature of 90.0°C, is added to the calorimeter. Calculate the final temperature of the mixture.

Given	Unknown	Basic equations
$m_C = 125\ g$ $c_C = 0.909\ J/g{\cdot}C°$ $m_W = 110.0\ g$ $c_W = 4.19\ J/g{\cdot}C°$ $T_W = 15.0°C$ $m_L = 200.0\ g$ $c_L = 0.128\ J/g{\cdot}C°$ $T_L = 90.0°C$	T_f	$Q_{Lost} = Q_{Gained}$ $Q = mc\Delta T$

Solution

Working equation:

$$\underbrace{m_L c_L \Delta T_L}_{\text{Heat lost}} = \overbrace{m_C c_C \Delta T_C + m_W c_W \Delta T_W}^{\text{Heat gained}}$$

$$m_L c_L (T_L - T_f) = m_C c_C (T_f - T_W) + m_W c_W (T_f - T_W)$$

$$T_f = \frac{m_L c_L T_L + m_C c_C T_W + m_W c_W T_W}{m_L c_L + m_C c_C + m_W c_W}$$

$$= \frac{(200.0\ g)(0.128\ J/g{\cdot}C°)(90.0°C) + (125\ g)(0.909\ J/g{\cdot}C°)(15.0°C) + (110.0\ g)(4.19\ J/g{\cdot}C°)(15.0°C)}{(200.0\ g)(0.128\ J/g{\cdot}C°) + (125\ g)(0.909\ J/g{\cdot}C°) + (110.0\ g)(4.19\ J/g{\cdot}C°)}$$

$$= \frac{(23\bar{0}0\ J) + (17\bar{0}0\ J) + (6910\ J)}{(25.6\ J/C°) + (114\ J/C°) + (461\ J/C°)}$$

$$= \frac{10\ 900}{601}\ °C$$

$$= 18.1°C$$

PRACTICE PROBLEMS **1.** An iron calorimeter has a mass of 150.0 g. It contains 300.0 g of water. The temperature of the calorimeter and the water is 21.5°C. A metal cylinder with a mass of 450.0 g and a temperature of 99.5°C is placed in the calorimeter. The final temperature of the mixture is 31.0°C. Calculate the specific heat of the metal cylinder. (Use Appendix B, Table 13, to obtain the necessary specific heats for your calculations.) *Ans.* 0.408 J/g·C°

2. A copper calorimeter with a mass of 170.0 g contains 145 g of water. The temperature of the calorimeter and water is 19.5°C. A 75.0-g mass of water, at a temperature of 90.5°C, is poured into the calorimeter. Calculate the final temperature of the mixture. *Ans.* 42.1°C

3. An aluminum calorimeter with a mass of 199.0 g contains 145 g of water. The temperature of both the calorimeter and the water is 20.2°C. A 32.0-g mass of water, at a temperature of 92.5°C, is poured into the calorimeter. Calculate the final temperature of the mixture. *Ans.* 30.7°C

QUESTIONS: GROUP A

1. What is meant by the expression *heat capacity?*
2. Which of the metals in Figure 8-13 has the greatest heat capacity?
3. The specific heat of a certain material is found to be 1.47 J/g·C°. Explain what this means.
4. How is the unit kJ/kg·K related to the following units: (a) J/g·C°; (b) cal/g·C°?
5. Why is water such an effective material to use in an automobile radiator?
6. Differentiate between endothermic and exothermic processes.

GROUP B

7. What data are required for the determination of the specific heat of an unknown substance by using the method of mixtures?
8. (a) What is a calorimeter? (b) How is a calorimeter used in determining the specific heat of any material? Be specific.
9. Upon what factors does the amount of heat energy that a material will absorb depend?

PROBLEMS: GROUP A

1. How much heat must be added to raise $25\bar{0}$ g of water at 10.0°C to its boiling point?
2. How much heat is given of by each block in Figure 8-13 if the mass off each is $10\bar{0}$ g and the temperature drops from 90.0°C to 40.0°C?
3. The addition of 1.36×10^5 J of energy raises the temperature of a block of aluminum from 345.0 K to 500.0 K. Calculate the mass (in grams) of the aluminum block.
4. If 25.0 g of water at a temperature of 10.0°C is mixed with 40.0 g of water at 80.0°C, what is the final temperature of the mixture?

GROUP B

5. A piece of brass with mass of 1750 g is heated. The brass is dropped into a 250.0-g aluminum calorimeter cup containing 600.0 g of water. The temperature of the water goes from 15.0°C to 80.0°C. What is the initial temperature of the brass?
6. Determine the specific heat of a material if 5640 J of energy will raise the temperature of 50.0-g amount of the material from a temperature of 9.00°C to a temperature of 86.0°C.
7. A 350-g glass beaker contains 500.0 g of water at a temperature of 10.0°C. If 400.0 g of ethyl alcohol at 35.0°C is subsequently poured into the beaker and the water and alcohol mixture is thoroughly stirred, what is its equilibrium temperature?
8. A sample of tin, with a mass of 225 g, is heated to a temperature of 100.0°C and dropped into 100.0 g of water at 10.0°C. If the equilibrium temperature for the tin and water mixture is a temperature of 20.0°C, what value does this experiment give for the specific heat of tin?

CHANGE OF PHASE

8.14 The Triple Point The phase and density of any pure substance are determined by its temperature and pressure. When temperature and pressure are controlled, it is possible to cause a pure substance to change from any one phase to either of the other two phases. The substance can make the phase change directly or by first passing through the third phase.

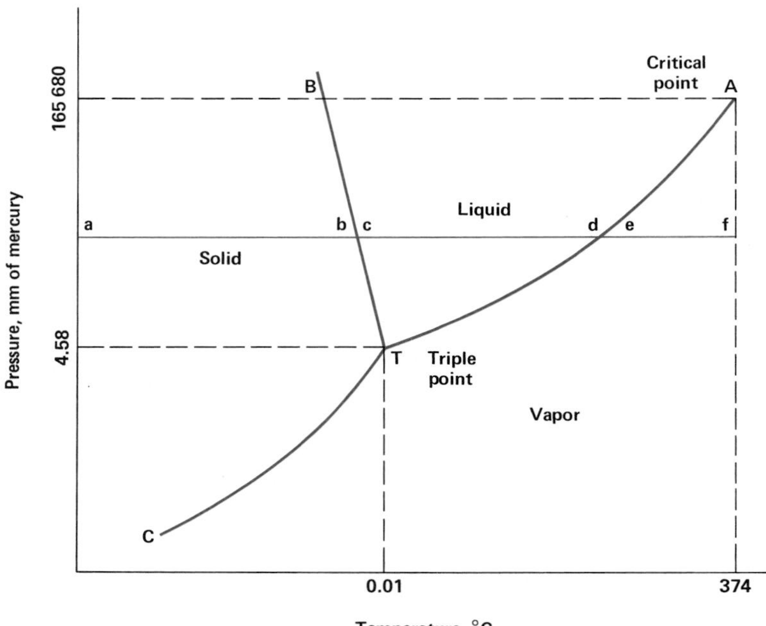

Figure 8-15. Temperature-pressure graph for water. The scales have been condensed in order to show the critical point.

Figure 8-15 is the temperature-pressure equilibrium graph for water. At the pressure and temperature of the triple point, **T,** the three phases of water can exist in equilibrium. It exists as a solid for temperature-pressure values lying in the area between the curves **TC** and **TB.** It is a liquid when these values lie in the area between **TB** and **TA.** In the area below the **ATC** curve it is a vapor. If the temperature-pressure values fall anywhere on the curve **TA,** water can exist with its vapor and liquid phases in equilibrium. Similarly, if these values fall on the curve **TB,** the solid and liquid phases can exist in equilibrium. For values on the curve **TC** the solid and vapor phases can exist in equilibrium. When there are two or more phases of a substance in equilibrium at any given temperature and pressure, there will always be interfaces separating the phases.

Figure 8-15 (with the solid-liquid curve **TB,** the liquid-vapor curve **TA,** and the solid-vapor curve **TC** plotted on a single graph) is typical of some pure crystalline materials. However, each substance has a different set of curves. This is because the temperature at which changes of phase occur (at a given pressure) are different for each substance.

Other features of the graph in Figure 8-15 will be explained in succeeding sections of this chapter. So it is important that you fully understand the meaning of the lines **TA, TB,** and **TC** in the figure.

8.15 Heat of Fusion At the temperature and pressure indicated by point **a** in Figure 8-15, water exists as ice, a solid. Let us assume that we keep the pressure constant as we apply heat at a uniform rate. The ice will be warmed from its initial temperature to the temperature at point **b.** At this temperature the ice will begin to melt. Melting is an endothermic process. The application of more heat will melt more ice, but the temperature will not rise until all the ice is melted.

In melting, a substance absorbs heat without a rise in temperature.

Following this change in phase, the temperature of the liquid water will rise. The horizontal line **abcdef** shows the temperature values as heat is applied first to the ice, then to the water, and finally to the vapor while the pressure on the system is held constant. This information can be shown more strikingly in another way. Assume that we start with a block of ice at −20°C and add heat at a constant rate while holding the pressure constant at a value indicated by the line **af.** A plot of temperature readings against time during which heat is being applied at a uniform rate gives us the graph in Figure 8-16. The line **ab** in each figure represents the warming of the ice without change of state; **bc** represents the heat required to change the solid to a liquid without a change of temperature (note that **b** and **c** of Figure 8-16 are the same point); **cd** represents the warming of the water. The addition of heat to a solid at its melting point produces a change of phase instead of a rise in temperature. All the heat energy is used to increase the potential energy of the particles. The average kinetic energy is unchanged. Thus the addition of heat to ice at a pressure of one atmosphere and a temperature of 0°C causes the ice to change into water at 0°C.

The amount of heat needed to melt a unit mass of a substance at its melting point is called its ***heat of fusion,*** L_F. L_F for ice is 334 J/g at 0°C,meaning that this quantity of heat must be added to each gram of ice at 0°C to convert it to water at 0°C. Heats of fusion of various substances are given in Appendix B, Table 13.

The heat of fusion of ice is 334 joules per gram at 0°C.

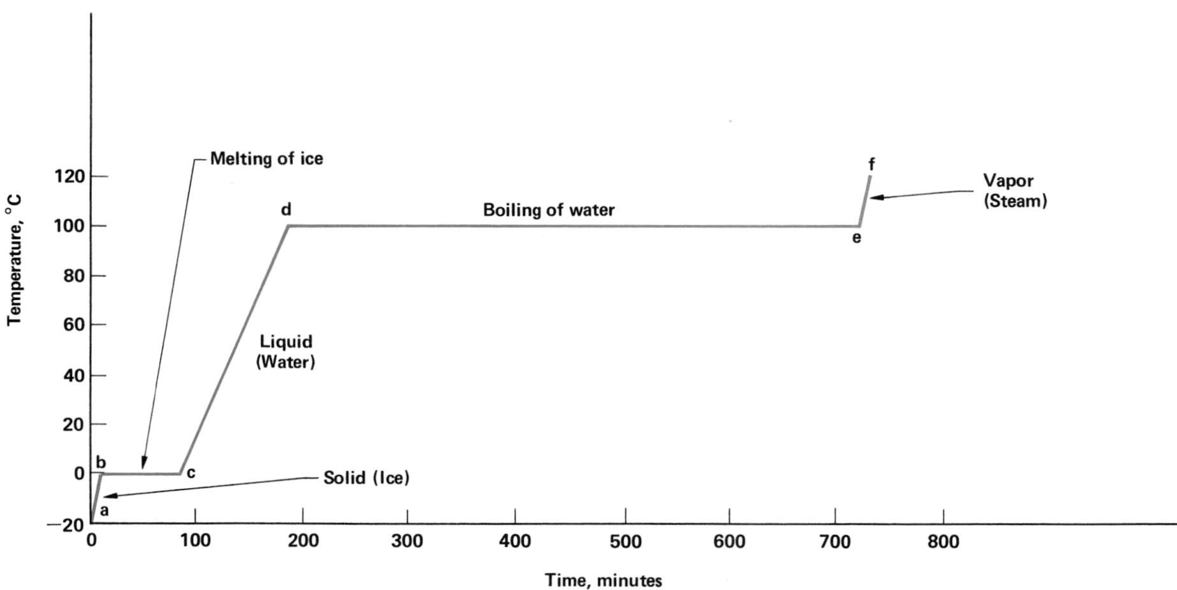

Figure 8-16. Time-temperature graph for the constant addition of heat to a mass of ice. The data for this graph are difficult to obtain experimentally.

The method of mixtures can be used to determine the heat of fusion of a solid. Suppose we wish to determine experimentally the heat of fusion of ice. Since hot water melts ice, we find the mass of ice at a known temperature that can be melted by a known mass of water at a known temperature. The following example illustrates this.

EXAMPLE An aluminum calorimeter, mass 150.0 g, contains 420.0 g of water at 35.0°C. The specific heat of the calorimeter is 0.909 J/g·C°. An 83.5-g amount of ice at 0.0°C is placed in the calorimeter and melts completely. The final temperature is 17.0°C. Calculate ice's heat of fusion.

Given	Unknown	Basic equations
$m_C = 150.0$ g	L_F	$Q_{Lost} = Q_{Gained}$
$c_C = 0.909$ J/g·C°		$Q = mc\Delta T$
$m_W = 420.0$ g		$Q = mL_F$
$T_W = T_C = 35.0$°C		
$m_I = 83.5$ g		
$T_I = 0.0$°C		
$T_f = 17.0$°C		

Solution

Working equation: $\overbrace{m_C c_C\ \Delta T_C + m_W c_W\ \Delta T_W}^{\text{Heat lost}} = \overbrace{m_I L_F + m_I c_I\ \Delta T_I}^{\text{Heat gained}}$

$$L_F = \frac{(m_C c_C + m_W c_W)(T_W - T_f) - m_I c_I (T_f - T_I)}{m_I}$$

$$= \frac{[(136\ J/C°) + (1760\ J/C°)](18.0°C) - (35\bar{0}\ J/C°)(17.0°C)}{83.5\ g}$$

$$= 337\ J/g$$

8.16 The Freezing Process The heat added to ice to make it melt increases the thermal energy of its molecules and changes it into water. This same amount of heat is evolved when water freezes because this process is a reversible energy change. Freezing is an exothermic process. Each gram of water at 0°C that forms ice at 0°C liberates 334 joules of heat. When the molecules of water return to their fixed positions in ice, they give up in the form of heat the energy that enabled them to slide over one another. The reversible energy change between 1 g of ice and 1 g of water at 0°C is shown in Figure 8-17. To change the ice to water, it is necessary to add 334 J; to change the water to ice, it is necessary to take away 334 J. A more precise value for the heat of fusion of ice is 333.7 J/g.

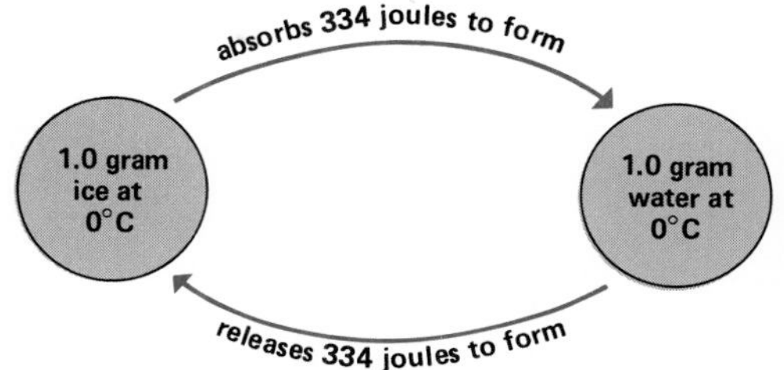

Figure 8-17. Heat of fusion. It takes 334 additional joules to change 1.0 g of ice at 0°C to water. Water at 0°C releases the same number of joules upon freezing.

If pure water is very carefully cooled without being disturbed, it can reach temperatures as low as −20°C without freezing. Water that is cooled below the normal freezing point is said to be *supercooled*. If a piece of ice or a speck of dust is added to such water, freezing takes place rapidly and the temperature rises to 0°C, the normal freezing point. The formation of ice takes place readily at 0°C if there is some dust or other foreign matter on which the first crystals of ice can form. Supercooling occurs when no such foreign matter is present. Supercooling is of particular interest in meteorology because the process of supercooling takes place in the formation of some clouds and when "freezing rain" turns to ice as it hits trees, telephone wires, and other surfaces.

8.17 The Boiling Process The equilibrium vapor pressure of a liquid is a characteristic of the liquid that depends on the temperature only. Appendix B, Table 12, gives the equilibrium vapor pressure of water at various temperatures. The vapor pressure curve for water is shown in Figure 8-18. This curve shows the relationship between the pressure and temperature of water and its saturated vapor. Any point on the curve or on **TA** (Figure 8-15) represents a definite temperature and pressure at which water is in equilibrium with its saturated vapor. Figure 8-18 is simply a portion of the curve **TA** of Figure 8-15

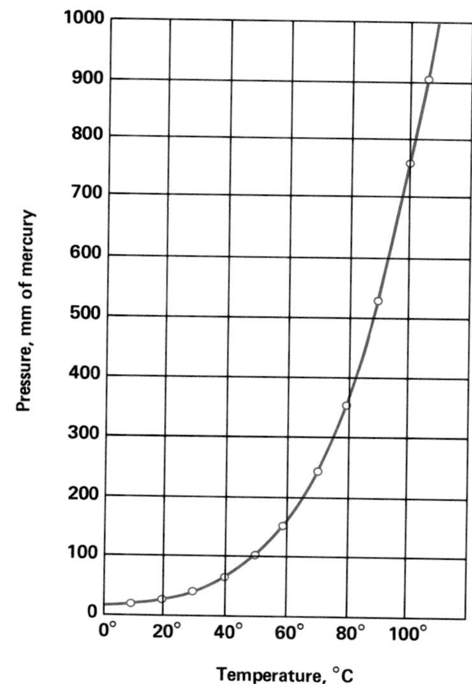

Figure 8-18. The equilibrium vapor pressure curve for water. The boiling point of water rises rapidly as the pressure is raised.

drawn to specific temperature and pressure scales. Other liquids show vapor pressure curves that are similar.

Solids, like liquids, exert a vapor pressure. The equilibrium vapor pressure of ice at 0°C is about 4.5 mm of mercury. The vapor pressure of solids is much less than that of liquids because solids sublime more slowly, if at all, at normal temperatures. Evaporation of both solids and liquids can occur at temperatures and pressures other than the equilibrium values that fall on the curves **TC** and **TA** of Figure 8-15. A solid at any temperature and pressure to the left of the curve **TC** can be evaporating to some degree depending upon the nature of the substance. Under these conditions the vapor pressure of the substance will be less than the pressure at saturation, however.

Equilibrium vapor pressure is defined in Section 7.21. Boiling point is defined in Section 7.22.

A non-SI unit of pressure is millimeters of mercury. Using this unit, one atmosphere equals 760 mm of mercury. The *boiling point* of a liquid is the temperature at which the vapor pressure of the liquid equals atmospheric pressure. The vapor pressure of water equals 760 mm of mercury at 100°C; therefore, 100°C is called the ***normal boiling point*** of water. If the air pressure is reduced to 525.8 mm of mercury, water boils at 90°C because the vapor pressure of water is 525.8 mm of mercury at that temperature.

At very high altitudes, it is difficult to prepare a hard-boiled egg in an open pan. Can you explain why?

Strong-walled pressure cookers, in which water is boiled at pressures up to about 2 atmospheres and at temperatures up to about 120°C, are useful for rapid cooking of foods. Special pans in which water is boiled at room temperature or slightly above are used in the production of sugar crystals.

If a mixture of ethyl alcohol and water is boiled, the boiling temperature is not the same as the boiling point of either liquid by itself. Alcohol boils at 78°C and water at 100°C; the boiling temperature of the mixture is between 78°C and 100°C, depending on the proportions of alcohol and water in the mixture. The boiling temperature of a mixture of two or more liquids each having different boiling points is different from that of any of the liquids used.

A liquid can be separated from a nonvaporizing disolved solid by *distillation*. The process of distillation involves evaporation followed by condensation of the resulting vapor in a separate vessel. Liquids that have different boiling points can be separated from one another by the process of *fractional distillation*. As the mixture boils, more of the component with the lower boiling point vaporizes and the boiling point of the resulting mixture rises. Samples from the mixture collected at different temperatures are then redistilled.

The heat of vaporization of water is 2260 joules per gram at 100°C.

8.18 Heat of Vaporization If a liter of water at 0°C is heated to 100°C, each gram of water will absorb 419 J. If heat is supplied at a constant rate, it takes more than five times as long to boil the water away as it did to heat it from 0°C to 100°C. (See Figure 8-16.) Each gram of water absorbs more than 2000 joules of heat as it is changed into steam. The temperature of the water remains constant during boiling, and the steam produced has the same temperature as the boiling water. *The heat required per unit mass to vaporize a liquid at its boiling point is called its **heat of vaporization,*** L_v.

Boiling is an endothermic process.

The heat required for vaporization gives the particles of liquid sufficient thermal energy to overcome the energy binding them to the liquid, and so enables them to separate from one another and move among the molecules of the gases above the liquid. As noted in Section 7.20, all the energy is used to increase the potential energy of the particles. Their average kinetic energy is not changed. Since the energy required for vaporization varies with temperature, the heat of vaporization varies with temperature. The heat of vaporization for water is 2260 J/g at 100°C. Water boiling under reduced pressure at a lower temperature has a heat of vaporization that is somewhat greater; at boiling temperatures above the normal boiling point, the heat of vaporization is smaller.

The method of mixtures is used to determine the heat of vaporization of water. As shown in Figure 8-19, a known mass of steam is passed into a known mass of cold water at a known temperature, and the increase in temperature is measured. To ensure that only steam enters the water in the calorimeter, a trap is used to catch any condensed water from the steam generator. The calculations for this method are given in the following example.

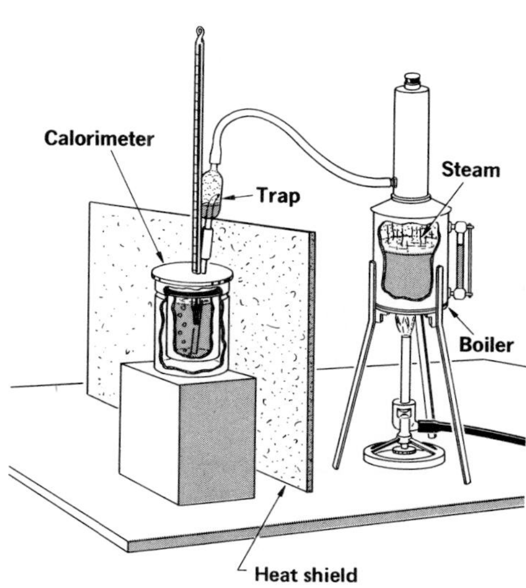

Figure 8-19. Laboratory apparatus that can be used for finding the heat of vaporization of water by the method of mixtures.

EXAMPLE An aluminum calorimeter, mass 130.4 g, contains 324.6 g of water at a temperature of 5.00°C. Steam at a temperature of 100°C is passed into the water, after which the total mass of the calorimeter and water is 473.0 g. What is the final temperature of the water? (Refer to Appendix B, Table 13.)

Given	Unknown	Basic equations
$m_C = 130.4$ g	T_f	$Q_{Lost} = Q_{Gained}$
$c_C = 0.909$ J/g·C°		$Q = mc\Delta T$
$m_W = 324.6$ g		$Q = mL_v$

$T_w = T_c = 5.00°C$
$m_s = 18.0$ g
$T_s = 100°C$
$L_v = 2260$ J/g

Solution

$$\text{Working equation:}\quad \overbrace{m_sL_v + m_sc_w(T_s - T_f)}^{\text{Heat lost}} = \overbrace{m_cc_c(T_f - T_w) + m_wc_w(T_f - T_w)}^{\text{Heat gained}}$$

$$T_f = \frac{(m_sL_v) + (m_sc_wT_s) + (m_cc_cT_w) + (m_wc_wT_w)}{(m_sc_w) + (m_cc_c) + (m_wc_w)}$$

$$= \frac{40\,700\text{ J} + 7540\text{ J} + 593\text{ J} + 6800\text{ J}}{75.4\text{ J/C°} + 119\text{ J/C°} + 1360\text{ J/C°}}$$

$$= 35.8°C$$

PRACTICE PROBLEM An aluminum calorimeter contains 420.0 g of water at a temperature of 15.0°C. The mass of the calorimeter is 152.0 g. How many grams of steam at 100.0°C are needed to raise the temperature of the water and the calorimeter to 75.0°C? (Use 0.909 J/g · C° as the specific heat of aluminum.) *Ans.* 48.2 g

---

Condensation is an exothermic process.

8.19 The Condensing Process Heat is absorbed during vaporization. This increase in the thermal energy of the molecules of a liquid enables them to break away from the liquid and become molecules of vapor. When the vapor condenses to a liquid, this thermal energy is evolved as heat. This reversible energy change, shown in graphic form in Figure 8-20, is useful in a steam-heating system. The heat of vaporization changes water to steam in the boiler. The steam passes into radiators where it gives up its heat of vaporization and condenses to a liquid.

Even though steam and boiling water may be at the same temperature, steam can produce a more severe burn. One reason is that steam at 100°C has acquired 2260 more joules of heat per gram than water at 100°C. When steam condenses, this heat of vaporization is given out. As it cools, the water that is formed gives out the same amount of heat that water at 100°C does during cooling.

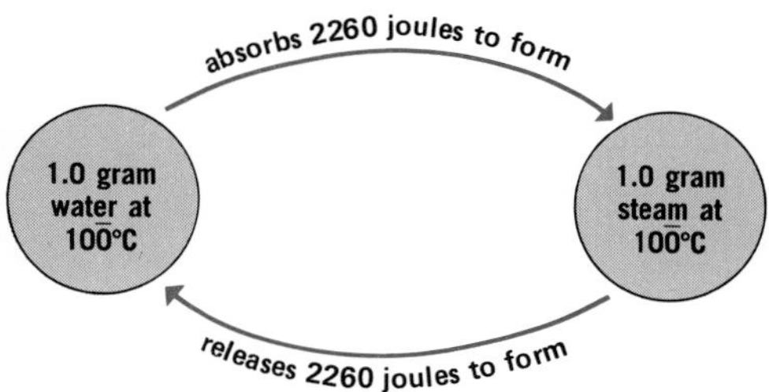

Figure 8-20. Heat of vaporization. It takes 2260 additional joules to change 1.0 g of water at 100°C to steam. Steam at 100°C releases the same number of joules upon condensing.

In Figure 8-16, the line **de** represents the heat required to change the liquid to a vapor (steam) without producing a change in temperature. Note that points **d** and **e** on Figures 8-15 and 8-16 represent the same values. After all the water is converted to steam, the addition of heat energy causes the temperature of the steam to rise as indicated by **ef** in both figures.

In order to keep the pressure constant while heat is applied to change ice from −20̄°C to steam at 120̄°C, the volume of the container must be greatly increased. In our previous discussions we noted that ice expands slightly as its temperature is raised to 0°C, it contracts as it is melted to water at 0°C, the water formed contracts to its minimum volume at 4°C, and above 4°C the water expands as its temperature is raised to 100̄°C. At constant pressure, as the water at 100̄°C is changed to steam at 100̄°C by the addition of heat, the expansion is about 1700 times. As the vapor (steam) is heated further, it continues to expand.

8.20 The Critical Point The vaporization curve **TA** in Figure 8-15 is not unlimited in extent. The lower limit is the temperature and pressure of the triple point. The upper limit is the *critical point*. The temperature and pressure of the critical point are called the *critical temperature* and the *critical pressure*. A substance cannot exist as a liquid at a temperature above its critical temperature; no matter how great the pressure, it cannot be condensed to the liquid state. At the critical point the densities of the liquid and the vapor are equal and the heat of vaporization is zero. At temperatures above the critical temperature, a substance is usually called a gas, while at temperatures below the critical temperature it is called a vapor. The critical temperature of water is 374°C, and the critical pressure is 218 atmospheres. This means that the temperature of liquid water cannot be raised to 374°C unless it is under a pressure of 218 atmospheres, and that at any higher temperature water can exist only in its gaseous phase, no matter how high the pressure.

The lower limit of the vaporization curve of a substance is the triple point. The upper limit is the critical point.

Gaseous substances, such as oxygen and nitrogen, have very low critical temperatures. The critical temperature of oxygen is −119°C and of nitrogen is −174°C. These gases must first be cooled to their low critical temperatures before they can be liquefied. Helium has the lowest critical temperature, −268°C.

8.21 Summary of Phase Changes We have described several effects of addition or loss of heat on water. In Figure 8-21 these are summarized in a somewhat different manner than has been shown previously. The graph shows the relationship among temperature, thermal energy, and phase as ice at −20°C is heated to steam at 120̄°C. Since the specific heat of ice is 2.2 J/g·C°, 1.0 g of ice absorbs 44 J in being warmed to 0.0°C. As this ice melts there

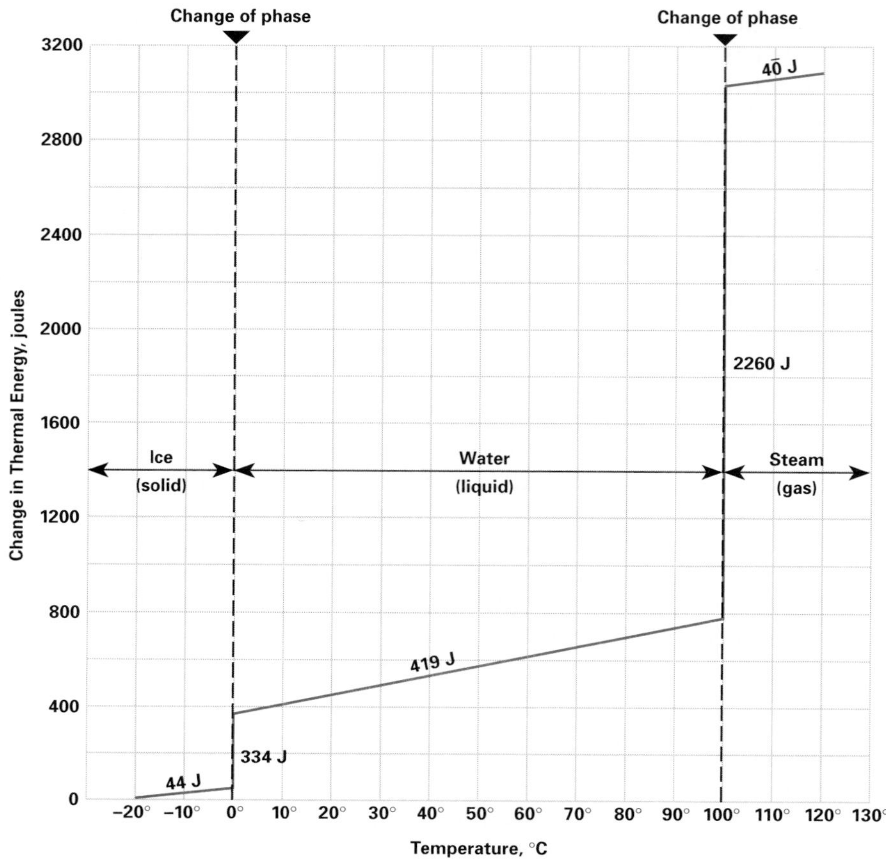

Figure 8-21. Temperature-energy graph for 1 g of water.

is *no temperature change* while 334 J of heat is being absorbed. There is a phase change only. As heating continues, the next 419 J increases the temperature to 10$\bar{0}$°C. As the water boils, 2260 J of heat converts the water into steam at 10$\bar{0}$°C; during this change of phase there is *no temperature change.* If the steam is under one atmosphere pressure, its specific heat is 2.0 J/g·C° and 4$\bar{0}$ J is needed to heat it to 120°C. Thus 3$\bar{0}$00 J is required to change 1.0 g of ice at −2$\bar{0}$°C into 1.0 g of steam at 12$\bar{0}$°C. Conversely, 3$\bar{1}$00 J is evolved if 1.0 g of steam at 12$\bar{0}$°C is changed into 1.0 g of ice at −2$\bar{0}$°C.

QUESTIONS: GROUP A

1. (a) What process is referred to as "fusion"? (b) What is the heat of fusion?

2. Heat of fusion is sometimes called *latent* heat, from a Latin word meaning "hidden." Is this term appropriate? Why?

3. (a) What is supercooling? (b) Under what circumstances does supercooling occur?

4. Give examples of endothermic and exothermic processes.

5. Why is the vapor pressure of solids lower than that for liquids?

6. Use Figure 8-18 to determine the boiling temperature of water at the following pressures. (a) 150 mm Hg (b) 1000 mm Hg (c) 760 mm Hg
7. (a) Explain how a liquid is separated from a dissolved solid. (b) How can two liquids, like water and ethyl alcohol, be separated?
8. What happens to the heat added to a material already at its boiling point?

GROUP B

9. What is the critical point for a material?
10. Since boiling is a rapid form of evaporation, it could be considered a "cooling" process. Explain.
11. Why does steam at 100°C produce a more severe burn to the skin than the same mass of water at 100°C?
12. At 150°C, oxygen and nitrogen would be called "gases," while water in the gaseous state is a "vapor." Why?
13. The temperature of the air above an ice-covered lake is −10°C. Predict the temperature of (a) the surface of the ice, (b) the water just below the ice, and (c) the water at the bottom of the lake.
14. Using Figure 8-15, determine the range of temperatures and pressures where water can exist as (a) a solid and a liquid, (b) a solid and a vapor, and (c) a liquid and a vapor.
15. Why is the heat of vaporization of water so much higher than its heat of fusion?
16. Why does the value of the heat of vaporization of water vary with the boiling temperature?
17. If you had a sample of pure water at its triple point, what would happen if (a) the pressure increased while the temperature remained constant, (b) the temperature increased while the pressure remained constant, (c) the pressure decreased while the temperature remained constant, and (d) the temperature decreased while the pressure remained constant?

PROBLEMS: GROUP A

1. How many joules of energy will be absorbed by 1.50 kg of ice at 0.0°C as it melts?
2. To what temperature must a 2270-g iron ball be heated so that it can completely melt a 1150-g piece of ice at 0.0°C?
3. A 500.0-g aluminum block is heated to 350°C. How many grams of ice at 0.0°C will the aluminum block melt as it cools?
4. A copper calorimeter has a mass of 220.0 g. It contains 450 g of water at a temperature of 21.0°C. How many grams of ice at 0.00°C must be added to reduce the temperature of the mixture to 5.00°C?
5. To what temperature must a 500.0-g brass cube be heated to convert 60.0 g of ice at a temperature of −20.0°C to water at 20.0°C?
6. How many joules of heat are released by 50.0 g of steam at 100.0°C when it condenses?
7. How many joules of energy are used to vaporize 20.0 kg of water at a temperature of 100.0°C?
8. Calculate the number of joules evolved when 4.00 kg of steam at a temperature of 100°C is condensed, cooled, and then changed to ice at 0.00°C.
9. How many grams of mercury can be vaporized at its boiling point, 356.58°C, by the addition of 4.19×10^3 J?
10. What is the final temperature attained by the addition of 11.4 g of steam at 100.0°C to 681.0 g of water at a temperature of 25.0°C in an aluminum calorimeter having a mass of 182.0 g?

GROUP B

11. What is the final temperature of a mixture of 50.0 g of ice at 0.00°C and 50.0 g of water at 80.0°C?
12. A calorimeter, specific heat 0.500 J/g·C°, mass 200.0 g, contains 300.0 g of water at 40.0°C. If 50.0 g of ice at 0.00°C is dropped into the water and stirred, the temperature of the mixture when the ice has melted is 23.8°C. Calculate the heat of fusion of ice.
13. A block of silver, with a mass of 500.0 g, temperature 100.0°C, is put in a calorimeter with 300.0 g of water, temperature 30.0°C. The mass of the calorimeter is 50.0 g, and its specific heat is 0.500 J/g·C°. A 50.0-g mass of ice at −10.0°C is put in the calorimeter. Calculate the final temperature.
14. What is the final temperature attained when 900.0 g of ice at 0.00°C is dropped into 3400.0 g of water at 93.3°C in a calorimeter having a mass of 1350 g and a specific heat of 0.400 J/g·C°?
15. To determine the heat of vaporization of water, 15.0 g of steam at 100.0°C is added to 150.0 g of water at a temperature of 20.0°C in a calorimeter. The mass of the calorimeter is 75.0 g; its specific heat is 0.500 J/g·C°. The equilibrium temperature of the mixture is 73.9°C. Calculate the heat of vaporization.
16. A mixture of ice and water, mass 200.0 g, is in a 100.0-g calorimeter, specific heat 0.800 J/g·C°. When 40.0 g of steam is added to the mixture, the temperature is 60.0°C. How many grams of ice were originally in the calorimeter?
17. An aluminum cylinder, mass 50.0 g, is placed in a 100.0-g brass calorimeter with 250.0 g of water at 20.0°C. What equilibrium temperature is reached after the addition of 25.0 g of steam at 120.0°C?
18. A copper ball with a mass of 4.54 kg is removed from a furnace and dropped into 1.36 kg of water, temperature 22.0°C. After the water stops boiling, the combined mass of the ball and water is 5.45 kg. What is the furnace temperature?

SUMMARY

The thermal energy of a material is the potential and kinetic energy of its particles. Heat is the thermal energy that is absorbed, given up, or transferred from one material to another. Thus heat is a form of energy. Temperature is the physical property that determines the direction in which heat flows between substances adjacent to each other.

Temperature is measured in degrees. The Celsius and Kelvin temperature scales are based on the triple point of water, which is the temperature at which the solid, liquid, and vapor phases of water can coexist. Absolute zero (0 K) is the temperature at which the kinetic energy of the molecules of a substance is at a minimum. Heat is measured in calories. A calorie is equivalent to 4.19 joules of energy.

The change in unit length of a solid when its temperature is changed one degree is its coefficient of linear expansion. The expansion of most liquids is proportional to their increase in temperature. The expansion of water is abnormal. Water reaches its maximum density at 4°C. Gases expand uniformly except at very high pressures and very low temperatures. According to Charles' law, the volume of a dry gas is directly proportional to its Kelvin temperature, provided the pressure is constant. Boyle's law states that if the temperature is constant, the volume of a dry gas is inversely proportional to the pressure. The universal gas constant relates the pressure, volume, Kelvin temperature, and number of moles of an ideal gas. Since the behavior of a real gas is similar to that of an ideal gas except at extreme pressures and low temperatures, the gas laws can be used with reasonable exactness with real gases.

The heat capacity of a body is the quantity of heat needed to raise its temperature one degree. Its specific heat is the ratio of its heat capacity to its mass. The heat given off by hot materials equals the heat received by cold materials.

The heat of fusion is the amount of heat needed to bring about fusion in a unit mass of a substance at its melting point. The heat required to vaporize a unit mass of liquid at its boiling point is the heat of vaporization. The temperature to which any gas must be cooled before it can be liquefied by pressure is called its critical temperature; the pressure needed to liquefy a gas at this temperature is called its critical pressure.

VOCABULARY

Boyle's law
calorie
calorimeter
Celsius scale
Charles' law
coefficient of area expansion
coefficient of cubic expansion
coefficient of linear expansion
critical point
critical pressure
critical temperature
endothermic
exothermic
heat
heat capacity
heat of fusion
heat of vaporization
Kelvin scale
law of heat exchange
mole
pascal
specific heat
standard pressure
standard temperature
temperature
thermal energy
triple point
universal gas constant

Heat Engines

thermodynamics (thur-mo-di-Na-miks) n.: the study of the quantitative relationships between heat and other forms of energy.

OBJECTIVES

- Discuss the relationship between heat and work.
- Differentiate between adiabatic and isothermal processes.
- State and apply the two laws of thermodynamics.
- Describe the principles and characteristics of various external and internal combustion engines.

HEAT AND WORK

9.1 Mechanical Equivalent of Heat Since heat is a form of energy, it can be produced by mechanical work. For example, you can warm your hands by briskly rubbing them together. You can heat up a nail by pounding it into a board with a hammer. Even as you walk, a certain amount of heat is given off as your shoes touch the floor.

As we noted in Section 1.13, Count Rumford and James Prescott Joule studied the relationship between heat and work in the 19th century. In experiments conducted between 1842 and 1870, Joule found that the heat produced by a given amount of mechanical work is always the same.

The study of the quantitative relationships between heat and other forms of energy is called ***thermodynamics.*** In this chapter we shall be concerned with the relationship of heat energy to mechanical energy. In discussing energy transformations, we shall also use the term *internal energy.* ***Internal energy*** *is the total available potential and kinetic energy of the particles of a substance.* These particles include molecules, ions, and atoms. When heat is added to a substance, its internal energy increases.

Work and heat are interchangeable.

Joule used an apparatus similar to that shown in Figure 9-2 to determine the relationship between mechanical energy and heat energy. The system on which work was done was a mass of water in an insulated vessel designed to reduce the escape of heat from the vessel to a minimum. A set of movable paddles on a shaft was turned by a falling mass connected to the shaft by a cord. The paddles moved past fixed vanes, churned the water, and increased its temperature from T_i to some final value T_f. The work done on the water equals the loss of potential energy of mass m as it falls through a distance d, less the kinetic energy it possesses as it reaches the bottom. The loss in potential energy is mgd. The final kinetic energy of the mass is $\frac{1}{2}mv^2$. (In these experiments, v was small.) Hence, the amount of work, W, done on the system by the falling mass is

$$W = mgd - \frac{1}{2}mv^2$$

The amount of heat that would be needed to produce the observed temperature change in the water is

$$Q = m_W c_W (T_f - T_i)$$

where Q is the heat energy in joules, m_W is the mass of the water, c_W is the specific heat of water, and T_f and T_i are the final and initial temperatures of the water, respectively. By showing that W and Q are equal, Joule proved that heat is a form of energy.

If heat is measured in calories, Joule's experiment shows that a calorie equals 4.19 J (as pointed out in Section 8.3).

Just as in the case of the hot brass chips in Count Rumford's cannon-boring experiments, it is possible to produce internal energy indefinitely in the Joule apparatus provided we continue to supply mechanical energy to it. There is further evidence that heat is a form of energy and not a substance.

Figure 9-1. James Prescott Joule determined the quantitative relationship between heat and mechanical energy.

9.2 First Law of Thermodynamics We can now broaden the statement of the conservation of energy in Section 1.13 to include internal energy as well as mechanical energy. This generalization is known as the ***first law of thermodynamics:*** *The quantity of energy supplied to any isolated system in the form of heat is equal to the work done by the system plus the change in internal energy of the system.* Thus the energy input to an isolated system equals the energy gained by the system plus the energy output in the form of work. We may also state the first law of thermodynamics as follows: *When heat is converted to another form of energy or when other forms of energy are converted to heat, there is no loss of energy.*

As an application of the first law of thermodynamics, let us assume that an amount of heat, Q, is added to a substance whose total internal energy is E_i. Generally we would find that the addition of this heat energy increases the internal energy of the substance to E_f and also causes

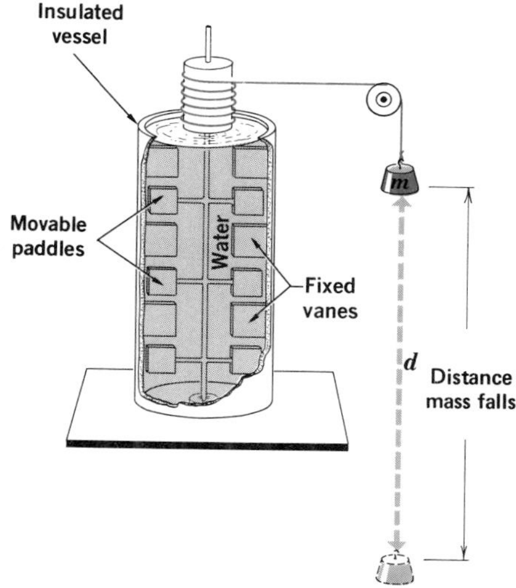

Figure 9-2. A simplified diagram of Joule's apparatus for observing the conversion of mechanical energy into internal energy.

the substance to do a quantity of work, W, on its surroundings. This can be stated algebraically as

$$Q = (E_f - E_i) + W$$

Q is positive when heat is added to the substance, and W is positive when the body does work on surrounding objects. All quantities must, of course, be expressed in the same units.

In situations when no work is done by or on the substance, the change in internal energy equals the quantity of heat added to or removed from the substance. Adding heat to water that undergoes no change of phase will cause each gram of water to increase in temperature by one Celsius degree for each 4.19 J of heat added. Here, we neglect the very small amount of work involved in changing the volume of the water. We know that the internal energy of the water is changed because the temperature of the water is changed.

Figure 9-3. What part of this activity is an adiabatic process?

A process in which no heat is added to or removed from a substance is called an ***adiabatic process.*** In such a case, $Q = 0 = (E_f - E_i) + W$. Thus $E_f - E_i = -W$. An example is the Joule experiment for determining the relationship between mechanical energy and heat energy. When the vessel containing the water permits no heat to enter or leave during the churning process, the work done on the water equals the change in its internal energy.

In another example let us assume that we have a quantity of air trapped in an *insulated* cylinder with a tight-fitting but freely moving piston. When the air is compressed by pushing the piston into the cylinder, the change in the internal energy of the air must equal the work done on it. This relationship is shown by the change in volume, pressure, and temperature of the air.

9.3 Isothermal Expansion In nearly all situations involving gases, the gases are confined by barriers on which they exert a force and which, in turn, exert an equal and opposite force on the gases. A gas may be under pressure in a storage tank or in a cylinder above a movable piston of an internal combustion engine. A quantity of air (an air mass) in the atmosphere is under pressure exerted by the air around it. In any of these cases, expansion of the gas requires it to do work against an external force; thus work is done on the external medium. On the other hand, the quantity of gas may be compressed by the action of an outside force; work may be done on the gas.

To calculate the work done by a gas in expanding, let us imagine the gas enclosed in a cylinder with a tight-fitting piston, as illustrated in Figure 9-4. The piston rod is connected to a device on which it may exert a force. The force, F, acting on the piston due to the pressure, p, exerted by the gas on area A of the piston head is

$$F = pA$$

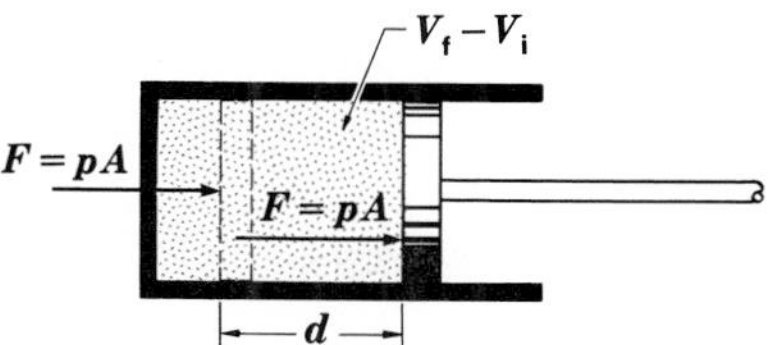

Figure 9-4. Expansion of a gas at constant pressure.

Suppose the piston is moved a distance d by the expanding gas in the cylinder. Suppose further that some heat is applied to the gas while the pressure remains constant during the expansion. The work done by the expanding gas in moving the piston will be

Pressure = force per unit area

$$W = Fd = pAd$$

where the quantity Ad is the change in volume of the expanding gas. Thus the work done by the gas expanding at constant pressure is

$$W = p(V_f - V_i)$$

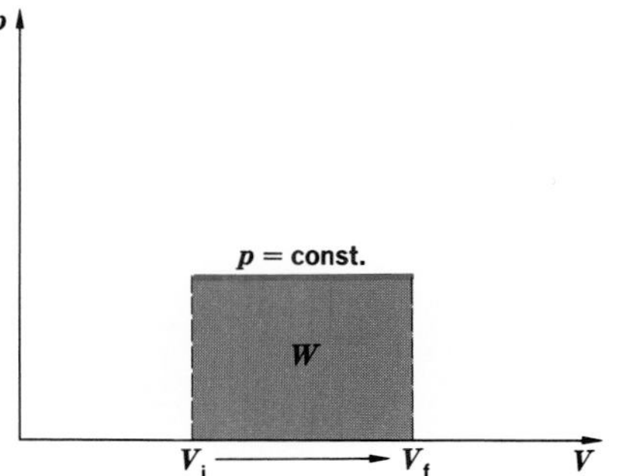

Figure 9-5. Graph of work done by a gas expanding at constant pressure.

where V_f and V_i are the final and initial volumes, respectively, of the confined gas. We may conveniently show the work done by the gas expanding at constant pressure by using a graph, as shown in Figure 9-5. On this graph volumes are plotted as abscissas and pressures as ordinates. The expansion at constant pressure is represented by a blue horizontal line extending from V_i to V_f. The work, W, done by the gas in expanding is given by the area between this line and the x axis and between V_i and V_f (the blue portion shown in Figure 9-5). This graph is for the *expansion* of a gas at constant pressure. If the gas is *compressed* at constant pressure, the work done is represented in the same manner. It is considered negative, however, because the volume is decreasing as a result of work being done on the gas.

If during the expansion the pressure of the gas changes, the calculation becomes somewhat more complicated. Let us consider a situation in which sufficient heat is supplied to keep the temperature constant during an expansion. For a dry gas, the relationship between pressure and volume is given by Boyle's law

$$pV = \text{constant}$$

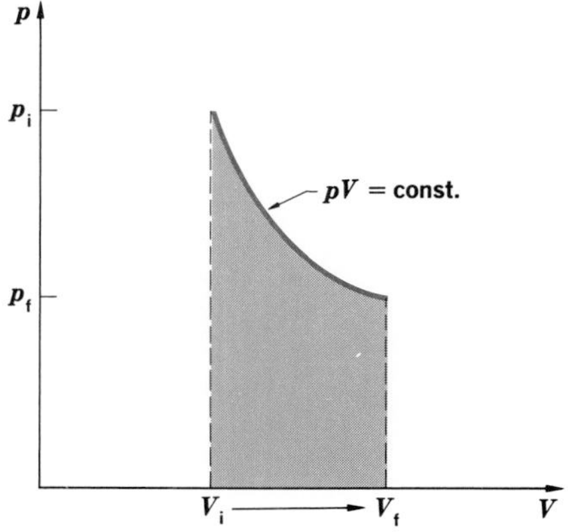

Figure 9-6. Graph of work done during the expansion of a gas kept at constant temperature.

This relationship can be shown graphically, as in Figure 9-6. The area under the pV curve between V_f and V_i represents the work involved. This work can be computed by dividing the total volume change into a number of very small volume changes. Each of these volume changes must be so small that the pressure remains practically constant for the change. The work associated with each minute change in volume is determined by multiplying the change in volume by the pressure at the volume at which the change takes place. The sum of all these quantities of work equals the total work involved. *A process that takes place at constant temperature is known as an* ***isothermal process.***

Isothermal: constant temperature

If the volume of an ideal gas increases at constant temperature, its pressure decreases. Similarly, if the volume of an ideal gas decreases at constant temperature, its pressure increases. Since the internal energy of an ideal gas determines the temperature, there can be no change in the internal energy of an ideal gas during isothermal processes. However, work is done by a gas during expansion. Since the volume occupied by the gas becomes greater, work is done by the gas molecules against external pressure. (In a real gas, some work is also done on the gas molecules, increasing their potential energy by moving them farther apart.) The heat equivalent of the work done isothermally *by* an ideal gas during expansion *must be absorbed from its surroundings.* In a similar manner, work must be done *on* an ideal gas during isothermal compression and the equivalent amount of heat *must be transferred to its surroundings.*

A sealed plastic bag containing air will increase in volume as the atmospheric pressure drops during an approaching storm, even if the room temperature remains constant. The air in the plastic bag pushes out against the atmosphere, yet there are no temperature changes.

9.4 Adiabatic Expansion

During an adiabatic expansion of an ideal gas, work is done just as it is during an isothermal expansion. Now, however, the equivalent amount of heat is not withdrawn from the surroundings but is obtained at the expense of the thermal energy of the gas. Thus during adiabatic expansion the temperature as well as the pressure of the ideal gas is lowered. Similarly, when work is done on a gas as it is adiabatically compressed, the heat equivalent of the work is not lost to the surroundings but is used to increase the internal energy of the gas. Hence, during adiabatic compression both the pressure and the temperature of the gas increase.

Adiabatic processes do occur in fast compressions when there is not enough time for the resulting heat to escape. An example of this process is the diesel engine, which will be described later in this chapter.

9.5 Specific Heats of Gases While the specific heat of a solid or liquid is a fixed value to several significant figures for the substance at a given temperature, this is not true for gases. When heat is added to a mass of gas, its pressure, volume, or both may change, and its temperature may rise.

When the volume of a unit mass of gas is held constant as heat is added, the quantity of heat required to change the temperature of this unit mass of gas by 1 C° is called its *specific heat at constant volume,* c_V.

When the pressure of the same unit mass of gas is held constant as heat is applied to raise its temperature 1 C°, the quantity of heat required is called its *specific heat at constant pressure,* c_p.

The numerical value of c_V differs from that of c_p even though the same mass of the same gas is taken through the same temperature interval. The internal energy of the unit mass of gas must be changed by the same amount in each instance. Where the pressure is held constant, added energy must be provided to do the work required to produce the volume increase. That is, work must be done on the movable part of the gas container or the surrounding atmosphere. This added work equals the pressure times the change in volume. As a result c_p is greater than c_V. In addition to increasing the internal energy, as evidenced by the increased temperature of the gas, some of the added heat energy is transformed into mechanical energy as a result of the expansion.

Refer to Section 8.12 for a definition of specific heat.

The same situation exists for liquids and solids. However, the amount of expansion is so small that the two specific heats are numerically the same to the number of significant figures with which we work, even in most research situations.

The values of c_p and c_V for some common gases at room temperature and atmospheric pressure are given in Table 9-1.

Table 9-1
SPECIFIC HEATS OF GASES

Gas	c_p $\left(\frac{J}{g \cdot C^\circ}\right)$	c_V $\left(\frac{J}{g \cdot C^\circ}\right)$
air	1.01	0.725
ammonia	2.19	1.67
carbon dioxide	0.838	0.645
hydrogen	14.2	10.1
nitrogen	1.04	0.737
oxygen	0.913	0.654

QUESTIONS: GROUP A

1. Define thermodynamics.
2. Describe Joule's experiment for determining the relationship between mechanical energy and heat.
3. (a) Make a flow chart of the energy changes in Joule's experiment. (b) How did Joule's experiment show that heat is a form of energy?
4. (a) What is the mechanical equivalent of heat? (b) What is its value?
5. (a) Give the two statements for the first law of thermodynamics. (b) How is it related to the law of conservation of mechanical energy?
6. (a) What happens to the internal energy of a substance during an adia-

batic process? (b) What happens to the substance's temperature?

7. How is the work done on a gas calculated?
8. (a) What is an isothermal process? (b) What happens in an isothermal expansion?
9. (a) How could you use a pressure-volume graph of a gas to determine the amount of work done by the expanding gas? (b) What would be different in your calculations if the gas was being compressed rather than expanding?
10. (a) Does Figure 9-6 illustrate an isothermal or an adiabatic process? (b) What is the shape of the curve pV = constant?

GROUP B

11. What physical law(s) governing the behavior of gases applies to (a) the adiabatic expansion and compression of an ideal gas and (b) the isothermal expansion and compression?
12. (a) What are the two types of specific heat for gases? (b) Describe each one.
13. (a) Which type of specific heat of gases is greater? (b) Why?
14. Why is there only one value for the specific heat of a solid?
15. Give an example of (a) an adiabatic expansion of a gas and (b) an isothermal expansion.
16. When an ideal gas expands isothermally, it does work on its surroundings. Describe the changes in energy.

PROBLEMS: GROUP A

1. How many joules of work must be done to produce 6500 calories of heat?
2. An apple contains about 80.0 kcal of chemical potential energy. If it were possible, how many joules of work could be produced?
3. How much work is done by a gas when the pressure is kept at 35.0 Pa while the volume goes from 5.00 to 7.50 m^3?
4. A 1350-kg car moving at a speed of 45.0 km/h is brought to a stop. If all the energy needed to stop the car were changed to heat, how many joules would this be?
5. A 150-kg wooden box is pushed a distance of 35.0 m across a horizontal wooden floor at constant velocity. How much heat is developed?

GROUP B

6. The water going over Niagara Falls drops 50.6 m. (a) Trace the energy changes of the water. (b) If all the potential energy at the top becomes heat, how much warmer is the water at the bottom?
7. The natural gas burned in a gas turbine has a heating value of 1.80×10^6 J/g. If 2.00 g of gas is burned in the turbine each second and its efficiency is 38%, calculate (a) the work input each second and (b) the power output in kilowatts.
8. What must be the speed in m/s of a snowball at a temperature of 0°C if it is completely melted when it strikes a stone wall?
9. A worker is driving 0.500-kg railroad spikes by striking them with a 2.50-kg sledge hammer. If the hammer hits each spike with a speed of 65.0 m/s and one third of the energy becomes heat, what is the increase in temperature of the iron hammer and the iron spike?

PHYSICS ACTIVITY

Strike a large nail repeatedly with a heavy hammer. Feel the nail. What has happened to its temperature? What energy transformations have taken place?

HEAT TRANSFER MECHANISMS

9.6 Efficiency of Ideal Heat Engines A device that converts heat energy into mechanical energy is called a ***heat engine.*** In studying heat engines, we shall use a theoretical model to represent the actual processes that take place in such engines. The model will include the necessary transformation of heat and work. We need not be concerned here with the fact that there are many different types of heat engines utilizing a variety of working substances such as steam, a mixture of fuel and air, or a mixture of fuel and oxygen.

In an engine, the working substance is taken through a series of operations known as a *cycle.* The result is that some of the heat supplied to the substance from a high-temperature source is converted into work that is delivered to an external object. For example, high-temperature steam drives a turbine that in turn does work on an electric generator. Experimental and theoretical evidence indicates that not all the heat supplied to the engine can be converted into work. The heat that is not converted into work is delivered by the engine to some external reservoir at lower temperature. Such a reservoir is called a *heat sink.* It is a system that absorbs the exhausted heat, preferably without a significant increase in its own temperature.

A heat sink is the opposite of a heat source.

In an ideal heat engine, the working substance is a gas that is returned at the end of the cycle to its original pressure, volume, and temperature conditions. There is no permanent loss or gain of internal energy; the internal energy of the working substance remains unchanged.

The operation of an ideal heat engine is schematically shown in Figure 9-7. A quantity of heat, Q_1, is delivered to the engine during the beginning of a cycle. This heat comes from a high-temperature heat source. The engine performs an amount of work, W, on some outside object and exhausts an amount of heat, Q_2, to a low-temperature heat sink. Low temperature in this case means any temperature below that of the heat source, and ideally much lower. Applying the first law of thermodynamics to this cycle,

$$W = Q_1 - Q_2$$

The thermal efficiency, e, of the heat engine is defined as

$$e = \frac{\text{work done during one cycle}}{\text{heat added during one cycle}}$$

or

$$e = \frac{W}{Q_1}$$

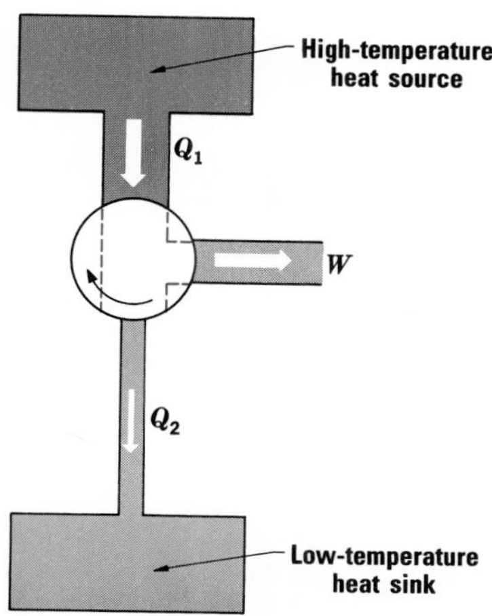

Figure 9-7. Diagram of the operation of a heat engine.

Since

$$W = Q_1 - Q_2$$

then

$$e = \frac{Q_1 - Q_2}{Q_1} = 1 - \frac{Q_2}{Q_1}$$

The efficiency of a heat engine is always less than 100%.

From this equation we see that the thermal efficiency of an operating heat engine must always be less than 100%.

It can be shown that the ratio Q_2/Q_1 is always equal to the ratio of the absolute temperatures T_2/T_1, provided that the engine is considered frictionless. Therefore we can write

$$e = 1 - \frac{T_2}{T_1}$$

From this equation we can see that the efficiency of a heat engine may be increased by making the temperature of the heat source as high as possible and the temperature of the heat sink as low as possible.

From experience we know that gasoline and steam engines expel some heat during the exhaust part of the cycle. The efficiency of any real heat engine will be less than the efficiency of an ideal heat engine because of heat loss from engine parts and friction.

It is interesting to note that the efficiency of steam engines has increased from 0.17% for the first steam engines of the seventeenth century to over 40% for the turbines used in modern power plants.

Figure 9-8. Rudolf Clausius formulated the second law of thermodynamics that relates heat transfer and differences in temperature.

9.7 Second Law of Thermodynamics It was noted in the previous section that not all the heat supplied to a heat engine can be converted into mechanical work. This fact is true for all types of heat engines and is the basis for a far-reaching generalization known as the ***second law of thermodynamics,*** which was first formulated by the German physicist Rudolf Clausius (1822–1888). *It is not possible to construct an engine whose sole effect is the extraction of heat from a heat source at a single temperature and the conversion of this heat completely into mechanical work.* Thus if a heat engine takes a quantity of heat from a heat source at a high temperature, it must transfer some of this heat to a heat sink at a lower temperature.

One interpretation of the second law of thermodynamics is that it is impossible to attain the absolute zero of temperature. There is simply no place to which heat can be transferred in order to cool a substance to exactly 0 K.

Values only a few thousandths of a degree above absolute zero have been attained. The experimental difficulties increase enormously as still lower temperatures are sought. The unattainability of absolute zero is sometimes called the third law of thermodynamics.

Some have considered extracting heat from the internal energy of the ocean and using this to operate the engines of a ship. If the water surrounding the ship is considered a heat source at a single temperature, the second law of thermodynamics indicates that it would be impossible to operate the engines of the ship from the internal energy of the sea water.

It also follows from the second law of thermodynamics that transfer of internal energy from a low-temperature heat source to a high-temperature heat sink requires work. This type of internal energy transfer takes place in refrigerators, air conditioners, and heat pumps.

Work is required to make heat flow from an object with a low temperature to an object with a higher temperature.

In order to express the second law of thermodynamics in quantitative form, we must be able to measure the amount of mechanical work a system can do. For example, consider placing a hot and a cold body in thermal contact. After a time they will reach thermal equilibrium; the two bodies will come to the same temperature. There is no loss of energy in the process, but the system as a whole loses its capacity for doing work. (The temperature of the environment is disregarded in this discussion.) A heat engine connected between the two bodies before the bodies were placed in contact could have done useful work. After contact of the two bodies, when a temperature difference no longer exists between them, work cannot be done by an engine connected between the two.

9.8 Entropy *The amount of energy that cannot be converted into mechanical work is related to* ***entropy.*** Entropy is a measurable property as important in the study of thermodynamics as is energy. As with potential energy or internal energy, it is the *difference* in entropy that is significant rather than the actual value of entropy. If an amount of heat, ΔQ, is added to a system that is at a Kelvin temperature, T, the *change in entropy*, ΔS, is

$$\Delta S = \frac{\Delta Q}{T}$$

If heat is removed from the system, the quantity ΔQ is negative and the change in entropy is also negative. Note that this equation defines the *change* in entropy and not entropy itself. The units for denoting the change in entropy are joules per kelvin.

EXAMPLE It takes 8.50×10^5 J of heat to melt a given sample of a solid. In the process, the entropy of the system rises by $200\bar{0}$ J/K. Find the melting point of the solid in °C.

Given	Unknown	Basic equations
$\Delta Q = 8.50 \times 10^5$ J	T_c	$\Delta S = \dfrac{\Delta Q}{T}$
$\Delta S = 200\bar{0}$ J/K		°C = K − 273°

Solution

$$\textit{Working equation: } T_c = \frac{\Delta Q}{\Delta S} - 273°\text{C}$$

$$= \frac{8.50 \times 10^5 \text{ J}}{2.000 \times 10^3 \text{ J/K}} - 273°\text{C}$$

$$= 152°\text{C}$$

Let us again refer to the heat transfer taking place between two bodies of different temperature that are brought into thermal contact. Assume that these bodies are insulated from their surroundings and the initial temperature, T_1, of the first body is greater than the initial temperature, T_2, of the second body. The moment they touch, a small quantity of heat, ΔQ, will be transferred from body 1 to body 2. The entropy change is $+\ \Delta Q/T_2$ for body 2 and $-\ \Delta Q/T_1$ for body 1. The net change in entropy for the system is $\Delta Q/T_2 - \Delta Q/T_1$. Since $T_1 > T_2$, the change in entropy is positive; the entropy of the system increased. This is an example of an irreversible process in which a loss of capacity for work results in an increase in entropy for the system. It can be shown that for any transformation occurring in an isolated system, the entropy of the final state can never be less than that of the initial state. Only in cases where the transformation is reversible will the system undergo no change in entropy.

The available energy of the universe is diminishing.

The *second law of thermodynamics* is ***the law of entropy:*** *A natural process always takes place in such a direction as to increase the entropy of the universe. In the case of an isolated system, it is the entropy of the system that tends to increase.*

All natural processes are irreversible and involve increases in entropy. Thus the second law of thermodynamics is equivalent to the statement that the entropy of the universe is increasing, and the first law of thermodynamics is equivalent to the statement that the total energy of the universe is constant.

Heat can sometimes be considered disordered energy. An example of ordered energy is a flying rifle bullet that has kinetic energy. When the bullet is stopped suddenly by impact, the energy of its motion is transformed to random motion of atoms. This disordered energy is evidenced by the heating of both the bullet and the area of impact.

Work involves orderly motion. But when work is done against friction and is dissipated into internal energy, the disorderly motion of molecules is increased. There is an accompanying increase in disorder.

There are many examples in nature where energy processes are observed to go toward a state of greater disorder. The melting of an isolated crystalline solid is an example. When heat is added, the system goes from a well-ordered array of molecules in a crystal to a less well-ordered array of molecules in a liquid without a change in the temperature of the sample. Entropy may thus be thought of as a measure of the order-disorder in the system.

Figure 9-9. As heat flows from the air and water into the ice, the ice melts and the entropy of the system increases.

The concept of entropy has important implications about the available energy resources of the universe and the ways in which we use them. The first law of thermodynamics states that the total amount of energy in the universe does not change, but the second law reminds us that the availability of that energy constantly decreases. An example of this is the conversion of the hydrogen fuel in the sun into disordered heat and other radiations. On the earth, some of this solar energy drives biological processes and can be used in heat engines before it is "lost."

9.9 Steam Engines The earliest known devices in which heat is converted to work were described by Hero of Alexandria in the first century A.D. In the so-called Hero's fountain, heated air was used to expel water vertically out of a container. In another device, known as the aeolipile, steam generated in a boiler passed through tubes into a sphere and out through nozzles that were mounted tangentially to the sphere, as shown in Figure 9-10. The reaction force of the steam leaving the nozzles made the sphere rotate. Neither of these devices had much practical value, however.

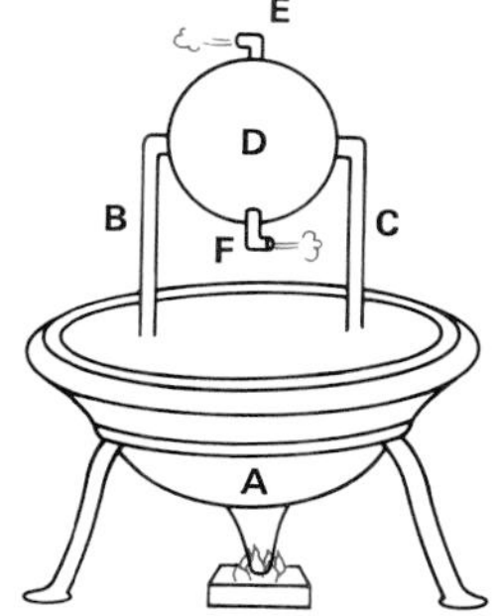

Figure 9-10. Hero's aeolipile. Steam produced in the boiler (A) passed through the hollow tubes (B and C), into the sphere (D), and out through the nozzles (E and F). The reaction force of the escaping steam made the sphere rotate.

The first practical steam engine was built in England in 1712 by Thomas Newcomen (1663–1729). In the design of the Newcomen engine, steam was directed against a movable piston inside a cylinder. A rod that was connected to the moving piston activated a water pump or some

It was noted in Section 6.5 that the metric unit of power is named in honor of James Watt.

other mechanical device. James Watt and others made important improvements on the steam engine, which became a major impetus for the Industrial Revolution.

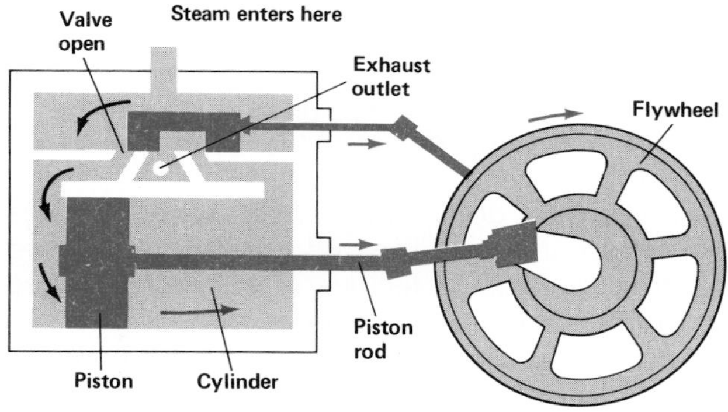

Figure 9-11. A double-acting steam engine. The sliding valve above the cylinder directs the steam first to one side and then the other side of the piston.

Figure 9-11 is a cross-sectional view of a double-acting steam engine. As the name implies, steam enters the cylinder through a sliding valve and pushes the piston first in one direction and then in the opposite direction.

The pascal is named for the French mathematician Blaise Pascal (1623–1662).

The SI unit of pressure is the pascal (Pa), which is equal to 1 N/m^2. Normal atmospheric pressure is about 10 kPa. The pressure in a typical steam engine is about 15 kPa. The efficiency of the steam engine can be increased by arranging several cylinders in tandem, so that the steam leaving the first cylinder is used to push the piston in a second cylinder, and so on. Some compound steam engines of this type in large ocean liners have power outputs of more than 11 000 kilowatts.

Steam engines have been largely replaced by other power sources, such as the steam turbine and the gasoline engine. But a small steam engine is more efficient than a small steam turbine, and where large supplies of steam are required for other purposes, it can be economically feasible to use a steam engine to drive pumps and compressors.

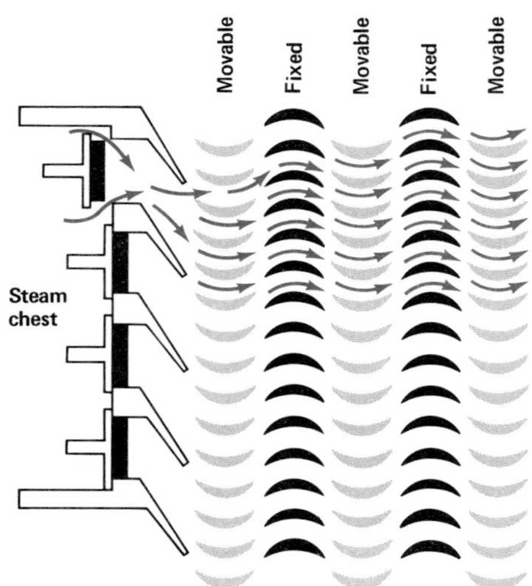

Figure 9-12. In a steam turbine, steam strikes a set of movable blades and exerts force against them. The steam is then deflected by a set of fixed blades to another set of movable blades, and so on.

9.10 Steam Turbines In one type of steam turbine, steam at a high temperature and high pressure is directed through nozzles against a set of cupped blades attached to a wheel. This arrangement is called an impulse turbine. The sideways component of the force of the steam against the blades causes the turbine wheel and the shaft on which it is mounted to rotate. Small impulse turbines may be made to move at rates of more than 10 000 revolutions per minute.

In a second type of steam turbine, high-pressure steam passes through alternating sets of movable and fixed vanes, as shown in Figure 9-12. This is called a reaction turbine. A single reaction turbine may contain forty or more pairs of movable and fixed sets of vanes. The fixed vanes are attached to the housing of the turbine, while the movable vanes are connected to a single shaft. The diameters of the turbine blades are made increasingly larger from the inlet to the outlet part of the engine, so that the energy of the expanding steam is transferred more efficiently. This arrangement can be seen in Figure 9-13.

Figure 9-13. A large steam turbine and electric generator during assembly. Notice that the turbine wheels have varying diameters.

Various combinations of steam turbines are used today for ship propulsion and for driving generators in electric power plants. In a large modern steam turbine, steam may enter at a pressure of several hundred atmospheres and a temperature of over 500°C. This obviously requires the use of special alloys that can withstand extreme conditions of temperature and pressure. As explained in Section 9.6, a large difference between the input and exhaust temperatures increases the efficiency of a heat engine. Such steam turbine-generator combinations with power outputs of more than 1 000 000 kilowatts are in use today.

9.11 Gasoline Engines Steam engines and steam turbines are examples of ***external combustion engines.*** That is, the fuel burns outside the engine and the heat is transferred to a cylinder or turbine chamber by means of steam.

In an ***internal combustion engine,*** the fuel burns inside the cylinder or engine chamber. This requires the use of a fuel that does not leave appreciable amounts of solid combustion products in the engine. Combustible vapors are therefore used in most internal combustion engines.

The gasoline engine, which is the most widely used of all heat engines, is an internal combustion engine. Figure 9-14 shows the sequence of events in a four-stroke gasoline engine. As the piston moves down, (A), air and gasoline vapor are brought into the cylinder through a valve. As the piston moves up again, (B), the valve closes and the air-vapor mixture is compressed to a fraction of its original volume. This compression causes the temperature of the mixture to rise. A spark then starts a chemical reaction between the gasoline vapor and the oxygen in the air, (C).

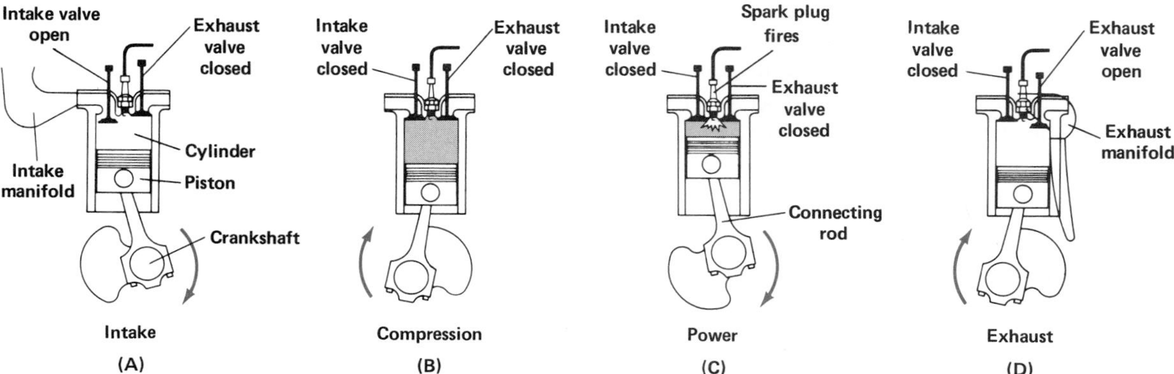

Figure 9-14. A four-stroke gasoline engine.

Heat produced by the chemical reaction increases the pressure of the gas in the cylinder. Changing some of the internal energy into mechanical energy causes the temperature of the gases to decrease, but not to a temperature as low as the temperature of the outside air. The heat that was added from the chemical reaction was partly converted to work and partly used to increase the internal energy of the gases that escape through the exhaust, (D).

The firing sequence for the cylinders in an automobile engine is not the same for each model.

The number of cylinders and their arrangement in a gasoline engine may vary, depending on the application. A lawn mower engine has only one cylinder, automobile engines have four or more cylinders, while some aircraft engines have as many as twenty-eight cylinders. In a V-8 gasoline engine, a row of four cylinders on one side of the engine is mounted opposite a second row of four cylinders, with the two rows of cylinders forming a V. This makes for compactness and freedom from vibration as the motion of the pistons is transferred to the crankshaft of the engine.

In two-stroke gasoline engines, ignition takes place during every revolution of the crankshaft and the intake and exhaust functions are carried out in a single operation. This increases the power-to-weight ratio of the engine.

The total piston displacement of an automobile engine is often given in liters.

The power of a gasoline engine depends on several factors. These include the compression ratio in the cylinder, piston size and displacement, and the amount of fuel the engine can burn in a given time period. In order to provide greater compression, especially in aircraft engines that are used at high altitudes, a turbine may be used to increase the pressure of the air before it enters the engine. This is called *turbocharging*. A turbocharged engine may have a power output as high as 2500 kilowatts.

Many types and sizes of gasoline engines have been developed for use in automobiles. One of these is called the Wankel engine, after its German inventor Felix Wankel (b. 1902). Instead of pistons, the Wankel engine has a triangular rotor that turns inside a specially shaped housing. See Figure 9-15. A Wankel engine has fewer moving parts and is quieter than a reciprocating engine. The power-to-weight ratio of a Wankel engine is also much higher than it is for a piston engine.

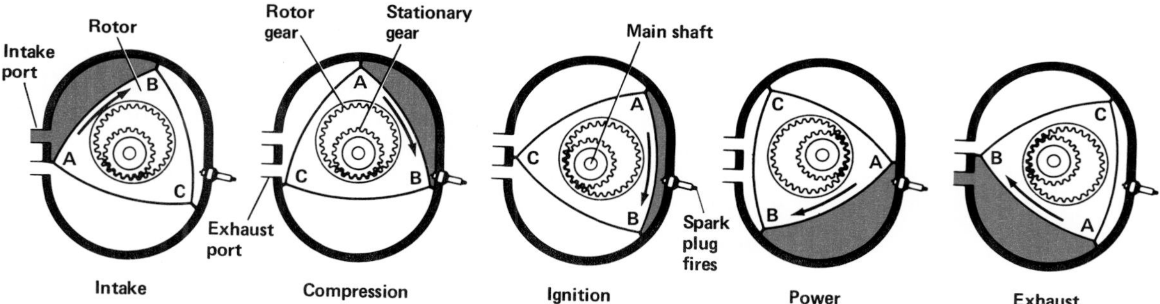

Figure 9-15. A Wankel engine.

9.12 Diesel Engines In 1895, the German inventor Rudolf Diesel (1858–1913) developed an internal combustion engine that does not require spark plugs to ignite the fuel. The heat of compression of the air inside the cylinder is sufficient to produce ignition. The fuel is injected at the time of maximum compression. The heating of the air during compression in a diesel engine is an example of an adiabatic process.

See Section 9.4.

A diesel engine has a low power-to-weight ratio, but it is more efficient and can burn a lower grade of fuel than its gasoline counterpart. The diesel engine is also known for its reliability and durability. A small, single-cylinder diesel engine may have a power output of less than five kilowatts. Larger versions are used in buses, trucks, tractors, and locomotives. A diesel engine used for ship propulsion or in an electric-generating plant may have twelve cylinders, a piston diameter of more than seventy centimeters, a stroke of over one hundred centimeters, and develop 7500 kilowatts.

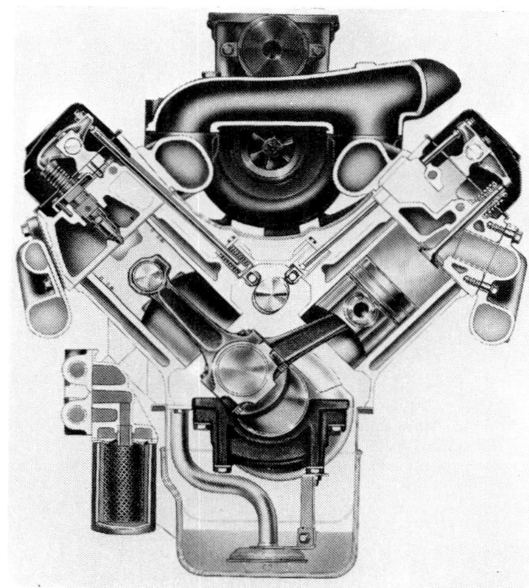

Figure 9-16. A diesel engine.

9.13 Gas Turbines The gas turbine is another type of internal combustion engine. In a gas turbine, a large volume of air is continually drawn into a compressor, where its pressure is increased. This compressed air then flows into a combustion chamber, where the fuel is injected and burns with a continuous flame. Only part of the compressed air is needed for complete combustion of the fuel. From the combustion chamber the hot combustion gases and heated air expand and move at great speed, first through a turbine and then through the exhaust nozzle.

Some of the energy imparted to the turbine is used to drive the compressor. The remaining energy may be taken from the shaft of the engine to run an electric generator or some other machine.

The efficiency of a gas turbine is rather low, partly because the temperature is not as high inside the engine as it is in a reciprocating engine. Very high temperatures would cause deterioration in the turbine blades. A gas turbine is also less efficient than a steam turbine, since more work is required to compress the air in a gas turbine than to pump the water needed for the steam in a steam turbine.

A gas turbine is most efficient at high velocities and under maximum load. Gas turbines are presently used to power fast ships and trains, to pump gas through long-distance pipelines, and to generate electricity. Used in conjunction with jet engines (which are described in the next section), gas turbines also power all but the lightest and slowest of today's aircraft.

Gas turbine automobiles are still in the developmental stage.

A practical gas turbine automobile engine still awaits development. Engineers estimate that good fuel efficiency will not be achieved in such an engine below an internal temperature of about 1350°C. The metal components of an engine lose strength at about 1050°C. Poor acceleration is another drawback of the gas turbine as a car engine.

While natural gas is the best fuel for a gas turbine, other fuels can also be used. In one test-model car, powdered coal is used as a fuel. Compressed air pushes the powdered coal into the turbine from a fuel tank over a front wheel. The fuel costs of this engine are comparatively low, but the emissions have not yet been reduced to acceptable levels.

9.14 Jet Engines Hero's aeolipile (Figure 9-10) was the first jet engine. As the steam leaves the nozzles of the aeolipile, the reaction force turns the sphere on which the nozzles are mounted. The reaction force in a jet engine is called *thrust*. (Thrust is *not* produced, as is often imagined, by the push of the escaping gases against the surrounding atmosphere.)

The simplest type of modern jet engine is the ramjet. A ramjet has no moving parts. See Figure 9-17. Air is forced at high pressure into the combustion chamber, where fuel is added and burns. The hot combustion gases and heated air rushing out through the jet nozzle produce the unbalanced reaction force that moves the engine forward.

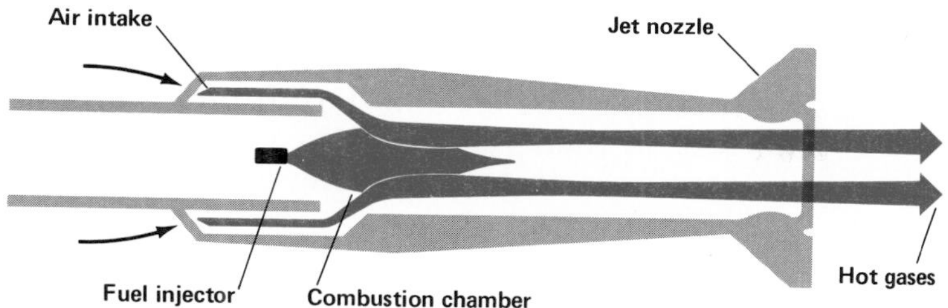

Figure 9-17. A ramjet. This engine has no moving parts and must be brought up to a certain speed before ignition.

Since a ramjet depends on its speed to compress the air necessary for combustion, it must have a high speed to operate efficiently. Therefore a ramjet must be boosted to the proper speed for ramjet operation. Rockets (which are described in the next section) are usually used for this purpose. Ramjets are used today mainly to propel guided missiles and are operated at speeds of up to five times the speed of sound.

The speed of sound is called Mach 1, in honor of the Austrian physicist Ernst Mach (1838–1916).

A turbojet engine is, as the name implies, a combination gas turbine and jet engine. As shown in Figure 9-18, the turbine turns a compressor at the front end of the engine. The compressor draws air into the engine and compresses the air before it reaches the combustion chamber.

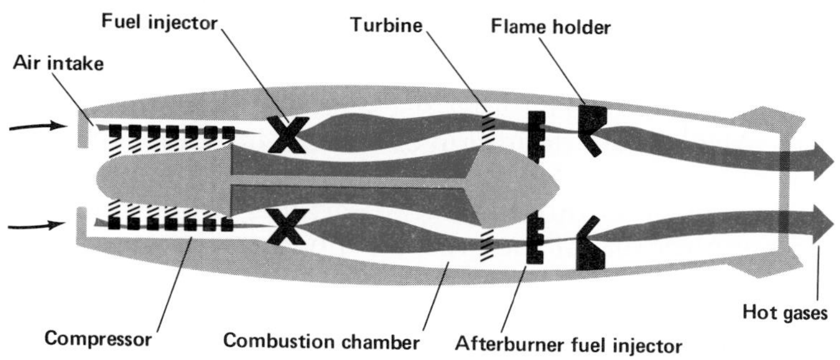

Figure 9-18. A turbojet engine. Part of the energy of the escaping gases is used to turn a turbine which, in turn, operates a compressor at the front of the engine.

Fuel, usually kerosene, is injected into the air stream just behind the compressor. Here it burns very rapidly at a high temperature, producing great pressure. The hot, high-pressure gases pass through the turbine that drives the compressor, and the unbalanced reaction force exerted by the gases as they escape through the nozzle pushes the engine forward at great speed.

The shaft in a turbojet engine rotates at about 12 000 revolutions per minute and internal temperatures can be as high as 1000°C. The amount of thrust depends on the mass and velocity of the exhaust gases. A medium-sized turbojet engine may use as much as 4000 kilograms of air and 70 liters of fuel per minute! The exhaust velocity is about 6000 meters per second. Each engine on a large jet aircraft may produce 180 000 newtons of thrust.

Auxiliary power must be used to start turning a turbojet engine. The fuel is then ignited by a spark plug. The fuel mixture continues to burn as long as the engine is running.

A rocket engine carries an oxidizer as well as fuel.

9.15 Rockets A rocket operates on the same principle as a jet engine. However, a rocket carries both fuel and the oxidizer needed to burn it, while a jet engine carries only fuel and uses oxygen from the air as the oxidizer. Thus a rocket can go beyond the earth's atmosphere. Rockets can move better through outer space than through air, because there is no friction.

Rocket engines can be classified as solid-propellant or liquid-propellant engines. The gunpowder rockets made by the Chinese in the thirteenth century were the first solid-propellant rockets. The first successful launch of a liquid-propellant rocket was made in 1926 by the American physicist Robert Goddard (1882–1945).

The performance of a rocket propellant is described in terms of *specific impulse.* ***Specific impulse*** *is the product of a propellant's thrust-to-weight ratio and its burning time.* Specific impulse is measured in seconds. A propellant consisting of liquid hydrogen and liquid oxygen has a specific impulse of 335 seconds. In general, as the temperature in the rocket chamber rises, so does the specific impulse. Also, as the definition of specific impulse suggests, light propellants like hydrogen-oxygen generally have higher specific impulses than do heavier propellants.

To get a rocket off the ground, the thrust of the propellant must exceed the total weight of the vehicle on which the rocket is mounted. As mentioned in the previous section, thrust depends in part on the velocity of the exhaust gases. In a typical rocket, exhaust velocities of 21 000 meters per second and internal pressures of 35 kilopascals are produced.

Figure 9-19. Saturn V, the space vehicle that carried the first explorers to the moon. The rockets in the first stage of this vehicle were the most powerful ever built.

The most powerful rocket ever built was for Saturn V, which was used for the manned flights to the moon (Figure 9-19). The cluster of five engines in the first stage of Saturn V had a combined thrust of over 33 million newtons. The entire spacecraft was 132 meters tall and its fueled weight on the ground was slightly over 28 million newtons. The kerosene and liquid oxygen in the first stage burned at a rate of 14 100 liters per second for a total of 2.5 minutes. The maximum velocity attained by the spacecraft after the firing of its three stages was about eleven kilometers per second.

The Space Shuttle has three liquid-propellant engines that burn liquid hydrogen and liquid oxygen. At the time of launch, two solid-propellant booster rockets are also used. Each of these boosters weighs 5 750 000 newtons and has 12 900 000 newtons of thrust. The propellant is a mixture of powdered aluminum fuel and aluminum perchlorate oxidizer. The Shuttle boosters are the largest solid-propellant rockets ever built. Unlike previous space vehicles, the Space Shuttle and its booster chambers are reusable.

See Figure 6-16.

9.16 Heat Pumps A *heat pump* transfers heat from a low-temperature source to a high-temperature sink. This "uphill" flow of heat energy requires work, which is usually supplied in a modern heat pump by an electric motor. The principle of the heat pump was first proposed by Lord Kelvin in 1852, but it was not widely applied until electric refrigerators came into general use in the 1930's.

The fluid in a heat pump is a vapor that is easily condensed to a liquid when pressure is applied. The liquid, under pressure, gives up the heat developed during compression to the heat sink. The liquid is then released into a low-pressure zone where it is quickly evaporated, taking its heat of vaporization from the heat source. The vapor is then recompressed and the cycle is repeated.

In the winter, a house to be heated by a heat pump is made the high-temperature sink, as shown in Figure 9-20(A). The outside air serves as the low-temperature source. In the summer, Figure 9-20(B), these conditions

Figure 9-20. A heat pump. This heat transfer mechanism can be used to heat a home in the winter and cool it in the summer.

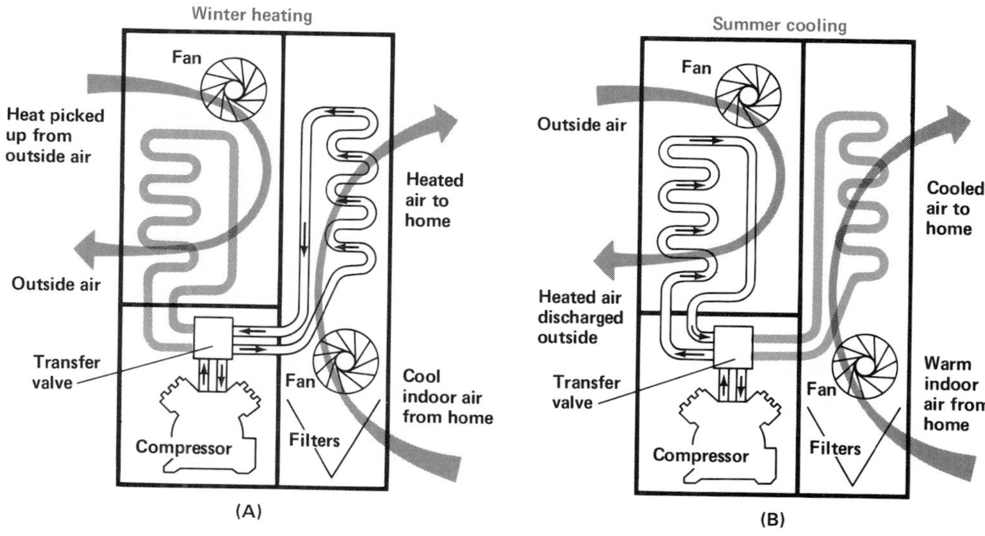

are reversed. The house is the low-temperature source and the outside air is the high-temperature sink.

Unlike other heating systems, a heat pump does not *generate* heat, but simply moves heat energy from one place to another. This makes the heat pump much more efficient than other electric heating systems and gas or oil furnaces. However, a heat pump may have to be augmented by some other heating device when the outside temperature drops to very low temperatures.

A refrigerator is a heat pump in which the freezing compartment is the low-temperature heat source and the air in the room is the high-temperature sink. Consequently, a refrigerator heats the room while at the same time cooling the food that is stored in the refrigerator.

QUESTIONS: GROUP A

1. (a) What is a heat source? A heat sink? (b) How are they related in an operating heat engine?
2. What is an ideal heat engine?
3. (a) How can you increase the efficiency of a heat engine? (b) Why is the efficiency of a heat engine always less than 100%?
4. (a) What is the second law of thermodynamics? (b) Is there a second way of describing it? Explain.
5. Describe heat in two ways.
6. Describe the SI unit of pressure.
7. What is the difference between an impulse turbine and a reaction turbine?
8. Describe the process of turbocharging.
9. (a) How do internal and external combustion engines differ? (b) Give an example of each.

GROUP B

10. What are the advantages and disadvantages of the diesel engine?
11. What can be done to increase the efficiency of a gas turbine?
12. How can an engine operate by moving heat from a low-temperature source to a high-temperature sink?
13. If the *reaction* force in a jet engine is the thrust, according to Newton's third law, what is the *action* force?
14. Is the compression/exhaust cycle an adiabatic or isothermal process? Explain.
15. What are some advantages and disadvantages of a steam engine as compared to a steam turbine?
16. Why is heat referred to as "disordered" energy?
17. Explain the following statement in light of the law of conservation of energy: "The available energy of the universe is diminishing."

PROBLEMS: GROUP A

1. What is the theoretical highest efficiency of a steam engine with an input temperature of 200.0°C and an exhaust temperature of 100.0°C?
2. Referring to Problem 1, assume the temperature of the heat sink to remain at 100.0°C. What input temperature would be required to increase the efficiency of this engine to 30.0%?
3. Determine the thrust-to-weight ratio for the space shuttle booster.
4. Using the data in Section 9.15, calculate the specific impulse of the Saturn V rocket.

SUMMARY

The study of the quantitative relationship between heat and other forms of energy is called thermodynamics. The first law of thermodynamics states that the quantity of heat supplied to a system is equal to the work done by the system plus its change in internal energy. The expansion of a gas may be an adiabatic or an isothermal process. Adiabatic processes are those in which no heat is added to or removed from a substance. Isothermal processes take place at constant temperatures. The specific heat of a gas varies with temperature.

The efficiency of an ideal heat engine increases as the difference between the temperatures of the heat source and heat sink increases. The second law of thermodynamics states that it is impossible to make an engine that will extract heat from a heat source at a single temperature and convert it completely to work.

The amount of energy that cannot be converted into mechanical work is related to entropy. The change in entropy of a system is equal to the amount of heat added to the system divided by its Kelvin temperature. The entropy of a system always tends to increase.

A heat engine converts heat energy into mechanical energy. Steam engines and steam turbines are external combustion engines. Gasoline engines, diesel engines, gas turbines, jet engines, and rockets are internal combustion engines. Unlike other engines, a rocket contains an oxidizer as well as fuel. The performance of rocket propellants is measured in terms of specific impulse. A heat pump transfers heat from a low-temperature source to a high-temperature sink. A heat pump is more efficient than most other heating systems.

VOCABULARY

adiabatic process
aeolipile
diesel engine
entropy
external combustion engine
first law of thermodynamics
gas turbine
gasoline engine
heat engine
heat pump
heat sink
heat source
internal combustion engine
internal energy
isothermal process
jet engine
mechanical equivalent of heat
ramjet
rocket
second law of thermodynamics
specific impulse
steam engine
steam turbine
thermodynamics
turbocharging
turbojet
Wankel engine

10

Waves

wave n.: a disturbance that propagates through a medium or space.

OBJECTIVES

- Describe energy transfer by waves.
- Define and describe characteristics of periodic waves.
- Define and differentiate among rectilinear propagation, reflection, refraction, and diffraction and interference of waves.
- Apply the superposition principle to wave trains.
- Describe the conditions necessary to produce standing waves.

THE NATURE OF WAVES

10.1 Energy Transfer Energy sources are very often far removed from the places where the energy is most needed. Clearly, some mechanism must be provided for the transport of energy from one place to another. One way of transporting energy is by the movement of materials or objects. Winds and projectiles in flight are well-known examples. When a baseball strikes a window, the glass is shattered by energy transferred from the ball during impact. In a hot air or hot water heating system, the heat energy is transferred from the furnace to the rooms of the house by the movement of a gas or a liquid. Even in carrying your textbook from your locker to your desk, the book's potential energy relative to the earth is transported from one location to another.

A more interesting but more complicated way of transporting energy involves *waves.* Several natural phenomena that typify waves can be recognized. These phenomena are studied collectively because of an important simplifying fact: the ideas and language used to describe waves are the same, regardless of the kinds of waves involved. The basic concept in the use of the term *wave* is that the wave involves some quantity or disturbance that *changes in magnitude with respect to time at a given location* and *changes in magnitude from place to place at a given time.* We shall see that some wave disturbances occur only in material media and others are not restricted in this way. In general, *a* ***wave*** *is a disturbance that propagates through a* *medium or space.* All kinds of waves are characterized by transfer of energy without the large-scale transfer of matter.

A stone, or a single drop of water, falling into a quiet pond produces the familiar wave pattern on the surface of the water shown in Figure 10-1. A sound is heard because a wave travels from the source through the intervening atmosphere. The energy released by a great explosion can shatter windows far from its source because a wave of compression moves out from the source in all directions. The "shock wave" of a sonic boom can have similar destructive effects. *These waves are disturbances that move through a material medium.*

(A)

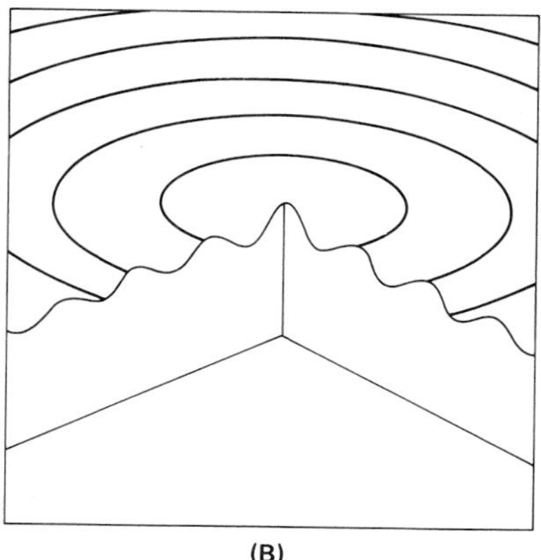

(B)

Figure 10-1. (A) A periodic circular wave train is generated by a drop of water falling on the surface of a quiet pool of water. (B) A cutaway diagram illustrates the succession of crests and troughs of the surface wave train as it expands outward from the point of disturbance.

Some properties of light can be explained by means of waves. Physicists have demonstrated that light waves, radio waves, infrared and ultraviolet waves, X rays, and gamma rays are fundamentally similar. They are *electromagnetic waves.* Although they can travel through matter, their transmission from one place to another does not require a material medium, that is, a medium composed of matter. The nature of electromagnetic waves will be discussed in detail in later chapters.

Figure 10-2. Shock waves are produced when air strikes an obstacle at supersonic speeds. In this wind-tunnel photo, the shock waves are visible as thin lines of high-density air.

For the present (through Section 10.5) we shall be concerned with waves that require a *material medium* for their propagation. These waves are called *mechanical waves.* If the motion responsible for the wave disturbance is *periodic,* a periodic *continuous wave,* or *wave train,* is produced. Such wave motion is related to harmonic motion. Harmonic motion is described in Section 5.10 in terms of a single *object* vibrating about its equilibrium position. Here, we must consider many *particles* vibrating about their respective equilibrium positions as the wave train travels through the medium. It is important to understand the behavior of waves because waves are very common and because the ideas of wave motion are involved in the study of subatomic particles.

As particles at some distance away from a source of vibrational energy are made to vibrate, their vibrations show that they have acquired energy. This energy has been transmitted to particles far from the energy source by the wave disturbance. Thus it is evident that waves provide a mechanism by which energy is transmitted from one location to another *without the physical transfer of matter between these locations.*

Elastic forces give rise to simple harmonic motion, in accordance with Hooke's law (Section 7.11).

10.2 Mechanical Waves *A **mechanical wave** is a disturbance in the equilibrium positions of matter, the magnitude of which is dependent on location and on time.* To generate mechanical waves, a *source of energy* is required to cause a disturbance and an ***elastic medium*** is required to transmit the disturbance. An elastic medium behaves as if it were an array of particles connected by springs with each particle having an equilibrium position. A simple model of such a medium is shown in Figure 10-3.

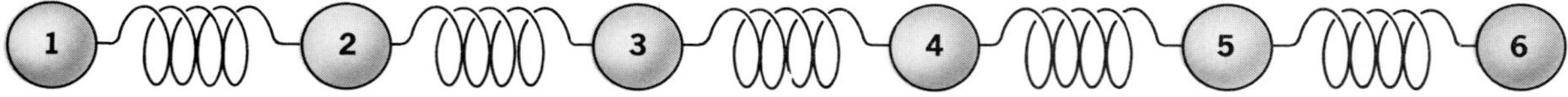

Figure 10-3. An elastic medium behaves as if it were an array of particles connected by springs, with each particle vibrating about an equilibrium position.

If particle **1** is displaced from its equilibrium position by being pulled away from particle **2,** it is immediately subjected to a force from particle **2** that attempts to restore particle **1** to its original position. At the same time, particle **1** exerts an equal but opposite force on particle **2** that attempts to displace it from its equilibrium position. Similar events occur but in opposite directions if particle **1** is displaced from its equilibrium position by being pushed toward particle **2.**

Suppose particle **1** is displaced by an energy source. Then particle **1** exerts a force that displaces particle **2.** Particle **2,** in turn being displaced from its equilibrium position, exerts a force on particle **3,** which is in turn displaced. In this way, the displacement travels along from particle to particle. Because the particles have inertia, the displacements do not all occur at the same time but successively as the disturbance affects particles farther and farther from the source. The energy initially imparted to particle **1** by the energy source is transmitted from particle to particle in the medium without motion of the medium as a whole.

10.3 Transverse Waves A long spiral spring stretched between two rigid supports can represent an elastic medium, as shown in Figure 10-4(A). A portion of the spring is displaced at point **2** to form a *crest,* or upward displacement, as in Figure 10-4(B). To force the spring into this shape, point **2** must be pulled up while points **1** and **3** are held in place. When the spring is released at points **2** and **3** simultaneously, point **2** accelerates downward and point **3** accelerates upward. The crest moves toward the right as in Figure 10-4(C). Similarly, as the spring in the region of point **3** moves downward, the spring in the region of point **4** is displaced upward. In this manner the crest travels along the spring as shown in Figure 10-4(D) and (E). A *trough,* or downward displacement, formed at **2** travels along the spring in a similar fashion.

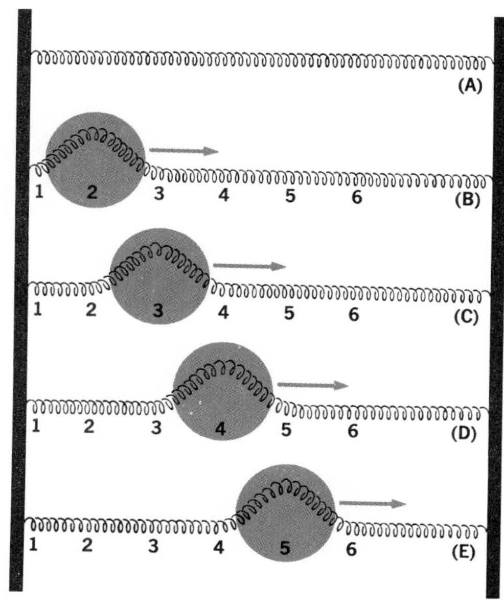

Figure 10-4. A transverse pulse traveling along an elastic medium.

A single nonrepeated disturbance such as a single crest or a single trough is called a single-wave pulse or, simply, a *pulse.* The displacement of the particles of the medium (the coiled spring) caused by the pulse is *perpendicular* to the direction in which the pulse travels. Such a pulse is said to be *transverse.* It is convenient to think of the crest as a *positive* pulse and the trough as a *negative* pulse.

If a pulse is produced in the middle of the spring, the disturbance will move in both directions.

When a periodic succession of positive and negative pulses is applied to the coiled spring, a series of crests and troughs, called a *continuous wave,* or *wave train,* travels through the medium. Again, the displacement of the particles of the medium is perpendicular to the direction the wave train travels. The wave disturbance is *transverse.* A ***transverse wave*** *is one in which the displacement of particles of the medium is perpendicular to the direction of propagation of the wave.*

10.4 Longitudinal Waves

Suppose a similar spiral spring is used, but instead of pulling the spring out of line, several coils are pinched closer together at one end, as in Figure 10-5(A). Such a distortion is called a *compression.* When these compressed coils are released, they attempt to spread out to their equilibrium positions. Thus, they compress the coils immediately to the right and the compression moves toward the right. See Figure 10-5(B)–(D).

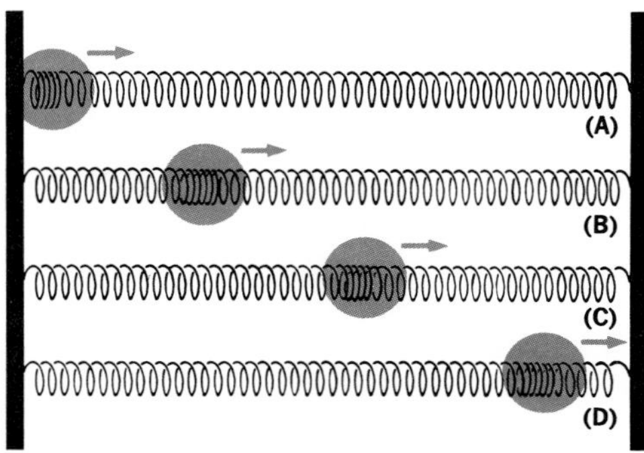

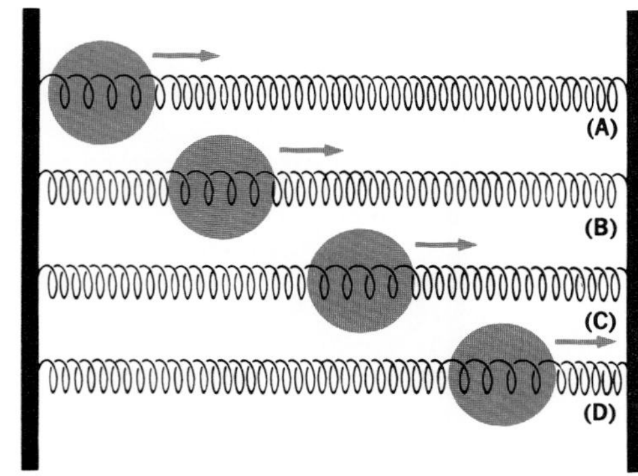

Figure 10-5. (Left) A compression pulse traveling along an elastic medium.

Figure 10-6. (Right) A rarefaction pulse traveling along an elastic medium.

Transverse displacements are across the path of propagation. Longitudinal displacements are along the path of propagation.

If the coils at the left end of the spring are stretched apart instead of compressed, a *rarefaction* is formed. When released, the rarefaction travels along the spring just as the compression did. This effect is shown in Figure 10-6.

These pulses propagate along the spring by displacing the particles of the spring in directions *parallel* to the direction the pulses are traveling. They are examples of *longitudinal pulses.* Similarly, a continuous wave disturbance of this type gives rise to *longitudinal wave motion.* Energy is transferred from particle to particle along the medium without motion of the medium as a whole. *A* ***longitudinal wave*** *is one in which the displacement of particles of the medium is parallel to the direction of propagation of the wave.*

10.5 Periodic Waves

We have examined the effect of a single, nonrecurring wave disturbance on one end of a long, rigidly mounted spring. Now let us consider what happens if similar disturbances are repeated periodically.

Suppose we attach the left end of a long spring to a mass suspended by a second spring as in Figure 10-7. Assume that the mass can move up and down without friction between its vertical guides. If the mass is pulled down

slightly and then released, it will vibrate within the guides with simple harmonic motion. Such a motion is *periodic.* That is, the mass repeats its motion once every certain time interval T, called the *period* of vibration. Since the vibrating mass is attached to the end of the long spring, it acts as a source of periodic disturbances. A *transverse wave train* is generated, which moves to the right along the spring. This wave is shown in Figure 10-7 at the instant the vibrating mass is at the upper limit of its excursion and again at the instant it is at the lower limit of its excursion.

This periodic transverse wave carries energy away from the vibrating mass. Unless energy is supplied to it, the mass loses amplitude and comes to rest. In order for a source to generate a continuous wave of uniform amplitude, energy must be supplied to the source at the same rate as the source transmits energy to the medium. Then

Figure 10-7. The oscillating mass generates a periodic transverse wave. The waves in color and the black waves represent a time difference of one-half period ($T/2$).

All simple harmonic motions are periodic; not all periodic motions are harmonic. For example, doing push-ups at a constant repetition rate is a periodic motion, but it is not simple harmonic motion.

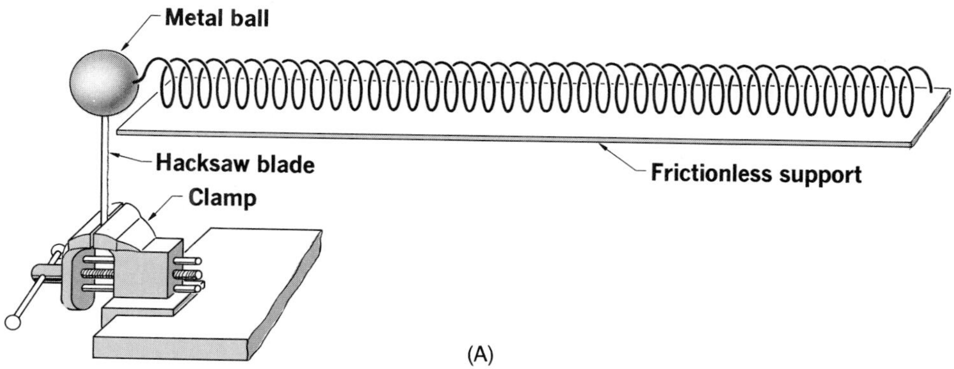

Figure 10-8. (A) An apparatus for generating longitudinal waves. (B) A periodic longitudinal wave in a section of the spring.

the successive wave disturbances will be identical. When the periodic wave is a simple harmonic wave, it gives each particle of the medium a simple harmonic motion.

A periodic longitudinal wave can be generated by the apparatus shown in Figure 10-8(A) on the preceding page. The long spring is attached to a metal ball fastened to one end of a hacksaw blade. The other end of the blade is rigidly clamped. If the metal ball is displaced slightly to one side and released, it vibrates with simple harmonic motion. This motion produces a series of periodic compressions and rarefactions in the spring, and therefore a periodic longitudinal wave. This is illustrated in Figure 10-8(B). In order for the successive wave disturbances to be identical, energy must be provided to the source at the same rate it is transmitted by the wave.

10.6 Characteristics of Waves The *characteristics* and *properties* of waves discussed in the remaining sections of this chapter are descriptive of *all* waves. An elastic medium is required for the propagation of mechanical waves, but not for that of electromagnetic waves. The propagation of electromagnetic waves in free space is dramatically illustrated by the transmission of information back to earth by space probes from the outer regions of our solar system and beyond. For the transmission of electromagnetic waves, periodic oscillations of accelerating electrons in a conductor (such as an antenna) give rise to periodically reversing electric and magnetic fields that propagate in free space away from their source at the speed of light.

The broad frequency range of the electromagnetic spectrum can be seen in Figure 12-8 of Chapter 12. Notice the locations of familiar regions of the electromagnetic spectrum such as X rays, visible light, ultraviolet light, and radio waves. The propagation of energy by means of electromagnetic waves is called *radiation*. Radiation can be detected as heat when it is absorbed. The direct radiation of heat by objects at ordinary temperatures occurs predominantly in the infrared region of the spectrum and is called *blackbody radiation*.

All waves have several common characteristics in addition to that of transferring energy without the transport of matter. As a periodic mechanical wave propagates through a medium, the particles of the medium vibrate about their equilibrium positions in identical fashion. Generally, however, the particles are in corresponding positions of their vibratory motion at different times. The *position* and *motion* of a particle indicate the *phase* of the wave. Particles that have the same displacement and are

The word "phase" has a different meaning here than it does in Chapters 7 and 8.

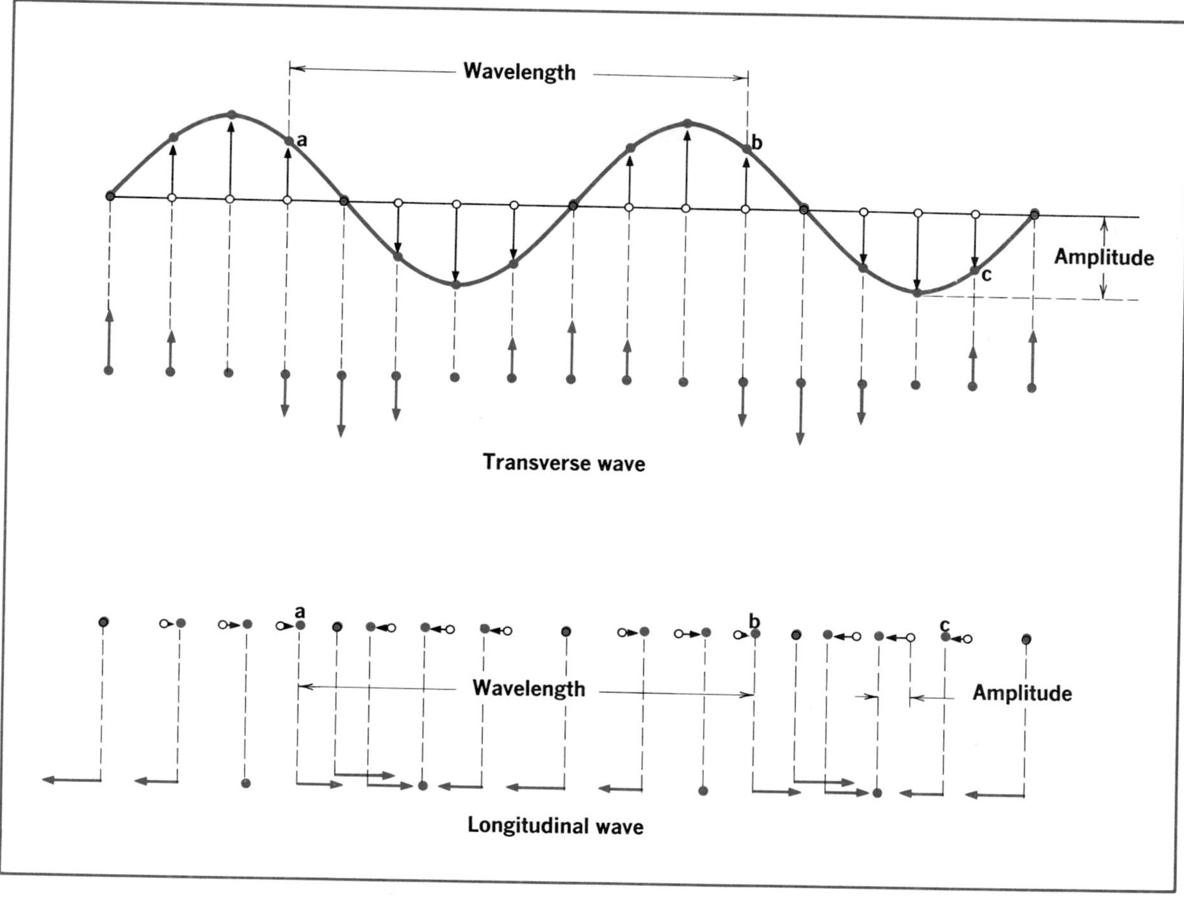

Figure 10-9. Characteristics of transverse and longitudinal waves. The black open circles show the equilibrium positions of the particles of the media. The dots in color show their displaced positions. The black arrows indicate the displacements of the particles from their equilibrium positions. The arrows in color are the velocity vectors for the particles above them.

moving in the same direction are said to be ***in phase.*** Particles **a** and **b** in both waves illustrated in Figure 10-9 are in phase. Those with opposite displacements and moving in opposite directions, such as particles **b** and **c** in both waves, are in *opposite phase,* or 180° out of phase.

The *frequency, f,* of a periodic wave is the number of crests (or troughs) passing a given point in unit time. It is the same as the frequency of the simple harmonic motion of the source and is conventionally expressed in terms of the vibrating frequency of the source. Thus the frequency of a motion is the number of vibrations, oscillations, or cycles per unit of time. The SI unit of frequency is the *hertz* (Hz). One Hz is equivalent to the expression "one cycle per second." Note that "cycle" is an event, not a unit of measure. Therefore it is not a part of the dimensional description of the hertz. The dimension of the hertz is simply s^{-1}. For example, a wave generated at 60 cycles per second has a frequency of 60 Hz, which is expressed dimensionally as 60/s.

$$\mathbf{f = 60 \text{ cycles per second} = 60 \text{ Hz} = 60/s}$$

The *period, T,* of a wave is the time between the passage of two successive crests past a given point. It is the same as the period of the simple harmonic motion of the source. Period is therefore the reciprocal of frequency.

See Section 1.8.

$$f = \frac{1}{T} \quad \text{and} \quad T = \frac{1}{f}$$

The *wavelength,* represented by the Greek letter λ (lambda), is the distance between any particle in a wave and the nearest particle that is in phase with it. The distance between particles **a** and **b** in either of the waves shown in Figure 10-9 is one wavelength. It is the distance advanced by the wave motion in one period, *T.*

An advancing wave has a finite *speed, v,* for a given transmitting medium. Wave speed may be quite slow, as that of water waves. It may be moderately fast, as that of sound waves that travel with speeds on the order of 10^2 to 10^3 m/s. It may be the speed of light or radio waves in a vacuum, 3×10^8 m/s. The speed of a wave depends primarily on the nature of the wave disturbance and on the medium through which it passes. In certain media, wave speed may also depend on wavelength.

When the speed of a wave depends on its wavelength (or frequency), the transmitting medium is said to be *dispersive.* A glass prism disperses (separates) a beam of white light into a color spectrum as the light of different wavelengths emerges along diverging paths. Glass is a dispersive medium for light waves. We can observe rainbows because water droplets in the atmosphere disperse sunlight in the same way. In fact, all electromagnetic waves propagating through matter, but not through a vacuum, show dispersion.

Rainbows are produced by the dispersion of light waves as they pass through raindrops.

Water is highly dispersive for surface waves, and the wave speed can change significantly with wavelength. Ripples produced by a very gentle wind or vibrations of high frequency have very short wavelengths. The wave speed is governed by the surface tension of the liquid, which is a measure of the tendency of the surface to resist stretching. In this case, the *shorter* the wavelength, the higher is the wave speed.

A strong steady wind may produce surface waves of very long wavelength. Here, the wave speed is governed by gravity, which provides a restoring force. Such waves are called *gravity waves*. In this case, the *longer* the wavelength, the higher is the wave speed.

Since a wave travels the distance λ in the time T for a complete vibration of the source, its speed v is given by the expression

$$v = \frac{\lambda}{T}$$

and because
$$f = \frac{1}{T}$$

then
$$v = f\lambda$$

The wave equation $v = f\lambda$ is true for all wave systems.

This equation is true for all periodic waves, whether transverse or longitudinal, no matter what the medium is.

The maximum displacement of the vibrating particles of the medium from their equilibrium positions is called the *amplitude* of the wave. It is related to the energy flow in the system.

10.7 Amplitude and Energy In all forms of traveling waves, energy is transmitted from one point to another. By expending energy, a vibrating source can cause a harmonic disturbance in a medium. The energy expended per unit of time depends on the amplitude, the frequency, and the mass of the particles of the medium at the source. Simple harmonic motion (vibrational motion) of a particle of the medium is characterized by an amount of energy, part kinetic and part potential, that the particle has at any instant. The disturbance is transferred through the medium because of the influence that a particle has on other particles adjacent to it. Thus the wave carries energy away from the source.

Assuming no losses in the system, the energy transported by the advancing wave during a given time is the same as the energy expended by the source during that time. The energy transported, or the energy expended, per unit of time is the *power* transmitted by the wave.

If the wave amplitude is doubled, the vibrational energy is increased fourfold. Doubling the vibrating frequency has the same effect. *The rate of transfer of energy, or the power transmitted by a wave system, is proportional to the square of the wave amplitude and also to the square of the wave frequency.*

Surface waves on water that emanate from a point source undergo a decrease in amplitude as they move away from their source. Each advancing wave crest is an expanding circle. The energy contained in each crest is a fixed amount, and the crest is expanding. The energy per unit length of crest must decrease. Thus the amplitude of

Figure 10-10. The collapse of the Tacoma Narrows Bridge in 1940. The vibrational energy supplied by even moderate winds of 70 km/h subjected the bridge to excessive torsional vibrations. These vibrations were amplified by the bridge's natural resonance frequencies and caused the roadway to collapse.

the wave must diminish. This effect is greater for sound waves since the advancing crests are expanding spheres.

We know from experience that the amplitude of a vibrating pendulum or spring gradually decreases with time and the vibrations eventually stop. Frictional and resistive effects that oppose the motion slowly remove energy from the vibrating system. Similarly in a wave system, energy is dissipated and the wave amplitude gradually diminishes. *The reduction in amplitude of a wave due to the dissipation of wave energy as it travels away from the source is called* ***damping.*** Damping effects may be quite small over relatively short distances.

QUESTIONS: GROUP A

1. (a) What is a wave? (b) What characteristic is common to all waves?

2. How does a mechanical wave differ from an electromagnetic wave?

3. Differentiate between a pulse and a periodic wave.

4. How is the direction of propagation related to the direction of vibration for (a) transverse wave and (b) a longitudinal wave?

5. Define (a) crest, (b) trough, (c) compression, and (d) rarefaction.

6. (a) How are the velocity, frequency, and wavelength of a wave related? (b) Does amplitude affect wave speed?

7. If you halve the frequency of a vibrating wave, what effect does this have on its period?

8. Classify the following waves as transverse or longitudinal; mechanical or electromagnetic; continuous or pulse: (a) wave caused by a pebble dropped into a pond; (b) infrared radiation from the sun; (c) microwaves in an oven; (d) compression wave caused by an explosion; (e) wave on a Slinky when the coils are pulled together and released.

9. If you call your friend with a louder voice, will your friend hear you faster? Explain.

10. How far does a wave travel in one period?

GROUP B

11. Why must a medium be "elastic" in order to transmit a mechanical wave?

12. Two tuning forks with frequencies of 256 Hz and 512 Hz are struck. Which of the sounds will move faster through the air?

13. (a) If you double the distance from a source of a water wave, what will happen to the amplitude? (b) What will happen to the amplitude of a sound wave in air under the same circumstances?

14. (a) Consider two transverse waves of equal wavelengths. If one wave is generating a crest at the same time the other is generating a trough, what is their phase relationship? (b) What if one is at equilibrium position beginning to form a crest and the other is at equilibrium position beginning to form a trough?

PROBLEMS: GROUP A

1. Your favorite radio station operates at 95.5 MHz. If the wavelength of the radio waves produced is 3.14 m, what is the speed of the wave?

2. Microwaves, with a frequency of 9.00×10^9 Hz, travel at the speed of light, 3.00×10^8 m/s. What is their wavelength?
3. (a) A tuning fork produces a sound with a frequency of 256 Hz and a wavelength of 1.35 m in air. What value does this give for the speed of sound in air? (b) What would be the wavelength of the wave produced by this same tuning fork in water, in which sound travels at an approximate speed of 1500 m/s?
4. You dip your finger into a pan of water twice each second, producing waves with crests that are separated by 0.15 m. Calculate (a) the wave frequency, (b) their period, and (c) their speed.
5. (a) Refer to Figure 10-8(B). If the distance between each compression and the adjacent rarefaction is 8.0 cm, what is the wavelength of this wave? (b) If the hacksaw blade takes 0.25 s to move back and forth once, how fast are the waves traveling along the spring?

WAVE INTERACTIONS

10.8 Properties of Waves In any study of wave behavior, the various properties common to all kinds of waves should be recognized. These common properties are: *rectilinear* (straight line) *propagation, reflection, refraction, diffraction,* and *interference.* Being familiar with the properties of waves helps to determine whether an observed phenomenon involves a wave or not.

Surface waves on water are particularly useful in studying wave properties. A surface wave is easily generated, its speed is quite slow, and the transparency of the water to light allows the wave pattern to be projected on a screen. A continuous wave can be established on the water surface by allowing a probe to dip into the surface periodically. An apparatus for observing wave motion on the surface of water is shown in Figure 10-11. It is called a ripple tank.

Figure 10-11. (Left) The ripple tank provides a means of observing the properties of waves in the laboratory.

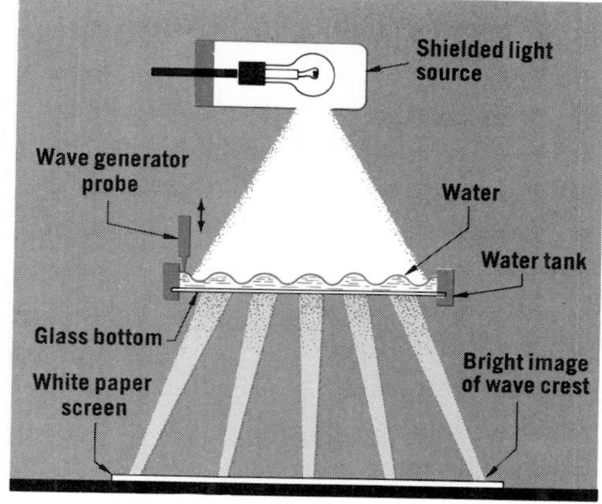

Figure 10-12. (Right) Diagram showing how crests in a ripple tank focus light onto a screen.

The convex surface of a wave crest serves to converge light.

The concave surface of a wave trough serves to diverge light.

The diagram in Figure 10-12 shows how images of surface wave crests can be projected on a screen. The ripple tank has a glass bottom that allows light to be projected down through the water and the glass onto a screen below the tank. The crests appear on the screen as bright regions, and troughs appear as dark regions.

10.9 Rectilinear Propagation We can generate a train of straight waves in a ripple tank by placing a long, straight edge, like that of a ruler, along the surface of the water and causing it to vibrate up and down. A photograph of the image of such a continuous wave is shown in Figure 10-13. The ***wave crests*** are the parallel white lines. The ***wave troughs*** are the dark spaces separating the crests. The portions of water surface in which particles are all in the same phase of motion are called ***wave fronts.*** The paths of the two points **a** and **b** show that *the direction of propagation of the advancing straight wave is perpendicular to the wave front.* This is called ***rectilinear propagation.***

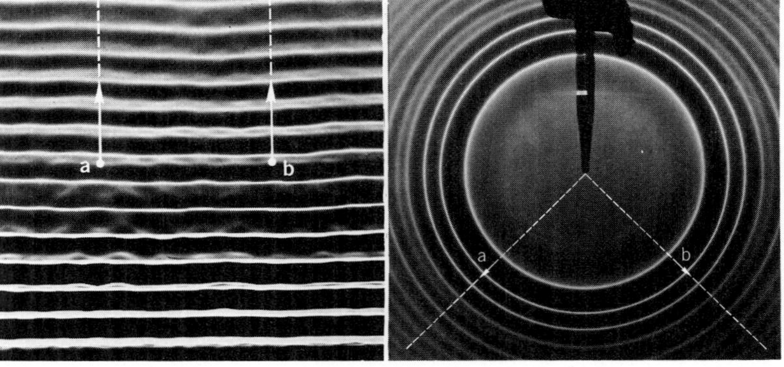

Figure 10-13. (Left) Periodic straight wave train moving across a ripple tank. The wavelength is the distance between adjacent crests or adjacent troughs.

Figure 10-14. (Right) Periodic circular wave train generated by a point source in a ripple tank. The wave train moves radially away from the disturbance point in ever-widening circles.

By changing from a straight to a pointed probe, such as the tip of a pencil, we can generate a train of circular waves. See Figure 10-14. Here the advancing wave crests are expanding circles; the wave fronts are moving out in all directions from the center of disturbance. The letters **a** and **b** locate two points on an expanding crest. Their paths show that *the directions of propagation of these two segments of the advancing circular wave lie along radial lines away from the center of disturbance.* This is true of all other points on any circular wave crest. The radial lines are perpendicular to the segments of the wave front through which they pass.

Light regions of a ripple-tank image relate to wave crests; dark regions relate to wave troughs.

From these experiments with the ripple tank we observe that *continuous waves traveling in a uniform medium propagate in straight lines perpendicular to the advancing wave fronts.* The uniform spacing between the wave crests suggests that the wave speed in the medium is constant. Considering the speed and direction of propagation of traveling waves, we can state that *the wave velocity at every point along the advancing wave front is perpendicular to the wave front.*

10.10 Reflection

Familiar examples of reflection are the sound echoes that return from a distant canyon wall, the reflection of light from a mirror, and the reflection of waves from the edge of a pool. Our image as seen in a mirror appears to be reversed, right for left. The crest of a water wave is reflected from the pool side as a crest traveling in a different direction. The echo of a sound retains the character of the original sound.

The illusion of reversal occurs because of the left-right symmetry of the body.

About the only inference we can draw from these casual observations is that *a wave is turned back, or* ***reflected,*** *when it encounters a barrier that is the boundary of the medium in which the wave is traveling.* A close examination of the boundary conditions can reveal much more about the nature of reflection.

Suppose we generate a single, straight pulse in a ripple tank that has a straight barrier placed across the tray parallel to the advancing wave front. When the pulse reaches the barrier, it is reflected back in the direction from which it came. No disturbance appears in the water behind the barrier. The paths of the incident pulse (approaching the barrier) and the reflected pulse lie perpendicular to the surface of the barrier.

The angle formed by the path of incidence and the perpendicular (normal) to the reflecting surface at the point of incidence is called the ***angle of incidence,*** i. The angle formed by the path of reflection and the normal is the ***angle of reflection,*** r. In this instance both the incident and reflected angles are 0°, and i is equal to r.

Certainly i and r equal 0° when the incident wave approaches the barrier along a line perpendicular to it. We may then ask whether the equality between i and r is coincidental in this situation, or whether it is characteristic of reflection in general. To investigate the relationship of i and r further, we need to change the position of the barrier so that it is no longer parallel to the wave fronts and then send more pulses against it. Figure 10-15 shows the reflection of a single pulse from a diagonal barrier.

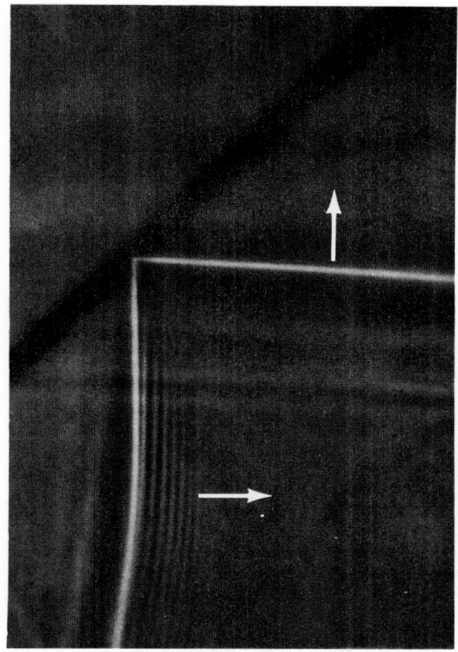

Figure 10-15. The image of a single, straight pulse being reflected from a diagonal barrier in a ripple tank. The incident segment of the pulse is moving toward the top of the picture. The reflected segment is moving toward the right.

Reasonably good measurements can be made of the angles that the incident and reflected pulses make with the barrier by placing suitable markers on the screen parallel to the projected images of the pulse. We shall call these angles i' and r' respectively. Measurements of i' and r' for several positions consistently indicate that these angles

Figure 10-16. The geometry of reflection. A straight pulse is reflected from a straight barrier in a ripple tank. The angle of incidence, *i*, is shown to be equal to the angle of reflection, *r*.

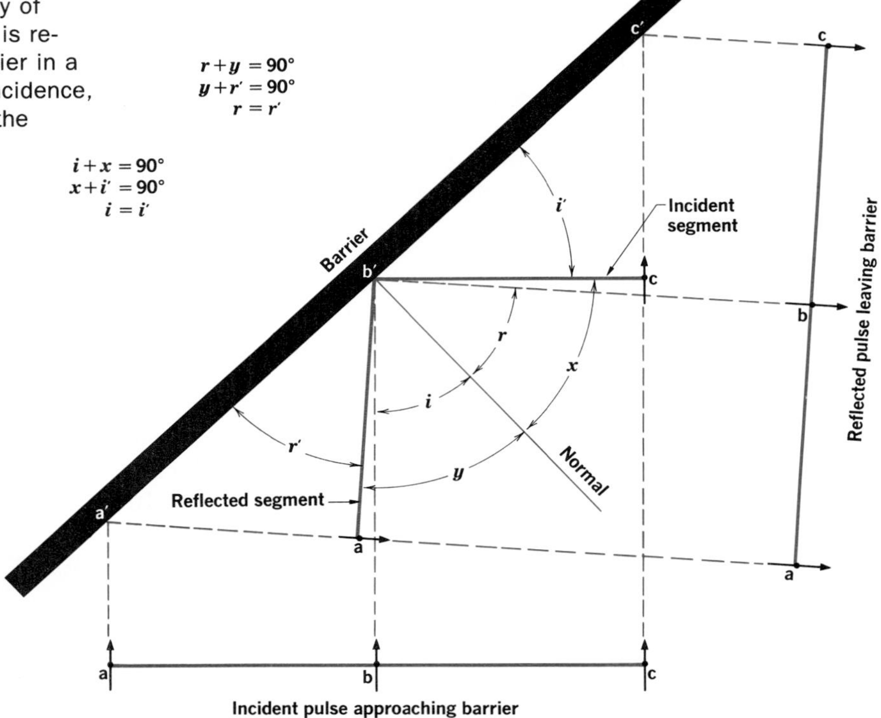

are equal. From Figure 10-16 it is evident that angles $i + x = 90°$ because the incident path **bb′** is perpendicular to the incident segment **b′c** of the pulse. Also, angles $x + i' = 90°$ because they form the angle between the barrier and its normal. Thus i and i' are equal angles. Similar reasoning shows that r and r' are also equal angles. From measurement, i' and r' are equal. Thus the angle of incidence, i, and the angle of reflection, r, are equal.

$$i = r$$

We can state this relationship as a ***law of reflection.*** *When a wave disturbance is reflected at the boundary of a transmitting medium, the angle of incidence, i, is equal to the angle of reflection, r.* A photograph of a periodic straight wave train reflected from a diagonally placed barrier is shown in Figure 10-17.

Figure 10-17. Reflection of a periodic straight wave train from a diagonal barrier. The incoming waves are generated by the black bar at the bottom and are moving toward the top. The left segment of each wave has already been reflected toward the right.

The reflection of a circular wave from a straight barrier is shown in Figure 10-18. Each segment of the expanding wave front reflected from the barrier surface rebounds according to the principle just stated. Observe that the reflected portions of the wave crests are arcs of circles. The apparent center of these reflected wave crests is as far behind the reflecting surface as the real center of disturbance is in front of it.

Reflection may be partial or complete depending on the nature of the reflecting boundary. If there is no boundary mechanism for extracting energy from the wave, all the energy incident with the wave is reflected back with it.

When a surface wave in the ripple tank encounters the straight barrier, which is a rigid vertical bulkhead, the vertical transverse component of the wave is unrestrained. The wave crest is reflected *as a crest,* and the trough is reflected *as a trough.* The surface wave is reflected without a change in phase. *A boundary that allows unrestrained displacement of the particles of a medium reflects waves with no change in the direction of the displacement, which is equivalent to no change in phase.* The bulkhead acts as a *free-end* or *open-end* termination to the wave medium.

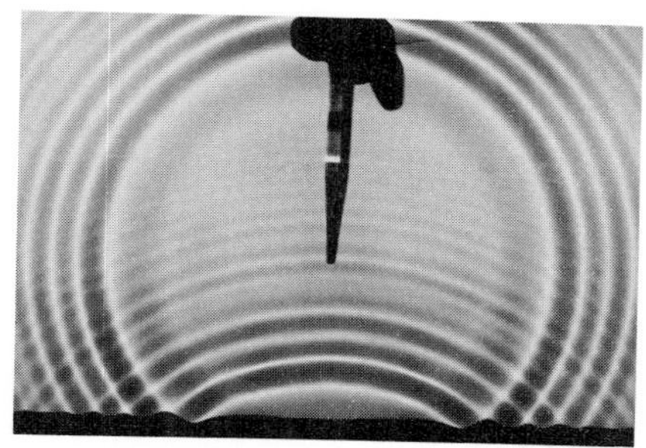

Figure 10-18. A periodic circular wave train generated at the center of the photograph is reflected from a straight barrier located at the bottom of the photograph.

Suppose we send a transverse pulse traveling along a stretched string of which the far end is free. Having such a string presents no problem in a "thought" experiment. In practice, the free-ended termination might be approximated as shown in Figure 10-19. To the extent that the termination is frictionless, the end of the string is free to move in the plane of the pulse.

After arriving at the free end of the string, the pulse is reflected as shown in Figure 10-19. Observe that it has not been inverted. Displacement of the particles of the string is in the same direction as that for the incident pulse. During the time interval that the pulse is arriving at the free end of the string and the reflected pulse is leaving it, the combined effect of the two pulses causes the free end to experience twice the displacement as from either pulse alone. *Reflection at the free-end termination of a medium occurs without a change in phase.*

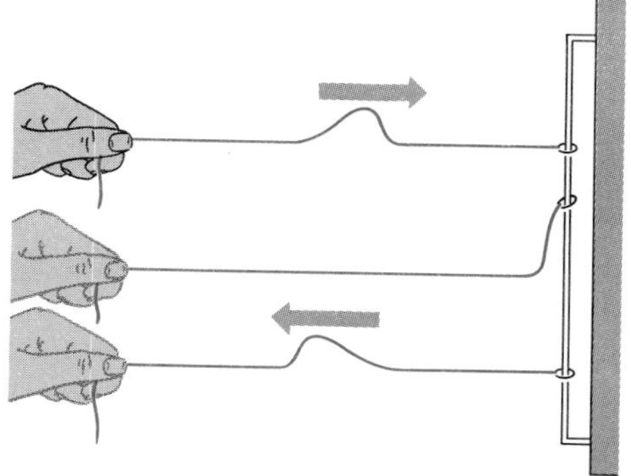

Figure 10-19. Reflection of a pulse at the "free" end of a taut string. The string is shown terminating in a "frictionless" ring that allows the end segment unrestrained displacement in the plane of the pulse.

Now suppose we clamp the far end of the string in a fixed position and send a transverse pulse toward this termination. See Figure 10-20. When the pulse arrives at the fixed end, it applies an upward force to the clamp. An equal but opposite reaction force is applied to the string generating a reflected pulse that is inverted with respect to the incident pulse. *Reflection at the fixed termination of a medium occurs with a reversal of the direction of the displacement, or a phase shift of 180°.* In general, we may conclude that *a boundary that restrains the displacement of the particles of a medium reflects waves inverted in phase.*

Some examples of reflecting terminations for different kinds of wave disturbances are given in Table 10-1. The first three systems listed summarize the phase relations discussed in this section.

Generally, something less than total reflection occurs at the termination of a transmitting medium. The "free" end

Table 10-1
REFLECTION BEHAVIOR

Type of wave	*Example*	*Termination*	
		For in-phase reflection	*For out-of-phase reflection*
transverse (mechanical)	stretched string	free-ended	fixed-ended
longitudinal (sound)	air motion in organ pipe	open pipe	closed pipe
liquid surface (vertical component)	water	solid bulkhead	(no simple analog)
optical (electromagnetic)	light	(no simple analog)	mirror
electric	transmission line	short circuit	open circuit

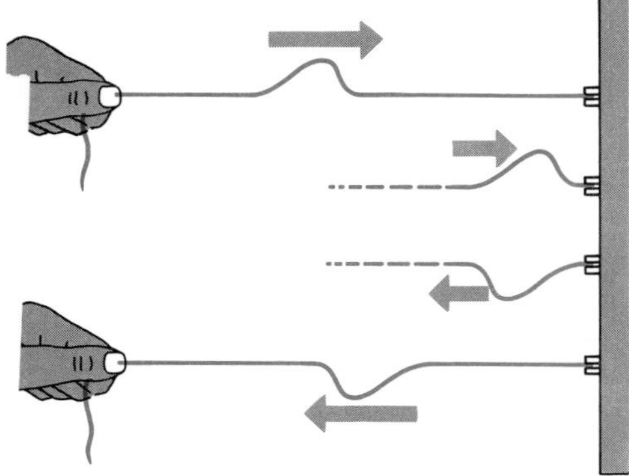

Figure 10-20. Reflection of a pulse at the clamped end of a taut string. The string is terminated in a fixed position and displacement of the end segment is restrained.

of a vibrating string, as well as the fixed end, must be supported in some manner, and some of the wave energy is transferred to the medium providing this support.

The speed of the wave disturbance depends on the properties of the transmitting medium. Waves often traverse the boundary between two media, the second of which can be thought of as the *terminating medium.* Thus waves in the string and in the terminating medium travel at different speeds. If this difference in speed is very great, the portion of wave energy reflected at the boundary will be very large. If the two wave speeds are very similar, little reflection will occur at the boundary. An intermediate effect is seen in Figure 10-22, where partial reflection is evident at the boundary of the deep and shallow water. Light is partially transmitted and partially reflected when it passes from air into glass because of the difference in the transmission speeds in the two media.

10.11 Impedance If we were to apply the same wave-producing force to a light string as to a heavy rope, two entirely different *displacement velocities* would result. The string would experience a very large displacement velocity compared with that of the rope. We could say that the ratio of the *wave-producing force* to the *displacement velocity* is quite small for the string and quite large for the rope. In the study of waves, this *ratio of the applied wave-producing force to the resulting displacement velocity is called the* ***impedance*** *of the medium.*

$$\textbf{Impedance} = \frac{\textbf{wave-producing force}}{\textbf{resulting displacement velocity}}$$

By this definition, the string is said to have a relatively low impedance and the rope a relatively high impedance.

The impedance of a wave medium denotes the ease or difficulty with which a wave disturbance of given amplitude can be launched in it.

The concept of impedance is quite general and can be applied to wave media of all kinds. For a stretched string, the wave-producing force is the transverse oscillatory force applied to the end of the string. The displacement velocity is the oscillating transverse velocity imparted to the particles of the first portion of the string. In an organ pipe, the ratio of the alternating compressional force applied to the molecules of the air to the longitudinal oscillating velocity acquired by these molecules defines the acoustic impedance of the air column.

"Impedance" is a common concept in alternating-current electricity and electronics.

In cases of *total* in-phase reflection discussed in Section 10.10, the impedance of the terminating medium may be considered to be zero. For *total* out-of-phase reflection, the impedance of the terminating medium is infinite. In systems having infinite or zero impedance terminations, reflection is complete and no wave energy is transferred to the terminating medium.

If the terminating impedance exactly matches that of the transmitting medium, the wave energy is completely transferred to the terminating medium and there is no reflection. What about systems that fall in between the extreme impedance mismatch, zero or infinite impedance terminations, and the perfect impedance match? In these cases some energy is transferred to the terminating medium and some is reflected back through the medium of the incident wave.

In practical energy-transfer systems, an abrupt impedance discontinuity between two wave propagation media causes unwanted wave reflection that wastes energy. Numerous devices and techniques are used to smooth over an impedance mismatch and reduce or prevent reflections. They are called *impedance transformers.*

An impedance transformer is an energy-transfer device.

A cheerleader's megaphone is a simple approximation of an acoustic impedance transformer. It effectively matches the impedance of the air column of the throat and mouth to the free air. The coating on an optical lens acts as an impedance transformer. It reduces reflection at the air-glass boundary and causes a greater portion of the total incident light energy to be transmitted through the lens. There are many applications of impedance transformers in electric and electronic systems. In general, whenever a device intended to receive wave energy efficiently cannot be designed to have the same impedance as the transmitting medium, it must be coupled to the medium by means of an impedance transformer.

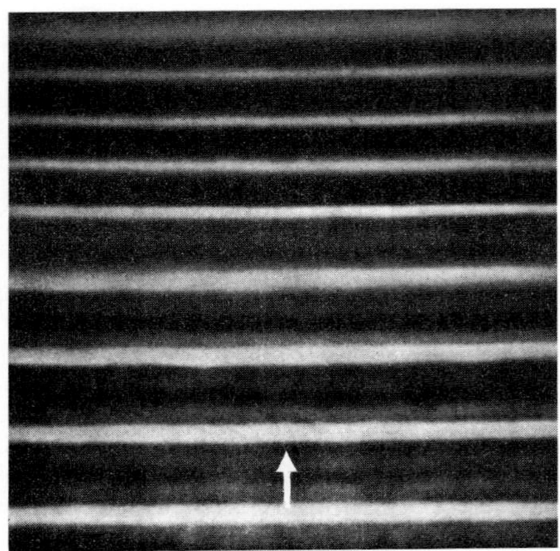

Figure 10-21. The passage of a surface wave from deep to shallow water. The shallow water is at the top of the picture.

10.12 Refraction

The properties of the medium through which a certain wave disturbance moves determine the propagation speed of that disturbance. It is not surprising therefore to find that traveling waves passing from one medium into another experience a change in speed *at the interface (boundary)* of the two media.

From the wave equation, $v = f\lambda$, we see that the wavelength, λ, for a wave disturbance of a given frequency, f, is a function of its speed, v, in the medium through which it is propagating. If the speed decreases when the wave enters a second medium, the wavelength is shortened proportionately. If the speed increases, the wavelength is lengthened proportionately.

A useful aspect of water waves is the relation of the speed of surface waves to the depth of the water. Such wave disturbances travel faster in deep water than in shallow water. For a given wave frequency, the wavelength in deep water is longer than it is in shallow water. Therefore water of two different depths acts just like two different transmitting media for wave propagation.

This difference is easily observed in the ripple tank by arranging the tray so that it has two different depths of water. Then surface waves are generated that travel across the boundary of the two regions. In Figure 10-21, the deep water representing the medium of higher propagating speed is in the lower portion of the photograph. The wave moves toward the shallow water in the upper portion. The boundary between the deep and shallow water is parallel to the advancing wave front. Thus each incident wave crest approaches the shallow region along the normal to the boundary.

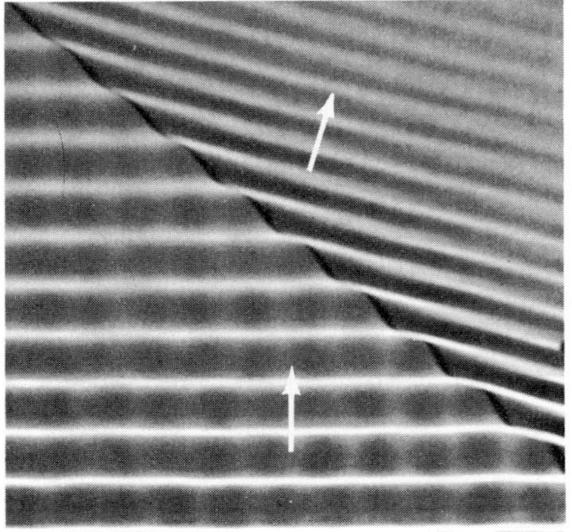

Figure 10-22. Refraction of a surface wave at the boundary of deep and shallow water. The shallow water is above the boundary in the picture. Observe the slight reflection of the incident wave to the left at the boundary.

Observe that the change in wavelength is abrupt. It occurs simultaneously over the entire wave front at the boundary of the deep and shallow water. All segments of the advancing wave front change speed at the same time, and the wave continues to propagate in its original direction.

Suppose we adjust the deep and shallow regions of the tray so that the boundary is no longer parallel to the advancing straight wave but cuts diagonally across its path. This arrangement is shown in Figure 10-22. Each advancing wave crest now approaches the boundary obliquely. Adjacent segments of the wave front pass from deep to shallow water successively rather than simultaneously. At the boundary, the direction of the wave front changes; the advancing wave has undergone *refraction*. ***Refraction** is the bending of the path of a wave disturbance as it passes obliquely from one medium into another of different propagation speed.*

The refraction of surface waves on water is pictured in Figure 10-23. As the wave passes into the shallow water where its speed is less and its wavelength is shorter, it is refracted *toward* the normal drawn to the deep-shallow boundary. The angle of refraction, r, is smaller than the angle of incidence, i. (The angle r is the angle between the path of the refracted wave and the normal.)

Had we generated the wave in the shallow region and directed it obliquely toward the boundary, it would have been refracted *away from* the normal on entering the deep water. Angle r would have been larger than i. The change in direction and the change in speed occur simultaneously. How are these two changes related? We shall investigate this question quantitatively later when we study optical refraction.

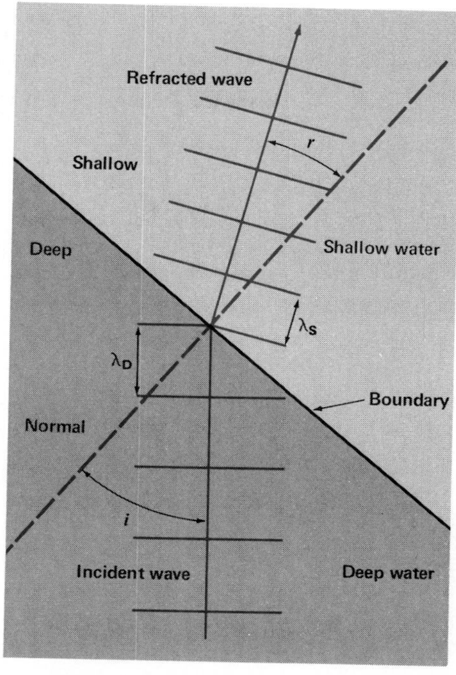

Figure 10-23. The geometry of refraction. Water of two depths in the ripple tank represents two media of different transmission speeds. Compare this diagram with Figure 10-22.

10.13 Diffraction Sounds can be heard even though they may originate around the corner of a building. However, the source of the sound disturbance cannot be seen around the corner. Rectilinear propagation appears to hold in this situation for light but not for sound. The building is an obstruction that prevents the straight-line transmission of light from the source of the sound disturbance to the observer.

Perhaps, where sound waves are concerned, the corner of the building creates a discontinuity in the transmitting medium that causes the waves to *spread*. Even if we find support for this idea, the question of the different behavior of light remains to be answered.

We will again use the ripple tank to observe wave behavior, this time with obstructions inserted into the transmitting medium. This may be done by placing two straight barriers across the tray on a line parallel with the straight-wave generator. An *aperture,* or opening, is left between them approximately equal to the wavelength of the wave to be used. When a periodic straight wave is sent out, the wave pattern beyond the barrier opening appears as shown in Figure 10-24. As a segment of each wave crest passes through the aperture, it clearly spreads into the region beyond the barriers. *This spreading of a wave disturbance beyond the edge of a barrier is called* ***diffraction.***

Sound waves spread around the edges of doorways in the same way that water waves spread around the edges of obstructions. Audible sounds have wavelengths in air ranging from a few centimeters to several meters. The doorway of a room or building has dimensions within this range. This suggests, but does not prove, that a discontinuity such as an aperture in the path of an advancing wave will diffract the wave if its dimensions are comparable with the wavelength of the oncoming wave.

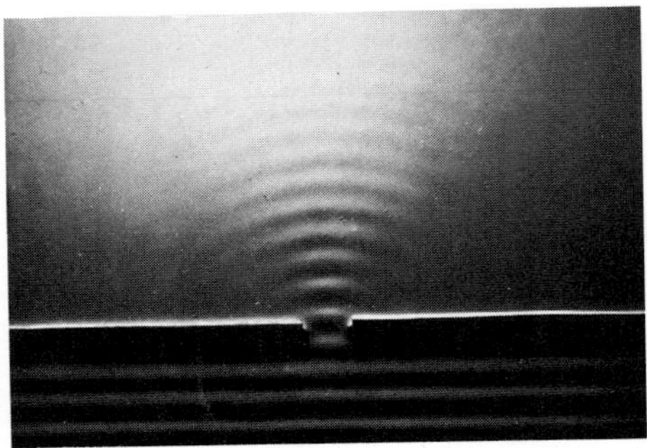

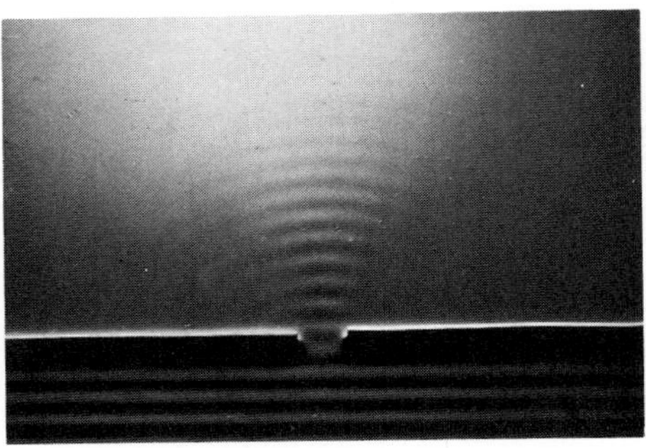

Figure 10-24. The diffraction of a periodic straight wave as it passes through a small aperture. Observe the decrease in the diffraction effect as the wavelength of the disturbance sent against the barrier is shortened.

An open doorway diffracts sound, but it does not diffract light.

We produced the diffraction pattern in the ripple tank with a barrier aperture approximately the same width as the wavelength of the surface wave used. Suppose we shorten the wavelength by increasing stepwise the frequency of the wave generator. We would observe a series of patterns with waves of decreasing wavelengths.

The spreading of the wave at the edges of the aperture diminishes as the wavelength of the wave sent against them is shortened. When the wavelength is a small fraction of the width of the opening between the barriers, the wave segments passing through show little tendency to spread into the shadow regions beyond the barriers. These observations suggest that if the wavelength is much smaller than the width of the aperture, no diffraction occurs. The part of the straight wave that passes through the opening will continue in straight-line propagation.

How are these observations related to the fact that light is not diffracted at the edges of a doorway as is sound? The wavelengths of visible light range from about 4×10^{-7} m for violet light to about 7×10^{-7} m for red light. These wavelengths are much smaller than the opening in a doorway. However, as we will see in Chapter 15, light waves *are* diffracted by very small openings, a phenomenon that has many important applications in science.

Light can also be diffracted by very small objects. The blue color of the daytime sky, for example, is partially due to the differential diffraction of sunlight by dust particles that have dimensions similar to those of light waves. And in crystallography, the arrangement of atoms can be studied by their diffracting effect on electromagnetic radiation.

10.14 The Superposition Principle We have described the passage of a single pulse and a continuous wave through a medium. It is possible for two or more wave disturbances to move through a medium at the same time. Sound waves from all the musical instruments of an orchestra move simultaneously through the air to our ears. Electromagnetic waves from many different radio and television stations travel simultaneously through regions of space to our receiving antennas. Yet, we still can listen to the sound of a particular musical instrument or select a particular radio or television station. These facts suggest that each wave system proceeds independently along its pathway as if the other wave systems were not present.

Let us consider the effects of two periodic transverse waves traveling along a taut string in the same direction. The displacements they produce *at a particular instant* are shown in Figure 10-25. These two waves have different amplitudes and frequencies. The displacements $\mathbf{y_1}$ produced by one wave are represented by the black solid curve. The displacements $\mathbf{y_2}$ produced by the other wave are represented by the black dashed curve. The resultant displacements **Y** all along the string at this instant are represented by the colored curve. These resultant displacements are determined by algebraically adding the displacements $\mathbf{y_1}$ and $\mathbf{y_2}$ for every point along the string.

In effect, the displacement of any particle of the medium by one wave at any instant is superimposed on the displacement of that particle by the other wave at that instant. The action of each wave on a particle is independent of the action of the other, and the particle displacement is the resultant of both wave actions. This phenomenon is known as *superposition.* The two waves are said to *superpose.*

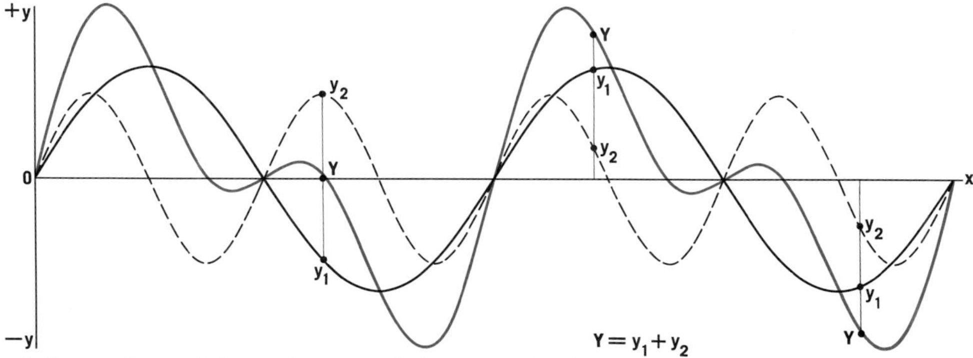

Figure 10-25. An instantaneous view of the superposition of two periodic transverse waves of different amplitudes and frequencies. The component waves y_1 and y_2 are shown in black; the resultant wave Y is in color.

The superposition principle: *When two or more waves travel simultaneously through the same medium, (1) each wave proceeds independently as though no other waves were present and (2) the resultant displacement of any particle at a given time is the vector sum of the displacements that the individual waves acting alone would give it.*

Providing the displacements are small, this principle holds for light and all other electromagnetic waves as well as for sound waves and waves on a string or on a liquid surface. For example, this principle does not hold for "shock waves" produced by violent explosions.

The component waves $\mathbf{y_1}$ and $\mathbf{y_2}$ in Figure 10-25 are simple sine waves. Observe that the frequency f_2 of wave $\mathbf{y_2}$ is twice the frequency f_1 of wave $\mathbf{y_1}$. That is,

$$f_2 = 2f_1$$

Either periodic wave alone would cause each particle of the string to undergo simple harmonic motion about its equilibrium point as the wave traveled along the string.

The resultant wave is also periodic. However, it has a complex, or complicated, wave form. The superposition principle makes it possible to analyze a complex wave in terms of a combination of simple component waves that are frequency multiples of the lowest frequency wave.

Actual wave disturbances generally have complex wave forms. For example, the sound waves produced by musical instruments may be very complicated compared with simple sine wave variations. Wave forms of musical tones from a French horn and a trumpet are shown in Figure 11-15 in Chapter 11. Each of these complex waves can be represented by a certain combination of simple waves that are whole number-multiples of the lowest frequency and that, by superposition, yield the complex wave form.

10.15 Interference The general term *interference* is used to describe the *effects* produced by two or more waves that superpose while passing through a given region. Special consideration is given to *waves of the same frequency*, particularly in the case of sound and light. Interference phenomena are exclusively associated with waves. In fact, the existence of interference in experiments with light first established the wave character of light.

Let us consider two waves of the same frequency traversing the same medium simultaneously. *Each particle* of the medium is affected by *both* waves. Suppose the displacement of a particular particle caused by one wave at any instant is in the same direction as that caused by the other wave. Then the total displacement of *that particle at that instant* is the *sum* of the separate displacements (superposition principle). The resultant displacement is *greater* than either wave would have caused separately. This effect is called *constructive interference.*

On the other hand, if the displacement effects of the two waves on the particle are in opposite directions, they tend to cancel one another. The resultant displacement of *that particle at that instant* is the *difference* of the two separate displacements and is in the direction of the larger (superposition principle). The resultant displacement is *less* than one of the waves would have caused separately. This effect is called *destructive interference.*

If two such opposite displacement effects are equal in magnitude, the resultant displacement is zero. The destructive interference is complete. The particle is not displaced at all but is in its equilibrium position *at that instant.*

Notice that we have been considering the effect of two waves on the position of just *one particle* of the medium at *one particular instant*. We shall continue to fix our attention on one instant in time as we consider the displacements of many *different* particles of the medium in which the two

(A)

(B)

(C)

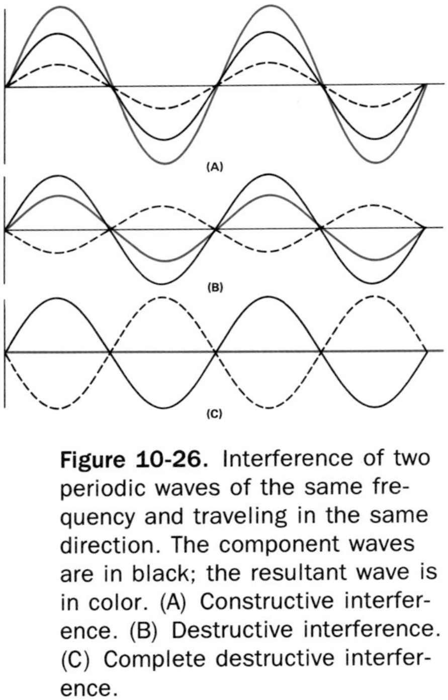

Figure 10-26. Interference of two periodic waves of the same frequency and traveling in the same direction. The component waves are in black; the resultant wave is in color. (A) Constructive interference. (B) Destructive interference. (C) Complete destructive interference.

waves are propagating. We find that the interference effects on different particles are different. There will appear constructive interference at some locations and destructive interference at others.

In Figure 10-26(A), two in-phase periodic waves having the same frequency are traveling in the same direction. They interfere constructively. The resultant periodic wave, shown in color, has the same frequency as the component waves, but has an amplitude equal to the sum of the component amplitudes.

In Figure 10-26(B) the two periodic waves have the same frequency but are opposite in phase, that is, 180° out of phase. The displacements of the two waves are opposite in sign, and they interfere destructively. In Figure 10-26(C) the amplitudes of the two waves are equal and the destructive interference is complete.

The arrangement of interference effects (called an *interference pattern*) depends on the relative characteristics of the interfering waves. Figure 10-27 is a photograph of an interference pattern produced in a ripple tank by two identical circular waves emanating from two points. The two waves are in phase at their sources. That is, the two probes that generate the waves by their vibrations are moving up and down together.

This interference pattern is diagrammed in Figure 10-28. Points **A** and **B** are the sources of two periodic circular waves of the same frequency and amplitude. The sources are acting in phase. Solid circular lines represent wave crests and dashed circular lines represent wave troughs.

At each point similar to those marked **C** the crest of a wave from one source is superposed with the crest of a wave from the other source. At points similar to **C′** the troughs are superposed. A half period later the crests are at **C′** and the troughs are at **C.** Constructive interference occurs along the lines **CC′** giving the resultant wave an amplitude twice the amplitudes of the individual waves.

At points similar to those marked **D** the crest of a wave from one source is superposed with the trough of a wave from the other source. Destructive interference occurs along the lines **DD′**, displacements being reduced to zero.

Points of zero displacement in the interference pattern are called *nodes*. The lines **DD′** along which they occur are *nodal lines*. Similarly, points of maximum displacement are called *antinodes* (or loops) and the lines **CC′** along which they occur are *antinodal lines*. Observe these nodal and antinodal regions in the photograph of the interference pattern shown in Figure 10-27.

At any nodal point where the destructive interference is complete, there is no motion of the medium and no energy. This does not mean that the energy of the two inter-

Figure 10-27. A photograph of the interference pattern of water waves from two point sources. Locate the nodal and antinodal regions.

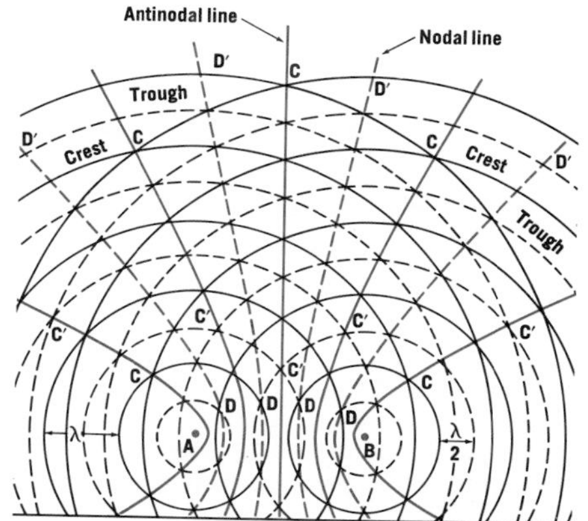

Figure 10-28. A diagram of the interference pattern of two periodic circular waves of the same frequency emanating in phase from points A and B.

fering waves has been destroyed or otherwise lost. Instead, it appears at points of constructive interference, such as points **C** in Figure 10-28. Recall from Section 10.7 that where the amplitude of the resultant wave crest is double that of a single interfering wave, the energy is four times greater. The total energy of the two wave systems remains unchanged, but the *energy distribution* resulting from the interference is different.

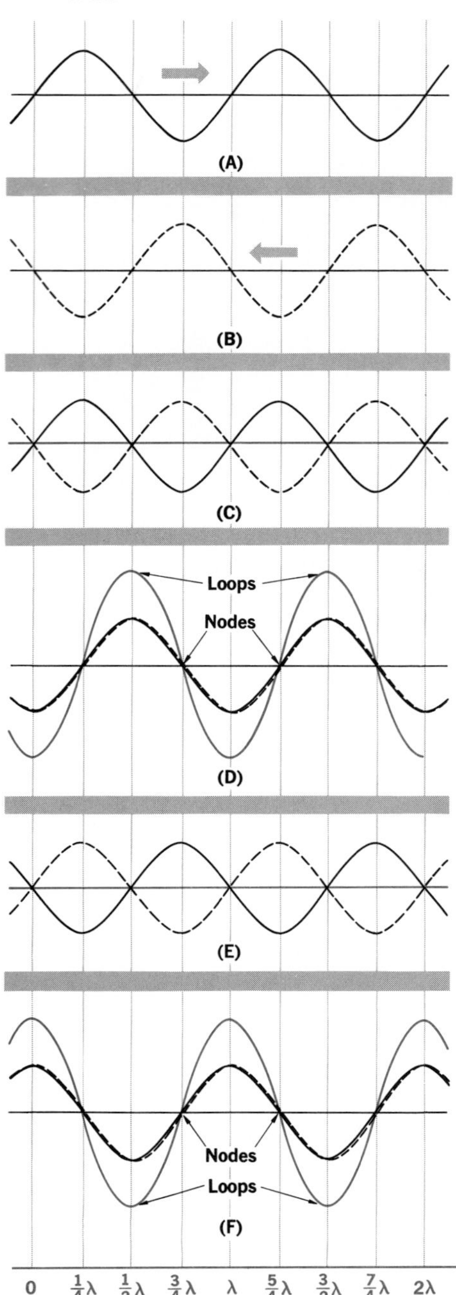

Figure 10-29. A standing wave (shown in color) produced by the interference of two periodic waves of the same frequencies and amplitudes traveling in opposite directions.

10.16 Standing Waves As seen in Section 10.10, a transverse pulse traveling on a taut string is inverted when reflected at the clamped end. This inversion upon reflection at a clamped end occurs also with a train of transverse waves. If no energy is lost in reflection at the fixed end, a continuous wave will be reflected back upon itself giving two wave trains of the same wavelength, frequency, and amplitude but traveling in opposite directions.

Portions of two such wave trains are shown at a certain instant in time in Figure 10-29(A) and (B). If their displacements are added at this instant, as in (C), the resultant displacement is zero. When the wave patterns have moved $\frac{1}{4}$ wavelength, each in its own direction of travel, superposition gives the resultant shown in (D). Another $\frac{1}{4}$ wavelength of movement gives the resultant in (E), and an additional movement of $\frac{1}{4}$ wavelength gives the resultant displacement in (F). Such a wave pattern is a *standing wave*. The particles in a standing wave vibrate in simple harmonic motion with the same frequency as each of the component waves. The amplitude of their motion is not the same for all points along the string. It varies from a minimum of zero amplitude at scale positions $\frac{1}{4}\lambda$, $\frac{3}{4}\lambda$, $\frac{5}{4}\lambda$, $\frac{7}{4}\lambda$, etc., to a maximum of twice the amplitude of one component wave at 0λ, $\frac{1}{2}\lambda$, 1λ, $\frac{3}{2}\lambda$, etc.

In the displacement shown in Figure 10-29(F) the motions of all the particles between $-\frac{1}{4}\lambda$ and $\frac{1}{4}\lambda$ are exactly in phase with one another and with the motion of particles between $\frac{3}{4}\lambda$ and $\frac{5}{4}\lambda$. They are 180° out of phase with the motion of particles between $\frac{1}{4}\lambda$ and $\frac{3}{4}\lambda$ and with the motion of particles between $\frac{5}{4}\lambda$ and $\frac{7}{4}\lambda$. When particles in a length of string equal to one-half wavelength are going up, the particles in the immediately adjacent parts of the string are going down.

Certain parts of a string vibrating with a standing wave pattern never move from their equilibrium positions. These parts are the *nodes*. Halfway between the nodes, where the amplitude of vibration is maximum, are the *loops*. The wavelength of the component periodic transverse waves that produce the standing wave is *twice* the distance between the adjacent nodes or the loops in a standing wave.

*A **standing wave** is produced by the interference of two periodic waves of the same amplitude and wavelength traveling in opposite directions.* No standing wave can be produced if the two waves have different wavelengths.

When a stretched string is clamped at both ends, a standing wave pattern can be formed only for certain definite wavelengths. In such a vibrating string both ends must be nodes. Four possible standing wave patterns in a stretched string are shown in Figure 10-30. Loops are indicated by the letter **L** and nodes by the letter **N.** The wavelength, λ, for each standing wave is expressed in terms of the length of the string, l. Figure 10-30(A) illustrates a time-exposure view showing the envelopes of the standing waves. Part (B) is a strobe-flash view showing positions of the strings at different instants of time.

In a standing wave, particles of the string at the nodes are continuously at rest. Energy is not transported along the string, but remains "standing" in the string. The energy of each particle remains constant as that particle executes its simple harmonic motion. When the string is straight, or undistorted, particles have their maximum velocities and the energy is all kinetic. The energy is all potential when the string has its maximum displacement and all particles are momentarily at rest.

Standing waves are established in water, in air, or in other elastic bodies, just as they are in taut strings. Standing waves can also be produced with electromagnetic waves. A standing wave pattern can occur in a circular ripple tank in which a periodic circular wave is generated at the center of the tank. Nodal rings appear where the surface of the water is at rest. The surface oscillates up and down between the nodal rings. Many vibrating objects vibrate normally in a way that establishes standing waves in the object. The strings or the air columns of musical instruments establish various modes of standing waves.

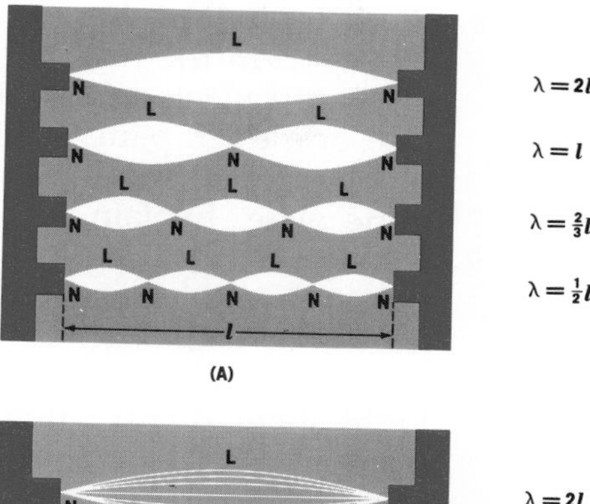

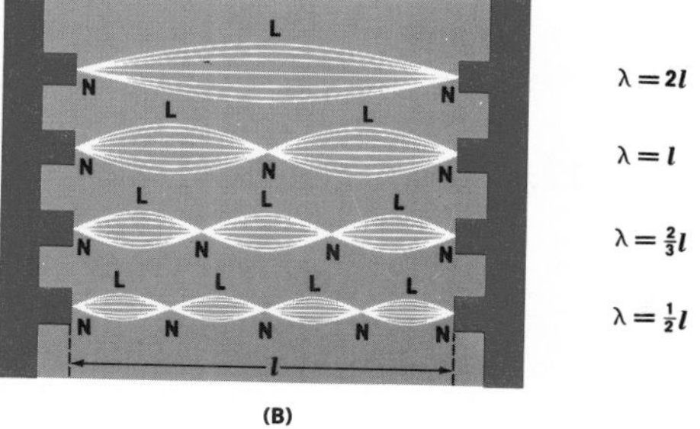

Figure 10-30. Standing wave patterns in a stretched string. (A) A time-exposure view. (B) A strobe-flash view. Wavelength λ is expressed in terms of the length l of the string. Labels L and N identify loops and nodes.

QUESTIONS: GROUP A

1. Does rectilinear propagation apply to circular wave fronts? Explain.
2. When is a wave reflected?
3. What determines whether reflection is partial or complete?
4. Differentiate between free-end and fixed-end reflection.
5. (a) What will happen when a transverse wave moves from a rope to a string attached to the end of the rope? (b) What property of waves causes this?
6. (a) What is an impedance transformer? (b) How could you construct one for the situation in Question 5?

GROUP B

7. (a) When will a standing wave be produced? (b) What types of waves can produce standing waves?

8. Predict whether there will be a phase inversion in the following cases: (a) light reflected from the front surface of a thick pane of glass; (b) light reflected from the back surface of the same plate. Explain.
9. Newton thought light was a particle, not a wave, because it was not diffracted by a doorway, as sound is. Comment on this.
10. Light slows down when it moves from air to glass. What will be the sizes of the angles of incidence and refraction?
11. A red spot light intersects a green spot light before shining on stage. Predict the individual color of each spot.
12. You are in an auditorium watching a play. The people around you are laughing at the dialogue, but you find it difficult to hear. What might be happening here?
13. On graph paper, draw two transverse waves of the same wavelength, different amplitudes, and same phase. Draw the wave pattern that results if these two waves are superimposed.
14. Now draw waves of equal amplitude, but one having twice the wavelength. Begin each wave at equilibrium position moving to its crest $\frac{1}{4}$ wavelength later.
15. Repeat Question 14, leaving the longer wave the same, but beginning the shorter wave at its crest, moving to rest position $\frac{1}{4}$ wavelength later.
16. What happens to the energy of the system when two waves interfere?
17. Which of the wave properties is the most characteristic of waves?

PROBLEMS: GROUP A

1. Electromagnetic waves travel at 3.0×10^8 m/s. The visible range has wavelengths ranging from about 4.0×10^{-7} m in the violet region to about 7.6×10^{-7} m in the red region. What is the frequency range?
2. A string stretched between two clamps is 2.0 m long. When plucked at the center, a standing wave is produced that has nodes at the clamped ends and a single loop at the center. The string completes 7.0 vibrations per second. (a) What is the wavelength of the component traveling waves? (b) What is the speed of the transverse waves in the string?
3. (a) If the oscillating mass shown in Figure 10-7 moves 3.00 cm from its lowest to its highest position, what is the amplitude of the wave being produced by the spring? (b) If the distance from one intersection of the spring positions pictured to the next intersection is a distance of 12.5 cm, what is the wavelength of these waves? (c) If the mass oscillates 5 times per second, what is the speed of the waves on the spring?

GROUP B

4. If a distance of 2.5 m separates a trough and an adjacent crest of water waves on the surface of a lake, and 33 crests pass in 30 s, what is the speed of these waves?
5. Two physics students were fishing from a boat anchored a distance of 24 m from the shore of a lake. One of the students observed that the boat rocked up and down through 11 complete oscillations in 19 s and that one wave crest passed the boat with each oscillation. The other student noted that each wave crest required 6.5 s to reach the shore. (a) What was the period of the surface wave? (b) What was the wavelength?

PHYSICS ACTIVITY

Put about ten centimeters of water in your bathtub. Drop a button into the water in order to form a surface wave. Measure the time it takes the wave crest to reach the end of the tub. Create a bigger wave and similarly time it. Is there a significant difference in the times? Explain.

SUMMARY

Energy can be transferred by the movement of materials or objects, by the motion of particles, by the flow of gases or liquids, and by waves. Energy is transferred by waves without the transfer of matter. Waves that require a material medium for their transmission are called mechanical waves. Electromagnetic waves can travel through matter but a material medium is not required for their transmission. A pulse is a single nonrecurring wave disturbance. A continuous wave results from a wave disturbance that recurs periodically.

Mechanical waves in which the particles of the transmitting medium are displaced perpendicular to the direction in which the wave travels are called transverse waves. Those in which the particles of the medium are displaced parallel with the wave direction are longitudinal waves.

Waves have common characteristics such as phase, frequency, period, wavelength, speed, and amplitude. A wave's velocity is the product of its frequency and its wavelength in the medium. Transmission media in which the speed of a wave is dependent on its frequency are said to be dispersive.

Some of the common properties of waves are rectilinear propagation, reflection, refraction, diffraction, and interference. Superposition occurs when two or more wave disturbances move through a medium at the same time. Standing waves are produced by the interference of two periodic waves of the same amplitude and wavelength and traveling in opposite directions.

VOCABULARY

amplitude
angle of incidence
angle of reflection
compression
constructive interference
damping
destructive interference
diffraction
dispersive medium
elastic medium
frequency
impedance
interference
longitudinal wave
mechanical wave
period
phase
rarefaction
rectilinear propagation
reflection
refraction
standing wave
superposition
transverse wave
wave
wave crest
wave front
wave trough
wavelength

Sound

Until recently, an entire orchestra would be brought into a recording studio as backup for a 3-piece rock band. This practice, however, has become more of a rarity since the introduction of digital sound sampling, which involves translating a sound into digits that represent that sound. This is accomplished by feeding a sound through a microphone and then into a computer. The computer measures various parameters of the sound, such as timbre and pitch. Each parameter is represented by a specific digital sequence. When a trigger, such as a drum pad or an electronic-piano key, is hit, the digits are translated back into sound that is emitted through a speaker. The sounds of musical instruments, such as a trumpet or violin, as well as nonmusical sounds such as the wind or a dog's bark, can be fed into the computer and stored. A composer can then combine, distort, modify or change the speed of the sampled sounds. By simply playing an electronic piano hooked up to the computer, a musician can use these samples to reproduce the

A Synclavier operator records the sound from a saxophone using a digital recorder.

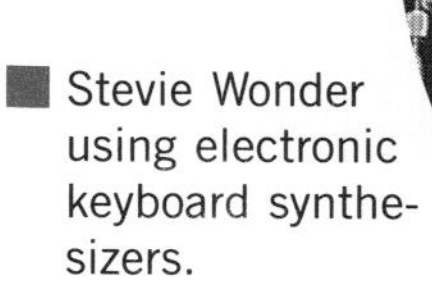

Stevie Wonder using electronic keyboard synthesizers.

Acoustical testing of the noise emission from a computer printer. Identification and control of noise sources carried out by an acoustical engineer.

sound of an entire orchestra or create special effects with a single sound. Drummers can use pads programmed to sample the sounds of a perfectly tuned drum set instead of setting up and tuning the actual drums each time they play.

Acoustical Physicists

Acoustical physicists and engineers study sound and its transmission and contribute to the design of acoustical systems. One area of this science involves developing acoustical equipment, such as stereos and digital sound synthesizers, as well as the studios where sounds are recorded. Some of these scientists help design concert halls. Others are involved in noise control.

A B.S. degree in physics or acoustical physics is necessary to pursue this field. A wide range of job opportunities is available in medical facilities, as architectural and engineering consultants, and in government and industrial laboratories.

Electronic-Piano Technicians

With the increasing variety and popularity of electronic pianos, an electronic-piano technician is very much in demand. These specialists install, repair, and test electronic pianos as well as other electronic musical instruments. They use manuals, wiring diagrams and electricians' tools to locate the source of malfunctions and repair electronic circuitry.

Those interested in becoming electronic-piano technicians should have excellent hearing and manual dexterity. Ability to play a musical instrument is helpful. Most technicians begin with on-the-job training in musical instrument repair and sales shops. Electronics training from a community college, technical, or vocational school can be useful. ■

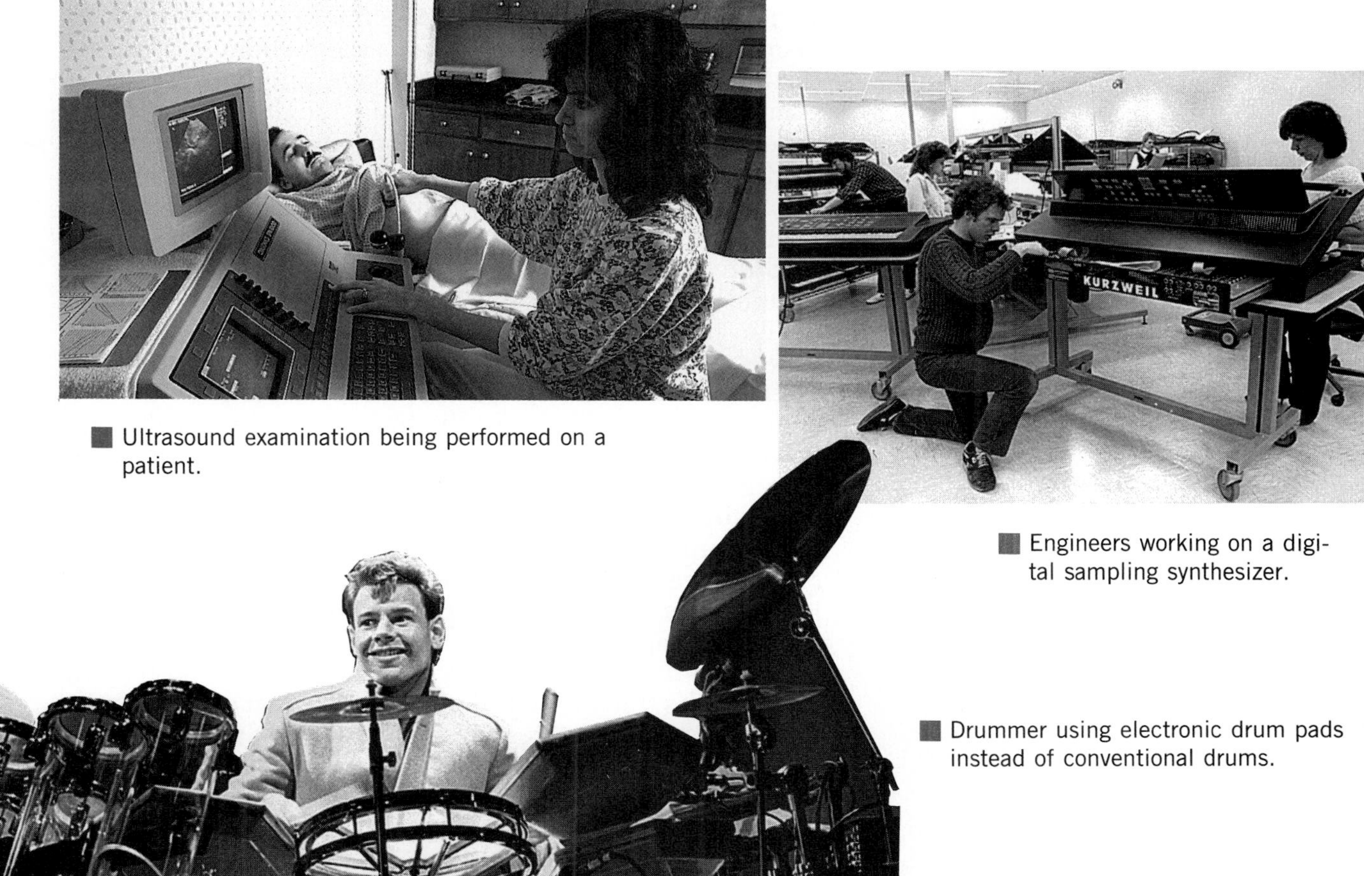

■ Ultrasound examination being performed on a patient.

■ Engineers working on a digital sampling synthesizer.

■ Drummer using electronic drum pads instead of conventional drums.

11 Sound Waves

sound n.: a range of compression-wave frequencies to which the human ear is sensitive.

OBJECTIVES

- Describe the spectrum of sound.
- Define and describe the properties of sound waves.
- Describe the measurement of sound levels.
- Define and describe the Doppler effect.
- Describe the nature of musical sounds.
- Apply the laws of vibrating strings.
- Describe forced vibrations and beats.

THE NATURE OF SOUND

11.1 The Sonic Spectrum The energy-transfer mechanism of compression waves is a major interest of physicists. In a medium having elastic and inertial properties, compression waves propagate as longitudinal disturbances. The disturbances consist of **compressions** and **rarefactions** that give rise to elastic forces in the propagating (transmitting) medium. Through these elastic forces the energy of the wave is transferred to the particles of the medium that are next in line. The energy exchange is continuous as particles receive wave energy and pass it along.

The frequency range over which such longitudinal waves occur is very large. It is called the **sonic spectrum.** There is no clearly defined lower limit of frequency of the sonic spectrum. However, the frequency of an earthquake wave may be a fraction of a cycle per minute and its wavelength may be measured in kilometers.

The upper limit of the sonic spectrum is well defined. For a constant-velocity condition, periodic waves of increasing frequency have proportionally decreasing wavelength. A wave is not transported by a medium in which the wavelength is small compared to the inter-particle spacing of the medium. For a gas the mean free path of the molecules is the limiting dimension. Thus at ordinary temperatures and pressures the upper range of sonic frequencies is of the order of 10^9 hertz in a gaseous medium. In liquids and solids the upper frequency limit is higher because of the smaller inter-particle spacing.

Within the sonic spectrum lies the region of ***sound,** a range of compression-wave frequencies to which the human ear is sensitive.* This audible range of frequencies, called the *audio range,* or the **audio spectrum** extends from approximately 20 to 20 000 hertz. Compression waves at frequencies above the audio range are referred to as **ultrasonic;** those below the audio range are referred to as **infrasonic.**

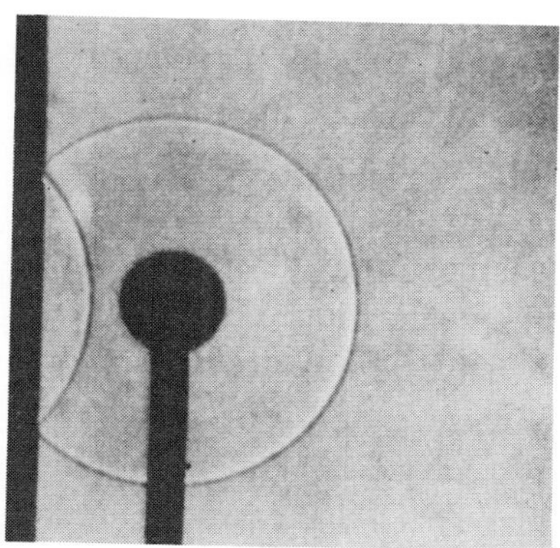

Figure 11-1. Photograph of a sound wave. The sound from a loud electric spark traveled outward from the black circle. The light from a second spark casts a shadow on the resulting shock wave, thereby making the wave visible as a dark circle. At the left, the wave is reflected from a plane surface.

11.2 The Production of Sound

A sound is produced by the initiation of a succession of compressive and rarefactive disturbances in a medium capable of transmitting these vibrational disturbances. Particles of the medium acquire energy from the vibrating source and enter the vibrational mode themselves. The wave energy is passed along to adjacent particles as the periodic waves travel through the medium.

Such vibrating elements as reeds (clarinet, saxophone), strings (guitar, vocal cords), membranes (drum, loudspeaker), and air columns (pipe organ, flute) initiate sound waves. Sound waves are transmitted outward from their source by the surrounding air. When they enter the ear, they produce the sensation of sound.

Suppose we clamp one end of a thin strip of steel in a vise to serve as a vibrating reed. When struck sharply, the free end vibrates to and fro. See Figure 11-2. If the reed vibrates rapidly enough, it produces a humming sound.

When the reed vibrates, its motion approximates simple harmonic motion. See Figure 11-3(A). A graph of its displacement with time produces the sine wave shown in Figure 11-3(B). As the vibrating reed moves from **a** to **b,** it compresses the gas molecules of the air immediately to the right. The reed transfers energy to the molecules in the direction in which the compression occurs. At the same time the air molecules to the left expand into the space behind the moving reed and become rarefied. Thus the movement of the reed from **a** to **b** produces a compression and rarefaction simultaneously with a net transfer of energy to the right, the direction of the compression.

As the reed moves in the reverse direction from **b** to **a,** it compresses the gas molecules to the left, while those to the right become rarefied. The combined effect of this simultaneous compression and rarefaction transfers energy to the left, again in the direction of the compression.

If we consider just the series of compressions and rarefactions produced to the right, it is apparent that the maximum compression occurs as the reed moves through

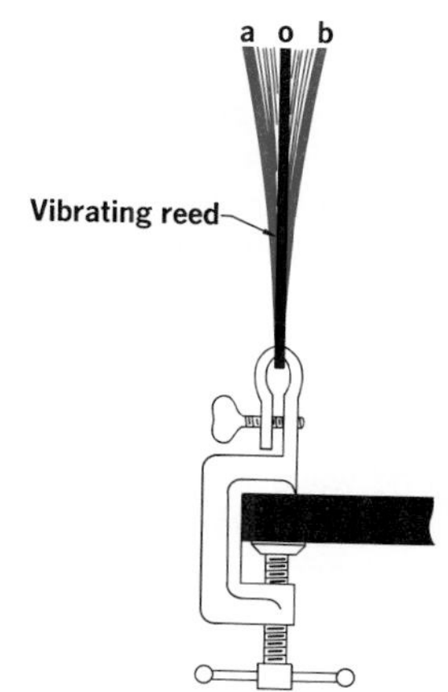

Figure 11-2. Sounds are produced by vibrating matter.

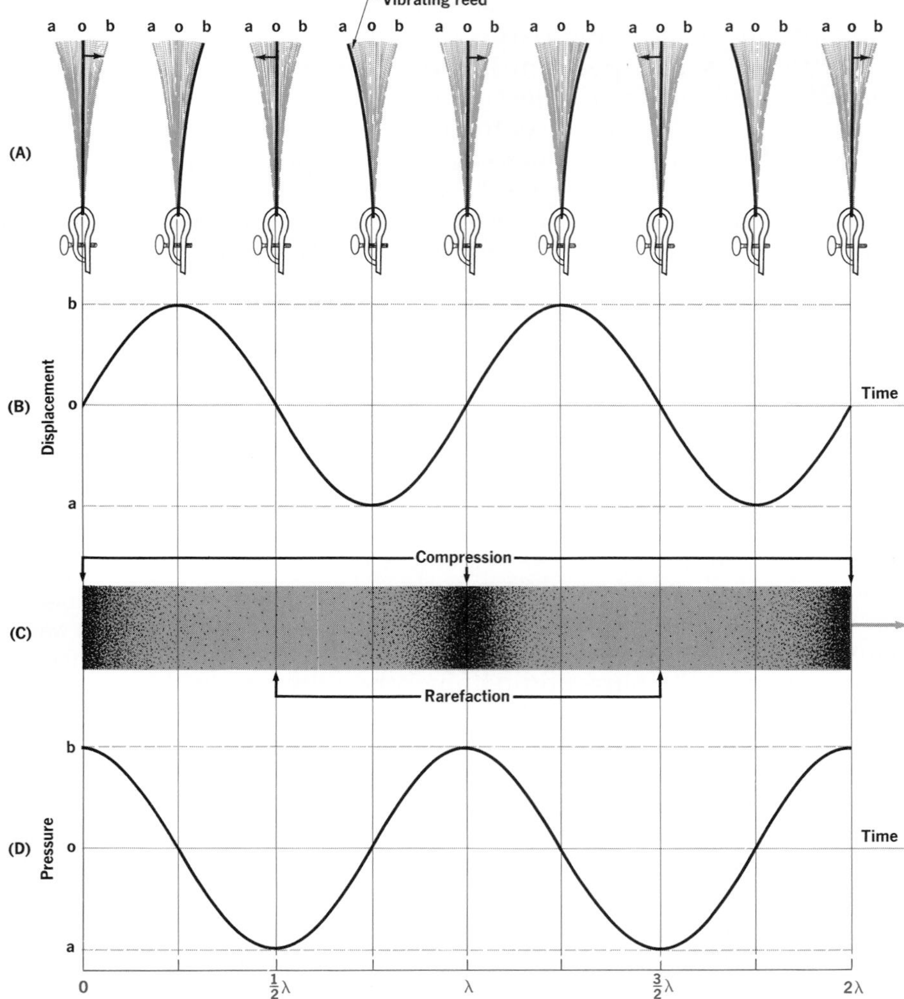

Figure 11-3. Variations of the vibrating reed in (A) produce the displacement variations with time in (B). The regions of compression and rarefaction of the longitudinal sound wave produced to the right of the reed are shown in (C). The pressure, or density variations, of the sound wave with time are plotted in (D). Observe that the displacement and pressure curves are 90° out of phase.

its equilibrium position **o** traveling from **a** to **b.** See Figure 11-3(C). The maximum rarefaction occurs as the reed moves through its equilibrium position **o** traveling from **b** to **a.** This is shown in (D) as a plot of the variation of pressure with time.

Simultaneously, of course, a corresponding series of rarefactions and compressions is produced to the left. The vibration of the reed thus generates longitudinal trains of waves. In these longitudinal waves, vibrating gas molecules move back and forth along the path of the traveling waves, receiving energy from adjacent molecules nearer the source and passing it on to adjacent molecules farther from the source. *Sound waves are longitudinal waves.*

Variations in the work done in setting the reed in vibration alter the amplitude of the vibration but not the frequency. The greater the energy of the moving reed, the greater is the amplitude of its vibrations and the amplitude of the resulting longitudinal waves.

Figure 11-4. In this specially designed room, only 0.1% of the sound is reflected. The acoustical conditions here are similar to those in the atmosphere 1.6 km above the earth's surface.

11.3 Sound Transmission To produce sound waves, we must have a source that initiates a mechanical disturbance and an elastic medium through which the disturbance can be *transmitted.* Most sounds come to us through the air, and it is the air that acts as the transmitting medium. At low altitudes, we usually have little difficulty hearing sounds. At higher altitudes, where the density of the air is lower, less energy may be transferred from the source to the air. Dense air is a more efficient transmitter of sound than rarefied air.

The following experiment provides evidence that a material is required for the transmission of sound. As shown in Figure 11-5, an electric bell under a bell jar is connected to a source of energy so that it rings. While it is ringing, air is removed from the bell jar. As the remaining air becomes less and less dense, the sound becomes fainter and fainter. When air is allowed to reenter, the sound becomes louder. These effects show that it is increasingly difficult to effect the transfer of sound energy to the air as the density of the air decreases. This demonstration suggests that if all the air could be removed from the bell jar, no sound would be heard. *Sound does not travel through a vacuum; it is transmitted only through a material medium.*

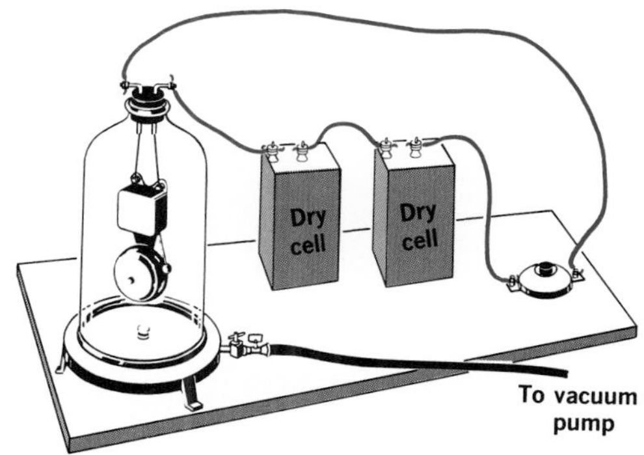

Figure 11-5. As air is pumped from the bell jar, the sound of the bell becomes fainter. Such a demonstration is evidence that a material medium is needed for the transmission of sound.

The sound of two rocks being struck together under water can be heard quite plainly by a swimmer submerged in the water. If the swimmer is near the source of the disturbance, the effect can be painful. Liquids are efficient transmitters of sound.

A loose faucet washer may vibrate when water is drawn from a water pipe. The sound of this vibration is carried to all parts of the house by water pipes and wood framework. Many solids are efficient sound transmitters. In long rods or pipes, the sound propagation may be restricted to one dimension instead of spreading in space.

The ratio of the speed of an object to the speed of sound in the same medium is called the Mach number, in honor of the Austrian physicist Ernst Mach (1838–1916).

11.4 The Speed of Sound During a thunderstorm, a distant lightning flash can be seen several seconds before the accompanying thunder is heard. The timer at the finish line during a track meet may see the smoke from the starter's gun before he hears the report. Over short distances, light travels practically instantaneously. Therefore, the time that elapses between a lightning flash being seen and the thunder being heard or between a gun being fired and the report being heard must be the time required for the sound to travel from its source to the listener. The speed of sound in air is 331.5 m/s at 0°C. This speed increases with temperature about (0.6 m/s)/C°.

The speed of sound in water is about four times the speed in air. In water at 25°C sound travels about 1500 m/s. In some solids, the speed of sound is even greater. In a steel rod, for example, sound travels approximately 5000 m/s—about 15 times the speed in air. In general the speed of sound varies with the temperature of the transmitting medium. For gases the change in speed is rather large. For liquids and solids, however, this change in speed is small and is usually neglected. Representative values for the speed of sound waves in various media are given in Table 11-1. (See also Appendix B, Table 17.)

Table 11-1
SPEED OF SOUND
(25°C)

Medium	***Speed*** (m/s)
air	346
hydrogen	1339
alcohol	1207
water	1497
glass	4540
aluminum	5000
iron (steel)	5200

11.5 The Properties of Sound A nearby clap of thunder is loud; a whisper is soft. A cricket has a shrill, high chirp; a bulldog has a deep growl. A violin string bowed by a virtuoso produces a note of fine quality and is far more pleasing to hear than that produced by the novice. Each of these sounds has characteristics clearly associated with it. Sounds differ from each other in several fundamental ways. We shall consider three physical properties of sound waves: *intensity, frequency,* and *harmonic content.* The effects of these properties on the ear are called, respectively, *loudness, pitch,* and *quality.*

11.6 Intensity and Loudness *The* ***intensity*** *of a sound is the time rate a : which the sound energy flows through a unit area normal to the direction of propagation.* Intensity thus has the dimension of power/area.

Recall that $P = \frac{E}{t}$ *(see Section* 6.5).

$$I = \frac{P}{A}$$

Where P is sound power in watts and A is area in square

meters, the sound intensity I is expressed in watts per square meter. The intensity of a sound wave of given frequency is dependent on its amplitude; it is proportional to the square of the amplitude.

Review Section 10.7.

Sound waves that emanate from a vibrator approximating a point source and that travel in a uniform medium spread out in a spherical pattern. Thus the area of the expanding wave front is directly proportional to the square of its distance from the source. Since the total power of the wave is constant, the intensity of the wave diminishes as it moves away from the source. The sound produced by a whistle is only one-fourth as intense at a distance of one kilometer as it is at a distance of half a kilometer from the source. The intensity of sound in a uniform medium is inversely proportional to the square of its distance from the point source.

*The **loudness** of a sound depends on an auditory sensation in the consciousness of a human listener.* In general, sound waves of higher intensity are louder, but the ear is not equally sensitive to sounds of all frequencies. Thus a high- or a low-frequency sound may not seem as loud as one of mid-range frequency having the same intensity.

The loudness of a sound is not directly proportional to its intensity.

An increase in the intensity of a sound of fixed frequency causes it to seem louder to the listener. Suppose a 1000-Hz tone is generated of such low intensity that it is barely audible. The intensity can be increased approximately 10^{12} times before the tone becomes so loud that it is painful to the listener. While the *intensity* of a sound can be directly measured with instruments, the *loudness* can not, as it depends on the ear and subjective judgment of the listener.

11.7 Relative Intensity Measurements Intensity is measured with acoustical instruments and does not depend on the hearing of a listener. At 1000 Hz, the intensity of the average faintest audible sound, called the **threshold of hearing,** is 10^{-12} W/m^2. When expressing the intensity of a sound, it is often compared with the intensity of the threshold of hearing, and this ratio of intensities yields an expression of the *relative intensity* of the sound. Because of the great range of intensities over which the ear is sensitive, a *logarithmic* rather than an arithmetic intensity scale is used. On this kind of scale, the relative intensity of a sound is given by the equation

$$\beta = 10 \log \frac{I}{I_0}$$

where β (the Greek letter beta) is the relative intensity in

Table 11-2
RELATIVE INTENSITIES OF SOUNDS (at 1000 Hz)

Type of sound	*Intensity* (dB)
threshold of hearing	0
whisper	10–20
very soft music	30
average residence	40–50
conversation	60–70
heavy street traffic	70–80
thunder	110
threshold of pain	120
jet engine	150

decibels (dB) of a sound of intensity I. I_0 is the intensity of the threshold of hearing. I and I_0 have the same dimensions, usually W/m^2. Therefore the decibel is a dimensionless unit. Relative intensities of several familiar sounds are listed in Table 11-2. The following example illustrates the use of the relative intensity equation.

EXAMPLE A small source uniformly emits sound energy at a rate of 2.0 W. Calculate the relative intensity at 34 m from the source.

Given	Unknown	Basic equations
$p = 2.0$ W $r = 34$ m $I_0 = 10^{-12}$ W/m^2	β	$I = \frac{P}{A}$ $\beta = 10 \log \frac{I}{I_0}$ $A = 4\pi r^2$

Solution

Working equation: $\beta = 10 \log \dfrac{P}{AI_0} = 10 \log \dfrac{2.0\text{ W}}{4\pi(34\text{ m})^2(10^{-12}\text{ W/m}^2)}$

$= 81$ dB

PRACTICE PROBLEM A jet plane is found to have a relative intensity upon takeoff of 110 dB. Calculate the intensity of the sound the jet makes. *Ans.* 0.10 W/m^2

Table 11-3
RELATIVE INTENSITIES

I/I_0	bel	dB
10	1	10
4		6
2		3
1	0	0
0.5		−3
0.25		−6
0.1	−1	−10

The **decibel** is the practical unit for the relative intensity of sound. It is equal to 0.1 **bel,** the original unit named in honor of Alexander Graham Bell (1847–1922), who invented the telephone.

$$\textbf{Relative intensity in bels} = \log \frac{I}{I_0}$$

In the human sensory response, a change in the sound power level of 1 dB is just barely perceptible. This change of 1 dB represents a change in sound intensity of 26%. Table 11-3 shows several power ratios and their decibel equivalents.

We stated in Section 11.1 that the audio spectrum extends from approximately 20 to 20 000 Hz. The lower frequency is called the *lower limit of audibility* and the higher frequency the *upper limit of audibility*. However, there is considerable variation in the ability of individuals to hear sounds of high or low frequencies.

The graph in Figure 11-6 shows the characteristics of audible sound waves. Since the graph is a composite of results obtained by testing many persons, it refers to the performance of the "average" ear.

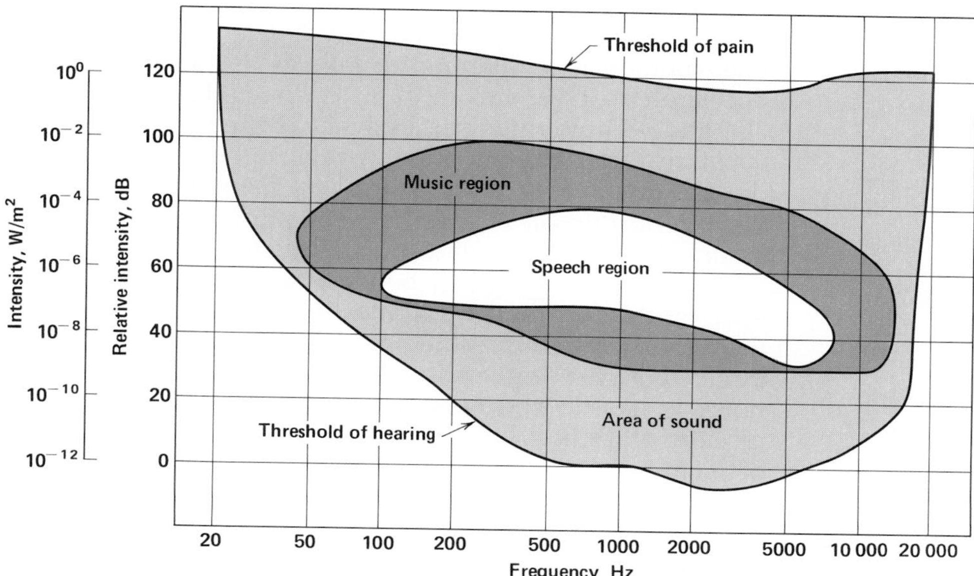

Figure 11-6. The range of audibility of the human ear.

The lower curve, the *threshold of hearing*, shows the minimum intensity level at which sound waves of various frequencies can be heard. Observe that the lowest intensity that will produce audible sound is for frequencies between 2000 and 4000 Hz. The ear is most sensitive in this frequency range. At the lower and upper limits of audibility, the intensities of sound waves must be greater to be heard.

The upper curve, the **threshold of pain,** indicates the upper intensity level for audible sounds. Sounds of greater intensity produce pain rather than hearing.

The graph shows the general frequency limits for the audio range to be between 20 and 20 000 Hz. The relative intensity limits in the region of maximum sensitivity are between 0 and 120 decibels, corresponding to intensities of from 10^{-12} W/m^2 to 10^0 W/m^2.

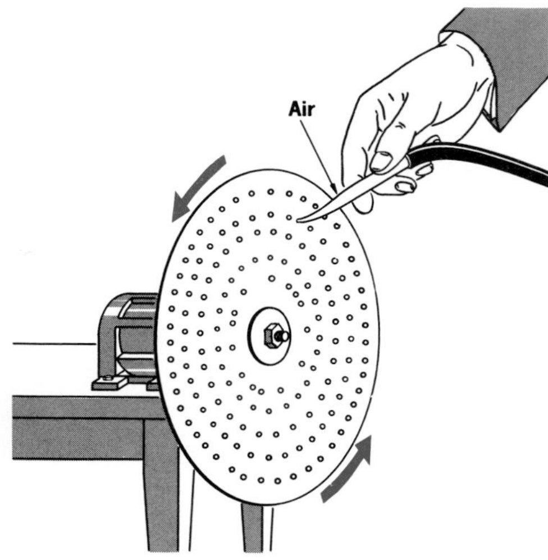

Figure 11-7. A siren disk perforated to demonstrate the relationship between pitch and frequency and the difference between musical tones and noise.

11.8 Frequency and Pitch A perforated disk, as shown in Figure 11-7, can be used to show that the frequency of a sound determines its *pitch*. The disk has five concentric rings of holes. The holes of the innermost ring are irregularly spaced. The holes in the other four rings are regularly spaced, and, respectively, there are 24, 30, 36, and 48 holes in them.

A stream of air directed against a ring of holes is interrupted by the metal between the holes as the disk rotates rapidly at constant speed. The interruptions give rise to a series of air pulses that produce a sound characteristic of the frequency of the pulses. A sound of constant pitch is heard if the holes are evenly spaced. The pitch rises as the disk rotation is accelerated; it falls as the disk rotation is decelerated. If the stream of air is directed at the 24-, 30-, 36-, and 48-hole rings successively as the disk rotates at a constant rate, tones of successively higher pitch are produced. ***Pitch** is the characteristic of a sound that depends on the frequency the ear receives.* It is the *pitch* that allows us to assign the sound its place in the musical scale.

We interpret the frequency of a sound in terms of its pitch.

Noise: sounds with nonperiodic waveforms.

We usually distinguish between a musical tone and a *noise* by the fact that the former is a pleasing sound, and the latter is a disagreeable one. When the stream of air is directed at the innermost ring of holes, an unpleasant noise is heard. The noise is characterized by a random mixture of frequencies and is not easily identified in terms of pitch.

The pitch of a tone produced by directing the air against one of the regularly spaced rings of holes is appropriately described in terms of the frequency of the sound produced. If the disk is rotating at the rate of 20 revolutions per second, the frequency of the tone produced by the ring with 24 holes is 480 hertz. The ring with 30 holes produces a tone having a frequency of 600 hertz; the 36-hole ring, a frequency of 720 hertz; and the 48-hole ring, 960 hertz.

Music: sounds with approximate periodic waveforms, unless their intensities are too high.

The two tones produced by the outermost and innermost of these four regularly spaced rings are an *octave* apart, their frequencies being in the ratio of 2:1. The four tones have frequencies in the ratios 4:5:6:8. When these tones are produced simultaneously, they comprise a musical sound known as a *major chord.*

Tone is associated with the pleasing quality of the sound that reaches the ear. Tone quality is enhanced by the complexity of the sound wave, that is, by the number and distribution of harmonics of the fundamental frequency superposed in the complex sound wave. *Pitch* is associated principally with the fundamental frequency of the complex sound wave. This distinction is important when considering the quality of musical sounds. For example, notes of the same pitch sounded by the pianist and the violinist are very different in tonal quality.

11.9 The Doppler Effect In our discussion of frequency and pitch in Section 11.8, both the source of the emission and the listener were assumed to be stationary. In that case, the pitch of the sound heard is characteristic of the frequency of the emission. If the source emits 1000 vibrations per second, the listener hears a 1000-Hz tone.

When there is relative motion between the source of sound and the listener, the pitch of the sound heard is not the same as that when both listener and source are stationary. Two situations are common: The source may be moving and the listener is stationary, and the listener may be moving and the source is at rest.

The first situation prevails when the listener stands at a railroad crossing while a train passes. The pitch of the warning horn of the locomotive drops abruptly as the train passes the listener's position. The second situation prevails when the listener rides in the train as it passes through a highway intersection. In this instance the pitch of the warning bell at the crossing drops abruptly as the listener passes through the intersection.

In both situations the steady pitch of the sound heard by the observer is *higher* than the actual frequency of the source would indicate as the distance between the listener and the source *decreases* at a constant rate. Also, the steady pitch of the sound heard is *lower* than the actual frequency of the source would indicate as the distance between the listener and the source *increases* at a constant rate. *The change in pitch produced by the relative motion of the source and the observer is known as the* ***Doppler effect.*** Of course, the frequency of the sound emitted by the source remains unchanged, as does the velocity of the sound in the transmitting medium. Figure 11-8 illustrates this principle.

The Doppler effect is named for the Austrian physicist Christian Johann Doppler (1803–1853).

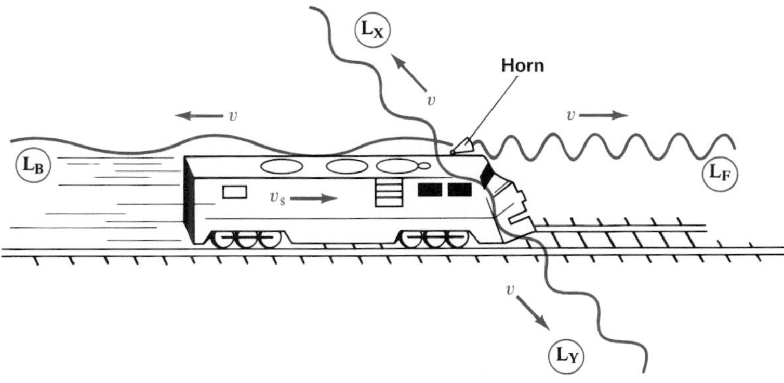

Figure 11-8. Doppler effect. The pitch of the horn on the speeding diesel locomotive sounds normal to the listeners L_X and L_Y, off to the sides, higher to the listener L_F, in front, and lower to the listener L_B, in back.

Suppose that a source of sound is stationed at **S** in Figure 11-9 and that one listener is at $\mathbf{L_F}$ and another is at $\mathbf{L_B}$. If the source emits a sound of frequency f_S and the velocity of sound in the medium is v, the wavelength λ is

$$\lambda = \frac{v}{f_S} \qquad \text{(Equation 1)}$$

The time t in this diagram is unit time of 1 second. Thus the distance $\mathbf{SL_F} = \mathbf{SL_B} = vt$, that is, the distance the sound travels in 1 second. Then there are f_S waves in both distances $\mathbf{SL_F}$ and $\mathbf{SL_B}$, each of wavelength λ.

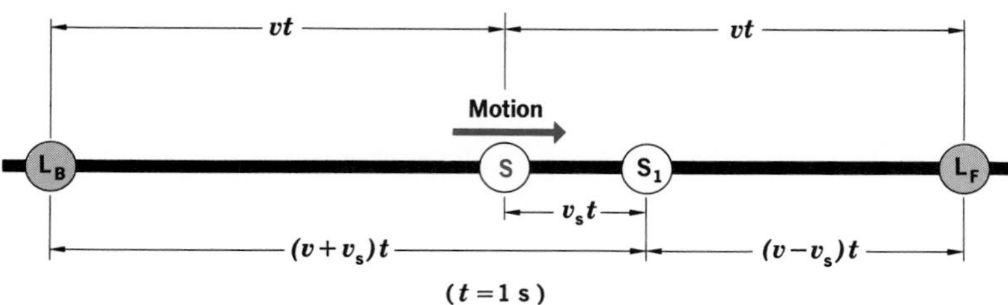

Figure 11-9. Sound waves emitted in unit time by the source **S** during its movement from **S** to $\mathbf{S_1}$ with speed v_S are contained within the distance $\mathbf{S_1L_F}$ in the direction of listener $\mathbf{L_F}$ in front of the source, and within the distance $\mathbf{S_1L_B}$ in the direction of the listener $\mathbf{L_B}$ behind the source.

Now suppose that the source is moving toward $\mathbf{L_F}$ and away from $\mathbf{L_B}$ with speed v_S and in 1 second travels from **S** to $\mathbf{S_1}$, a distance $v_S t$. During this time, f_S vibrations are emitted by the source. The first vibration is emitted at **S**. At the end of the 1-second interval this vibration is at $\mathbf{L_F}$ in front of the source, and also at $\mathbf{L_B}$ behind the source. Remember that the advancing wave fronts are spherical. The last vibration just emitted at the end of the 1-second interval is at $\mathbf{S_1}$. Therefore, the same number of vibrations, f_S, are in the front region $\mathbf{S_1L_F}$ and the back region $\mathbf{S_1L_B}$ in the 1-second interval. These two distances are respectively

$$\mathbf{S_1L_F} = (v - v_S)t \qquad \text{(Equation 2a)}$$

and

$$\mathbf{S_1L_B} = (v + v_S)t \qquad \text{(Equation 2b)}$$

The wavelength λ' of the sound reaching the stationary listener in front of the moving source is shorter than λ of Equation 1 and is given by

$$\lambda' = \frac{v - v_S}{f_S} \qquad \text{(Equation 3)}$$

The velocity of the sound in the medium, v, is independent of the motion of the source. Thus the frequency of the sound reaching the stationary observer in front of the moving source, f_{LF}, is

$$f_{LF} = \frac{v}{\lambda'} \qquad \text{(Equation 4)}$$

Substituting the expression for λ' given in Equation 3 into Equation 4 yields

$$f_{LF} = \frac{v}{\dfrac{v - v_S}{f_S}}$$

or

$$f_{LF} = f_S \frac{v}{v - v_S} \qquad \text{(Equation 5)}$$

Equation 5 shows us that *the frequency of the sound reaching the stationary listener in front of the moving source is higher than the frequency of the source.* Therefore, the sound that is heard has a higher pitch than that emitted by the source.

We shall now turn to the other listener, $\mathbf{L_B}$, in Figure 11-9 and show how the pitch of the sound heard is related to the frequency of the source. *The wavelength λ' of the sound reaching the stationary listener behind the moving source is longer than λ of Equation 1* and can be expressed as

$$\lambda' = \frac{v + v_S}{f_S} \qquad \text{(Equation 6)}$$

As in Equation 4 the frequency of the sound reaching the observer behind the moving source, f_{LB}, is

$$f_{LB} = \frac{v}{\lambda'} \qquad \text{(Equation 7)}$$

Substituting the expression for λ' given in Equation 6 in Equation 7, we get

$$f_{LB} = f_S \frac{v}{v + v_S} \qquad \text{(Equation 8)}$$

Equation 8 indicates that *the frequency of the sound reaching the stationary listener behind the moving source is lower than the frequency of the source.* Thus the sound that is heard has a lower pitch than that emitted by the source.

If the listener is moving toward a stationary source, the pitch of the sound heard is higher than f_S, but it is not exactly that indicated by Equation 5. If the listener has a closing velocity v_{LC}, then v_{LC}/λ waves per second are received in addition to the f_S emitted by the source. From Equation 1,

$$f_S = \frac{v}{\lambda}$$

so,

$$f_{LC} = \frac{v}{\lambda} + \frac{v_{LC}}{\lambda}$$

or

$$f_{LC} = \frac{v + v_{LC}}{\lambda} \qquad \text{(Equation 9)}$$

Substituting the expression for λ given in Equation 1 in Equation 9, we have

$$f_{LC} = f_S \frac{v + v_{LC}}{v} \qquad \text{(Equation 10)}$$

Equation 10 shows the frequency of the sound reaching the listener moving toward the stationary source. Thus a sound is heard that has a higher pitch than that emitted by the source but not the same pitch as given in Equation 5.

The listener moving away from the stationary source with an opening velocity v_{LO} will receive v_{LO}/λ waves per second fewer than f_S emitted by the source.

$$f_{LO} = \frac{v - v_{LO}}{\lambda} \qquad \text{(Equation 11)}$$

Substituting for λ as before,

$$f_{LO} = f_S \frac{v - v_{LO}}{v} \qquad \text{(Equation 12)}$$

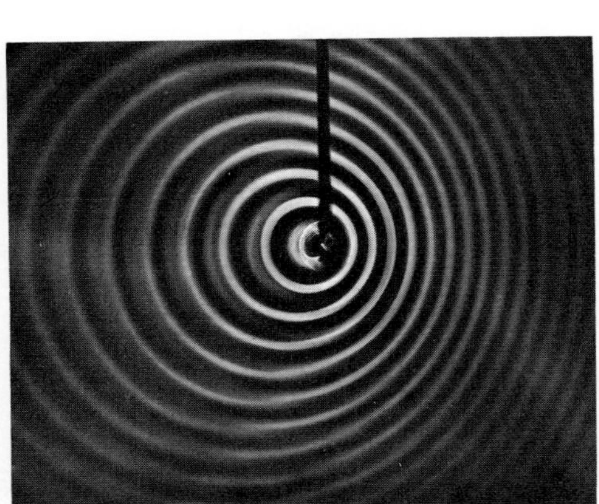

Figure 11-10. Doppler effect due to motion of the source. This effect is observed in a ripple tank in which the source of the disturbance is moving from left to right. Observe the crowding effect on the waves in front of the source and the spreading effect on those behind the source.

From Equation 12 we can determine the frequency of the sound reaching a listener moving away from the stationary source. Observe that the pitch is lower than that emitted by this source, but is not the same pitch as given in Equation 8.

We have considered the special cases in which the velocities of the source and the observer lie along the line common to both. If the medium has a velocity along this line joining the source and the listener, it must be considered also.

The Doppler effect has many important applications. Traffic police measure the speed of approaching and receding cars by observing the Doppler shift of radio waves from a stationary radar gun and then converting the shift into miles per hour (or kilometers per hour). Astronomers use the Doppler shift of light waves to measure the rotational velocities of planets and the radial velocity of celestial bodies. The large shift of spectral lines toward the red in distant galaxies is an important part of the Big Bang theory of the universe, which states that the universe is rapidly expanding.

QUESTIONS: GROUP A

1. What are the requirements for the production of a sound wave?
2. A dog can hear a sound produced by a dog whistle while his owner cannot. Explain.
3. How does the speed of sound vary with (a) the density of an elastic material and (b) the temperature of the material?

4. Distinguish between (a) intensity and loudness (b) frequency and pitch.
5. In an old western movie, the train robber is seen laying his ear against a rail. Why?
6. The relative intensity of sound in an average home is 40 dB, while that of street traffic is 80 dB. Is street traffic twice as loud as conversation?
7. Why is the threshold of hearing a curve, rather than a single point?
8. What is the difference between music and noise?
9. What determines the pitch of a complex musical sound?
10. (a) What is the range of frequencies to which the human ear is sensitive? (b) What is this range called?

GROUP B

11. (a) Look up the word *sonar*. Relate it to this chapter. (b) Why is sonar more practical under water than in the air?
12. An oboe and a French horn are being played at the same fundamental frequency. How can you tell them apart?
13. Why is the decibel scale logarithmic?
14. Why is there a well-defined upper limit to the frequency range of the sonic spectrum?
15. What is the frequency of a tone that is three octaves higher than 256 Hz?
16. You are at a street corner and hear an ambulance siren. Without looking at it, how can you tell when the ambulance passes by you?
17. You are listening to the soundtrack of race cars going around a track. Is there a way to distinguish the relative speeds of the cars?
18. You see a flash of lightning and hear the crash of the thunder a second later. Is the sound that reaches you loud or soft? Why?
19. (a) What name is given to the Doppler effect seen in light coming from distant stars? (b) Explain the significance of the name. (c) What information do we get about the motion of stars relative to the earth? (d) What does this tell us about what is happening to the universe?
20. (a) What is a chord? (b) How must the frequencies that make up a chord be related?
21. (a) How is the intensity of a sound related to the amplitude? (b) Why does intensity have this relationship to amplitude?

PROBLEMS: GROUP A

1. You look up and see a helicopter pass directly overhead: 3.10 s later you hear the sound of the engines. If the air temperature is 23.0°C, how high was the helicopter flying?
2. On a day when the outdoor air temperature is 10.5°C, you use a whistle of frequency 2.50×10^4 Hz to call your dog. What is the wavelength of the sound produced by the whistle?
3. You see a flash of lightning and hear the clap of thunder 4.00 s later. If the air temperature is 28.5°C, how far away is the storm?
4. At a track meet, you see the flash of the starter's pistol. If you are 155 m from the start and the air is at 20.0°C, how long will it be before you hear the report of the pistol?
5. What is the intensity range, in W/m^2, of a normal conversation?

GROUP B

6. A car moving at 72.0 km/h is sounding its horn as it approaches you. If the actual frequency of the horn is 5.50×10^2 Hz, what frequency will you hear at 25°C?
7. The relative intensity of the sound of a jet engine is 165 dB at a distance of 4.65 m from the plane. (a) What is its

intensity in W/m^2? (b) What is the intensity of the engine at a distance of 12.3 m from the plane?

8. A hobo is listening with his ear against the rail for an oncoming train. When the train is 1.65 km away, what is the time lag between the sound heard by the hobo and by his friend, who is standing nearby? (Assume the air temperature is 25.0°C.)
9. (a) What is the intensity of the sound produced by a 50.0-W stereo system at a distance of 2.85 m from the stereo? (b) What is the relative intensity in decibels?
10. On a day when the temperature is 25.0°C, a child drops a stone into a well that is 75.2 m deep. How much time passes before the child hears the stone hit the water?
11. A piece of machinery in a factory produces a relative intensity of 80.0 dB. What would be the relative intensity if the factory installed 12 more of these machines?
12. The frequency of a siren on a fire station is 3.50×10^3 Hz. If you are driving away from the fire house at a speed of 40.0 km/h, what frequency will you hear?

CHARACTERISTICS OF SOUND WAVES

11.10 Fundamental Tones Suppose a piano wire is stretched about a meter in length between two clamps and drawn tight enough to vibrate when plucked. If plucked in the middle, the wire vibrates *as a whole,* as shown in Figure 11-11(A). *A taut wire or string that vibrates as a single unit*

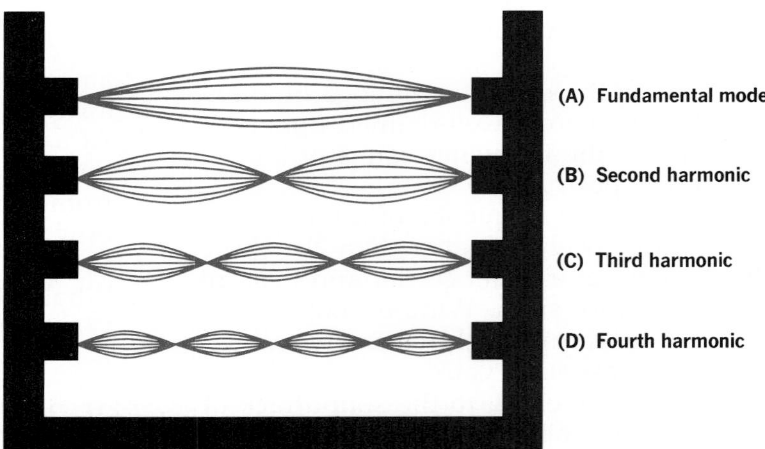

Figure 11-11. A strobe-flash view of some vibration modes of a single string.

produces its lowest frequency, called its ***fundamental.***

A vibrating string, having a small surface area, cuts through the air without disturbing it effectively. The string itself transfers very little of its vibrational energy to the air and therefore produces only a feeble sound. If, however, its vibrations are transferred to a larger surface, such as a sounding board, the air will be disturbed more effectively and a louder sound will be produced. The sounding board enables a substantial quantity of the vibrational energy to be transferred to the air. In this sense, the sounding board can be thought of as an impedance transformer (Section 10.11), which reduces the impedance discontinuity between the vibrational source and the air. These considerations are enlarged upon in Section 11.14.

The properties of vibrating strings and the sounds they produce can be studied using an instrument called a *sonometer*, which is shown in Figure 11-12. The sonometer may have two or more wires or strings stretched over a sounding board, which reinforces the sound produced by the wires. Because wires may vary in diameter, tension, length, and composition, a sonometer is useful for studying the effects of these variables on the vibrational characteristics of the wires.

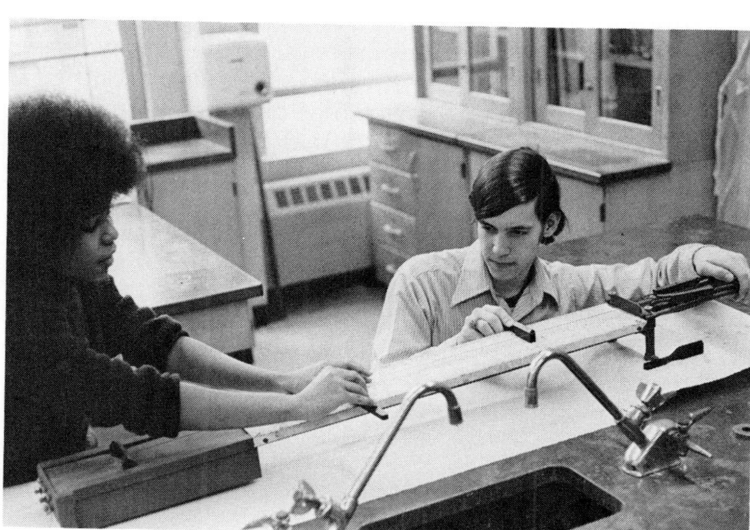

Figure 11-12. The sonometer can be used to study the vibrational characteristics of strings and wires.

By plucking a string, energy is transferred to the string and causes it to vibrate in transverse wave motion. Just as a vibrating reed transfers energy to molecules of air gases and causes them to exhibit longitudinal wave motion, so a vibrating string transfers energy to these gas molecules and causes them to vibrate and transmit a longitudinal wave. The frequency of the longitudinal vibration of the molecules, which is the frequency of the sound wave, is the same as the frequency of the transverse vibration of the string. Since the production of sound by a vibrating string dissipates energy, energy must be continuously supplied to the string by plucking or bowing to maintain the sound.

11.11 Harmonics A string may vibrate as an entire unit and may also vibrate in two, three, four, or more segments depending on the standing wave pattern that is established on the string. See Figure 11-11(B–D). When a string is plucked or bowed, not only the fundamental mode but other, higher modes of vibration may be present.

Suppose a string vibrates in two segments with a node in the middle, as in Figure 11-11(B). Its vibrating frequency is twice that of the string vibrating as a single segment. Doubling the frequency of a sound wave raises the pitch one *octave*. When the string vibrates in four segments, its vibrating frequency (and that of the sound waves produced) is four times its fundamental frequency. The pitch of the sound produced is then two octaves above the fundamental. *The fundamental and the vibrational modes having frequencies that are whole-number multiples of the fundamental are called* ***harmonics.*** Only these frequencies can be produced by plucking or bowing the string.

By this definition the fundamental is the *first harmonic* because it is *one* times the fundamental. The vibrational mode having a frequency twice that of the fundamental is the *second harmonic*. The vibrational mode of frequency *three* times that of the fundamental is the *third harmonic*, and so on. Oscillograms of a fundamental mode and several harmonics are shown in Figure 11-13.

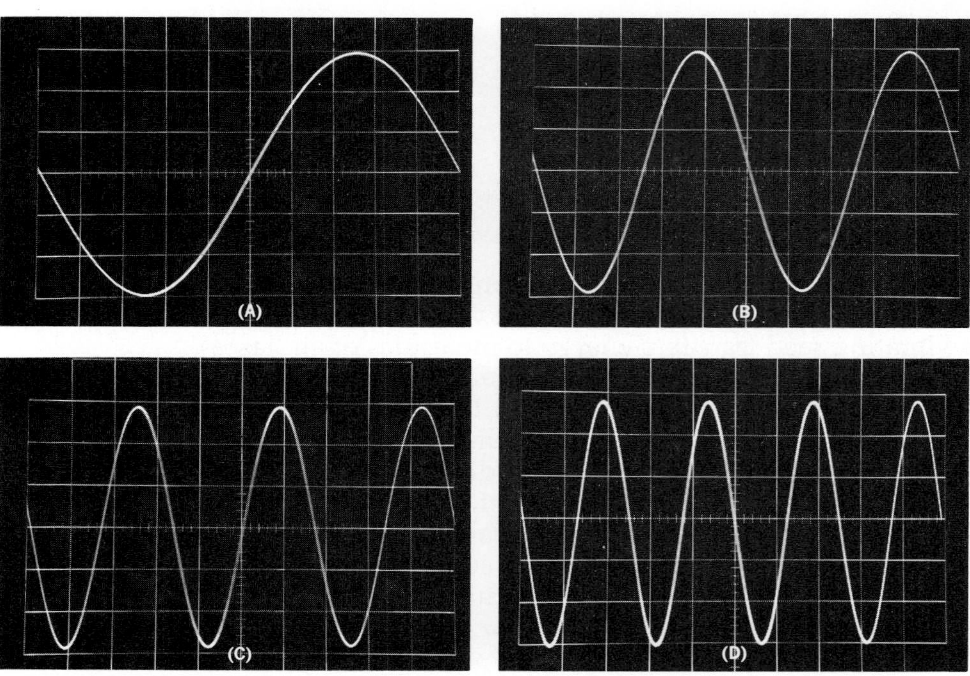

Figure 11-13. Oscillograms of (A) a fundamental, (B) the second harmonic, (C) the third harmonic, and (D) the fourth harmonic. These oscillograms are traces of light on the screen of a cathode-ray tube (similar to a television picture tube) that presents a graph of the instantaneous values of the amplitude of the sound wave as a function of time. The instrument that contains the cathode-ray tube is called an oscilloscope.

11.12 The Quality of Sound It is not difficult to pick out the sounds produced by different instruments in an orchestra, even though they may be producing the same tone with equal intensity. The difference is a property of sound called *quality*, or *timbre*.

If the string of a sonometer is touched very lightly in the middle while being bowed near one end, it will vibrate in two segments and as a whole at the same time, as Figure 11-14 illustrates. The fundamental is audible, but added to it is the sound of the second harmonic. The combination is richer and fuller; the quality of the sound is improved by the addition of the second harmonic to the fundamental. See Figure 11-15(A). *The* ***quality*** *of a sound depends on the number of harmonics produced and their relative intensities.* When stringed instruments are played, they are bowed, plucked, or struck near one end to enhance the production of harmonics that blend with the fundamental and give a richer sound. See Figure 11-15(B).

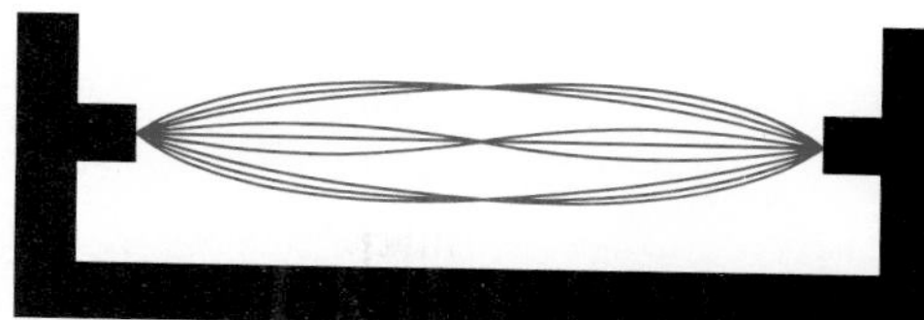

Figure 11-14. A strobe-flash view of a single string vibrating in its fundamental and second harmonic modes simultaneously.

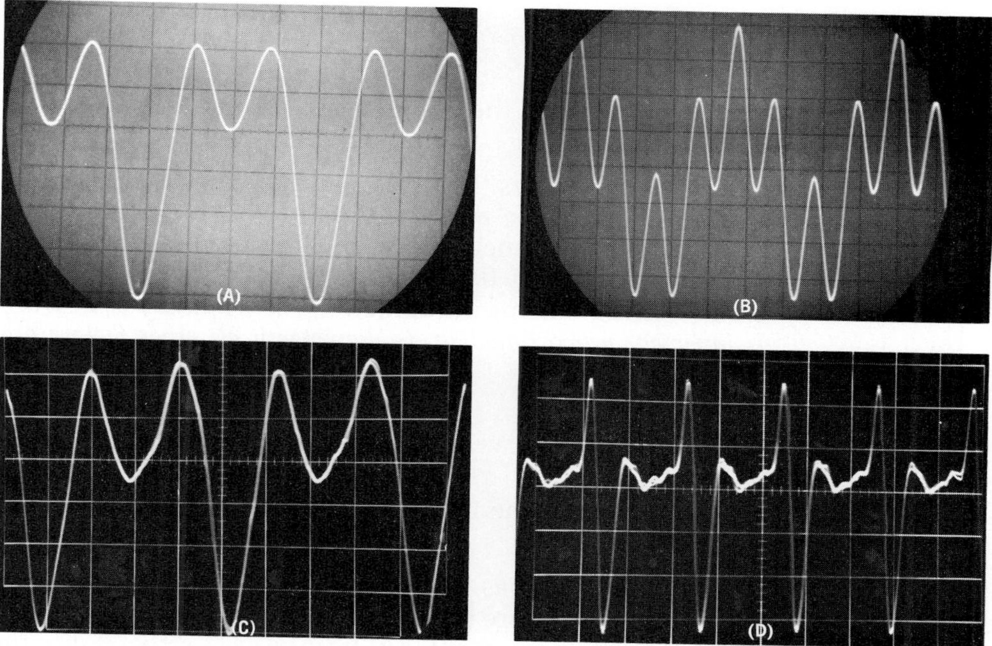

Figure 11-15. Oscillograms of (A) a fundamental and second harmonic, (B) a fundamental and fourth harmonic, (C) the sound of a French horn, and (D) the sound of a trumpet.

The quality of the tones produced by orchestral instruments varies greatly. For example, the tone produced by a French horn consists almost entirely of the fundamental and the second harmonic. This harmonic content can be recognized by comparing Figures 11-15(A) and 11-15(C). The tonal quality of the trumpet is due to the intensity of its high-frequency harmonics. See Figure 11-15(D).

We interpret the harmonic content of a sound in terms of its quality.

Figure 11-16. The laws of strings are used by the guitarist to produce a wide range of musical tones.

11.13 The Laws of Strings The frequency of a vibrating string is determined by its *length, diameter, tension,* and *density*. The strings of a piano produce tones with a wide range of frequencies. If we examine them we find that the strings that produce low-frequency tones are long, thick, and loose, while the strings that produce high-frequency tones are short, thin, and taut. One string produces a tone loud enough for the low-pitched notes, but three strings are needed to produce high-pitched notes of comparable loudness.

The conditions affecting the frequency of vibrating strings are summarized in four *laws of strings.*

1. *Law of lengths.* A musician may raise the pitch produced by a stringed instrument by shortening the vibrating length of the string. A violinist shortens the length of the A string about 2.5 cm to produce the note B. *The frequency of a string is inversely proportional to its length if all other factors are constant.*

$$\frac{f}{f'} = \frac{l'}{l}$$

Here f and f' are the frequencies corresponding to the lengths l and l'.

2. *Law of diameters.* In a piano and other stringed instruments like the cello and the guitar, the strings that produce higher frequencies have smaller diameters. A string with a diameter of 0.1 cm has twice the frequency of a similar string 0.2 cm in diameter. *The frequency of a string is inversely proportional to its diameter if all other factors are constant.*

$$\frac{f}{f'} = \frac{d'}{d}$$

Here f and f' are the frequencies corresponding to the diameters d and d'.

3. *Law of tensions.* When stringed instruments are tuned, the strings are tightened to increase their frequency or loosened to decrease it. *The frequency of a string is directly proportional to the square root of the tension on the string if all other factors are constant.*

$$\frac{f}{f'} = \frac{\sqrt{F}}{\sqrt{F'}}$$

Here f and f' are the frequencies corresponding to the tensions F and F'. See the following example.

4. *Law of densities.* The more dense a string is, the lower its frequency. Usually three of the strings on a violin are of plain gut; the fourth is wound with fine wire to increase its density so that it can produce the low-frequency tones. *The frequency of a string is inversely proportional to the square root of its density if all other factors are constant.*

$$\frac{f}{f'} = \frac{\sqrt{D'}}{\sqrt{D}}$$

Here f and f' are the frequencies corresponding to the densities D and D'.

EXAMPLE A violin string is 0.035 m long and is stretched with a tension of 27 N, so that it vibrates with a frequency of 256 Hz. What is the frequency when the length is 0.030 m and the tension is 32 N?

Given	Unknown	Basic equations
$l = 0.035$ m $f = 256$ Hz $l' = 0.030$ m $F = 27$ N $F' = 32$ N	f'	$\frac{f}{f'} = \frac{l'}{l}$ $\frac{f}{f'} = \frac{\sqrt{F}}{\sqrt{F'}}$

Solution

Working equation: $f' = f\frac{l\sqrt{F'}}{l'\sqrt{F}} = \frac{(256\text{ Hz})(0.035\text{ m})(32\text{ N})^{1/2}}{(0.030\text{ m})(27\text{ N})^{1/2}}$

$= 330$ Hz

11.14 Forced Vibrations When a tuning fork is struck with a rubber hammer, it vibrates at its fundamental frequency together with some low-order harmonics. The fundamental has a natural frequency that depends upon the fork length, thickness, and composition. When a key on a piano is struck, the piano string vibrates at its fundamental frequency and at harmonics of this frequency. The only external force that affects these natural rates of vibration is friction.

Suppose we strike a tuning fork and then press its stem against a table top. The tone becomes louder when the fork is in contact with the table because the fork *forces* the table top to vibrate with the same frequency. Since the table top has a much larger vibrating area than the tuning fork, these **forced vibrations** produce a more intense sound.

A vibrating violin string stretched tightly between two clamps does not produce a very intense sound. When the string is stretched across the bridge of a violin, however, the wood of the violin is forced to vibrate in response to the vibrations of the string; the intensity of the sound is increased by these forced vibrations. The sounding board of a piano acts in the same way to intensify the sounds produced by the vibrations of its strings.

11.15 Resonance The two tuning forks in Figure 11-17 have the same frequency. They are mounted on sounding boxes that increase the intensity of the sound through forced vibrations. One end of each sounding box is open.

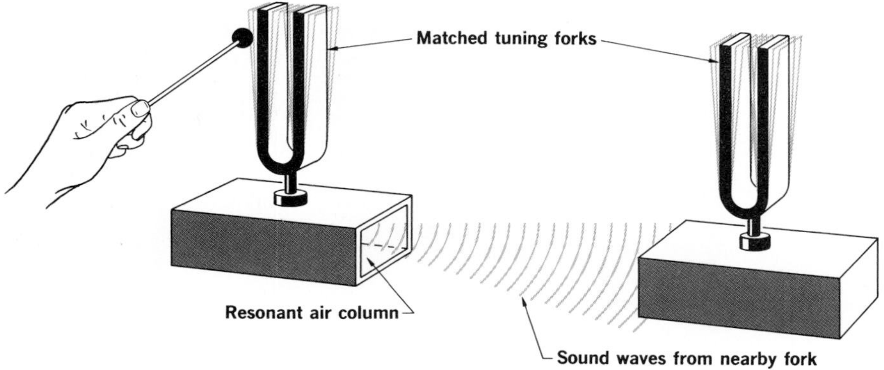

Figure 11-17. Resonance between two matched tuning forks.

Suppose we place these forks a short distance apart, with the open ends of the boxes toward each other. Now let us strike one fork and, after it has vibrated several seconds, touch its prongs to stop them. We find that the other fork is vibrating weakly. The compressions and rarefactions produced in the air by the first tuning fork act on the second fork in a regular fashion, causing it to vibrate. Such action is called **resonance,** or *sympathetic vibration.* A person who sings near a piano may cause the piano strings that produce similar frequencies to vibrate. Resonance occurs when the natural vibration rates of two objects are the same or when the vibration rate of one of them is equal to one of the harmonics of the other. If we changed the frequency of one of the tuning forks by adding mass to one of its prongs, we would alter the conditions of the experiment, and no resonance would be evident. Both forks should have the same natural frequency to produce this resonant effect.

Figure 11-18 illustrates a method of producing resonance between a tuning fork and a column of air. The vibrating fork is held above the hollow cylinder immersed in water. The length of the air column is gradually increased by raising the cylinder. As seen in Figure 11-18(A), there is a marked increase in the loudness of the sound. Here the column of air in the tube vibrates vigorously at the frequency of the tuning fork, and the two are in resonance.

This behavior is similar to the production of a standing wave on a stretched string. During resonance, a compression of the reflected wave unites with a compression of the direct wave, and a rarefaction of the reflected wave unites with a rarefaction of the direct wave. The constructive superposition of the two waves in phase amplifies the sound.

The resonant air column is simply a standing longitudinal wave system. The water surface in Figure 11-18(A) closes the lower end of the tube and prevents longitudinal displacement of the molecules of the air immediately adjacent to it. This termination effectively clamps the air column at the closed end and gives rise to a displacement node. Because the upper end of the tube is open, it provides a free-ended termination for the air column and gives rise to a displacement antinode, or loop. These are, respectively, shown in Figure 11-18(A) as **N** and **L.**

Because the air particles at the closed end of the tube are unable to undergo longitudinal displacement, they experience maximum changes in pressure. This position of a displacement node is also a pressure antinode, or a point of maximum pressure change. (Refer to Figure 11-3 for the phase relationship between displacement and pressure waves.) Compressions are reflected as compressions, and rarefactions are reflected as rarefactions. This behavior amounts to a change in phase upon reflection from a rigid termination in the sense that the displacement vector of the air particles is inverted. See Figure 11-19(A).

The open end of the tube is a displacement loop and, consequently, a pressure node. Here the pressure remains at the constant value of the outside atmosphere, and air rushes into and out of the tube as the column vibrates. It is this large movement of air at the open end of the tube that

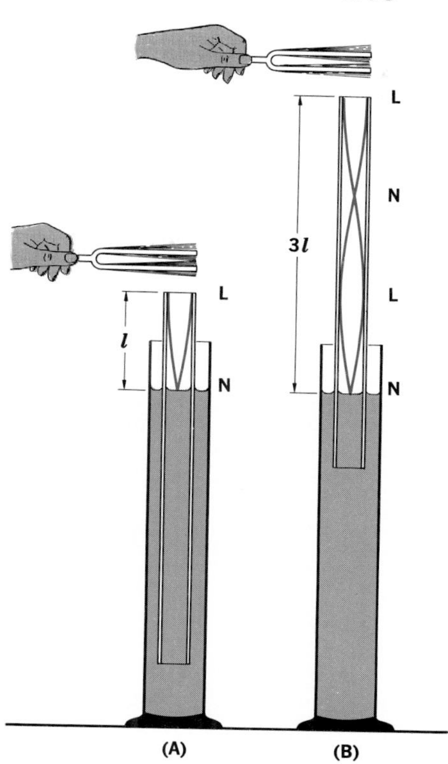

Figure 11-18. Resonance between a tuning fork and a vibrating air column. (A) and (B) give greater sound reinforcement than would any water-level positions in between.

Figure 11-19. A compression traveling from left to right is reflected at the closed end in (A) as a compression and at the open end in (B) as a rarefaction. The particle displacement vector in the closed tube is inverted by reflection, but it remains unchanged in the open tube because the displacement of air layers is in the same direction for a rarefaction traveling to the left as for a compression traveling to the right.

Particle displacement:
before reflection →
after reflection ←
(A) Closed tube

Particle displacement:
before reflection →
after reflection →
(B) Open tube

transfers energy to the atmosphere and provides the reinforcement of the sound produced by the tuning fork.

The fundamental frequency of the resonant column corresponds approximately to a displacement node at the closed end and an adjacent displacement loop at the open end, as shown in Figure 11-18(A). Since the distance separating a node and an adjacent loop of a standing wave is one-fourth wavelength, *the length of the closed tube is approximately one fourth the wavelength of its fundamental resonant frequency.*

$$\lambda \simeq 4l$$

By applying a small empirical correction proportional to the diameter of the tube (because the motion of molecules of the air at the open end of the tube is not strictly in one dimension), we can state the relationship more precisely as follows:

$$\lambda = 4(l + 0.4d)$$

λ is the wavelength of the fundamental resonant frequency, l is the length of the closed tube, and d is its diameter.

If the cylinder of Figure 11-18(A) is lifted higher out of the water, resonance will occur again when the length is three quarters of a wavelength. This length of air column also corresponds to a displacement node at the closed end and a loop at the open end and allows a standing wave to develop. See Figure 11-18(B). For a tube that is long enough, successively weaker resonance points could be found at $5/4\lambda$, $7/4\lambda$, etc. Thus *a closed tube is resonant at odd quarter-wavelength intervals.*

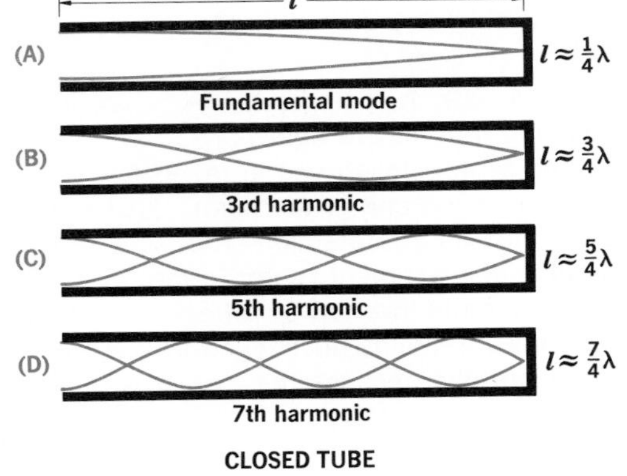

Figure 11-20. The normal modes of oscillation of a closed tube. The fundamental frequency is half that of an open tube of the same length, and only odd harmonics are produced as shown by the standing wave patterns.

The quarter-wave resonant column for the fundamental frequency of the tuning fork of Figure 11-18(A) is three quarters of a wavelength long for the third harmonic of this frequency. It is $5/4\lambda$ for the fifth harmonic and is $7/4\lambda$ for the seventh harmonic. Therefore *the resonant frequencies of a closed tube are harmonics, but only odd harmonics of the fundamental mode are present.* Figure 11-20 illustrates this.

Many musical instruments employ vibrating air columns open at one end and closed at the other *(closed tube)* or open at both ends *(open tube).* From the above discussion we have shown that the normal modes of oscillation of air columns are characterized by

1. *a displacement node (or pressure loop) at a closed end,* and
2. *a displacement loop (or pressure node) at an open end.*

Compressions traveling in an open tube are reflected at the open ends as rarefactions, and rarefactions are reflected as compressions. In such reflections, the longitudinal displacement vector of the air particles is *not* inverted, and in this sense there is no change in phase at the open termination of the air column. This condition is illustrated in Figure 11-19(B).

The fundamental frequency of a resonant air column in an open tube corresponds approximately to displacement loops at opposite ends and a displacement node in the middle, as shown in Figure 11-21(A). Since adjacent loops of a standing wave are one-half wavelength apart, *the length of an open tube is approximately one half the wavelength of its fundamental resonant frequency.*

$$\lambda \simeq 2l$$

Again, by applying a small empirical correction proportional to the diameter of the tube, we can state the relationship more precisely as follows:

$$\lambda = 2(l + 0.8d)$$

λ is the wavelength of the fundamental resonant frequency, l is the length of the open tube, and d is its diameter.

An open tube, which is a half-wave resonant column at its fundamental frequency, is a full wavelength long at the second harmonic of this frequency. It is $3/2\lambda$ at the third harmonic. In each case displacement loops occur at both open ends, and the column resonates in these modes. Therefore *the resonant frequencies of an open tube are harmonics, and all harmonics of the fundamental mode are present.* See Figure 11-21.

The quality of a musical tone is generally enhanced by its harmonic content. The open tube resonator has both odd and even harmonics of the fundamental mode present, whereas the closed tube resonator has only odd harmonics. Thus the quality of the sounds from open tubes and closed tubes is not the same.

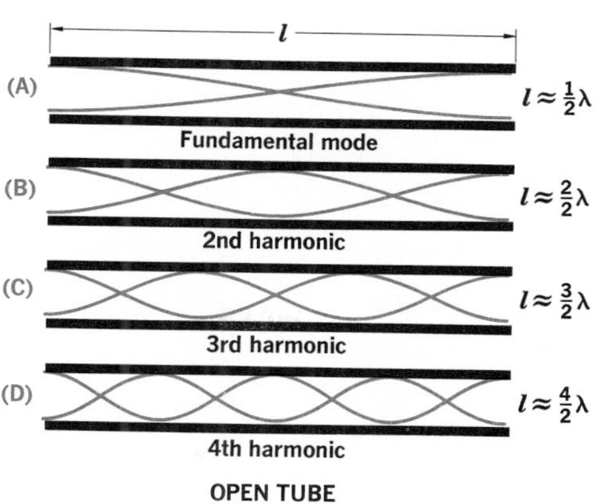

Figure 11-21. The normal modes of oscillation of an open tube. The fundamental frequency is twice that of a closed tube of the same length, and all harmonics are produced.

11.16 Beats

We have already discussed interference of transverse waves in strings and of water waves in the ripple tank in Chapter 10. We were concerned with standing waves in resonant air columns in the preceding section. There is abundant experimental evidence that two or more wave disturbances can travel through the same medium independently of one another. The superposition principle shows us that the displacement of a particle of a medium at any time is the vector sum of the displacements it would experience from the individual waves acting alone.

A standing wave is formed by two wave trains of the same frequency and amplitude traveling through a medium in opposite directions. We are generally concerned with the conditions or behavior of the space in which the standing wave exists. This generalization is true for resonant air columns used in musical instruments. The standing wave is characterized by an amplitude *that varies with distance or position in space.*

Two wave trains of slightly different frequencies traveling in the same direction through a medium will interfere in a different way. At any fixed point in the medium through which the waves pass, their superposition gives a wave characterized by an amplitude *that varies with time.*

In Figure 11-22 the resultant displacement of two wave trains of slightly different frequencies is plotted as a function of time. In curve **(A)** the frequency is 8 Hz, and in **(B)** it is 10 Hz. Curve **(C)** shows the combined effect of these two waves at a fixed point in their pathway. This resultant wave varies periodically in amplitude with time. Such amplitude pulsations are called **beats.**

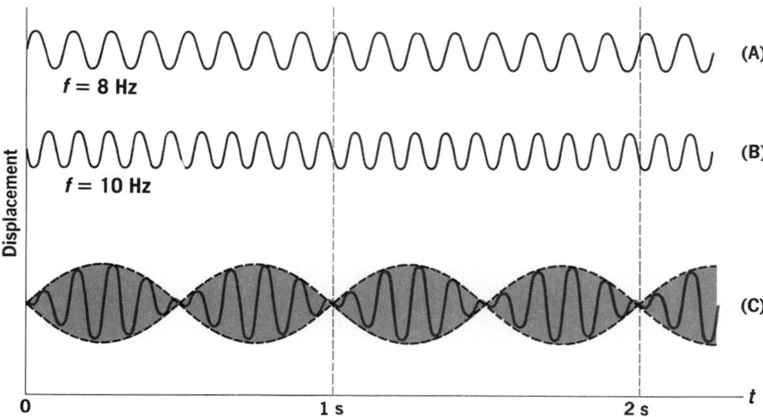

Figure 11-22. Beats. Waves **(A)** and **(B)** of slightly different frequencies combine to give a wave **(C)** that varies in amplitude with time.

Observe that waves **(A)** and **(B)** come into phase two times each second and out of phase the same number of times each second. The **beat frequency** can be described as *two beats per second. The number of beats per second equals the difference between the frequencies of the component waves.*

When the interfering frequencies are audible sounds, the amplitude variations, or beats, are recognized as variations in loudness. The average human ear can distinguish beats up to a frequency of approximately ten per second.

The beat phenomenon is frequently used to tune vibrating systems with great precision. Two vibrating strings can be tuned to the same frequency by adjusting the tension of one until the beats disappear. This procedure is called "zero beating." The piano tuner adjusts frequencies precisely by systematically beating harmonics of notes against each other.

To illustrate, assume that two vibrating strings have fundamental frequencies of 165 Hz and 325 Hz respectively. A person (who possesses a "trained" ear) may distinguish 5 beats per second as a consequence of the 325-Hz tone beating against the second harmonic of the 165-Hz tone.

A precision tuning fork vibrates only in its fundamental mode and produces a tone that is free of harmonics. Suppose a tuning fork of frequency 260 Hz is sounded together with one of 264 Hz. A listener will hear 4 beats each second of sound having an average frequency of 262 Hz.

If the lower frequency is represented by f_L and the higher frequency by f_H, the average frequency f_{av} can be expressed as follows:

$$f_{av} = 1/2(f_L + f_H)$$

The sound is a sine wave of average frequency f_{av} and varying amplitude that reaches a maximum $(f_H - f_L)$ times each second (the beat frequency). When the frequency separation of the two tones is too large for the beats to be distinguished by the listener, a difference *tone* of frequency $(f_H - f_L)$ may be heard if the difference frequency falls within the audio range.

QUESTIONS: GROUP A

1. (a) What is a fundamental frequency? (b) How are harmonics related to the fundamental?
2. A vibrating string disturbs a small amount of air. How can the sound produced be enhanced?
3. How does the frequency of the wave on the string of a musical instrument compare to that of the sound wave produced?
4. What factors influence the quality of a sound?
5. (a) What characterizes a sound as a *forced* vibration? (b) Give an example.
6. (a) Define resonance. (b) Under what conditions does resonance occur?
7. (a) To produce the best resonance, how should the length of an air column closed at one end be related to the wavelength of the sound? (b) How should length and wavelength be related for a column open at both ends?
8. What would happen to the frequency of a vibrating string under the following circumstances: (a) the density is tripled, (b) the length is halved, (c) the tension is quartered, (d) the diameter is doubled?
9. What distinguishes a *beat* from a *difference tone?*

GROUP B

10. Which harmonic is three octaves above the fundamental?
11. (a) In Figure 11-11(C), how many waves are shown? (b) If the distance between nodes in Figure 11-11(D) is 5.0 cm, what is the wavelength of this wave?
12. Marching soldiers always break step to walk across a bridge. Why?

13. What is the reason for wrapping some strings on a guitar with fine wire?
14. Sometimes the wind gusting across the top of a chimney will cause it to "sing." Explain what is happening.
15. Drivers are often cautioned not to sound car horns in tunnels. Why?
16. What conditions must exist for a singer to break a glass with her voice?
17. Research the "Tacoma Narrows Bridge Collapse." Explain what happened to the bridge in light of what you have learned in this chapter.
18. Do you think resonance is confined to sound waves? Elaborate.
19. Small radio speakers have poor reproduction properties for tones of low frequencies. Why are difference tones important here?

PROBLEMS: GROUP A

1. A child's whistle is 15.0 cm long with a diameter of 1.25 cm. What is the frequency of the sound that resonates through the whistle if the temperature is 20.0°C?
2. A physics student discovers that a tuning fork, $f = 3.00 \times 10^2$ Hz, will produce the best resonance with a 6.80-cm diameter tube closed at one end when the tube length is 26.0 cm. (a) What is the speed of the sound? (b) What is the air temperature in the lab that day?
3. Two tuning forks have frequencies of 256 Hz. When a bit of clay is placed on the prong of one tuning fork and both are struck simultaneously, 4 beats per second are heard. What is the frequency of the fork with the clay?
4. A whistle has two short pipes of unequal length. When the whistle is blown, one pipe emits a tone of 1.10×10^3 Hz and the other a tone of 1.294×10^3 Hz. What is the frequency of the difference tone produced?
5. A tension of 253 N on a guitar string gives its fundamental frequency of 452 Hz. If the guitarist wants to tune it to a frequency of 440.0 Hz, what should the tension be?
6. A violin string has a resonant frequency of 512 Hz. If the tension on the string is halved and the length is doubled, what will be the new frequency?
7. One string on an instrument has a density twice that of an adjacent string. If all other factors are equal, and the frequency of the first string is 330.0 Hz, what is the frequency of the second?
8. Two tuning forks of 324 Hz and 336 Hz are sounded simultaneously. (a) What frequency will a listener hear? (b) How many times per second will the amplitude of the sound reach a maximum?
9. A chimney, with the fireplace damper open, is a resonant tube open at both ends. If the chimney is 18.0 cm wide and 5.20 m long, what frequency will cause resonance? (Assume the air in the chimney has a temperature of 38.0°C)
10. A physics student blows across the top of a graduated cylinder that is 23.0 cm tall and 4.00 cm in diameter. If the temperature in the lab is 21.5°C, what is the frequency of the sound produced?
11. The frequencies of two notes in a chord produced by a guitar have a frequency ratio of 3:2. If the tension on the first string is 285 N, what is the tension on the second (assuming all other factors are equal)?
12. An organ pipe open at both ends is 1.23 m long and has a diameter of 10.0 cm. (a) What is its fundamental frequency when the air is 15.0°C? (b) What are the frequencies of the two lowest harmonics produced along with the fundamental tone?

PHYSICS ACTIVITY

Obtain a soda straw. Taper one end to a point with scissors. Chew on the tapered end to flatten it a little. Blow through the tapered end. You have made your own double-reed instrument! With a sharp pair of scissors, carefully snip pieces off the end of the straw as you blow through it. What happens to the pitch?

SUMMARY

Sound is a longitudinal disturbance consisting of a succession of compressions and rarefactions to which the ear is sensitive. The term also applies to similar disturbances above and below the normal range of hearing. A medium having elastic and inertial properties is required for its propagation. The intensity of a sound is the rate at which the wave energy flows through a unit area. The sensory response to sound intensity is loudness, the sensory response to frequency is pitch, and the sensory response to harmonic content is tonal quality.

The sensory response to sounds is nonlinear. Minimum intensity levels for hearing various sound frequencies define the threshold of hearing. Upper intensities that produce pain rather than increased loudness define a threshold of pain. The change in pitch heard when there is relative motion between an observer and a constant-frequency source is called the Doppler effect.

When set in motion, sound generators such as taut strings and air columns vibrate at their fundamental and various harmonic frequencies. The greater the harmonic content of a sound produced by a musical instrument, the higher is its quality and the more pleasing is the auditory response. The frequency of a vibrating string is determined by its length, diameter, tension, and density. The laws of strings enunciate the relationships between frequency and these physical factors.

The intensity of a sound generated by a vibrating source can be increased by a sounding board forced to vibrate at the sound frequency. When forced vibrations are at the natural frequency of the reinforcing body (resonance), maximum reinforcement occurs.

Resonant air columns in the form of open and closed tubes are standing longitudinal wave systems. The resonant length of a closed tube is approximately one-fourth the wavelength of its fundamental frequency, and it resonates at odd harmonics of the fundamental mode. The resonant length of an open tube is approximately one-half the wavelength of its fundamental frequency, and it resonates at all harmonics of the fundamental mode.

Two wave trains of slightly different frequencies traveling in the same direction in a medium interfere in a way characterized by an amplitude that varies with time at any fixed point in the medium. The amplitude pulsations are called beats. The number of beats per second is the difference between the frequencies of the interfering waves.

VOCABULARY

audio spectrum
beat frequency
beats (sound)
bel
decibel
Doppler effect
forced vibrations
fundamental (sound)
harmonics
infrasonic
intensity (sound)
loudness
pitch
quality (sound)
relative intensity
resonance
sonic spectrum
sonometer
threshold of hearing
threshold of pain
ultrasonic

12

The Nature of Light

light n.: the visible part of the electromagnetic spectrum, which also includes radio waves, microwaves, and X rays.

OBJECTIVES

- Describe the properties of light.
- Study the historical development of the wave theory of light.
- Describe the electromagnetic spectrum.
- Describe the photoelectric effect and its importance in the development of quantum theory.
- Describe the basic principle behind the laser.
- Define and describe photometry, the quantitative study of light.

WAVES AND PARTICLES

12.1 Properties of Light The general properties of waves are described in Chapter 10. They are restated and briefly summarized as follows:

1. *Propagation* within a uniform medium is along straight lines.

2. *Reflection* occurs at the surface, or boundary, of a medium.

3. *Refraction*, or bending, may occur where a change of speed is experienced.

4. *Interference* is found where two waves are superposed.

5. *Diffraction*, or bending around corners, takes place when waves pass the edges of obstructions.

These properties are easily recognized in the behavior of sound and water waves. Such disturbances occur in matter, the particles of the medium being set in motion about their equilibrium positions by the passing waves.

Matter is not required for the propagation of light. While light does pass readily through certain kinds of matter, its transmission is unhindered in interstellar space or through an evacuated vessel.

The regular reflection of light from smooth surfaces was known in the time of Plato, almost twenty-four hundred years ago. (Reflection is the subject of Chapter 13.)

The refraction of light at the interface (boundary) of two transparent media of different optical densities was observed by the Greeks as early as the second century A.D.

The Arabian mathematician Alhazen (965–1039) studied the refraction of light and disputed the ancient theory that visual rays emanated from the eye. He demonstrated the refractive behavior of light as it passed from one medium into another of greater optical density. See Figure 12-1. He believed that the angles of incidence and refraction are related, but was unable to determine how they are related. This relationship, now known as *Snell's law,* was established six hundred years later. (Refraction and Snell's law are discussed in Chapter 14.)

Figure 12-1. The Arabian mathematician Alhazen studied the refraction of light nearly a thousand years ago.

The shapes of shadows and the practical use of sight lines for placing objects in a straight line give evidence of the straight-line, or rectilinear, propagation of light. Greek philosophers were familiar with this property of light. Sir Isaac Newton conducted experiments on the separation of light into colors by means of a prism and was aware of the colors produced by thin films.

Rectilinear propagation, reflection, and refraction are the principal light phenomena that were familiar to observers in the seventeenth century. It must have been evident to Newton and his contemporaries that since the transmission of energy from one place to another was involved, only two general theories could explain these properties of light.

In general, energy can be propagated either by particles of matter or by wave disturbances traveling from one place to another. A window pane may be shattered by a moving object, such as a baseball thrown from a distance, or by the concussion ("shock wave") from a distant explosion.

The light phenomena known in Newton's time were rectilinear propagation, reflection, and refraction.

Arguments favoring both a particle (corpuscular) theory and a wave theory were plausible when applied to the properties of light observed in the seventeenth century. The principal advocate of the corpuscular theory was Newton, whose arguments were supported by the French mathematical physicist and astronomer Laplace (1749–1827). The wave theory was upheld principally by Christian Huygens (*hi*-ganz) (1629–1695), a Dutch mathematician, physicist, and astronomer. He was supported by Robert Hooke of England. Because of the plausibility of both theories, a scientific debate concerning the nature of light developed between the followers of Newton and the followers of Huygens and continued unresolved for more than a century. What are some of the arguments that support each of these classical theories of light?

12.2 The Corpuscular Theory Sir Isaac Newton believed that light consists of streams of tiny particles, which he called "corpuscles," emanating from a luminous source. Let us examine the arguments used by those who believed the particle theory best explained the various light phenomena known to them.

1. *Rectilinear propagation.* A ball thrown into space follows a curved path because of the influence of gravity. Yet if the ball is thrown with greater and greater speed, we know that its path curves less and less. We can easily imagine minute particles traveling at such enormous speed that their paths essentially form straight lines.

Newton experienced no difficulty in explaining the rectilinear propagation of light by means of his particle model. In fact, this property of light provided the supporters of the corpuscular theory with one of their strongest arguments against a wave theory. How, they asked, could waves travel in straight lines? A sound can easily be heard around the corner of an obstruction, but a light certainly cannot be seen from behind an obstruction. The former is unquestionably a wave phenomenon. How can the latter also be one?

The simple and direct explanation of rectilinear propagation provided by the particle model of light, together with the great prestige of Sir Isaac Newton, was largely responsible for the preference shown for the corpuscular theory during the seventeenth and eighteenth centuries.

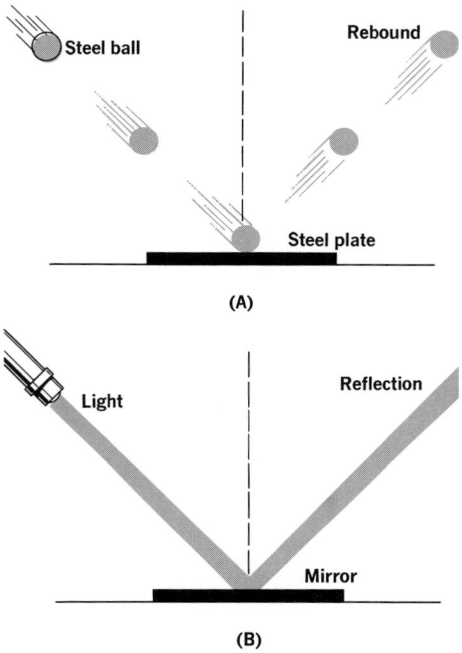

2. *Reflection.* Where light is incident on a smooth surface, such as a mirror, we know that it is regularly reflected. How do particles behave under similar circumstances? Steel ball bearings thrown against a smooth steel plate rebound in much the same way light is reflected. Perfectly elastic particles rebounding from a resilient surface, then, could provide a suitable model for the reflection of light. See Figure 12-2.

Steel ball
Rebound
Steel plate
(A)
Light
Reflection
Mirror
(B)

Figure 12-2. The rebound of a steel ball from a resilient surface resembles the reflection of light from a mirror surface. This example was used as an early argument to "prove" that light rays were streams of tiny particles.

3. *Refraction.* Newton was able to demonstrate the nature of refraction by means of his particle model. We can duplicate this experimentally by arranging two level surfaces, one higher than the other, with their adjacent edges joined by an incline, as shown in Figure 12-3. A ball may be rolled across the upper surface, down the incline, and across the lower surface. Of course, it will experience an acceleration due to the force of gravity while rolling down the incline and will move across the lower surface at a higher speed than it had initially.

Suppose the ball is set rolling on the higher surface toward the incline at a given angle with the normal to the edge. At the incline, the accelerating force exerts a pull on the ball causing it to roll across the lower surface at a smaller angle with the normal to its edge. Now if we think of the upper surface as representing air, the lower surface as an optically more dense medium like water, and the incline as the interface of the two transmitting media, the rolling ball behaves as particles of light being redirected, or refracted, as they pass from air into water.

By varying the grade of the incline while maintaining both constant rolling speed and constant angle with the normal on the upper surface, the refractive characteristics of different transmitting media can be illustrated by the rolling-ball model.

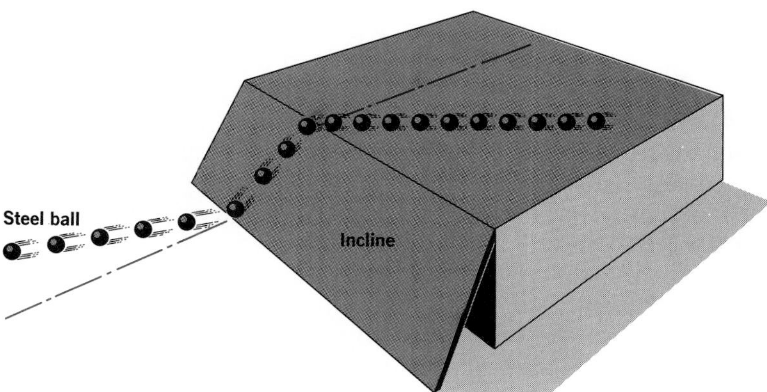

Figure 12-3. A ball bearing rolling from a higher to a lower surface illustrates one aspect of the refraction of light. In what way does this rolling-ball model of refraction fail?

Newton believed that water attracted the approaching particles of light in much the same way gravity attracts the rolling ball on the incline. The rolling-ball experiments imply, as they did to Newton, that light particles accelerate as they pass from air into a medium of greater optical density like water or glass. The corpuscular theory required that the speed of light in water be higher than the speed of light in air. Newton recognized that if it should ever be determined that the speed of light in water is lower than the speed of light in air, his corpuscular theory would have to be abandoned. This uncertainty about the speed of light in water remained unresolved for one hundred twenty-three years after Newton's death. In 1850 the French physicist Jean Foucault (foo-*koh*) (1819–1868) demonstrated experimentally that the speed of light in water is indeed lower than the speed of light in air. This condition was predicted by the wave theory.

Physicist Foucault is recognized primarily for his pendulum experiment, which demonstrates that the earth rotates on its axis in space. The Foucault pendulum experiment is now repeated daily in science museums throughout the world.

12.3 The Wave Theory Christian Huygens is generally considered to be the founder of the wave theory of light. Although somewhat different in its modern form, Huygens' basic concept is still very useful to us in predicting and interpreting the behavior of light. Let us recall a familiar characteristic of water waves as an introduction to this important principle.

If a stone is dropped into a pool of quiet water, it creates a disturbance in the water and a series of concentric waves travels out from the disturbance point. The stone quickly comes to rest on the bottom of the pool, so its action on the water is of short duration. However, wave disturbances persist for a considerable time thereafter and cannot reasonably be attributed to any activity on the part of the stone. It must be that the disturbances existing at all points along the wave fronts at one instant of time generate those in existence at the next instant.

Huygens recognized this logical deduction as a basic aspect of wave behavior and devised a geometric method of finding new wave fronts. His concept, published in 1690 and now recognized as ***Huygens' principle,*** may be stated as follows: *Each point on a wave front may be regarded as a new source of disturbance.*

According to this principle, a wave front originating at a source **S** in Figure 12-4(A) arrives at the position **AB.** Each point in this wave front may be considered as a secondary source sending out wavelets. Thus from points **1, 2, 3,** etc., a series of wavelets develops simultaneously. After a time t these wavelets have a radius equal to vt, where v is the velocity of the wave.

The principle further states that the surface **A′B′**, tangent to all the wavelets, constitutes the new wave front. It is apparent from Figure 12-4 that spherical wave fronts are

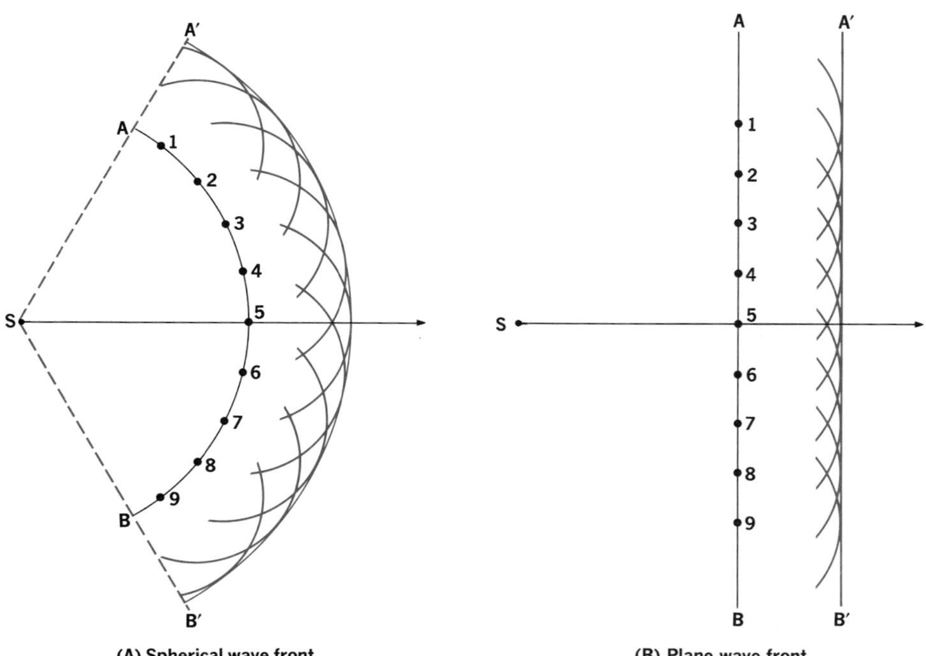

Figure 12-4. By Huygens' principle, every point on an advancing wave front is regarded as a source of disturbance.

propagated from spherical wavelets and planar wave fronts from planar wavelets.

The wave theory treats light as a train of waves having wave fronts perpendicular to the paths of the light rays. In contrast to the particle model discussed earlier, the light energy is considered to be distributed uniformly over the advancing wave front. Huygens thought of a ray merely as a line of direction of waves propagated from a light source.

The supporters of the wave theory were able to satisfactorily explain reflection and refraction of light. The explanation of refraction required that the speed of light in optically dense media, such as water and glass, be *lower* than the speed of light in air. They had trouble, however, explaining rectilinear propagation. This was the primary reason Newton rejected the wave theory.

Before the nineteenth century, interference of light was unknown and the speed of light in such media as water and glass had not been measured. Diffraction fringes or shadows had been observed as early as the seventeenth century. In the absence of knowledge of interference, however, neither Newton nor Huygens attached much significance to this diffraction phenomenon.

In 1801 the interference of light was discovered. This was followed in 1816 by the explanation of diffraction based on interference principles, as discussed in Chapter 15.

These two phenomena imply a wave character and cannot be satisfactorily explained by the behavior of particles. Thus despite the great prestige of Sir Isaac Newton, the corpuscular theory was largely abandoned in favor of the wave theory. The final blow to the corpuscular theory came when Foucault found that the speed of light in water was lower than the speed of light in air. Through the remainder of the nineteenth century the wave concept supplied the basic laws from which came remarkable advances in optical theory and technology.

The corpuscular theory was largely abandoned in favor of the wave theory following the discovery of interference and diffraction of light. Note that the speed of light in water had not yet been measured.

12.4 The Electromagnetic Theory Hot objects transfer heat energy by radiation. If the temperature is high enough, these objects radiate light as well as heat. If a light source is blocked off from an observer, its heating effect is cut off as well. For this reason, a cloud that obscures the sun's light cuts off some of the sun's heat at the same time.

The English physicist Michael Faraday (1791–1867) became concerned with the transfer of another kind of energy while investigating the attraction and repulsion of electrically charged bodies. In 1831 these experiments led him to the principle of the electric generator.

Figure 12-5. Before he was fifteen years old, James Clerk Maxwell (1831–1879), the Scottish theoretical physicist, wrote papers that were recognized for their scientific value. His laws of electromagnetic waves (Maxwell's equations) have the same role in electromagnetism that Newton's laws of gravitation and motion have in mechanics.

Faraday's practical mind required a model to interpret and explain physical phenomena. It was difficult for him to visualize electrically charged objects attracting or repelling each other at some distance with nothing taking place in the intervening space. Thus he conceived a space under stress and visualized *tubes of force* between charged bodies.

Faraday was not an astute mathematician and so did not put his model for this "transmission of electric force" into abstract mathematical form.

The Scottish mathematical physicist James Clerk Maxwell (1831–1879) set out to determine the properties of a medium that would transmit the energies of heat, light, and electricity. By the year 1865 he had developed a series of mathematical equations from which he predicted that all three are propagated in free space at the speed of light as *electromagnetic disturbances.* This unification, the *electromagnetic theory,* brought into common focus the various phenomena of radiation. Maxwell determined that the energy of an electromagnetic wave is equally divided between an electric field and a magnetic field, each perpendicular to the other, and both perpendicular to the direction of propagation of the wave. In Section 10.1 we defined a wave as a periodic disturbance that propagates itself through a medium or space. In this sense, *an* ***electromagnetic wave*** *is a periodic disturbance involving electric and magnetic forces.* A model of an electromagnetic wave at a given instant is shown in Figure 12-6.

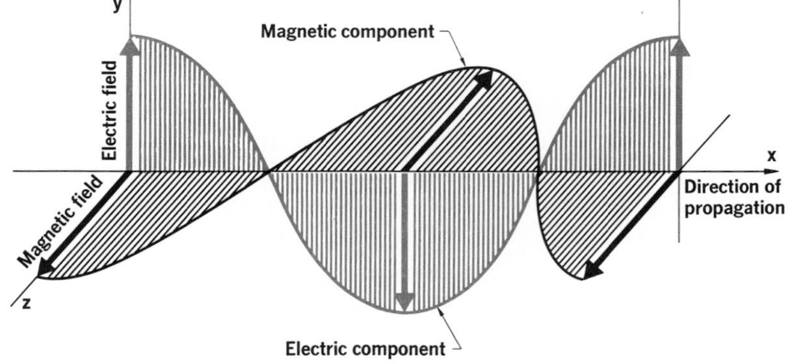

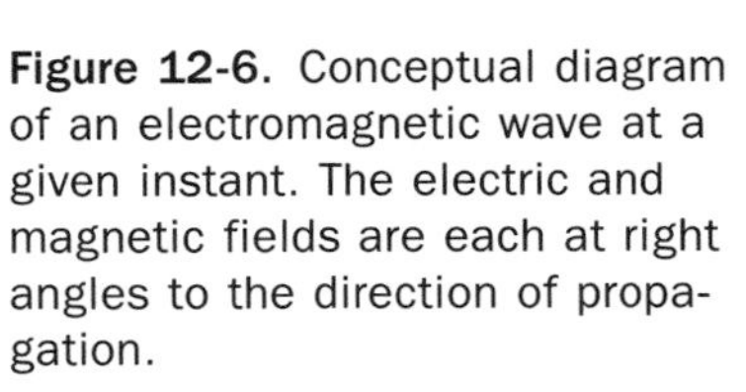

Figure 12-6. Conceptual diagram of an electromagnetic wave at a given instant. The electric and magnetic fields are each at right angles to the direction of propagation.

By 1885, experimental confirmation of the electromagnetic theory was achieved by the German physicist Heinrich Rudolf Hertz (1857–1894). Hertz showed that light transmissions and electrically generated waves are of the same nature. Of course, many of their properties are quite different because of great differences in frequency.

Maxwell's theory of electromagnetic waves seemed to provide the final architecture for optical theory; all known optical effects could now be fully explained. Many physicists felt at this time that all the significant laws of physics had been discovered and that there was little left to do other than develop new and more sophisticated techniques for measuring everything more accurately. In this connection Hertz stated, "The wave theory of light is, from the point of view of human beings, a certainty." Ironically, it was Hertz who was soon to discover a most important phenomenon having to do with the absorption of light energy, one that would create a dilemma involving the wave theory and at the same time set the stage for the *new physics* that was to emerge in the early years of the twentieth century.

Figure 12-7. Heinrich Hertz, the German experimental physicist whose research provided the experimental confirmation of the electromagnetic waves predicted by Maxwell.

12.5 The Electromagnetic Spectrum Electromagnetic energy can be detected and measured by physical means only when it is intercepted by matter and changed into another form of energy such as thermal, electric, kinetic, potential, or chemical energy. Today, the electromagnetic spectrum is known to consist of a tremendous range of radiation frequencies extending from about 10 Hz to more than 10^{25} Hz. All electromagnetic radiations travel in free space with the constant velocity of 3×10^8 meters per second. From the wave equation ($v = f\lambda$) of Section 10.6, it is evident that the wavelengths (λ) of these radiations are inverse functions of their respective frequencies (f). Therefore, the range of the electromagnetic spectrum in terms of radiation wavelengths is from about 3×10^7 meters in the low-frequency region to less than 3×10^{-17} meter in the high-frequency region. See Figure 12-8. In terms of the angstrom (Å), a unit frequently used to express the wavelengths of electromagnetic radiations, the range is from about 3×10^{17} Å to 3×10^{-7} Å, an angstrom being equal to 10^{-10} meter.

Eight major regions of the electromagnetic spectrum are commonly recognized. These regions are based on the general character of the radiations and are identified in Figure 12-8. All kinds of electronic transmissions are accommodated in the *radio-wave* region. Commercial electricity falls within the *power* region. Observe the very small region occupied by the **visible spectrum.**

The optical spectrum includes those radiations, commonly referred to as light, that can be detected visually. Their wavelengths range from approximately 7600 Å to 4000 Å. Accordingly, ***light*** *may be defined as radiant energy*

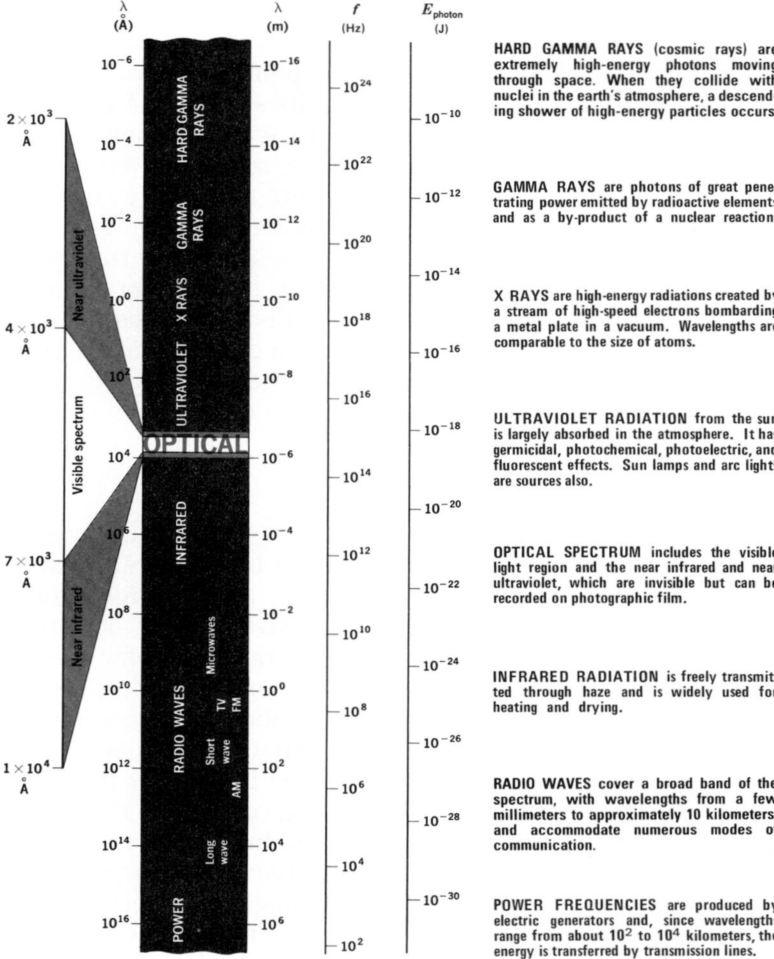

Figure 12-8. The electromagnetic spectrum.

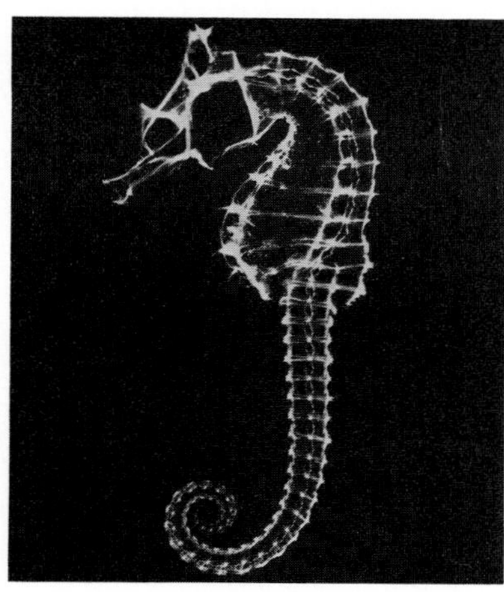

Figure 12-9. X-ray photograph of a sea horse. Locate the X-ray region of the electromagnetic spectrum in Figure 12-8.

that a human observer can see. The optical spectrum also extends into the near infrared and into the near ultraviolet. Although our eyes cannot see these radiations, they can be detected by means of photographic film.

12.6 The Photoelectric Effect As mentioned in Section 12.4, Hertz conducted experiments that led to confirmation of the electromagnetic theory. While studying the radiation characteristics of oscillatory discharges, he observed that a spark discharge occurred more readily between two charged spheres when they were illuminated by another spark discharge. At about the same time, other investigators found that negatively charged zinc plates lost their charge when illuminated by the ultraviolet radiations from an arc lamp. Positively charged plates were not discharged when similarly illuminated.

Observations of the peculiar effects of ultraviolet radiation on metal surfaces led to the discovery of the photoelectric effect, a phenomenon that defied explanation based on the electromagnetic wave theory of light.

Figure 12-10 shows two freshly polished zinc plates **A** and **B** that are sealed in an evacuated tube having a quartz window and are connected externally to a battery and galvanometer (a sensitive current-indicating meter). The quartz window transmits ultraviolet radiation, which does not pass through glass. The galvanometer indicates a small current in the circuit when ultraviolet light falls on the negative plate **A.** If a sensitive electrometer circuit is substituted for the battery, it may be shown that the plate exposed to the ultraviolet light acquires a positive charge.

The results of these experiments imply that the action of the light on the zinc plate causes it to lose electrons. The German physicist Philipp Lenard (1862–1947) published the results of the first quantitative studies of the photoelectric phenomenon in 1902. By measuring the charge-to-mass ratio of the negative electricity derived from an aluminum plate illuminated by ultraviolet light, he was able to prove that electrons were ejected from the metal surface. Such electrons are called **photoelectrons.** Subsequent investigations have shown that all substances exhibit photoemission of electrons. *The emission of electrons by a substance when illuminated by electromagnetic radiation is known as the* ***photoelectric effect.***

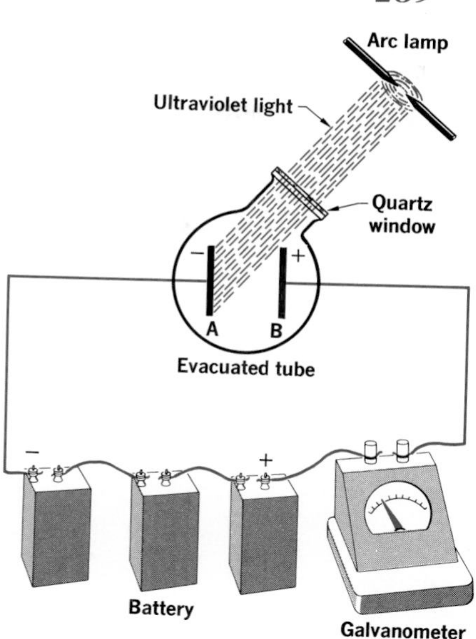

Figure 12-10. Apparatus for demonstrating the photoelectric effect.

12.7 Laws of Photoelectric Emission

Plate **B** of Figure 12-10, having a positive potential *V* with respect to the emitter plate **A,** acts as a collector of photoelectrons ejected from the emitter. If the positive potential is increased enough, all photoelectrons are collected by plate **B** and the photoelectric current *I* reaches a certain limiting, or *saturation,* magnitude.

Curve **a** of Figure 12-11 is a graph of photoelectric current as a function of collector plate potential for a given source of light. Curve **b** shows the result of doubling the intensity of the light. Observe that the magnitude of the saturation current is doubled. This means, of course, that the rate of emission of photoelectrons is doubled. Here we have evidence of the ***first law of photoelectric emission:*** *The rate of emission of photoelectrons is directly proportional to the intensity of the incident light.*

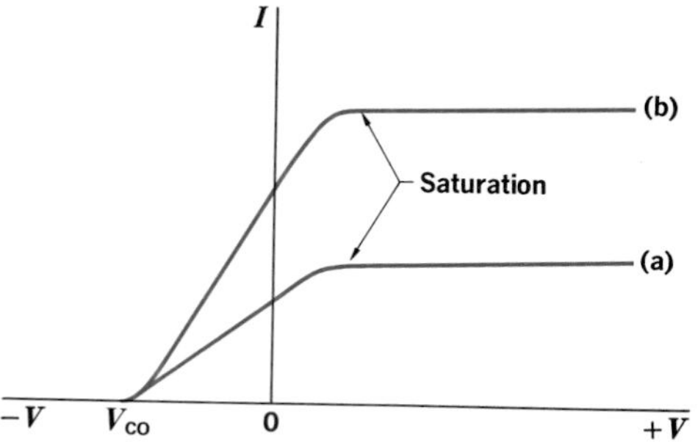

Figure 12-11. Photoelectric current as a function of collector plate potential. Curve (b) shows the effect of doubling the intensity of the same incident light.

For an electron to escape through the surface of a metal, work must be done against the forces that bind it within the surface. This work is known as the *work function.* The photoelectrons must acquire from the incident light radiation the energy needed to overcome this surface barrier. If electrons acquire less energy than the work function of the metal, they cannot be ejected. On the other hand, if they acquire more energy than is required to pass through the surface, the excess appears as kinetic energy and consequently as velocity of the photoelectrons.

We may assume that light penetrates a few atom layers into the metal and that the photoelectric effect will occur at varying depths beneath the surface. Photoelectrons ejected from atom layers below the surface will lose energy through collisions in reaching the surface and must then give up energy equal to the work function of the metal in escaping through the surface. Photoelectrons ejected from the surface layer of atoms lose only the energy necessary to overcome the surface attractions. Thus in any photoelectric phenomenon we should expect photoelectrons to be emitted at various velocities ranging up to a maximum value possessed by electrons having their origin in the surface layer of atoms.

We can test the logic of these deductions by experimenting further with the photoelectric cell of Figure 12-10. A positive potential of a few volts on the collector plate **B** produces a saturation current as shown in Figure 12-11. If the collector plate potential is lowered towards zero, the photoelectric current decreases slightly, but at zero potential may still be close to the saturation magnitude.

As the collector plate potential is made slightly negative with respect to the emitter, the photoelectric current decreases also. By increasing this negative potential a value is reached where the photoelectric current drops to zero. This is called the *stopping,* or *cutoff,* potential V_{co} and is shown in Figure 12-11.

The cutoff potential "cuts off" the photoelectric current at the collector plate.

A negative potential on the collector plate repels the photoelectrons, tending to turn them back to the emitter plate. Only those electrons having enough kinetic energy and velocity to overcome this repulsion reach the collector. As the cutoff potential is approached, only those photoelectrons with the highest velocity reach the collector. These are the electrons with the maximum kinetic energy that are emitted from the surface layer of the metal. At the cutoff potential even these electrons are repelled and turned back to the emitter.

Thus the negative collector potentials, which repel rather than attract photoelectrons, reveal something about the kinetic energy distribution of the ejected electrons. As this potential is made more negative, a nearly linear decrease in photoelectron current shows us that the photoelectrons do have a variety of velocities. The cutoff potential measures the kinetic energy of the *fastest* photoelectrons.

The cutoff potential measures the maximum kinetic energy of photoelectrons.

From curves **a** and **b** shown in Figure 12-11, observe that the photoelectric currents produced by different intensities of incident light from a certain source reach zero at the same collector potential. Thus the cutoff potential for a given photoelectric system and *the velocity of the electrons expelled* are independent of the intensity of the light source. From these observations we can formulate a ***second law of photoelectric emission:*** *The kinetic energy of photoelectrons is independent of the intensity of the incident light.*

The second law of photoelectric emission makes trouble for the wave theory.

The American physicist Robert A. Millikan (1868–1953) performed many experiments with various emitters and light sources. By illuminating emitters made of sodium metal with light radiations of different frequencies, Millikan found that the cutoff potential had different values for the various frequencies of incident light. By plotting the cutoff potentials V_{co} as a function of light frequencies f, the straight line shown in Figure 12-12 is obtained. Millikan demonstrated that *the cutoff potential depends only on the frequency of the incident light.*

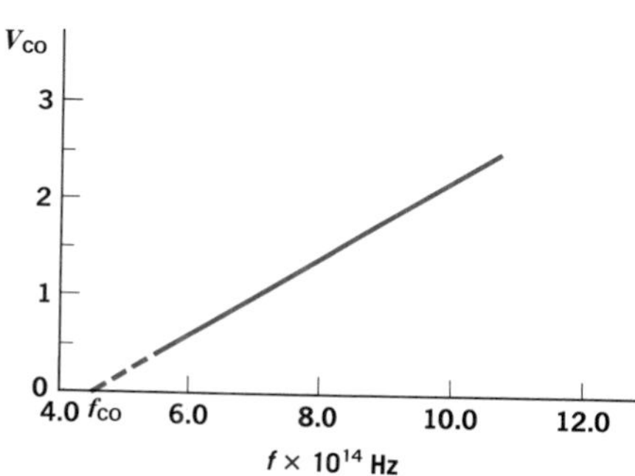

Figure 12-12. A plot of cutoff potentials for sodium as measured by Millikan at various frequencies of illumination. The cutoff frequency of sodium is 4.4×10^{14} Hz.

Recalling that the cutoff potential measures the kinetic energy of the fastest photoelectrons ejected in a particular photoelectric system, we must conclude that the *maximum kinetic energy of photoelectrons increases with the frequency of the light illuminating the emitter.*

Other physicists, notably A. L. Hughes at the Cavendish Laboratory in England and K. T. Compton at Princeton University, obtained similar results in numerous experiments. All these experiments further showed that *for each kind of surface there is a characteristic threshold or cutoff frequency f_{co} below which the photoelectric emission of electrons ceases* no matter how intense the illumination.

Only a few elements, in particular the alkali metals, exhibit photoelectric emission with ordinary visible light. The cutoff frequency of cesium is in the infrared region of the electromagnetic spectrum. Millikan determined the f_{co} of sodium to be 4.4×10^{14} Hz, which is red light with a wavelength of about 6800 Å. The cutoff frequency of potassium is in the visible green region. Those of copper and platinum are in the ultraviolet and deep ultraviolet.

The number of materials available as photoelectric emitters with visible light is limited.

For photoelectric emission from any surface, the incident light radiation must contain frequencies higher than the cutoff frequency characteristic of that surface. The maximum kinetic energy of photoelectrons emitted from any surface can be increased only by raising the frequency of the light illuminating the surface. These facts provide us with a ***third law of photoelectric emission:*** *Within the region of effective frequencies, the maximum kinetic energy of photoelectrons varies directly with the difference between the frequency of the incident light and the cutoff frequency.*

12.8 Failures of the Wave Theory The experiments of Hertz, together with those that followed, confirmed that Maxwell's electromagnetic theory correctly describes the transmission of light and other kinds of radiation. This wave theory requires the radiated energy to be distributed uniformly and continuously over a wave front. The higher the intensity of the radiating source, the greater should be the energy distributed over the wave front.

The first law of photoelectric emission does not imply any deviation from the electromagnetic theory since the magnitude of the photoelectric current is proportional to the incident light intensity. The surprising thing is that the velocity of the photoelectrons is not raised with an increase in the intensity of the illumination on the surface of an emitter, contrary to what the wave theory suggests. Measurements of emission cutoff potentials over very large ranges of intensities show that the maximum kinetic energy of photoelectrons is independent of the light intensity.

According to the wave theory, light of any frequency should cause photoelectric emission provided it is intense enough. Experiments show, however, that all substances have characteristic cutoff frequencies below which emission does not occur, no matter how intense the illumination. On the other hand, a very feeble light containing frequencies above the threshold value causes the ejection of photoelectrons.

Again, from the wave theory we could argue that given enough time an electron in an area illuminated by a very feeble light would "soak up" enough energy to escape from the surface. No such time lag between the illumination of a surface and the ejection of a photoelectron has ever been measured. If emission of photoelectrons occurs at all, it begins simultaneously with the illumination of a photosensitive surface.

The wave theory does not provide an explanation for the photoelectric effect.

The wave theory, which serves so well in the explanation of radiation transmission phenomena, is incapable of describing the processes of radiation absorption observed in the photoelectric effect. In this instance a particle model of light appears to be useful. Had Newton perhaps been on the right track after all? The classic particles of Newton's theory could explain the relation between photoelectron emission rate and intensity of illumination but not that between photoelectron velocities and frequency of illumination.

The discovery of the photoelectric effect, coming as it did from the experiments that established the correctness of the electromagnetic wave theory, presented a paradox to twentieth-century physicists. Was there no theory to explain these widely divergent phenomena?

12.9 The Quantum Theory Aside from the disturbing implications that the photoelectric effect introduced into twentieth-century physics, physicists were much concerned about discrepancies between their experimental studies of absorption and emission of radiant energy and the requirements derived from the electromagnetic radiation theory.

The first great step toward the resolution of these discrepancies was taken by the German theoretical physicist Max Planck (1858–1947). See Figure 12-13. In 1900, Planck was investigating the spectral distribution of electromagnetic radiation from a hot body. Classic electromagnetic theory predicted that the emission intensity would increase continuously as the radiation frequency increased. The spectral distribution of radiated energy measured experimentally was very different from that predicted by theory at the higher frequencies. It approached zero rather than infinity as the frequency increased without limit. Clearly something was wrong in this prediction that the radiated energy would keep on increasing with frequency without limit.

Figure 12-13. Max Planck, the German physicist, developed the initial assumptions on which the quantum theory is based while studying radiation phenomena.

Planck found that he could bring radiation theory and experiment into agreement by assuming the energy emitted by the radiating sources to be an integral multiple of a fundamental quantity hf, where h is a universal constant now known as **Planck's constant,** and f is the frequency of these radiating sources. To explain why this bold assumption worked, he postulated that light is radiated and absorbed in indivisible packets, or *quanta*. We now call these packets, or quanta, ***photons.***

The amount of energy comprising a photon is determined by the frequency of the radiation. It is directly proportional to this frequency since

$$E = hf$$

When f is given in hertz and h in joule-seconds ($h = 6.63 \times 10^{-34}$ J·s), the energy, E, of a photon is expressed in joules.

Planck published his quantum hypothesis in 1901 and, although not immediately accepted, it was destined to profoundly influence the *new physics* emerging with the twentieth century. Certainly there is nothing in the classical physics of Newton to suggest that certain values of energy should be allowed and others should not.

Albert Einstein recognized the value of Planck's quantum hypothesis in connection with the photoelectric effect. He reasoned that, since emission and absorption of light radiation occur discontinuously, certainly the transmission field should be discontinuous. Proceeding with this hypothesis, in 1905 Einstein published a very simple and straightforward explanation of the photoelectric effect. A few years later, the accurate experimental work of Millikan, Hughes, and Compton established the correctness of Einstein's explanation. The bold extension of Planck's quantum ideas by Einstein firmly proved *the* ***quantum theory,*** *which assumes that the transfer of energy between light radiations and matter occurs in discrete units called quanta, the magnitude of which depends on the frequency of the radiation.*

12.10 Einstein's Photoelectric Equation According to Einstein, light illuminating an emitter surface consists of a stream of photons. When a photon is absorbed by the emitter, its quantum of energy hf is transferred to a single electron within the surface. If the acquired energy is sufficient to overcome the surface barrier and the electron is moving in the right direction, the electron will escape from the surface. In penetrating the surface, the electron must give up a certain energy w, the work function of the substance. If the energy hf imparted by the photon is greater than w, the electron will have kinetic energy as indicated by its velocity after leaving the surface.

The maximum kinetic energy possessed by photoelectrons ejected from an emitter illuminated by light of frequency f was given by Einstein as

$$\tfrac{1}{2}mv^2_{\mathrm{max}} = hf - w$$

where m and v are the mass and velocity, respectively, of the photoelectrons, h is Planck's constant, f is the frequency of the impinging light radiation (the product hf is the energy of the photon), and w is the work function of the emitting material.

Now, recalling the failures of the wave theory to explain photoelectric emission, we find no such difficulties in the application of Einstein's photon hypothesis. A lower light intensity means fewer photons impinging on the emitter surface in a given period of time and fewer photoelectrons ejected. However, as long as the frequency of the incident light remains unchanged, each photon transfers the same energy hf to the electron when a collision occurs.

Einstein's photoelectric equation shows clearly why light of too low a frequency will not cause photoelectric emission from a given material no matter how intense the illumination. Photon energy is a linear function of the frequency of the light since it equals the product hf. Now if $\frac{1}{2}mv^2_{max} = 0$, it follows that

$$hf_{co} = w$$

Recall that f_{co} for an emitting surface is the illumination frequency below which the photoemission of electrons ceases.

Here the photon can impart to the electron just enough energy for it to penetrate the surface barrier. No energy is left over to appear as electron velocity. The frequency in this situation is the cutoff frequency f_{co}. Illumination of any lower frequency on this emitter will consist of photons with energy $hf < w$, and photoelectric emission cannot occur no matter how many photons there are.

If photons have enough energy to eject electrons, no "soaking up" time is required before a feeble light can start the photoelectric emission process. This is true because the ejection energy for each photoelectron is delivered in a single concentrated bundle.

Light exhibits a dual character: that of waves and of particles.

The quantum theory, which provides a particle model of light, meets every objection raised when the electromagnetic wave theory is employed to interpret photoelectric emission phenomena. Yet many experiments have proved the correctness of the wave theory. When we consider the full range of radiation phenomena, we are persuaded that light must have a dual character. In some circumstances it behaves like waves, and in other circumstances it behaves like particles. *The modern view of the nature of light recognizes this dual character:* **Radiant energy** *is transported in photons that are guided along their path by electromagnetic waves.*

12.11 The Quantized Atom

Striking evidence favoring the quantum theory was given by Niels Bohr (1885–1962), who in 1913 devised a model of the atom based on quantum ideas. Ernest Rutherford (1871–1937), the English physicist, had conducted experiments (described in Section 23.7) that in 1911 led him to postulate the nuclear atom with its planetary arrangement of electrons. This model of Rutherford's immediately encountered difficulties when subjected to the classical laws of physics, that is, the known laws of mechanics and electromagnetism as formulated by Newton and Maxwell.

Figure 12-14. Niels Bohr, the Danish physicist, demonstrating a point on his blackboard. Bohr received the Nobel Prize in 1922 for his work in atomic physics.

The electromagnetic theory predicts that charged particles undergoing acceleration must radiate energy. Indeed,

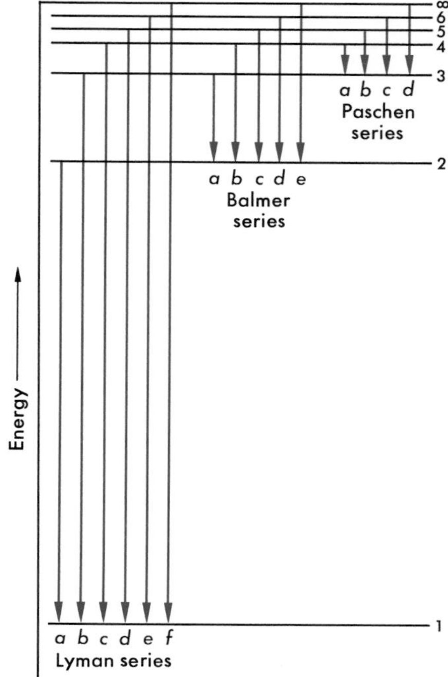

Figure 12-15. This electronic energy-level diagram for hydrogen shows some of the transitions that are possible in this atom.

Figure 12-16. Representative lines in the hydrogen spectrum. The small letter below each line indicates which of the energy-level transitions shown in Figure 12-15 produces it.

this accounts for the radiation of energy from a radio transmitting antenna. A planetary electron moving about the nucleus of an atom experiences acceleration and accordingly should radiate energy. As energy is drained from the electron, it should spiral in toward the nucleus causing the atom to collapse. Of course atoms do not collapse, and so this classical model cannot be correct.

Bohr assumed that an electron in an atom can move about the nucleus in certain discrete orbits without radiating energy. Such orbits represent "allowable" *energy levels* in the Planck concept. The level closest to the nucleus corresponds to the orbit of lowest energy. This postulate could account for the stability of an atom. However, atoms do radiate energy. Bohr assumed further than an electron may "jump" from one discrete orbit to another of lower energy. In the process a photon is emitted. Its energy represents the difference between the energies of the levels involved in the transition. The frequency of the emission is proportional to this energy difference, the proportionality factor being Planck's constant h. Thus the frequency of the photon emission depends on the magnitude of this energy change ΔE and can be expressed as $\Delta E/h$.

The return of excited atoms to their stable state results in the emission of ***bright line spectra.*** (See Color Plate VII, Chapter 14.) Excited hydrogen atoms produce the simplest such atomic spectra. Hydrogen lines appear in several series in and beyond the visible region of the radiation spectrum. These include the *Balmer series* in the visible region, the *Lyman series* in the ultraviolet region, and the *Pashen series* in the near infrared region. See Figures 12-15 and 12-16.

Numerous attempts had been made to account for the discrete frequencies of the bright lines observed in these spectra. It was Bohr who first associated spectral lines

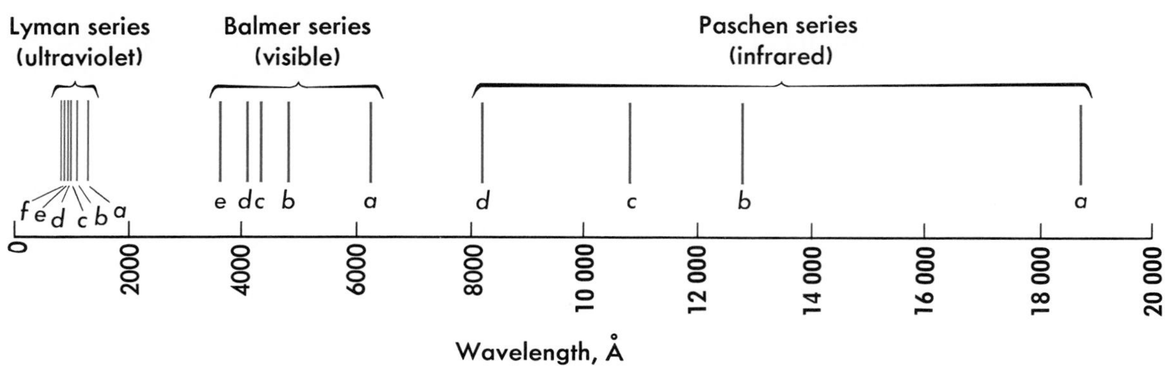

with pairs of energy levels, thus marking the initial step in the development of the *quantum mechanics* of atomic structure.

Bohr's atom model provided no information about the mechanics of photon emission when an electron passes from a higher to a lower energy level, but his concept of energy levels provided a great stimulus for theoretical studies in this field. In 1924 the French physicist Louis de Broglie (1892–1987) suggested that the dual particle-and-wave nature of light provided evidence of the wave nature of particles as well. Accepting Einstein's idea of the equivalence between mass and energy, he postulated that *in every mechanical system, waves are associated with matter particles*. The application of this concept of matter waves to the study of the structure of matter is known as *wave mechanics*. Just as the ordinary laws of mechanics are essential to any explanation of the behavior of objects of large dimensions, *wave mechanics* is necessary in dealing with objects that have atomic and subatomic dimensions.

A photon has energy that is the product of Planck's constant and its radiation frequency.

$$E = hf$$

In Einstein's equation $E = mc^2$ for mass-energy equivalency, m is the mass of the particle while in motion. When at rest, its mass m_0, known as the rest mass, is smaller than m.

If we consider the energy of the photon to be equivalent to that of a moving particle, then

$$hf = mc^2$$

and we could solve for m. However, this solution would also imply a certain rest mass for the photon. But a photon is never at rest; it always moves with the speed c. Photons do have momentum and can transfer it to any surface on which they impinge. Calculating this momentum mc, we get

$$mc = \frac{hf}{c}$$

Since

$$c = f\lambda$$

by substitution,

$$mc = \frac{h}{\lambda}$$

or

$$\lambda = \frac{h}{mc}$$

The wavelength of the photon may be expressed in terms of its momentum *mc*. Similarly, the wavelength of any matter particle having a velocity *v* may be expressed in terms of its momentum *mv*.

$$\lambda = \frac{h}{mv}$$

There is abundant evidence today of the wave nature of subatomic particles. Accelerated electrons have been found to behave like X rays. The electron microscope is an application of electron waves. The development of the quantum theory and the system of wave mechanics have provided physicists with their most powerful means of studying the structure and properties of matter. Matter waves will be discussed more extensively in Chapter 25.

Incoherent waves differ in frequency or phase, or both.

Coherent waves have the same frequencies and a constant phase relationship.

12.12 Coherent Light: The Laser Common light sources such as fluorescent and incandescent lamps radiate light in all directions over a wide range of frequencies with random phase relationships. Even groups of waves of the same frequency have random phase relationships. These disordered light waves are said to be *incoherent*.

A stationary interference pattern is described in Section 10.15 and represented in Figures 10-27 and 10-28. To produce such a pattern, it is necessary to employ two wave trains with identical wavelengths and a constant phase relationship. Such wave trains have *coherence;* they are said to be *coherent* waves.

Figure 12-17. A laser used in conjunction with an aircraft instrument landing.

It is not difficult to obtain coherence from two sources of sound waves. However, a stationary interference pattern from light waves is ordinarily produced in a high school laboratory by means of a subtle arrangement using a single source of incoherent light.

Coherent light is produced by a remarkable instrument called a *laser.* It is a light amplifier that functions because of stimulated emissions of radiation. The term **"laser"** is an acronym derived from the description of its function: **L**ight **A**mplification by **S**timulated **E**mission of **R**adiation. The laser output is a single-frequency (monochromatic) radiation of parallel waves with a constant phase relationship. An example of a laser application is shown in Figure 12-17.

The first successful laser produced a pulsed coherent red beam from a ruby crystal. The ruby laser was followed by gaseous lasers which produced continuous coherent beams. Today there are many different types of lasers using gases, liquids, crystals, semiconductors, glass, and plastics. These lasers produce coherent beams of different spectral colors. Laser research is continuing on a broad front. Industry, engineering, medicine, communications, photography, and the military are areas in which lasers are being used. Scientists and engineers are constantly finding new applications.

The underlying principle of laser technology can be illustrated with a model of the basic ruby laser system shown in Figure 12-18. The model consists of a pencil-shaped ruby crystal and an external flash lamp with its power source. The ends of the ruby crystal are optically flat, silvered reflecting surfaces, one end being partly silvered to allow some light to emerge. The flash lamp surrounds the ruby crystal as illustrated. When this lamp flashes, photons enter the ruby crystal and initiate the laser action.

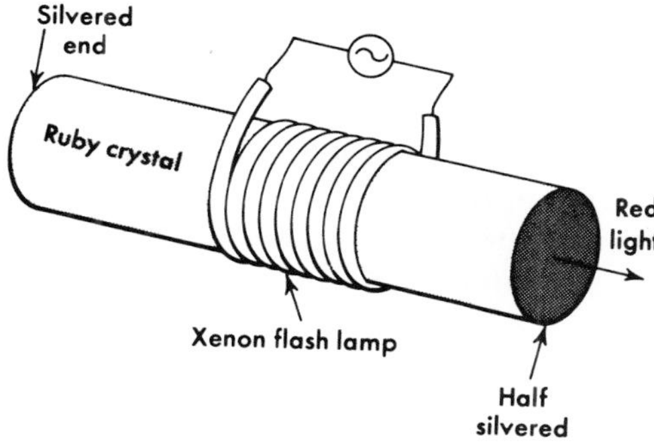

Figure 12-18. Simplified model of a ruby laser.

The effect of the flash lamp is to "pump" energy into the crystal. Some atoms of the crystal become excited by absorbing photons and are elevated to a higher energy state. They may then emit some of the energy and make a transition to an intermediate state. From this intermediate state an atom may return to its stable state either by spontaneous or stimulated emission of a photon of light energy equal to its energy level difference.

Suppose one atom emits a photon spontaneously. If traveling in the proper direction, the photon is reflected at the end of the ruby crystal. As it travels back through the crystal, the photon stimulates another atom in the intermediate state to emit a photon. This process continues, and the intensity of the photon beam in the crystal is *amplified*.

With the reflective ends of the crystal the proper distance apart, a standing wave of single-frequency red light is set up in the ruby rod. This particular standing light wave is analogous to the standing sound wave in an organ pipe. As the photon beam is amplified, photons pass through the partially reflective end of the rod in the form of intense pulses of coherent red light. Gaseous lasers produce a continuous but less intense beam of coherent light. The laser provides practical evidence of the quantized energy levels of atoms as proposed by Bohr in 1913. See Section 12.11.

12.13 Production of X Rays In the photoelectric effect, photons of frequency f cause the ejection of electrons from atoms of a substance with a maximum velocity given by the equation

$$\tfrac{1}{2}mv^2_{\mathbf{max}} = hf - w$$

It is not surprising that the converse of this phenomenon can be produced also. High-speed electrons with velocity v projected against the surface of a target cause the emission of photons of radiation having a maximum frequency given by the photoelectric equation

$$\tfrac{1}{2}mv^2 = hf_{\mathbf{max}} - w$$

The work function w of target material is negligible compared to the photon energy at X-ray frequencies and is usually omitted from the above equation.

This phenomenon is sometimes referred to as the *inverse photoelectric effect*. An example is the production of accelerated electrons. Each electron that strikes the target loses its energy, and this energy reappears as photons of radiation. Neglecting the work-function energy, as previously indicated, no more than the full kinetic energy of the electron can appear in the X-ray photons as the energy hf. Since h is constant, the maximum frequency f_{max} of the X-ray radiation is determined by the velocity v of the electron.

The penetrating properties of X rays and their many practical uses stemming from these properties are well known. X rays lie deep in the high-frequency end of the electromagnetic spectrum beyond the ultraviolet region. The energy hf of X-ray photons is very high, being of the order of 10^4 times the energy of photons in the visible region. As a consequence of this high photon energy and the very short wavelengths, X rays are used extensively in numerous areas of research. Table 12-1 gives photon energies for important regions of the spectrum.

Table 12-1
PHOTON ENERGIES

Radiation	*Typical value of f* (Hz)	*hf* (J)
gamma rays	3.0×10^{19}	2.0×10^{-14}
X rays	3.0×10^{18}	2.0×10^{-15}
visible light	6.0×10^{14}	4.0×10^{-19}
heat waves	3.0×10^{13}	2.0×10^{-20}
radio waves	3.0×10^{6}	2.0×10^{-27}

12.14 The Pressure of Light The electromagnetic theory of Maxwell predicted that incident light should exert pressure in the direction of the radiation. The conservation of momentum requires that the pressure on a totally reflecting surface should be twice that on a totally absorbing surface illuminated by the same radiation.

In 1901, Nichols and Hull in the United States and Lebedev in Russia succeeded experimentally in measuring the pressure of light. Their results were in such close agreement with theory as to provide proof of the existence of light pressure.

What does the particle model tell us about the pressure of light? Our common experiences with pressure are those involving collisions of particles in motion. Since particle collisions are momentum problems, conservation of momentum applies.

A perfectly elastic ball rebounding from an object transfers twice the momentum that a perfectly inelastic ball having the same mass and velocity would transfer. In Section 12.11 we expressed the momentum of a photon in terms of its frequency as

$$mc = \frac{hf}{c}$$

When a beam of photons is incident on a totally absorbing surface, the pressure exerted depends on the rate of change of photon momentum per unit area of illuminated surface.

$$p = \frac{F}{A} = \frac{\Delta(hf/c)}{\Delta t \times A} = \frac{\Delta hf}{c\Delta tA}$$

The pressure, p, has the units

$$\frac{(\text{J}\cdot\text{s}) \times (\text{Hz})}{(\text{m/s}) \times (\text{s}) \times (\text{m})^2} = \frac{\text{J}\cdot\text{s}\cdot\text{s}^{-1}}{\text{m}^3} = \frac{\text{J}}{\text{m}^3} = \frac{\text{N}}{\text{m}^2}$$

If the photon energy hf is totally reflected from the surface of a body, the momentum change is $2hf/c$ and the pressure of light is twice that for the totally absorbing surface illuminated by the same beam.

The pressure of light is exceedingly small in comparison to pressures we commonly experience. The pressure of sunlight on the earth is approximately 4×10^{-11} standard atmosphere. Comet tails always point away from the sun, due to the pressure of sunlight and solar particles on the extremely diffuse matter in them.

QUESTIONS: GROUP A

1. (a) Who was the principal proponent of the corpuscular theory of light? (b) of the wave theory of light?
2. What assumption did the corpuscular theory make in order to explain refraction?
3. Why was the wave theory rejected before the nineteenth century?
4. The acceptance of what theory led physicists to believe that all significant laws of physics had been discovered?

5. Which band of the electromagnetic spectrum has the (a) lowest frequency (b) shortest wavelength (c) highest energy per photon?
6. Define *photon.*
7. What is the modern theory about the nature of light?

GROUP B

8. (a) What is the photoelectric effect? (b) What practical applications does it have? (c) What important implications did it have for theories about the nature of light?
9. Given a monochromatic (single frequency) light source to illuminate a photocell, how can you explain the fact that photoelectrons are ejected at various velocities ranging up to a maximum value?
10. (a) What is meant by the term *quantized?* (b) Are there physical phenomena other than light that are quantized?
11. If all particles exhibit some wave properties, why is your car not diffracted as you drive under a highway overpass?
12. What property, also exhibited by sound waves, results in the amplification of the light within a laser?
13. Research some practical uses for the laser. Share your findings with your classmates.
14. How does an X-ray technician in a hospital depend on Einstein's photoelectric equation?
15. A beam of light shines on the blackened bulb of a thermometer. When a piece of ordinary glass is placed in front of the thermometer, the reading drops 2 degrees. When a piece of quartz glass is used, the reading returns to its original value. What does this suggest about the nature of the light source?
16. Is it correct to say that light is both a particle and a wave? Explain.
17. What difficulties did Rutherford's model of a quantized atom face with respect to the ideas of classical physics?

ILLUMINATION

12.15 Luminous and Illuminated Objects Nearly all the natural light we receive comes from the sun. Distant stars account for an extremely small amount of natural light. Of course, moonlight is sunlight reflected from the surface of the moon.

We produce light from artificial sources in several ways. Materials may be heated until they glow, or become incandescent, as in an electric lamp. Molecules of a gas at reduced pressure may be bombarded with electrons to produce light, as in a neon tube. Most of the visible light from fluorescent tubes results from the action of ultraviolet radiations on phosphors that coat the inside surface of the glass tubes. The firefly produces light by means of complex chemical reactions.

Most artificial sources of light are hot bodies that radiate energy in the infrared region of the spectrum as well as visible light. The energy of the *thermal radiation* emitted by hot bodies depends upon the temperature of the body and the nature of its surface.

At a temperature of 300°C, the most intense radiation emitted by a hot body has a wavelength of about 5×10^{-4} cm, well down in the infrared region of the electromagnetic spectrum. As the temperature is raised to 800°C, enough radiation is emitted in the visible region to cause the body to appear "red hot" although the bulk of the energy radiated is still in the infrared region. If the temperature of the body is raised to 3000°C, the most intense emission remains in the near infrared region; however, there is enough blue visible radiation to cause the body to appear "white hot."

Visible spectrum: 7.6×10^{-5} cm to 4.0×10^{-5} cm.

This last temperature is near that of the filament of an incandescent lamp. Hence such lamps have low efficiencies as producers of visible radiation. Generally the efficiency improves as the filament temperature is raised.

The white-hot filament of an incandescent lamp is said to be *luminous*. It is visible primarily because of the light it emits. *An object that gives off light because of the energy of its accelerated particles is said to be* ***luminous.*** The sun and the other stars are luminous objects.

Just as radiant heat may be reflected, light may be reflected from the surfaces of objects. Mirrors reflect a beam of light in a definite direction. Other surfaces scatter the light that is incident on them in all directions. *An object that is seen because of the light scattered from it is said to be* ***illuminated.*** The moon is illuminated, for it reflects radiant energy from the sun. Some of the light energy arriving at the surface of an object is reflected, some of it is transmitted, and some of it is absorbed.

Any dark-colored object absorbs light, but a black object absorbs nearly all the light it receives. When the rays of the sun strike a body of water vertically, most of them are either absorbed or transmitted. Most of the rays are reflected, however, when they strike the water at an oblique angle. For this reason the image of the sun can be seen in the water without discomfort when the sun is directly overhead, but not when the sun is near the horizon.

Air, glass, and water transmit light readily and are said to be *transparent*. Other substances transmit light but scatter or diffuse it so that objects seen through them cannot be identified. Such substances are *translucent*. Typical examples are frosted electric lamps and parchment lampshades. *Opaque* substances do not transmit light at all.

From a luminous point source, light waves travel outward in all directions. If the medium through which they pass is of the same nature throughout, the light progresses along the normal to the wave front. A single line of light from a luminous point is called a *ray;* a group of closely spaced rays forms a *beam* of light. Small beams are referred to as *pencils.* See Figure 12-19. Light coming from the sun is in rays so nearly parallel that it may be considered as a parallel beam. When several rays of light come from a point, they are called a *diverging* pencil, while rays proceeding toward a point form a *converging* pencil. When the sun's rays pass through a magnifying glass, they converge at a point called a *focus.*

Figure 12-19. The formation of beams and pencils of light.

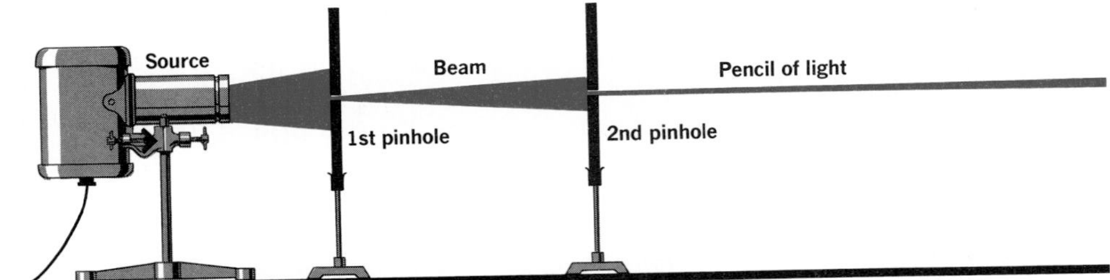

Because an opaque object absorbs light, it casts a shadow in the space behind it. When the source of light is a point, as in Figure 12-20(A), an opaque ball, **B,** cuts off all the rays that strike it and produces a shadow of uniform darkness on the screen, **S.** If the light comes from an extended source, the shadow varies in intensity, as shown in Figure 12-20(B). The part from which all the rays of light are excluded is called the **umbra;** the lighter part of the shadow is the **penumbra.** Within the region of the penumbra the luminous source is not entirely hidden from an observer. The umbral and penumbral regions of the moon's shadow cast during a solar eclipse are shown in Figure 12-21.

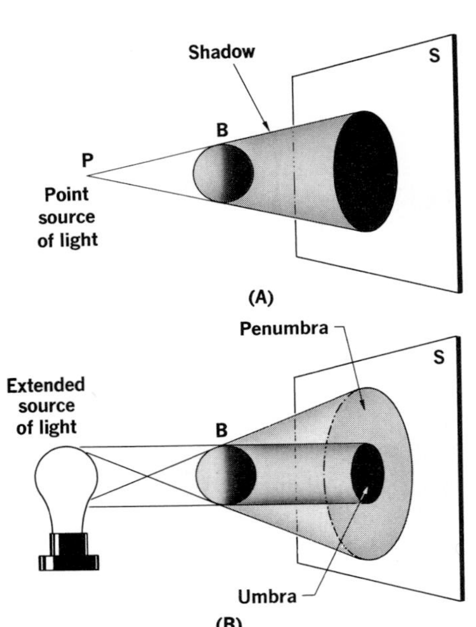

Figure 12-20. An obstruction in the path of light from a point source (A) casts a shadow of uniform density. If the light is from an extended source (B), the shadow is of varying density.

12.16 The Speed of Light The speed of light is one of the most important constants used in physics, and the determination of the speed of light represents one of the most precise measurements ever achieved. Before 1675 light propagation was generally considered to be instantaneous, although Galileo had suggested that a finite time was required for it to travel through space. In 1675 the Danish astronomer Olaus Roemer (1644–1710) determined the first value for the speed of light to be 140 000 miles per second. This is approximately equivalent to 225 000 kilometers per second. He had been puzzled by a variation in his calculations of the time of eclipse of one of Jupiter's satellites as seen from different positions of the earth's orbit about the sun. Roemer concluded that the variation was due to differences in the distance that the light traveled to reach the earth.

Very precise modern measurements of the speed of light are made using laboratory methods. The most notable experiments were performed by Albert A. Michelson (1852–1931), a professor of physics at the University of Chicago, who measured the speed of light in air and in a vacuum with extraordinary precision.

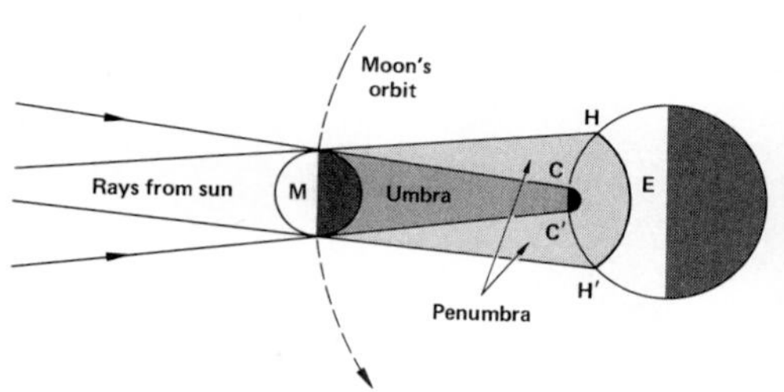

Figure 12-21. A diagram of a total solar eclipse showing the umbral and penumbral regions of the moon's shadow. CC′ is the region of totality on the earth; CH and C′H′ are the regions from which a partial eclipse is seen. (Sizes and distances are not to scale.)

1. *The speed of light in air.* Michelson measured the speed of light over the accurately determined distance between Mt. Wilson and Mt. San Antonio, California. His method is illustrated in principle by Figure 12-22. The light source, octagonal mirror, and telescope were located on Mt. Wilson and the concave mirror and plane mirror were located on Mt. San Antonio, 35.4 km (approximately 22 mi) away. The octagonal mirror, **M,** could be rotated rapidly under controlled conditions and was timed very accurately.

Foucault first used a rotating mirror in 1850 to compare the speed of light in water and in air.

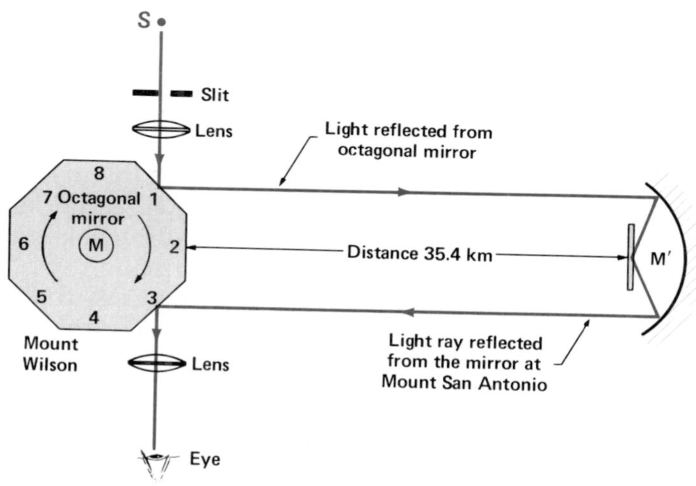

Figure 12-22. A diagram of Michelson's octagonal-mirror method for measuring the speed of light.

With mirror **M** stationary, a pencil of light from the slit opening was reflected by $\mathbf{M_1}$ to the distant mirror $\mathbf{M'}$, from which it was returned to $\mathbf{M_3}$. The image of the slit in the mirror at the $\mathbf{M_3}$ position could be observed accurately through the telescope. The octagonal mirror was then set in motion and the speed of rotation brought up to the value that moved $\mathbf{M_2}$ into the position formerly occupied by $\mathbf{M_3}$ during the time required for the light to travel from $\mathbf{M_1}$ to Mt. San Antonio and return. The slit image was again seen in the telescope precisely as it was when the octagon was stationary. The light traveled twice the optical path of 35.4 km in one eighth of the time of one revolution of the octagonal mirror. Thus

$$c = \frac{2MM'}{t}$$

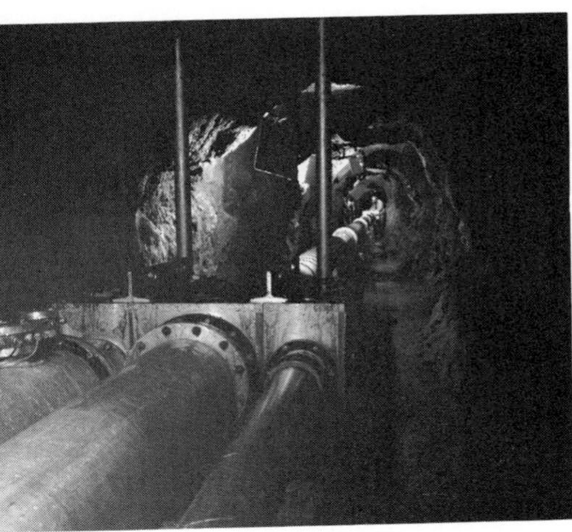

Figure 12-23. A laser beam is directed down a 30-meter vacuum tube in an experiment to measure the speed of light. The apparatus is installed in an abandoned gold mine and is so sensitive that it records "tides" in the earth's crust caused by the sun and moon.

where MM' is the optical path, t is the time of one-eighth revolution of the octagon, and c is the speed of light in air.

Michelson's investigation of the speed of light in air required several years to complete and extremely high precision was attained in making the necessary observations. The optical path was measured by the U.S. Coast and Geodetic Survey and found to be 35 385.5 meters, accurate to about one part in seven million. The rate of the revolving mirror was measured by stroboscopic comparison with an electric signal of standard frequency. The average of a large number of determinations yielded a speed of light in air of 299 729 km/s.

2. *The speed of light in a vacuum.* Michelson conducted similar experiments using an evacuated tube one mile long to eliminate the problems of haze and variations in air density. In these investigations he determined the speed of light to be 299 796 km/s, which he believed to be accurate to within 1 km/s.

Modern laboratory methods of measuring the speed of light require very complex apparatus and are considered to be more accurate than the methods used by Michelson. In some experiments, electromagnetic waves much longer than light waves have been used and good agreement has been found between their speed and that of visible light. These findings provide experimental confirmation of Maxwell's electromagnetic theory, which requires that all electromagnetic waves throughout the electromagnetic spectrum have the same speed c in free space. Thus we must consider the speed of light within the larger framework of the speed of electromagnetic radiations in general.

As Michelson's figures show, the speed of light is slightly higher in a vacuum than in air. In 1983, the speed of light in free space was fixed arbitrarily as

$$\mathbf{c = 2.997\,924\,58 \times 10^8\ m/s}$$

and became the basis for the definition of the fundamental unit of length (as explained in Section 2.2). Future refinements of the speed of light will, therefore, result in a change in the definition of the meter.

12.17 Light Measurements

The quantitative study of light is called ***photometry.*** Three quantities are generally measured in practical photometry: the *luminous intensity* of the source; the *luminous flux,* or light flow, from a source; and the *illuminance* on a surface.

1. *Luminous intensity.* In order to derive a set of photometric units it is necessary to introduce an additional fundamental unit into our measurement system. The unit for luminous intensity, I, is the arbitrary choice. The SI unit of luminous intensity is the ***candela*** (cd).

The original standard candle was one made of spermaceti (wax obtained from the oil of sperm whales) that burned 120 grains per hour.

Originally the light from a certain type of candle was used as a standard, but this has been replaced by a more readily reproducible source of luminous intensity. *The candela is the luminous intensity, in a given direction, of a source that emits monochromatic radiation of frequency 5.40×10^{14} Hz and that has an intensity in that direction of 1/683 watt per steradian.* In practice it is convenient to use incandescent lamps that have been rated by comparison with the standard.

Incandescent lamps used for interior lighting generally have an intensity ranging from a few candelas to several hundred candelas. The common 40-watt lamp has an intensity of about 35 candelas; a 100-watt lamp gives about 130 candelas. Note that a 40-watt fluorescent lamp has an intensity of about 200 candelas. The intensity of any lamp depends on the direction from which it is measured; the average luminous intensity in all directions in space is often given.

2. *Luminous flux.* Not all of the energy radiated from a luminous source is capable of producing a visual sensation. Most of the radiation is in the infrared region, and a small amount is in the ultraviolet region. The rate of flow of visible radiation is called *luminous flux,* which is a power quantity. ***Luminous flux,*** *F, is that part of the total energy radiated per unit of time from a luminous source that is capable of producing the sensation of sight.* The unit of luminous flux is the *lumen,* lm.

Suppose a standard light source of 1 candela is placed at the center of a hollow sphere having a radius of 1 meter, as illustrated in Figure 12-24. The luminous source is presumed to radiate light equally in all directions and to have such small dimensions that it may be termed a "point source." The area of the surface of a sphere of radius r is equal to $4\pi r^2$. Since the radius of the unit sphere is 1 meter, its surface area is $4\pi \times (1\ \text{m})^2$. One lumen of flux is radiated by the 1-candela source to each square meter of inside surface of the sphere. *The* ***lumen,*** *lm, is the luminous flux on a unit surface all points of which are at unit distance from a point source of one candela.* The lumen is not a measure of a total *quantity* of luminous energy but a *rate* at which luminous energy is being emitted, transmitted, or received.

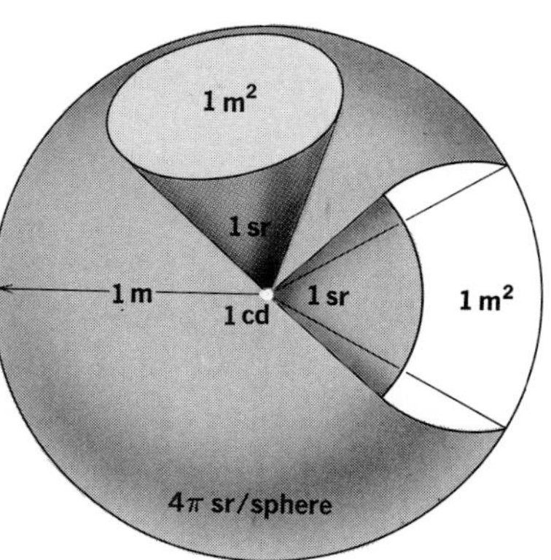

Figure 12-24. A 1-candela point source radiates luminous flux at the rate of 4π lumens.

The unit surface area of the unit sphere is intercepted by a solid angle of 1 *steradian,* sr. The unit surface area of a sphere of radius r is intercepted by a solid angle of $1/r^2$ steradians. *The ratio of the intercepted surface area of a sphere to the square of the radius is the measure of the solid angle in steradians.*

$$\textbf{Angle} = \frac{A}{r^2} \qquad \textbf{(in sr)}$$

Observe that the steradian is dimensionless.

The luminous flux of the standard light source of 1 candela is 1 lumen per steradian.

Since the unit sphere has 4π unit areas of surface, there are 4π steradians per sphere. The total luminous flux emitted by a point source is therefore 4π lumens per candela of luminous intensity.

The equation for the area of a sphere is $A = 4\pi r^2$

$$F = \frac{4\pi \text{ lm}}{\text{cd}} \times I \text{ cd}$$

$$F = 4\pi I \qquad \textbf{(in lumens)}$$

Thus a luminous source having an intensity of 1 candela emits light at the rate of 4π lm, or 12.57 lumens. In fact, light sources are usually rated in terms of the total flux emitted, with 12.57 lumens being radiated by 1 candela. A 40-watt incandescent lamp is rated at about 450 lumens and a 40-watt fluorescent lamp at about 2600 lumens.

3. *Illuminance.* In Figure 12-24 it is evident that as the intensity of the source is increased, the luminous flux transmitted to each unit area of surface and the flux on each unit area are similarly increased. The *illuminance* on the surface is said to be increased. ***Illuminance** is the density of the luminous flux on a surface.* When the surface is uniformly illuminated, the illuminance, E, is the quotient of the flux on the surface divided by the area of the surface and is expressed as the luminous flux per unit area.

$$E = \frac{F}{A}$$

Since F is in lumens and A in square meters, E is in lm/m^2. One lm/m^2 is called a lux (lx).

Suppose the radius of the unit sphere shown in Figure 12-24 is increased to 2 meters. The surface area is $4\pi(2 \text{ m})^2$, or approximately 50 m^2, and is *four* times the area of the unit sphere. Similarly, a radius of 3 meters gives an area of $4\pi(3 \text{ m})^2$, which is *nine* times the area of the unit sphere. Thus as the radial distance from the luminous source is increased, the area illuminated is increased in proportion to the *square* of the distance.

If a point source of constant intensity is located at the center of the sphere, *the illuminance decreases as the square of the distance from the source.* See Figure 12-25. If the intensity of the source is doubled, of course the luminous flux transmitted to the surface is doubled and the illuminance is doubled. Thus the illuminance E on a surface perpendicular to the luminous flux falling on it is dependent on the intensity I of the source and its distance r from the source.

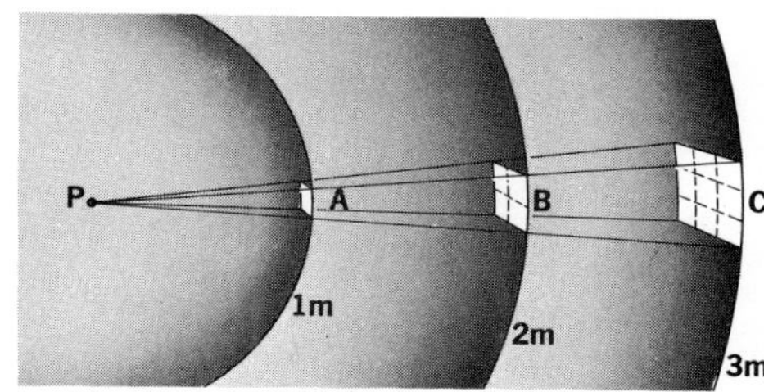

Figure 12-25 The illuminance on a surface varies inversely as the square of the distance from the luminous point source.

The distance r from the point source is the radius of a spherical surface of area $4\pi r^2$ illuminated by the source I at its center. Thus the proper dimensions of E become

E depends on distance; ***I*** *does not.*

$$E = \frac{F}{A} = \frac{\cancel{4\pi} I}{\cancel{4\pi} r^2}$$

$$E = \frac{I}{r^2}$$

If r is expressed in meters and I in candelas, then

$$E = \frac{\cancel{4\pi}\ \text{lm}/\cancel{\text{cd}} \times I\ \cancel{\text{cd}}}{\cancel{4\pi} r^2\ \text{m}^2}$$

$$E = \frac{I}{r^2} \qquad \textbf{(in lm/m}^2\textbf{)}$$

The **inverse square law** is used to calculate the illuminance from an individual point source on planes perpendicular to the beam. In practice, if the dimensions of the source are negligible compared with its distance from the illuminated surface, it is considered a point source. For long fluorescent tubes, the illuminance varies inversely as the distance, but only for distances somewhat smaller than the length of the tube.

When a surface is not perpendicular to the beam of light illuminating it, the luminous flux spreads over a greater area and the level of illuminance is reduced. In Figure 12-26 the perpendicular surface **ABC** illuminated by the beam is square, having an area equal to **AB × BC.** As the surface is tilted away from the source, the illuminated area becomes rectangular, the width remaining the same and the length increasing to **BD.** In the right triangle **ABD**

$$\mathbf{BD} = \frac{\mathbf{AB}}{\cos\theta}$$

where θ equals the tilt angle of the surface from its perpendicular position with respect to the beam. (It is evident that θ also represents the angle that the light beam makes with the perpendicular to the illuminated surface.)

This equation is similar to the one that describes the work done by a force that is applied at an angle (Section 6.1)

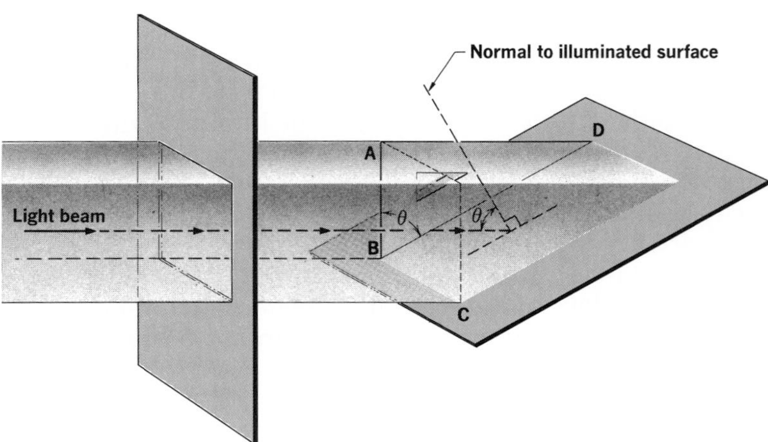

Figure 12-26. Illuminance varies directly with the cosine of the angle between the luminous flux and the normal to the surface.

As θ reaches 60°C, cos θ equals 0.5 and the length **BD** becomes twice the width **AB.** Thus the area is doubled and, since the same flux is spread over twice the area, the illuminance is reduced to half the original level. In the general case,

This equation is related to the relative heating effect of the sun in summer and in winter.

$$E = \frac{I \cos \theta}{r^2}$$

Illuminance on a surface varies inversely with the square of the distance from the luminous source and directly with the cosine of the angle between the luminous flux and the normal to the surface. A surface that is perpendicular to the luminous flux is simply a special case of the general expression for illuminance in which the angle θ becomes zero and cos $\theta = 1$. See the following example.

EXAMPLE A table is illuminated by a 2000.0-lumen incandescent lamp. The perpendicular distance from the lamp to the table is 3.00 m. Calculate the illuminance at a spot on the table that is 1.00 m to one side of the point directly below the lamp.

Given	Unknown	Basic equations
$F = 2000.0$ lm $r = 3.00$ m $d = 1.00$ m	E	$F = 4\pi I$ $\tan \theta = \frac{d}{r}$ $E = \frac{I \cos \theta}{r^2}$

Solution

Working equation: $E = \dfrac{F\,[\cos(\arctan d/r)]}{4\pi r^2} = \dfrac{(2000.0 \text{ lm})(0.949)}{4(3.14)(3.00 \text{ m})^2}$

$= 16.8 \text{ lm/m}^2$

PRACTICE PROBLEMS **1.** The light from a 1500.0-lumen incandescent lamp strikes a horizontal surface. What is the illuminance at a spot that is 2.50 m directly below the lamp? *Ans.* 19.1 lm/m^2

2. What is the illuminance in Problem 1 at a spot that is 2.00 m to one side of the original spot? *Ans.* 14.9 lm/m^2

3. Suppose that the surface in Problem 1 is tilted so that the normal to the surface makes an angle of 30.0° with the luminous flux. Now what is the illuminance at the original spot? *Ans.* 16.5 lm/m^2

12.18 The Intensity of a Source The luminous intensity of a light source can be measured by comparing its intensity with that of a standard light source having the same color quality. This is done by using an instrument called a **photometer.** See Figure 12-27. Photometric measurements are made in a darkened room.

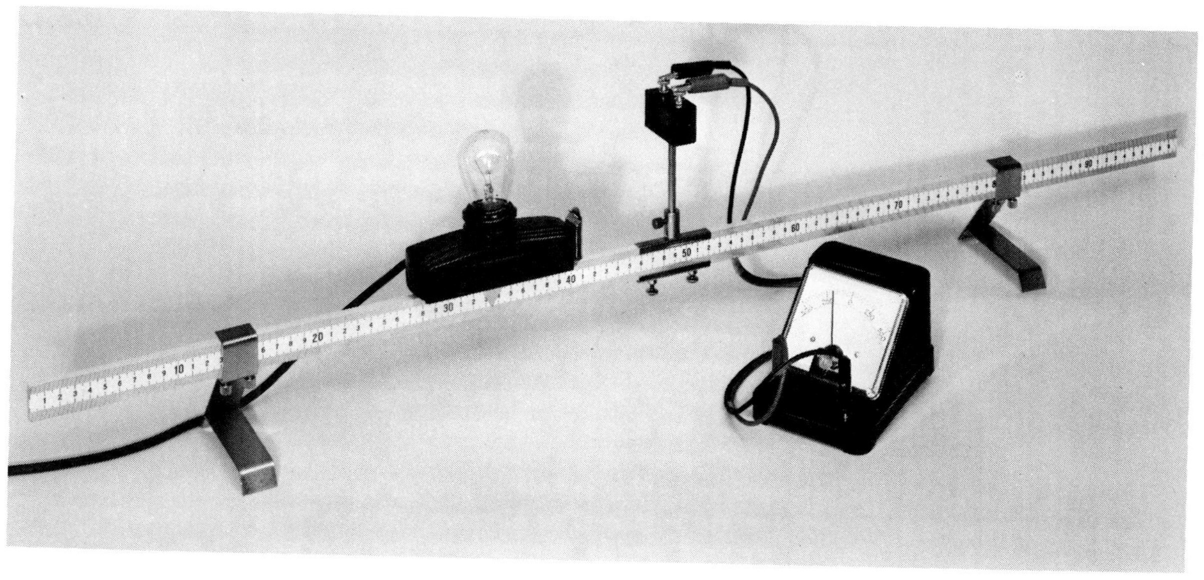

Figure 12-27. A photoelectric photometer.

1. *Bunsen photometer.* The Bunsen photometer is sometimes called the grease-spot photometer. If a disk of white paper with a grease spot in the center is held toward a light, the grease spot will appear lighter than the rest of the paper because it transmits more light. On the other hand, because it is a poorer reflector of light than the paper, the grease spot will appear darker than the paper when held away from a light. In that case it is seen by reflected light. A laboratory model of the Bunsen photometer consists of such a paper screen supported on a meter stick between a standard lamp and a lamp of unknown intensity. The paper screen is moved back and forth along the meter stick until it is equally illuminated on both sides. At this point the grease spot seems to disappear. When the screen is correctly positioned, illuminances E_1 and E_2 of the two sides of the screen are equal.

$$\text{Since} \qquad E_1 = E_2 \qquad \text{then} \qquad \frac{I_1}{r_1^2} = \frac{I_2}{r_2^2}$$

I_1 and I_2 are the intensities of the two sources producing illuminances E_1 and E_2, *and* r_1 and r_2 are their respective distances from the screen. Each distance r is a radial distance from a source.

2. *Joly photometer.* The Joly photometer gives more satisfactory results than the grease-spot photometer. It consists of two blocks of plastic separated by a thin sheet of metal. The light from either side is transmitted by the plastic but is stopped by the metal. By looking at the edges of the blocks of plastic it is easy to adjust the photometer so that both sides are equally illuminated. Then distances from the sheet of metal to the lamps are measured and calculations made as with the Bunsen photometer.

3. *Photoelectric photometer.* The Bunsen or Joly photometer head can be replaced by a photoelectric cell connected to a meter suitable for measuring the photoelectric current of the cell. The photoelectric cell is then placed a given distance from a standard lamp of known luminous intensity, and the photocurrent is measured. An unknown lamp is then substituted for the standard lamp, and the position of the photocell is adjusted to give the same photocurrent as for the standard lamp. The luminous intensity of the unknown lamp can then be calculated as with the Bunsen photometer.

4. *Spherical photometer.* The light source being tested is placed in the center of a large sphere that is painted white on the inside. The luminous flux is received by a photocell located inside the sphere but shielded from direct light

from the source. The light on this cell is equal to that received by any other similar portion of the sphere interior, due to cross-reflections, and is therefore proportional to the total light emitted by the test source. A meter outside the sphere is connected to the cell and is calibrated to read the mean spherical intensity or lumens directly. This accurate photometer is used commercially.

QUESTIONS: GROUP A

1. Define and give an example of (a) a luminous object and (b) an illuminated object.
2. What is the difference between an incandescent and a fluorescent light?
3. How does the frequency of the light emitted from an incandescent source vary with temperature?
4. (a) Polaris is a blue-white star, while our sun appears yellow. What does this say about their relative temperatures? (b) Betelgeuse is a "cool" star. Predict its color.
5. (a) What are the three effects that an object can have on incident light? (b) Which of the three effects are predominant in objects that are transparent? translucent? opaque?
6. (a) What is the quantitative study of light called? (b) What are the three quantities measured? (c) Which is the fundamental quantity?

GROUP B

7. Relate your answer to Question 3 to the temperatures and colors of a bunsen burner flame.
8. (a) What experiment provides evidence that light is an electromagnetic wave? (b) What simple piece of astronomical evidence could you present to show light is an electromagnetic wave?
9. What happens to the relative size of the umbra and penumbra as the moon moves farther from the earth? Refer to Figure 12-21.
10. Galileo performed an experiment to measure the speed of light by timing how long it took light to go from a lamp he was holding to an assistant on a nearby mountain top and back again. Why was this experiment inconclusive?
11. Transparency is dependent on thickness as well as the nature of the material. What might be happening internally to prevent light from passing through a very thick glass plate?
12. What property of light allows us to draw light rays as straight lines?
13. (a) What is an *umbra?* (b) Can you think of another English word with the same root? Elaborate.
14. What is the relationship among the intensity of a source, the luminous flux transmitted, and the illuminance?
15. What happens to the illuminance provided by a light source if the distance from the source is tripled?
16. What is a photometer?
17. A member of a television studio audience in New York sits 30.0 m from the performer. A viewer, sitting 2.0 m from his TV set, observes the performance at home in Chicago, 1.20×10^3 km away. Who hears the performer first? Why?

PROBLEMS: GROUP A

1. A *lightyear* is the distance that light travels in one year. About how many kilometers is this?
2. The radius of the earth's orbit is, on average, 1.5×10^{11} m. If the sun suddenly enlarged and became a super-

nova, how long would it be before we knew about it?

3. It takes light from Alpha Centauri, the nearest star beyond the sun, 4.3 years to reach us. How far is Alpha Centauri?
4. What is the illuminance on your physics book if it is 58 cm below a reading lamp with an intensity of 130 cd?
5. The amount of illuminance on a photometer screen is the same when two lamps are 4.0 m and 2.5 m from the screen. If the intensity of the first lamp is 55 cd what is the intensity of the second lamp?
6. At a distance of 15.0 cm, a lamp provides an illuminance of 530 lm/m^2. What will be the illuminance provided by this lamp on an object 40.0 cm away?
7. An incandescent lamp of 30.0 cd is placed at the 0.0-cm mark on a meter stick, and a lamp of 20.0 cd is placed at the other end. Where must the screen of a photometer be placed so that both sides are equally illuminated?

Physics Activity

Cut a circle from a piece of heavy paper or cardboard. Make a screen from a piece of white paper and tape it to the wall. Using a flashlight or a slide projector as your light source, see what happens to the umbra and penumbra of the shadow on the screen as you change the relative distances between the screen and the object and between the object and the source.

SUMMARY

A particle model and a wave model were developed in the seventeenth century to explain the then known properties of light. The particle model failed after additional properties of light were discovered. The wave model prevailed, and with the experimental confirmation of the electromagnetic theory the wave model of light was thought to be complete. The discovery of the photoelectric effect, the failure of the electromagnetic theory, and the success of the quantum theory in explaining photoemission established the dual character of light.

An object is luminous if it is visible because of the light it emits. An object is illuminated if it is visible because of the light it reflects. An object may reflect, absorb, or transmit the light it receives. Opaque objects cast shadows when illuminated. Shadows may consist of umbras and penumbras.

Photometry is the quantitative study of light. Three quantities measured in practical photometry are luminous intensity, luminous flux, and illuminance. The intensity of a light source can be measured by means of a photometer.

VOCABULARY

candela
coherent light
corpuscular theory
electromagnetic spectrum
electromagnetic wave
first law of photoelectric emission
Huygens' principle
illuminance
illuminated object
inverse square law
laser
lumen
luminous flux
luminous object
penumbra
photoelectric effect
photoelectron
photometer
photometry
photon
Planck's constant
quantum theory
radiant energy
second law of photoelectric emission
third law of photoelectric emission
umbra
visible spectrum

13 Reflection

reflection (re-FLEK-shun) n.: the turning back of a wave meeting a boundary of a medium.

13.1 Reflectance A light beam passing through any material medium will become progressively weaker due to two effects. Part of the light beam's energy will be *absorbed* by molecules of the medium and part will be *scattered* in all directions by the molecules. A light beam striking the boundary between two media can be partly *transmitted* and partly returned to the first medium. See Figure 13-1. Light returned to the first medium is said to be *reflected* at the boundary. Reflection is described as a wave property in Section 10.10.

Part of the reflected light proceeds in a common direction but part can be reflected in all possible directions as scattered light. See Figure 13-2 for these effects. The fact that a portion returns in a common direction provides us

OBJECTIVES

- Describe regular and diffused reflection of light.
- Describe the two laws of reflection.
- Analyze the images formed by plane mirrors.
- Define the terminology of curved mirrors.
- Describe the location of image points formed by concave and convex mirrors.
- Do problem solving with the mirror equation.

Figure 13-1. A diagram showing the effects on light rays of a material medium and a boundary between two media.

with a method for controlling the reflection of a light beam.

The amount of light reflected at the surface of an object, whether by scattering or by unidirectional reflection, depends on the kind of material underlying the surface, the smoothness of the surface, and the angle at which the light strikes it. *The ratio of the light reflected from a surface to the light falling on the surface is called* ***reflectance.*** This ratio is commonly expressed as a percentage. Materials differ widely in their reflectance. The material of highest reflectance is magnesium oxide, a white chalky substance that reflects about 98% of incident light with practically complete scattering. The reflectance of a smooth surface of silver is about 95% with negligible scattering. Some black surfaces have reflectances of 5% or less.

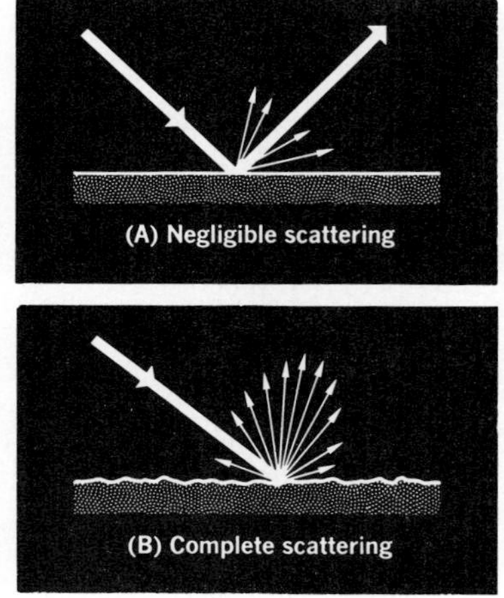

Figure 13-2. Reflecting surfaces vary in the extent to which reflected rays are scattered.

13.2 Regular and Diffused Reflection

The reflection of sunlight by a mirror produces a blinding glare. The rays of the sun reaching the earth are practically parallel and thus have the same angle of incidence. They remain practically parallel after being reflected from a plane mirror. Such reflection, in which scattering of the reflected rays is negligible, is called **regular** ***reflection.*** Polished, or **specular,** surfaces causes regular reflection and the image of the luminous source is sharply defined. The regular reflection of a narrow beam of light is in one direction with no appreciable loss of definition or intensity. The nature of regular reflection is shown in Figure 13-3.

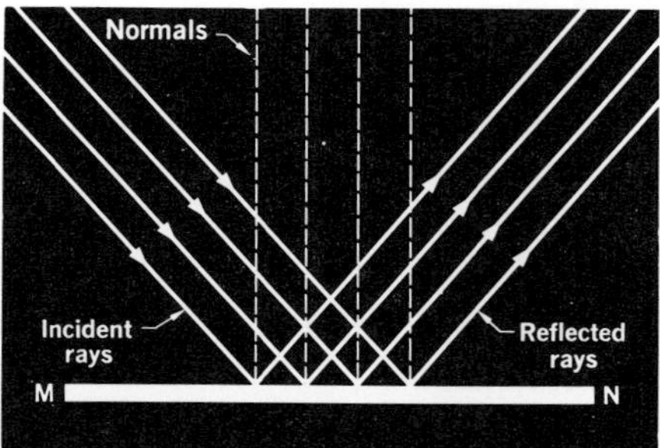

Figure 13-3. Regular reflection of light from a specular surface.

Regular reflection from highly polished surfaces provides for relatively accurate control of light rays. Searchlights, beacons, and automobile spotlights use concentrated light sources of high intensity and highly-polished regular reflectors to redirect light rays in the desired direction.

In Figure 13-4 we observe what happens to a beam of light incident on an irregular surface. The laws of reflection hold true for each particular ray of light, but the normals to the surface are not parallel and the light is reflected in many directions. Such ***scattering,*** or **diffusion,** of light is extremely important.

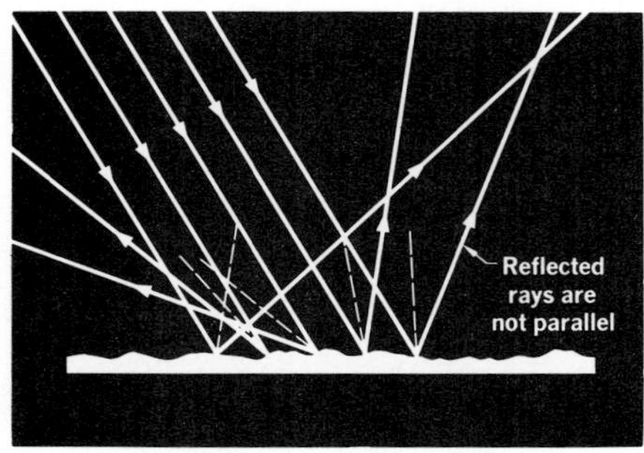

Figure 13-4. Diffused reflection promotes the diffusion of light.

If the sun's rays were not diffused by rough and irregular surfaces and by dust particles in the air, the corners of a room and the spaces under shade trees would be in almost total darkness and the glare would be dazzling in sunlit areas. Astronauts have found that it is possible for them to see in the shadows on the lunar surface where there is no air, but they cannot see as easily as we can on the surface of the earth.

13.3 Mirrors as Reflectors The reflection of light is similar to the reflection of sound or to the rebound of an elastic ball. The line **MN** in Figure 13-5(A) represents a reflecting surface; **AD** is a ray of light incident upon the reflector at **D; DB** is the path of the reflected ray. The line **CD,** perpendicular to the reflecting surface at the point of incidence, is called a **normal.** In Figure 13-5(B) the normal is perpendicular to the tangent to the curved reflecting surface at the point of incidence. Recall from Section 10.10 that the **angle of incidence** (i) is the angle between the incident ray and the normal at the point of incidence [angle **ADC** in Figure 13-5(A)]. In Chapter 10 we examined the nature of reflection by sending water waves against a barrier in the ripple tank. At that time we recognized *the **first law of reflection:** The angle of incidence, i, is equal to the angle of reflection, r.* The **angle of reflection,** r, is the angle between the reflected ray and the normal at the point of incidence [angle **BDC** in Figure 13-5(A)].

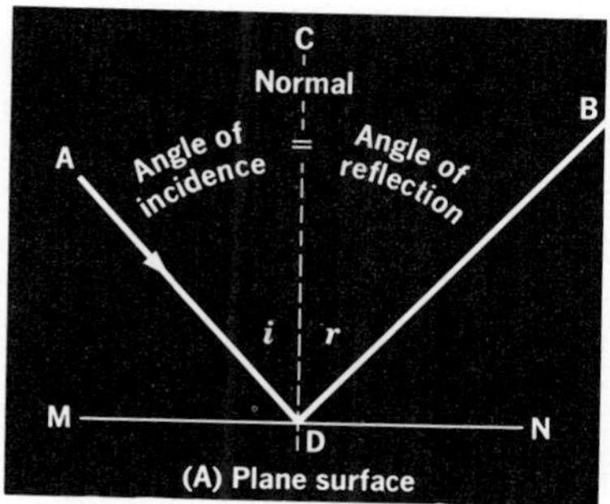

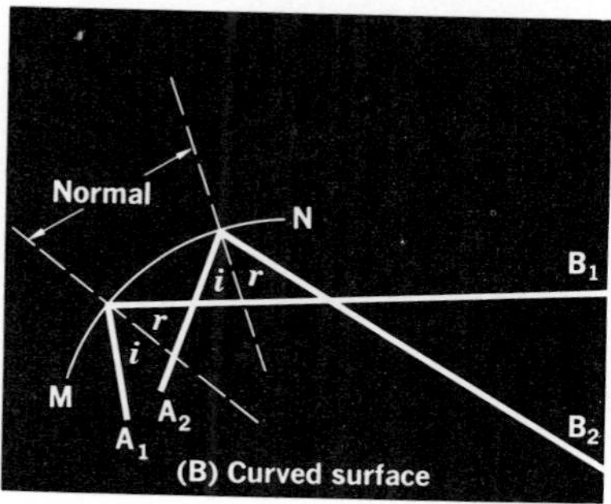

Figure 13-5. Reflections from plane and curved surfaces.

The relationships between an incident ray of light and the reflected ray and between the angles they form with the normal are easily determined in the laboratory. Critical observation of the specular reflection of light reveals a ***second law of reflection:** The incident ray, the reflected ray, and the normal to the reflecting surface lie in the same plane.* As mentioned in Section 10.10, these laws are true for all forms of wave propagation.

The reflective coating is commonly placed on the back surface of a plane mirror to protect the coating from damage.

Any highly polished surface that forms images by the regular reflection of light can act as a mirror. A **plane mirror** of plate glass is silvered on one surface to reflect the light efficiently. Its reflectance may be nearly 100%. The plane parallel surfaces of the plate glass allow an essentially distortion-free image to be observed.

A **spherical mirror** is a small section of the surface of a

Figure 13-6. Spherical mirrors are circular sections of spheres.

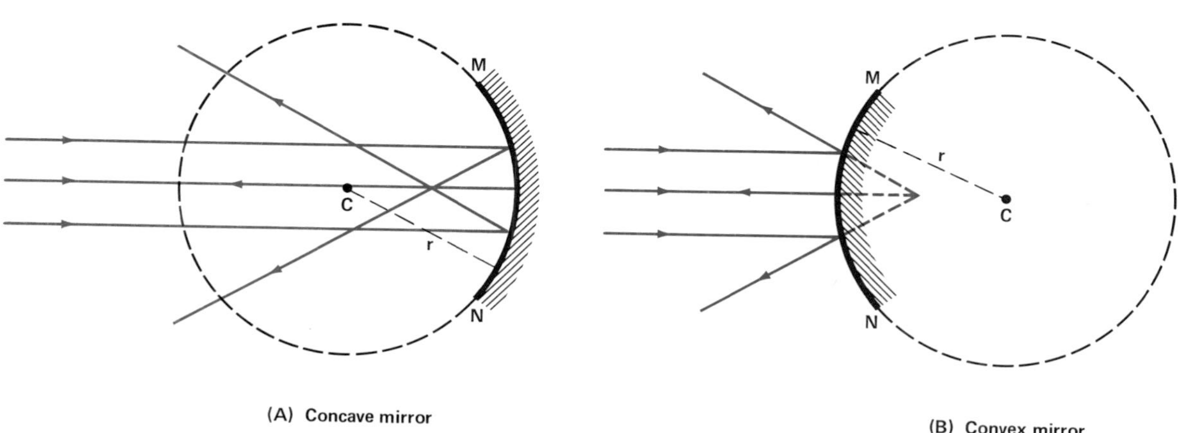

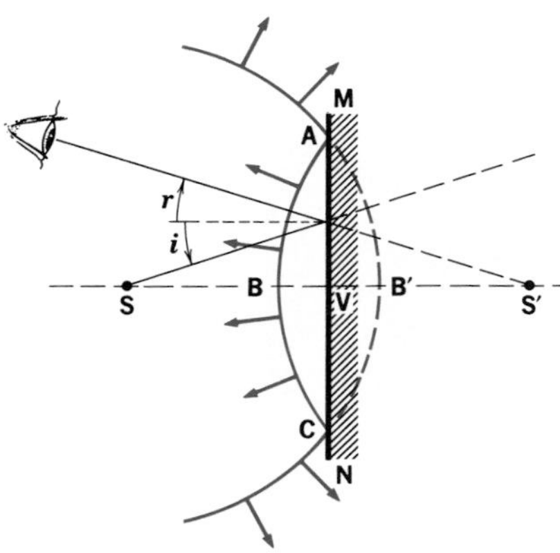

Figure 13-7. A wave-front diagram of reflection of light from a point source by a plane mirror.

sphere. One side of the section is polished or coated with a reflective material. When viewed from the *inside,* the mirror is called a **concave mirror,** as indicated in Figure 13-6(A). When viewed from the *outside,* it is called a **convex mirror,** as indicated in Figure 13-6(B).

The laws of reflection hold for both spherical mirrors and plane mirrors. However, the size and position of the images formed by spherical mirrors are quite different from the size and position of images formed by plane mirrors.

13.4 Images by Reflection Light rays reflected from a concave mirror may meet in front of the mirror and form an image of the object from which the light comes. Such an image can be projected on a screen placed at the image location. *An image formed by converging rays of light actually passing through the image point is called a **real image.*** Real images are inverted relative to the object and can be larger or smaller than the object.

Your image observed in a plane mirror seems to be behind the mirror, although the rays forming it actually are reflected forward from the mirror surface. *This image is called a **virtual image.** Rays of light that appear to have diverged from the image point do not actually pass through that point.* Virtual images cannot be projected on a screen. They are erect with respect to the object and can be enlarged or reduced in size.

In Figure 13-7 light traveling from a point source in spherical wave fronts is reflected from a plane mirror. At the instant that points **A** and **C** on the wave front reach the mirror, point **B** has been reflected back towards the source at **S,** having traveled the distance **VB.** The reflected wave front **ABC** has an apparent source **S′** at an apparent distance **VS′** behind the mirror. Note that **VS′** is equal to the distance **VS** of the source in front of the mirror. The reflected wave front approaches an observer, the eye, as though its source were **S′**, **S′** being the virtual image of the source **S.** The real image of an object is composed of many image points, each being the point image of the corresponding point on the object.

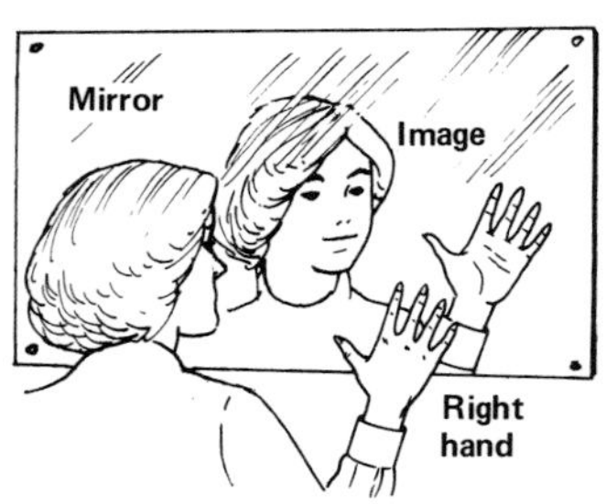

Figure 13-8. The image formed by a plane mirror appears to be reversed right and left. Observe that the image of the right hand looks like the left hand.

13.5 Images Formed by Plane Mirrors The image formed by a plane mirror is neither enlarged nor reduced but is always virtual, erect, and appears to be as far behind the mirror as the object is in front of it. The image differs from the object in that right and left appear to have been interchanged. In Figure 13-8 the image of the right hand has the appearance of a left hand. If an object is spinning clockwise about its axis, its image appears to be spinning counterclockwise about its image axis.

The ray diagram in Figure 13-9 shows a simple graphical method of locating the image formed by a plane mirror. The triangle **ABC**, drawn on paper, is used as the object in front of a mirror, **MN**. A pin is placed at each vertex of the triangle. Rays $\mathbf{AO_1}$ and $\mathbf{AO_2}$ from pin **A** strike the mirror so as to be reflected to the two eyes, $\mathbf{E_1}$ and $\mathbf{E_2}$, of the observer. When the sight lines $\mathbf{E_1O_1}$ and $\mathbf{E_2O_2}$ are produced (extended) behind the mirror, they intersect at **A′**, which is the location of the image of pin **A**. A comparison of triangles $\mathbf{AO_2O_1}$ and $\mathbf{A'O_2O_1}$ makes it apparent that **A′** is as far behind the mirror as **A** is in front of it.

The images of **B** and **C** are located in similar fashion at **B′** and **C′**. In each instance *the image appears to be as far behind the mirror as the object is in front of it*. When the image points **A′B′C′** are joined, it becomes evident that *the image is the same size as the object*.

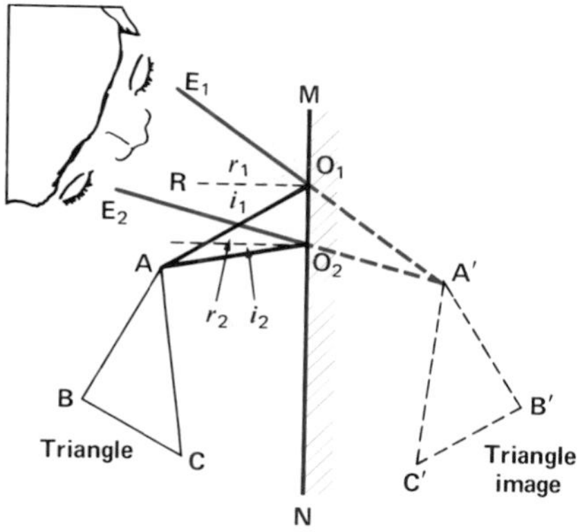

Figure 13-9. A ray diagram used to construct an image formed by a plane mirror.

This method of constructing images is based on the laws of reflection. The light ray $\mathbf{AO_1}$ is reflected at point $\mathbf{O_1}$ and is sighted at $\mathbf{E_1}$. By drawing a normal to the mirror at $\mathbf{O_1}$, the angle $\mathbf{AO_1E_1}$ is bisected and the angle of reflection $\mathbf{E_1O_1R}$ is equal to the angle of incidence $\mathbf{AO_1R}$. The incident ray $\mathbf{AO_1}$, the reflected ray $\mathbf{O_1E_1}$, and the normal $\mathbf{O_1R}$ all lie in the same plane.

Sound waves bound and rebound between parallel cliffs or walls and produce multiple echoes. In a similar fashion, light reflects back and forth between parallel mirrors or mirrors set at an acute angle and forms multiple images. The image formed in one mirror appears to act as the object forming the image in the next mirror. Because some light energy is absorbed and some is scattered at each mirror, the succeeding images become fainter. A thick plate-glass mirror produces multiple reflections, as shown in Figure 13-10, because some light is reflected each time a boundary is encountered.

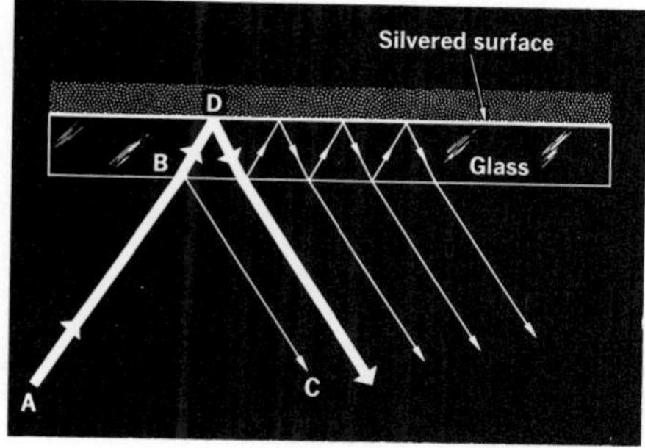

Figure 13-10. Multiple reflections that result from the use of a back-silvered mirror.

Multiple reflections can be avoided by aluminizing or silvering the front surface of a glass mirror. Mirrors for astronomical telescopes and other precision instruments are usually coated with aluminum in this manner. Such reflecting surfaces are exposed and can be easily damaged.

13.6 Curved-Mirror Terminology In order to discuss image formation by spherical mirrors, several terms associated with this process must be identified. In this discussion, reference is made to the diagrams of mirrors shown in Figures 13-11 and 13-12.

The arc **MVN** is the portion of the spherical surface separated from the sphere by a cutting plane represented by the line **MN**.

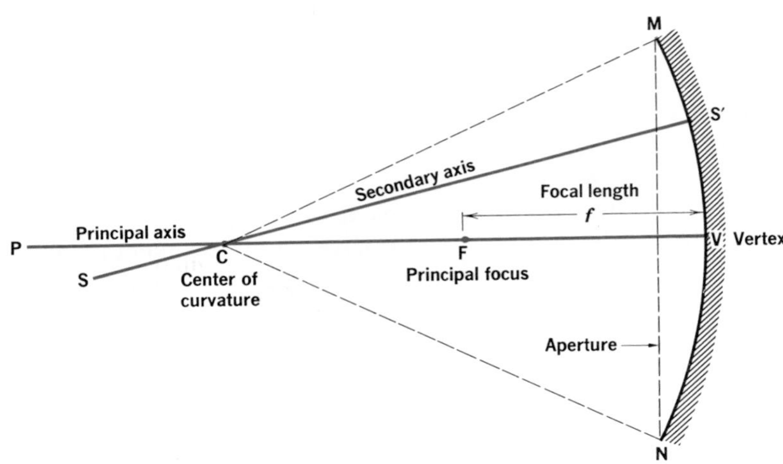

Figure 13-11. A plane diagram for defining terms used with curved mirrors.

1. The **center of curvature, C,** is the center of the original sphere. The radius of the sphere is also the *radius* **CV** of the spherical mirror.

2. The **vertex, V,** is the center of the mirror.

3. The **principal axis** of the mirror is the line **PV** drawn through the center of curvature and the vertex.

4. A **secondary axis** is any other line drawn through the center of curvature to the mirror: **SS′**, for example.

5. A **normal** to the surface of a spherical mirror is a radius drawn from the point of incidence of a light ray. Therefore the normal is perpendicular to the mirror surface at the incidence point. (Normals are illustrated in Figure 13-12.)

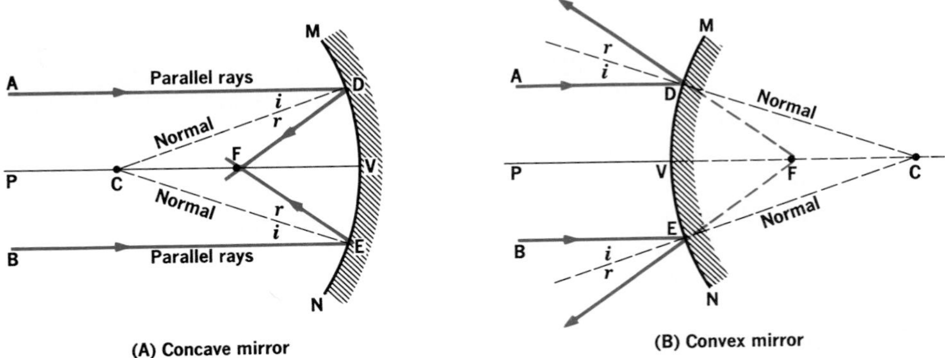

Figure 13-12. Locating the principal focus of spherical mirrors.

The aperture size relative to focal length in mirror diagrams is exaggerated for the sake of clarity.

6. The **principal focus, F,** is the point on the principal axis where light rays close to and parallel to the principal axis converge, or from which they appear to diverge.

7. The **focal length,** *f*, is the distance from the vertex of the mirror to its principal focus.

8. The **aperture** determines the amount of light intercepted by the mirror. The length **MN** is called the **linear**

aperture of the mirror, and the angle **MCN** is called the **angular aperture.** For a spherical mirror to be capable of forming sharp images, its aperture must be small compared with its focal length.

13.7 Rays Focused by Spherical Mirrors In Figure 13-12(A) the rays **AD** and **BE,** parallel to the principal axis, are shown incident on a concave spherical mirror at points **D** and **E** respectively. The normals **CD** and **CE** are drawn to the points of incidence. By the second law of reflection, the reflected rays converge on the principal axis at **F,** the principal focus. Because the incident rays converge upon being reflected, *concave mirrors* are also known as **converging mirrors**. The distance **FV** is the *focal length* of the mirror. For spherical mirrors of small aperture, the focal length is one-half the radius of curvature.

Parallel rays incident on the surface of a convex spherical mirror are shown in Figure 13-12(B). The normals **CD** and **CE** produced beyond the mirror surface show the reflected rays to be divergent along the lines **FD** and **FE** produced, **F** being the principal focus of the mirror. Rays parallel to the principal axis appear to diverge from the principal focus of a convex mirror. Because incident rays diverge on being reflected, *convex mirrors* are known as **diverging mirrors.**

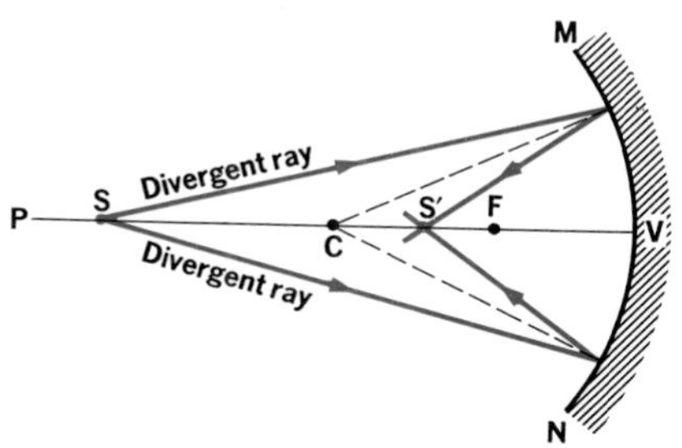

Figure 13-13. The reflection of diverging rays by a concave mirror.

Diverging rays from a point **S** incident on a concave mirror, Figure 13-13, converge to a focus at some point **S′** beyond the principal focus of the mirror. If **S** is located on the principal axis, the image **S′** is also on the principal axis. Suppose the source of diverging rays is now moved to position **S′**. The image formed by the mirror will be at **S.**

Converging rays are also focused by concave mirrors. As shown by Figure 13-14, the point of focus is between the vertex and the principal focus. In every instance the direction of a reflected ray is determined by the laws of reflection.

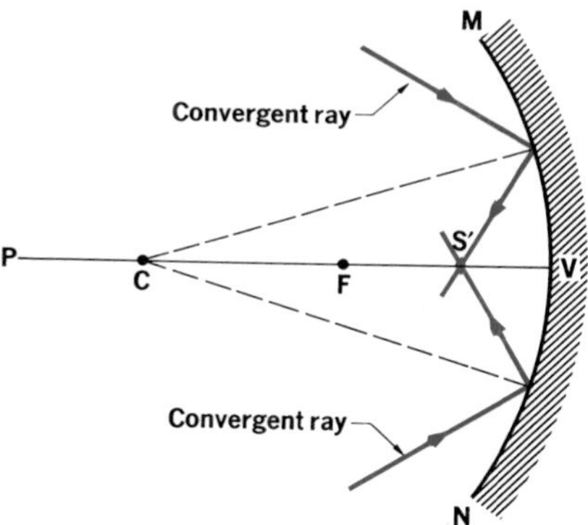

Figure 13-14. The reflection of converging rays by a concave mirror.

As observed in Figure 13-12(A), the incident parallel rays that are also parallel to the principal axis of the concave mirror are reflected and converge at the principal focus. The image is ideally a point of light at the principal focus. If the parallel rays are not parallel to the principal axis but are incident at a small angle with the principal axis, they do not converge at the principal focus. Instead they are focused at a point in a *plane* that contains the principal focus and is perpendicular to the principal axis. This plane is called the **focal plane** of the mirror. Figure 13-15 illustrates the reflection of a bundle of parallel light rays incident on a concave mirror. In Figure 13-15(A) the rays are parallel to the principal axis of the mirror, and in

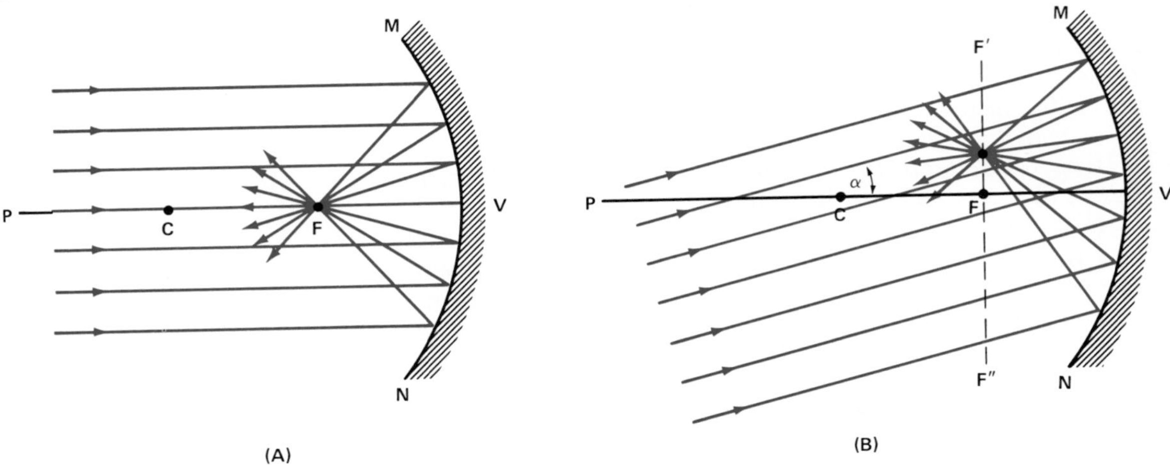

Figure 13-15. In (A) parallel rays are shown incident on a concave mirror parallel to its principal axis. The reflected rays converge at the principal focus. In (B) the parallel rays are incident at a small angle α (alpha) with the principal axis. Here the reflected rays are focused at a point in the focal plane F′F″ of the mirror.

Figure 13-15(B) the rays are at a small angle α (the Greek letter alpha) with the principal axis.

If the aperture of a spherical mirror is large, the parallel rays of light striking the mirror near its edge are not reflected through the principal focus but are focused at points nearer the mirror, as shown in Figure 13-16(A). Only those parallel rays that are incident on the mirror near its vertex are reflected to the principal focus. This characteristic of spherical mirrors of large aperture is known as **spherical aberration** and results in the formation of fuzzy images.

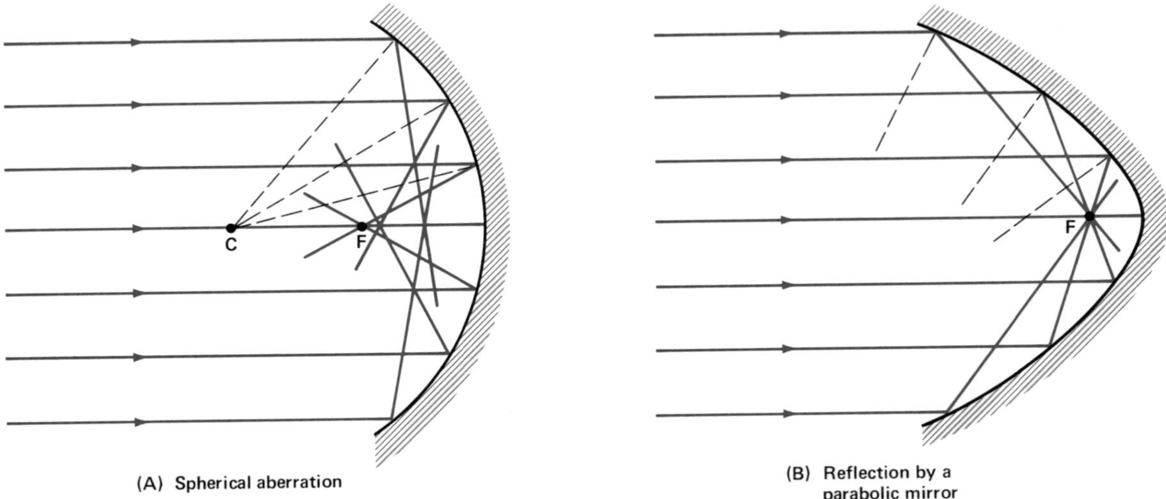

Figure 13-16. Spherical aberration is a characteristic of spherical mirrors with large apertures (A). This defect is avoided by the use of parabolic mirrors (B).

In order to reduce spherical aberration, the aperture of the spherical mirror must be small. For angular apertures no larger than about 10°, the distortion of the image is negligible. Spherical aberration can be avoided for parallel rays if the mirror surface is made parabolic instead of spherical. See Figure 13-16(B). The parabolic mirror shown in Figure 13-17 reflects light rays originating at the focus as parallel rays. Parabolic reflectors are used in automobile headlights, microwave communication relays, radar, and radio telescopes. Parabolic mirrors are also used in large reflecting telescopes to collect and focus light rays from distant objects.

Parabola: the curve of intersection of a plane parallel to a straight line in the surface of a right circular cone.

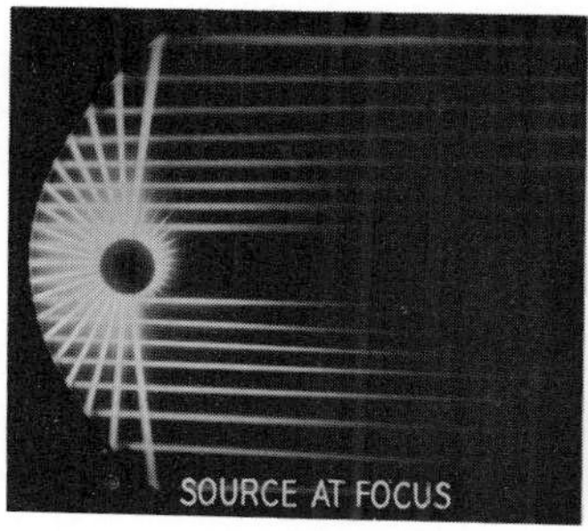

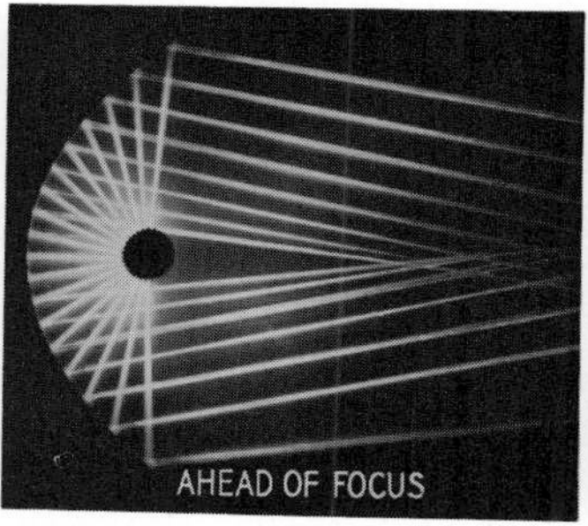

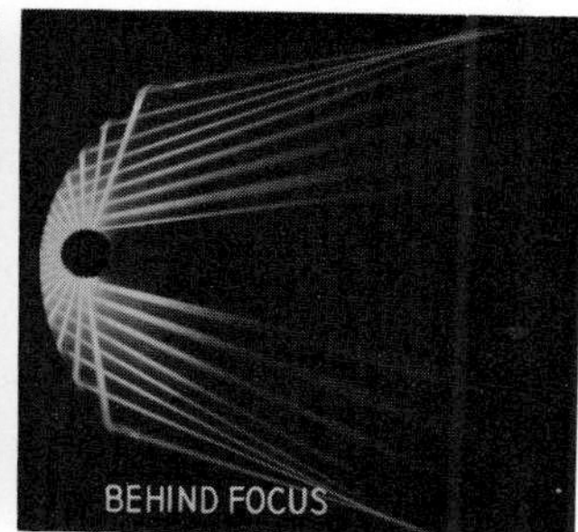

Figure 13-17. Reflections from a parabolic mirror.

13.8 Constructing the Image of a Point In Figure 13-18 the image of point **S** formed by a concave mirror is to be located. First the principal axis **PV** is drawn through **C.** Because light is given off in all directions from point **S,** *any two lines* can be drawn from **S** to represent rays of light incident on the mirror. The point of intersection of the *reflected* rays locates the image of **S** at **S′.**

The construction is greatly simplified if incident rays are selected for which the directions of the reflected rays are known from the geometry of the system. Three such special rays are shown in Figure 13-18. Ray **SA,** *parallel to the principal axis,* is reflected through the principal focus. Ray **SB** *passes through the principal focus* and is reflected parallel to the principal axis. Ray **SC** *lies along the secondary axis.* It is reflected back along its path of incidence because the

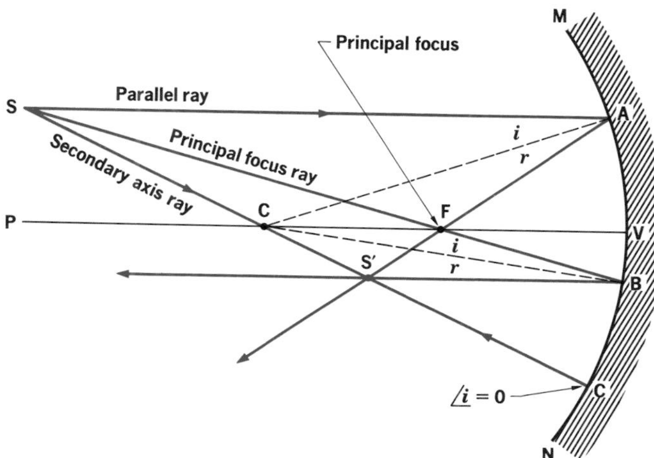

Figure 13-18. Locating an image formed by a concave mirror.

angle of incidence is zero. *Note that the convergence of any two of these three reflected rays at a common point* **S′** *forms the image of* **S.**

Conjugate points: any two points so situated that light from one is concentrated at the other.

If the object were located at **S′** the image would be formed at **S**. The object and image are thus interchangeable, or *conjugate*. The positions of the object and its image, **S** and **S′**, are called *conjugate points*.

13.9 Images Formed by Concave Mirrors The image formed by a concave mirror can be constructed by locating the images of an adequate number of object points. Suppose the object is an arrow. The entire image can be located by constructing the image points of the head and the tail of the arrow. Concave mirror images can be grouped into six cases based on the position of the object in front of the mirror.

Case 1. Object at an infinite distance. Rays emanating from an object at an infinite distance from a mirror are parallel at the mirror. If the rays are parallel to the principal axis, they are reflected through the principal focus. See Figure 13-19(A). We therefore conclude that *the image formed by a concave mirror of an object an infinite distance away is a point at the principal focus.* The rays of the sun reaching the earth are practically parallel and provide a simple method for finding the approximate focal length of a concave mirror.

Case 2. Object at a finite distance beyond the center of curvature. In Figure 13-19(B) the image of the object arrow can be located using the method illustrated in Section 13.8. Two known rays, **1** and **2,** emanating from the head of the arrow, intersect after reflection to locate the image of the arrowhead. The image of the tail lies on the principal axis as shown earlier in Figure 13-13. *The image in this case is real, inverted, reduced in size, and located between the center of curvature and the principal focus.* The nearer the object approaches the center of curvature, the larger the image becomes and the nearer the image approaches **C.**

Case 3. Object at the center of curvature. When the object arrow is at the center of curvature, as shown in Figure 13-19(C), the image of the arrowhead is found inverted at **C.** *When the object is at the center of curvature, the image is real, inverted, the same size as the object, and located at the center of curvature.*

Case 4. Object between the center of curvature and principal focus. This case is the converse of Case 2 and is shown in Figure 13-19(D). *The image is real, inverted, enlarged, and located beyond the center of curvature.* The nearer the object approaches the principal focus of the mirror, the larger the image becomes and the farther it is beyond **C.**

Case 5. Object at principal focus. This case is the converse

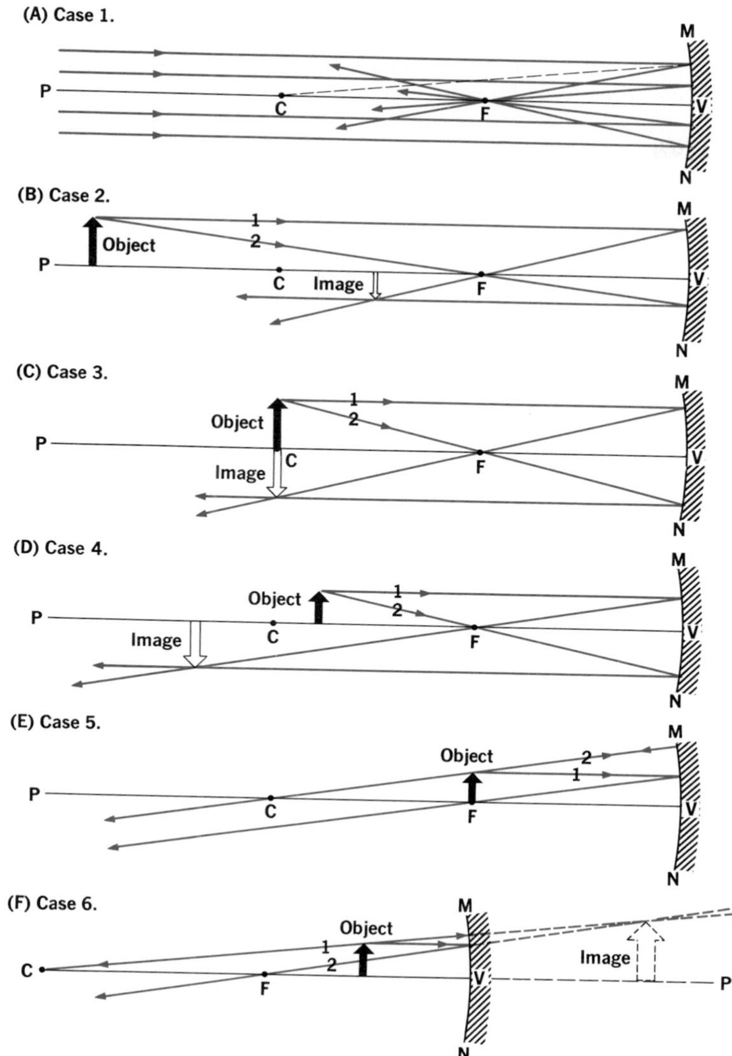

Figure 13-19. Ray diagrams of image formation by concave mirrors. The eye sees the light rays that are directed toward the left in each case.

of Case 1; all rays originating from the same point on the object are reflected from the mirror as parallel rays. See Figure 13-19(E). *When the object is at the principal focus, no image is formed.* If the object is a point source, all reflected rays are parallel to the principal axis.

Case 6. Object between principal focus and mirror. The reflected rays from any point on the object are divergent; they can never meet to form a real image. They appear to meet behind the mirror, however, to form a virtual image as shown in Figure 13-19(F). In this case, *the image is virtual, erect, enlarged, and located behind the mirror.*

13.10 Images Formed by Convex Mirrors A convex mirror forms an erect image of reduced size. The diagram shown in Figure 13-20 can be used to show how such an image is formed. The arrow **AB** represents the object. The secondary axes and the normals at the points of incidence of the parallel rays are *radii produced.* The parallel rays are reflected from the surface of the mirror as divergent rays.

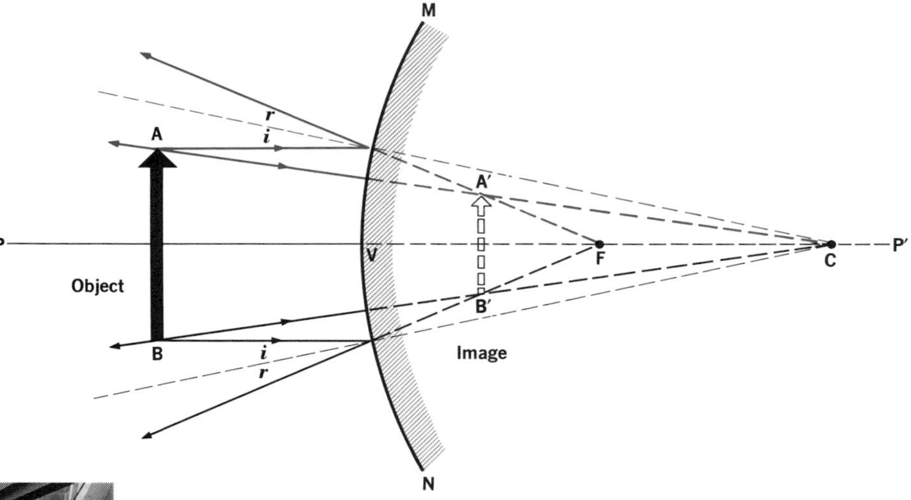

Figure 13-20. The only image formed by a convex mirror is virtual, erect, reduced, and located behind the mirror. The mirror and image are being viewed from the left.

When produced behind the mirror, the reflected parallel rays *appear* to meet at the principal focus. The image is formed where parallel and secondary-axis rays *emanating from the same object point* appear to intersect behind the mirror. In a convex mirror, *all images are virtual, erect, smaller than the object, and located behind the mirror between the vertex and the principal focus.* The size of the image increases as the object is moved closer to the mirror, but it can never become as large as the object itself.

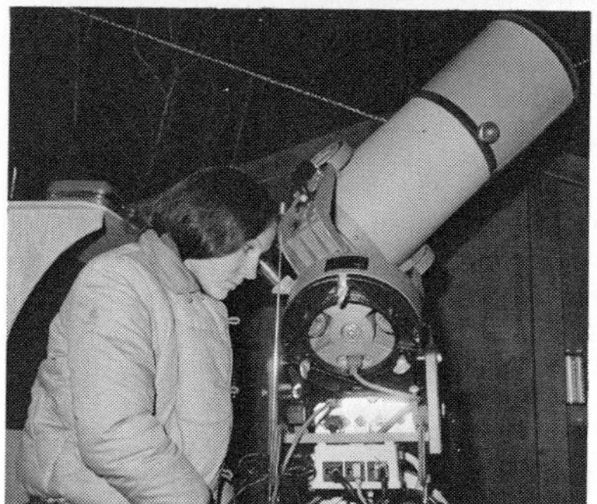

Figure 13-21. (A) The 200-inch mirror of the Mt. Palomar, California, reflecting telescope. (B) An 8-inch reflecting telescope ideally suited for amateur astronomers.

13.11 The Reflecting Telescope The largest optical telescopes are reflecting telescopes. Large concave parabolic mirrors are used to collect and focus light rays. See Figure 13-21. Reflecting telescopes can be made much larger than refracting (lens) telescopes. It is difficult to obtain a large piece of glass of sufficiently high optical quality for a lens, to grind and polish two surfaces with precision, and to support it appropriately.

When objects being viewed with a reflecting telescope are at a finite distance beyond the center of curvature of the mirror, Case 2 (Section 13.9) applies. The image is real, inverted, and located near the principal focus of the mirror. This real image can be reflected out of the path of the rays of light incident on the mirror so that it can be viewed or photographed from a position outside the telescope.

The inversion of the image is not objectionable when reflecting telescopes are used for astronomical viewing.

13.12 Object-Image Relationships A simple relationship exists among the distance of an object d_o from a curved mirror, the distance of its image d_i from the mirror, and the focal length f of the mirror. This relationship, known as the **mirror equation,** can be derived by considering the diagram shown in Figure 13-22, in which a real image of the object **AB** is formed by a concave mirror. (In this derivation we will consider as insignificant the very slight error introduced by the curvature **MV.**)

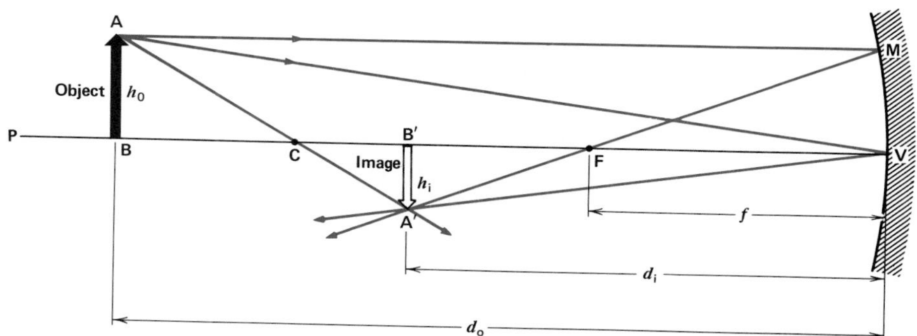

Figure 13-22. A diagram for the derivation of the mirror equation for curved mirrors.

In similar triangles **A′B′F** and **MVF,**

$$\frac{\mathbf{A'B'}}{\mathbf{MV}} = \frac{\mathbf{B'F}}{\mathbf{VF}} \qquad \text{(Equation 1)}$$

Taking **MV** = **AB, VF** = f, and **B′F** = $d_i - f$, we have

$$\frac{\mathbf{A'B'}}{\mathbf{AB}} = \frac{d_i - f}{f} = \frac{d_i}{f} - 1 \qquad \text{(Equation 2)}$$

In similar triangles **ABV** and **A′B′V,**

$$\frac{\mathbf{A'B'}}{\mathbf{AB}} = \frac{\mathbf{B'V}}{\mathbf{BV}} \qquad \text{(Equation 3)}$$

Taking **BV** = d_o and **B′V** = d_i,

$$\frac{\mathbf{A'B'}}{\mathbf{AB}} = \frac{d_i}{d_o} \qquad \text{(Equation 4)}$$

Substituting in Equation 2

$$\frac{d_i}{d_o} = \frac{d_i}{f} - 1$$

Dividing by d_i,

$$\frac{1}{d_o} = \frac{1}{f} - \frac{1}{d_i} \qquad \text{(Equation 5)}$$

Or,

$$\frac{1}{f} = \frac{1}{d_o} + \frac{1}{d_i} \qquad \text{(Equation 6)}$$

The mirror equation also applies to image formation by refraction (Section 14.11).

where d_o is the distance of the object from the mirror, d_i the image distance, and f the focal length of the mirror.

Equation 6, the mirror equation, is applicable to all concave and convex mirrors. When an object is located less than one focal length in front of a concave mirror, the image is virtual. It appears to be located behind the mirror, and the image distance d_i is a *negative* quantity. In the case of convex mirrors, the radius of curvature is negative and the image is virtual. Therefore, both the focal length f and the image distance d_i are *negative* quantities. For concave mirrors, the radius of curvature and the focal length are *positive.* The image distance can be either positive or negative—being *positive* for real images and *negative* for virtual images. The significance of these negative quantities is illustrated in the following example.

EXAMPLE A concave mirror has a focal length of 20.0 cm. An object is located 10.0 cm in front of the mirror. What is the location of the image?

Given	Unknown	Basic equations
$f = 20.0$ cm $d_o = 10.0$ cm	d_i	$\frac{1}{f} = \frac{1}{d_o} + \frac{1}{d_i}$

Solution

Working equation: $d_i = \dfrac{fd_o}{d_o - f} = \dfrac{(20.0\text{ cm})(10.0\text{ cm})}{10\text{ cm} - 20.0\text{ cm}}$

$= -20.0$ cm (behind the mirror)

The relative heights of an object and its image formed by a curved mirror depend on their respective distances from the center of curvature of the mirror. This can be seen from similar triangles **ABC** and **A′B′C** of Figure 13-22.

We can demonstrate that for mirrors of small aperture the relative heights of the object and image also depend on their respective distances from the vertex of the mirror. If the height of object **AB** in Figure 13-22 is represented by h_o and the height of the image **A′B′** is represented by h_i, Equation 4 of Section 13.12 becomes

$$\frac{h_i}{h_o} = \frac{d_i}{d_o}$$

The equations on this page apply to virtual images as well as to real images.

The sign conventions for spherical mirror computations using the mirror equation are summarized as follows:

1. The numerical value of the focal length f of a concave mirror is positive.
2. The numerical value of the focal length f of a convex mirror is negative.
3. The numerical value of the real object distance d_o is positive.
4. The numerical value of the real image distance d_i is positive.
5. The numerical value of the virtual image distance d_i is negative.

QUESTIONS: GROUP A

1. (a) What is reflectance? (b) Which materials have the highest reflectance?
(c) What astronomical phenomenon is believed to have 0% reflectance?

2. Are the laws of reflection true for diffuse and specular reflection?

3. (a) What is a real image? Which type(s) of mirror produce(s) such an image? (b) What is a virtual image? Which mirror(s) can give you virtual images?

4. (a) Define spherical aberration.
(b) How can it be reduced and/or avoided?

5. Why are the largest telescopes reflectors rather than refractors?

6. What is the relationship between the principal axis and the mirror surface at their intersection (the vertex)?

7. (a) What does a negative value for a mirror's focal length tell you about the mirror? (b) What does it mean if the image distance is also negative?

8. Could an object be a regular reflector for some electromagnetic waves, yet diffuse for others?

9. What happens to the image as you bring an object from very far away towards the focus of a concave mirror?

10. One of the mirrors mounted on the side of your car has this warning: "Objects in mirror are closer than they appear." (a) What type of mirror is this? (b) Why do objects appear closer?

11. Leonardo da Vinci kept many notebooks that could only be read if viewed in a plane mirror. How did he write them?
12. You look in the rearview mirror of your car and read the word "Police" written on the front of the van behind you. What will the word look like if you turn around and look at it directly?
13. Suppose you write the word "light" on a mirror with white paint. When the mirror is placed in the sunlight, the reflection consists of a bright area where the word "light" appears dark. How does the property of reflectance explain what is happening here?
14. With a compass and ruler, construct a diagram of a concave spherical mirror with a shallow curvature. Label the center of curvature, focal point, vertex, and principal axis. Illustrate and label the three principal rays.
15. Construct ray diagrams locating and identifying the image of an object placed (a) beyond the center of curvature of a concave mirror (b) between the focus and vertex of a concave mirror (c) in front of a convex mirror.

PROBLEMS: GROUP A

1. What is the focal length of a mirror whose radius of curvature is 32 cm?
2. An object placed 15 cm from a curved mirror produces a virtual, enlarged image. (a) What type of mirror is it? (b) What is the minimum value for the radius of curvature of this mirror?
3. How far from a concave mirror, focal length 6.0 cm, would a candle have to be placed to look like it was burning at both ends?
4. An object and its images are the same height when the object is 0.42 m from a curved mirror. What is (a) the radius of curvature of the mirror? (b) its focal length?
5. (a) To what point would a concave mirror converge the light from a star that is very far away? (b) If the radius of curvature of this mirror is 2.75 m, how far from the mirror should a camera be placed to receive this image?
6. (a) An object is placed 15.0 cm from a mirror whose focal length is 5.0 cm. Determine the distance of the image from the mirror. (b) If the object is 6.0 cm high, how big is the image?

GROUP B

7. Where would you place an object in front of a concave mirror (focal length 8.50 cm) so that the image is twice as far from the mirror as the object?
8. A candle, 3.5 cm high, placed 28 cm from a curved mirror forms an image on a screen at a distance of 12 cm from the mirror. (a) Find the focal length of the mirror. (b) How high is the image? (c) What type of mirror is this?
9. A 5.00-cm arrow stands at the 0.0-cm mark on a meter stick. At the 50.0-cm mark is a convex mirror whose radius of curvature is 45.0 cm. (a) How far from the mirror is the image? (b) Is it virtual or real? (c) How tall is it?
10. The image of the moon is formed by a concave telescope mirror whose radius of curvature is 4.00 m at a time when the distance to the moon is 3.80×10^8 m. What is the diameter of the image if the diameter of the moon is 3.48×10^6 m.
11. A spherical concave mirror of small aperture has a radius of curvature of 40.0 cm. When your face is 15.0 cm from the mirror, (a) what type of image will be formed? (b) what is the magnification of the image?

PHYSICS ACTIVITY

Measure the length of your face from the top of your forehead to your chin. Look at yourself in the bathroom mirror. Predict how big the mirror would have to be to show your whole face. With a piece of soap, mark the top and bottom of the image. Measure it and compare it to the length of your face. Now back away from the mirror, keeping the top of your forehead even with the top mark. What happens to the image size?

SUMMARY

The amount of incident light reflected from the surface of an object depends on the composition of the object, the character of its surface, and the angle at which the light rays strike the surface. The ratio of the reflected light to the incident light is known as reflectance. The angle of reflection and the angle of incidence are equal, and both lie in the same plane.

Images may be either real or virtual. Real images are inverted with respect to the object and can be projected onto a screen. Virtual images are erect and cannot be projected onto a screen. Images formed by plane mirrors are virtual, erect, and the same size as the object. Images formed by plane mirrors appear to be the same distance behind the mirror as the object is in front of the mirror.

Concave mirrors cause reflected rays of light to converge. These mirrors can form real images. The size and position of an image are related to the size of the object and its position relative to the mirror. A virtual image is formed by a concave mirror if the object is less than a focal length away from the mirror. Convex mirrors cause reflected rays of light to diverge. They form virtual images of objects irrespective of the object distance from the mirror.

Spherical mirrors of large aperture form indistinct images. The spherical-mirror defect responsible for fuzzy images is known as spherical aberration. Parabolic mirrors do not show this aberration defect. When mirrors of large aperture are required, parabolic surfaces rather than spherical surfaces are used to reflect incident light.

VOCABULARY

angle of incidence
angle of reflection
angular aperture
aperture
concave
conjugate points
converge
convex
diffusion (light)
diverge
focal length
focal plane
linear aperture
mirror equation
normal (surface)
plane mirror
principal axis
principal focus
real image
reflectance
regular reflection
scattering (light)
secondary axis
specular (surface)
spherical aberration
spherical mirror
vertex (mirror)
virtual image

14 REFRACTION

refraction (re-FRAK-shun) n.: the bending of the path of a wave disturbance as it passes obliquely from one medium into another of different propagation speed.

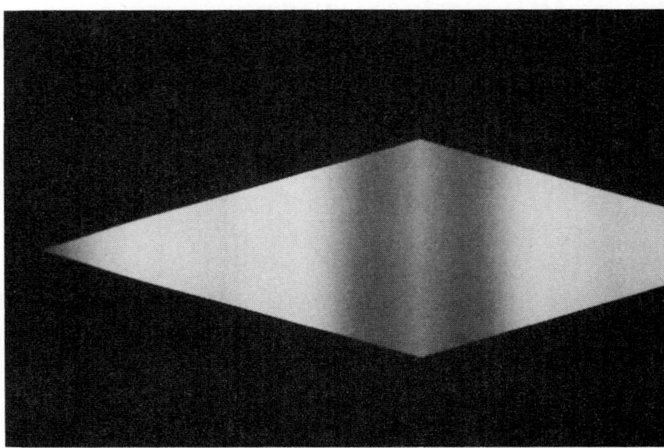

OBJECTIVES

- Describe the relationship between optical refraction and the wave character of light.
- Show the effect of refraction on the speed of light.
- Describe the control of light beams with lenses.
- Analyze the formation of images by ray diagrams.
- Solve object-image problems.
- Study the magnification of images.
- Describe the dispersion of light by prisms.
- Define color as a property of light.
- Discuss primary and secondary colors.

OPTICAL REFRACTION

14.1 The Nature of Optical Refraction In Section 10.12 we discussed refraction as a property of waves. Now we shall examine the refractive behavior of light and relate this behavior to its wavelike nature.

When aiming a rifle at a target, one relies on the common observation that light travels in straight lines. It does so, however, only if the transmitting medium is of the same *optical density* throughout. ***Optical density*** *is a property of a transparent material that is an inverse measure of the speed of light through the material.*

Consider a beam of light transmitted through air and directed onto the surface of a body of water. Some of the light is reflected at the interface (boundary) between the air and water; the remainder enters the water and is transmitted through it. Because water has a higher optical density than air, the speed of light is reduced as the light enters the water. This change in the speed of light at the air-water interface is diagrammed in Figure 14-2.

A ray of light that strikes the surface of the water at an *oblique* angle (less than 90° to the surface) changes direction abruptly as it enters the water because of the change in speed. The reason for this change in direction with a change in speed can be illustrated if we redraw the wavefront diagram of Figure 14-2(A) to make the angle of the incident ray oblique, as in Figure 14-2(B). When interpreting this diagram, you should remember that a light ray indicates the direction the light travels and is perpendicular to the wave front. We have already defined refraction as a bending of a wave disturbance. (See Section 10.12.) This bending of a light ray is called *optical refraction.* ***Optical refraction*** *is the bending of light rays as they pass obliquely from one medium into another of different optical density.*

Because of refraction, a fish observed from the bank appears nearer to the surface of the water than it actually is. A teaspoon in a tumbler of water appears to be bent at the surface of the water. A coin in the bottom of an empty teacup that is out of the line of vision of an observer may become visible when the cup is filled with water.

Figure 14-1. Parallel beams of light incident upon a rectangular glass plate. The beams are mainly refracted at both surfaces. Some reflection occurs at each surface, however.

14.2 Refraction and the Speed of Light Line **MN** of Figure 14-3 represents the surface of a body of water. The line **AO** represents a ray of light through the air striking the water at **O.** Some of the light is reflected along **OE.** Instead of continuing in a straight line along **OF,** the light ray entering the water is bent as it passes from air into water, taking the path **OB.**

The incident ray **AO** makes the angle **AOC** with the normal. Angle **AOC** is the **angle of incidence,** *i.* Recall that the angle of incidence is defined as the angle between the incident ray and the normal at the point of incidence. The refracted ray **OB** makes the angle **DOB** with the normal produced. *The angle between the refracted ray and the normal*

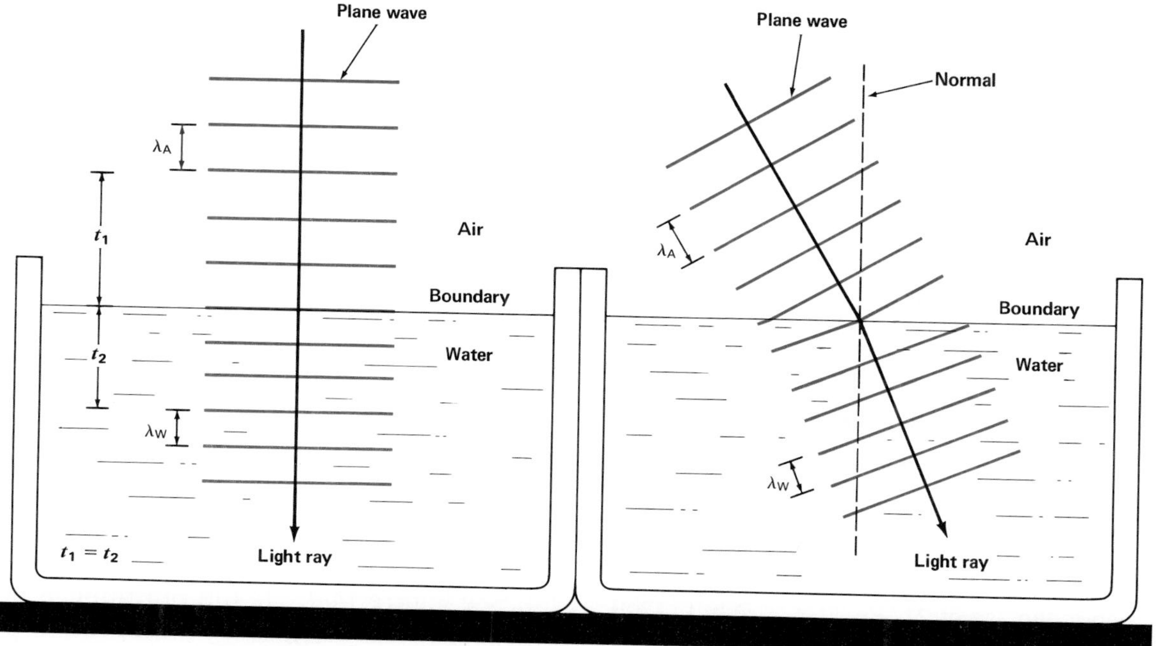

Figure 14-2. Wave-front diagrams illustrating (A) the difference in the speed of light in air and in water and (B) the refraction of light at the air-water boundary.

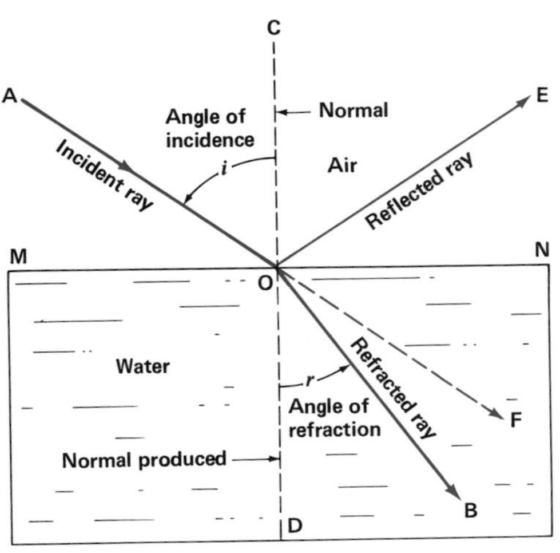

Figure 14-3. A ray diagram of refraction showing the angle of incidence and the angle of refraction.

at the point of refraction is called the ***angle of refraction,*** *r.*

In the examples given so far, the light rays have passed from one medium into another of *higher* optical density, with a resulting reduction in speed. When a ray enters the denser medium normal to the interface, no refraction occurs. When a ray enters the denser medium at an oblique angle, refraction does occur and the ray is bent *toward* the normal.

What is the nature of the refraction if the light passes obliquely from one medium into another of *lower* optical density? Suppose the light source were at **B** in Figure 14-3. Light ray **BO** then meets the surface at point **O,** and angle **BOD** is the angle of incidence. Of course some light is reflected at this interface. However, on entering the air the refracted portion of the light takes the path **OA.** The angle of refraction in this case is angle **COA.** It shows that the light is bent away from the normal. When a light ray enters a medium of lower optical density at an oblique angle, the ray is bent *away* from the normal. Had the light ray entered this less dense medium normal to the interface, no refraction would have occurred.

14.3 The Index of Refraction The speed of light in a vacuum is approximately 300 000 kilometers per second. The speed of light in water is approximately 225 000 kilometers per second, or just about three fourths of that in a vacuum. The speed of light in ordinary glass is approximately 200 000 kilometers per second or about two thirds of that in a vacuum. *The ratio of the speed of light in a vacuum to its speed in a substance is called the* ***index of refraction*** *for that substance.* For example:

$$\textbf{Index of refraction (glass)} = \frac{\textbf{speed of light in vacuum}}{\textbf{speed of light in glass}}$$

Using the approximate values just given, the index of refraction for glass would be about 1.5 and for water about 1.3. The index of refraction for a few common substances is given in Appendix B, Table 18. The speed of light in air is only slightly different from the speed of light in a vacuum. Therefore, with negligible error, we can use the speed of light in a vacuum for cases where light travels from air into another medium.

The fundamental principle of refraction was discovered by the Dutch mathematician and astronomer Willebrord Snell (1580–1626). See Figure 14-4. He did not publish his discovery, but his work was taught at the University of Leyden, where he was a professor of mathematics and

Figure 14-4. Willebrord Snell became a professor of mathematics and physics at the University of Leyden at the age of 21. He discovered the law of refraction now known as Snell's law.

physics. The French mathematician and philosopher René Descartes (1596–1650) published Snell's work in 1637.

Snell's discoveries about refraction were not stated in terms of the speed of light. The speed of light in empty space was not determined until 1676, and the speed in water was not measured until 1850. From his observations, however, Snell defined the index of refraction as the ratio of the sine of the angle of incidence to the sine of the angle of refraction. This relationship is known as **Snell's law.** If n represents the index of refraction, i the angle of incidence, and r the angle of refraction,

$$n = \frac{\sin i}{\sin r}$$

The sines of angles from 0° to 90° are given in Appendix B, Table 6.

The relationship between Snell's law and the ratio of the speeds of light in air and a refracting medium can be recognized from Figure 14-5. Rays of light travel through the first medium (air) with a speed v_1 and enter the refracting medium in which their speed is v_2.

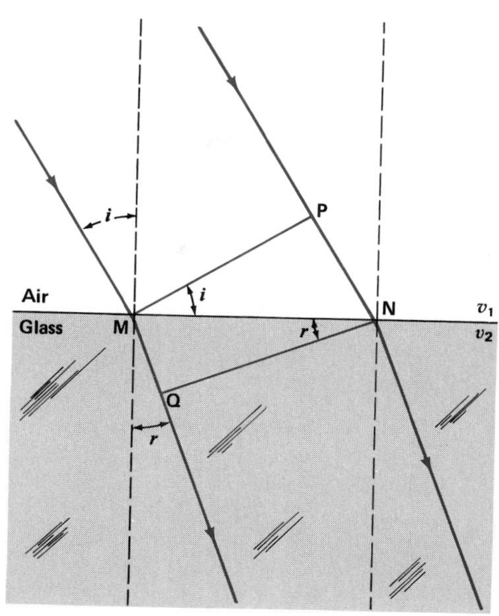

Figure 14-5. A wave-front diagram of refraction.

A wave front **MP** approaches the refracting interface **MN** in the first medium at an incident angle i. The wave front at **P** travels to **N** in the time t with a speed v_1. Simultaneously the wave front at **M** travels in the second medium to **Q** in the same time t but with a speed v_2. The new wave front is **NQ,** which travels forward in the second medium at a refractive angle r. The distances **PN** and **MQ** are respectively $v_1 t$ and $v_2 t$, so that

$$\frac{\mathbf{PN}}{\mathbf{MQ}} = \frac{v_1 t}{v_2 t} = \frac{v_1}{v_2}$$

In triangle **MNP**

$$\sin i = \frac{\mathbf{PN}}{\mathbf{MN}}$$

In triangle **MNQ**

$$\sin r = \frac{\mathbf{MQ}}{\mathbf{MN}}$$

The ratio

$$\frac{\sin i}{\sin r} = \frac{\mathbf{PN/MN}}{\mathbf{MQ/MN}} = \frac{\mathbf{PN}}{\mathbf{MQ}} = \frac{v_1}{v_2}$$

From Snell's law

$$n = \frac{\sin i}{\sin r}$$

Therefore

$$n = \frac{v_1}{v_2}$$

These relationships are illustrated in the following example.

EXAMPLE A ray of light passes from air into water, striking at an angle of 65.0° with the boundary between the two media. The index of refraction of water is 1.33. Calculate the angle of refraction of the ray of light.

Given	Unknown	Basic equation
$\theta = 65.0°$ $i = 90.0° - \theta$ $n_w = 1.33$	r	$n_w = \dfrac{\sin i}{\sin r}$

Solution

Working equation: $\sin r = \dfrac{\sin(90° - \theta)}{n_w}$

$$r = \arcsin\left(\frac{0.423}{1.33}\right)$$

$$= 18.5°$$

PRACTICE PROBLEMS **1.** Using the information given in the Example and the fact that the speed of light in air is 3.00×10^8 m/s, calculate the speed of light in water. *Ans.* 2.26×10^8 m/s

2. A ray of light passes from air into a gemstone at an angle of incidence of 40.0°. The angle of refraction is measured to be 15.4°. Using the information in Appendix B, Table 18, determine whether the gemstone is a diamond. *Ans.* Yes

---

When a piece of glass is immersed in water, the edges of the glass are visible because of the difference between the index of refraction of glass and that of water, even though both substances are transparent. The difference in the indices of refraction of air and water also make the top surface of the water visible.

But when a piece of crown glass ($n = 1.517\,2$) is immersed in benzene ($n = 1.501$), the crown glass seems to disappear in the benzene because the nearly identical indices of refraction make the edges of the glass virtually invisible.

The index of refraction of a homogeneous substance is a constant quantity that is a definite physical property of the substance. Consequently, the identity of such a substance can be determined by measuring its index of refraction with an instrument known as a **refractometer.** For example,

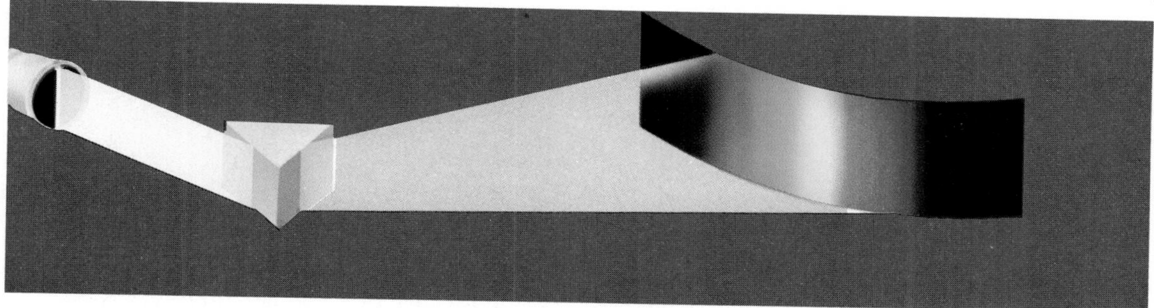

A. A prism spreads out a narrow beam of white light into the visible spectrum; light with higher frequencies (shorter wavelengths) is bent more than light with lower frequencies (longer wavelengths).

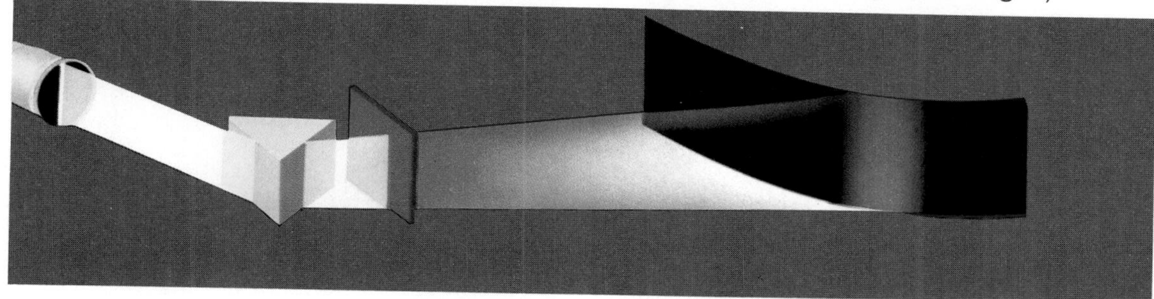

B. A red filter between the prism and the screen allows only light with lower frequencies (longer wavelengths) to pass.

C. A green filter allows only light from the middle region of the spectrum to pass.

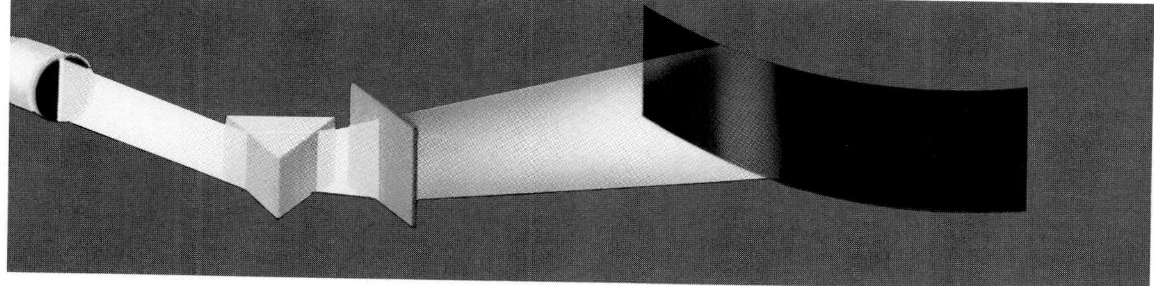

D. A blue filter passes only light with higher frequencies (shorter wavelengths).

(Adapted from COLOR AS SEEN AND PHOTOGRAPHED, Eastman Kodak)

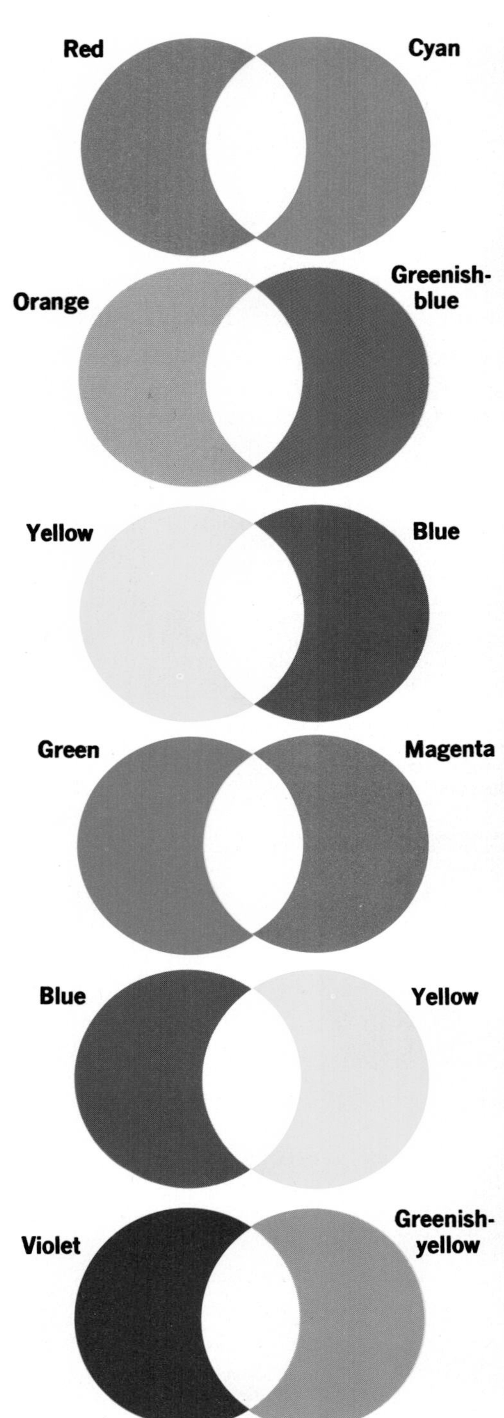

A. Complementary colors.

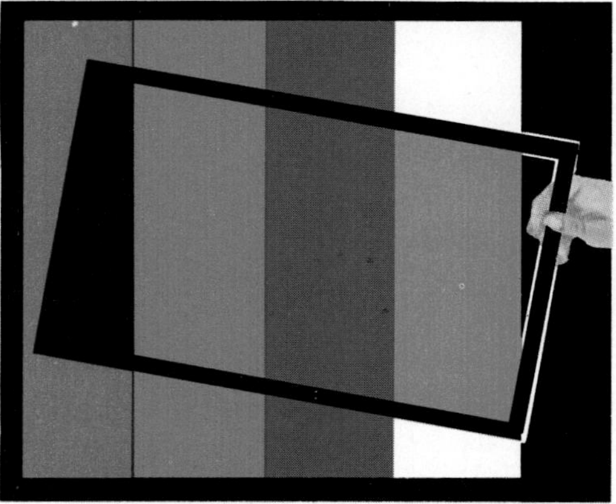

B. A cyan filter absorbs its complement, red light, and transmits blue and green light.

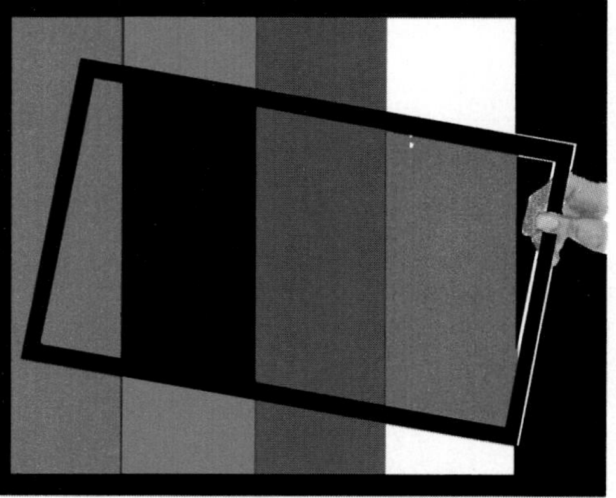

C. A magenta filter absorbs its complement, green light, and transmits red and blue light.

D. A yellow filter absorbs its complement, blue light, and transmits red and green light.

(Adapted from COLOR AS SEEN AND PHOTOGRAPHED, Eastman Kodak)

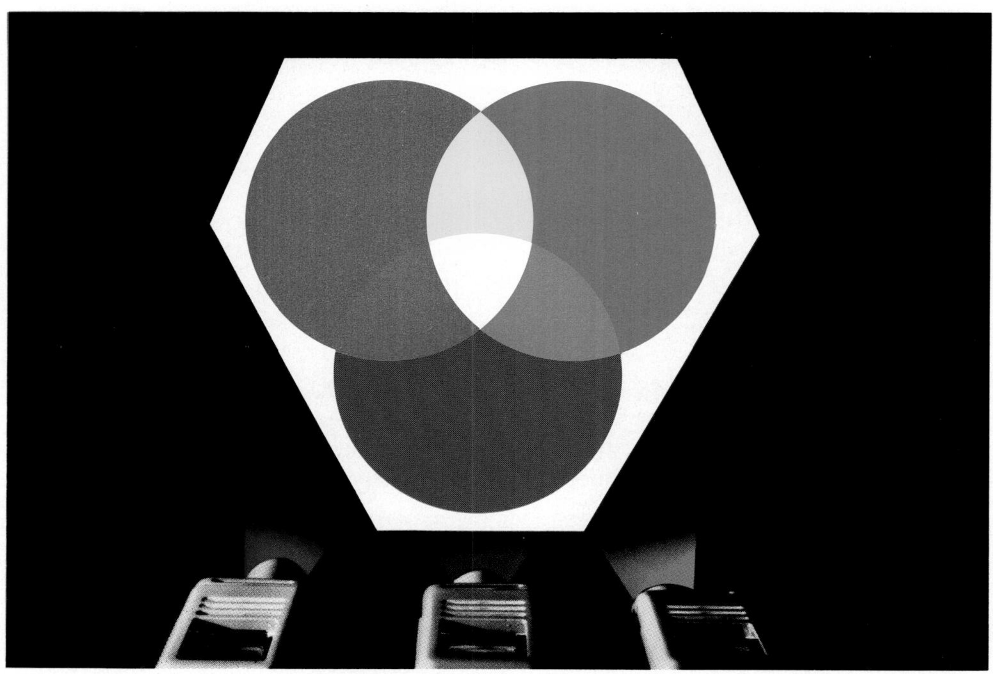

A. Primary colors. The additive mixture of red, green, and blue lights produces white light. Combined in pairs, two primaries give the complement of the third: cyan, magenta, and yellow. (Adapted from COLOR AS SEEN AND PHOTOGRAPHED, Eastman Kodak)

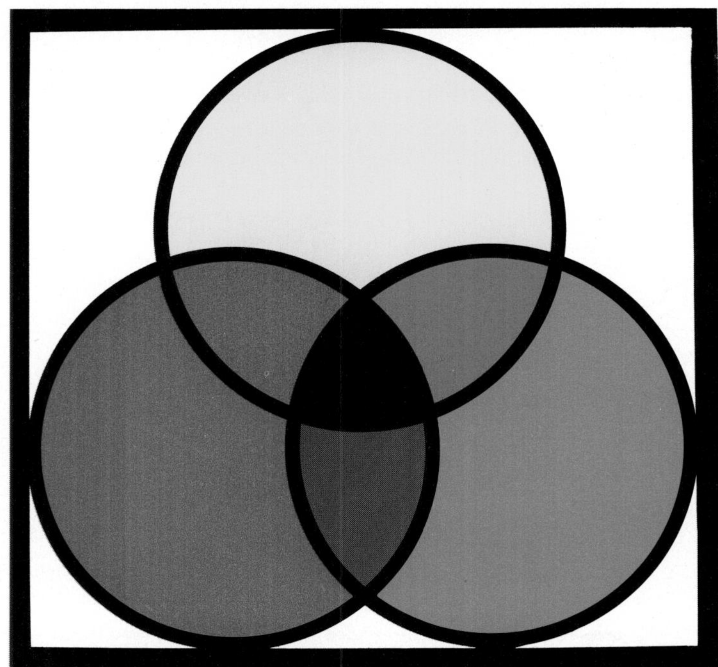

B. Primary pigments. The subtractive combination of cyan, magenta, and yellow filters transmits no light. Combined in pairs, two primary pigments give the complement of the third by subtraction: red, green, and blue.

A. A continuous spectrum is produced by incandescent solids, liquids, and gases under high pressure. The very dense incandescent gases in the main body of the sun and stars produce continuous spectra.

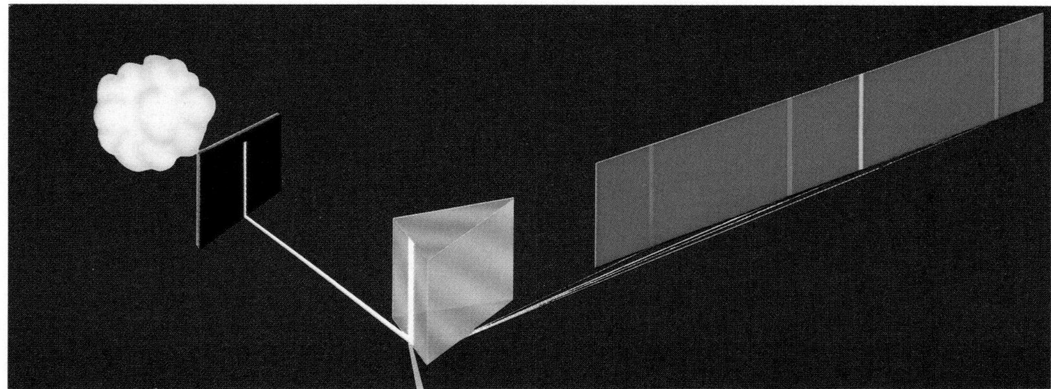

B. A bright-line spectrum is produced by incandescent gases of low density. Each chemical substance yields a characteristic pattern of lines that differ from all others.

C. A dark-line (absorption) spectrum is produced by a low-density (cooler) gas in front of the source of a continuous spectrum. The gas absorbs and reradiates some of the energy from the continuous spectrum. Since only a small amount of the reradiated energy strikes the prism, the resulting spectrum contains lines that are darker than the rest of the spectrum.

Plate IV

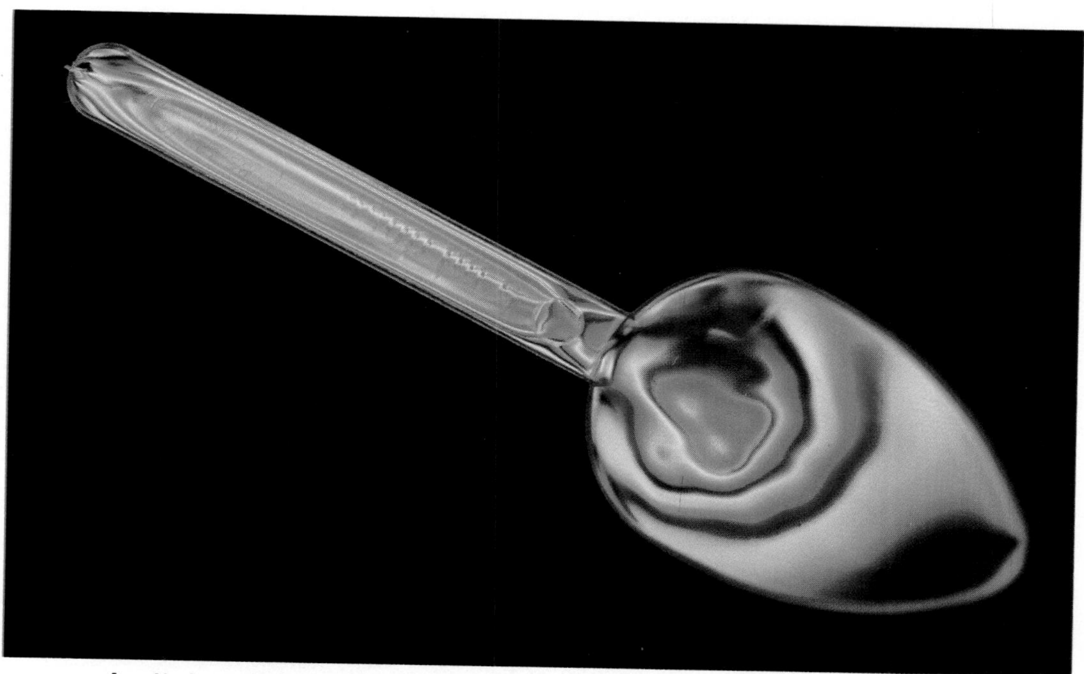

A. Under polarized light, strain patterns become visible in photoelastic materials. The closer the line pattern, the greater the stress distribution.

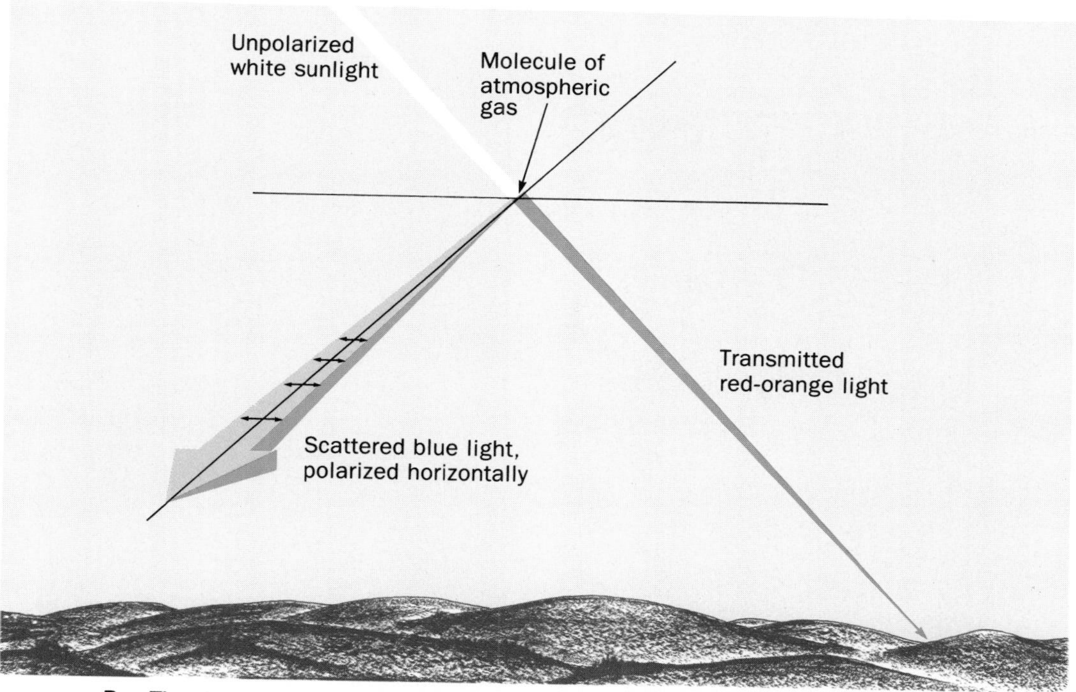

B. The daytime sky looks blue because the gas molecules of the atmosphere scatter the shorter wavelengths of sunlight more than the longer wavelengths.

A. Total internal reflection is displayed by fiber optics. The angle of incidence of the light inside the glass fiber exceeds the critical angle. Therefore the light cannot escape and is reflected down the length and out the end of the fiber.

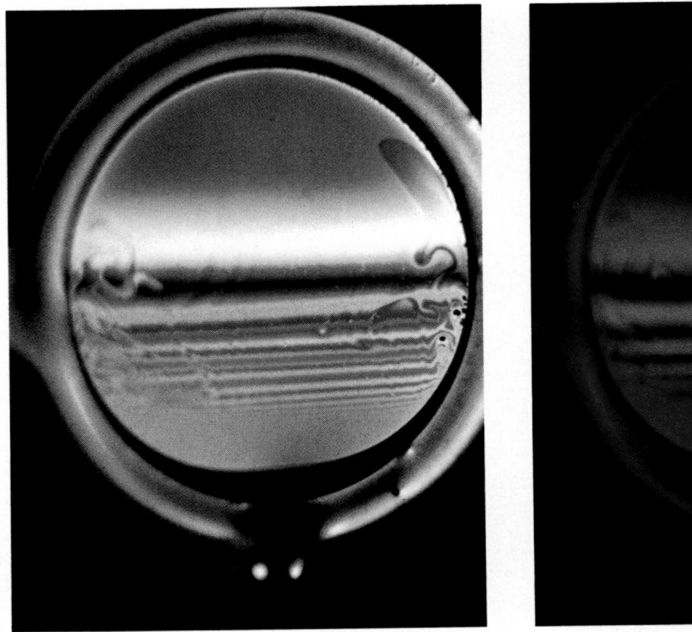

B. Interference produced by reflecting white light from a soap film. The picture on the right shows the pattern produced by red light.

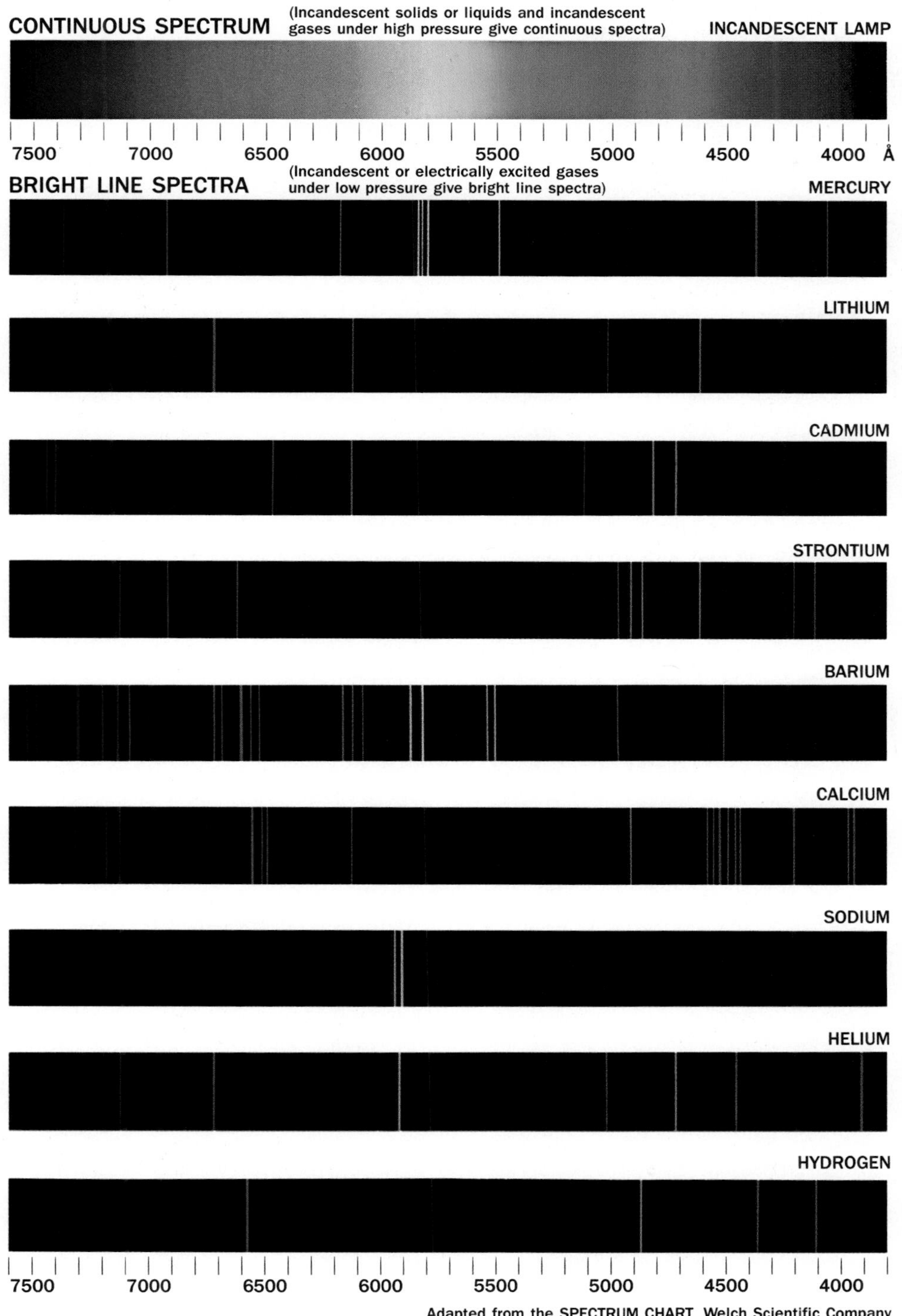
EMISSION SPECTRA
CONTINUOUS SPECTRUM
(Incandescent solids or liquids and incandescent gases under high pressure give continuous spectra)
INCANDESCENT LAMP
7500 7000 6500 6000 5500 5000 4500 4000 Å
BRIGHT LINE SPECTRA
(Incandescent or electrically excited gases under low pressure give bright line spectra)
MERCURY
LITHIUM
CADMIUM
STRONTIUM
BARIUM
CALCIUM
SODIUM
HELIUM
HYDROGEN
7500 7000 6500 6000 5500 5000 4500 4000
Adapted from the SPECTRUM CHART, Welch Scientific Company

A. Interference pattern produced by light of a single spectral color passing through two narrow slits.

B. Passing white light through a single slit produces this diffraction pattern.

Red light passed through the same slit produces this pattern.

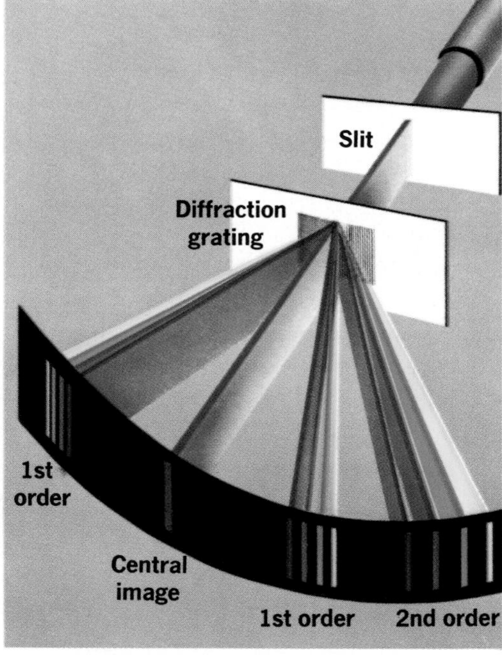

C. Mercury light through a diffraction grating.

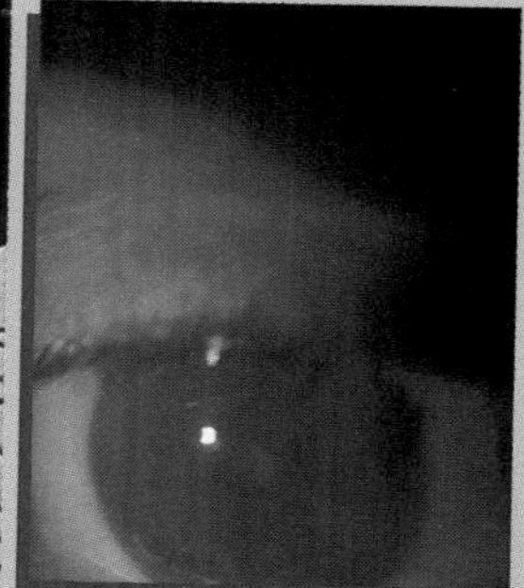

Modern Physics Intra-Science

How the Sciences Work Together

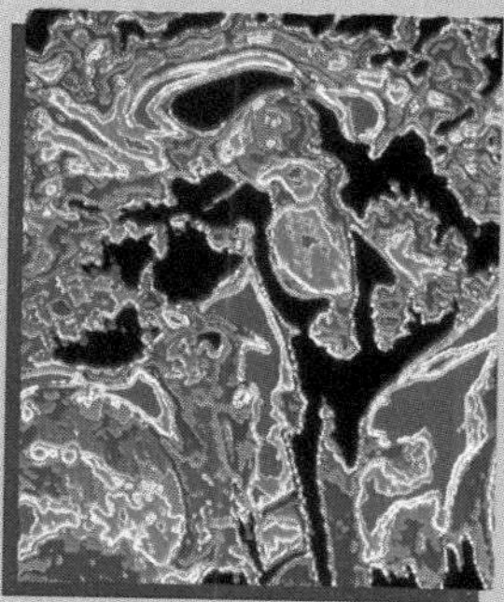

The Principle of Intra-Science

Technological achievements cut across scientific disciplines. Few are the work of a single science. For example, how did humans get to the moon? Sir Isaac Newton, a physicist, set forth the principle of rocketry more than 300 years ago in his Third Law of Motion. But the science of physics alone was not enough to put us on the moon. Without chemistry to develop the fuel, the rocket could not have gotten off the ground. Without astronomy to find the launch window, it would have drifted into outer space or been pulled into the sun. All three sciences were essential. Others—geology, medicine, metallurgy, exobiology—also contributed. It is this intermeshing of sciences toward one end that we call Intra-Science.

The Question in Physics: What's Next for Fiber Optics?

Fiber optics is a little like the wheel: simple in concept, rather useless by itself, but full of potential.

The wheel did not become important until it was put to use. Someone connected an axle to two wheels, built a box over it, and had a cart. Next came the four-wheeled wagon. Then, with the advent of advanced metallurgy, came the bicycle. Finally, with the invention of steam and gasoline engines, came the locomotive and the automobile. Always, the limiting factor for the growth of the wheel was not the design itself, but the ancillary technology.

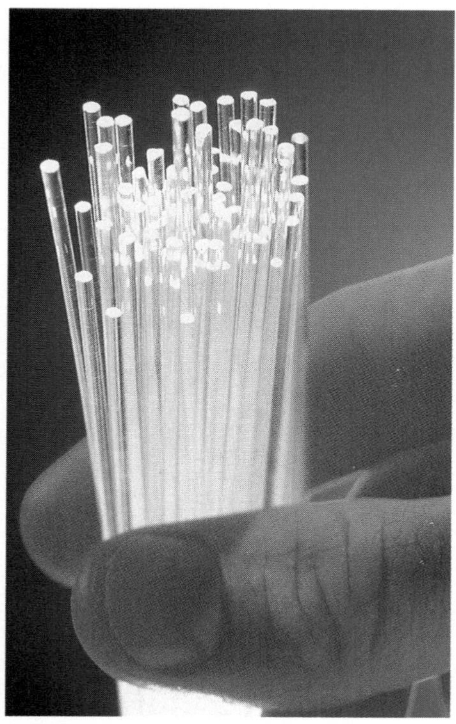

Optical fibers

The same may be said of fiber optics. An exceedingly simple concept, fiber optics involves sending light through strands of glass. The potential applications of this process are so enormous that we find ourselves in the midst of a "fiber optics revolution." Scientists are developing the technology to transmit 10 million telephone conversations on a single optical fiber. Physicians can peer inside a living human body and perform laser surgery using optical fibers.

Optical fibers are thin, flexible, transparent strands through which light can be transmitted with little loss of intensity. Most artificial fibers consist of a solid glass fiber core measuring from 5–250 micrometers (μm) in diameter, surrounded by a solid fiber cladding that is from 10 to 150μm wide.

Telecommunications is one of the most important applications of fiber optics. Sound is converted into pulses of light emitted by a laser, and the pulses are sent through an optical fiber to a photodetector. The photodetector transforms the pulses of light into electrical signals and finally back into sound. Optical-fiber lasers now in use are set to flash about 100 million pulses per second—enough for nearly 10 000 simultaneous telephone conversations on a single line. The lasers could work much faster, but the switchers that sort the signals out would be overloaded. So, better switchers are being developed.

If costs can be cut, fiber optics may be used to link computers with the buttons, switches, and engines of automobiles. Unlike wires, optical fibers are not subject to radio interference.

Optical fibers are also used to connect parts of stereos. There is no signal loss with a fiber-optic link because sound is converted to digitally pulsed light, then back to sound in the amplifier.

Another application of fiber optics is in the field of robotics. Manufacturers use laser beams that

Robot containing optical fibers

have been passed through optical fibers to control robots that perform cutting, drilling, and other precision operations.

The Connection to Health Sciences

Fiberscopes are fiber optic bundles containing thousands of thin (5–10 μm in diameter) optical fibers. The fibers are rigidly bound together at the ends so that the relative positions of the fibers are the same at both ends of the bundle. This facilitates the transmission of a clear image.

Physicians can view internal organs with endoscopes, instruments that combine fiberscopes with light guides. Light guides are fiber-optic bundles that carry high-intensity light, often produced by a laser. Endoscopes can be inserted into the body through natural openings or small incisions. Some endoscopes contain additional channels that allow the physician to take biopsy samples or to clear blood from the area being viewed. Even with all these features, an endoscope usually has a diameter no greater than 15 mm.

Angioscopes are very thin fiberscopes (2 mm in diameter) that are guided through blood vessels into the heart, where they are used to examine heart valves, and into coronary arteries to probe for fatty plaques.

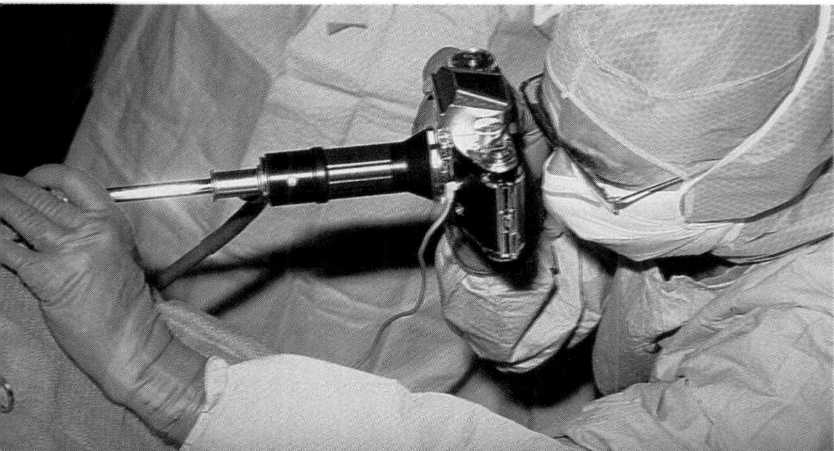

sician using an endoscope

The Connection to Other Sciences

The attention focused on artificial optical fibers has led to the discovery that natural optical fibers may play an important role in biology. It has been found, for example, that natural fiber-optic bundles in a germinating seedling transmit light throughout the seedling. This light, although faint, is absorbed by a pigment called phytochrome, which regulates the development of the seedling according to its absorption of light.

Chemically, ordinary window glass is an amorphous solid formed by melting together a mixture of 75 percent sand (silica, SiO_2), 15 percent soda (sodium carbonate, Na_2CO_3), and 10 percent lime (calcium oxide, CaO). Although window glass could be used for short optical fibers, it is not sufficiently transparent for the long optical fibers used in telecommunications. These fibers are formed from a special optical glass that is almost pure silica with traces of boron oxide (B_2O_3) and germanium oxide (GeO_2). Special manufacturing techniques are required because of the much higher melting point of almost pure silica than that of the silica-soda-lime mixtures used to make ordinary glass.

The highest quality optical fibers are prepared by a technique called chemical vapor deposition. This process involves the building up of successive layers of silica mixed with boron oxide or with germanium oxide on the inside of a hollow silica tube. These added inner layers are formed by the condensation of vapor mixtures produced by heating gaseous mixtures of silicon tetrachloride ($SiCl_4$) and boron trichloride(BCl_3), or germanium tetrachloride ($GeCl_4$) with oxygen. When these mixtures are heated, the oxygen displaces the chlorine, which is drawn off in the exhaust, leaving the condensible vapor mixture of oxides. After the proper layers have been deposited, the tube assembly is heated sufficiently to shrink the tube to the point of producing a solid rod. This rod is then heated and shaped to make the desired thin optical fibers.

The Connection to Careers

The telecommunications industry is the biggest user of fiber optics. Mechanical engineers design the photodetectors that receive light signals. Laser scientists develop narrow-frequency lasers in an effort to increase the number of beams that can be carried on a single fiber. Chemical engineers search for more transparent materials for fibers. Technicians install and maintain the optical fiber equipment. Manufacturers use fiber optics to control robots in factories. Audio engineers link stereo equipment with optical fibers to improve sound quality. Automotive designers employ fiber optics to link electronic equipment. Increasingly, computer technicians use optical fibers in computers. Physicians use fiber optics to diagnose and treat patients.

Fiber-optics research

The Process in Physics: Laser, Tool of All Trades

When the first working laser was built in 1960, scientists weren't quite sure what to do with it. Some called it ''a solution in search of a problem.'' Today, the laser—a device that produces intense, focused light beams—is the solution to hundreds of problems. It reads price labels at supermarket checkouts. It cuts through the hardest materials, including diamonds. It cleans valuable paintings and welds joints. It plays video and audio discs. With incredible accuracy, it can analyze chemical reactions, send thousands of simultaneous telephone conversations through a single line, perform delicate surgery, measure the movement of continents and the distance to the moon. It even casts spectacular images into the nighttime sky in laser shows.

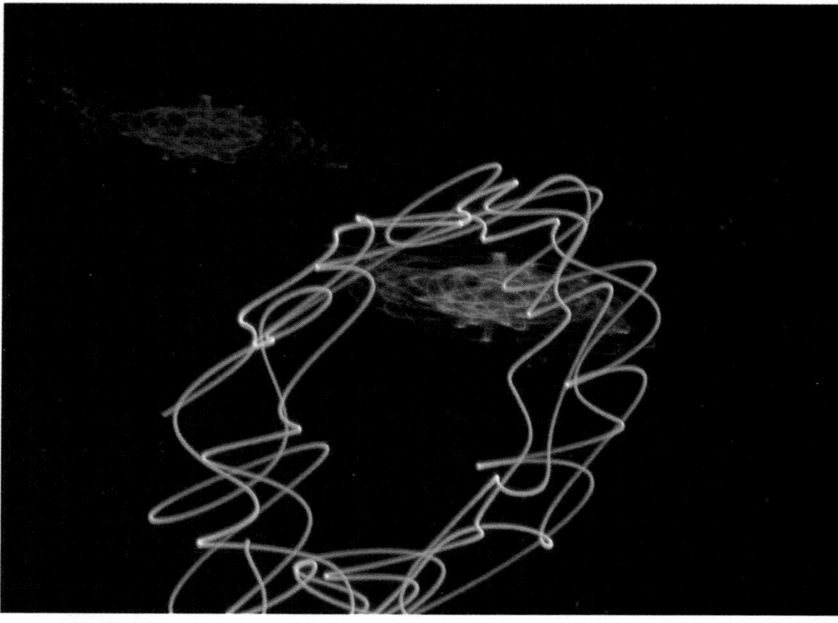

Laser light show

The future will bring more applications of laser technology. Plans for the ''Star Wars'' defense system included lasers to detect and destroy enemy missiles. Lasers have an enormous potential for generating electrical power. Today's nuclear plants produce power by fission (the breaking apart of heavy elements). A laser plant would use the opposite process – fusion, which is cleaner and safer. One of the methods under study is to heat the nuclei of two different forms of hydrogen with lasers to a temperature of 100 000 000°C, six times the temperature at the center of the sun. At this point they would fuse and release energy.

Laser barcode scanner

The Connection to Health Sciences

Laser beams can be aimed through the pupil of an eye onto a detached retina, forming a tiny scar that welds the retina back into place. No surgical incision need be made. Certain birthmarks called port-wine stains can be removed with little scarring by laser treatment. These birthmarks are actually dilated blood vessels. Laser light heats and is absorbed by the red pigment in the blood, causing the vessels to shrink and the marks to fade.

Lasers are frequently used as ''light knives'' to perform delicate types of surgery. Carried by a hair-thin optical fiber that is inserted into the body through a natural opening or small incision, the laser beam is aimed at cancerous tissue and destroys it. A side benefit of this technique is that the laser seals the lymphatic system—the network of vessels that returns fluid from tissue to the bloodstream—closing a door through which the cancer might otherwise spread. The cauterizing effect of lasers reduces blood loss and the risk of infection in surgery. Lasers are used to close bleeding ulcers, unclog coronary arteries, and remove brain tumors.

The Connection to Earth Science

In 1969 a laser pulse sent from the earth was bounced off a reflector left on the moon by astronauts. By measuring the time it took the pulse to make its return trip, scientists were able to calculate the distance to the moon with an accuracy of a few inches.

The same technique, called laser ranging, is used to keep track of the drift of continents. Great sections of the earth's crust called tectonic plates float above the semimolten interior. The plates move quite slowly—often no more than 10 cm per year. Where two plates crunch or slide along each other, the pressure creates a constant threat of earthquakes. The precise monitoring of such plate movements may someday help geologists predict quakes.

Laser research

The Connection to Other Sciences

The focused beam of the laser permits the marking of a perfectly straight line over long distances. In the past, long tunnels cut through mountains often deviated slightly from the straight path intended. This was particularly embarrassing to work crews when the tunnel was bored from each end and crews that were supposed to meet in the middle didn't. Now, machines used for boring are kept on errorless alignment by laser beams.

In chemistry, laser pulses enable the measuring of chemical reactions that occur too quickly for conventional techniques. A laser pulse is shot into a chemical mixture, inciting a reaction. Then, a tiny fraction of a second later, a second pulse is flashed. By measuring the amount of energy drawn from the second pulse by the reacting chemicals, scientists can determine the progress of the reaction. Lasers used for this purpose emit pulses measured in femtoseconds. (There are as many femtoseconds in a second as there are seconds in 30 million years.) Laser spectroscopy is used to analyze substances chemically and identify particles of air pollution.

pulse measurement

Another type of laser, called the excimer laser, holds promise for communicating with submerged submarines, which ordinary radio waves cannot reach.

The entertainment industry has put the laser to other uses. The purity of its color and the narrowness of its beam make it possible to project finely detailed images. Lasers can also produce three-dimensional images called holograms.

The Connection to Careers

Although physicists developed the laser, people in many sciences and professions find it to be an extremely useful tool. Chemists analyze substances and initiate, alter, and monitor chemical reactions with lasers. Biochemists use lasers to identify the steps in biochemical reactions such as photosynthesis. Molecular biologists and geneticists employ a laser beam to alter chromosomes or other molecules. Surgeons, ophthalmologists, and dermatologists use laser devices to treat patients.

Geologists and astronomers have used the laser as a measuring device. In the manufacturing industries, lasers cut and weld.

Artists use lasers to produce holograms and light shows. And the combined technologies of fiber optics and lasers hold the promise of a revolution in communications. A single optical fiber can carry many channels by using a number of very narrow laser beams of different frequencies.

The Question in Physics: How Can We See in the Dark?

Many people feel ill at ease, even vulnerable, in the dark. We usually have artificial light around us, but some people, such as soldiers on night missions and naturalists studying nocturnal animals, must work in the dark. How do people see in the dark? What techniques are available to enhance our night vision?

In total darkness, unaided vision is impossible—we need at least a little light. Visible light is electromagnetic radiation ranging from violet to red light. In dim light, we can make out only general forms through our night-vision system. However, light amplifiers can help us view an object using our day-vision system, which provides detailed images.

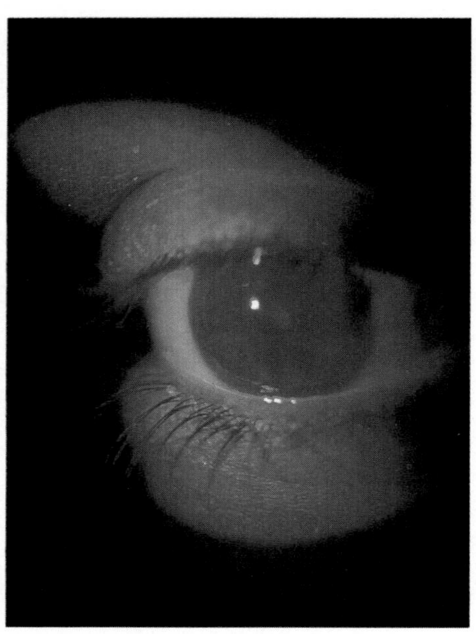

Human eye at night

Image-intensifier tubes are devices that make dim images brighter. The image strikes the front side of a photocathode, causing the other side to emit electrons. An electron-optical system intensifies the electron image and focuses it onto a fluorescent screen, which converts it into a visible light image.

Light striking the rod cells is absorbed by them, causing a nerve impulse to be transmitted to the brain. We do not see colors with rods but only shades of gray. In the periphery of the retina, hundreds of rods are connected to a single nerve. This has the effect of concentrating light and aids in seeing dim objects. At the same time, it prevents the rod cells from sending a detailed visual image to the brain. Just as fine-grained film is required to provide finely detailed photographs, finely detailed night vision would require each tiny rod cell to be able to send a message to the brain that is independent of the message sent by adjacent rod cells.

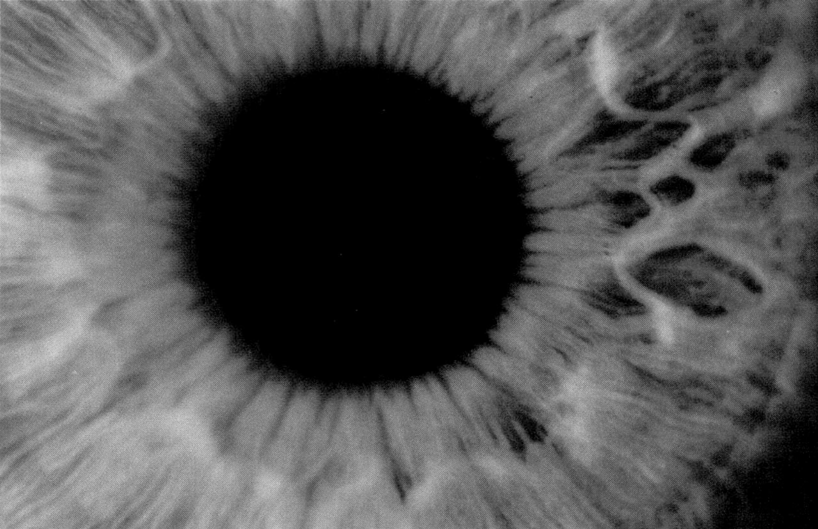

Pupil and iris of human eye

Infrared-sensitive film and infrared-imaging devices can be used to observe things in the dark from the invisible infrared radiation that they emit or reflect. Without being aware that they are observed, people or animals can be seen in the dark with the use of infrared light.

The Connection to Biology

Humans are relatively rare in their ability to see both in bright daylight and in the darkness of night. Human night vision involves the retinal absorption of light emitted or reflected by the object being seen. This light is focused by the eye's cornea and lens onto the retinal light-absorbing cells called rod cells or rods. Because rods are concentrated at the periphery of the retina, we can see objects at night best if they are off to the sides of our field of vision.

Day vision is color vision, and it involves the absorption of light by retinal cells, called cone cells or cones. These cells are concentrated in the center of the retina; each is connected to the brain by a separate nerve. As a result, this visual system gives us detailed color images in the center of our visual field. When we use a light amplifier at night, it produces an image bright enough to be observed by our day-vision system and thus enables us to see detailed images. The image is monochromatic, the color depending on that produced by the fluorescing screen in the light amplifier. In many cases, this fluorescence is a green color similar to that produced by many luminous watch dials.

Rattlesnakes and other members of the pit viper family have pit organs below each eye that are sensitive to infrared light. At night, warm-blooded animals are typically warmer than their surroundings. So, even in the complete absence of visible light, rattlesnakes can see their prey because the prey's infrared radiation differs from that emitted by the rest of the environment.

The Connection to Chemistry and Health Sciences

Vision is fundamentally a photochemical process. The rods contain a substance called rhodopsin which consists of retinal, a form of vitamin A that is chemically bonded to a protein called opsin. The absorption of light breaks rhodopsin into the retinal and opsin molecules. This breakdown depolarizes the rod cell and generates a nerve impulse.

In bright light, all of the rhodopsin is broken down quickly. The process of resynthesizing rhodopsin is relatively slow. This is why we are temporarily blind when we move from a bright to a dark environment. As our eyes adjust, dark adaptation occurs as rhodopsin molecules are resynthesized. Red light breaks down rhodopsin more slowly than other wavelengths. Aviators and other people who must preserve night vision often use red lighting.

Retinitis pigmentosa is a hereditary disease that results in the degeneration of the eye's light-sensitive pigments. The rods are usually affected first, making the person blind at night. Such night blindness can be alleviated with scopes that make an image bright enough to be detected by the day-vision system.

A lack of vitamin A in the diet may also result in night blindness. If this state is prolonged, rod cells, which cannot regenerate, may be destroyed.

Scientist using night vision scope to observe animal behavior

dar specialist scanning airspace

The Connection to Careers

Eventually, we will be able to improve our most important sense—sight—with the contributions of people in various scientific fields. Physicists study the properties of different wavelengths of light; engineers apply this information to the design of night-vision scopes that are used by people in many occupations. Morphologists study the anatomy of the eye and optic nerve. Physiologists describe the function of the rods and cones of the eye. Biochemists examine the light-sensitive pigments and chemical reactions that generate nerve impulses. Ophthalmologists treat people who have impaired night vision.

The Process in Physics: Finding New Alloys

Exciting new ideas and scientific discoveries cannot be converted into practical technology until the right materials are created. Just as past advances in semiconductor, transistor, and laser technology depended on the development of new alloys, the current revolution in superconductivity is also contingent upon the development of new materials.

Today it is difficult to imagine a world without alloys. Almost every manufactured item contains an alloy of some kind.

An alloy used to be defined as a mixture of two or more metals. Today, some alloys are made of metals that are mixed with nonmetallic "whiskers" (hairlike crystals), plastic fibers, or tiny ceramic spheres. Such alloys are called composite alloys.

To design an alloy with the chemical and physical properties needed for a specific job, physicists must first understand the atomic and molecular structure of materials. Microscopy and x-ray diffraction are used to observe the crystalline structure of various substances.

The need for strong, stable alloys that are resistant to high temperatures has kept physicists busy designing hundreds of new ceramic-metal-fiber materials. Strong lightweight alloys of aluminum (Al), magnesium (Mg), and copper (Cu) are now being developed for aircraft construction.

Alloys are used in devices that convert heat into electricity. These are called thermoelectric devices. The same principle is used to extract power from a nuclear furnace in outer space and to use waste heat from kitchens to run appliances. In recent experiments, thermoelectric cells were implanted in humans to convert body heat into energy to run watches, hearing aids, pocket computers, and heart pacemakers.

Not all alloy research is devoted to high-tech applications. We also need materials that do not deteriorate—cars that don't rust, spoons that don't tarnish, and bridges that don't collapse. Over a quarter of the world's steel production is needed just to replace rusted iron products. The addition of chromium (Cr) helps stop corrosion, but not if the alloy is exposed to severe heat or saltwater conditions.

Several years ago, a manufacturer used one of the strongest alloys known—an aluminum-copper alloy used commonly in aircraft construction—to build small boats. The boats fell apart within a year, however, because of saltwater corrosion.

The human body is another very corrosive environment. Inert, corrosion-resistant alloys are needed to construct artificial limbs and organs. Artificial heart valves, for example, must function without interruption in blood, a liquid that is both corrosive and electrically conducting.

Another application of alloys is in noise control. During World War II, the military needed quieter ships and submarines to prevent enemy electronic detection. The navy tried, unsuccessfully, to silence the ships with coatings and insulation. They had to build a quiet alloy with the damping capacity of lead. Such an alloy, iron manganese, was found to be an effective muffler that also had

Aluminum alloy under a scanning electron microscope

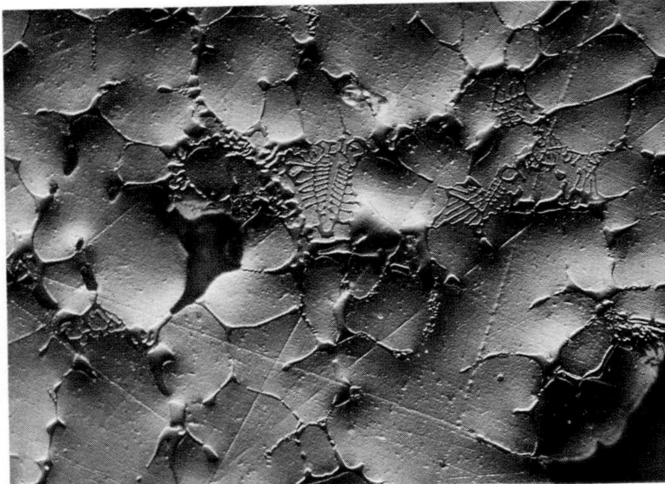

Interference contrast micrograph of an aluminum alloy

the strength and mechanical properties needed for ship construction.

The Connection to Other Sciences

Soon materials will cease to be classified as metals, ceramics, polymers, or glasses, but will be classed according to their application, such as electrical, nuclear, chemical, or magnetic. It will not matter what a spacecraft's protective nose cone is made of, as long as the material meets stress and structure requirements.

e shuttle's protective nose cone made of carbon and silicone compounds

The constant search for higher-volume, faster, and less expensive semiconductors has greatly increased production of rare-earth elements. Elements such as yttrium (Y), indium (In), tellurium (Te), and selenium (Se) are used in alloys for computers and other advanced applications. A dilute indium alloy, for instance, is used as a coating on glass to block ultraviolet and infrared radiation. And a rare-earth alloy composed mostly of neodymium is now being used to manufacture permanent magnets. This new alloy can be used to make motors that are more compact and lighter than conventional motors.

The more common metals and alloys are also being put to new uses. The research necessary to develop nickel-cadmium batteries led to the development of cadmium telluride as a major component of solar photovoltaic cells. Aluminum-lithium alloys used in the redesign of airliners led to the first lithium battery ever marketed.

The Connection to Careers

The demand for alloys is so great that mining engineers, geologists, and mineralogists continue to search for additional sources of minerals. Even biotechnologists entered the mining field by developing workable techniques for the separation of minerals by biological action. For example, they use bacteria to remove gold from iron sulfide crystals and create an acid solution that dissolves the gold. Physicists and electrical engineers built a superconducting magnetic separator that removes impurities from the extracted minerals.

Aeronautical engineers work with chemists, physicists, and metallurgists to develop refractory oxide alloys—high-temperature alloys of tungsten (W), tantalum (Ta), and iridium (Ir), among other elements. Their goal is to develop materials for spacecraft that will withstand the heat created by air resistance when traveling through the earth's atmosphere.

Mineralogist at work

The Process in Physics: Infrasound and Ultrasound

As noisy as this world is at times, it would be much noisier if we could hear *all* the sounds around us. Many sounds that can be detected by instruments or animals with wide hearing ranges are totally inaudible to humans. The dog whistle is an example. It emits a trill so high-pitched that humans cannot hear it, but dogs can.

Scientists have found numerous applications for sound. Sound waves are used to explore previously hidden worlds—the seafloor, a fetus in its mother's womb, and the inner structure of the sun.

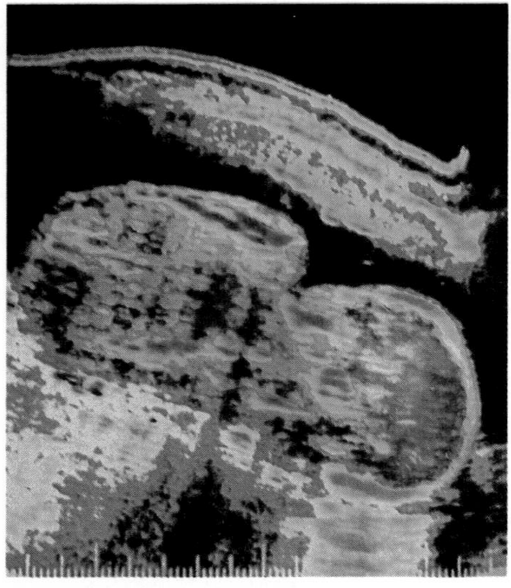

Ultrasound image of a fetus

Sound is a wave of energy that moves through a medium. When an object vibrates, its kinetic energy causes the surrounding medium to alternately compress and expand. The compressions and expansions are transmitted as sound waves in all directions from the vibrating object. The speed of transmission depends on the density and elasticity of the medium. Air, for example, transmits sound at a speed of 331.5 m/s (1193 km/h) at a temperature of 0°C.

Artist's conception of sound waves

The average person can hear waves produced by objects vibrating from approximately 20 times a second to approximately 20 000 times a second. A frequency of 1 per second is called a frequency of 1 hertz (Hz). Wave frequencies of less than 20 Hz are called infrasound. Those above 20 000 Hz are called ultrasound.

Ultrasonic waves are produced by a sound source called a transducer that changes electrical energy into the mechanical energy of vibration. One type of transducer sends an alternating electrical current through a crystal. The crystal expands and contracts, creating sound waves that have the same frequency as that of the oscillating electrical field.

The Connection to Earth Science

Many processes in the earth's crust, its oceans, and its atmosphere produce infrasound. An earthquake sends out waves in all directions from its focus. From their knowledge of how the direction and speed of sound changes when it passes from one medium into another, scientists use seismographic recordings of earthquakes to gain evidence about the nature of the earth's mantle and inner and outer cores.

Volcanic eruptions, the aurora borealis, and tornadoes produce infrasound waves that travel undiminished in intensity for thousands of miles through the atmosphere. When the Indonesian volcano Krakatoa exploded in 1883, the resulting infrasound waves could still be detected after circling the globe several times.

The inner structure of the sun is being explored using infrasound. This new field, helioseismology, is based on the fact that sound waves generated within the sun cause a pattern of peaks and valleys on the sun's surface. Because the movement of sound through a medium depends on factors such as elasticity and density, helioseismologists can learn about processes inside the sun by analyzing the surface patterns.

Sonar (*SO*und *N*avigation *A*nd *R*anging) is a transducer that sends out sound waves ranging from 5000 to 25 000 Hz. When the waves hit an object, they bounce back and are detected by a receiver. A sonar device called Gloria is being used to provide detailed maps of the seafloor. As Gloria is towed behind a ship, it sends out a fan of sonar waves that are reflected by the seafloor and converted into pictures by a computer. Huge volcanoes, canyons, and alluvial fans have been discovered using this technique.

The Connection to Biology

Even in complete darkness, bats can catch flying insects and avoid obstacles by hearing the echo of ultrasonic radiation emitted from their mouths and reflected from any object in the path of the sound. The bats emit a precisely directed beam of high-frequency sound (up to 30 000 Hz). Bats can precisely locate their prey by measuring the time an emitted sound takes to return and the direction from which it returns.

using ultrasound to locate prey

A variety of insects seem to signal each other by using ultrasound. Dogs and cats can hear frequencies as high as 30 000 Hz. Sea mammals such as the killer whale use a sonar system similar to that of bats to find prey and to locate obstacles in the darkness of deep water. This system is called echolocation.

The Connection to Other Sciences

Sound has wide-ranging application in the health sciences. Ultrasound can be used to pasteurize milk by creating air bubbles that destroy tiny organisms in the milk, and to homogenize milk by breaking up fat globules.

Another recent application of ultrasound in the health sciences is diagnostic ultrasonography, which produces images of internal organs from reflected ultrasound waves. Ultrasonography is used to detect the number and position of fetuses in the uterus and provides warning of fetus development problems. The technique has also been used to detect problems in the lymph nodes, heart, kidneys, and spleen.

In chemistry, the application of sound has led to a science called sonochemistry, which explores the interaction of sound energy with matter. Irradiation by high-intensity ultrasound causes tiny gas bubbles to grow and collapse within a solution. This is called acoustic cavitation. Very high temperatures (near 5000 °C) can be created within the exploding bubbles. When some solutions are irradiated by high-intensity ultrasound, they emit light. Each solution emits a characteristic spectrum of light. By comparing the spectrum emitted from an unknown solution to known spectra, the chemical composition of the solution can be determined.

Acoustic cavitation is also used to break up the solid particles in a suspension. The smaller pieces remain uniformly suspended in the liquid instead of gradually settling to the bottom. This process is used in making alloys from different molten metals, in preparing photographic emulsions that yield film of extremely small grain size, and in breaking up polymers into smaller molecules.

The Connection to Careers

Geologists study the inner structure of the earth by monitoring the transmission of sound waves through the earth. They use sonar to map the seafloor. Helioseismologists use infrasounds produced by the sun to examine its inner structure. The fishing industry relies on sonar to locate schools of fish. Sonar is also used by the military and the shipping industries for navigation. Chemists can use ultrasound to analyze the chemical composition of a substance. Manufacturers use ultrasound to check for cracks or flaws in equipment, clean precision instruments, and weld materials together. Ultrasound is a tool used by physicians to diagnose and treat patients.

One of the most fascinating fields of study involving sound is animal communication. We know that birds, primates, sea mammals, and other animals utter complex and varied audible sounds. But relatively little is known about animal use of infrasound and ultrasound. It is possible that further study will reveal even more complex patterns of animal communication.

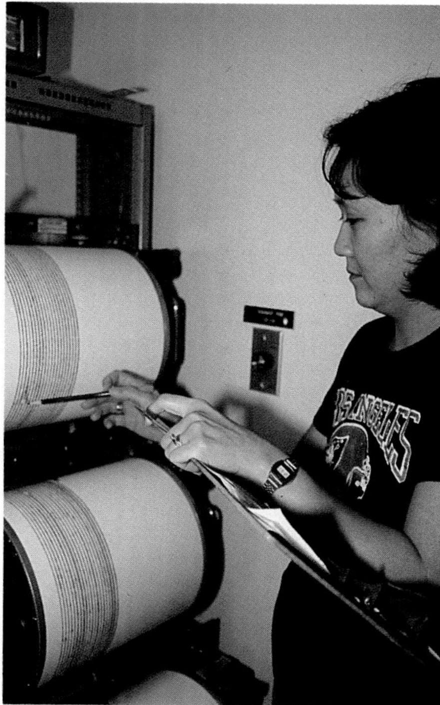

Geologist checking seismographic recording

The Question in Physics: How to Make Cold Our Ally

The U.S. space shuttle *Challenger* exploded in 1986, killing all seven aboard, because unusually cold weather had caused a critical seal to become brittle. The same year, the Japanese built a model of a levitating 500-km/h supertrain, which was the result of research into the effects of cold.

Throughout human history, the effects of extreme cold have been more destructive than beneficial. Farmers in particular have been hurt by freezing temperatures. But science and technology are beginning to change that.

Dramatic discoveries are being made in cryogenics, the study of the effects of extreme cold. One of the recent advances in this field is superconductivity, in which a number of substances—mostly ceramics—allow the flow of electricity without resistance when they are cooled to temperatures near absolute zero, –273.15°C.

The 1987 Nobel Prize in physics went to K. Alex Müller and J. Georg Bednorz, who produced superconductivity at a transition temperature of –238°C, about 12 degrees higher than the temperatures at which the phenomenon occurred previously. Ching-Wu

Space shuttle Challenger *at the point of exploding*

Frozen orange crop

Chu of the University of Houston used a different version of the Müller-Bednorz superconductor and lost all resistance at –181°C. Early in 1988, Chu announced the development of a new material that has a transition temperature of –159°C. It is made of bismuth (Bi), aluminum (Al), strontium (St), calcium (Ca), copper (Cu), and oxygen (O). The higher this temperature is pushed, the more economical superconductors will become.

The first experiments were made with liquid helium (He), which boils at –269°C. Now, it is possible to use liquid nitrogen (N_2), which remains liquid up to –196°C and is cheaper, more abundant, and easier to work with than liquid helium. Machines using liquid nitrogen need less insulation, which means that they are more compact than those using liquid helium.

With superconducting wires, electricity can be transmitted much more efficiently than with copper wires, since no energy is lost in overcoming resistance. The same technology may also produce electric power station generators and storage systems, smaller and faster computer chips, more efficient electric-powered cars, and safer nuclear reactors that use fusion instead of fission.

Particle accelerators use superconducting magnets chilled with liquid helium to power subatomic particles to nearly the speed of light.

The experimental supertrain built in Japan uses superconducting magnets to levitate above its tracks and travel almost friction-free.

ɔanese levitating supertrain

Through a process called cryopumping, vacuums similar to those in outer space can be created. They are frequently used in high-energy physics research. First, the air is pumped out of an enclosure. Then, surfaces within the chamber are chilled to extremely low temperatures to freeze the leftover gases, leaving a near-perfect vacuum.

The Connection to Biology

Low temperatures retard enzyme activity and development of fungi and bacteria. Refrigerators and freezers allow us to preserve food for long periods of time.

Cold weather restricts biological activity in nature. Hundreds of insect species and other invertebrates, as well as some frogs, freeze rock-solid during winter. Unlike mammals, they are not subject to the destructive formation of ice crystals in their blood, tissues, and organs. With the first spring thaw, they begin to move again. Some warm-blooded mammals, such as ground squirrels and brown bats, go into a coma-like hibernation at the onset of winter. Their metabolism slows so they can survive cold weather.

The bacterium *Pseudomonas syringae* has a protein in its cell membranes that causes water molecules to form ice crystals. A mutant of this bacterium that does not have the protein has been sprayed on plants to retard frost formation. Water molecules do not align into ice crystals, as they would without the mutant bacteria. The bursting of plant cells and the severing of plant tissues that normally result from freezing are thereby prevented. This process holds promise for protecting fruit and vegetable crops against frost.

The Connection to Other Sciences

The health and earth sciences are also concerned with the effects of cold.

Exposure to cold may cause frostnip, frostbite, or hypothermia. Frostnip is the freezing of outer skin, and frostbite is the freezing of deep tissues. Hypothermia is a decline in body temperature below 37°C, which may result in constriction of surface blood vessels, disorientation, hallucinations, and even death.

In medicine, cryosurgeons use freezing probes that constrict blood vessels and minimize bleeding during surgery. The flow of liquid nitrogen (N_2) through the probe can be electronically controlled so the temperature can be varied. Only the tip of the probe is used to destroy tissue. Warts, tonsils, and cataracts can be removed with liquid-nitrogen probes. Liquid nitrogen is also used to preserve blood, bone marrow, semen, and ova. The next challenge is to freeze whole organs without damaging them.

In earth science, studies of permafrost (the permanently frozen subsoil of extremely cold regions) are vital in the development of polar areas. For example, heated structures built on permafrost could heat the ice below and sink. The Trans-Alaska pipeline was elevated above the permafrost areas because the warm oil would have heated the frozen ground and caused shifting.

The Connection to Careers

The race is on among physicists to develop materials that superconduct at higher and higher temperatures. The applications of superconductivity will provide jobs in power, communications, transportation, and computers. Cryoengineers develop techniques for producing, storing, and handling materials at very low temperatures. Chemical engineers are needed to develop durable materials that can withstand extreme temperatures.

In the health sciences, medical technologists are needed to develop cryosurgical tools for the physicians who use them. Cryobiologists are looking for ways to preserve organs. Geneticists and molecular biologists working in the field of genetic engineering are developing new strains of organisms that will withstand cold.

Cryobiologist monitoring cell samples

The Process in Physics: The Multitalented Magnet

The magnet provides an excellent demonstration of Intra-Science because it shows clearly how a single discovery can benefit several branches of learning.

The rudimentary applications of magnets are well known. Refrigerator doors and kitchen cabinets are held shut by bar magnets. Gift shops sell the same type of magnet for use in tacking notes to a metallic surface. More sophisticated applications, ranging from household appliances to huge particle accelerators used in research, involve electromagnets.

Bending a bar magnet creates the familiar horseshoe magnet. This produces a magnetic field between its poles and is used for small motors, such as those used in slot cars. Circular magnets in electronic computers store information. Disc magnets in radio speakers help create sound from electric impulses.

Magnetism is a force of attraction or repulsion caused by the arrangement of electrons in atoms. Magnetism and electricity are related. Essentially, they are two separate expressions of the force of electromagnetism.

People learned to harness electromagnetism only about 150 years ago, but today its applications are all around us. If you visit a large hydroelectric facility, you will see rows of generators, each one as big as a small house. These generators convert mechanical energy into electricity by using the principle of electromagnetic induction. Electric motors perform the opposite function of generators. Their design is similar to that of generators, but the roles of electricity and magnetism are reversed; motors convert electrical energy to magnetic energy, which is converted to mechanical energy.

Electromagnets are used when magnetism is needed on demand, rather than continuously. They can be found in doorbells, in magnetic vibrators, and in wrecking yards. The power of electromagnets is demonstrated by junkyard cranes that can pick a 2000-kg load of scrap metal with a disk-shaped magnet and then drop it the moment the electricity is turned off.

A more advanced application of magnetic fields results from their ability to control the direction of high-energy particles in particle accelerators. Perhaps the most spectacular use of electromagnetic forces lies in high-speed trains that are suspended almost magically above the rails. The Japanese came up with a design in which the repulsion of a magnetic field lifts the train 10 cm above the rails. This eliminates not only friction but also the noise associated with trains. Propulsion is also achieved with magnetic power. Magnets in the track continuously reverse polarity in sequence to push or pull on the train's magnets and move it forward or backward.

Hydroelectric facility at Hoover Dam

The Connection to Earth Science

The earth itself is a magnet, but a weak one. The magnetism of the earth causes the needle of a compass to align in a north-south direction. If a compass reading were taken on the magnetic equator, though, the needle would remain horizontal. The magnetic equator is an imaginary line circling the globe halfway between the magnetic poles. It is not far from the geographic equator.

When the earth's magnetic field is disturbed by solar flares or other radiation emitted by the sun, we experience magnetic storms. They are not storms in the conventional sense—we cannot feel them. But, if listening to a short-wave radio, we can hear their effect because they interfere with the reception, and we can see their effect in the auroras (called northern lights or aurora borealis in this hemisphere).

ra borealis at Fairbanks, Alaska

The magnetism of other planets and stars helps us understand their structure and history. Neutron stars have extremely powerful magnetic fields, a billion times stronger than that of the earth. Galaxies have exceedingly weak magnetic fields, about one millionth that of the earth.

The Connection to Other Sciences

Magnetic resonance imaging (MRI) is a technique used in the health sciences to detect tumors and other abnormalities of tissues and organs. A magnetic field causes the nuclei of some elements in the human body, such as hydrogen (H) and phosphorus (P), to line up, and a high-frequency radio wave causes the nuclei to change direction. The change in direction creates signals that can be detected by a nuclear scanner and turned into an image by a computer.

MRI is particularly useful for viewing the brain. In addition to finding tumors, studies of magnetic fields generated by electrical currents in the brain allow physicians to locate and remove areas of the brain that cause epileptic seizures.

Nuclear magnetic resonance is used extensively in chemical analysis to assess the magnetic properties of atomic nuclei. Chemists in industry and university research laboratories determine the structure of new compounds with nuclear magnetic resonance.

The Connection to Careers

Because magnetism is not completely understood, physicists continue to probe the inner workings of the atom in an effort to explain the phenomenon. They use magnets in particle accelerators to study subatomic particles. The revolution in superconductivity has spurred the development of new types of magnetic materials. Engineers apply electromagnetism to the development of electric generators, motors, and storage devices.

Physicians and medical technologists use magnetism in their diagnoses. Chemists analyze substances using nuclear magnetic resonance. Microbiologists employ similar techniques to identify the structure of biological molecules such as proteins and nucleic acids.

Earth scientists study the earth's magnetic field, and astronomers learn about neutron stars and galaxies by examining their magnetic fields.

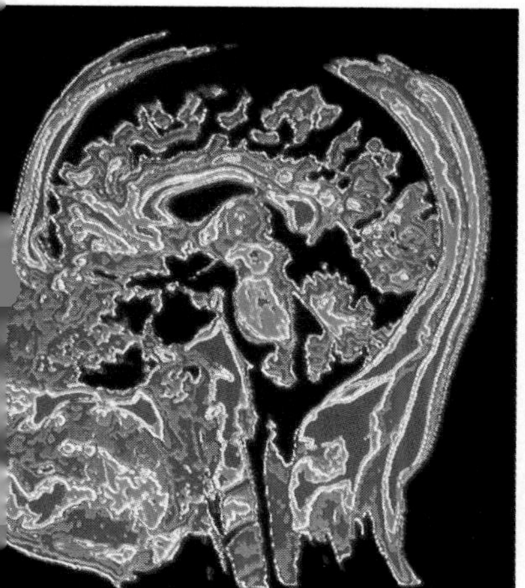

scan of human brain

Magnets being checked for polarity at the Stanford Linear Accelerator

The Hubble Space Telescope

How many sciences can you name that are involved in the exploration of the universe?

butterfat and margarine have different indexes of refraction. One of the first tests made in a food-testing laboratory to determine whether butter has been mixed with margarine is the measurement of the index of refraction. The high index of refraction of a diamond furnishes one of the most conclusive tests for its identification.

Because light travels very slightly faster in outer space than it does through air, light from the sun or the stars is refracted when it enters the earth's atmosphere obliquely. Since the atmosphere is denser near the earth's surface, a ray of light from the sun or a star striking the atmosphere obliquely follows a path suggested by the curve shown in Figure 14-6. There is no abrupt refraction such as that which occurs at the interface between two media of different optical densities.

Atmospheric refraction prevents the sun and stars from being seen in their true positions except when they are directly overhead. In Figure 14-6, the sun appears at **S′** instead of **S,** its true position. Since the index of refraction from outer space to air is only 1.000 29, the diagram is greatly exaggerated to show the bending. Refraction of sunlight by the earth's atmosphere causes the sun, when geometrically on the horizon, to appear about one diameter ($\frac{1}{2}°$) higher than it really is. Around 2 minutes are required for the earth to rotate through $\frac{1}{2}°$ of arc. Thus, we gain about 4 minutes of additional daylight each day because of atmospheric refraction at sunrise and sunset.

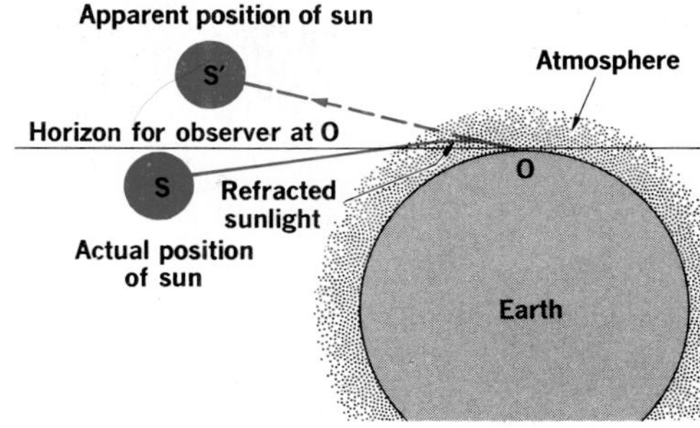

Figure 14-6. The sun is visible before actual sunrise and after actual sunset because of atmospheric refraction.

14.4 The Laws of Refraction

If the index of refraction of a transparent substance is known, it is possible to trace the path that a ray of light will take in passing through the substance. In Figure 14-7 rectangle **ABCD** represents a piece of plate glass with parallel surfaces. The line **EO** represents a ray of light incident upon the glass at point **O.** From point **O** as a center, two arcs are drawn. The radii of these arcs are in the ratio 3/2 based on the index of refraction of the glass being 1.5 and that of the air being 1. The normal **OF** is drawn. Then a line is drawn parallel to **OF** through the point **H** where the incident ray intersects the *smaller* arc. This line intersects the *larger* arc at point **G.** The line **OP,** determined by points **G** and **O,** marks the path of the refracted ray through the glass.

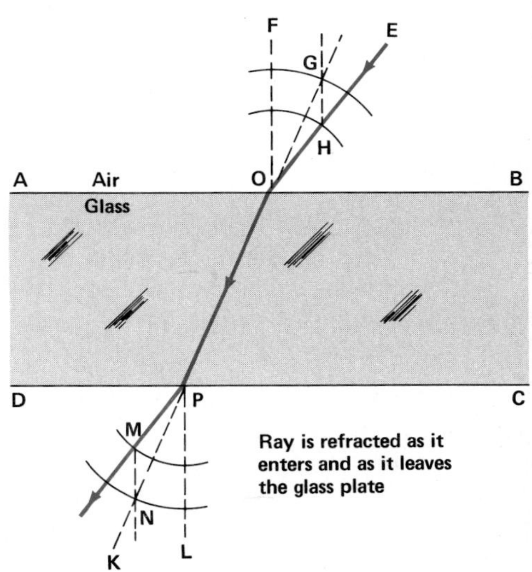

Figure 14-7. A method of tracing a light ray through a glass plate.

If the ray were not refracted as it leaves the glass, it would proceed along the line **PK.** To indicate the refraction at point **P,** this point is used as a center and arcs are drawn having the same ratio, 3/2, as before. The normal **PL** and a line **MN** parallel to the normal are drawn. The parallel line is drawn this time through point **N** where the

Test the validity of this construction using Figure 14-5 and Snell's law.

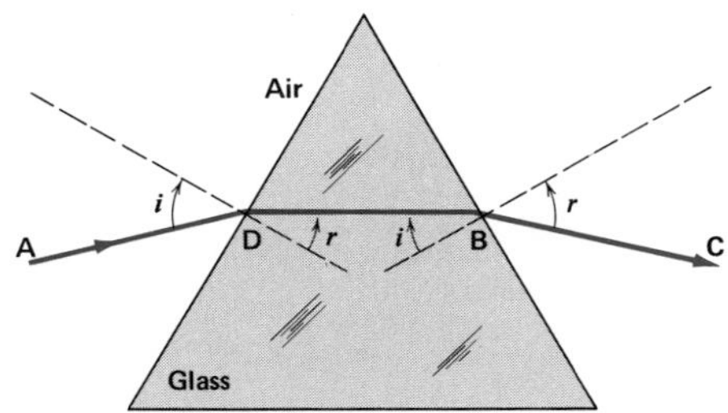

Figure 14-8. The path of a light ray through a prism. As the ray **AD** passes from the optically less dense to the optically more dense medium at **D**, its path is refracted toward the normal. As the ray passes from the optically more dense to the optically less dense medium at **B**, its path is refracted away from the normal.

larger arc is intersected by the extension of refracted ray **OP.** This line intersects the smaller arc at point **M.** Points **P** and **M** determine the path of the refracted ray as it enters the air.

In Figure 14-8 line **AD** represents a light ray incident upon a triangular glass prism at point **D.** Passing obliquely from air into glass, the ray is refracted toward the normal along line **DB.** Observe that at point **D** angle i in air is larger than angle r in glass. As light ray **DB** passes obliquely from glass into air at point **B,** it is refracted away from the normal along line **BC.** At point **B** angle i in glass is smaller than angle r in air.

Refraction of light can be summarized in ***three laws of refraction:***

1. *The incident ray, the refracted ray, and the normal to the surface at the point of incidence are all in the same plane.*
2. *The index of refraction for any homogeneous medium is a constant that is independent of the angle of incidence.*
3. *When a ray of light passes obliquely from a medium of lower optical density to one of higher optical density, it is bent toward the normal to the surface. Conversely, a ray of light passing obliquely from an optically denser medium to an optically rarer medium is bent away from the normal to the surface.*

Figure 14-9. (Left) The critical angle i_c is the limiting angle of incidence in the optically denser medium that results in an angle of refraction of 90°.

Figure 14-10. (Right) Total reflection at the water-air interface occurs when the angle of incidence exceeds the critical angle.

14.5 Total Reflection

Suppose an incident ray of light, **AO,** passes from water into air and is refracted along the line **OB,** as shown in Figure 14-9. As the angle of incidence, i, is increased, the angle of refraction, r, also increases. When this angle approaches the limiting value, $r_L = 90°$, the refracted ray emerges from the water along a path that gets closer to the water surface. As the angle of incidence continues to increase, the angle of refraction finally equals 90° and the refracted ray takes the path **ON**

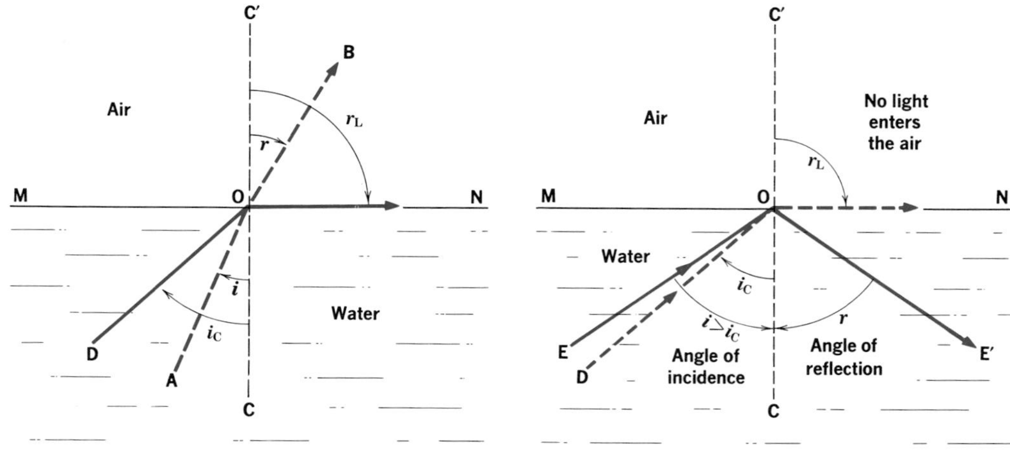

along the water surface. *The limiting angle of incidence in the optically denser medium that results in an angle of refraction of 90° is known as the* ***critical angle,*** i_C. The critical angle for water is reached when the incident ray **DO** makes an angle of 48.5° with the normal; the critical angle for crown glass is 42°, while that for diamond is only 24°.

Figure 14-11. The meter stick appears to be bent at the surface of the water as a result of the refraction of light. Can you account for the second image of the end of the meter stick at the lower left?

In Section 14.3 we defined the index of refraction of a material as the ratio of the speed of light in a vacuum (air) to the speed of light in the material or as the ratio $\sin i/\sin r$. In Figure 14-9 the light passes from the optically denser water to the air. Thus in the form of Snell's law, the roles of the angles of incidence and refraction are reversed. Here the angle of refraction, r, is related to the speed of light in air. The angle of incidence, i, is associated with the speed of light in water. The index of refraction of the water in this instance is

$$n = \frac{\sin r \text{ (air)}}{\sin i \text{ (water)}}$$

At the critical angle, i_C, r is the limiting value r_L, which equals 90°. Thus

$$n = \frac{\sin r_L}{\sin i_C} = \frac{\sin 90°}{\sin i_C} = \frac{1}{\sin i_C}$$

Therefore, in general

$$\sin i_C = \frac{1}{n}$$

where n is the index of refraction of the optically denser medium relative to air and i_C is the critical angle of this medium.

If the angle of incidence of a ray of light passing from water into air is increased beyond the critical angle, no part of the incident ray enters the air. The incident ray is totally reflected from the water interface. In Figure 14-10 **EO** represents a ray of light whose angle of incidence exceeds the critical angle, the angle of incidence **EOC** being greater than the critical angle **DOC.** The ray of light is reflected back into the water along the line **OE′**, a case of simple reflection in which the angle of incidence **EOC** equals the angle of reflection **E′OC.** *Total reflection* always occurs when the angle of incidence exceeds the critical angle.

Total internal reflection is the basis for fiber optics, which is used in modern telecommunications.

A diamond is a brilliant gem because its index of refraction is exceedingly high and its critical angle is therefore correspondingly small. Very little of the light that enters the upper surface of a cut diamond passes through the diamond; most of the light is reflected internally (total reflection), finally emerging from the top of the diamond. The faces of the upper surface are cut at such angles as to ensure that the maximum light entering the upper surface is reflected back to these faces. See also **Plate VI(A).**

QUESTIONS: GROUP A

1. What three conditions must be met for refraction to occur?
2. Does the fact that light refracts violate the principle of rectilinear propagation? Explain.
3. Snell's law relates angles of incidence and refraction. Did he know why refraction occurs? Explain.
4. Why is the index of refraction a characteristic of a material?
5. (a) How could you distinguish between a diamond and a piece of glass cut the same way? (b) Why does this happen?
6. What is the relationship between the velocity of light and the index of refraction of a transparent substance?

GROUP B

7. Why do we gain about four minutes of daylight each day?
8. Research "optical fibers" and relate their operation to this section.
9. (a) Most of what you see outside your classroom window consists of images of the actual objects. Explain. (b) Under what circumstances would you see the actual object?
10. A friend throws a coin into a pool. You dive toward the spot where you saw it from the edge of the pool in order to retrieve it. Explain what will happen.
11. (a) In which of the gases listed in Table 18 in Appendix B, does light travel the slowest? (b) Which material has the highest optical density?
12. According to the definition, what is the index of refraction of air?

PROBLEMS: GROUP A

1. (a) What is the refractive index of a material in which the speed of light is 1.85×10^8 m/s? (b) Using Table 18, identify this material.
2. Determine the speed of light in each of the following materials: (a) a sapphire (b) ice (c) glycerol.
3. Light passes from air into water at an angle of incidence of 42.3°. Determine the angle of refraction in the water.
4. The angles of incidence and refraction for light going from air into an optically more dense material are 63.5° and 42.9°, respectively. What is the index of refraction of this material?

GROUP B

5. A man in a boat shines a light at a friend under the water. If the beam in the water makes an angle of 36.2° relative to the normal, what was the angle of incidence?
6. Calculate the critical angle for light going from glycerol into air.
7. A ray of light passes from water into Lucite. If the angle relative to the normal in the Lucite is 45.0°, what was the angle in the water?
8. Light moves from flint glass into water at an angle of incidence of 28.7°;
   (a) What is the angle of refraction?
   (b) At what angle would the light have to be incident to give an angle of refraction of 90.0°?

LENS OPTICS

14.6 Types of Lenses A lens is any transparent object having two nonparallel curved surfaces or one plane surface and one curved surface. The curved surfaces can be spherical, parabolic, or cylindrical. Lenses are usually made of glass but can be made of other transparent materials. There are two general classes of lenses based upon their effects on incident light.

1. **Converging lenses.** Cross sections of converging lenses are shown in Figure 14-12(A). All are thicker in the middle than at the edge. The concavo-convex lens is known as a meniscus lens.

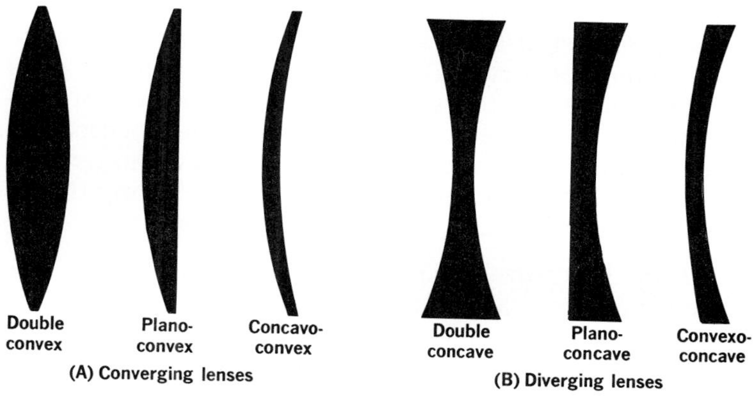

Figure 14-12. Common lens configurations.

Recall that light travels more slowly in glass and other lens materials than it does in air. Light passing through the thick middle region of a converging lens is retarded more than light passing through the thin edge region. Consequently, a wave front of light transmitted through a converging lens is bent as shown in Figure 14-13(A). The plane wave in this diagram is incident on the surface of the converging lens parallel to the lens plane. The wave is refracted and converges at the point **F** beyond the lens.

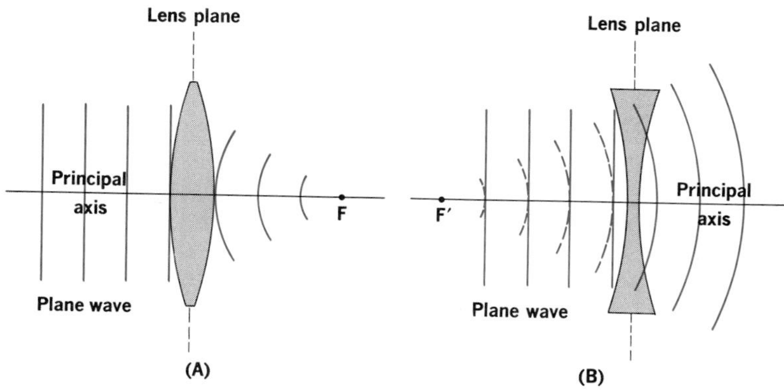

Figure 14-13. Refraction of a plane wave (A) by a converging lens and (B) by a diverging lens. The incident wave is perpendicular to the lens axis.

2. **Diverging lenses.** Lenses that are thicker at the edges than in the middle are diverging lenses. Their cross sections are shown in Figure 14-12(B). The convexo-concave lens is a meniscus lens.

Light passing through the thin middle region of a diverging lens is retarded less than light passing through the thick edge region. A wave front of light transmitted through a diverging lens is bent as shown in Figure 14-13(B). Observe that the plane wave in this diagram is also incident on the surface of the lens parallel to the lens plane. In this case, however, the refracted wave diverges in a way that makes it appear to come from the point **F′** in front of the lens.

A spherical lens usually has two centers of curvature—the centers of the spheres that form the lens surfaces. These centers of curvature determine the principal axis of the lens. The radii of curvature of the two surfaces of double convex and double concave lenses are not necessarily equal, although they are so drawn in the ray diagrams in this chapter. The geometry of lens surfaces having different radii of curvature is shown in Figure 14-14.

14.7 Ray Diagrams

The nature and location of the image formed by a lens is more easily determined by a ray diagram than by a wave-front diagram. The plane waves of Figure 14-13 can be represented by light rays drawn perpendicular to the wave fronts. Since these wave fronts

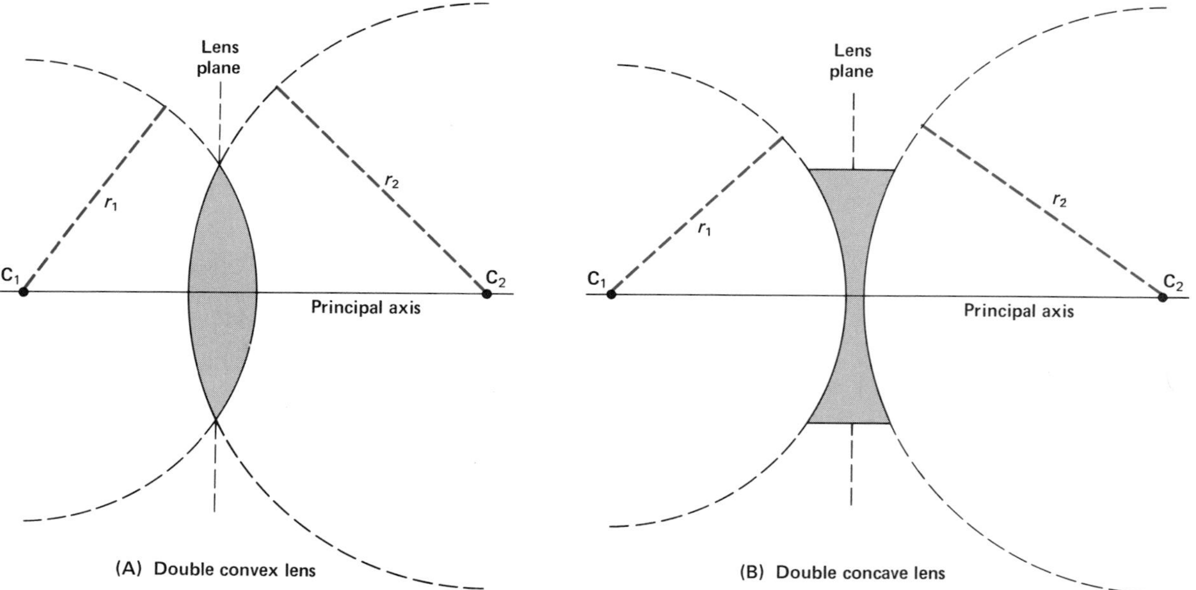

Figure 14-14. The geometry of lens surfaces that have different radii of curvature. Observe that the surfaces of the double convex lens are sections of intersecting spheres.

approaching the lens are shown parallel to the lens plane, their ray lines are parallel to the **principal axis** of the lens. Observe that the principal axis passes through the center of the lens and is perpendicular to the lens plane.

In Figure 14-15(A) these parallel rays (which are also parallel to the principal axis) are shown incident on a converging lens. They are refracted as they pass through the lens, and they converge at a point on the principal axis that locates the **principal focus F** of the lens. At point **B** the incident ray is refracted toward the normal drawn to the front surface of the lens. At point **D** the incident ray is refracted away from the normal drawn to the back surface of the lens. Because the rays of light actually pass through the principal focus **F**, it is called the **real focus. Real images** are formed on the same side of the lens as the real focus. Had these rays approached the lens from the right parallel

Real images can be projected on a screen.

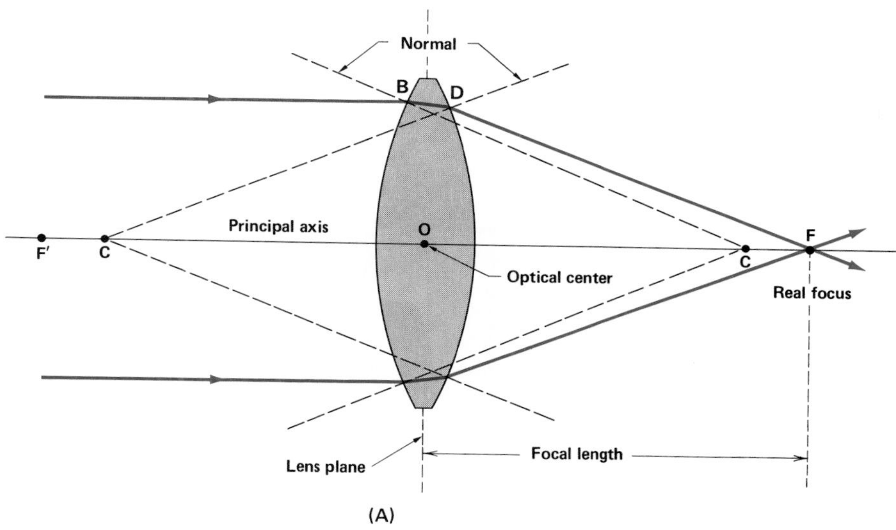

(A)

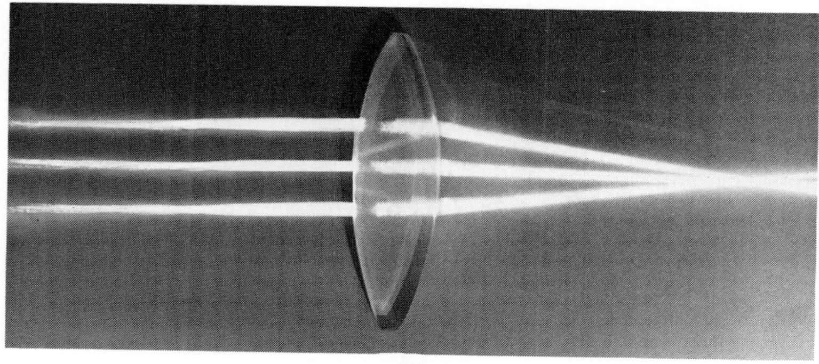

(B)

Figure 14-15. The converging lens. (A) A ray diagram that illustrates the refractive convergence of incident light rays parallel to the principal axis of the lens. (B) A photograph using light pencils shows a similar convergence.

to the principal axis, they would have converged at a point on the principal axis that located the principal focus **F′**.

Rays parallel to the principal axis are shown incident on a diverging lens in Figure 14-16(A). The rays of light are refracted as they pass through the lens and diverge as if they had originated at the principal focus **F′** located in front of the lens. Because the rays do not actually pass through this principal focus, it called a **virtual focus. Virtual images** are formed on the same side of the lens as the virtual focus.

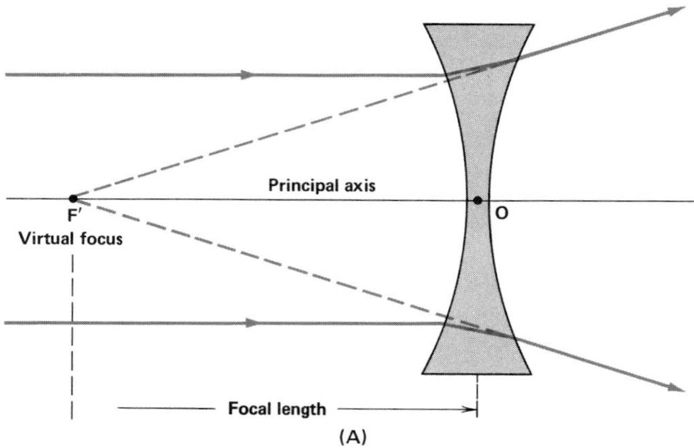

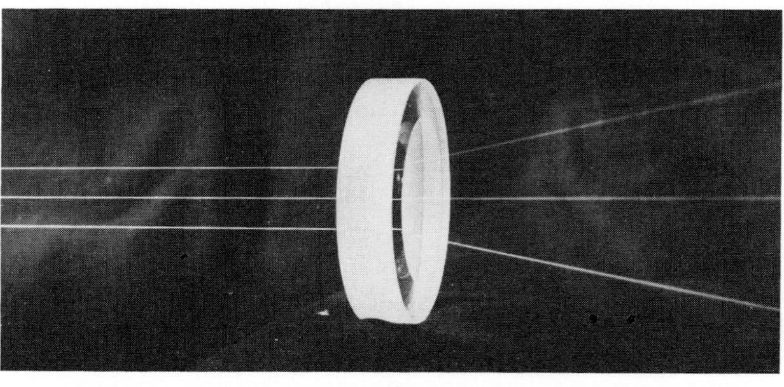

Figure 14-16. The diverging lens. (A) A ray diagram illustrates the refractive divergence of incident light rays parallel to the principal axis of the lens. (B) A photograph using light pencils shows a similar divergence.

In general, lenses refract rays that are parallel to the principal axis and the refracted rays either converge at the real focus behind the (converging) lens or diverge as if they originated at the virtual focus in front of the (diverging) lens. However, these foci are not midway between the lens and the center of curvature as they are in spherical mirrors. *The positions of the foci on the principal axis depend on the index of refraction of the lens.* A common double convex lens of crown glass has principal foci and centers of curvature that practically coincide and a focal length that is approximately equal to its radius of curvature. The **focal length,** *f*, of a lens is the distance between the optical center of the lens and the principal focus. The focal length of any lens depends on its index of refraction and the curvature of its surfaces. The higher its index of refraction, the shorter is its focal length. The longer its radius of curvature, the longer is its focal length.

The focal length of a lens is the same for light traveling in either direction even if the two lens surfaces have different radii of curvature.

A "bundle" of parallel light rays incident on a converging lens parallel to its principal axis is refracted and converges at the principal (real) focus behind the lens. The image is ideally a point of light at the real focus. This refraction is shown in Figure 14-17(A).

Bundles of parallel rays are not always incident on a lens parallel to its principal axis. If such rays are incident on a converging lens at a small angle with the principal axis, they converge at a point in the plane that contains the principal focus of the lens and is perpendicular to the principal axis. This plane is called the **focal plane** of the lens. A bundle of parallel rays incident on a converging lens at a small angle α (alpha) with its principal axis is shown in Figure 14-17(B).

Bundle of rays: the essentially parallel rays in a beam or a pencil of light.

The focal plane of a concave mirror is described in Section 13.7.

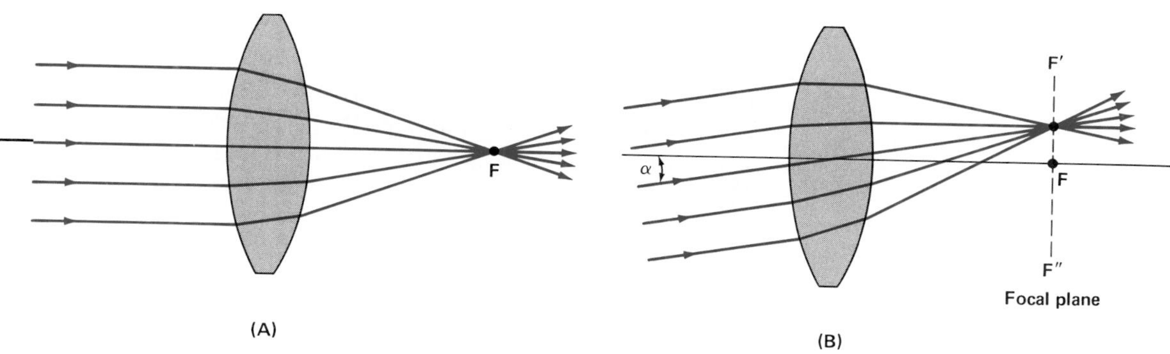

Figure 14-17. In (A) parallel rays are shown incident on a converging lens parallel to its principal axis. The refracted rays converge at its principal focus. In (B) the rays are incident at a small angle α with the principal axis. Here the refracted rays are focused at a point in the focal plane F′F″ of the lens.

14.8 Images by Refraction

The image of a point on an object is formed by intersecting refracted rays emanating from the object point. In Section 13.8 we recognized that the task of locating graphically an image formed by reflection is simplified if we select incident rays that have known paths after reflection. These rays, called **principal rays,** are shown with their *reflected* paths in Figure 13-18. Their paths after refraction are also known. They are shown with their *refracted* paths in Figure 14-18.

The image **S′** of point **S** (the object in Figure 14-18) is formed by the converging lens when the refracted rays from point **S** intersect behind the lens. Ray **1,** *parallel* to the principal axis, is refracted through the real focus, **F.** See Figure 14-15. Ray **2,** along the **secondary axis,** passes

Review Figure 13-18.

through the optical center of the lens, **O**, without being appreciably refracted. Ray **3** passes through the principal focus, **F′**, and is refracted parallel to the principal axis. The three refracted rays intersect at the image point **S′**. Observe that any two of the principal rays *emanating from the same object point* are able to locate the image of that point.

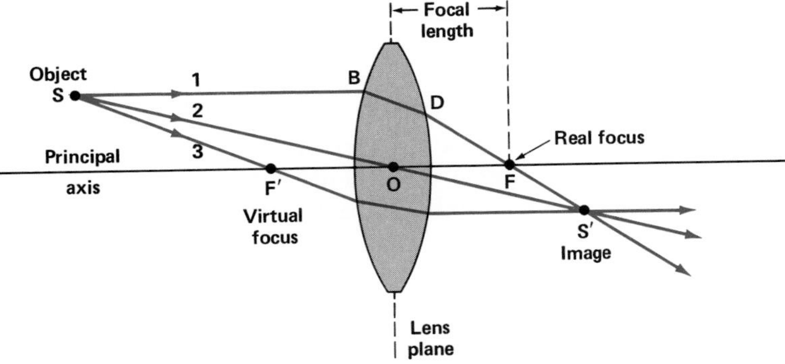

Figure 14-18. The principal rays used in ray diagrams. Any two of these three rays from the same point on the object locate the image of that point.

Lenses and mirrors differ in several ways.

1. Secondary axes pass through the optical center of a lens and not through either of its centers of curvature.
2. The principal focus is usually near the center of curvature, depending on the refractive index of the glass from which the lens is made. Thus the focal length of a double convex lens is about equal to its radius of curvature.
3. Since the image produced by a lens is formed by rays of light that actually pass through the lens, a *real* image is formed on the side of the lens opposite the object. *Virtual* images formed by lenses appear to be on the same side of the lens as the object.
4. Convex (converging) lenses form images in almost the same manner as concave mirrors, while concave (diverging) lenses are like convex mirrors in the manner in which they form images.

The lens principles discussed in this chapter apply to thin lenses. Ray diagrams become more complicated when they are applied to thick lenses.

Spherical lenses, like spherical mirrors, have aberration defects. See Figure 14-19. When such lenses are used with large apertures, images formed by rays passing through the central zones of the lens are generally sharp and well-defined, while images formed by rays passing through the edge zones are fuzzy. This defect of lens images is called **spherical aberration.**

Similarly, rays of light coming from an object point not on the principal axis are not brought to a sharp focus in the image plane. This defect of lenses is known as *lens astigmatism*. By using a combination of lenses of suitable refractive indexes and focal lengths, lens makers produce *anastigmatic* lenses. Such lenses give good definition over the entire image even when used with large apertures.

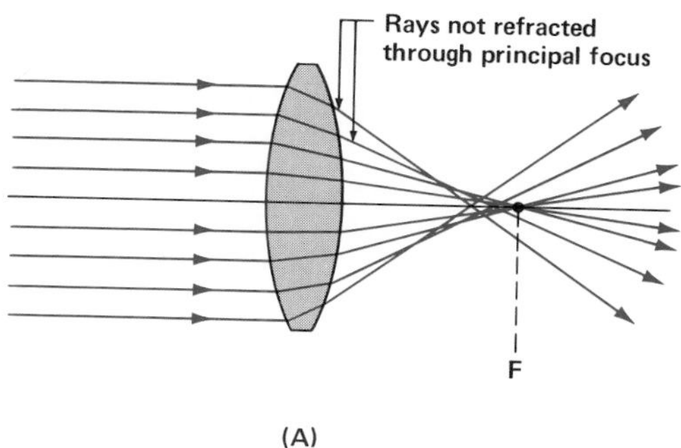

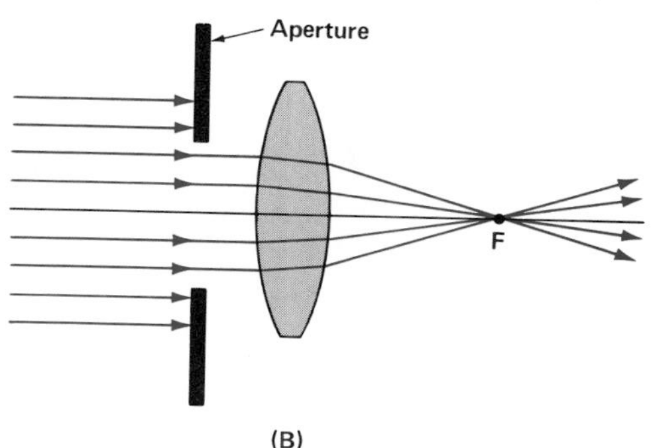

Figure 14-19. (A) Rays parallel to the principal axis but near the edge of a converging lens are not refracted through the principal focus. (B) An aperture can be used to block these rays from the lens.

A lens of short focal length that can be used with a large aperture has a large light-gathering capability. It is said to be a "fast" lens. The light-gathering power of a camera lens is given in terms of its **f-number.** This number is determined by the focal length of the lens and its effective diameter. The effective diameter is the diameter of the camera aperture (diaphragm) that determines the useful lens area. The light-gathering power, or "speed," of a lens is expressed as the ratio of its focal length to its effective diameter. If the speed of a lens is given as f/4, it means that its focal length is 4 times its effective diameter.

Because the useful area of a lens is proportional to the square of its effective diameter, the light-gathering power of a lens increases four times when its effective diameter is doubled. An f/4 lens is 4 times as fast as an f/8 lens, and is 16 times as fast as an f/16 lens. It follows that the required time of exposure increases as the square of the f-number.

The area of a circle:
$A = \pi r^2 = \frac{\pi d^2}{4}$. *Thus,* $A \propto d^2$.

It should be recognized that all lenses having the same f-number give the same illumination in the image plane, regardless of their individual diameters. Therefore, a lens having twice the diameter of another lens of the same f-number will have four times the light-gathering power, but this light will be spread over an image having four times the area and will give the same image brightness.

14.9 Images Formed by Converging Lenses We shall consider six different cases of image formation. These cases are illustrated in Figure 14-20.

Case 1. Object at an infinite distance. The use of a small magnifying glass to focus the sun's rays upon a point approximates this first case. While the sun is not at an infinite distance, it is so far away that its rays reaching the earth are nearly parallel. When an object is at an infinite distance and its rays are parallel to the principal axis of the lens, *the image formed is a point at the real focus.* See Figure 14-20(A). This principle can be used to find the focal length of a lens by focusing the sun's rays on a white

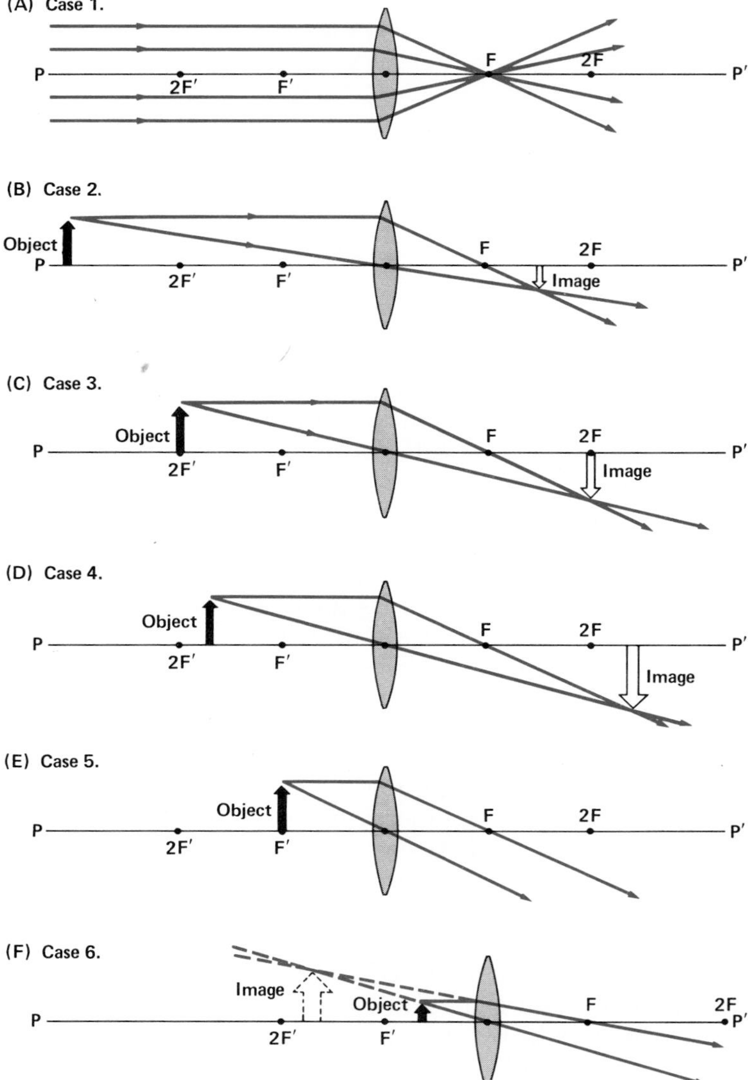

Figure 14-20. Ray diagrams of image formation by converging lenses.

screen. The distance from the screen to the optical center of the lens is the focal length of the lens.

Case 2. Object at a finite distance beyond twice the focal length. Case 2 is illustrated in (B). Rays parallel to the principal axis, along the secondary axis, and emanating from a point on the object are used to locate the corresponding image point. *The image is real, inverted, reduced, and located between F and 2F on the opposite side of the lens.* The lenses of the eye and the camera, and the objective lens of the refracting telescope are all applications of this case.

Case 3. Object at a distance equal to twice the focal length. The construction of the image is shown in (C). *The image is real, inverted, the same size as the object, and located at 2F on the opposite side of the lens.* An inverting lens of a field telescope, which inverts an image without changing its size, is an application of Case 3.

Case 4. Object at a distance between one and two focal lengths away. This is the converse of Case 2 and is shown in (D). *The image is real, inverted, enlarged, and located beyond 2F on the opposite side of the lens.* The compound microscope, slide projector, and motion picture projector are all applications of a lens used in this manner.

Case 5. Object at the principal focus. This case is the converse of Case 1. *No image is formed,* since the rays of light are parallel as they leave the lens (E). The lenses used in lighthouses and searchlights are applications of Case 5.

Case 6. Object at a distance less than one focal length away. The construction in (F) shows that the rays are divergent after passing through the lens and cannot form a real image on the opposite side of the lens. These rays appear to converge behind the object to produce *an image that is virtual, erect, enlarged, and located on the same side of the lens as the object.* The simple magnifier and the eyepiece lenses of microscopes, binoculars, and telescopes form images as shown in Case 6.

14.10 Images Formed by Diverging Lenses

The only kind of image of a real object that can be formed by a diverging lens is one that is *virtual, erect, and reduced in size.* Diverging lenses are used to neutralize the effect of a converging lens, or to reduce its converging effect to some extent. The image formation is shown in Figure 14-21.

Figure 14-21. The image of an object formed by a diverging lens.

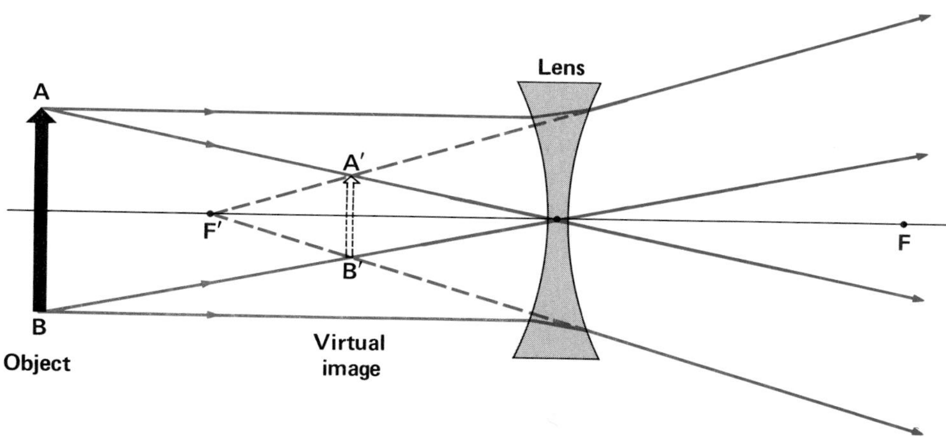

14.11 Object-Image Relationships For thin lenses, the ratio of object size to image size equals the ratio of the object distance to image distance. This rule is the same as the rule for curved mirrors. Thus

$$\frac{h_i}{h_o} = \frac{d_i}{d_o}$$

Observe that the lens equations are the same as the mirror equations of Section 13.12.

where h_o and h_i represent the heights of the object and the image respectively, and d_o and d_i represent the respective distances of the object and image from the optical center of the lens.

The equation used to determine the distances of the object and image in relation to focal length for curved mirrors applies also to lenses. It can be restated here as

$$\frac{1}{f} = \frac{1}{d_o} + \frac{1}{d_i}$$

The lens equation is valid providing these sign conventions are followed:

d_o *is* [*positive for real objects* / *negative for virtual objects*]

d_i *is* [*positive for real images* / *negative for virtual images*]

f *is* [*positive for converging lenses* / *negative for diverging lenses*]

where d_o represents the distance of the object from the lens, d_i the distance of the image from the lens, and f the focal length.

The numerical value of the focal length f is *positive* for a converging lens and *negative* for a diverging lens. For real objects and images, the object and image distances d_o and d_i have *positive* values. For virtual objects and images, d_o and d_i have *negative* values.

14.12 The Simple Magnifier A converging lens of short focal length can be used as a simple magnifier. The lens is placed slightly nearer the object than one focal length and the eye is positioned close to the lens on the opposite side. This is a practical example of Case 6; the image is virtual, erect, and enlarged as shown in Figure 14-22. A reading glass, a simple microscope, and an eyepiece lens of a compound microscope or telescope are applications of simple magnifiers.

Magnification *M is simply the ratio of the image height to the object height.*

$$M = \frac{h_i}{h_o}$$

But,

$$\frac{h_i}{h_o} = \frac{d_i}{d_o}$$

So,

$$M = \frac{d_i}{d_o}$$

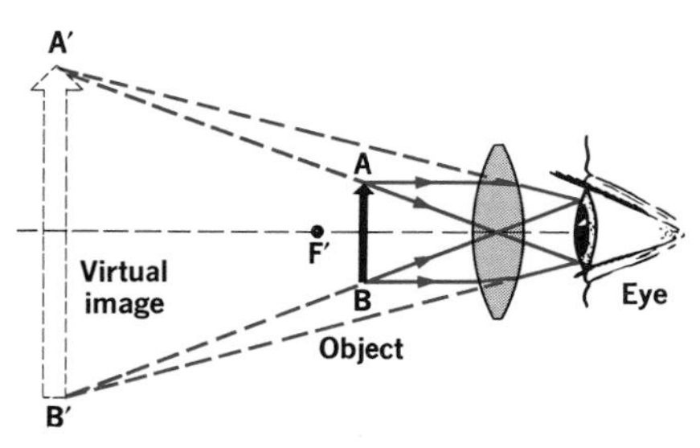

Figure 14-22. The simple magnifier.

Suppose an object is viewed by the unaided eye. As it is moved closer and closer to the eye, the image formed on the retina becomes larger and larger. Eventually, a *nearest* point is reached for the object at which the eye can still form a clear image. This minimum distance for distinct vision is approximately 25 cm from the eye. Although this nearest point varies among individuals, 25 cm is taken as the standard distance for most distinct vision; it is called the **near point.** As a person grows older the muscles of the eye, which thicken the lens and thus increase its convergence (shorten its focal length), gradually weaken. Consequently the near point moves out with aging.

If a converging lens is placed in front of the eye as a simple magnifier, the object can be brought much closer and the eye focuses on the virtual image. When the lens is used in this way, the object is placed just inside the principal focus ($d_o \simeq f$). The image is then formed approximately at the near point. Magnification, shown above to be equal to the ratio d_i/d_o, can now be expressed for a simple magnifier as the ratio of the distance for most distinct vision to the focal length of the lens. When f is given in centimeters, the magnification becomes approximately

$$M = \frac{25 \text{ cm}}{f}$$

Magnifiers are labeled to show their magnifying power. Thus a magnifier with a focal length of 5 cm would be marked 5X. One with a focal length of 2.5 cm would be marked 10X, etc. Observe that the shorter the focal length of a converging lens, the higher is its magnification.

14.13 The Microscope

The compound microscope, thought to be invented in Holland by Zacharias Janssen about 1590, uses a lens, the **objective,** to form an enlarged image as in Case 4. This image is then magnified, as in Case 6, by a second lens, called the **eyepiece.**

In Figure 14-23 a converging lens is used as the objective, with the object **AB** just beyond its focal length. At **A′B′**, a distance greater than twice the focal length of the objective lens, an enlarged, real, and inverted image is formed. The eyepiece lens acts as a simple magnifier to enlarge this image.

The magnifying power of the objective is approximately equal to the length of the tube, l, divided by the focal length, f_o, of the objective, or l/f_o. The magnifying power of the eyepiece, acting as a simple magnifier, is approximately 25 cm/f_E. The total magnification is the product of

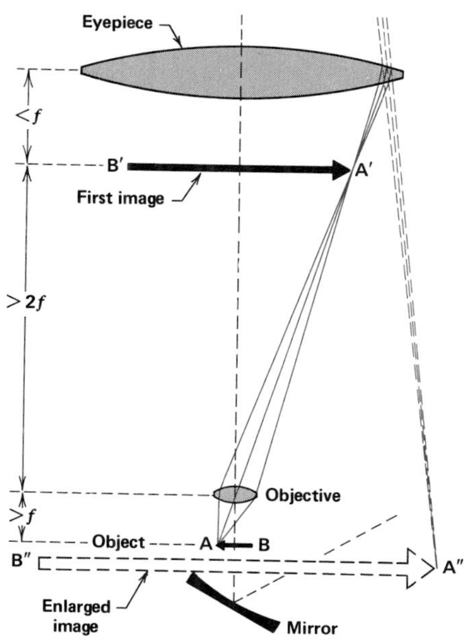

Figure 14-23. Image formation by a compound microscope.

the two lens magnifications. (The equation is again an approximation.)

$$M = \frac{25 \text{ cm} \times l}{f_E \times f_o}$$

14.14 Refracting Telescopes A refracting astronomical telescope has two lens systems. The *objective lens* is of large diameter so that it will admit a large amount of light. The objects to be viewed in telescopes are always more distant than twice the focal length of the objective lens. As a consequence the image formed is smaller than the object. The *eyepiece lens* magnifies the real image produced by the objective lens. The magnifying power is approximately equal to the focal length of the objective, f_o, divided by the focal length of the eyepiece, f_E, or f_o/f_E.

The lenses of a *terrestrial,* or field, telescope form images just as their counterparts do in the refracting astronomical telescope. Since it would be confusing to see objects inverted in a field telescope, another lens system is used to reinvert the real image formed by the objective. This additional inverting lens system makes the final image erect, as shown in Figure 14-24. The inverting lens system does not magnify the image because the lens system is placed exactly its own focal length from the image formed by the objective.

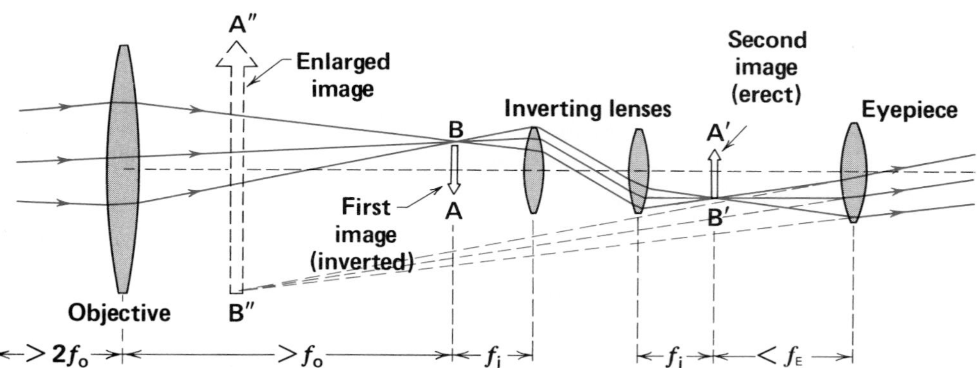

Figure 14-24. Image formation by a terrestrial telescope.

The *prism binocular* is actually a double field telescope that uses two sets of totally reflecting prisms instead of a third lens system to reinvert the real images formed by the objective lenses. This method of forming final images that are upright and correctly oriented also has the effect of folding the optical path, thus making binoculars more compact and easier to use than telescopes.

The prism binocular customarily has descriptive markings stamped on its case, such as 7 × 35, 8 × 50, etc. The first number gives its magnification, and the second number gives the diameters (in mm) of its objective lenses.

A viewing device with one eyepiece and one objective lens is called a monocular.

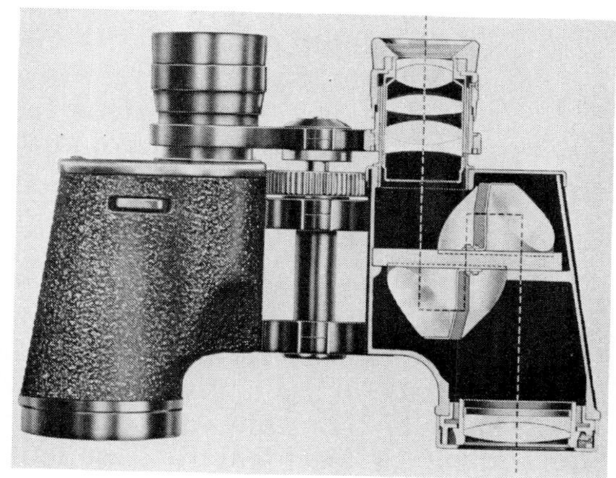

Figure 14-25. A pair of totally reflecting prisms reinvert the image in binoculars.

QUESTIONS: GROUP A

1. (a) What is the difference between a convex and a concave lens? (b) What effect does each have on light?
2. How does the relationship between the focal length and radius of curvature of a lens differ from that of a curved mirror?
3. (a) What factors determine the focal length of a lens? (b) How do they affect the focal length?
4. How many principal rays are needed to locate the image of an object when using a lens?
5. Why is no image formed when the object is at the focal point of a convex lens?
6. What is the relationship between the focal length of a converging lens and its magnification?
7. Which optical instrument most clearly approximates the human eye?
8. What property of real images explains why you have to load a slide projector with the slide upside down?
9. (a) How many lenses does a refracting telescope have? (b) Name and describe them.
10. What must be true about the relationship between the index of refraction of the air and the index of refraction of the lens material in order for a convex lens to be a converging lens?

GROUP B

11. What additional property of real images would help you distinguish between a real and a virtual image?
12. Describe how you would determine the focal length of a converging lens.
13. If there are two converging lenses in a compound microscope, why is the image still inverted?
14. (a) What are you doing when you focus your camera? (b) How does this differ from the way the eye focuses?
15. What is happening to your granddad's eyes that makes him say

"my arms are too short to read the newspaper"?

16. Why is an extra set of lenses needed in a terrestrial telescope?
17. A student uses a lens to focus an inverted, reduced image of a candle. (a) What kind of lens is it? (b) Where is the candle located?

Problems: Group A

1. A converging lens with a focal length of 15.0 cm is placed 53.0 cm from a light bulb. Where would you place a screen to focus an image of the object?
2. An object is 32.5 cm from a converging lens with a focal length of 12.0 cm. (a) Locate and describe the image using a ray diagram. (b) Calculate the distance of the image from the lens.
3. A convex lens of focal length 25.0 cm is placed 5.50 m from a screen. (a) Where should you place a candle to form a sharp image? (b) If the candle flame is 1.85 cm high, how high will its image be?
4. An object 30.0 cm from a converging lens forms a real image 60.0 cm from the lens. (a) Find the focal length of the lens. (b) If the object is 9.75 cm high, how high is the image?
5. What is the magnifying power of a simple magnifier whose focal length is 15 cm?
6. A camera lens has a 5.10-cm focal length. How far must the lens be from the film to take a clearly focused picture of your friend, 6.50 m away?
7. The objective lens of a compound microscope has a focal length of 0.500 cm and the eyepiece has a focal length of 2.00 cm. If the tube of the microscope is 15.0 cm long, what is the magnifying power of this microscope?

Group B

8. You set up a slide projector 3.50 m from the screen to get an image 1.35 m high. (a) If the slide is 3.50 cm tall, how far from the lens is the slide? (b) What is the focal length of this lens?
9. A camera, equipped with a lens of focal length 4.80 cm, is to be focused on a tree that is 10.0 m away. (a) How far must the lens be from the film? (b) How much would the lens have to be moved to take a picture of another tree that is only 1.75 m away?
10. The distance from the front to the back of your eye is approximately 1.90 cm. If you are to see a clear image of your physics book when it is 35.0 cm from your eye, what must be the focal length of the lens/cornea system?
11. Suppose you look out the window and see your friend, who is standing 15.0 m away. To what focal length must your eye muscles adjust your lens so that you may see your friend clearly? (See Problem 10)
12. When a 5.0-cm object is placed 12 cm from a converging lens, an image is produced on the same side of the lens as the object but 60.0 cm away from the lens. (a) What type of image is this? (b) Find the focal length of the lens. (c) Calculate the image size.

Physics Activity

Borrow a pair of glasses from a near-sighted friend, unless you have a pair yourself. Hold them about 12 cm from your eye and look at different objects through them. Describe what you see. If possible, use lenses with different correction factors. How does the correction factor affect what you see? Try this with the glasses of a far-sighted person, too.

DISPERSION

14.15 Dispersion by a Prism Suppose a narrow beam of sunlight is directed onto a glass prism in a darkened room. If the light that leaves the prism falls on a white screen, a band of colors is observed, one shade blending gradually into another. *This band of colors produced when sunlight is dispersed by a prism is called a* ***solar spectrum.*** The dispersion of sunlight was described by Newton, who observed that the spectrum was "violet at one end, red at the other, and showed a continuous gradation of colors in between."

We can recognize six distinct colors in the visible spectrum. These are *red, orange, yellow, green, blue,* and *violet.* Each color gradually blends into the adjacent colors giving a *continuous* spectrum over the range of visible light. A continuous spectrum is shown in **Plate VII** of the color insert between pages 336 and 337. *Light consisting of several colors is called* ***polychromatic light;*** *light consisting of only one color is called* ***monochromatic light.*** All colors of the spectrum are present in the incident beam of sunlight. White light is a mixture of these colors.

The dispersion of light by a prism is shown in **Plate I** of the color insert. It is evident that the refraction of red light by the prism is not as great as that of violet light; the refractions of other colors lie between these two. Thus the index of refraction of glass is not the same for light of different colors. If we wish to be very precise in measuring the index of refraction of a substance, monochromatic light must be used and the monochrome color must be stated. Some variations in the index of refraction of glass are given in Table 14-1.

Table 14-1
VARIATION OF THE INDEX OF REFRACTION

Color	*Crown glass*	*Flint glass*
red	1.515	1.622
yellow	1.517	1.627
blue	1.523	1.639
violet	1.533	1.663

14.16 The Color of Light A hot solid radiates an appreciable amount of energy that increases as the temperature is raised. At relatively low temperatures, a small amount of energy is radiated in the infrared region. As the temperature of the solid is raised, some of the energy is radiated at higher frequencies. These frequencies range into the red portion of the visible spectrum as the body becomes "red hot." At still higher temperatures, the solid may be "white hot" as the major portion of the radiated energy shifts toward the higher frequencies.

Suppose we have a clear-glass tungsten-filament lamp connected in an electric circuit so that the current in the filament, and thus the temperature of the filament, can be controlled. A small electric current in the filament does not change the filament's appearance. As we gradually increase the current, however, the filament begins to glow with a dark red color. To produce this color, electrons in the atoms of tungsten must have been excited to sufficiently high-energy levels so that upon de-excitation they emit energy with frequencies of about 3.9×10^{14} Hz. Even before the lamp filament glows visibly, experiments show that it radiates infrared rays that can be detected as heat. As the current is increased further, the lamp filament gives off orange light in addition to red; then the filament adds yellow, and finally at higher temperatures it adds enough other colors to produce white light.

A photographer's tungsten-filament flood lamp operates at a very high temperature. If the white light of such a lamp is passed through a prism, a band of colors similar to the solar spectrum is obtained. Figure 14-26 shows the distribution of radiant energy from an "ideal", or blackbody, radiator for several different temperatures.

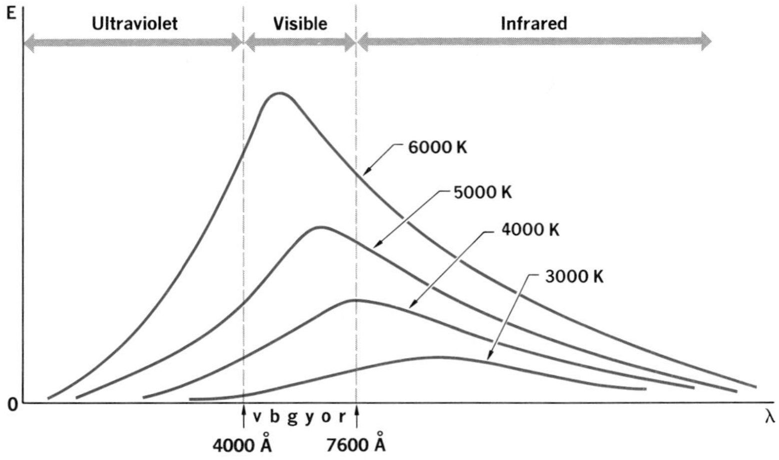

Figure 14-26. Distribution of radiant energy from an ideal radiator.

Considering the wavelengths of various colors shown in **Plate I,** it is evident that our eyes are sensitive to a range of frequencies equivalent to about one octave. The wavelength of the light at the upper limit of visibility (7600 Å) is about twice the wavelength of the light at the lower limit of visibility (4000 Å). We use the word *color* to describe a psychological sensation through the visual sense related to the physical stimulus of light. The color perceived for monochromatic light depends on the frequency of the light. For example, when light of about 4.6×10^{14} Hz (6500 Å) enters the eye, the color perceived is red.

14.17 The Color of Objects Color is a property of the light that reaches our eyes. Objects may absorb certain frequencies from the light falling upon them and reflect other frequencies. For example, a cloth that appears blue in sunlight appears black when held in the red portion of a solar spectrum in a darkened room. A red cloth held in the blue portion of the solar spectrum also appears black. *The color of an opaque object depends upon the frequencies of light it reflects.* If all colors are reflected, we say it is *white*. It is *black* if it absorbs all the light that falls upon it. It is called *red* if it absorbs all other colors and reflects only red light. The energy associated with the colors absorbed is taken up as heat.

Actually, in order to be reflected the red light must interact with the object. The interaction consists of a resonant absorption by electrons of the object's atoms. The energy is immediately re-emitted or "reflected."

A piece of blue cloth appears black in the red portion of the spectrum because there is no blue light there for it to reflect and it absorbs all other colors. For the same reason, a red cloth appears black in the blue portion of the spectrum. *The color of an opaque object depends on the color of the light incident upon it.*

Ordinary window glass, which transmits all colors, is said to be colorless. Red glass absorbs all colors but red, which it transmits. The stars of the United States flag would appear red on a black field if viewed through red glass. *The color of a transparent object depends upon the color of the light that it transmits.*

14.18 Complementary Colors Because polychromatic light can be dispersed into its elementary colors, it is reasonable to suppose that elementary colors can be combined to form polychromatic light. There are three ways in which this can be done.

In physics, the term "color" refers to light. In art, the term "color" often refers to the shade of a pigment.

1. A prism placed in the path of the solar spectrum formed by another prism will recombine the different colors to produce white light. Other colors can be compounded in the same manner.

2. A disk that has the spectral colors painted on it can be rotated rapidly to produce the effect of combining the colors. The light from one color forms an image that persists on the retina of the eye until each of the other colors in turn has been reflected to the eye. If pure spectral colors are used in the proper proportion, they will blend to produce the same color sensation as white light.

Combining colored lights is an additive process.

3. Colored light from the middle region of the visible spectrum combined with colored light from the two end regions produce white light. This method is described in Section 14.19.

Two prisms are used as described above, but with the red light from the first prism blocked off from the second prism. The remaining spectral colors are recombined by the second prism to produce a blue-green color called

Become familiar with the six elementary colors and their complements illustrated in ***Plate II.***

cyan. Red light and cyan should therefore combine to produce white light, and a rotating color wheel shows this to be true. *Any two colors that combine to form white light are said to be* ***complementary.***

In similar fashion it can be shown that blue and yellow are complementary colors. White fabrics acquire a yellowish color after continued laundering. A blue dye added to laundry detergents neutralizes the yellow color and the fabrics appear white. Iron compounds in the sand used for making glass impart a green color to the glass. Manganese gives glass a *magenta,* or purplish-red, color. However, if both these elements are present in the right proportion, the resulting glass will be colorless. Green and magenta are complementary colors. The complements of the six elementary spectral colors are shown in **Plate II.**

14.19 The Primary Colors

The six regions of color in the solar spectrum are easily observed by the dispersion of sunlight. Further dispersion within a color region fails to reveal any other colors of light. We generally identify the range of wavelengths comprising a color region by the color of light associated with that region. These are the six **elementary *colors*** of the visible spectrum; they combine to produce white light. However, the complement of an elementary color is not monochromatic but is a mixture of all of the elementary colors remaining after the one elementary color has been removed.

Experiments with beams of different colored lights have shown that most colors and hues can be described in terms of three different colors. Light from one end of the visible spectrum combined with light from the middle region in various proportions will yield all of the color hues in the half of the spectrum that lies in between them. Light from the opposite end, when combined with light from the middle region, will also yield all hues in the half of the spectrum that lies in between them. Colored light from the two end regions and the middle region can be combined to match most of the hues when mixed in the proper proportions. The three colors that can be used most successfully in color matching experiments of this sort are *red, green,* and *blue.* Consequently these have been called the ***primary colors.***

Suppose we project the three primary colors onto a white screen as shown in **Plate III(A).** The three beams can be adjusted to overlap, producing additive mixtures of these primary colors. Observe that green and blue lights combine to produce cyan, the complement of red; green and red lights combine to produce yellow, the complement of blue; and red and blue lights combine to produce magenta, the complement of green. Thus two primary colors combine to produce the complement of the third primary color. Where the three primary colors overlap, white light is produced.

Combining primary colors is an additive process.

14.20 Mixing Pigments When the complements blue light and yellow light are mixed, white light results by an additive process. If we mix a blue pigment with a yellow pigment, a green mixture results. This process is subtractive since each pigment subtracts or absorbs certain colors. For example, the yellow pigment subtracts blue and violet lights and reflects red, yellow, and green. The blue pigment subtracts red and yellow lights and reflects green, blue, and violet. Green light is the only color reflected by both pigments; thus the mixture of pigments appears green under white light.

Combining primary pigments is a subtractive process.

The subtractive process can be demonstrated by the use of various color filters that absorb certain frequencies and transmit others from a single white-light source.

When pigments are mixed, each one subtracts certain colors from white light, and the resulting color depends on the frequencies that are not absorbed. *The **primary pigments** are the complements of the three primary colors.* They are cyan (the complement of red), magenta (the complement of green), and yellow (the complement of blue). When the three primary pigments are mixed in the proper proportions, all the colors are subtracted from white light and the mixture is black. See **Plate III(B).**

*Compare the additive combinations of primary colors in **Plate III(A)** with the subtractive combinations of primary pigments in **Plate III(B).***

14.21 Chromatic Aberration Because a lens configuration has some similarity to that of a prism, some dispersion occurs when light passes through a lens. Violet light is refracted more than the other colors and is brought to a focus by a converging lens at a point nearer the lens than the other colors. Because red is refracted the least, the focus for the red rays is farthest from the lens. See Figure 14-27(A). Thus images formed by ordinary spherical lenses are always fringed with spectral colors. *The nonfocusing of light of different colors is called **chromatic aberration.*** Sir Isaac Newton developed the reflecting telescope to avoid the objectionable effects of chromatic aberration that occur when observations are made through a refracting telescope.

The English optician John Dollond (1706–1761) discovered that the fringe of colors could be eliminated by means of a combination of lenses. A double convex lens of crown glass used with a suitable plano-concave lens of flint glass

Figure 14-27. (A) Chromatic aberration is caused by unequal refraction of the different colors. (B) A two-lens combination of crown and flint glass corrects chromatic aberration.

corrects for chromatic aberration without preventing refraction and image formation. A lens combination of this type is shown in Figure 14-27(B). It is called an *achromatic* (without color) lens.

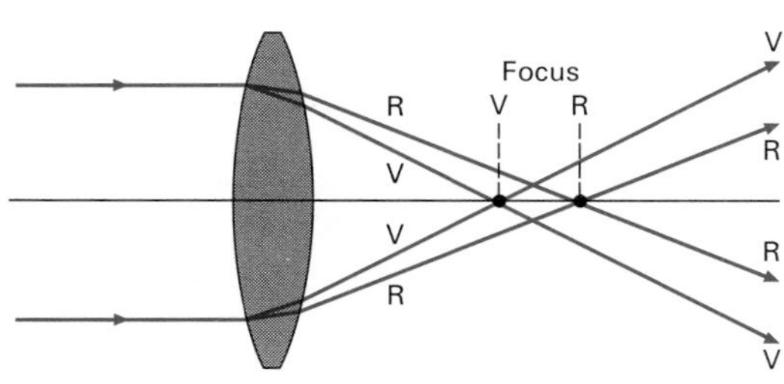

(A) Chromatic aberration

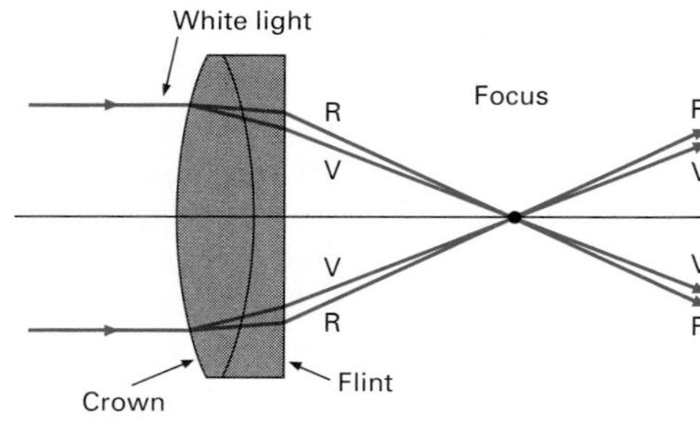

(B) Achromatic Lens

QUESTIONS: GROUP A

1. What are the six colors of the spectrum?
2. Use Table 14.1 to answer these questions: (a) Which of the two materials is optically less dense? (b) How does the index of refraction vary with the wavelength of the light? (c) Which color changes speed the most?
3. (a) What are the three primary additive colors? (b) What happens when you mix them?
4. You have a magenta opaque object. What color would it look under the following colors of light? (a) white (b) red (c) cyan (d) green (e) yellow
5. (a) What are complementary colors with reference to light? (b) Name the three pairs of complementary colors.
6. Explain what could happen when you mix the following: (a) cyan and yellow pigment (b) blue and yellow light (c) spectral blue and yellow pigments.
7. A student, operating a spotlight for the school show, points out that there are colored fringes around the edges of the white spot from the light. Explain what is happening here.
8. (a) What is meant by the term *dispersion?* (b) Who first explained this phenomenon?
9. How does the range of frequencies that our eyes perceive as visible light compare to the range of vibrations our ears perceive as sound?
10. If a black body absorbs all visible radiation incident on its surface, how can we see it?

SUMMARY

The laws of refraction describe the behavior of light rays that pass obliquely from one medium into another of different optical density. The index of refraction of any transparent material is defined in terms of the speed of light in a vacuum and the speed of light in the material. The index of refraction is also expressed in terms of Snell's law. Total reflection is explained on the basis of the critical angle of a material and the limiting value of the angle of refraction.

Converging lenses have convex surfaces and form images in a manner similar to that of concave mirrors. Diverging lenses have concave surfaces and form images in a manner similar to that of convex mirrors. The general lens equations correspond to those for curved mirrors. These equations relate object and image sizes with their respective distances from the lens and object and image distances with the focal length of the lens.

A converging lens forms either real or virtual images of real objects depending on the position of the object relative to the principal focus of the lens. A diverging lens forms only virtual images of real objects. Lens functions in common types of refractive instruments such as microscopes and telescopes can be analyzed by considering each lens separately.

Sunlight is composed of polychromatic light that undergoes dispersion when refracted by a prism. Six elementary colors are recognized in dispersed white light. The visual perception of color is related to the frequency of visible light.

The removal of an elementary color from white light leaves a polychromatic color that is the complement of the color removed. Addition of complementary colors produces white light. White light is produced by adding the primary colors. Any two of the three primary colors will combine to produce the complement of the third. The primary pigments are complements of the primary colors. Their combination is considered to be a subtractive process.

VOCABULARY

achromatic lens
angle of refraction
chromatic aberration
complementary color
converging lens
critical angle
diverging lens
elementary color
eyepiece
f-number
focal length
focal plane
index of refraction
lens equation
monochromatic
near point
objective
optical density
optical refraction
polychromatic
primary color
primary pigment
principal axis
principal focus
principal ray
real focus
real image
refraction
refractometer
secondary axis
Snell's law
solar spectrum
spherical aberration
total reflection
virtual focus
virtual image

15 Diffraction and Polarization

interference (in-tur-FEER-ens) n.: the mutual effect of two wave trains that meet, resulting in a loss of intensity in certain regions and a reinforcement of intensity in other regions.

OBJECTIVES

- Solve problems based on the relationship between wavelength and diffraction angle.
- Show that polarization confirms that light is a transverse wave.
- Describe the different ways that light becomes plane-polarized.
- Discuss double refraction.
- Describe polarized light and strain patterns.

Figure 15-1. A method of observing double-slit interference.

INTERFERENCE AND DIFFRACTION

15.1 Double-Slit Interference The superposition of two identical wave trains traveling in the same or opposite direction illustrates the phenomenon of *interference.* Sound waves of the same amplitude and wavelength projected in the same direction by two loudspeakers provide an interference pattern in which alternate regions of reinforcement and cancellation can be found. Two identical waves traveling in opposite directions on a stretched string interfere and produce a standing wave pattern.

The interference of two water waves of identical wavelength is easily observed in the laboratory using the ripple tank. More precise studies of such wave behavior are made by photographing light reflections from the surface of a mercury ripple tank. Interference and superposition topics can be reviewed by referring to Sections 10.14 through 10.16.

The superposition principle holds for these wave disturbances in material media and for light waves and other electromagnetic waves in free space. Thus light shows ***interference:*** *the mutual effect of two beams of light that results in a loss of intensity in certain regions* **(destructive interference)** *and a reinforcement of intensity in other regions* **(constructive interference).**

The English physician and physicist Thomas Young (1773–1829) first demonstrated interference of light in 1801 and showed how this phenomenon supports the wave theory of Huygens. Two narrow slits, as shown in Figure 15-1, about 1 mm apart in a piece of black paper are used to observe a narrow source of light, **S,** placed 2 or 3 meters away. A series of narrow bands alternately dark and light is seen in the center of a fairly wide band of light. See Figure 15-2. If a red filter is placed between the source and the slits so that a single spectral color is the light source, a series of red and black bands is seen. A double-slit interference pattern from a monochromatic light is shown in **Plate VIII(A).**

Figure 15-2. A double-slit interference pattern.

A wave-front diagram of double-slit interference with a monochromatic light source is shown in Figure 15-3. Slits $\mathbf{S_1}$ and $\mathbf{S_2}$ are equidistant from the source **S.** As light from **S** reaches $\mathbf{S_1}$ and $\mathbf{S_2}$, each slit serves as a new light source producing new wave fronts in phase with each other. These waves travel out from $\mathbf{S_1}$ and $\mathbf{S_2}$ producing bright bands of light due to reinforcement where constructive interference occurs and dark bands due to cancellation where destructive interference occurs.

Compare the diagram in Figure 15-3 with the photograph in Figure 10-27.

Figure 15-3. A wave-front diagram of double-slit interference.

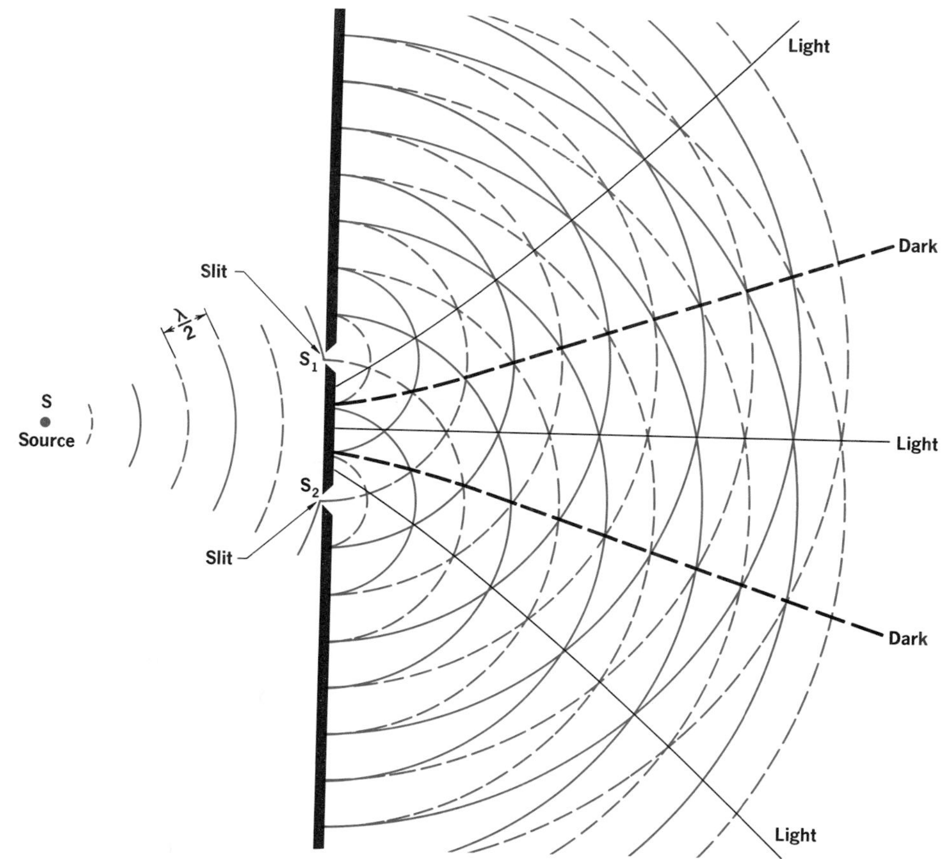

15.2 Interference in Thin Films References to thin films have been found on 3000-year-old clay tablets. The ancient Assyrians used the color patterns produced by oil films on water for fortune telling. Later, Isaac Newton devised experiments to measure the film thicknesses corresponding to specific colors. But Newton's corpuscular explanation of the partial reflection and partial refraction of light in thin films is inadequate in view of modern theories of light.

When viewed by reflected white light, thin transparent soap films, oil slicks, and wedge-shaped films of air show varying patterns of colors. When illuminated by monochromatic light, alternate bright and dark regions are observed, and the positions of the bands shift as the color of the monochromatic light is changed. Thomas Young explained this thin-film phenomenon in terms of the interference of light waves.

A ray of monochromatic light is incident on a thin transparent film at a small angle i, as shown in Figure 15-4.

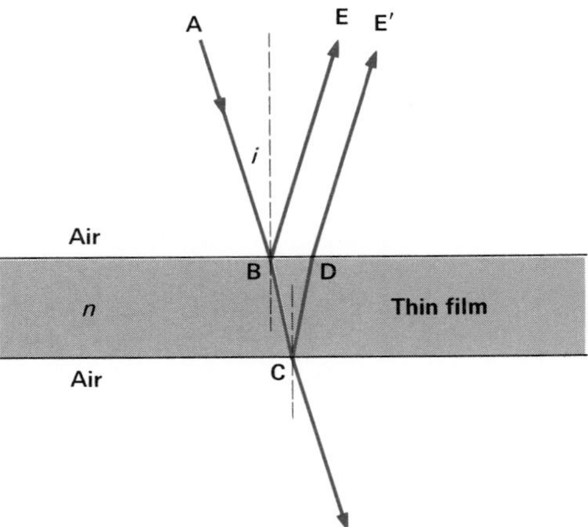

Figure 15-4. Partial reflection and refraction of light in a thin film results in interference.

Some light is reflected at **B** and some is refracted toward **C.** The index of refraction, n, of the transparent film is higher than that of air. The refracted light is then partially reflected at **C,** emerging from the film along **DE′** parallel to **BE.** If the wave fronts traveling along **BE** and **DE′** reach the eye, interference is observed. This is because the paths **ABE** and **ACE′** differ in length and the partial path **BCD** of **ACE′** is in a medium of different optical density. The nature of the interference depends on the phase relation between the waves arriving at the eye.

The two reflected waves, BE and DE′, are coherent because they both originate from the same point A on the monochromatic source.

For the monochromatic light that is used, assume that the incidence angle i is very small and the film at **B** has an optical thickness of a quarter wavelength. Thus the optical path *in the thin film* from **B** to **C** is one-quarter wavelength. From **C** to **D** is another one-quarter wavelength. Light emerging from the film at **D** is therefore one-half wavelength behind that reflected at **B** and will be expected to interfere destructively causing the film to appear dark.

The wavelength referred to is the wavelength of the monochromatic light in the thin film. Since the index of refraction of the film is higher than that of air, the wavelength of the light is shorter in the film than in the air.

Now if we assume the film has an optical thickness of a half wavelength, light reflected from the lower surface of the film will emerge at **D** a whole wavelength behind that reflected from the upper surface at **B.** The waves will be expected to interfere constructively causing the film to appear bright at this point. *Observation of reflected light from thin films shows that a reverse effect actually occurs.* Thus, the quarter-wave point appears *bright* and the half-wave point appears *dark.*

Camera lenses are often coated with transparent thin films to reduce stray reflected light and increase image contrast.

This anomalous situation prevails whenever air is on both sides of the film. It is easily observed in soap films. If a soap film is supported vertically, as it drains its upper region will become quite thin and a very small fraction of a wavelength in thickness just before it breaks. If observed by reflected monochromatic light at this moment, this upper region appears dark. See **Plate VI(B).**

Applying the same line of argument to this film of almost negligible thickness, we would expect the light reflected from the front surface and that reflected from the back surface to reach the eye so nearly in phase that this upper region of the film would appear bright by reflected light. Here is a clear-cut discrepancy between observation and theory. How can it be reconciled?

Thomas Young resolved the discrepancy by suggesting that *one of the interfering waves undergoes a phase change of 180° during reflection.* This phase change is in addition to the phase change that results from the unequal lengths of their optical paths. The phase inversion occurs during one reflection and not the other because the two reflections are opposite in kind. One reflection takes place at an interface at which the medium beyond has the higher index of refraction. This reflection corresponds to that at point **B** in Figure 15-4. *The reflected wave is 180° out of phase.* The other reflection occurs at an interface at which the medium beyond has the lower index of refraction. This reflection corresponds to that at point **C** in Figure 15-4. Here, *the reflected wave experiences no change in phase.*

See Section 10.10 for a review of incident and reflected waves at a reflecting interface.

Because of the phase-inverting effect of one of the two reflecting interfaces, the rules for constructive and destructive interference for thin films *bounded on both sides with a medium of lower index of refraction* are just the opposite from those we might expect. *Maximum constructive interference occurs with such thin films if the optical path difference is an odd number of half wavelengths*. This happens where the film thickness is an *odd* number of quarter wavelengths. *Maximum destructive interference occurs if the optical path difference is a whole number of wavelengths*. Here the film thickness is an *even* number of quarter wavelengths.

The vertically supported soap films of **Plate VI(B)** gradually increase in thickness toward the bottom as they drain. A vertical cross section of such a film would have a wedge-like appearance. As the thickness increases, the odd $\lambda/4$ requirement for the reinforcement of different colors and the even $\lambda/4$ requirement for their cancellation are met at successive intervals down the film. The soap film shown on the left of **Plate VI(B)** therefore reflects white light as a succession of colors at intervals down the film.

An air film between two optically flat glass plates produces a regular pattern of interference fringes. Irregular surfaces produce irregular patterns of interference fringes. Using interference patterns, inspection techniques have been developed that make extremely high precision measurements possible. Test plates can be polished optically flat with a tolerance of approximately 5×10^{-7} cm. Interference techniques are used to establish standards of measurement in many mechanical processes. The **interferometer** is an instrument that uses interference in the measurement of distance in terms of known wavelengths of light or in the measurement of wavelengths in terms of a standard of length.

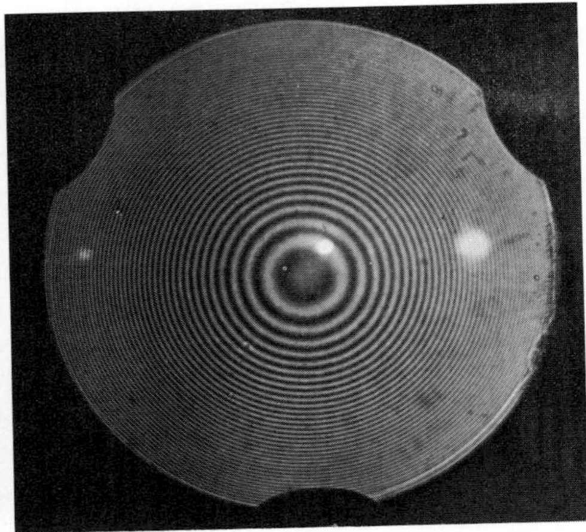

Figure 15-5. Interference patterns. (Left) Straight interference fringes from an optical flat. (Right) An interference pattern obtained from a spherical mirror being tested in an interferometer. The white spots are reflections from the interferometer light source.

15.3 Diffraction of Light According to the wave theory, light waves should bend around corners although our common experience with light shows that it travels in straight lines. Under some conditions light waves do bend out of their straight paths. When light waves encounter an obstruction with dimensions comparable with their wavelengths, the light spreads out and produces spectral colors due to interference. *The spreading of light into a region behind an obstruction is called* ***diffraction.*** A slit opening, a fine wire, a sharp-edged object, or a pinhole can serve as a suitable obstruction in the path of a beam of light from a point source. It is possible to see diffraction fringes if one peers between two fingers at a distant light source. With the fingers held close to the eye and brought together to form a slit opening, dark fringes will be seen just before the light is shut out.

Review diffraction as observed in the ripple tank in Section 10.13.

Diffraction effects, such as indistinct edges of shadows and shadow fringes, are known to have been observed as early as the seventeenth century. However, before the discovery of interference in 1801, neither the wave theory nor the corpuscular theory could offer a suitable explanation for these diffraction effects. In 1816 the French physicist A. J. Fresnel (fray-*nel*) (1788–1827) demonstrated that the various diffraction phenomena are fully explained by the interference of light waves.

Very useful diffraction patterns can be produced by illuminating an optical surface, either plane or spherical concave, that has many thousands of straight, equally-spaced parallel grooves ruled on it. These ruled surfaces are known as **diffraction gratings.** Light is diffracted when it is transmitted through or reflected from the narrow spaces between the ruled lines.

Figure 15-6. A master diffraction grating being ruled by a diamond-pointed scribe.

Standard gratings can have as many as 12 000 ruled lines per centimeter of grating surface. X rays are investigated with ordinary diffraction gratings set with their surface at a low angle to the rays. The gratings thus have the equivalent of many lines per centimeter. Gratings are generally superior to optical prisms for displaying the length or spread of spectra. However, grating spectra tend to be less intense than those formed by prisms.

15.4 Wavelength by Diffraction

A transmission grating placed in the path of plane waves disturbs the wave front because the ruled lines are opaque to light and the narrow spacings between the lines are transparent. These spaces provide a large number of fine, closely-spaced transmission slits. New wavelets generated at these slits interfere in such a way that several new wave fronts are established. One wave front travels in the original direction and the other wave fronts travel at various angles from this direction depending on their wavelength.

Recall Huygens' principle explained in Section 12.3: Each point on a wave front may be regarded as a new source of disturbance.

Suppose a single narrow slit is illuminated by white

light and viewed through a transmission grating. A white image of the illuminated slit is seen directly in line with the slit opening. In addition, pairs of continuous spectra are observed, each pair equally spaced on opposite sides of the principal image.

If we illuminate the slit with monochromatic light, successive pairs of slit images of decreasing intensity will appear on opposite sides of the principal image. The two images forming the first pair are known as **first-order *images,*** those forming the second pair as **second-order *images,*** etc.

In Figure 15-7, **A** and **B** are parallel spaces between the ruled lines on a diffraction grating. They act as adjacent transmission slits, being uniformly separated by the distance d, called the *grating constant*. Monochromatic light from a distant illuminated slit traveling normal to the grating surface produces secondary wavelets simultaneously at **A** and **B.** A new wave front of these wavelets proceeds along **MN** and produces the principal image of the distant slit at **N.**

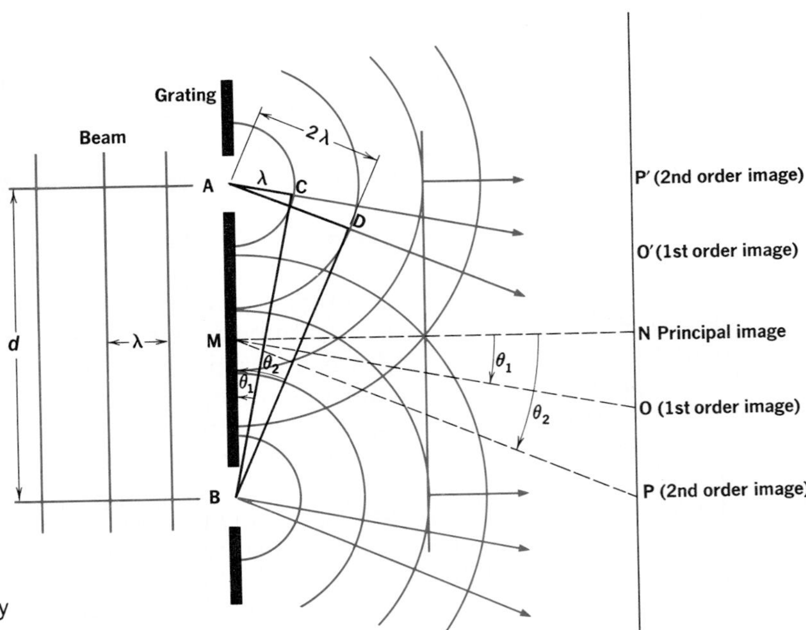

Figure 15-7. The optical geometry of a transmission grating.

A given wavelet from **B** and the first preceding wavelet from **A** produce a wave front **CB** that travels along **MO** and gives a *first-order* image of the slit at **O.** Similarly, a given wavelet from **B** and the second preceding wavelet from **A** give a wave front **DB** that yields a *second-order* image at **P,** etc. Of course, corresponding images appear at **O′, P′,** etc., on the other side of **N.**

In the right triangle **ABC,** side **AC** equals the wavelength λ of the incident light and angle θ_1 is the angle of the first-order diffracted wave front from the grating plane, the **diffraction angle.**

It is evident that

$$\boldsymbol{\lambda = d \sin \theta_1}$$

Side **AD** of triangle **ABD** is equal to 2λ, and θ_2 is the second-order diffraction angle. Thus for second-order images,

$$\boldsymbol{\lambda = \frac{d \sin \theta_2}{2}}$$

In the general case, for any order n, the grating equation becomes

$$\lambda = \frac{d \sin \theta_n}{n}$$

The diffraction angle θ can be determined experimentally. Knowing the order of image n observed, the diffraction angle θ, and the grating constant d, the wavelength of the light can be calculated from this equation. Of course, if the grating constant of a particular diffraction grating is not known, it can be calculated from an experimental determination of θ_n in which a monochromatic light of known wavelength is used. In the example, the grating equation is used to determine the wavelength of a monochromatic light from experimental data.

Spectra produced by diffraction depend on differences in wavelength; spectra produced by prisms rely on differences in refractive index (Section 14.15).

EXAMPLE A diffraction grating is ruled with 6.50×10^3 lines per centimeter. The grating produces a second-order image of a monochromatic light source at a diffraction angle of 55.0°. Calculate the wavelength of the light source in nanometers.

Given	Unknown	Basic equations
$d = 1/(6.50 \times 10^3)$ cm $n = 2$ $\theta = 55.0°$	λ	$\lambda = \dfrac{d \sin \theta_n}{n}$

Solution

Working equation: $\lambda = \dfrac{d \sin \theta_n}{n} = \dfrac{(\sin 55.0°)\text{ cm}}{(6.50 \times 10^3)(2)}(10^7 \text{ nm/cm})$

$= 6.30 \times 10^2$ nm

PRACTICE PROBLEMS 1. A diffraction grating produces a second-order image at a diffraction angle of 32.0° when used with a 500-nm monochromatic light source. Calculate the grating constant of the diffraction grating. *Ans.* 1.89×10^{-6} m

2. A source of monochromatic light shines through a slit and is viewed through a transmission grating 1.00 m from the slit. A first-order image appears 43.0 cm from the slit on each side. The grating constant is 1.67×10^{-4} cm. What is the color of the light? *Ans.* Red

15.5 Single-Slit Diffraction According to Huygens' principle, every point on an advancing wave front can be regarded as a new source of disturbance from which secondary waves spread out as spherical wavelets. If a barrier with a narrow slit opening is placed in the path of advancing plane waves, the disturbance will be transmitted by the slit to the region beyond the barrier. See Figure 15-8(A). When the width of the slit opening is reduced to only a few wavelengths of the light, a broad region **MN** is illuminated by the slit. Experiments show that the central portion of this region is always brighter than the remote portions that reveal diffraction fringes of diminishing intensity. Examples of single-slit diffraction patterns are shown in **Plate VIII(B).**

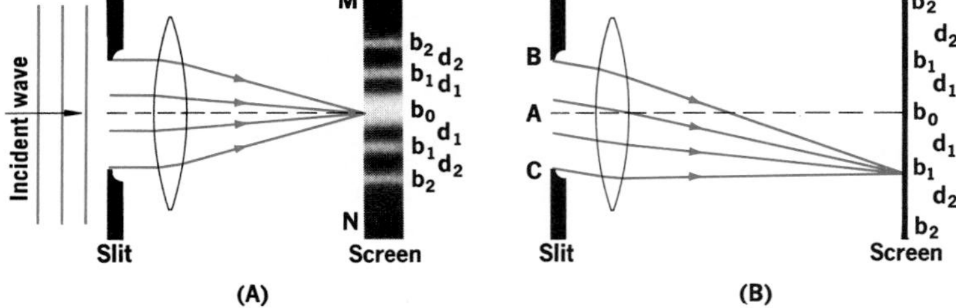

Figure 15-8. Single-slit diffraction. Slit size is exaggerated relative to the lines and screen.

In Figure 15-8(B) the point $\mathbf{b_0}$ lies on the perpendicular bisector of the slit and is equidistant from points **B** and **C.** Because the distance from the slit to the screen is very large compared with the width of the slit, point $\mathbf{b_0}$ is essentially equidistant from all points along the line **BAC.** Thus the wavelets originating simultaneously from all points along **BA** and from all points along **AC** will reach $\mathbf{b_0}$ in phase and the screen in this region will be bright.

At points $\mathbf{d_1}$, above and below $\mathbf{b_0}$, where the distance from **B** and **C** is different by a whole wavelength, the distance from **A** and **C** is different by a half wavelength. For every point along **CA** there is a corresponding point along **BA** that is a half wavelength different in distance from $\mathbf{d_1}$.

Therefore wavelets arriving at $\mathbf{d_1}$ from one half of the slit will be annulled by the wavelets arriving from the other half of the slit. The region of $\mathbf{d_1}$ will be dark.

Beyond points $\mathbf{d_1}$ there are points $\mathbf{b_1}$ where the distance from **B** and **C** is different by 1.5 wavelengths. Now we can think of the slit opening as being divided into three equal parts. Wavelets from one part arrive at $\mathbf{b_1}$ a half wavelength behind wavelets from the second part. Wavelets from these parts of the slit opening interfere destructively, while wavelets arriving from the third part produce brightness at $\mathbf{b_1}$. However, the illumination at $\mathbf{b_1}$ will be lower than that at $\mathbf{b_0}$ since only one-third of the slit opening contributes to the brightness of these regions.

By the same reasoning the regions of $\mathbf{d_2}$, where the difference in distance from **B** and **C** is 2 wavelengths, will be dark. Similarly, the regions of $\mathbf{b_2}$ will be bright. Thus a series of alternate bright and dark regions appears on either side of $\mathbf{b_0}$. The intensity of these regions decreases as the distance from $\mathbf{b_0}$ increases.

QUESTIONS: GROUP A

1. Interference in sound is recognized by differences in the loudness of the sound; how is interference in light recognized?
2. What will be the result if two light waves interfere (a) constructively? (b) destructively?
3. For constructive interference to occur in thin films, how must the wavelength of the light be related to the thickness of the film?
4. Which theory about the nature of light is able to explain the properties of diffraction and interference?
5. Why is diffraction more readily observed for sound waves than light waves?
6. Which properties of light account for the formation of colors in soap bubbles?
7. Which part of Young's double-slit experiment depended on diffraction and which part depended on interference?
8. How does the width of the central region of a single-slit diffraction pattern change as the wavelength of the light increases?
9. Make a generalization about the indices of refraction of two materials and what happens to the phase of light waves reflected at their interface.
10. Why does light diffract and show interference patterns when coming through a single slit as well as through a double slit?
11. Why is polychromatic light separated into a whole spectrum of colors when it is passed through a diffraction grating?
12. What data would you need to collect to be able to correctly calculate the wavelength of light in a diffraction experiment?
13. Radio waves have very long wavelengths. Are they diffracted more or less than visible light?

PROBLEMS: GROUP A

1. A transmission grating with 5800 lines/cm is illuminated by monochromatic light with a wavelength of 4920 Å. What is the diffraction angle for the first-order image?
2. In Problem 1, the perpendicular distance from the grating to the image screen was 34.5 cm. How far from the principal image was the first-order image found?
3. Monochromatic light illuminates a grating having 5900 lines/cm. The diffraction angle for a second-order image is 38.0°. (a) What is the wavelength of the light in angstroms? (b) What is its color?
4. White light falls on a grating that has 3500 lines/cm. The perpendicular distance from the grating to the image screen is 50.0 cm. (a) Find the distance of the near edge of a first-order image from the principal image on the screen. (b) Find the distance of the far edge of the same image.
5. In the experiment of Problem 4, a marker was placed in the first-order image on the screen 9.25 cm from the principal image. (a) Determine the wavelength (in angstroms) at the marker position. (b) What color corresponds to this wavelength?

POLARIZATION

15.6 Polarization of Transverse Waves The general wave properties of rectilinear propagation in a homogeneous medium—reflection, refraction, interference, and diffraction—are clearly recognized in the behavior of light as well as in sound and water disturbances.

In Chapter 12 we considered a particle model of light that at one time was as successful as the wave model in explaining rectilinear propagation, reflection, and refraction. Only interference and diffraction failed to accommodate the particle model. They required a wave model for explanation. Thus interference and diffraction give the best evidence that light has wavelike characteristics.

Before the introduction of the electromagnetic theory, light was assumed to be a longitudinal wave disturbance. Electromagnetic theory predicts that light is a transverse wave. Sound waves are longitudinal disturbances and water waves have both transverse and longitudinal characteristics. From interference or diffraction experiments, we can infer nothing concerning the transverse nature of light waves since both sound and water waves show these properties. What experimental basis do we have that supports the theoretical prediction that light waves are transverse?

Fresnel observed that a beam of light falling on a calcite crystal was separated into two beams that were incapable of producing interference fringes. Young suggested that this could be explained by assuming that the light consisted of transverse waves that were separated into component waves having oscillating planes at right angles to each other. He called this *a plane-polarization effect.*

Many experiments with calcite and other similar materials have demonstrated the correctness of Young's polarization hypothesis. ***Plane-polarized light*** *is light in which the oscillations are confined to a single plane that includes the line of propagation.* Transverse waves but not longitudinal waves can be polarized. This is because the oscillations of transverse waves are independent and randomly oriented. Longitudinal waves vibrate along only one line of direction, and that direction is parallel to its propagation.

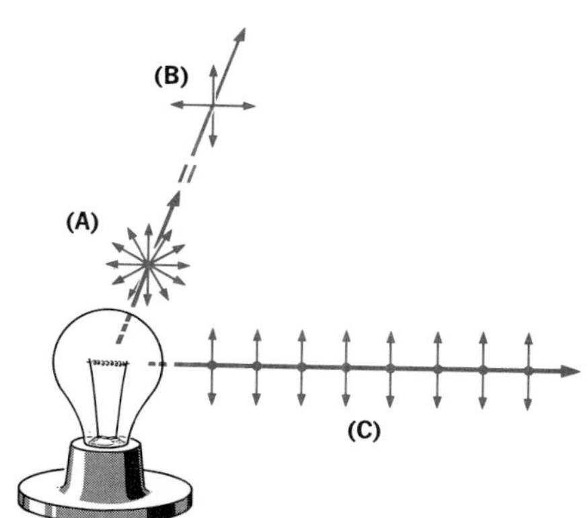

Figure 15-9. The vectors representing random oscillation planes of unpolarized light (A) may be arbitrarily resolved into vertical and horizontal component vectors as in (B) for an end-on view and as in (C) for a side view.

Light radiated by ordinary sources is unpolarized since the primary radiators, the atoms and molecules of the light source, oscillate independently. Beams of unpolarized light are made up of independent wave trains with oscillation planes oriented in a random manner about the line of propagation. Diagrammatically, we can represent unpolarized light being propagated in space by a system of *light vectors* as shown in Figure 15-9(A). It is customary to resolve these vectors into vertical and horizontal components as at (B), which is a convenient but entirely arbitrary orientation. Figure 15-9(C) represents a side view of (B) in which the horizontal component vectors are perpendicular to the page. It is customary for the light vectors of Figure 15-9 to represent the oscillating *electric components* of the electromagnetic (light) waves.

Review Section 12.4.

Ordinary light can become plane-polarized through interactions with matter. The scattering effect of small particles (see Section 15.11) is accompanied by a polarization effect. Polarization can also result from reflection of light from various surfaces, from refraction of light through some crystals, and from selective absorption of light in some crystals.

A simple mechanical model can be used to illustrate the polarization concept. Suppose we set up transverse waves in a rope passed through a slot as shown in Figure 15-10. We can readily see that vibrations are transmitted beyond the slot only when the vibrating plane of the rope and the plane of the slot are aligned. We shall call this frame of slots the *polarizer.* A second slot, parallel to the first, will transmit the waves. See Figure 15-10(A). If the second slot is perpendicular to the first slot, it obstructs them. See

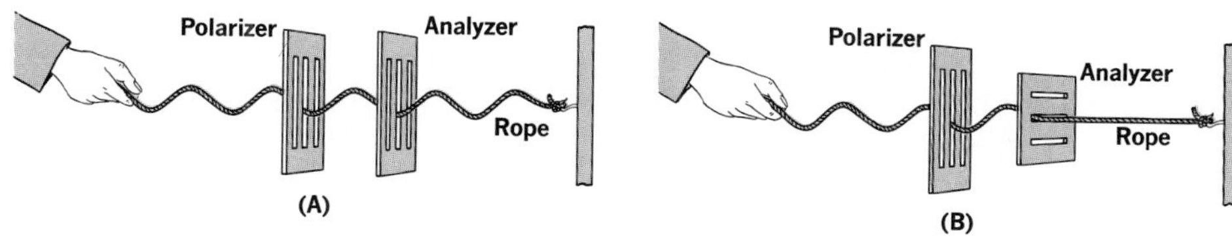

Figure 15-10. A mechanical analogy of polarization.

Polaroid sunglasses are perhaps the most familiar optical polarizing device.

Figure 15-10(B). We shall call this second frame of slots the *analyzer*.

If the rope is replaced by a long coiled spring, longitudinal waves set up in the spring will pass through both slots regardless of their orientation. *Polarization is a property of transverse waves.*

15.7 Selective Absorption It is known that certain crystalline substances transmit light in one plane of polarization and absorb light in other polarization planes. Tourmaline is such a material. Unpolarized light incident on a tourmaline crystal emerges as green, plane-polarized light of low intensity. This property of crystals in which one polarized component of incident light is absorbed and the other is transmitted is called **dichroism.** See Figure 15-11.

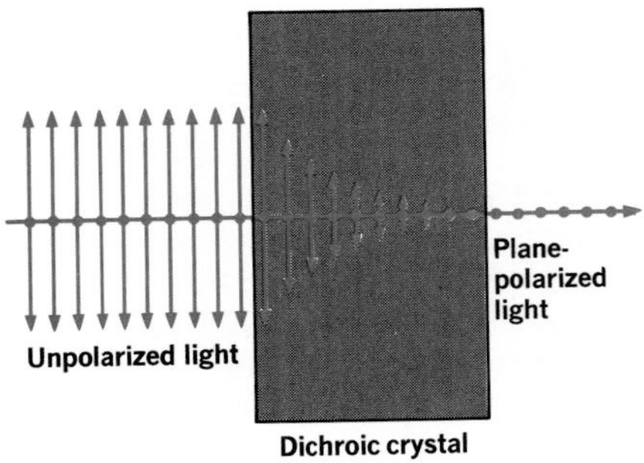

Figure 15-11. Polarization by selective absorption.

Dichroic crystals of quinine iodosulfate transmit plane-polarized light very efficiently but the crystals are too small for practical use. In 1935 Edwin H. Land developed a method of imbedding these crystals in cellulose film so that the dichroic properties of the crystals were retained. This highly efficient polarizing film is known commercially as Polaroid. Improved polarizing sheets have now been developed in which polarizing molecules, rather than crystals, are imbedded in appropriate films.

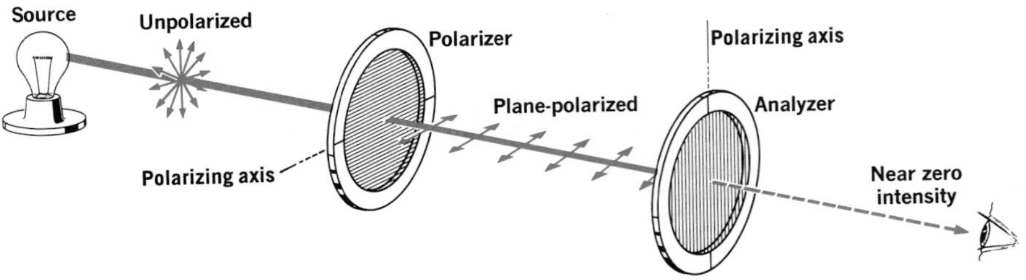

Figure 15-12. Polarizing disks in use. One acts as a polarizer and the other as an analyzer. Notice that the polarizing axis of each disk is marked on the rim.

15.8 Polarization by Reflection Sunlight reflected from the surface of calm water or from a level highway can be quite objectionable to the observer. Sunglasses made of polarizing films reduce the intensity of these reflections. By rotating the lenses slightly from side to side, the reflections are seen to pass through a minimum, indicating that these rays are partially polarized.

Polarization by reflection does not occur for metals and other good conductors. Polarizing glasses can reduce glare from painted surfaces of automobiles, but not from chrome-plated bumpers.

Ordinary light incident obliquely on the surface of a glass plate is partly reflected and partly refracted. Both the transmitted and the reflected beams are partly polarized. This can be verified by observing the light through a polarizing disk used as an analyzer. By rotating the disk in a plane perpendicular to the light axis, one can determine the direction of the dominant light vectors in each beam.

The component of the incident light lying in the plane parallel to the surface of the glass is largely reflected. The component lying in the plane perpendicular to the surface is largely refracted. A particular angle of incidence at which polarization of the reflected light is complete, known as the *polarization angle,* can be found experimentally. Polarization by reflection is shown in Figure 15-13.

At the polarizing angle the reflected beam is of low intensity and the reflectance is about 15%. The refracted beam, which is not completely polarized, is bright. The intensity of the plane-polarized reflected beam can be increased by combining reflections as a result of stacking several plates. The combined refracted beam becomes less intense but more completely plane-polarized.

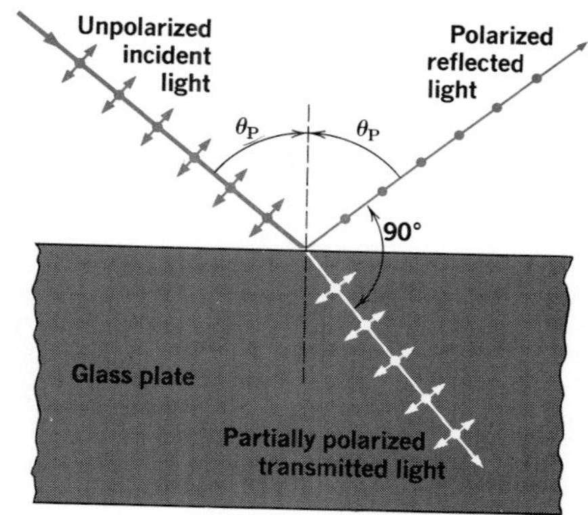

Figure 15-13. At the polarizing angle the reflected light is completely polarized but of low intensity.

15.9 Polarization by Refraction If a thick glass plate is placed on a printed page, the print viewed through the glass may appear displaced because of refraction. A natural crystal of calcite placed on the page shows *two* refracted images of the print, as shown in Figure 15-14. Calcite and many other crystalline materials exhibit this property of *double refraction.*

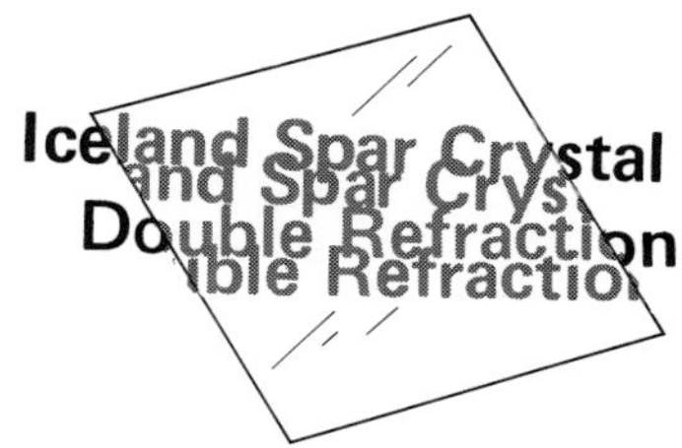

Figure 15-14. Double refraction in calcite.

Upon entering a doubly refracting crystal such as calcite, a beam of unpolarized light can divide into two beams at the crystal surface. This separation is shown diagrammatically in Figure 15-15. Analysis of these separate beams with polarizing disks reveals that they are plane-polarized with their planes of polarization perpendicular to each other.

Experimentally, one of the polarized beams can be shown to follow Snell's law; the other beam does not. Using monochromatic sodium light of 5893 Å and measuring the angles of incidence and refraction, we find that one beam yields a constant index of refraction of 1.66. However, the other beam shows an index of refraction that varies from 1.49 to 1.66 depending on the angle of incidence. This difference suggests that the light energy is propagated through the crystal at different speeds that are determined by the orientation of the light planes with the crystal lattice.

Calcite crystals are sometimes polished, cut through, and cemented back together in such a way that one of the polarized beams is totally reflected at the cemented face. Such a crystal, known as a **Nicol prism,** can be used to produce a beam of completely polarized light.

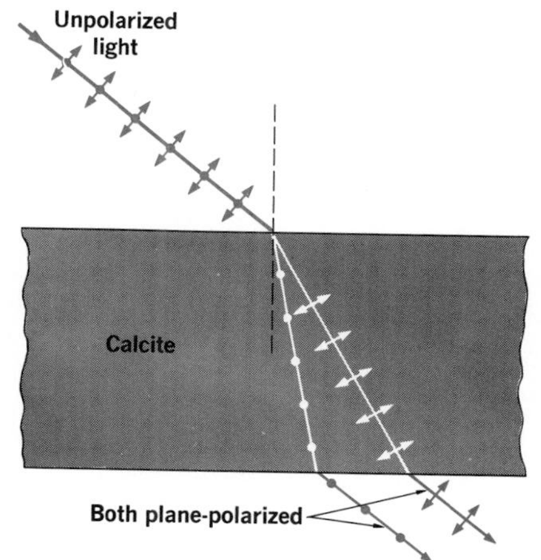

Figure 15-15. Double refraction. Light in one polarization plane conforms to Snell's law; light in the other plane does not.

15.10 Interference Patterns The two plane-polarized beams of light that emerge from a doubly refracting crystal cannot be made to interfere with each other even though they are transmitted through the crystal at different speeds. This is because their planes of polarization are perpendicular to each other. It follows that *if two light waves interfere, their oscillations (or components of these oscillations) must lie in the same plane.*

If *polarized light* is incident on a doubly refracting crystal at the proper angle, the emerging beams *are* found to interfere as they pass through an analyzer disk. The analyzer passes only the components of these perpendicularly polarized rays that lie in its transmission plane. Those waves from the two beams having a phase difference of an odd number of half wavelengths interfere destructively, the corresponding color is removed, and its complement is observed.

Materials, such as glass and Lucite, that become doubly refracting when subjected to mechanical stress are said to be *photoelastic*. When a photoelastic material is placed between polarizing and analyzing disks, the strain patterns (and thus the stress distributions) are revealed by interference fringes, as shown in Figure 15-16. See **Plate V(A).**

Figure 15-16. Strain patterns in a cylindrical disk subjected to diametrical compression.

15.11 Scattering A beam of light traveling in dust-free air cannot be seen even in a darkened room. If, however, there is dust in the path of the beam, it becomes visible by reflection of light from the surface of the particles.

Suppose a beam of white light travels in a medium containing suspended particles that have diameters smaller than the mean wavelength of visible light. When observed at right angles to its path, the beam has a bluish cast.

When observed in its path, the remaining transmitted beam has a red-orange cast. With certain concentrations of suspended particles, the bluish light sent off radially becomes quite intense and the transmitted light acquires a more intense red-orange color. An excess of shorter wavelengths is emitted at right angles to the path of the beam and an excess of longer wavelengths is transmitted along the path of the beam. This phenomenon, known as *scattering,* occurs when a beam of light encounters suspended particles with dimensions that are small compared to the wavelengths of the light.

If we look at blue sky light through a polarizing disk, we observe that it is plane polarized. By selecting a point in the blue sky near the horizon with the sun overhead, we find the scattered light to be horizontally polarized.

In the late afternoon near the time of sunset, sunlight must travel a maximum distance through the atmosphere to reach an observer. The setting sun has a yellow to red hue. The sunlight reflected to an observer from clouds near the sight line of the observer has the red-orange hue characteristic of the sunset. Much of the energy in the blue region of the sunlight has been removed by the scattering effect of the atmospheric gas molecules. Under these conditions the transmitted light is largely red-orange. This accounts for the blue appearance of the sky and the reddish appearance of the setting sun. See **Plate V(B).**

Atmospheric scattering accounts for our colorful sunsets and our blue sky.

Figure 15-17. Scattering of light by particles in the air brightens the daytime sky on earth (left). An observer on the lunar surface, however, sees a dark daytime sky (right) because the moon has almost no atmospheric particles.

If the earth had no envelope of atmospheric gases, the sky would appear black to an observer on the surface of the earth since there would be no scattering of sunlight passing overhead. Astronauts have observed the blackness of outer space while in flight beyond the earth's atmosphere.

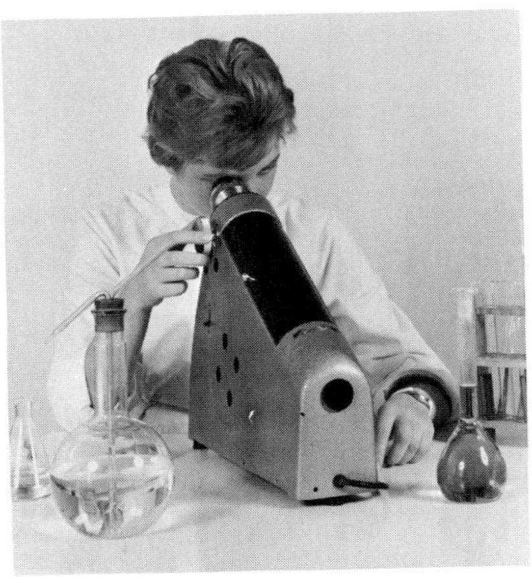

Figure 15-18. A modern polarimeter in use.

Scattering is responsible for certain other natural color phenomena known as **structural colors** (in contrast to the pigment colors that we have already considered). The colors of many minerals are structural. The blue color of a bluejay's feathers results from scattering of light by tiny bubbles of air dispersed through the feather structure. The blue color in the eyes of a person is also a structural color; there is no blue pigment in the irises of blue eyes.

15.12 Optical Rotation A number of substances, such as quartz, sugar, tartaric acid, and turpentine, are said to be **optically active** because they rotate the plane of polarized light. A water solution of cane sugar (sucrose) rotates the plane of polarization to the right. For a given path length through the solution, the angle of rotation is proportional to the concentration of the solution.

In analytical procedures chemists find numerous applications of this property of optically active substances. Instruments for measuring the angle of rotation under standard conditions are known as **polarimeters;** those used specifically for sugar solutions are called **saccharimeters.**

QUESTIONS: GROUP A

1. Which property of electromagnetic waves makes them transverse?
2. What is the phenomenon of "scattering"?
3. What are three different ways in which light can be polarized?
4. What is a structural color?
5. What is a dichroic crystal?
6. (a) Explain why a polarizing disk used as an analyzer blocks the beam of light from a first polarizing disk when it is properly oriented. (b) What is its orientation?

GROUP B

7. Why do the irises of an eye have different colors at different angles?
8. How should a manufacturer orient the polarizing molecules in his sunglasses to provide the greatest glare reduction?
9. Why are two images seen through calcite?
10. A civil engineer may make a mockup of a new structure in Lucite and then use its property of photoelasticity to run some tests on it. (a) How would she do this? (b) What is she hoping to find?
11. (a) Why is the daytime sky blue? (b) Why is the daytime sky black when viewed from moon? (c) Why does the sky near the earth's horizon look red or orange sometimes at sunset?
12. Since the angle of rotation of polarized light is proportional to the concentration of an optically active solution, describe a possible commercial application of this optical activity.

PHYSICS ACTIVITY

Pop the lenses out of an old pair of polarizing sunglasses to see what happens to the intensity of light. Look through one of the lenses at the sunlight reflected from a body of water. Rotate the lens. Explain what is happening.

SUMMARY

Interference of light can be compared with interference in other wave disturbances. It is interpreted in terms of the superposition principle. Thomas Young first demonstrated interference of light and related double-slit interference and thin-film interference to Huygens' wave theory.

The analysis of thin-film interference requires an understanding of the nature of wave reflections at interfaces where the medium beyond has the higher or the lower index of refraction.

The various diffraction phenomena are explained by the interference of light waves. Diffraction gratings produce diffraction patterns by the reflection or transmission of incident light. The wavelength of light can be determined experimentally from measurements of grating spectra.

Polarization of light depends on the transverse nature of light waves. Polarization is a property of transverse waves. Light may be plane polarized by the selective absorption of certain crystals, by reflection from certain smooth surfaces, and by doubly refracting crystals. Photoelastic substances become doubly refracting when subjected to mechanical stress.

The scattering of sunlight accounts for the blue appearance of sky light and the reddish appearance of sunset and sunrise. Substances that rotate the plane of polarized light are said to be optically active.

VOCABULARY

constructive interference
destructive interference
dichroism
diffraction
diffraction angle
diffraction grating
first-order image
grating constant
interference (light)
interferometer
Nicol prism
optically active substance
plane-polarized light
polarimeter
saccharimeter
scattering (light)
second-order image
structural color

16

Electrostatics

electricity (e-lek-TRI-si-tee) n.: the study of phenomena involving charged objects at rest (electrostatics) or in motion (electrodynamics).

OBJECTIVES

- Describe the two kinds of electrostatic charge and how each is transferred.
- Define Coulomb's law.
- Describe potential difference.
- Show electric-field mapping, along with a quantitative discussion.
- Describe the properties of conductors and nonconductors.
- Show how capacitors store electric charge.
- Describe the relationship between potential difference, capacitance, and amount of charge.

ELECTRIC CHARGE

16.1 Charges at Rest We can observe about us a number of physical effects that are sometimes produced by rubbing pieces of dry matter. Sometimes an annoying shock is felt when the door handle of an automobile is touched after one slides over the plastic-covered seat. We may feel a shock after we walk on a woolen carpet and then touch a doorknob or other metal object. The slight crackling sound heard when dry hair is brushed and the tendency of thin sheets of paper to resist separation are other common observations of these physical effects.

When an object shows effects of the type we have described, we say that it has an *electric charge. The process that produces electric charges on an object is called* ***electrification.*** Electrification is most apparent when the air is dry. An object that is electrically charged can attract small bits of cork, paper, or other lightweight particles. Because the electric charge is confined to the object and is not moving, it is called an *electrostatic charge.* Thus ***static electricity*** *is stationary electricity in the form of an electric charge at rest.* Static electricity is commonly produced by friction between two surfaces in close contact.

16.2 Two Kinds of Charge We can detect the presence of an electrostatic charge by means of an instrument called an **electroscope.** The simplest kind of electroscope is a small ball of wood pith or Styrofoam suspended by a silk thread. This electroscope is more sensitive if the pith ball is coated with aluminum or graphite. Such an instrument is shown in Figure 16-1.

A pith ball coated with aluminum or graphite is not appreciably more massive, but it can acquire a higher electric charge.

Suppose a hard rubber or Bakelite rod is charged by stroking it with flannel or fur. If the end of the charged rod is then held near a simple electroscope, the pith ball is attracted to the rod.

If the pith ball is allowed to come in contact with the charged rod, it immediately rebounds and is then repelled by the rod. We can reasonably assume that some of the *charge* has been transferred to the pith ball so that both the rod and the ball are now similarly charged.

Now suppose a glass rod is charged by stroking it with silk. If the glass rod is held near the charged electroscope, the pith ball is attracted rather than repelled as it was with the charged rubber rod. These effects of attraction and repulsion can be explained if we assume that there are two kinds of electric charge.

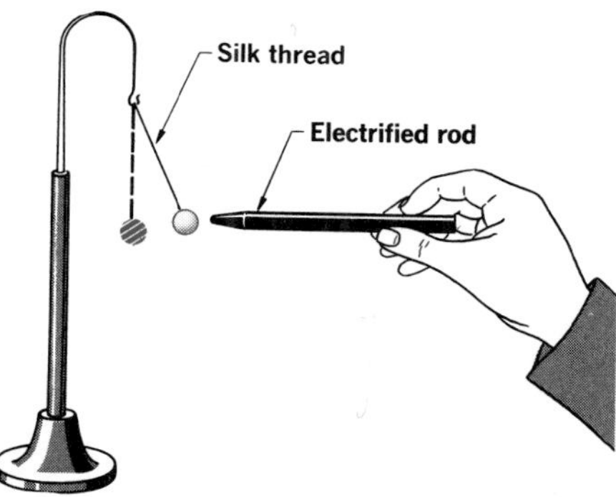

Figure 16-1. The pith-ball electroscope can be used to detect an electrostatic charge.

The electric charge produced on the rubber rod when the rod is stroked with flannel or fur is called a *negative* charge. The rod is said to be charged negatively. The electric charge produced on the glass rod when the rod is stroked with silk is called a *positive* charge. The rod is said to be charged positively.

From a study of atomic structure it is known that all matter contains both positive and negative charges. For simplification, however, the diagrams that follow show only the excess charge. If an object has an excess negative charge, the *net* charge will be indicated by negative (−) signs. If it has an excess positive charge, the *net* charge will be indicated by positive (+) signs. A neutral object will have no sign because it has the same amount of each kind of charge and therefore *a net charge of zero.*

Objects that are electrically neutral have zero net charge.

Two pith balls that are negatively charged by contact with a charged rubber rod repel each other. Similarly, two pith balls that are positively charged by contact with a charged glass rod also repel each other. However, if a negatively charged pith ball is brought near a positively charged pith ball, they attract each other. See Figure 16-2. These observations are summarized in ***a basic law of electrostatics:*** *Objects that are similarly charged repel each other; objects that are oppositely charged attract each other.*

16.3 Electricity and Matter As an understanding of the nature of static electricity requires a knowledge of the basic concepts regarding the structure of matter, we will briefly discuss these concepts now. (These concepts will be discussed more fully in Chapter 23.) All matter is composed of atoms, of which there are many different kinds.

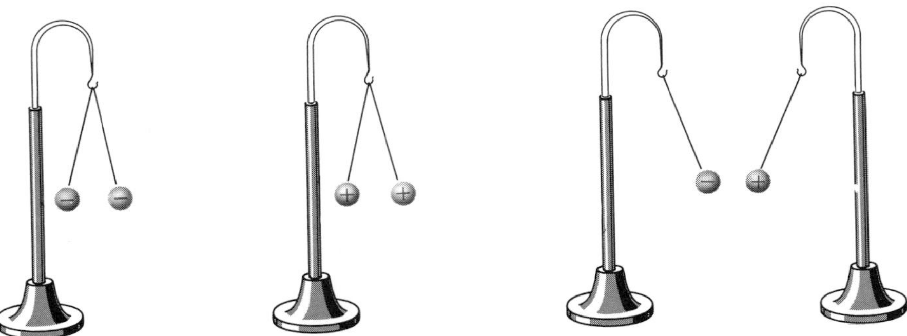

Figure 16-2. Like charges repel each other and unlike charges attract each other.

Each atom consists of a positively charged nucleus surrounded by negatively charged electrons.

Protons and neutrons are tightly packed into the very dense nucleus. Because each proton possesses a single unit of positive electric charge and neutrons, as their name suggests, are neutral particles, the nucleus is positively charged. This charge is determined by the number of protons the nucleus contains.

Objects that are electrically neutral have equal numbers of protons and electrons.

All electrons surrounding the nucleus are alike. They all carry the same amount of negative electric charge—one unit of negative charge per electron. Because an atom is electrically neutral, we know that the nucleus has a positive charge equal in magnitude to the total negative electronic charge. Thus the number of electrons of a neutral atom equals the number of protons.

The rest mass of an electron is $9.109\,4 \times 10^{-31}$ kg. The mass of a proton is $1.672\,6 \times 10^{-27}$ kg and that of a neutron is $1.67\,49 \times 10^{-27}$ kg. Both the proton and the neutron have masses that are nearly 2000 times greater than the mass of an electron. The mass of an atom is almost entirely concentrated in the nucleus.

Protons and neutrons are bound together within the nucleus by strong forces acting through very short distances. By comparison, the repulsions between protons that are due to their similarity of charge are weak forces.

The electrons are retained in the atom structure by the electric attraction exerted by the positive nuclear charge. In general the outermost electrons of higher energies are held less firmly in the atom structure than inner electrons of lower energies. The outer electrons of the atoms of metallic elements in particular are loosely held and are easily influenced by outside forces.

A net negative charge results from electrons being deposited on an isolated neutral body.

When two appropriate materials are in close contact, some of the loosely held electrons can be transferred from one material to the other. If a hard rubber rod is stroked with fur, some electrons can be transferred from the fur to the rod. The rubber rod becomes negatively charged because it has a *net excess of electrons,* and the fur becomes positively charged because it has a *net deficiency of electrons.*

Similarly, when a glass rod is stroked with silk, some electrons are transferred from the glass to the silk. The glass is positively charged because it has a *net deficiency of electrons,* and the silk is negatively charged because it has a *net excess of electrons.* All these charged states result from *the transfer of electrons.*

A net positive charge results from electrons being removed from an isolated neutral body.

Electric charge is a scalar quantity. The net charge on an object is the sum of its positive charges minus the sum of its negative charges.

16.4 The Electroscope Two common types of electroscopes that are more sensitive than the simple pith-ball device are shown in Figure 16-3.

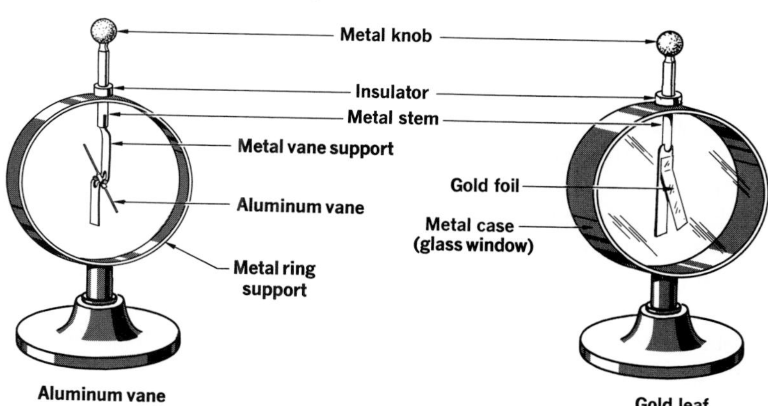

Figure 16-3. Two types of sensitive electroscopes.

The *vane electroscope* consists of a light aluminum rod mounted by means of a central bearing on a metal support that is insulated from its metal stand. When charged, the vane is deflected at an angle by electrostatic repulsion. The angle of deflection depends on the magnitude of the charge.

The *leaf electroscope* consists of very fragile strips of gold leaf suspended from a metal stem that is capped with a metal knob. The leaves are enclosed in a metal case with glass windows for their protection, and the metal stem is insulated from the case. When electrified, the leaves diverge because of the force of repulsion due to their similar charge. Good sensitivity is realized because of the very low mass of the gold leaf.

A **proof plane** is frequently used with an electroscope to test or transfer charges. The proof plane is a small metal disk with an insulating handle. One can easily be made by cementing a small coin to a glass rod. When the proof plane is used to transfer a charge, the metal disk is brought in contact with the charged object and then with an electroscope.

When used to test a charge, the charged disk of the proof plane is brought in contact with an electroscope previously electrified with a known charge. If the charge on the electroscope increases, the test charge is of the same sign as that on the electroscope. If the charge on the electroscope decreases, the test charge is of the opposite sign as that on the electroscope.

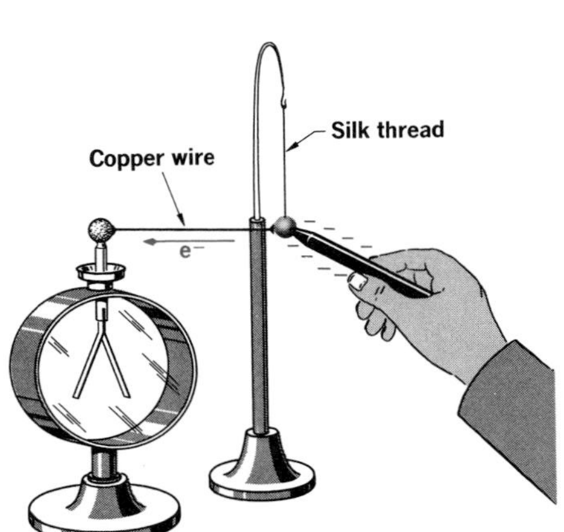

Figure 16-4. A charge is conducted to the electroscope by the copper wire.

16.5 Conductors and Insulators Suppose that an aluminum-coated pith ball is suspended by a silk thread and the ball is connected to the knob of a leaf electroscope by means of a copper wire, as shown in Figure 16-4. When a charge is placed on the pith ball, the leaves of the electroscope diverge. Apparently the charge on the ball is transferred to the electroscope by the copper wire.

Now suppose a silk thread is substituted for the copper wire. When a charge is placed on the pith ball, the leaves of the electroscope do not diverge. The charge has not been conducted to the electroscope by the silk thread.

*A **conductor** is a material through which an electric charge is readily transferred.* Most metals are good conductors. At normal temperatures, silver is the best solid conductor; copper and aluminum follow in that order.

*An **insulator** is a material through which an electric charge is not readily transferred.* Good insulators are such poor conductors that for practical purposes they are considered to be nonconductors. Glass, mica, paraffin, hard rubber, sulfur, silk, dry air, and many plastics are good insulators.

Liquid solutions and confined gases conduct electricity in a different way than solids. At this time we are interested in the conductivity of solids. We can use our knowledge of the structure of matter in describing why some materials are conductors and why some are insulators.

A few grams of matter contain a very large number of atoms and, except for hydrogen, an even larger number of electrons. For example, 27 g of aluminum consists of 6.02×10^{23} aluminum atoms containing 7.83×10^{24} electrons. This amounts to one electron for every 3.45×10^{-24} g, or 2 atomic mass units, of aluminum. (The atomic mass unit is defined in Section 7.3.)

Metals have close-packed crystal structures. The crystal lattice consists of positively charged particles surrounded by a cloud of **free electrons.** This cloud of free electrons is commonly referred to as the **electron gas.** The binding force in such structures is the attraction between the positively charged metal ions and the electron gas. The loosely held outermost electrons of the metal atoms have been "donated" to the electron gas and belong to the crystal as a whole. These electrons are free to migrate throughout the crystal lattice. Their migration gives rise to the high electric conductivity commonly associated with metals.

A good conductor contains a large number of free electrons whose motions are relatively unimpeded within the material. Since like charges repel, the free electrons spread throughout the material in order to relieve any local concentration of charge. If such a material is in contact with a charged body, the free electrons surge in a common direction. If the charged object is deficient in electrons (positively charged), this surge is in the direction of the object. If the charged object has an excess of electrons (negatively charged), the surge is away from the object. See Figure 16-5. In either case a transfer of electric charge continues until the repulsive forces between the free electrons are in equilibrium throughout the entire system.

An insulator is characterized by a lack of free electrons because even the outermost electrons are rather firmly held within the atom structure. Thus the transfer of charge through an insulator is usually negligible. If an excess of electrons is transferred to one particular region of such a material, the extra electrons remain in that region for some time before they gradually leak away.

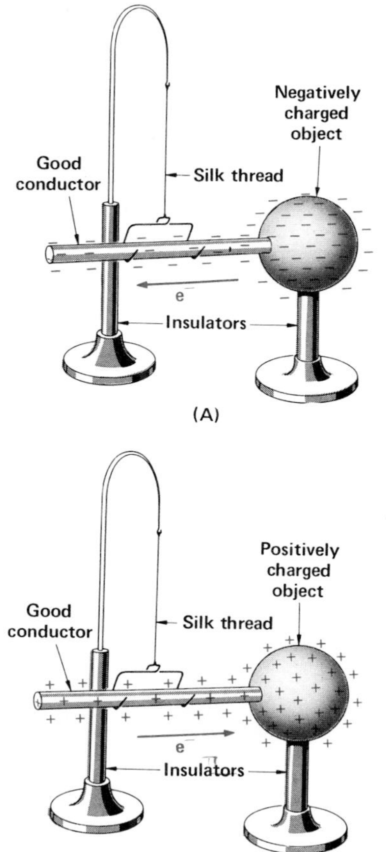

Figure 16-5. Free electrons of a conductor surge in the direction that reduces the net charge.

16.6 Transferring Electrostatic Charges Suppose a rubber rod is charged negatively by stroking it with fur. If the rod is brought near the knob of an electroscope, the leaves diverge. If the rod is removed, the leaves collapse; no charge remains on the electroscope. The charge that makes the leaves diverge is called an **induced *charge.*** The electroscope is said to be charged temporarily by *induction.*

We can reason that the negative charge on the rod when brought near the metal knob of the electroscope repels free electrons in the knob and metal stem and forces them down to the leaves. See Figure 16-6(A). The force of repulsion of the extra electrons on the leaves causes them to diverge. The knob is then deficient in electrons and is positively charged. As soon as the force of repulsion exerted by the charged rod is withdrawn, the excess free electrons on the leaves scatter throughout the stem and knob, restoring the normal uncharged state throughout the electroscope.

Similarly, a glass rod that has been stroked with silk temporarily induces a positive charge on the leaves of an electroscope by attracting electrons up through the stem to the knob. See Figure 16-6(B).

Electrostatic experiments are best performed in dry air.

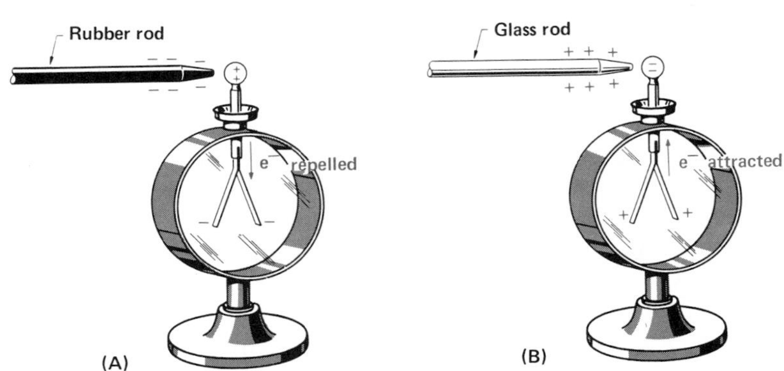

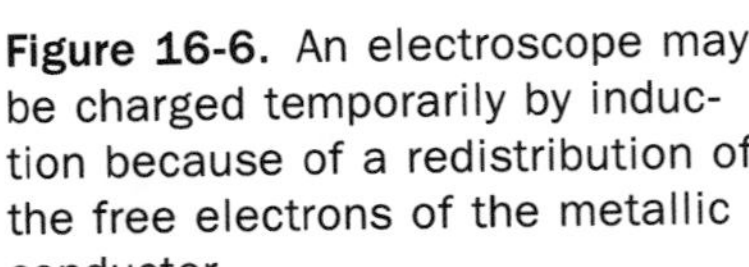

Figure 16-6. An electroscope may be charged temporarily by induction because of a redistribution of the free electrons of the metallic conductor.

In moist air an invisible film of water condenses on the surfaces of objects, including those of charged insulators. Dissolved impurities in the film make these surfaces conductive, and an isolated charge cannot be maintained for any length of time.

Any conducting object, when properly isolated in space, can be temporarily charged by induction. The region of the object nearest the charged body will acquire a charge of the *opposite* sign; the region farthest from the charged body will acquire a charge of the same sign. See Figure 16-7. This statement is consistent with the basic law of electrostatics stated in Section 16.2.

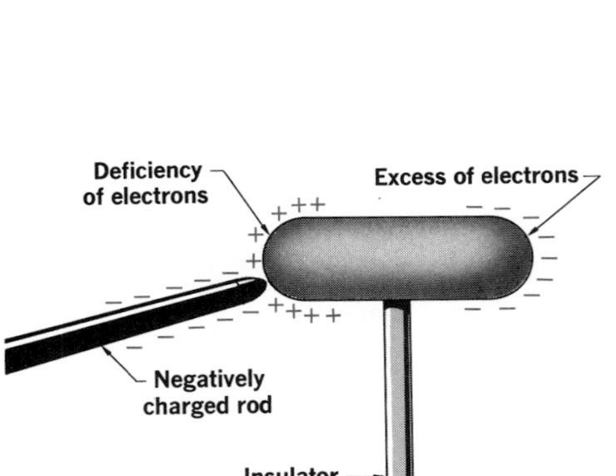

Figure 16-7. A charged rod brought near an isolated conductor induces electric charge of the same sign on the far end of the conductor.

A charge can be transferred by *conduction.* If we touch the knob of an electroscope with a negatively charged rubber rod, the leaves diverge. When the rod is removed, the leaves remain apart, indicating that the electroscope retains the charge. How can we determine the nature of this residual charge on the electroscope?

We can reason that some of the excess electrons on the rod have been repelled onto the knob of the electroscope. The electrons are freed from the rod by the ionization around tiny sparks of electricity that are produced at the end of the rod just before it touches the conducting electroscope. Any free electrons thus transferred to the electroscope, together with other free electrons of the electroscope itself, would be repelled to the leaves by the excess electrons remaining on the parts of the rod not in contact with the electroscope.

When the rod is removed, and with it the force of repulsion, the electroscope is left with a residual negative charge of a somewhat lower density. This deduction can be verified by bringing a positively charged glass rod near the electroscope to induce a positive charge on the leaves. The leaves collapse and diverge again when the glass rod is removed. See Figure 16-8. *Any conducting object, properly isolated in space and charged by conduction, acquires a* **residual charge** *of the same sign as that of the body touching it.*

An electroscope with a known residual charge can be used to identify the nature of the charge on another object. This second object merely needs to be brought near the knob of this charged electroscope.

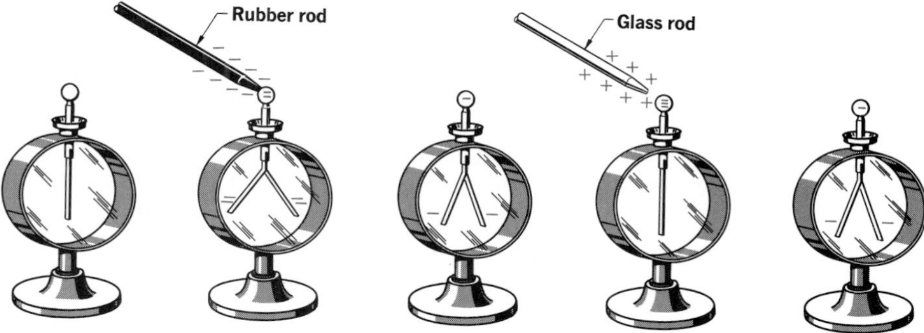

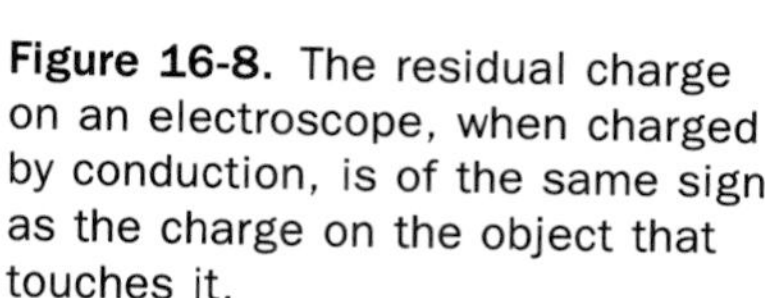
Figure 16-8. The residual charge on an electroscope, when charged by conduction, is of the same sign as the charge on the object that touches it.

16.7 Residual Charge by Induction When a charged rubber rod is held near the knob of an electroscope, there is no transfer of electrons between the rod and the electroscope. If a path is provided for electrons to be repelled from the electroscope while the repelling force is present, free electrons escape. Then if the escape path is removed before removing the repelling force, the electroscope is left with a deficiency of electrons, giving it a residual positive charge. We can verify this conclusion by bringing a positively charged glass rod near the knob of the charged electroscope. The leaves of the electroscope diverge even more. When the charged rod is withdrawn, the leaves fall back to their original divergence and remain apart. The steps in placing a residual charge on an electroscope by induction are shown in Figure 16-9.

Figure 16-9. Steps in placing a residual charge on an electroscope by induction.

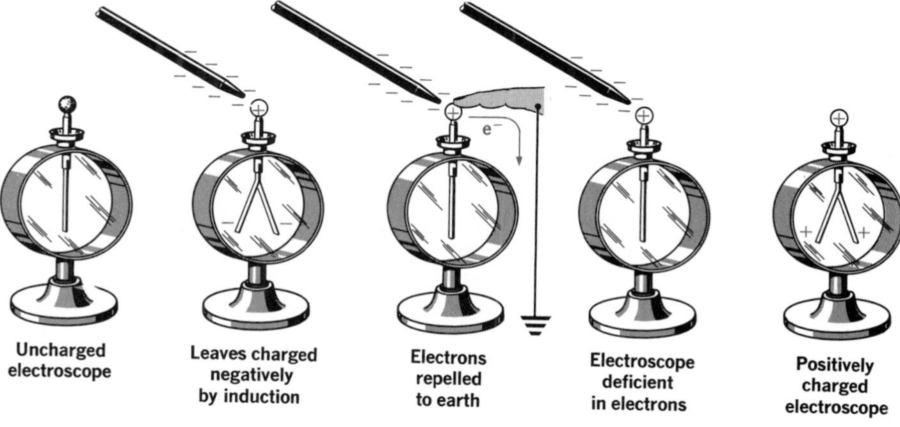

Similarly, we can induce a residual negative charge on an electroscope using a positively charged glass rod. *When an isolated conductor is given a residual charge by induction, the charge is opposite in sign to that of the object inducing it.*

16.8 The Force Between Charges From the basic law of electrostatics stated in Section 16.2, we recognize that like charges repel and unlike charges attract. If a charge is uniformly dispersed over the surface of an isolated sphere, its influence on another charged object some distance away is the same as if the charge were concentrated at the center of the sphere. Thus the charge on such an object is considered to be located at a particular point and is called a **point charge.**

The quantity of charge on a body, represented by the letter Q, is determined by the number of electrons in excess of (or less than) the number of protons. The quantity of charge is measured in ***coulombs*** (C), named for the French physicist Charles Augustin de Coulomb (1736–1806).

$$\textbf{1 coulomb = the charge on } 6.25 \times 10^{18} \textbf{ electrons}$$

Thus the charge on one electron, expressed in coulombs, is the reciprocal of this number and the sign of Q is $-$.

$$e^- = 1.60 \times 10^{-19}\ \mathrm{C}$$

Similarly, the charge on one proton is 1.60×10^{-19} coulomb and the sign of Q is $+$.

The coulomb is a very large unit of charge for the study of electrostatics. Frequently it is convenient to work with a fraction of this unit called the *microcoulomb* (μC).

$$1\ \mu\mathrm{C} = 10^{-6}\ \mathrm{C}$$

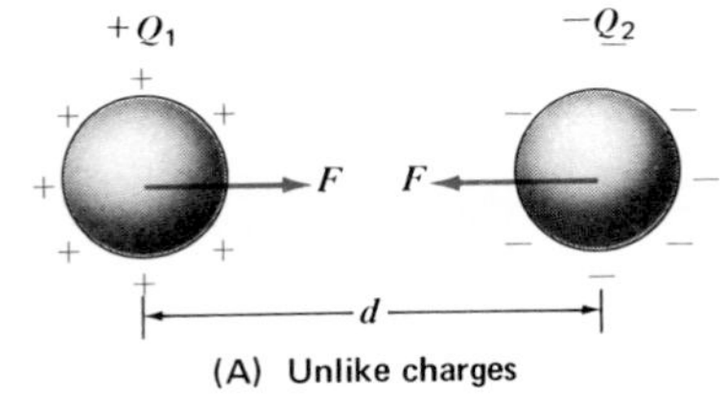

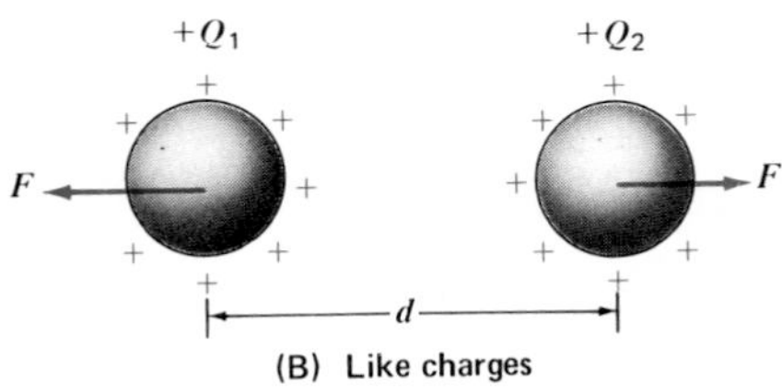

Figure 16-10. When Q_1 and Q_2 are of opposite sign, F is negative to indicate it is a force of attraction. When Q_1 and Q_2 are of the same sign, F is positive to indicate it is a force of repulsion.

Coulomb's many experiments with charged bodies led him to conclude that the forces of electrostatic attraction and repulsion obey a law similar to Newton's law of universal gravitation. We now recognize his conclusions as ***Coulomb's law of electrostatics:*** *The force between two point charges is directly proportional to the product of their magnitudes and inversely proportional to the square of the distance between them.* Charged bodies approximate point charges if they are small compared to the distances separating them. See Figure 16-10.

Coulomb's law can be expressed as

$$F \propto \frac{Q_1 Q_2}{d^2}$$

Note that Coulomb's law of electrostatics has the same form as Newton's law of universal gravitation, Section 3.11.

If we use a proportionality constant that takes into account the properties of the medium separating the charged bodies and that has the proper dimensions, Coulomb's law becomes

$$F = k\frac{Q_1 Q_2}{d^2}$$

The proportionality constant k has the numerical value 8.987×10^9 for vacuum and 8.93×10^9 for air. The dimensions of k are $N{\cdot}m^2/C^2$. The point charges Q_1 and Q_2 are in coulombs and are of proper sign to indicate the nature of each charge. As d is distance in meters, F is expressed in newtons of force.

In Figure 16-10(A) the charges Q_1 and Q_2 have opposite signs and the force F acts on each charge to move it toward the other. In Figure 16-10(B), charges Q_1 and Q_2 have the same sign and the force F acts on each charge to move it away from the other. The force between the two charges is a vector quantity that acts on each charge. The two charges will accelerate toward or away from each other.

Suppose the objects in Figure 16-10(B) are charged to 0.01 coulomb each and placed 10 meters apart. Since the charges are of like sign, the force between them is one of repulsion. Using k for air (to one significant figure it is $9 \times 10^9\ N{\cdot}m^2/C^2$), the Coulomb's law expression for this force becomes

$$F = k\frac{(+Q)(+Q)}{d^2}$$

$$F = 9 \times 10^9\ \frac{\mathbf{N{\cdot}m^2}}{\mathbf{C^2}} \times \frac{(10^{-2}\ \mathbf{C})(10^{-2}\ \mathbf{C})}{(10\ \mathbf{m})^2}$$

$$F = 9 \times 10^9\ \frac{\mathbf{N{\cdot}m^2}}{\mathbf{C^2}} \times \frac{10^{-4}\ \mathbf{C^2}}{10^2\ \mathbf{m^2}}$$

$$F = 9 \times 10^3\ \textbf{N of repulsive force}$$

EXAMPLE Two electrostatic charges of 60.0 μC and 50.0 μC exert a repulsive force on each other of 175 N in air. Calculate the distance between the two charges.

Given	Unknown	Basic equations
$Q_1 = 60.0\ \mu C$ $Q_2 = 50.0\ \mu C$	d	$F = k\dfrac{Q_1Q_2}{d^2}$

Solution

$$d = \sqrt{\left(k\frac{Q_1Q_2}{F}\right)}$$

$$= \sqrt{\frac{(8.93 \times 10^9\ \text{N}\cdot\text{m}^2/\text{C}^2)(60.0 \times 10^{-6}\ \text{C})(50.0 \times 10^{-6}\ \text{C})}{175\ \text{N}}}$$

$$= 0.391\ \text{m}$$

PRACTICE PROBLEM An electrostatic charge of 17.0 μC is placed at a distance in air of 15.0 cm from a second charge. The force of attraction between the two charges is 21.4 N. Calculate the magnitude of the second charge. *Ans.* $-3.17\ \mu$C

The idea of a field of force was introduced by Faraday.

16.9 Electric Fields The concept of a field of force will be helpful as we consider the region surrounding an electrically charged body. A second charge brought into this region experiences a force according to Coulomb's law. Such a region is an *electric field. An **electric field** is said to exist in a region of space if an electric charge placed in that region is subject to an electric force.*

Let us consider a positively charged sphere $+Q$ of Figure 16-11(A) isolated in space. A small positive charge $+q$, which we shall call a *test charge,* is brought near the surface of the sphere. Since the test charge is in the electric field of

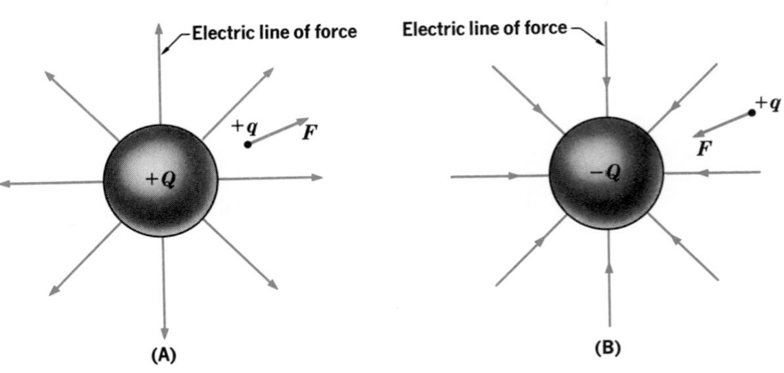

Figure 16-11. The electric field surrounding a charged sphere isolated in space.

the charged sphere and the charges are similar, it experiences a repulsive force directed radially away from $+Q$. Were the charge on the sphere negative, as in Figure 16-11(B), the force acting on the test charge would be directed radially toward $-Q$.

An electric ***line of force*** *is a line so drawn that a tangent to it at any point indicates the orientation of the electric field at that point.* We can imagine a line of force as the path of a test charge moving slowly in a very viscous medium in response to the force of the field. By convention, electric lines of force *originate* at the surface of a positively charged body and *terminate* at the surface of a negatively charged body, each line of force showing the direction in which a positive test charge would be accelerated in that part of the field. A line of force is *normal* to the surface of the charged conducting body where it joins that surface.

The direction of an electric line of force is opposite to that of the flow of free electrons.

The *intensity*, or strength, of an electrostatic field, as well as its direction, can be represented graphically by lines of force. *The electric field intensity is proportional to the number of lines of force per unit area normal to the field.* Where the intensity is high, the lines of force will be close together. Where the intensity is low, the lines of force will be more widely separated in the graphical representation of the field.

In Figure 16-12(A), electric lines of force are used to show the electric field near two equally but oppositely charged objects. At any point in this field the resultant force acting on a test charge $+q$ can be represented by a vector drawn tangent to the line of force at that point.

The electric field near two objects of equal charge of the same sign is shown by the lines of force in Figure 16-12(B). The resultant force acting on a test charge $+q$ placed at the midpoint between these two similar charges would be zero.

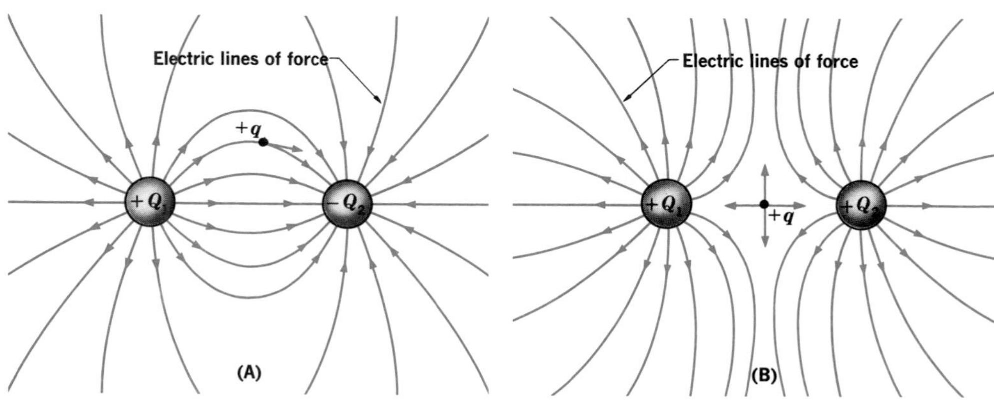

Figure 16-12. Lines of force show the nature of the electric field near two equal charges of opposite sign (A), and near two equal charges of the same sign (B).

The **electric field intensity,** *E, at any point in an electric field is the force per unit positive charge at that point.* The electric field intensity has the dimensions *newton per coulomb.* Thus,

$$E = \frac{F}{q}$$

where E is the electric field intensity and F is the force in newtons acting on the test charge q in coulombs.

The following example will illustrate the use of this equation.

EXAMPLE An electrostatic charge is placed in an electric field that has an intensity of 1.50×10^5 N/C. The charge experiences a force of 2.10 N. Calculate the magnitude of the charge in μC.

Given	Unknown	Basic equations
$E = 1.50 \times 10^5$ N/C $F = 2.10$ N	q	$E = \frac{F}{q}$

Solution

Working equation: $q = \frac{F}{E} = \frac{2.10\text{ N}}{1.50 \times 10^5\text{ N/C}}$

$= 14.0\ \mu\text{C}$

QUESTIONS: GROUP A

1. What is meant by "static" charges?
2. What do we mean when we say an object is "charged." Don't *all* objects have charges?
3. Describe the basic law of electrostatics.
4. (a) Which particle moves when an object acquires a net positive or negative charge? (b) Why does the particle of opposite charge not move?
5. (a) Describe an electroscope. (b) What is its function?
6. How do conductors differ from insulators?
7. Name three different ways to charge an uncharged object.
8. Define the term *electric field.*
9. Describe the results of touching a charged metal sphere to an uncharged, identical sphere.
10. (a) Describe Coulomb's law. (b) Compare it to Newton's law of gravitation.
11. Upon what factors does the strength of an electric field around a charged object depend?

GROUP B

12. Why are items of low mass used to detect charges?
13. Which effect proves more conclusively that an object is charged, its

attraction to or repulsion from another charged object?

14. After you rub a balloon on your sweater, it will stick to the wall. Why?
15. What do we mean when we say an electroscope has a residual charge?
16. What conditions must exist for a conductor to be able to hold an electrostatic charge?
17. If water is a nonconductor, why is it hard to build up a charge on a wet day?
18. (a) Mom forgot the fabric softener. As you pull apart your socks from the dryer, you hear a crackling sound. What happened? (b) Why does it take work to separate the socks?
19. Why should you remain in your car during a lightning storm?
20. All electrons are charged negatively and protons are charged positively. What would happen if suddenly all the polarities were reversed?
21. Why does the repulsive force between protons not destroy the atomic nucleus?
22. Explain how it is possible to give an object a residual negative charge with a positively charged rod.
23. A negatively charged rod is brought near the knob of a charged electroscope. The leaves first collapse and then as the rod is brought nearer they again separate. (a) What is the residual charge on the electroscope? (b) Explain why the leaves act this way.

PROBLEMS: GROUP A

1. A small pith ball carries a charge of $-2.5\ \mu C$. How many excess electrons does this represent?
2. What is the charge on an oxygen ion (oxidation number $= -2$) in coulombs?
3. A pith ball with an excess charge of $+6.0\ \mu C$ is placed 12 cm from another pith ball which carries a charge of $-4.3\ \mu C$. (a) What is the magnitude of the force between them? (b) Is it an attractive force or a repulsive force?
4. If the distance between the pith balls in Problem 3 is decreased to 4.0 cm, what is the magnitude of the force acting on each ball?
5. The pith balls in Problem 3 are allowed to touch and are then separated until they are 5.5 cm apart. (a) What is the magnitude of the force between them? (b) What type of force is it?
6. (a) Calculate the electrostatic force between an electron and a proton that are 1×10^{-10} m apart. (b) Calculate the gravitational force of attraction between them. (c) Compare the forces.
7. The two balls on a pith ball electroscope (as shown in Figure 16-2) are given equal charges. The balls separate so that the angle between them is 20.0°. If the mass of each ball is 0.15 g, what is the force acting on each ball? (Hint: Draw a free-body diagram showing all the forces acting on each ball.)
8. If the silk thread from which each ball in Problem 7 is suspended is 15.0 cm long, (a) how far apart are the balls? (b) what is the magnitude of the charge on each ball?
9. An object with a charge of $12.3\ \mu C$ experiences a force of 6.4 N from an electric field. What is the field's strength at this point?
10. How much electrostatic force would a magnesium ion (oxidation number $+2$) experience if it was placed in the electric field in Problem 9?
11. An electron, moving through an electric field, experiences an acceleration of $6.3 \times 10^{3}\ m/s^2$. (a) How much electrostatic force is acting? (b) What is the strength of the electric field?

POTENTIAL DIFFERENCE

16.10 Electric Potential Let us consider the work done by gravity on a wagon coasting down a hill. The wagon is within the gravitational field of the earth and experiences a gravitational force causing it to travel downhill. Work is done by the gravitational field, so the energy expended comes from within the gravitational system. The wagon has less potential energy at the bottom of the hill than it had at the top; in order to return the wagon to the top, work must be done on it. However, in this instance, the energy must be supplied from an outside source to pull against the gravitational force. The energy expended is stored in the system, imparting to the wagon more potential energy at the top of the hill than it had at the bottom.

Similarly, a charge in an electric field experiences an electric force according to Coulomb's law. If the charge moves in response to this force, work is done by the electric field. Energy is removed from the system. If the charge is moved against the coulomb force of the electric field, work is done on it using energy from some outside source, this energy being stored in the system.

If work is done as a charge moves from one point to another in an electric field or if work is required to move a charge from one point to another, these two points are said to *differ in electric potential. The magnitude of the work is a measure of this difference of potential.* The concept of potential difference is very important in the understanding of electric phenomena. *The* ***potential difference,*** *V, between two points in an electric field is the work done per unit charge as a charge is moved between these points.*

Figure 16-13. A strand of copper wire is completely vaporized as a potential difference of 5 000 000 volts is placed across it.

$$\textbf{potential difference } (V) = \frac{\textbf{work } (W)}{\textbf{charge } (q)}$$

The unit of potential difference is the *volt* (V). *One* ***volt*** *is the potential difference between two points in an electric field such that 1 joule of work is done in moving a charge of 1 coulomb between these points.*

$$1 \text{ volt} = \frac{1 \text{ joule}}{1 \text{ coulomb}}$$

Small differences of potential are commonly expressed in *millivolts* (mV) or *microvolts* (μV). Large differences of potential are measured in *kilovolts* (kV) and *megavolts* (MV).

$$1\ \mu\text{V} = 10^{-6}\ \text{V}$$
$$1\ \text{mV} = 10^{-3}\ \text{V}$$
$$1\ \text{kV} = 10^{3}\ \text{V}$$
$$1\ \text{MV} = 10^{6}\ \text{V}$$

Suppose the potential difference between two points in an electric field is 6.0 V. The work required to move a charge of 3.00×10^2 μC between these points can be determined as follows.

$$V = \frac{W}{q}$$

$$W = Vq = 6.0 \text{ V} \times 3.00 \times 10^2 \text{ } \mu\text{C} \times \text{C}/10^6 \text{ } \mu\text{C}$$
$$W = 6.0 \text{ V} \times 3.00 \times 10^{-4} \text{ C}$$
$$W = 1.8 \times 10^{-3} \text{ J}$$

Since a joule of work is a force of 1 newton applied through a distance of 1 meter, it follows that

$$\text{V} = \frac{\text{J}}{\text{C}} = \frac{\text{N}\cdot\text{m}}{\text{C}}$$

and

$$\frac{\text{V}}{\text{m}} = \frac{\text{N}}{\text{C}}$$

In Section 16.9 it was shown that E, the electric field intensity, is expressed in newtons per coulomb. Thus,

$$E = \frac{\text{N}}{\text{C}} = \frac{\text{V}}{\text{m}}$$

The electric field intensity is commonly expressed in terms of *volts per meter* and can be referred to as the *potential gradient. The* ***potential gradient*** *of an electric field is the change in potential per unit of distance.*

We can consider the earth to be an inexhaustible *source* of electrons, or limitless *sink* into which electrons can be "poured" without changing its potential. For practical purposes, the potential of the earth is arbitrarily taken as *zero.* Any conducting object connected to the earth must be at the same potential as the earth; that is, the potential difference between them is zero. Such an object is said to be *grounded.*

The potential at any point in an electric field is the potential difference between the point and earth taken as zero. This potential can be either positive or negative depending on the nature of the charge producing the electric field.

16.11 Distribution of Charges Michael Faraday performed several experiments to demonstrate the distribution of charge on an isolated object. He charged a conical silk bag like that shown in Figure 16-14 and found that the charge was on the outside of the bag. By pulling on the

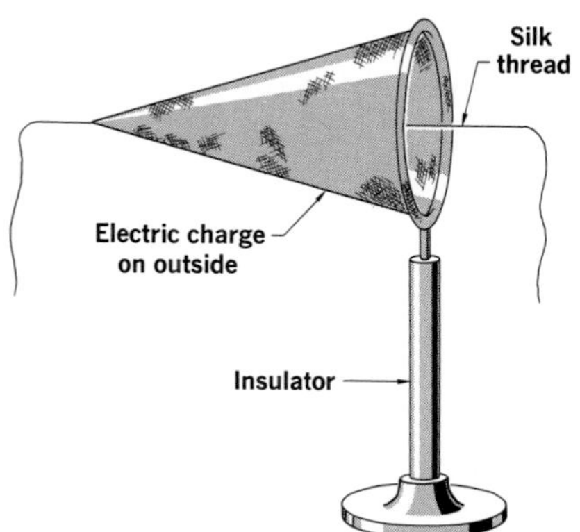

Figure 16-14. The type of conical silk bag used by Faraday to demonstrate that electric charges reside on the outside.

silk thread, he turned the bag inside out and found that the charge was again on the outside. The inside of the bag showed no electric charge in either position.

Faraday connected the outer surface of an insulated metal pail to an electroscope by means of a conducting wire, as shown in Figure 16-15(A). He then lowered a positively charged metal ball supported by a silk thread into the pail. The leaves of the electroscope diverged (B), indicating a charge by induction. The positive charge on the ball attracted free electrons in the pail to the inner surface, leaving the outside of the pail and the electroscope positively charged. When the ball was allowed to come in contact with the inside of the pail (C) and was then removed, the leaves remained apart without change. It was found that the ball no longer had a charge after its removal. The fact that the electroscope remained unchanged after the ball was removed (D) indicated (1) that there was no redistribution of positive charge on the outside surface of the pail and (2) that the outside of the pail (and the electroscope) had acquired a net charge equal to the charge originally placed on the ball.

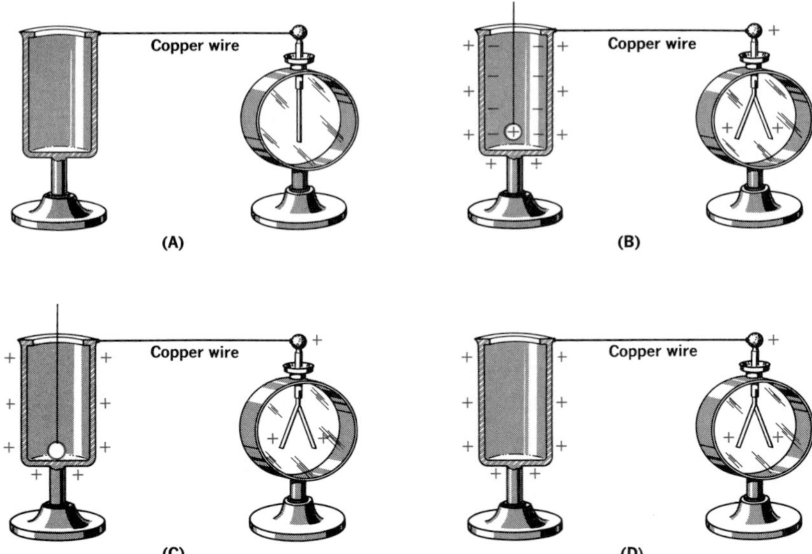

Figure 16-15. Faraday's ice-pail experiment.

From these and other experiments with isolated conductors we can conclude that:

1. *All the static charge on a conductor lies on its surface.* Electrostatic charges are at rest. If the charge were beneath the surface so that an electric field existed within a conductor, free electrons would be acted upon by the coulomb force

of this field. Work would be done, the electrons would move because of a difference of potential, and energy would be given up by the field. Since movement is not consistent with a static charge, the charge must be on the surface and the electric field must exist only externally to the surface of a conductor. See Figure 16-16.

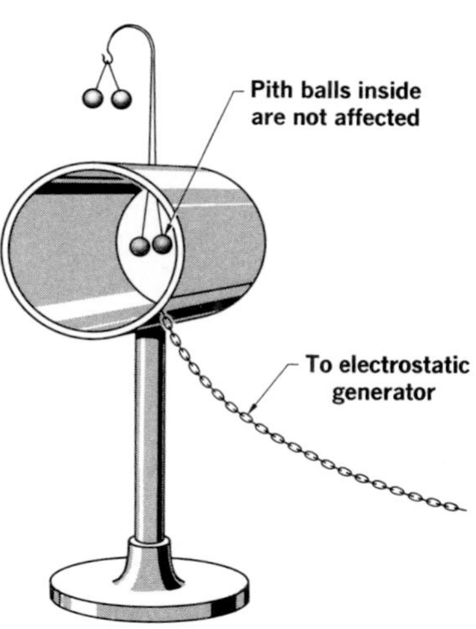

Figure 16-16. When the metal cylinder is charged by an electrostatic generator, the pith balls outside diverge while those inside are not affected.

2. *There can be no potential difference between two points on the surface of a charged conductor.* A difference of potential is a measure of the work done in moving a charge from one point to another. As no electric field exists within the conductor, no work is done in moving a charge between two points on the same conductor. No difference of potential can exist between such points.

3. *The surface of a conductor is an equipotential surface.* All points on a conductor are at the same potential and no work is done by the electric field in moving a charge residing on a conductor. If points of equal potential in an electric field near a charged object are joined, an *equipotential line* or *surface* within the field is indicated. No work is done when a test charge is moved in an electric field along an equipotential surface.

4. *Electric lines of force are normal to equipotential surfaces.* A line of force shows the direction of the force acting on a test charge in an electric field. It can be shown that there is no force acting normal to this direction. Thus no work is done when a test charge is moved in an electric field normal to the lines of force. See Figure 16-17.

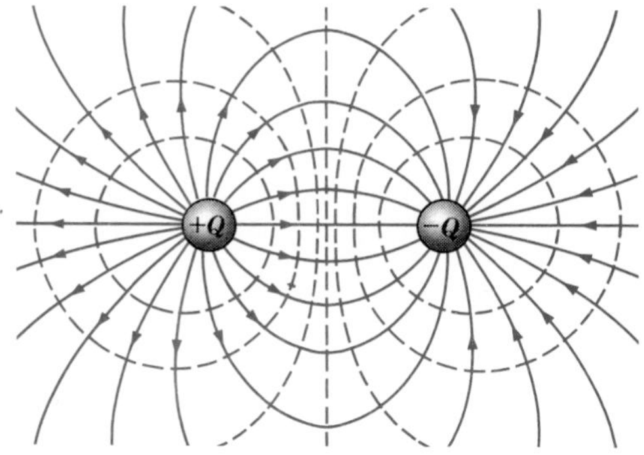

Figure 16-17. Lines of force (solid lines) and equipotential lines (dashed lines) define the electric field near two equal but opposite charges.

5. *Lines of force originate or terminate normal to the conductive surface of a charged object.* Since the surface of a conductor is an equipotential surface, lines of force must start out perpendicularly from the surface. For the same reason, a line of force cannot originate and terminate on the same conductor.

16.12 Effect of the Shape of a Conductor A charged spherical conductor perfectly isolated in space has a uniform charge density, or charge per unit area, over the outer surface. Lines of force extend radially from the surface in all directions and the equipotential surfaces of the electric field are spherical and concentric. Such symmetry is not found in all cases of charged conductors.

A charge acquired by a nonconductor such as glass is confined to its original region until it gradually leaks away. The charge placed on an isolated metal sphere quickly spreads uniformly over the entire surface. If the conductor is not spherical, the charge distributes itself according to the surface curvature, concentrating mainly around points. The pear-shaped conductor illustrated in Figure 16-18 shows the charge more concentrated on the curved regions and less concentrated on the straight regions. If the small end is made more pointed, the charge density will increase at that end.

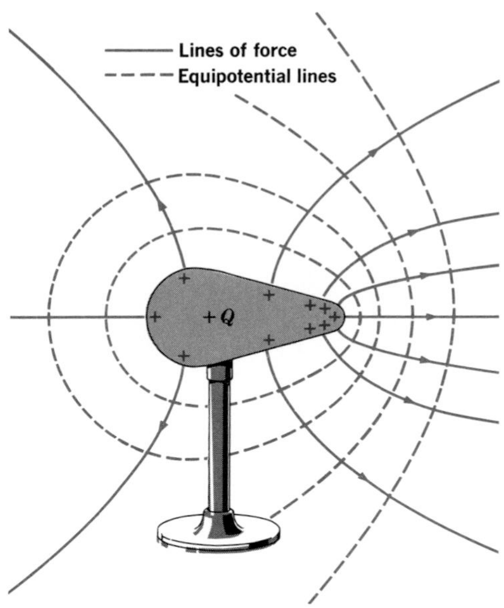

Figure 16-18. The charge density is greatest at the point of greatest curvature.

16.13 Discharging Effect of Points In Figure 16-18 the lines of force and equipotential lines are shown more concentrated at the small end of the charged conductor. This geometry indicates that the intensity of the electric field, or potential gradient, in this region is greater than elsewhere around the conductor. If this surface is reshaped to a sharply pointed end, the field intensity can become great enough to cause the gas molecules in the air surrounding the pointed end to *ionize.* Ionized air consists of gas molecules from which an outer electron has been removed, thus allowing both the positively charged *ion* and the freed electron to respond to the electric force. When the air is ionized, the point of the conductor is rapidly discharged.

There are always a few positive ions and free electrons present in the air. The intense electric field near a sharp point of a charged conductor will set these charged particles in motion such that the electrons are driven in one direction and the positive ions in the opposite direction. Violent collisions with other gas molecules will knock out some electrons and produce more charged particles. In this way air can be ionized quickly when it is subjected to a sufficiently large electric stress.

In dry air at atmospheric pressure, a potential gradient of 30 kV/cm between two charged surfaces is required to ionize the intervening column of air. When such an air gap is ionized, a *spark discharge* occurs. There is a rush of free electrons and ionized molecules across the ionized gap, discharging the surfaces and producing heat, light, and sound. Usually the quantity of static electricity involved is quite small and the time duration of the spark discharge is very short. Atmospheric lightning, however, is a spark discharge in which the quantity of charge is great.

A lightning bolt transfers an average of about 1 coulomb of charge between a cloud and the earth.

The intensity of an electric field near a charged object can be sufficient to produce ionization at sharp projections or sharp corners of the object. A slow leakage of charge can occur at these locations producing a *brush* or **corona discharge.** A faint violet glow is sometimes emitted by the ionized gases of the air. A glow discharge known as St. Elmo's fire can sometimes be observed at night at the tips of ship masts and at the trailing edges of wing and tail surfaces of aircraft. The escape of charges from sharply pointed conductors is important in the operation of electrostatic generators and in the design of lightning arresters.

Lightning is a gigantic electric discharge in which electric charges rush to meet their opposites. The interchange can occur between clouds or between a cloud and the earth. See Figure 16-19. According to the Lightning Protection Institute, one hundred bolts of lightning strike the earth every second, each bolt initiated by a potential difference of millions of volts.

Figure 16-19. Streaks of lightning brighten the nighttime sky. Each flash requires a potential difference of millions of volts.

There are no known ways of preventing lightning. However, there are effective means of protection from its destructiveness. Lightning rods, invented by Benjamin Franklin, are often used to protect buildings made of nonconducting materials from lightning damage. Sharply pointed rods are strategically located above the highest projections of the building, and by their discharging effect, they normally prevent the accumulation of a dangerous electrostatic charge. Lightning rods are thoroughly grounded. If lightning strikes, the rods provide a good conducting path into the ground. Being well grounded, the steel frames of large buildings offer excellent protection from lightning damage.

Television receiving antennas, even though equipped with lightning arresters, do not protect a building from lightning. Without lightning arresters they are a distinct hazard because they are not grounded.

16.14 Capacitors Any isolated conductor is able to retain an electrostatic charge to some extent. If we place a positive charge on such a conductor by removing electrons, the potential is raised to some positive value with respect to ground. Conversely, a negative charge placed on the conductor results in a negative potential with respect to ground. By increasing the charge, we increase the potential of the conductor since the potential of an isolated conductor is a measure of the work done in placing a charge on the conductor. It is evident that we can continue to increase the charge until the potential with respect to ground or other conducting surface becomes so high that corona or spark discharges occur. However, if the conductor is in an evacuated space, it could be raised to a much higher potential by continuing the addition of charge.

Suppose a charged conductor is connected to an electroscope, as shown in Figure 16-20(A). The leaves of the electroscope will diverge indicating the potential of the charged conductor. The electroscope is now a part of the

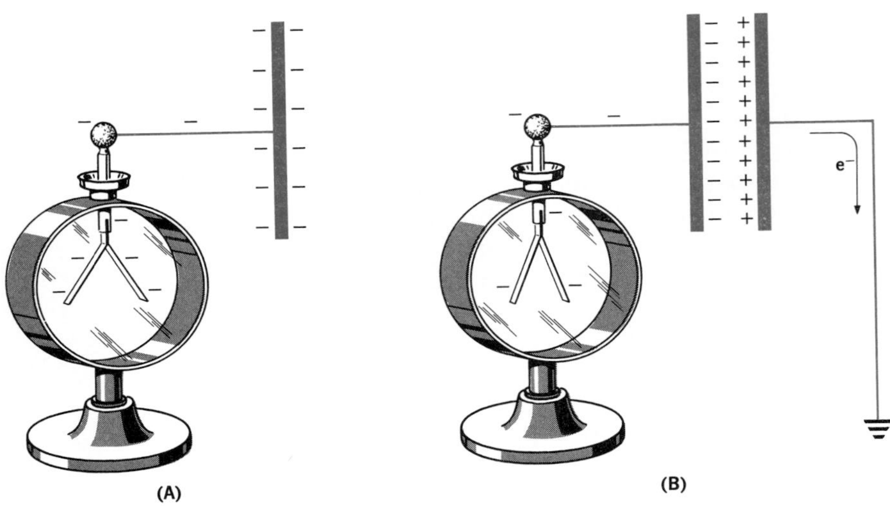

Figure 16-20. The principle of operation of a capacitor.

conducting surface and there can be no difference of potential between different regions of a single charged conducting surface.

Now suppose a grounded conductor is brought near the charged conductor, as shown in Figure 16-20(B). A positive charge is induced on this second conductor as free electrons are repelled to ground by the negative field of the first conductor. The leaves of the electroscope partially collapse. The closer we move the grounded plate to the charged conductor, the more pronounced is this effect. Because of the attractive force of the induced charge on the grounded plate, less work is required to place the same negative charge on the first conductor. Its potential is consequently reduced accordingly. A greater charge can now be placed on this conductor to raise the potential back to the initial value.

A combination of conducting plates separated by an insulator that is used to store an electric charge is known as a ***capacitor.*** The area of the plates, their distance of separation, and the character of the insulating material separating them determine the charge that can be placed on a capacitor. The larger the charge, the greater is the potential difference between the plates of a capacitor. The ratio of charge Q to the potential difference V is a *constant* for a given capacitor and is known as its *capacitance, C.* ***Capacitance*** *is the ratio of the charge on either plate of a capacitor to the potential difference between the plates.* Thus

$$C = \frac{Q}{V}$$

where C is the capacitance of a capacitor, Q is the quantity

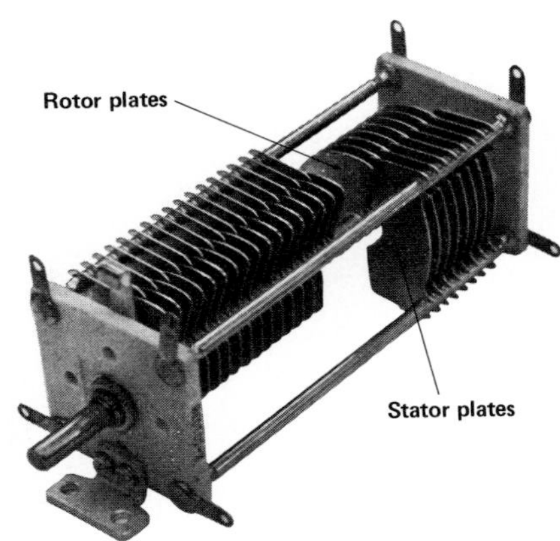

Figure 16-21. A two-section, air-dielectric variable capacitor. The capacitance of each section is varied by rotating the shaft to which the rotor (movable) plates are attached, thereby altering the area of rotor plates that engage the stator (fixed) plates.

of charge on either plate, and V is the potential difference between the conducting plates.

The unit of capacitance is the *farad* (F), named in honor of Michael Faraday. *The capacitance is 1* ***farad*** *when a charge of 1 coulomb on a capacitor results in a potential difference of 1 volt between the plates.* Capacitors used in electronic devices have capacitances on the order of *microfarads* (μF) or *picofarads* (pF).

$$1\ \mu\text{F} = 10^{-6}\ \text{F}$$
$$1\ \text{pF} = 10^{-12}\ \text{F}$$

16.15 Dielectric Materials Faraday investigated the effects of different insulating materials between the plates of capacitors. He constructed two capacitors with equal plate areas and equal plate spacing. Using air at normal pressure in the space between the plates of one and an insulating material between the plates of the other, he charged both to the same potential difference.

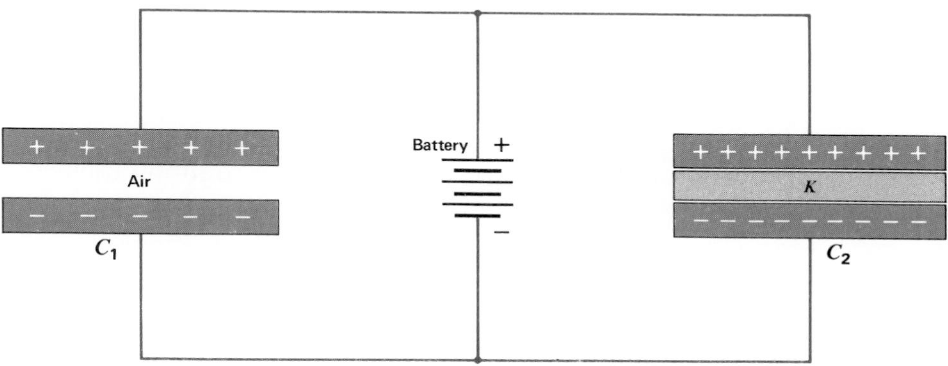

Figure 16-22. Capacitors C_1 and C_2 have identical dimensions and are charged to the same potential difference by the battery. C_2 accumulates the larger quantity of charge.

These capacitors are shown as C_1 and C_2 respectively in Figure 16-22. Faraday measured the quantity of charge on each capacitor and found that C_2 had a greater charge than C_1 by a factor K.

$$Q_2 = KQ_1$$

Except for the material separating the plates, the two capacitors are identical and have the same potential difference across their plates. The factor K by which the charge on C_2 exceeds the charge on C_1 must be due to a property of the insulating material of C_2. The ratio Q_2/V is larger than Q_1/V by this factor K.

$$\frac{Q_2}{V} = K\frac{Q_1}{V}$$

Thus the capacitance of C_2 is larger than the capacitance of C_1 by the factor K.

$$C_2 = KC_1$$

Many materials such as mica, paraffin, oil, waxed paper, glass, plastics, and ceramics can be used instead of air in the space between the plates of a capacitor. For each material, the resulting capacitance Q/V will have a different value.

Materials used to separate the plates of capacitors are known as **dielectrics.** *The ratio of the capacitance with a particular material separating the plates of a capacitor to the capacitance with a vacuum between the plates is called the* ***dielectric constant,*** *K, of the material.* Dry air at atmospheric pressure has a dielectric constant of 1.000 6. In practice this is taken as *unity* (the same as vacuum) for dielectric constant determinations. Dielectric constants are dimensionless numbers ranging from 1 to 10 for materials commonly used in capacitors. The dielectric constant, K, of the dielectric material used in Figure 16-22 is

$$K = \frac{C_2}{C_1}$$

Typical dielectric constants for some common dielectric materials are given in Table 16-1.

Table 16-1
DIELECTRIC CONSTANTS

Dielectric material	*Dielectric constant* K	*Proportionality constant* k $N \cdot m^2/C^2$
air	1.0	8.9×10^9
paper (oiled)	2.0	4.5×10^9
paraffin	2.2	4.1×10^9
polyethylene	2.3	3.9×10^9
polystyrene	2.5	3.6×10^9
hard rubber	2.8	3.2×10^9
mica	6.0	1.5×10^9
glass	8.0	1.1×10^9

Instead of charging the identical capacitors of Figure 16-22 to the same potential difference, suppose we place the same charge Q on them. The experiment now shows that the potential difference across C_2 is smaller than that across C_1 by the factor $1/K$.

$$V_2 = \frac{V_1}{K}$$

This leads us to our previous conclusion about the influence of the dielectric material separating the plates of a capacitor. Since $C = Q/V$ and Q_1 and Q_2 are equal,

$$\frac{Q_2}{V_2} = \frac{Q_1}{V_1/K} \quad \text{and} \quad C_2 = KC_1$$

16.16 The Effect of Dielectrics The molecules of some dielectrics have a permanent separation of their positive and negative centers of charge. This property is described as a permanent *electric dipole moment*. These molecules are called *polar* molecules, or *dipoles*. When placed in an electric field, polar molecules tend to become aligned with the external field to a degree characteristic of the molecules.

Other dielectrics are composed of molecules that are essentially *nonpolar* and that show no permanent electric dipole moments. When placed in an electric field, these molecules can acquire a temporary polar character by *induction*. While in the electric field, they have induced electric dipole moments.

The space between the plates of a charged capacitor is permeated by a uniform electric field. If a dielectric slab is inserted into such an electric field denoted by the potential gradient E_0, the slab as a whole becomes polarized by induction. The surface near the positive plate of the capacitor acquires a negative charge and the surface near the negative plate acquires a positive charge. See Figure 16-23. These surface charges result from the dipole moments of the dielectric molecules and not from the transfer of electrons as in the case of metallic conductors.

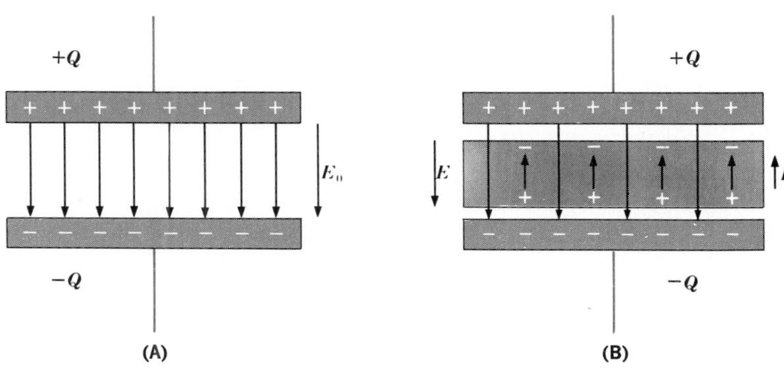

Figure 16-23. A dielectric slab placed in an electric field tends to weaken the field within the dielectric. What should be the effect on the potential difference across the plates?

The electric field established in the dielectric slab by the surface charge *opposes* the external field E_0 of the capacitor. This opposing field is shown in Figure 16-20(B) as E_d. Its effect is to weaken the original field within the dielectric. The net electric field E is the vector sum of the two fields E_0 and E_d and is always a weaker field in the direction of E_0.

$$E = E_0 + E_d$$

The net effect of a dielectric between the plates of a charged capacitor is to lower the potential gradient of the electric field. If a parallel-plate capacitor with an air dielectric is charged and then isolated, a potential difference of Q/C volts remains across the plates. The introduction of a dielectric slab between the plates causes a decrease in this potential difference and reveals its weakening effect on the electric field of the capacitor.

If the charged spheres of Figure 16-10 are immersed in some dielectric medium, the medium would also become polarized by induction. Clustered around each of the charged spheres would be opposite charges from the medium. The result would be an effective reduction in the values of Q_1 and Q_2. A rigorous analysis shows that the force between the two spheres is reduced in value by the factor K, which is the dielectric constant of the medium. Thus in the equation for Coulomb's law

In Table 16-1 observe that the value of k for air divided by K for another dielectric material yields k for that material.

$$F = k\frac{Q_1Q_2}{d^2}$$

the value of k is $8.987 \times 10^9\ \text{N}\cdot\text{m}^2/\text{C}^2$ for vacuum but is reduced by the factor K for material media. Its value for various materials is given in the last column of Table 16-1.

Capacitors find wide application in all kinds of electronic circuits, ignition systems, and telephone and telegraph equipment. Large capacitances are achieved by using large plate areas, insulators with high dielectric constants, and small separation of the plates. Physical size limits the plate area while cost limits the choice of dielectric. *Dielectric strength* of the insulator limits the reduction in spacing between the plates.

Dielectric strength should not be confused with the dielectric constant of a material. The dielectric strength defines the quality of the material as an insulator; that is, the potential gradient it will withstand without being punctured by a spark discharge. Some typical values are given in Table 16-2.

Table 16-2
DIELECTRIC STRENGTHS

Dielectric material	***Dielectric strength*** (kV/cm to puncture)
air	30
oil	75
paraffin	350
paper (oiled)	400
mica	500
glass	1000

16.17 *Combinations of Capacitors* Suppose three capacitors of capacitances C_1, C_2, and C_3 are connected in *parallel,* that is, with one plate of each capacitor connected to one conductor while the other plate is connected to a second conductor. See Figure 16-24(A).

The plates connected to the + conductor are parts of one conducting surface. Those connected to the − conductor form the other conducting surface. If the three capacitors are charged, it is apparent that they must have the same difference of potential, V, across them. The quantity of charge on each must be respectively

$$Q_1 = C_1V,\ Q_2 = C_2V,\ \text{and } Q_3 = C_3V$$

The total charge, Q_T, must be the sum of the separate charges on the three capacitors,

$$Q_T = Q_1 + Q_2 + Q_3$$

then

$$Q_T = C_1V + C_2V + C_3V$$

Because the total charge is equal to the product of the total capacitance, C_T, and the potential difference, V, it is evident that

$$Q_T = C_TV$$

Substituting,

$$C_TV = C_1V + C_2V + C_3V$$

and

$$C_T = C_1 + C_2 + C_3$$

For capacitors connected in parallel, the total capacitance is the sum of all the separate capacitances.

Now suppose we connect the three capacitors in *series,* as shown in Figure 16-24(B). A positive charge placed on C_1 from the + source induces a negative charge on the second plate as the electrons are attracted away from the plate C_2, which is connected to C_1. A positive charge of the same magnitude is left on the + plate of C_2. Similarly the plates of C_3 acquire the same magnitude of charge. Thus

$$Q = Q_1 = Q_2 = Q_3$$

The negative plate of C_1 must be at the same potential with respect to ground as the positive plate of C_2 since they are connected and are parts of the same conducting surface. Similarly the negative plate of C_2 must be at the same potential as the positive plate of C_3. The total difference of potential, V_T, across the three series capacitors

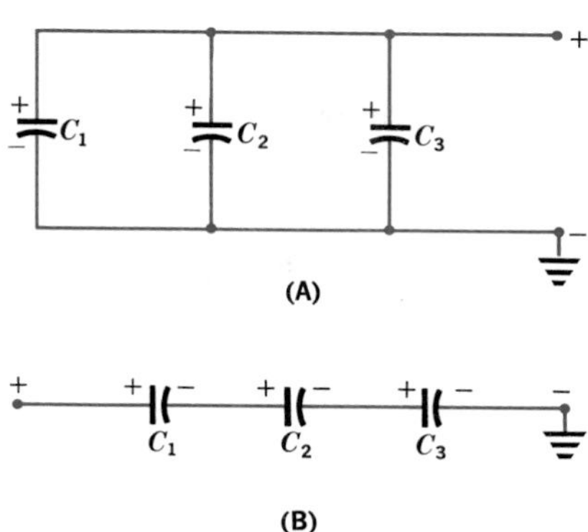

Figure 16-24. Capacitors connected in parallel (A) and in series (B).

must be equal to the sum of the separate potential differences across each capacitor: V_1, V_2, and V_3.

$$V_T = V_1 + V_2 + V_3$$

Since the charge received from the source is Q, then by definition

$$V_T = \frac{Q}{C_T}$$

and

$$V_1 = \frac{Q}{C_1}, \quad V_2 = \frac{Q}{C_2}, \quad V_3 = \frac{Q}{C_3}$$

Substituting,

$$\frac{Q}{C_T} = \frac{Q}{C_1} + \frac{Q}{C_2} + \frac{Q}{C_3}$$

or

$$\frac{1}{C_T} = \frac{1}{C_1} + \frac{1}{C_2} + \frac{1}{C_3}$$

For capacitors connected in series, the reciprocal of the total capacitance is equal to the sum of the reciprocals of all the separate capacitances. See the following example.

EXAMPLE Four capacitors are connected in parallel and charged to a potential difference of 150.0 V. The values of the capacitors are 0.300 μF, 0.300 μF, 0.400 μF, and 0.400 μF. Calculate the charge on one of the 0.400-μF capacitors.

Given	Unknown	Basic equations
$C_1 = 0.300\ \mu F$ $C_2 = 0.300\ \mu F$ $C_3 = 0.400\ \mu F$ $C_4 = 0.400\ \mu F$	Q_3	$V = V_1 = V_2 = V_3 = V_4$ $C = \frac{Q}{V}$

Solution

Working equation: $Q_3 = C_3V_3$
$= (0.400\ \mu F)(150.0\ V)$
$= 60.0\ \mu C$

EXAMPLE The four capacitors in the previous example are discharged and then connected in series. A charge of 60.0 μC is placed on the ungrounded terminal. Calculate the potential difference across one of the 0.300 μF capacitors.

Given	Unknown	Basic equations
$C_1 = 0.300\ \mu F$ $C_2 = 0.300\ \mu F$ $C_3 = 0.400\ \mu F$ $C_4 = 0.400\ \mu F$	V_1	$Q = Q_1 = Q_2 = Q_3 = Q_4$ $C = \dfrac{Q}{V}$

Solution

Working equation: $V_1 = \dfrac{Q_1}{C_1}$

$$= \frac{60.0\ \mu C}{0.300\ \mu F}$$

$$= 200\ V$$

PRACTICE PROBLEMS **1.** Two capacitors with values of 0.25 μF and 0.50 μF, respectively, are connected in parallel across an 80.0-V circuit. Calculate the total charge stored in the capacitors. *Ans.* $6\overline{0}\ \mu C$

2. The capacitors in Problem 1 are fully discharged and then connected in series. A charge of 25 μC is transferred to the ungrounded terminal. Calculate the potential difference across the two capacitors in series. *Ans.* 150 V

QUESTIONS: GROUP A

1. How is potential gradient related to electric field strength?
2. (a) Why is the earth considered a "ground" in electric terms? (b) Can any other object act as a ground?
3. Differentiate between electric potential energy and electric potential.
4. What does a line of force in an electric field diagram indicate?
5. Describe the charge distribution on a sharp-edged conductor?
6. What causes a spark discharge?
7. (a) What is a lightning rod? (b) Who invented it?
8. What is the difference between dielectric strength and dielectric constant?
9. How would you connect the plates of a set of capacitors so that they are (a) in parallel and (b) in series?

GROUP B

10. What happens to the work done in moving a positive charge against an electric field?

11. How did Faraday's icepail experiment help us understand why a person in a car is safe in a lightning storm?
12. Why can a charge move along a conductor without any work being done?
13. Two charged metal spheres are 5 cm apart. How great a potential difference will ionize the air between them?
14. Explain St. Elmo's fire.
15. How is the electric field of the dielectric between the plates of a capacitor related to the external field?
16. Why is Coulomb's constant *not* the same for all dielectrics?
17. (a) Why is the knob of an electroscope usually round? (b) Why is its charge never really "permanent"?
18. In Figure 16-18, how does the diagram indicate the strength of the electric field around the insulated conductor?

PROBLEMS: GROUP A

1. What is the potential difference between 2 points if 6.5×10^{-3} J is required to move a charge of 3.2 μC between the points?
2. How much work must a 1.5-V battery do if it is to move 0.25 C per second through a wire for 13 min?
3. A 12.0-V storage battery is connected to a 6.0-pF parallel-plate capacitor. What is the charge on each plate?
4. A potential difference of 100.0 V exists across the plates of a capacitor when the charge on each plate is 400.0 μC. What is the capacitance?
5. What is the equivalent capacitance of a 2.00-μF and 6.00-μF capacitor connected (a) in series and (b) in parallel?
6. Calculate the total charge when the capacitors in Problem 5(b) are connected to a 9.0-V battery.

GROUP B

7. A proton is accelerated through a potential difference of 4.5×10^6 V. (a) How much kinetic energy has the proton acquired? (b) If it started at rest, how fast is it moving?
8. A 3750-pF capacitor carries a charge of 1.75×10^{-8} C. (a) What is the potential difference across the plates? (b) If the plates are 6.50×10^{-4} m apart, what is the strength of the electric field between them?
9. A pair of parallel plates is 5.00×10^{-2} cm apart. Their capacitance with air between them is 75.0 pF. (a) What is the capacitance when air is replaced with paraffin? (b) If the potential difference between the plates is 25.5 V, what is the charge on each plate with the paraffin? (c) What is the equivalent capacitance of two pairs of these plates connected in series?
10. You have three 2.00 μF capacitors. (a) Draw all possible circuit diagrams. (b) Calculate the equivalent capacitance for each combination.
11. A force of 4.30×10^{-2} N is needed to move a charge of 56.0 μC a distance of 20.0 cm in an electric field. What is the potential difference?

SUMMARY

Electric charges are of two kinds, negative and positive. Like charges repel and unlike charges attract. Electrification is the process that produces electric charges on an object. The process involves the transfer of free electrons. An electroscope may be used to indicate the presence of an electric charge and to provide a rough measurement of its magnitude.

Substances are classified as conductors or insulators based on their ability to conduct an electric charge. Most metals are good conductors. Most nonmetals are good insulators because they are poor conductors of electric charges. The conduction of electric charge by solids is related to the availability of free electrons in their structures. Isolated bodies with conducting surfaces can be charged by either induction or conduction.

The force of attraction or repulsion between point charges is enunciated in terms of Coulomb's law. Charged bodies approximate point charges if they are small in comparison with the distance separating them. The region about a charged body in which the coulomb force exists is an electric field. The path along which a positive test charge is driven by an electric field is called a line of force. Lines of force are used to represent both the direction and intensity of the field.

The difference in potential between two points in an electric field is defined in terms of work expended in moving a charge between two points and quantity of charge moved between these points.

Experiments with isolated conductors have shown that (1) the residual charge on a conductor resides on its surface; (2) there can be no difference of potential between two points on the surface of a charged conductor; (3) the surface of a conductor is an equipotential surface; (4) electric lines of force are normal to equipotential surfaces; and (5) lines of force originate and terminate normal to the conductive surface of a charged object.

Charge density on the conductive surface of a charged body varies with the surface curvature. The intensity of the electric field near sharp points may be great enough to ionize the surrounding air, thus discharging the body.

An arrangement of conducting surfaces separated by insulators and used to store electric charge is called a capacitor. The capacity for storing electric charge is its capacitance. The dielectric constant of an insulating material is determined by comparing the material's effect on capacitance with the effect of air. Capacitors may be connected in series or in parallel. The total capacitance of the capacitor network is dependent upon the kind of connection used.

VOCABULARY

ampere
capacitance
capacitor
conductor (electric)
corona discharge
coulomb
Coulomb's law of electrostatics
dielectric
dielectric constant
electric field
electric field intensity
electrification
electron gas
electroscope
electrostatics
farad
free electron
induced charge
insulator (electric)
line of force (electric)
point charge
potential difference
potential gradient
proof plane
residual charge
static charge
volt

17 Direct-Current Circuits

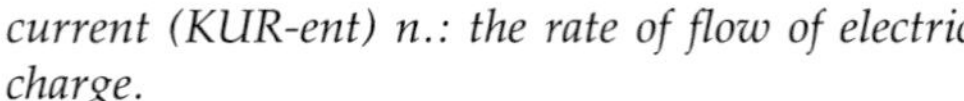

current (KUR-ent) n.: the rate of flow of electric charge.

OBJECTIVES

- Discuss the nature of electric current in terms of charge.
- Study the dry cell.
- Analyze series and parallel circuits.
- Define Ohm's law.
- Discuss internal resistance and emf.
- Define the measurements of voltage, current, and resistance in series and parallel circuits.
- Analyze the Wheatstone bridge circuit and its use.

SOURCES OF DIRECT CURRENT

17.1 Electric Charges in Motion The quantity of charge on a capacitor is indicated by the potential difference between the plates.

$$Q = CV$$

If the capacitance, C, for a given capacitor is a constant, then the charge, Q, is proportional to the potential difference, V, across it.

$$Q \propto V \quad (C \text{ constant})$$

A potential difference across a charged capacitor, and therefore a charge on the capacitor, can be indicated by an electroscope connected across the capacitor plates. See Figure 17-1. Suppose the two plates of the capacitor are now connected by a heavy copper wire as in Figure 17-2(A). The leaves of the electroscope immediately collapse, indicating that the capacitor has been discharged rapidly. Free electrons flow from the negative plate of the capacitor and from the electroscope through the copper wire to the positive plate. This action quickly establishes the normal distribution of electrons characteristic of an uncharged capacitor.

Considering the charged capacitor of Figure 17-1 again, suppose the plates are connected by means of a long, fine wire made of nichrome, which has few free electrons compared with copper. The leaves of the electroscope collapse more gradually than before. See Figure 17-2(B). This delay means that a longer time is required to discharge the capacitor completely. When any conductor connects the plates of the capacitor, electrons move from the negative plate through the conductor to the positive plate. This transfer decreases the charge on each plate and the difference of potential between the plates.

The charge, Q, on either plate of a capacitor is proportional to the potential difference, V, between the plates. The *rate* at which the charge decreases, in C/s, is proportional to the rate at which the potential difference decreases, in V/s. Now, the rate of decrease of charge on the capacitor must represent the rate of flow of the charge, in C/s, through the conductor. Thus the rate at which the leaves of the electroscope collapse indicates the rate of flow of the charge through the conductor. During the time the capacitor is discharging, a *current* is said to exist in the conducting wire. *An* ***electric current,*** *I, in a conductor is the rate of flow of charge through a cross section of the conductor.*

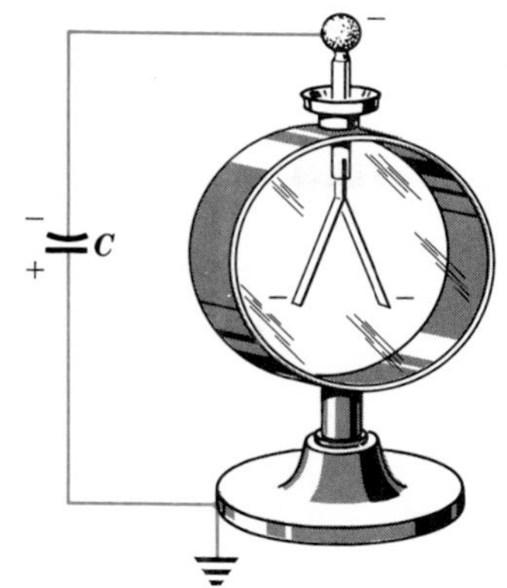

Figure 17-1. The charge on a capacitor can be indicated by an electroscope.

$$\textbf{current } (I) = \frac{\textbf{charge } (Q)}{\textbf{time } (t)}$$

The unit of current is the *ampere* (A), named for the French physicist André Marie Ampère (1775–1836). *One ampere is a current of 1 coulomb per second.*

$$1\text{ A} = \frac{1\text{ C}}{\text{s}}$$

Recall that 1 coulomb is the charge of 6.25×10^{18} electrons or a like number of protons. Small currents can be expressed in *milliamperes* (mA) or *microamperes* (μA).

$$\mathbf{1\ mA = 10^{-3}\ A}$$
$$\mathbf{1\ \mu A = 10^{-6}\ A}$$

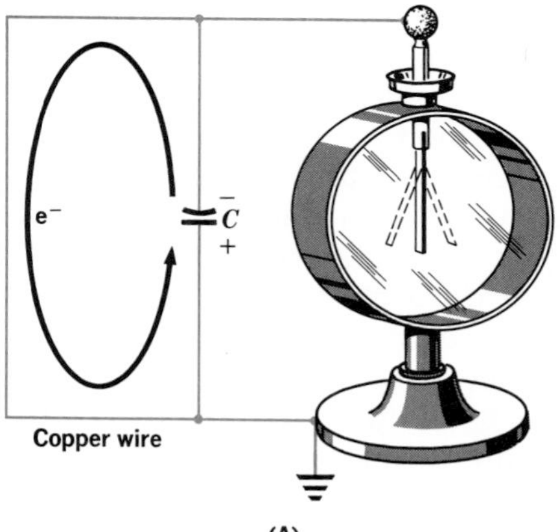

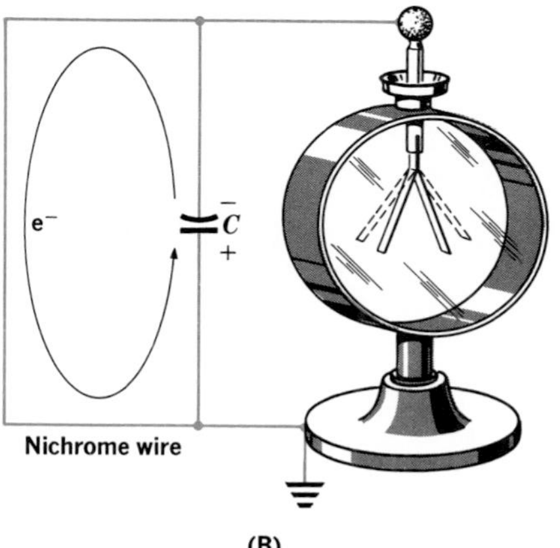

Figure 17-2. The rate of discharge of a capacitor depends on the conducting path provided.

A moving charge is basic for an electric current. The character of a conducting substance determines the nature of the charge set in motion in an electric field. In solids the charge carriers are electrons; in gases or electrolytic solutions the charge carriers can be positive ions, negative ions, or both. *An* ***electrolyte*** *is a substance whose aqueous solution conducts an electric current.*

A negative charge moving in one direction in an electric circuit is equivalent to a positive charge moving in the opposite direction insofar as external effects are concerned. Negative charges departing from a negatively charged surface and arriving at a positively charged surface leave the former surface less negative and the latter surface less positive. Positive charges departing from the positively charged surface and arriving at the negatively charged surface would have a similar effect, leaving the positively charged surface less positive and the negatively charged surface less negative.

If both positive and negative charges are set in motion in a conducting medium, the current in the medium is the sum of the currents due to the motion of the positive charges in one direction and the negative charges in the opposite direction.

The electric current in metallic conductors is an electron flow from negative to positive.

In an electric circuit, the directional sense of electron flow through an electric field in a metallic conductor is from negative to positive. The directional sense of a positive ion flow in an ionized gas is from positive to negative. In our study of electric circuits we will be concerned generally with current in metallic conductors and will deal mainly with electron flow in which the directional sense is from negative to positive. In any case, confusion about current direction can be avoided by indicating the directional sense of the charge carriers that are in motion.

The above reference to an electric field in a conductor does not contradict the assertions of Section 16.11, which require the electric field in a conductor to equal zero. In this earlier discussion we were concerned with static electric charges on an isolated conductor. The charges reside on the surface of such a conductor and there is no net charge motion on this equipotential surface. We are now dealing with moving charges and with conductors across the ends of which potential differences are deliberately maintained. Thus this earlier restriction does not apply.

Free electrons in empty space are accelerated by an electric field. The effect of an electric field on the free electrons of a metallic conductor is quite different. An acceleration of extended duration of these electrons is not realized because of their frequent collisions with the fixed particles of the conductor, the metallic ions. With each collision the electron loses whatever velocity it had acquired in the direction of the accelerating force, transfers energy to the fixed particle, and makes a fresh start.

The kinetic energy transferred in these collisions increases the vibrational energy of the conductor particles and heat is given up to the surroundings. The free electrons move in the direction of the accelerating force with an average velocity called the **drift velocity.** This drift velocity has a constant average value for a given conductor carrying a certain magnitude of current. It is much smaller than the speed with which *changes in the electric field* propagate through the conductor, which is approximately 3×10^8 m/s.

It has been estimated that the free electrons of a copper conductor of 1-mm^2 cross section have an average drift velocity of about 1 mm/s when a current of 20 A is in the conductor. The situation is somewhat analogous to that of a ball rolling down a flight of stairs. With the proper conditions, the successive collisions of the ball with the steps effectively cancel the acceleration it acquires between steps and it rolls down with a constant average speed.

As we saw, collision between the free electrons and the fixed particles of a conductor is an energy-dissipating process that opposes the flow of charge through the conductor. *This opposition to the electric current is called* ***resistance, R.*** The practical unit of resistance is the *ohm,* symbolized by the Greek letter **Ω** (omega). The laws of electric resistance are discussed in Section 17.11.

17.2 Continuous Current The current from a discharging capacitor persists for a very short interval of time; it is known as a *transient* current. The effects of such a current are also transient, but those effects are very important in certain types of electronic circuits. In the general applications of electricity, more continuous currents are required.

The effects of current electricity are quite different from those of static electricity. The electric current is one of our most convenient means of transferring energy. A *closed-loop* conducting path is needed if energy is to be utilized outside the source; this conducting loop is known as an *electric circuit.* The device or circuit component utilizing the electric energy is called the *load.* The basic components of an electric circuit are illustrated in three different ways in Figure 17-3. Conventional symbols used in schematic (circuit) diagrams are shown in Figure 17-4.

A capacitor would be useful as a source of continuous current over a prolonged period of time only if some

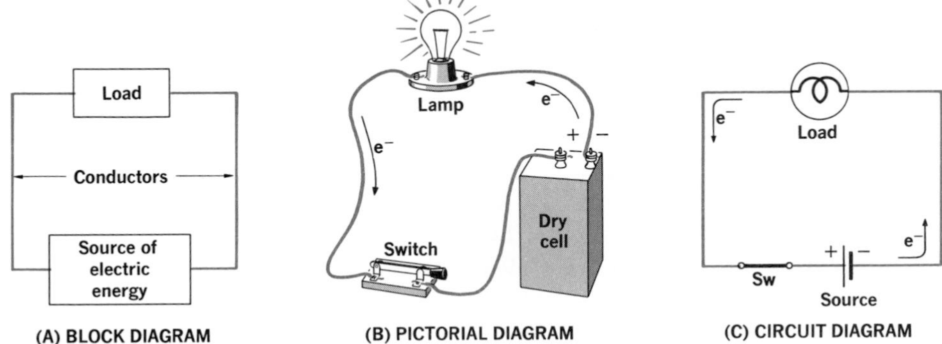

Figure 17-3. An electric circuit is a conducting loop in which a current can transfer electric energy from a suitable source to a useful load.

means were available for keeping it continuously charged. We would need to supply electrons to the negative plate of the capacitor as rapidly as they were removed by the current in the conducting loop. Similarly, we would need to remove electrons from the positive plate of the capacitor as rapidly as they were deposited by the current. We must

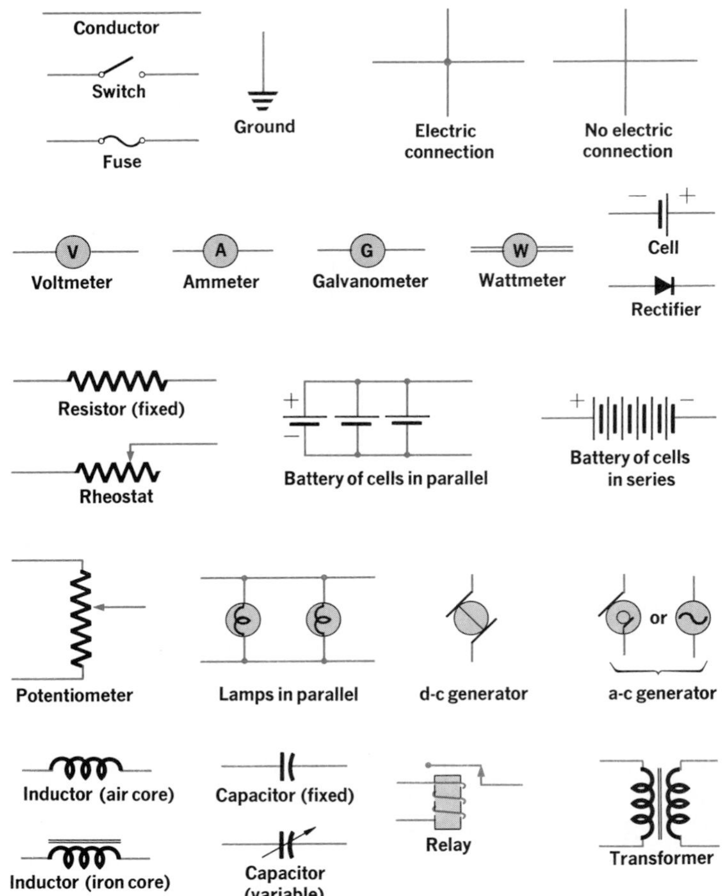

Figure 17-4. Conventional symbols used in schematic diagrams of electric circuits.

maintain the potential difference across the source during the time a continuous current is in the circuit.

Maintaining a potential difference across a source of current in an operating circuit is an energy-consuming process; work is done on the electrons in opposition to the force of an electric field. In the process the potential energy of electrons is raised, and this energy is available to do useful work in the external circuit. What is needed is some kind of "electron pump." Several practical energy sources are available to maintain a difference of potential across an operating electric circuit.

17.3 Sources of Continuous Current The transformation of available energy into electric energy is accomplished by one of two basic methods: *dynamic conversion* or *direct conversion*. Dynamic converters are rotating machines, and it is these machines that supply most of our requirements for electric energy today. Direct conversion methods produce energy transformations without moving parts. We shall consider the following sources of continuous current:

1. *Electromagnetic.* The major source of electric current in practical circuits today depends on the principle of *electromagnetic induction*. If a conducting loop is rotated in a magnetic field, the free electrons of the conductor are forced to move around the loop and thus constitute a current. This is the basic operating principle of electric generators. It is an example of the transformation of mechanical energy into electric energy, a dynamic conversion process. Electromagnetic induction and the electric generator are discussed in Chapter 20.

The energy supply for the dynamic conversion process comes mainly from burning fossil fuel, from falling water, and from nuclear fission reactions.

2. *Photoelectric.* The **photoelectric effect** was first observed by Heinrich Hertz. He noticed that a spark discharge occurred more readily when charged metal spheres were illuminated by another spark discharge.

Electrons are emitted from the surface of a metal illuminated by light of sufficiently short wavelength. The alkali metals are photosensitive to light in the near ultraviolet region of the radiation spectrum. Two of these metals, potassium and cesium, emit photoelectrons when exposed to ordinary visible light. In the case of most metals, the incident light for photoemission corresponds to radiations deep in the ultraviolet region. When a photon of the incident light is absorbed by the metal, its energy is then transferred to an electron. If the motion of this electron is toward the surface and its energy at the surface is sufficient to overcome the forces that bind it within the surface of the metal, it is emitted as a photoelectron.

Review the laws of photoelectric emission in Section 12.7.

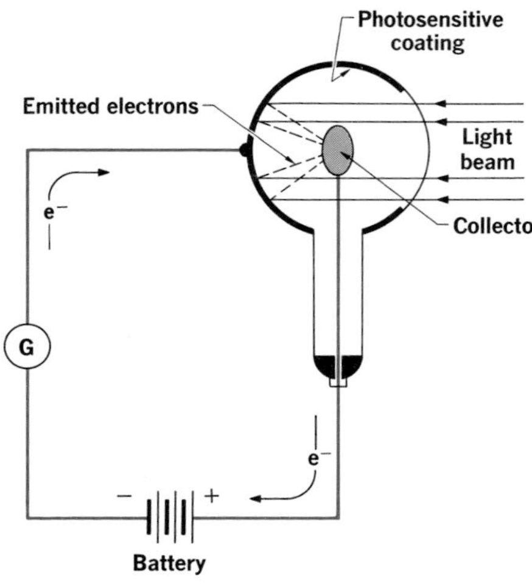

Figure 17-5. A circuit that includes a photoelectric cell.

If properly isolated, the metal surface acquires a positive charge as photoelectrons are emitted. As the charge increases, other electrons are prevented from escaping from the metal and the action stops. If, however, the metal is placed in an evacuated tube and made part of a circuit, as in Figure 17-5, it can act as a source of current whenever light is incident upon its surface. The metal thus serves as the electron pump mentioned in Section 17.2. The evacuated tube containing the metallic emitter is called a *photoelectric cell.*

The photoelectric cell shown in Figure 17-5 has a coating of potassium metal deposited on the inner surface of a glass tube, leaving an aperture through which light can enter. A metallic ring near the focus of the emitter surface acts as a collector of photoelectrons. The emitter is connected externally to the negative terminal of a battery, and the collector is connected to the positive terminal. The collector is thus maintained at a positive potential with respect to the emitter. The force of the resulting electric field acts on the emitted electrons, driving them to the collector. The small photoelectron current can thus be maintained in the external circuit as long as light photons are incident on the emitter surface. A **galvanometer,** an electric meter sensitive to very feeble currents, can be placed in the circuit to indicate the relative magnitude and direction of the current.

A bright light of effective frequency causes the ejection of many electrons from the surface of the photosensitive metal, while light of the same frequency but of a lower intensity produces fewer photoelectrons. Thus the current produced by the photoelectric cell is determined by the intensity of the incident light; this is a case in which radiant energy is transformed into electric energy. Some form of photosensitive cell is used in devices that are controlled or operated by light.

3. *Thermoelectric.* Suppose we form a conducting loop circuit that consists of a length of iron wire, a length of copper wire, and a sensitive galvanometer, as shown in Figure 17-6. One copper-iron junction is put in a beaker of ice and water to maintain a low temperature; the other junction is heated by a gas flame. As the second junction is heated, the galvanometer indicates a current in the loop. An electric circuit incorporating such junctions is known as a **thermocouple.**

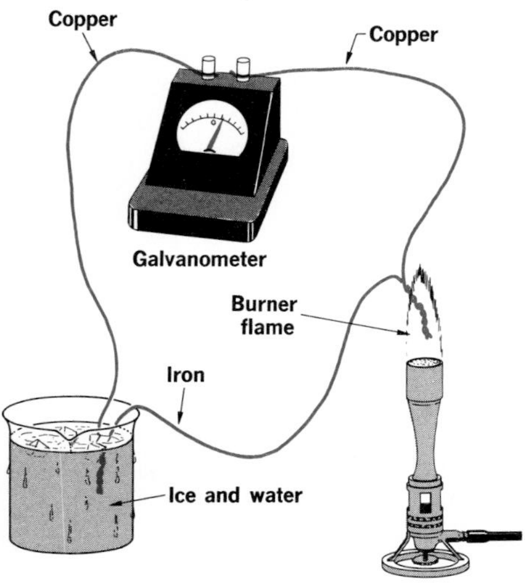

Figure 17-6. A thermocouple.

The magnitude of the current is related to the nature of the two metals forming the junctions and the temperature difference between the junctions. A thermocouple can be used as a sensitive thermometer to measure radiant energy. The iron-copper junction is useful for temperatures up to about 275°C. Junctions of copper and constantan, a copper-nickel alloy, are widely used in lower temperature ranges. Junctions of platinum and rhodium are used to measure temperatures ranging up to 1600°C.

Since heat energy is transformed directly into electric energy, the thermocouple is a thermoelectric source of current. If a sensitive galvanometer is properly calibrated, temperatures can be read directly. The hot junction of the thermocouple can be placed in a remote location where it would be impossible to read an ordinary thermometer.

Recall that the first law of thermodynamics tells us that heat can be converted into other forms of energy, but energy can neither be created nor destroyed. Thus in a closed thermodynamic system no energy will be gained or lost. Thermoelectric converters are fundamentally heat engines and, as for all other heat engines, are limited to the efficiency of the conversion process. The actual efficiency with which heat is converted to electricity is a function of the temperature difference within the operating system.

The heating effect of an electric current in the resistance of a conductor is an irreversible phenomenon. On the other hand, the basic thermoelectric effects discussed in this section are reversible. The best known of these effects is described in this section: when the junctions of two dissimilar metals are subjected to a difference in temperature, a current is set up in the loop. Known as the **Seebeck effect,** it is named after its discoverer who first observed the phenomenon in 1821. See Figure 17-7(A).

The **Peltier effect** is produced when a direct current is passed through a junction formed by two dissimilar metals. The junction becomes warmer or cooler depending on the direction of the current, as shown in Figure 17-7(B).

Another thermoelectric effect, called the *Thomson effect,* is observed in certain homogeneous materials. If a single solid is subjected to a temperature gradient between opposite faces, a difference of potential is established across it. This potential difference can be used to supply an electric current in an external circuit connected to the solid.

Certain semiconductor materials show thermoelectric properties and perform the heat-to-electricity conversion more efficiently than metals. In addition to the application of semiconductor converters as electric generators, the Peltier effect makes possible their use as cooling devices.

Thermocouples in which heat from radioactive isotopes

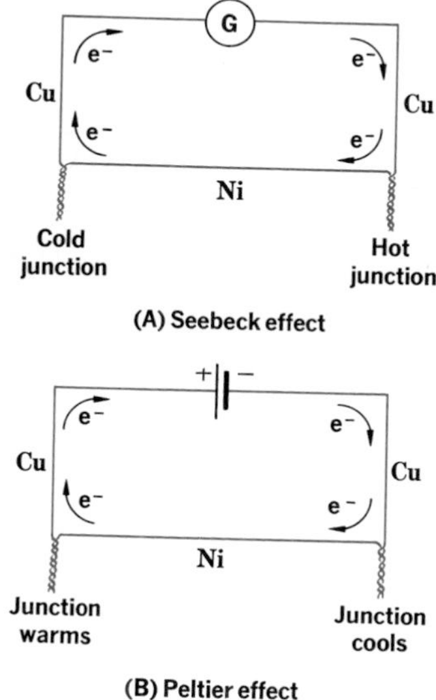

Figure 17-7. The Seebeck and Peltier effects for copper and nickel. One is the inverse of the other.

is converted to electric energy are used in some unmanned satellites. Nuclear heat sources that produce an electric current by thermoelectric conversion were used by astronauts on the surface of the moon to provide power for their lunar experiments. These energy converters (Figure 17-8) used plutonium-238 as the heat source.

Figure 17-8. A thermoelectric generator on the surface of the moon. Heat is supplied by nuclear energy. The shadow of an astronaut is seen at the right.

4. *Piezoelectric.* When certain crystals, such as Rochelle salt and quartz, are subjected to a mechanical stress, the opposite surfaces will become electrically charged. The difference of potential between the stressed surfaces is proportional to the amount of stress applied to the crystal. Reversing the stress reverses the charges also.

If the crystal is suitably supported and a conducting circuit is connected between surfaces having a difference of potential, electricity will flow through the circuit. Thus mechanical energy is transformed into electric energy. This transformation is known as the **piezoelectric effect,** and the crystal with its supporting mechanism is called a *piezoelectric cell.*

Crystal microphones use the piezoelectric effect to transform the alternating pressure of sound waves into an electric current that varies as the pressure of the sound waves varies. A crystal phonograph pickup responds in a similar manner to the varying stress caused by the needle riding in the groove of the record.

The piezoelectric effect is reversible; that is, the crystal experiences a mechanical strain when subjected to an electric stress. An alternating difference of potential properly placed across the crystal will cause mechanical vibrations proportional to the variations of the voltage. Crystal headphones use this reciprocal effect.

5. *Chemical.* There are certain chemical reactions that involve a transfer of electrons from one reactant to the other. The reactions that occur *spontaneously* (that is, of and by themselves) can be used as sources of continuous current. During the chemical reaction, electrons are removed from one reactant. A like number of electrons is added to the other reactant. If the reactants are separated in a conducting environment, the transfer of electrons will take place through an external circuit connecting the reactants. This electron transfer will continue for as long as reactants are available. Such an arrangement is known as an *electrochemical cell.* Chemical energy is transformed into electric energy during the chemical reaction.

A storage cell is reversible but it does not store electric energy.

Electrochemical cells include primary, storage, and fuel cells. A ***primary cell,*** a flashlight cell for example, is ordinarily replaced when its reactants are used up. *Storage cells,* as in an automobile battery, are reversible. Electric energy is initially stored as chemical energy, and the cell can be repeatedly "recharged" with energy. If the reactants in a primary cell are supplied continuously, the arrangement is called a *fuel cell*. Fuel cells have been used as the source of electric current in spacecraft.

17.4 The Dry Cell The essential components of a primary cell are an *electrolyte* and two *electrodes* of unlike materials, one of which reacts chemically with the electrolyte. Let us examine the carbon-zinc dry cell as a common example of the primary cell. A cut-away diagram of this cell is shown in Figure 17-9. The electrolyte is a moist paste of ammonium chloride. One electrode is a carbon rod. The other electrode, a zinc cup, is consumed as the cell is used.

When an external circuit is connected across the cell, zinc atoms react with the electrolyte and it loses electrons, which remain with the zinc cup. This reaction leaves the zinc electrode *negatively charged*. We shall refer to the negative electrode as the ***cathode.*** *It is the electron-rich electrode.*

Simultaneously, the electrolyte removes a like number of electrons from the carbon electrode. This reaction leaves the carbon electrode positively charged. We shall call the positive electrode the ***anode.*** *It is the electron-poor electrode.*

The electrons move through the external circuit from cathode to anode because of the difference in potential between the electrodes. Thus the cell acts as a kind of electron pump. See Figure 17-10. It removes electrons with low potential energy from the anode and supplies electrons with high potential energy to the cathode while current is in the external circuit. Electrons flow through the external circuit from cathode to anode. This unidirectional current is an example of *direct current*, dc.

If we open the circuit and stop the current, the *maximum* potential difference is quickly established across the cell. The magnitude of this **open-circuit** potential difference is dependent solely on the materials that make up the electrodes. For the carbon-zinc cell it is about 1.5 volts.

The open-circuit potential difference can be referred to as the *emf*, or the electromotive force, of the cell. The symbol for emf is $\mathscr{E}$ and the unit is the *volt* (V). The cell itself must supply a definite amount of energy for each unit of charge that is stored on either electrode; *the* ***emf*** *of a source is the energy per unit charge supplied by the source.* Thus the potential difference across the cell on an *open* circuit is equal to the emf of the cell.

$$V_{oc} = \mathscr{E}$$

As shown in Section 17.6, the potential difference, V, across a source of emf on a closed circuit is less than the emf, $\mathscr{E}$.

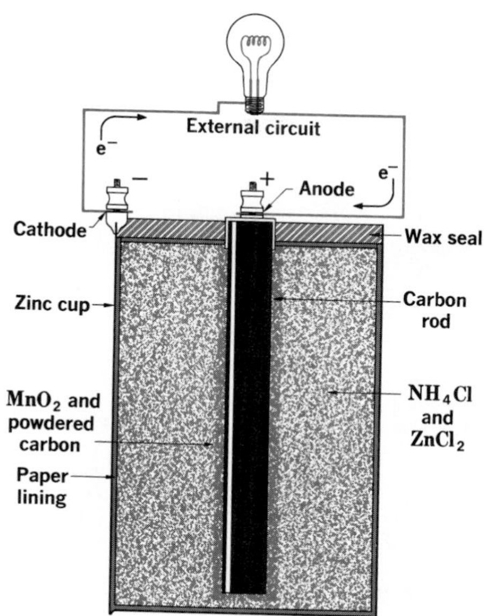

Figure 17-9. The dry cell is a practical source of direct current in the physics laboratory.

The definitions for electrodes are sometimes based on oxidation-reduction reactions rather than on their electric charge. In that case their names in primary cells become the reverse of those used here.

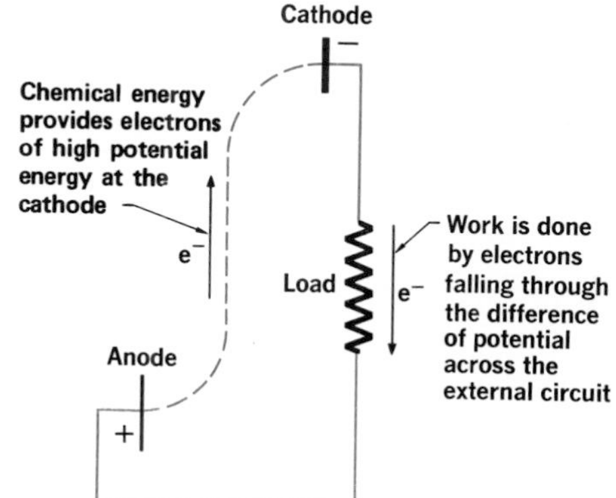

Figure 17-10. A voltaic cell provides electrons of high potential energy, the result of chemical reactions within the cell.

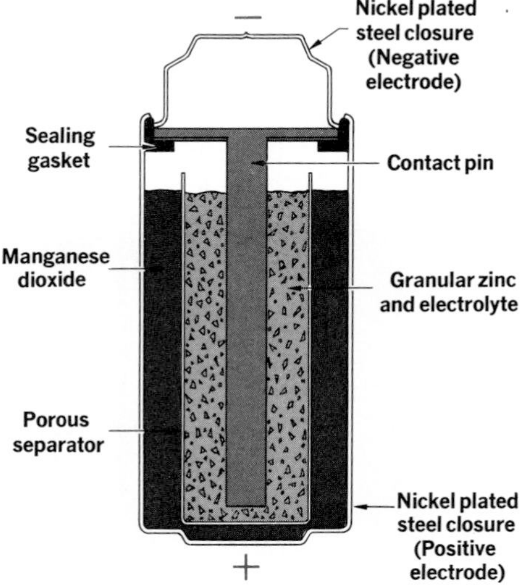

Figure 17-11. Cross section of an alkaline-manganese cell. An outer jacket (not shown) and appropriate terminal connections provide the conventional appearance of the carbon-zinc cell.

The carbon-zinc cell has an emf of 1.5 volts, regardless of the size of the cell. However, the larger the electrode surfaces, the larger the current the cell can deliver. The standard No. 6 dry cell is capable of supplying a *momentary* current up to 35 amperes. It should not be required to furnish more than 0.25 ampere of continuous current, however. The internal resistance can vary from less than 0.1 ohm as in the No. 6 cell to over 1 ohm as in small special-purpose cells.

Alkaline-manganese cells can be substituted for carbon-zinc cells in many applications. While more expensive to produce, alkaline cells have lower internal resistance, higher efficiency, and the ability to deliver from 50% to 100% more current than their carbon-zinc counterparts.

A sectional view of an alkaline-manganese cell is shown in Figure 17-11. The positive electrode consists of manganese dioxide in conjunction with a steel can that serves as the electrode terminal. The negative electrode is granulated zinc mixed with a highly conductive potassium hydroxide electrolyte. The nominal potential difference across the cell is 1.5 volts.

17.5 Combinations of Cells Energy-consuming devices that form the loads of electric circuits are designed to draw the proper current for their operation when a specific potential difference exists across their terminals. For example, a lamp designed for use in a 6-volt electric system of an automobile would draw an excessive current and burn out if placed in a 12-volt system. A lamp that is intended for use in a 12-volt system, on the other hand, would not draw sufficient current at 6 volts to function as intended.

Electrochemical cells are sources of emf. Each cell furnishes a certain amount of energy to the circuit for each coulomb of charge that is moved. It is often necessary to combine cells in order to provide either the proper emf or an adequate source of current. Groups of cells can be connected in *series,* in *parallel,* or in *series-parallel* combinations. Two or more electrochemical cells connected together form a *battery.* The emf of a battery depends on the emf of the individual cells and the way in which they are connected.

A *series* combination of cells is shown in Figure 17-12. Observe that the positive terminal of one cell is connected to the external circuit, and the negative terminal is connected to the positive terminal of a second cell. The negative terminal of this cell is connected to the positive terminal of a third cell, and so on. The negative terminal of the

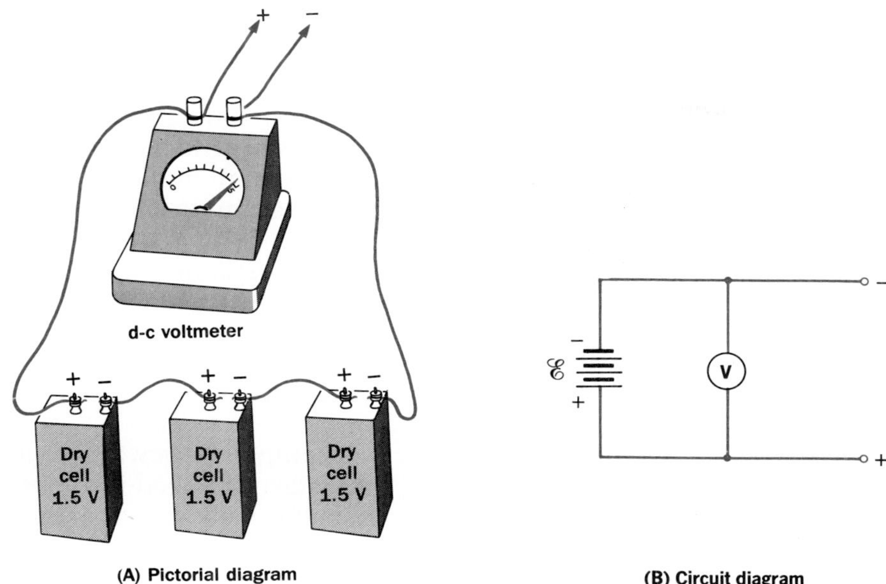

Figure 17-12. Dry cells connected in series to form a battery. The same quantity of electric charge flows through each cell.

last cell in the series is connected to the load to complete the circuit. Thus the cells are joined end to end and *the same quantity of electricity must flow through each cell.*

Suppose the battery is thought of as a group of electron pumps connected in *series* (adding). Each electron pump (cell) removes electrons of low potential energy from its anode and supplies electrons of high potential energy to its cathode. The first pump lifts electrons to a certain level of potential energy, the second pump takes them from this level to the next higher level, the third lifts them to a still higher potential energy level, and so on. Similarly, the emfs of the cells are added to give the emf of the battery.

A battery made up of cells connected in *series* has the following characteristics:

1. *The emf of the battery is equal to the sum of the emfs of the individual cells.*
2. *The current in each cell and in the external circuit has the same magnitude throughout.*
3. *The internal resistance of the battery is equal to the sum of the internal resistances of the individual cells.*

Resistances in series are treated analytically in Section 17.8.

In a *parallel* combination of cells, all negative terminals are connected together. One side of the external circuit is then connected to any one of these common terminals. The positive terminals are connected together and the other side of the external circuit can be connected to any one of these common terminals to complete the circuit. See Figure 17-13. This arrangement has the effect of increasing the total cathode and anode surface areas. The battery is then roughly the equivalent of a single cell having greatly enlarged electrodes in contact with the electrolyte and a lower internal resistance. Each cell of a group in parallel merely furnishes its proportionate share of the total circuit current. Of course, cells should not be connected in parallel unless they have equal emfs.

A battery of *identical* cells connected in *parallel* has the following characteristics.

1. *The emf is equal to the emf of each separate cell.*
2. *The total current in the circuit is divided equally among the cells.*
3. *The reciprocal of the internal resistance of the battery is equal to the sum of the reciprocals of the internal resistances of the cells.*

Resistances in parallel are treated analytically in Section 17.9.

The lead-acid storage battery for automobiles with 6-volt electric systems consists of three cells connected in series. The emf is approximately 6.6 volts. Six cells are connected in series to make up the battery for 12-volt systems.

As mentioned previously, a No. 6 dry cell should not be required to deliver more than 0.25 ampere of continuous current. Suppose we have an electric device designed to perform with a 0.75-ampere continuous current in a 3-volt circuit. How should the 1.5-volt dry cells be grouped to make up the battery for operating the device?

Certainly, two dry cells in series would provide an emf of 3 volts for the circuit. Three cells in parallel would require no more than 0.25 ampere from each cell. Thus a series-parallel arrangement of dry cells, as shown in Figure 17-14, is appropriate.

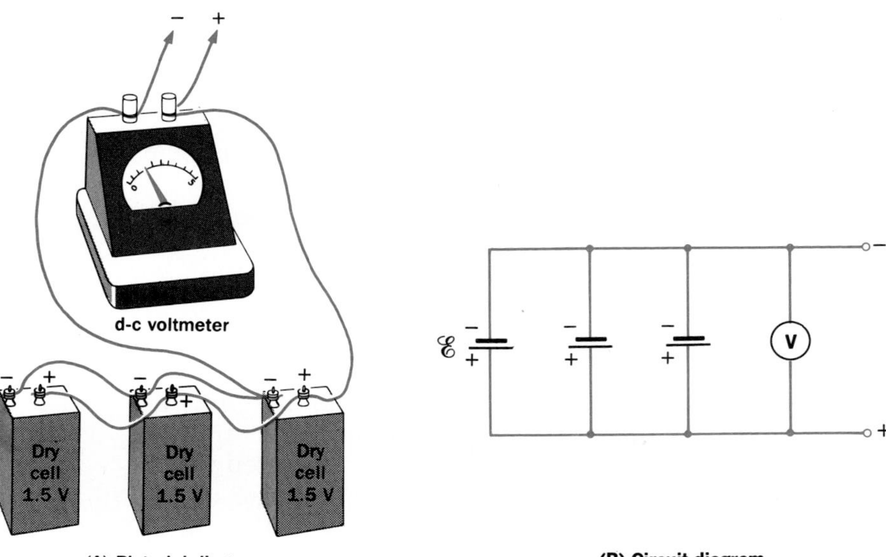

Figure 17-13. Dry cells connected in parallel to form a battery.

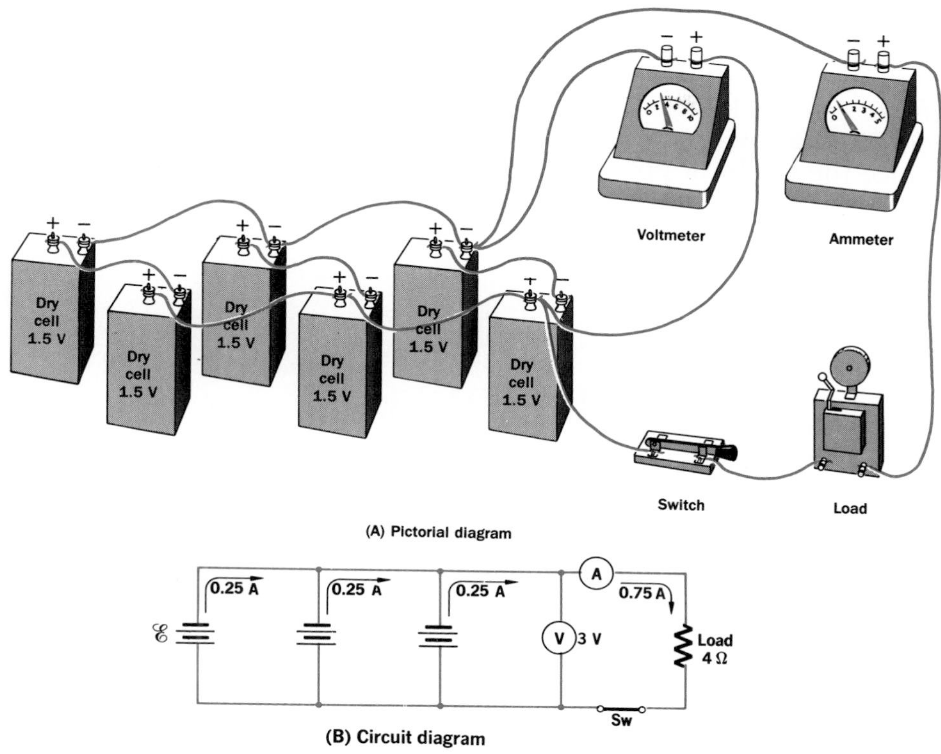

Figure 17-14. A circuit using a battery that consists of a series-parallel arrangement of cells.

QUESTIONS: GROUP A

1. What happens when a conductor connects the plates of a charged capacitor?
2. (a) Define *electric current.* (b) What is the SI unit for electric current?
3. What is an electrolyte?
4. What is the difference between a transient current and a continuous current?
5. Name and describe the two basic methods of producing a continuous current.
6. (a) What is the electron-rich electrode in a dry cell called? (b) What is the electron-poor electrode called?
7. (a) What is meant by the term *emf?* (b) Define "emf."
8. What is an electric battery?

GROUP B

9. What is meant by "drift velocity"?
10. Why do conductors normally offer resistance to a flow of electric current?
11. Why must energy be continuously pumped into a circuit to keep an electric current flowing?
12. For each of the following devices, indicate the type of energy that is being converted to electrical energy: (a) storage battery (b) electric genera-

tor (c) piezoelectric cell (d) thermocouple (e) fuel cell (f) photoelectric cell.

13. You have three dry cells and a small lamp. How would you arrange the cells (a) to produce the brightest possible light (b) to keep the light lit for the longest possible time?
14. What determines the efficiency of a thermoelectric converter?
15. A car battery is rated at 5 ampere-hours. What is this telling you?

SERIES AND PARALLEL CIRCUITS

17.6 Ohm's Law for d-c Circuits Resistance, R, is defined as the opposition to the flow of electric charge. Metallic substances in general are classed as good conductors of electricity. Silver and copper, because of their abundance of free electrons, are excellent conductors; yet even these metals offer some opposition to the current in them. Through its normal range of operating temperatures, every conductor has some inherent resistance. Thus every device or component in an electric circuit offers some resistance to the current in the circuit.

Copper wire is used to connect various circuit components in electric circuits because ordinarily its resistance is low enough to be neglected. Certain metallic alloys offer unusually high resistance to the flow of electricity; *nichrome* and *chromel* are notable examples. Spools wound with wire made from these alloys can be used in an electric circuit to provide either *fixed* amounts of resistance or *variable* resistance in definite amounts.

Carbon granules can be mixed with varying amounts of clay and molded into cylinders having a definite resistance. These devices, known as *carbon resistors,* are commonly used in electronic circuits.

Figure 17-15. Georg Simon Ohm, the German physicist for whom the unit of resistance is named.

The German physicist Georg Simon Ohm (1789–1854) discovered that *in a* **closed circuit** *the ratio of the emf of the source to the current in the circuit is a constant.* This constant is the *resistance* of the circuit. The above statement, known as ***Ohm's law of resistance,*** describes a basic relationship in the study of current electricity. It can be stated mathematically:

$$\frac{\mathscr{E}}{I} = \mathbf{R} \quad \text{or} \quad \mathscr{E} = IR$$

The units are the volt, ampere, and ohm respectively. The *ohm* (Ω) has the following dimensions:

$$\Omega = \frac{\text{V}}{\text{A}} = \frac{\text{J/C}}{\text{C/s}}$$

$$\Omega = \frac{\text{J}\cdot\text{s}}{\text{C}^2} = \frac{\text{N}\cdot\text{m}\cdot\text{s}}{\text{C}^2}$$

$$\Omega = \frac{\text{kg}\cdot\text{m}^2\cdot\text{s}}{\text{s}^2\cdot\text{C}^2} = \frac{\text{kg}\cdot\text{m}^2}{\text{C}^2\cdot\text{s}}$$

Suppose there is an emf of 12 V across an electric circuit that has a 33-Ω resistance. The current in the circuit is

$$\mathscr{E} = IR \quad \text{so} \quad I = \mathscr{E}/R$$

$$I = \frac{12\text{ V}}{33\ \Omega} = 0.36\text{ A}$$

Now suppose the circuit resistance is doubled. In this situation the opposition to the current is doubled but the emf remains the same, and so we should expect the current in the circuit to be reduced to one-half the former value.

$$I = \frac{12\text{ V}}{66\ \Omega} = 0.18\text{ A}$$

The emf, $\mathscr{E}$, applied to a circuit equals the drop in potential across the *total* resistance of the circuit, but Ohm's law applies equally well to *any part* of a circuit that does not include a source of emf. Suppose we consider the circuit shown in Figure 17-16. Points **A** and **B** are the terminals of the battery that supply an emf of 12.0 volts to the circuit. The total resistance of the circuit is 12.0 ohms. Of this total, 0.20 ohm is **internal resistance,** r, of the battery and is conventionally represented as a resistor in series with the battery inside the battery terminals. The remaining 11.8 ohms of resistance consists of a **load resistor,** R_L, in the external circuit.

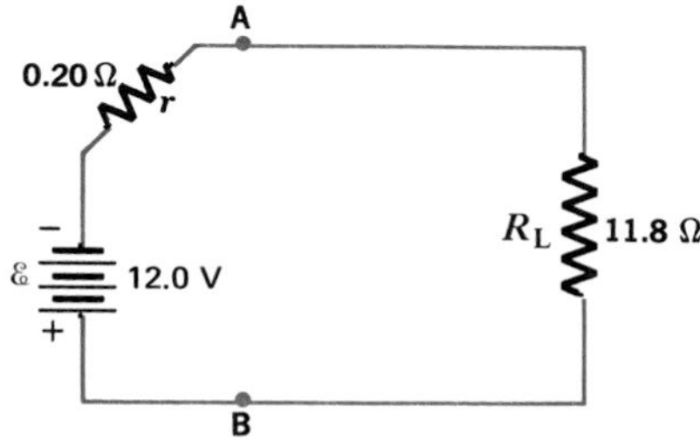

Figure 17-16. Ohm's law applies to an entire circuit or to any part of the circuit.

For the entire circuit, the total resistance, R_T, is equal to the sum of internal and external resistances.

$$R_T = R_L + r$$

The current in the entire circuit is

$$I = \frac{\mathscr{E}}{R_T} = \frac{\mathscr{E}}{R_L + r}$$

$$I = \frac{12.0\text{ V}}{11.8\ \Omega + 0.20\ \Omega} = 1.00\text{ A}$$

When we apply Ohm's law to a part of a circuit that does not include a source of emf, we are concerned with the potential difference, V, the current, I, and the resistance, R, that apply only to that part of the circuit. Thus Ohm's law for a part of the circuit becomes

$$V = IR$$

The drop in potential across the external circuit of Figure 17-16 must be

$$V_L = IR_L$$
$$V_L = 1.00\ \text{A} \times 11.8\ \Omega = 11.8\ \text{V}$$

The drop in potential across the internal resistance of the battery due to current in the battery can be found by a similar application of Ohm's law.

$$V_r = Ir$$
$$V_r = 1.00\ \text{A} \times 0.20\ \Omega = 0.20\ \text{V}$$

This is a drop in potential across the battery resistance that removes energy from the electrons. It is, therefore, opposite in sign to the emf of the battery. The potential difference across the battery terminals of a closed circuit with a current I is determined by subtracting this drop in potential from the emf.

$$V = \mathscr{E} - Ir$$

The closed-circuit potential difference across the terminals of a source of emf that is applied to the external circuit will always be less than the emf by the amount Ir volts. This is not an important difference for a source of emf of low internal resistance unless an excessive current is in the circuit. (In the discussion of Sections 17.8–17.10, the internal resistance of a source of emf will be neglected and the emf will be assumed to be applied to the external circuit.)

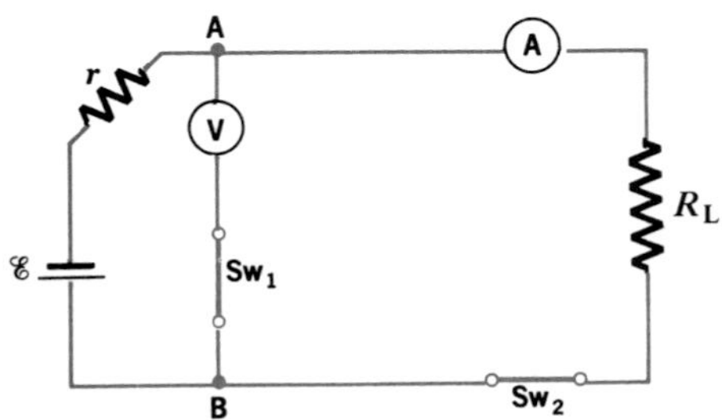

Figure 17-17. A circuit for measuring the internal resistance of a cell.

17.7 Determining Internal Resistance A good quality voltmeter is an instrument of very high resistance. When such a voltmeter is connected across the terminals of a source of emf, even a source with low internal resistance, negligible current is drawn from the source. The meter registers an "open circuit" voltage that, for all practical purposes, is the emf of the source. Such a measurement of the emf of a cell or battery provides a convenient method of determining its internal resistance. The following example, based on the circuit shown in Figure 17-17, illustrates this method of determining the internal resistance, r, of a dry cell.

EXAMPLE A dry cell gives an open-circuit voltmeter reading of 1.5 V with **Sw_1** closed and **Sw_2** open. The voltmeter, **V**, is then removed from the circuit by opening switch **Sw_1**, and an external load of 2.8 Ω, R_L, is connected across the source of emf by closing switch **Sw_2**. The ammeter, **A**, in the external circuit reads 0.50 A. Calculate the internal resistance, r, of the cell.

Given	Unknown	Basic equations
$\mathscr{E} = 1.5$ V $I = 0.50$ A $R_L = 2.8\ \Omega$	r	$\mathscr{E} = IR_T$ $R_T = R_L + r$

Solution

Working equation: $\mathscr{E} = I(R_L + \ )$

$$r = \frac{\mathscr{E} - IR_L}{I} = \frac{1.5\ \text{V} - (0.50\ \text{A})(2.8\ \Omega)}{0.50\ \text{A}}$$

$$= 0.20\ \Omega$$

PRACTICE PROBLEMS **1.** In the previous example, what would the voltmeter have read if it had been left connected while the current was in the load? *Ans.* 1.4 V

2. A resistance load of 7.0 Ω is connected across a battery with an emf of 4.5 V. The current in the external circuit is 0.60 A. Calculate the internal resistance of the battery. *Ans.* 0.50 Ω

3. A resistance load of 3.7 Ω is connected across a battery with an internal resistance of 0.30 Ω. The current in the circuit is 4.0 A. Calculate the potential difference across the load. *Ans.* 15 V

17.8 Resistances in Series If several electric devices are connected end to end in a circuit, electricity must flow through each in succession. There is a single path for the moving charge. The same current must be in each device or else there would be an accumulation of charge at different points around the conducting circuit. We know that a charge can be accumulated on a conductor only if it is isolated. *An electric circuit with components arranged to provide a single conducting path for current is known as a* ***series circuit.***

All elements connected in series in a circuit have the same magnitude of current.

Series-circuit operation has several inherent characteristics. If one component of a series circuit fails to provide a conducting path, the circuit is opened. Each component of the circuit offers resistance to the current, and this resistance limits the current in the circuit according to Ohm's law. When components are connected in series, their resistances are cumulative. The greater the number of resistive components, the higher is the total resistance and the smaller is the current in the circuit for a given applied emf. Obviously, since the current must be the same at all points in the circuit, *all devices connected in series must be designed to function at the same current magnitude.*

From Ohm's law, the drop in potential across each component of a series circuit is the product of the current in the circuit and the resistance of the component. Electrons lose energy as they fall through a difference of potential; in a series circuit these losses occur in succession and are therefore cumulative. *The drops in potential across successive components in a series circuit are additive, and their sum is equal to the potential difference across the whole circuit.*

These observations are summarized by stating three cardinal rules for resistances in series:

1. *The current in all parts of a series circuit has the same magnitude.*

$$I_T = I_1 = I_2 = I_3 = \text{etc.}$$

2. *The sum of all the separate drops in potential around a series circuit is equal to the applied emf.*

$$\mathscr{E} = V_1 + V_2 + V_3 + \text{etc.}$$

3. *The total resistance in a series circuit is equal to the sum of all the separate resistances.*

$$R_T = R_1 + R_2 + R_3 + \text{etc.}$$

These rules have been applied to a circuit consisting of several resistors connected in series as shown in Figure 17-18. For simplicity, the internal resistance of the source is assumed to be negligible and is omitted from the circuit diagram. The computations are shown at the right of the

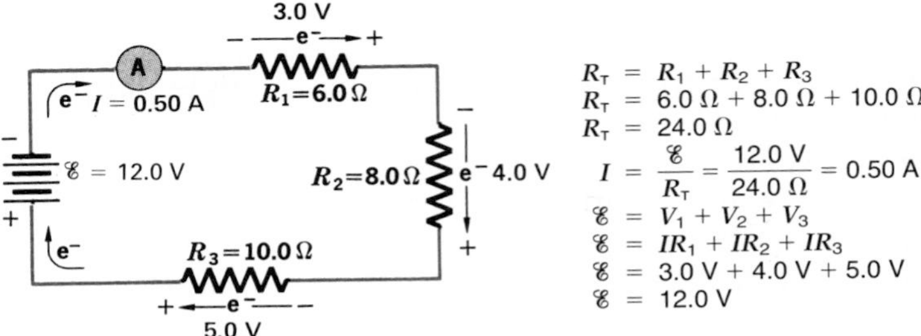

Figure 17-18. A series circuit with Ohm's law relationships calculated.

circuit. Observe that the "positive" side of R_1 is 3.0 volts *positive with respect to the cathode of the battery* and is 9.0 volts *negative with respect to the anode of the battery*. The three voltage drops around the external circuit are in series, and their sum equals the applied emf in magnitude, but their signs are opposite.

Stated in another way, *the algebraic sum of all the changes in potential occurring around the complete circuit is equal to zero*. This is a statement of ***Kirchhoff's second law,*** credited to Gustav Robert Kirchhoff (1824–1887), an eminent German physicist. Since potential difference is an expression of work (or energy) per unit charge involved in transporting a charge around a closed circuit, Kirchhoff's second law can be recognized as an application of the law of *conservation of energy* to electric circuits.

Kirchhoff's first law is given in Section 17.9.

17.9 Resistances in Parallel Some of the characteristics of series operation of electric devices were mentioned in the preceding section. It would be quite disconcerting to have all the lights go out in your home each time one lamp were turned off. For this reason each lamp is on a separate circuit so that each is operated independently. Such lamps are connected in parallel. *A **parallel circuit** is one in which two or more components are connected across two common points in the circuit to provide separate conducting paths for current.*

Since there can be only one difference of potential between any two points in an electric circuit, *electric devices or appliances connected in parallel should usually be designed to operate at the same voltage.* The currents in the separate branches must vary inversely with their separate resistances since the same potential difference exists across each branch of the parallel circuit.

By Ohm's law, which can be applied to any part of a circuit, the current in each branch is

$$I = \frac{V}{R}$$

But V is constant for all branches in parallel. Then

$$I_{\text{(for each branch)}} \propto \frac{1}{R_{\text{(of that branch)}}}$$

In Figure 17-19 it is apparent that more electric charge can flow between points **A** and **B** with the two parallel paths R_1 and R_2 in the circuit than with either path alone. The current I entering junction **A** must be the same as the current $(I_1 + I_2)$ leaving junction **A.** Similarly, the current $(I_1 + I_2)$ entering junction **B** must be the same as the current I leaving junction **B.** Then I must be equal to $I_1 + I_2$.

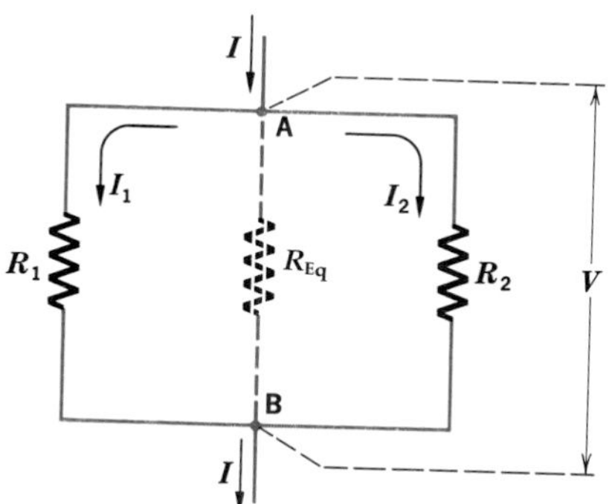

Figure 17-19. Two resistances in parallel and the single equivalent resistance, R_{Eq}, that can replace them with no change in the circuit parameters.

All elements connected in parallel in a circuit have the same voltage across them.

Let us consider in another way the currents entering and leaving junctions **A** and **B**. Electric current is the rate of transfer of charge, $I = Q/t$. In the steady-state operation of the circuit, the charge of **A** can neither increase nor decrease. Current I carries charge toward this junction and I_1 and I_2 carry charge away from it. Thus

$$I - I_1 - I_2 = 0$$

Similarly, currents I_1 and I_2 carry charge toward junction **B** and I carries charge away from it. Using the same sign convention for current direction as before,

$$I_1 + I_2 - I = 0$$

These expressions tell us that *the algebraic sum of the currents at any circuit junction is equal to zero.* This is known as ***Kirchhoff's first law.*** It states the law of *conservation of charge* in electric circuits. Kirchhoff's two laws define the basic principles we must employ in our interpretation of electric circuits: *that energy and charge are conserved.*

In Figure 17-19, assume that $R_1 = 1\bar{0}\ \Omega$, $R_2 = 15\ \Omega$, *and* $V = 3\bar{0}$ V. From Ohm's law, the current in R_1 is

$$I_1 = \frac{V}{R_1} = \frac{3\bar{0}\ \text{V}}{1\bar{0}\ \Omega} = 3.0\ \text{A}$$

The current in R_2 is

$$I_2 = \frac{V}{R_2} = \frac{3\bar{0}\ \text{V}}{15\ \Omega} = 2.0\ \text{A}$$

and

$$I = I_1 + I_2 = 5.0\ \text{A}$$

According to Ohm's law, the current from **A** to **B**, which is I, is the quotient of the potential difference across **A-B** divided by the effective resistance between **A** and **B.** It is convenient to think of the effective resistance due to resistances in parallel as a single **equivalent resistance,** R_{Eq}, that, if substituted for the parallel resistances, would provide the same current in the circuit.

An equivalent resistance is shown in Figure 17-19.

$$I = \frac{V}{R_{Eq}}$$

Then,

$$R_{Eq} = \frac{V}{I} = \frac{3\bar{0}\ \text{V}}{5.0\ \text{A}} = 6.0\ \Omega$$

The equivalent resistance for the parallel combination of

R_1 and R_2 is *less than* either resistance present. As the current approaching junction **A** *must be larger* than either branch current, clearly the equivalent resistance of the parallel circuit *must be smaller* than the resistance of any branch.

Since $$I = \frac{V}{R_{Eq}}, \; I_1 = \frac{V}{R_1}, \; I_2 = \frac{V}{R_2}$$

and $$I = I_1 + I_2$$

then, $$\frac{V}{R_{Eq}} = \frac{V}{R_1} + \frac{V}{R_2}$$

Dividing by V, we get

$$\frac{1}{R_{Eq}} = \frac{1}{R_1} + \frac{1}{R_2} \qquad \text{(Equation 1)}$$

Equation 1 shows that the sum of the reciprocals of the resistances in parallel is equal to the reciprocal of their equivalent resistance. Then solving Equation 1 for R_{Eq},

$$R_1R_2 = R_{Eq}(R_1 + R_2)$$

and $$R_{Eq} = \frac{R_1R_2}{R_1 + R_2}$$

$$R_{Eq} = \frac{10\ \Omega \times 15\ \Omega}{10\ \Omega + 15\ \Omega} = \frac{150\ \Omega^2}{25\ \Omega}$$

$$R_{Eq} = 6.0\ \Omega$$

These observations are summarized by stating three cardinal rules for resistances in parallel.

1. *The total current in a parallel circuit is equal to the sum of the currents in the separate branches.*

$$I_T = I_1 + I_2 + I_3 + \text{etc.}$$

2. *The potential difference across all branches of a parallel circuit must have the same magnitude.*

$$V = V_1 = V_2 = V_3 = \text{etc.}$$

3. *The reciprocal of the equivalent resistance is equal to the sum of the reciprocals of the separate resistances in parallel.*

$$\frac{1}{R_{Eq}} = \frac{1}{R_1} + \frac{1}{R_2} + \frac{1}{R_3} + \text{etc.}$$

17.10 Resistances in Simple Networks Practical circuits can be quite complex compared to the series and parallel circuits we have considered. There can be resistances

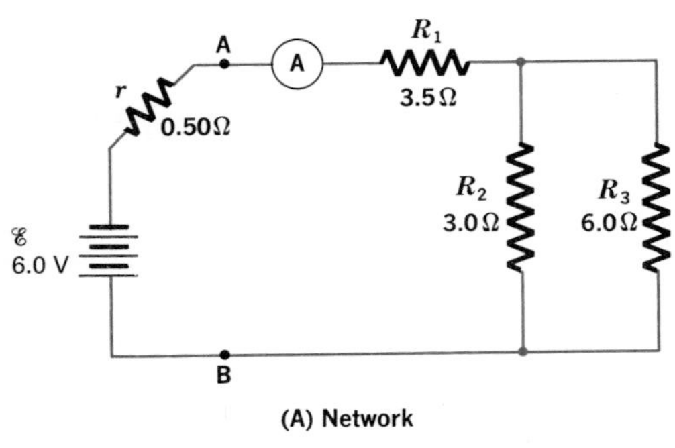

(A) Network

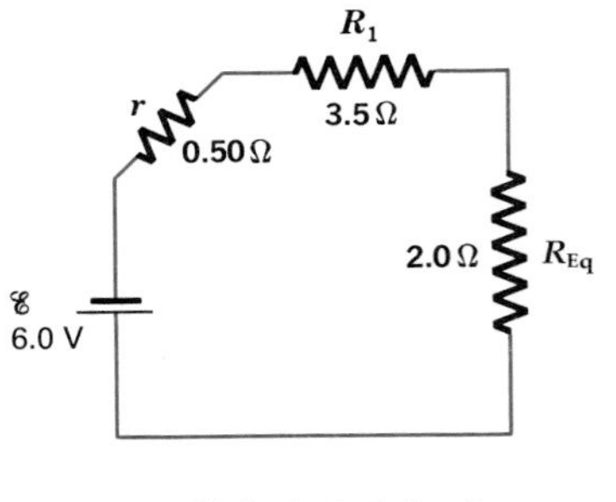

(B) Equivalent circuit

Figure 17-20. A simple network and its equivalent circuit.

in series, other resistances in parallel, and different sources of emf. Such complex circuits are commonly called *networks*.

A simple network is shown in Figure 17-20(A). There is a single source of emf and one resistor in series with the combination of two resistors in parallel. We will use this circuit to demonstrate the steps taken to simplify a network and reduce circuit problems to simple relationships.

We are given the emf and internal resistance of the battery and the resistance of each component resistor in the circuit. The first step is to find the total current in the circuit. Since the emf, $\mathscr{E}$, is known, the total resistance, R_T, must also be known in order to calculate the total current, I_T. To find the total resistance, we must first reduce the parallel resistances to an *equivalent* value.

$$\frac{1}{R_{Eq}} = \frac{1}{R_2} + \frac{1}{R_3} \quad \text{and} \quad R_{Eq} = \frac{R_2 R_3}{R_2 + R_3}$$

$$R_{Eq} = \frac{3.0\ \Omega \times 6.0\ \Omega}{3.0\ \Omega + 6.0\ \Omega}$$

$$R_{Eq} = 2.0\ \Omega\text{, the equivalent resistance}$$

A simpler *equivalent* circuit can now be drawn showing this equivalent resistance in series with R_1 and r. See Figure 17-20(B). The total resistance is now readily seen to be 6.0 ohms.

$$R_T = R_1 + R_{Eq} + r = 6.0\ \Omega$$

And

$$I_T = \frac{\mathscr{E}}{R_T} = \frac{6.0\ \text{V}}{6.0\ \Omega} = 1.0\ \text{A}$$

The drop in potential across R_1 is

$$V_1 = I_T R_1 = 1.0\ \text{A} \times 3.5\ \Omega$$

$$V_1 = 3.5\ \text{V}$$

Since R_2 and R_3 are in parallel, the same potential drop must appear across each. This is found by considering the total current in the equivalent resistance of the parallel segment.

$$V_{Eq} = I_T R_{Eq} = 1.0\ \text{A} \times 2.0\ \Omega$$

$$V_{Eq} = 2.0\ \text{V}$$

The potential difference across the terminals of the battery is the difference between the emf and the $I_T r$ drop.

$$V = \mathscr{E} - I_T r = 6.0\ \mathrm{V} - (1.0\ \mathrm{A} \times 0.50\ \Omega)$$

$$V = 5.5\ \mathrm{V}$$

This result is verified by the fact that V must equal $V_1 + V_{Eq}$.

The current in R_2 may be found since the potential difference, V_{Eq}, is known.

$$I_2 = \frac{V_{Eq}}{R_2} = \frac{2.0\ \mathrm{V}}{3.0\ \Omega}$$

$$I_2 = 0.67\ \mathrm{A}$$

The current in R_3 is

$$I_3 = \frac{V_{Eq}}{R_3} = \frac{2.0\ \mathrm{V}}{6.0\ \Omega}$$

$$I_3 = 0.33\ \mathrm{A}$$

This is verified by the fact that I_3 must equal $I_T - I_2$.

A more complex network is shown in Figure 17-21(A).

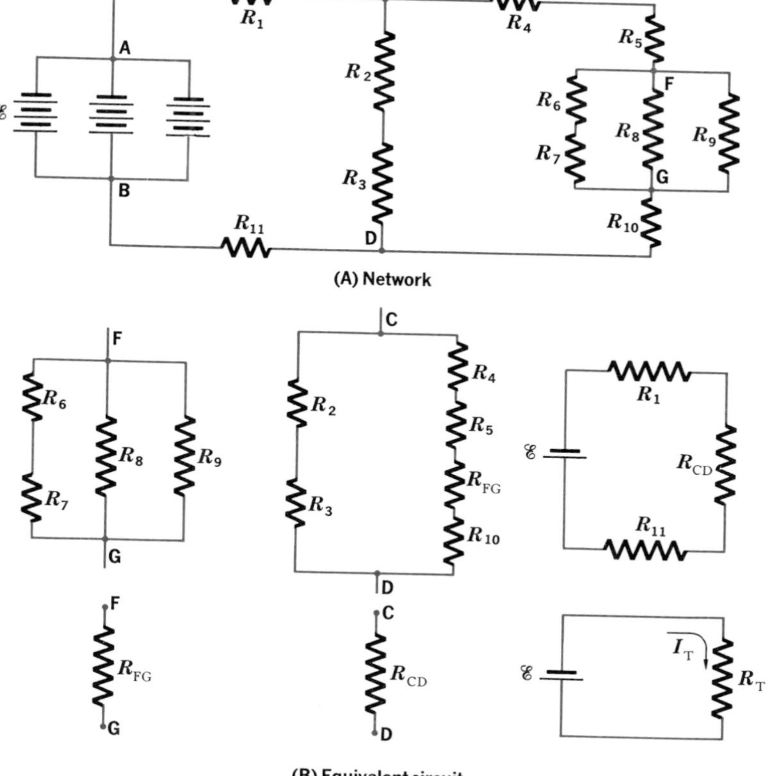

Figure 17-21. A complex network can be reduced stepwise to its simple equivalent circuit by applying the equivalent-circuit analysis techniques.

In simplifying a complex circuit, it is best to work from the "inside" out.

The battery is a series-parallel arrangement of similar cells, and so the magnitude of the emf will depend on the number of cells in *series*. The total current must be known. The approach is the same as that in our earlier example. First, reduce the segment **F-G** to its equivalent value. We will label this R_{FG}. Now R_{FG} is in series with R_4, R_5, and R_{10} to form one branch of the segment **C-D.** When the two branches of **C-D** are reduced to their equivalent, R_{CD}, we can consider R_1, R_{CD}, and R_{11} to be in series. This reduction provides R_T, and I_T may be calculated. See the diagrams in Figure 17-21(B).

Suppose we need to know the magnitude of current in R_7 and the drop in potential across this resistor. Now the current in R_{CD} must be I_T and the drop in potential across **C-D** must be $I_T R_{CD}$, or V_{CD}. The current in the branch of **C-D** containing R_{FG} is the quotient of V_{CD} divided by the sum $(R_4 + R_5 + R_{FG} + R_{10})$. This current (we will label it I_{FG}) is in R_{FG}, and the potential drop, V_{FG}, across R_{FG} is the product $I_{FG}R_{FG}$.

Since R_6 and R_7 are in series, the resistance of this branch of **F-G** is equal to their sum. The current in R_7 is V_{FG} divided by $(R_6 + R_7)$. With I_7 known, the drop in potential across R_7 is the product I_7R_7. Observe that we have depended on Ohm's law and the six extensions of Ohm's law listed as cardinal rules for series and parallel circuits in Sections 17.8 and 17.9.

17.11 The Laws of Resistance Various factors affect the resistance of a conductor.

1. *Temperature.* The resistance of all substances changes to some degree with changes in temperature. *In the case of pure metals and most metallic alloys, the resistance increases with a rise in temperature.* On the other hand, carbon, semiconductors, and many electrolytic solutions, show a decrease in resistance as their temperature is raised. A few special alloys, such as constantan and manganin, show very slight changes in resistance over a large range of temperatures. For practical purposes, the resistance of these alloys can be considered independent of temperature.

Thermal agitation of the particles composing a metallic conductor increases as the metal is heated. As the temperature is lowered, thermal agitation diminishes. The electric resistance of metals is related to this thermal agitation; in some substances, the resistance drops to zero at low temperatures. The graph of resistance as a function of temperature of a metallic conductor is essentially linear over the normal temperture range of the solid state of a metal. Thus when the electric resistance of a conductor is stated, the temperature of the material at which this resistance applies should be indicated. See the graph in Figure 17-22.

2. *Length.* Resistances in series are added to give the total resistance of the series circuit. From this fact we can conclude that 10 meters of a certain conductor should have a resistance 10 times that of 1 meter of the same conductor. Experiments prove this relationship to be true: *The resistance of a uniform conductor is directly proportional to the length of the conductor, or*

$$R \propto l$$

where l is the length of the conductor.

3. *Cross-sectional area.* If three *equal* resistances are connected in parallel in an electric circuit, the equivalent resistance is one-third the resistance of any one branch. The equal currents in the three branches add to give the total current in the circuit.

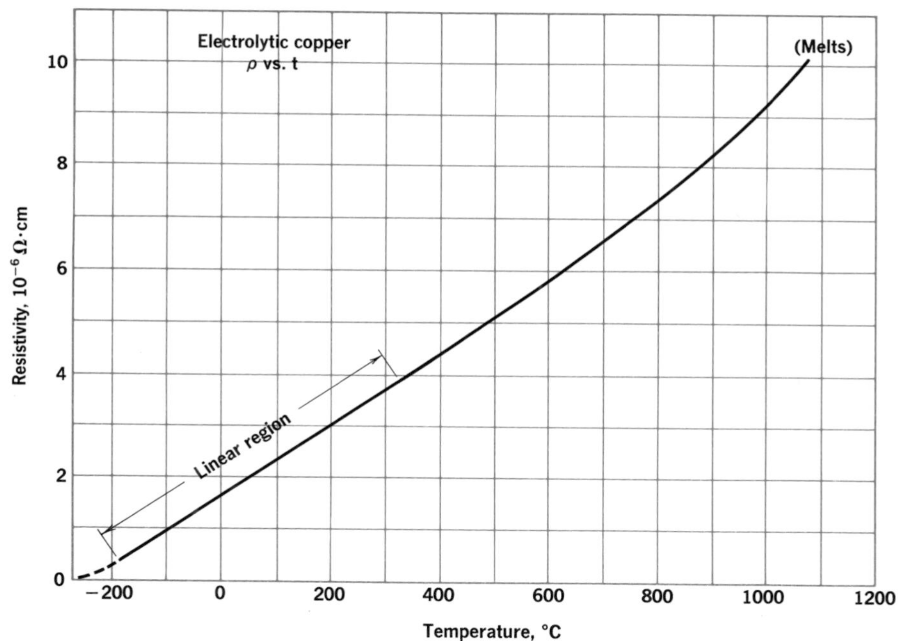

Figure 17-22. The resistivity of copper as a function of temperature.

Assuming the three resistances to be three similar wires 1 meter in length and each having a cross-sectional area of 0.1 cm^2, we can think of the equivalent resistance as a similar wire of the same length but having a cross-sectional area of 0.3 cm^2. This deduction is reasonable since a wire

Current density is the current per unit area of the cross-sectional area of a conductor.

carrying a direct current of constant magnitude has a constant *current density* throughout its cross-sectional area. If the parallel resistances are designated R_1, R_2, and R_3, then

$$\frac{1}{R_{Eq}} = \frac{1}{R_1} + \frac{1}{R_2} + \frac{1}{R_3}$$

But

$$R_1 = R_2 = R_3$$

So

$$\frac{1}{R_{Eq}} = \frac{3}{R_1}$$

And

$$R_1 = 3R_{Eq}$$

We can conclude that the resistance of the large wire, R_{Eq}, is one-third the resistance of the small wire, R_1, and experiments prove this conclusion to be true. *The resistance of a uniform conductor is inversely proportional to its cross-sectional area, or*

$$R \propto \frac{1}{A}$$

where A is the cross-sectional area of the conductor.

4. *Nature of the materials.* A copper wire having the same length and cross-sectional area as an iron wire offers about one-sixth the resistance to an electric current. A similar silver wire presents even less resistance. *The resistance of a given conductor depends on the material of which it is made.*

Taking these four factors into account, several useful conclusions become apparent. For a uniform conductor of a given material and temperature, the resistance is directly proportional to the length and inversely proportional to the cross-sectional area.

$$R \propto \frac{l}{A}$$

Suppose we introduce a proportionality constant, called *resistivity,* represented by the Greek letter ρ (rho). The laws of *resistance* for wire conductors can then be summarized by the following expression:

$$R = \rho\frac{l}{A}$$

Resistivity depends on the material composing the wire and its temperature, as shown in Figure 17-22. If R is expressed in ohms, l in centimeters, and A in centimeters^2, then ρ has the dimensions *ohm-centimeter* ($\Omega\cdot$cm).

$$\rho = \frac{RA}{l}$$

$$\rho = \frac{\Omega \cdot \text{cm}^2}{\text{cm}} = \Omega \cdot \text{cm}$$

Resistivity, *ρ, is equal to the resistance of a wire 1 cm long and having a uniform cross-sectional area of 1 cm*^{2}. (The resistivities of various materials are listed in Appendix B, Table 20. The wire gauge scale and the dimensions of wire by gauge number are given in Table 21.) If the resistivity of a material is known, the resistance of any other wire composed of the same material can be calculated. See the following example.

EXAMPLE A spool of 20-gauge copper wire at 20.0°C has a total resistance of 0.655 Ω. Calculate the length of the wire in meters.

Given	Unknown	Basic equation
Gauge No. = 20 $R = 0.655$	l	$R = \rho \frac{l}{A}$

Solution

The cross-sectional area, A, of a 20-gauge wire is given in Appendix B, Table 21, as 0.517 6 mm^2. The resistivity, ρ, of copper is given in Appendix B, Table 20, as $1.724 \times 10^{-6}\ \Omega\cdot\text{cm}$ at 20.0°C.

Working equation: $l = RA/\rho$

$$= \frac{(0.655\ \Omega)(5.176 \times 10^{-1}\ \text{mm}^2)(10^{-2}\ \text{cm}^2/\text{mm}^2)}{1.724 \times 10^{-6}\ \Omega \cdot \text{cm}}$$

$$= 1.97 \times 10^3\ \text{cm} = 19.7\ \text{m}$$

PRACTICE PROBLEMS **1.** A spool of 28-gauge constantan wire is $1\bar{0}$ meters long. Calculate the resistance of the wire at a temperature of 20.0°C. *Ans.* $6\bar{0}\ \Omega$

2. A spool of wire 64.0 meters long and 0.255 mm in diameter is found to have a resistance of 1440 Ω at 20°C. Determine the probable composition of the wire by calculating its resistivity and consulting Appendix B, Table 20. *Ans.* nichrome

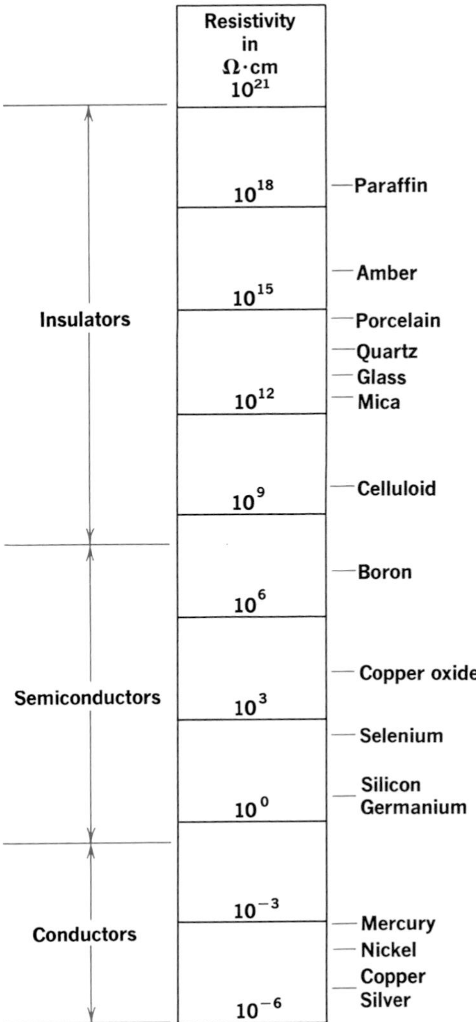

Figure 17-23. A resistivity spectrum. Observe the enormous range of resistivities between the best conductor and the best insulator.

17.12 Range of Resistivities The enormous range of resistivities of common materials is shown in Figure 17-23. The resistance of the best insulator can be greater than that of the best conductor by a factor of approximately 10^{25} Insulators and semiconductors can decrease in resistivity with increasing temperature and with increasing potential gradient. We can interpret this latter behavior to mean that these materials do not follow Ohm's law.

A small group of **semiconductors** has assumed considerable importance because of unusual variations in resistance with changes in temperature and potential gradient. Copper oxide *rectifiers* will pass a current when a potential difference is applied in one direction but not in the opposite direction. Silicon and germanium *transistors* are substituted for vacuum tubes in many electronic circuits.

At extremely low temperatures, some metals exhibit the unusual property of zero resistance. The production of very low temperatures and the study of the properties of materials at low temperatures are called **cryogenics.** Cryogenic temperatures are generally considered to be those below the boiling point of liquid oxygen, which is 80 K (−193°C). Ultra-low temperature research began with the liquefaction of helium in 1908 by the Dutch physicist Heike Kamerlingh Onnes (1853–1926). The boiling point of liquid helium is about 4 K (−269°C).

We have learned that the resistivity of a conductor is a function of its temperature. We would expect that the resistivity would become less and less as the temperature of a conductor drops, until the conductor offers no resistance at all to the passage of electrons. In 1911 Kamerlingh Onnes found that resistivity did indeed continue to drop with falling temperature. However, instead of gradually approaching zero resistivity, some materials had a specific temperature at which the resistivity dropped suddenly to zero. This temperature is called the *transition temperature. The condition of zero resistivity below the transition temperature of a material is called* ***superconductivity.***

Until 1986, the highest known temperature for a superconductor was 23.2 K (in a niobium-germanium alloy). Then, it was found that superconductivity can be achieved in ceramic conductors at much higher temperatures. By 1988, transition temperatures of up to 125 K had been reached. Superconductivity at room temperature may even become possible. Among the many possible applications of this remarkable technological breakthrough are high-speed magnetic trains (Figure 17-24), smaller and better computers, improved medical diagnostic tools, and more powerful "atom smashers."

17.13 Measuring Resistance There are several methods of measuring the resistance of a circuit component. Two methods utilizing apparatus normally available in the laboratory will be discussed here. These are the *voltmeter-ammeter* method and the *Wheatstone bridge* method.

1. *The voltmeter-ammeter method.* Ohm's law applies equally well to any part of an electric circuit that does not contain a source of emf and to an entire circuit to which an emf is applied. For a part of a circuit,

$$V = IR \quad \text{and} \quad R = \frac{V}{I}$$

If the potential difference, V, across a resistance component is measured with a voltmeter and the current, I, in the resistance is measured with an ammeter, the resistance, R, can be computed.

Suppose we wish to measure the resistance of R_1 in the circuit shown in Figure 17-25. The ammeter is connected in series with the resistance R_1 to determine the current in R_1. The voltmeter is connected in *parallel* with R_1 to determine the potential difference across R_1.

If the resistance of R_1 is very low, it can have an excessive loading effect on the source of emf, causing an excessive current in the circuit. This difficulty can be avoided by placing a variable resistance, or *rheostat,* in series with R_1. The rheostat provides additional series resistance and limits the current in the circuit. The rheostat can be adjusted to provide an appropriate magnitude of current in the circuit. See Figure 17-26. Observe that the voltmeter is connected across that part of the circuit for which resistance is to be determined.

Figure 17-24. A prototype magnetic train. Superconducting electromagnets suspend the maglev (*mag*netically *lev*itated train) and propel it at speeds of more than 500 kilometers per hour.

The voltmeter-ammeter method of measuring resistance is convenient but not particularly precise. A voltmeter is a high-resistance instrument. However, as it is connected in parallel with the resistance to be measured, there is a small current in it. The ammeter reading is actually the sum of the currents in these two parallel branches. Precision is limited by the extent to which the resistance of R_1 exceeds the equivalent resistance of R_1 and the voltmeter in parallel.

Of course the ammeter can be placed in the R_1 branch in series with R_1. In this location the ammeter reading is the current in R_1. However, the voltmeter reading is now the sum of the voltage drop across the ammeter and R_1 in series.

2. *The Wheatstone bridge method.* A precise means of determining resistance utilizes a simple *bridge* circuit known

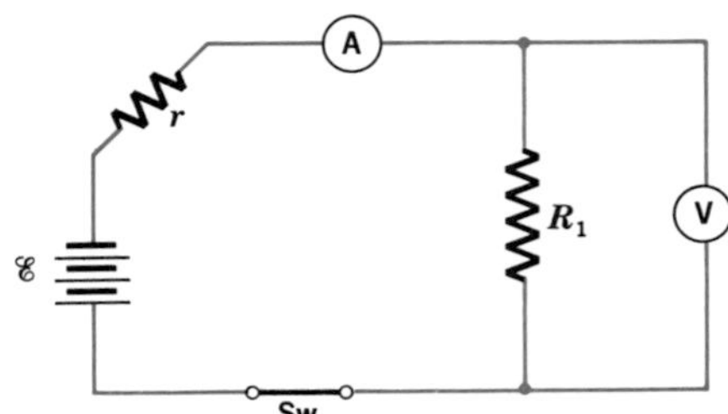

Figure 17-25. Determining resistance by the voltmeter-ammeter method.

as the **Wheatstone bridge.** It consists of a source of emf, a galvanometer, and a network of four resistors. If three are of known resistance, the resistance of the fourth can be calculated.

The circuit diagram of a Wheatstone bridge is shown in Figure 17-27, in which an unknown resistance, R_x, may be balanced against known resistances R_1, R_2, and R_3. A galvanometer is *bridged* across the parallel branches **ADB** and **ACB.** By adjusting R_1, R_2, and R_3, the bridge circuit can be balanced and the galvanometer will show zero current.

When the bridge is balanced in this manner, there can be no potential difference between **C** and **D,** since the galvanometer indicates zero current. Thus,

$$V_{AD} = V_{AC}$$

and

$$V_{DB} = V_{CB}$$

The current in the branch **ACB** is I_1, and the current in **ADB** is I_2. Then

$$I_2R_x = I_1R_1$$

and

$$I_2R_3 = I_1R_2$$

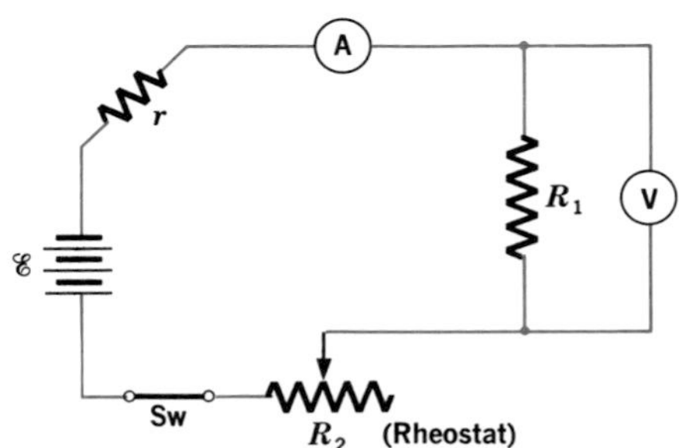

Figure 17-26. A rheostat provides a means of limiting the current in a circuit.

Dividing one expression by the other, we have

$$\frac{R_x}{R_3} = \frac{R_1}{R_2}$$

or

$$R_x = R_3\frac{R_1}{R_2}$$

In the laboratory form of the Wheatstone bridge, resistances R_1 and R_2 usually consist of a uniform resistance wire, such as constantan, mounted on a meter stick. See Figure 17-28. The length of wire comprising R_1 and the length comprising R_2 are determined by the position of the contact **C.** If l_1 is the length of wire forming R_1 and l_2 is the length of wire forming R_2, it is apparent that

$$\frac{l_1}{l_2} = \frac{R_1}{R_2}$$

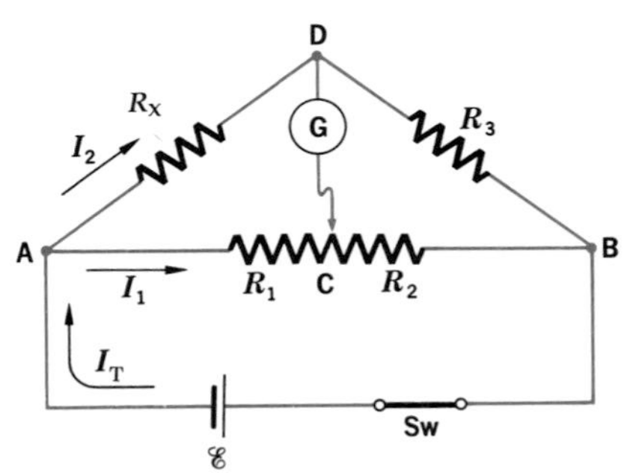

Figure 17-27. The circuit diagram of the Wheatstone bridge set up to measure the resistance R_x.

We do not need to know the resistances of R_1 and R_2, since the lengths of wire l_1 and l_2 can be taken directly from the meter stick. Our Wheatstone bridge expression

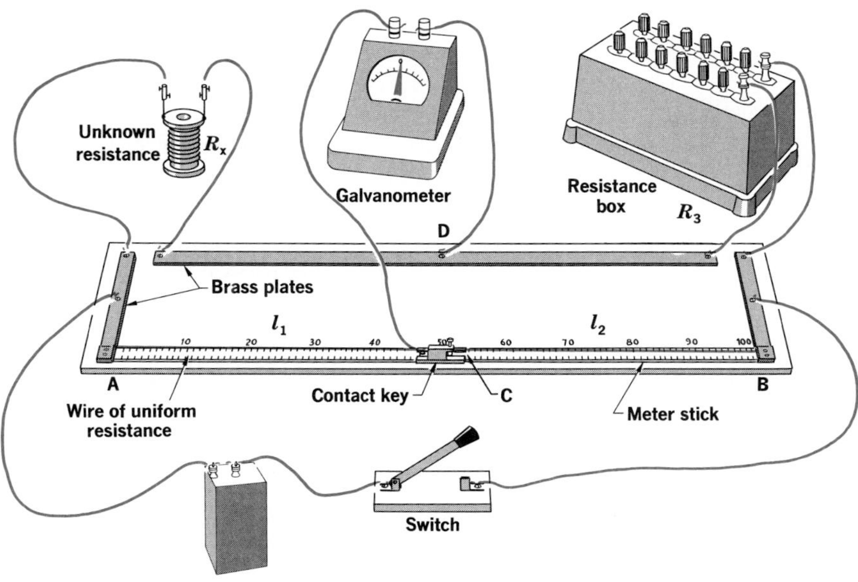

Figure 17-28. Pictorial diagram of the Wheatstone bridge circuit for measuring resistance R_x shown in Figure 17-27.

then becomes

$$R_x = R_3\frac{l_1}{l_2}$$

A precision resistance box of the type shown in Figure 17-29 is often used to provide a suitable known resistance R_3 in the bridge circuit. The internal arrangement is shown in the diagram. Resistance coils are wound with manganin or constantan wire to minimize changes in resistance with temperature. The free ends of each resistance coil are soldered to heavy brass blocks that can be electrically connected by inserting brass plugs between them. Each plug in place provides a low-resistance path across a resistance coil and thus effectively removes the coil from the circuit. Each plug removed places a coil in the circuit. The resistance between the terminals of the box is the sum of the resistances of all coils whose plugs are removed.

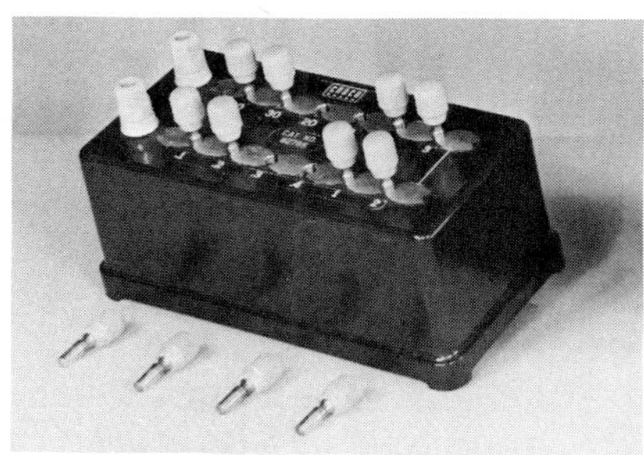

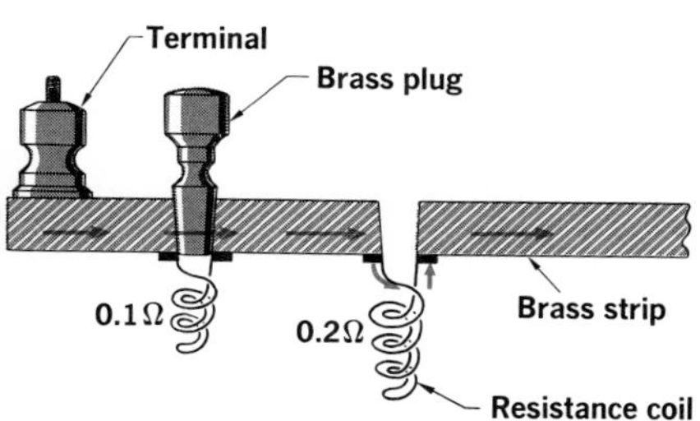

Figure 17-29. A plug-type resistance box that provides resistance between 0.1 ohm and 111.0 ohms in 0.1-ohm steps. The total resistance across the terminals will depend on which plugs are removed.

QUESTIONS: GROUP A

1. Define resistance.
2. Why do most metals exhibit low resistance?
3. (a) For a constant resistance, how are voltage and current related? (b) If voltage across a circuit is constant, how is current dependent on resistance?
4. State Ohm's law (a) verbally (b) mathematically.
5. To which part(s) of a circuit does Ohm's law apply?
6. Why does a source of emf have an "internal" resistance?
7. What is a series circuit?
8. Why is current the same through all parts of a series circuit?
9. (a) What is a parallel circuit? (b) What remains constant in it?
10. What is meant by *equivalent resistance?*
11. How would you mathematically determine the total resistance of three resistors (a) in series (b) in parallel?
12. Differentiate between resistance and resistivity.
13. Upon what factors does the resistance of a wire depend?
14. What is meant by *cryogenics?*
15. (a) What is a superconductor? (b) What is meant by "transition" temperature?

GROUP B

16. Is current "used up" in a circuit?
17. Decorative lights used to be connected in series but now are wired in parallel. What is the advantage of this type of connection?
18. If the drift velocity of electrons in a wire is so low, why does a light seem to turn on almost immediately when you flip the switch?
19. A 5-Ω, a 10-Ω, and a 15-Ω resistor are connected in series. (a) Which resistor receives the most current? (b) Which experiences the greatest voltage drop?
20. (a) A graph of a wire's resistance as a function of its length would have what shape? (b) What would a graph of resistance vs. diameter look like?
21. How must an ammeter be connected to correctly read circuit current?
22. Should an ammeter be a high- or a low-resistance device? Why?
23. How must a voltmeter be connected to correctly read the potential drop across a resistor?
24. Should a voltmeter have a very high or a very low resistance? Explain.
25. Two light bulbs, 10-Ω and 20-Ω respectively, are connected in series. (a) Which bulb is brighter? (b) Which would be brighter in a parallel circuit?
26. You switch on the electric coffeemaker and notice that the toaster and electric can opener do *not* automatically start working. What does this tell you about the wiring in your house?
27. Can equal lengths of copper and nichrome wire also have equal resistances?
28. If you draw an analogy between a circuit and water being pumped through pipes, which part of the circuit would represent (a) the pump (b) the pipes of different diameters (c) the flowing water?
29. Can you create an analogy similar to that in Problem 28, based on the human circulatory system?
30. Two small lamps and a battery are connected in series. What will happen to the brightness of the lamp(s) if (a) one of them burns out (b) you add two more lamps to the circuit?
31. Use Figure 17-22 to find the resistivity of copper at the following temperatures: (a) 200°C (b) 600°C (c) 1000°C.

PROBLEMS: GROUP A

1. A 1.5-V dry cell is connected to a small light bulb with a resistance of 3.5 Ω. How much current flows through the bulb?

2. A current of 6.25 A flows through a microwave oven. If the resistance of the circuitry in the oven is 17.6 Ω, what is the voltage drop across the oven?

3. A hair dryer draws 8.33 A of electric current when it is connected to a 110-V line. What is the resistance of the heating coil in the hair dryer?

4. Three 1.50-V dry cells are connected in series. How much current will flow through the circuit when they are connected to a resistance of 25.0 Ω?

5. A 12.0-V storage battery is connected to three resistors, 6.75 Ω, 15.3 Ω, and 21.6 Ω respectively. The resistors are joined in series. (a) Draw a circuit diagram. (b) Calculate the total resistance. (c) What is the current?

6. A series combination of two resistors, 7.25 Ω and 4.03 Ω, is connected to a battery of six 1.5-V dry cells. (a) Draw the circuit diagram. (b) Calculate the total resistance of the circuit and the total current. (c) What is the voltage drop across each resistor?

7. A small lamp is designed to draw 300.0 mA in a 6.0-V circuit. (a) How much current (C/s) flows past a point in the circuit? (b) How many electrons per second?

8. Three resistors, 14.0 Ω, 4.00 Ω and 6.00 Ω, respectively, are connected in series to a voltage source. A voltmeter across the 4.00-Ω resistor shows a voltage drop of 2.00 V. Calculate
(a) the current through the circuit
(b) the voltage output of the source
(c) the potential difference across each of the other resistors.

9. Resistances of 2.00 Ω and 4.0 Ω are connected in series across a battery composed of 1.5-V dry cells. An ammeter in the circuit shows a current of 0.50 A. (a) What is the total potential difference applied across the circuit by the battery? (b) If the maximum current that can be supplied by each cell is 0.252 A, how many cells are used and how are they arranged in the battery? (c) Draw a diagram of the circuit.

10. You have three 12-Ω resistors. What is the combined resistance if they are (a) in series and (b) in parallel? (c) Draw a diagram for each.

11. A 5.50-m piece of wire connected to two 1.5-V dry cells draws 0.250 A of current. What current would 1.75 m of this same wire draw if it is hooked up across a 4.5-V source?

12. A wire 6.24 m long with a diameter of 2.00 mm has a current of 0.500 A when it is connected to a battery of five 1.50-V dry cells in series. Calculate the wire's (a) resistance (b) resistivity.

13. An 18.0-Ω, 9.00-Ω, and 6.00-Ω resistor are connected in parallel to a voltage source. A 4.00-A current flows through the 9.00-Ω resistor. (a) Draw the circuit diagram and calculate R_{Eq}. (b) What is the voltage output of the source? (c) Calculate the current through the other resistors.

14. A 100.0-Ω resistance spool is made of No. 24 gauge German silver wire. At 20.0°C, what is the length of wire wrapped on the spool?

15. (a) Find the resistance of 100.0 m of No. 18 gauge copper wire at 20.0°C. (b) What is the resistance of 250.0 m of this wire?

16. (a) Determine the resistance of 35.5 m of No. 2 gauge aluminum wire at 20.0°C. (b) What length of No. 10 gauge aluminum wire is needed to provide the same resistance?

GROUP B

17. Prove algebraically that if R_1 and R_2 are connected in parallel, $R_{Eq} = R_1R_2/(R_1 + R_2)$.

18. Solve the following expression for R_{Eq}: $1/R_{Eq} = 1/R_1 + 1/R_2 + 1/R_3$.

19. A battery consists of three cells connected in series, each with an emf of

2.5 V and an internal resistance of 0.25 Ω. If a 6.50-Ω load is connected across the battery, (a) what will the current in the circuit be? (b) Calculate the potential difference across the load. (c) Draw a circuit diagram.

20. A 30.0-Ω resistance is connected in parallel to a 15.0-Ω resistor. These are joined in series to a 5.00-Ω resistor and a source with an emf of 30.0 V. (a) Draw a circuit diagram. Calculate (b) the total resistance (c) the voltage drop across each resistor (d) the current through each resistor.

21. Complete the chart of voltage, current, and resistance for the following circuit.

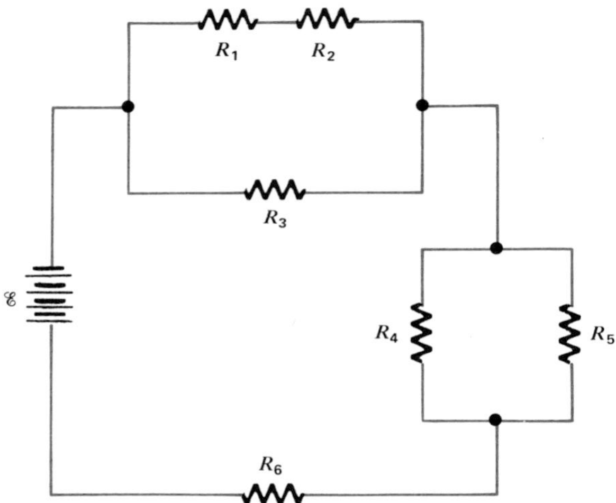

	$\mathscr{E}$	I	R
Source		2.0 A	
R_1			5.0 Ω
R_2	3.5 V		
R_3		1.5 A	
R_4	4.0 V		
R_5		1.0 A	
R_6			2.0 Ω

22. Complete the chart for the following circuit.

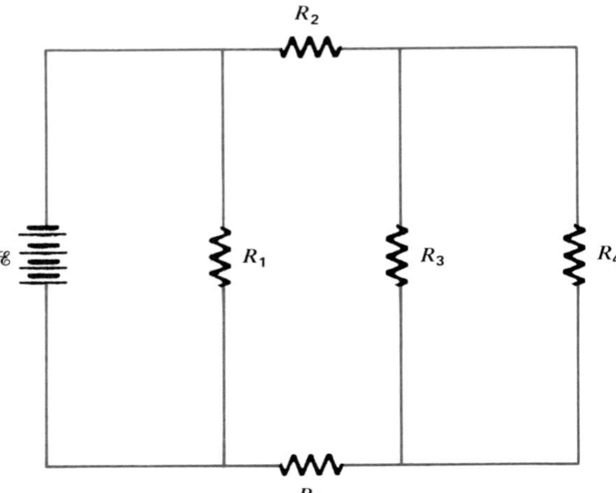

	$\mathscr{E}$	I	R
Source	45 V		
r			0.50 Ω
R_1		8.0 A	
R_2		2.0 A	
R_3	10.0 V		
R_4			10.0 Ω
R_5			9.0 Ω

Physics Activity

Make your own "chemical cell" by pushing a piece of copper wire and an unbent paper clip into a lemon. Hold the lemon and touch the ends of the wires to your tongue. The slight tingling sensation you feel and salty taste result because your saliva provides an electrolytic solution through which the current flows.

SUMMARY

Electric current is the rate of flow of charge in a circuit. The charge carriers in metallic conductors are electrons. The current in these conductors is an electron current and its directional sense is from negative to positive. A continuous current requires a source of emf. There are several sources of emf and continuous current, each of which transforms energy of some form to electric energy.

The dry cell is a common source of emf that supplies direct current to a circuit. It is a primary cell that is a form of voltaic cell in which chemical energy is converted to electric energy. Electric cells can be combined in series, in parallel, or in series-parallel to form batteries of suitable characteristics for a given circuit.

All circuit components offer some resistance to the current in a closed circuit. The relationship between resistance, current, and potential difference is expressed by Ohm's law. Ohm's law applies to an entire circuit or to any part of a circuit.

Resistances can be connected in series, in parallel, or in series-parallel combinations called networks. Networks can be reduced to simple equivalent circuits in which those components in parallel are reduced to their series equivalents. Equivalent circuit problems are solved by the application of Ohm's law for series circuits.

The resistance of a metal conductor depends on the temperature, length, cross-sectional area, and identity of the metal. From a knowledge of these factors, the laws of resistance are formulated and a property of materials called resistivity is defined. Resistivity depends on only the identity of the material and its temperature.

The resistance of a circuit element can be measured directly by the voltmeter-ammeter method or by the Wheatstone bridge method. The voltmeter-ammeter method is rapid, and the Wheatstone bridge method is more precise.

VOCABULARY

anode
battery
cathode
closed circuit
cryogenics
drift velocity
electric current
electrochemical cell
electrode
electrolyte
emf
equivalent resistance
internal resistance
Kirchoff's first law
Kirchoff's second law
load resistor
network
Ohm's law of resistance
open circuit
parallel circuit
Peltier effect
photoelectric effect
piezoelectric effect
primary cell
resistance
resistivity
Seebeck effect
semiconductor
series circuit
superconductivity
thermocouple
Wheatstone bridge

Superconductors

Most superconductors are metals that lose resistance to electricity when cooled below a certain temperature. Because they conduct electricity with no energy loss, the transfer of electricity is less costly. Unfortunately, these superconductors must be cooled to temparatures below 70 K.

However, new superconducting materials are being developed that conduct electricity at much higher temperatures than conventional superconductors. Temperatures nearing 130 K have been attained with some of the new compounds.

In the future, room-temperature superconductors may be used to produce more efficient electrical transmission cables, electrical motors, and high-performance electromagnets.

Sensors, called squids, use superconductors to detect weak magnetic fields, such as those emitted by the human brain. Images made of these fields indicate the areas of specific brain functions, abnormal brain activity, and possible tissue scarring.

The use of the new high-temperature superconductors in

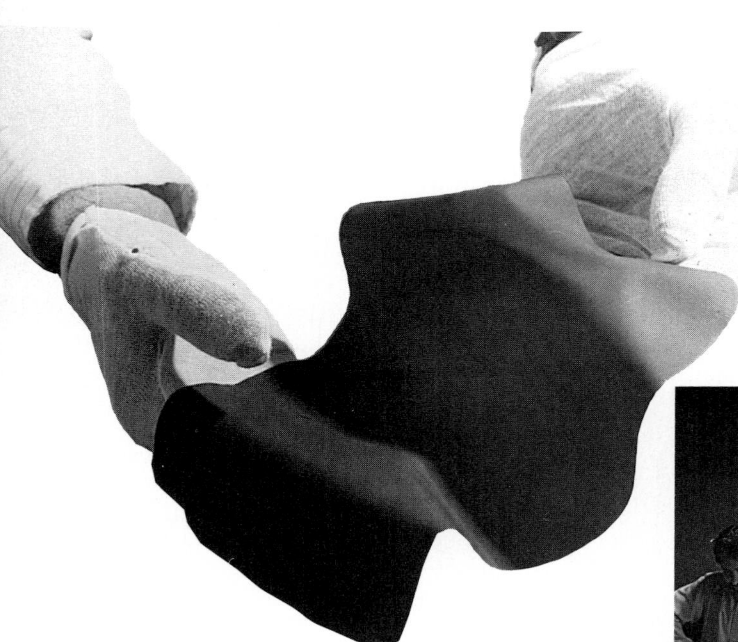

A tape of high-temperature superconducting ceramic is flexible enough to be shaped before firing, but must be fired before it becomes a superconductor. Once fired, it also becomes extremely brittle.

Ceramic superconductor manufacturing: grinding, heating, cooling, and coating.

squids requires less coolant than the old type. More squids can be packed into a smaller space. Thus a more maneuverable device can be developed and more accurate magnetic images can be produced.

Ceramic Engineers

Superconductors are only one of the many uses of ceramics, or nonmetal inorganic materials that require high-temperature processing. Ceramic engineers develop new ceramic materials and methods of producing useful products from the materials. Many of these engineers specialize in making certain types of ceramic, such as those used in aeronautics, electronics, or optics.

Ceramic engineers usually have a bachelor's degree in ceramics or materials engineering. A strong background in mathematics, chemistry, and physics, plus specialized courses such as electronics, is required. Ceramic engineers often are employed in the glass, steel, electronics, clay, and chemical industries.

Materials-Science Technicians

The production and testing of newly developed materials such as alloys, ceramics, and plastics are carried out by materials-science technicians. These technicians also may develop new equipment or methods for producing the materials. Some materials-science technicians operate spectrographs and other equipment to determine the material's internal structure and locate defects. Others use devices to test such properties as hardness, strength, and general usefulness of the new product.

A materials-science technician must have an interest in science and at least average mathematical skills. These technicians need a high school diploma. Others go on to receive a two-year degree in materials-science technology from a community college or technical institute. Materials-science technicians are employed in industrial and government laboratories and production plants. ■

■ A cube of superconducting material is suspended above a magnet. Superconductors are repelled by magnetic fields.

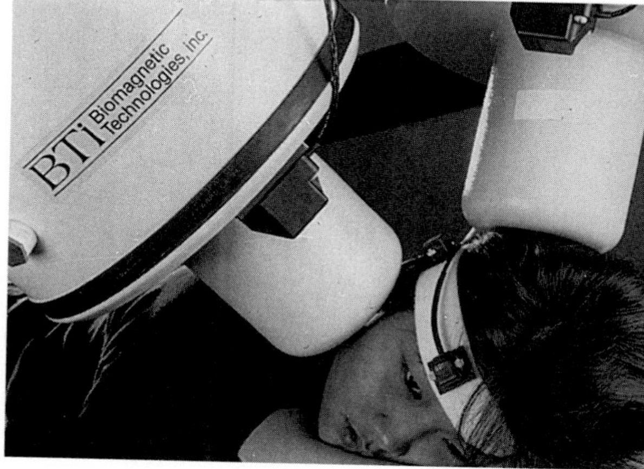

■ Patient undergoes a brain scan using a 14-channel Neuromagnetometer, an instrument that precisely measures the electrical activity of the brain without surgically entering the skull. The process, called magnetoencephalography (MEG), detects the magnetic fields generated by electrical currents within the brain. The device is used to seek the causes of epilepsy, Alzheimer's disease, Parkinson's disease, and other brain disorders.

■ A master ceramist inspects fired powder of yttrium-barium-copper oxide. The material loses all resistance to electricity when cooled to 94 K (−290°F).

18

Heating and Chemical Effects

power n.: the rate at which electric energy is delivered to a circuit.

OBJECTIVES

- Discuss the transfer of energy in an electric circuit.
- Study the relationship between heat energy and resistance.
- Discuss power dissipation in an electric circuit.
- Define the distinction between electrolytic and electrochemical cells.
- Discuss Faraday's laws of electrolysis.

HEATING EFFECTS

18.1 Energy of an Electric Current The relationships among objects, time, and space are basic concerns of physics. Work, power, and energy are important physical concepts involved in these relationships. We have described energy as the ability to do work and have measured it in units of work and heat. To maintain an electric current, energy must be supplied by the source of emf. Chemical energy, which can be thought of as the potential energy of chemical bonds, is transformed into electric energy in an **electrochemical cell.** Thus a charge acquires energy within the cell or other source of emf and expends this energy in the electric circuit.

In Section 16.10 we recognized that one joule of work is done when one coulomb of charge is moved through a potential difference of one volt.

$$W = qV$$

One coulomb of charge transferred in one second constitutes one ampere of current.

$$I = \frac{q}{t} \quad \text{and} \quad q = It$$

Therefore

$$W = VIt \qquad \text{(Equation 1)}$$

Within a source of emf, the charge is moved in opposition to the potential difference, and work, W, is done on the electrons *by the source of emf*. As the charge moves through the external circuit in response to the potential difference across the circuit, work, *W*, is done *by the electrons* on the components of the circuit. It is this electric energy, expended in the external circuit, that is available to perform useful work.

We can observe the work done by an electric current in a variety of ways. If a lamp is in the circuit, part of the work appears as light; if an electric motor is in the circuit, part of the work appears as rotary motion; if an electroplating cell is in the circuit, part of the work promotes chemical action in the cell. In all cases, *part* of the work, *W*, appears as *heat* because of the inherent resistance of the circuit components. If an ordinary resistor comprises the circuit, *all* the work goes to the production of heat since no mechanical work or chemical work is done. See Figure 18-2.

18.2 Energy Conversion in Resistance Suppose short lengths of No. 30 gauge copper wire and German silver wire are connected in series and placed in the circuit shown in Figure 18-3. When the switch is closed, the German silver wire may become red hot and melt. The copper wire will become warm but is less likely to melt. The melting point of German silver is about 30 C° higher than that of copper, but its resistivity is nearly 20 times higher than that of copper. Since the two wires were in series, the same magnitude of current was in each. We may infer that *a greater quantity of electric energy was converted to heat in the wire having the higher resistance.*

Figure 18-1. The electric energy expended in the lamp filament by the electron charge carriers heats the filament to incandescence, and hence the name "incandescent lamp."

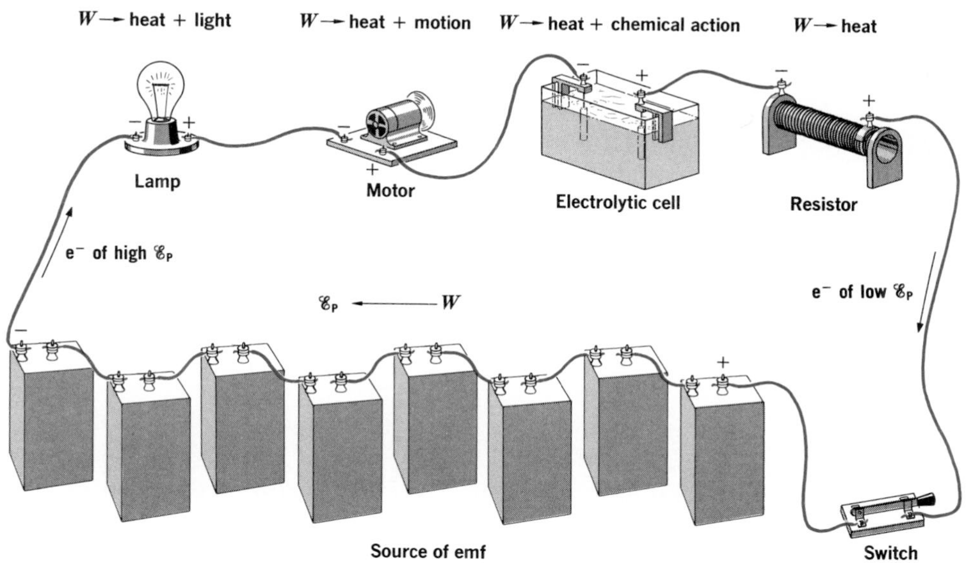

Figure 18-2. Work is done when a charge is moved through a potential difference.

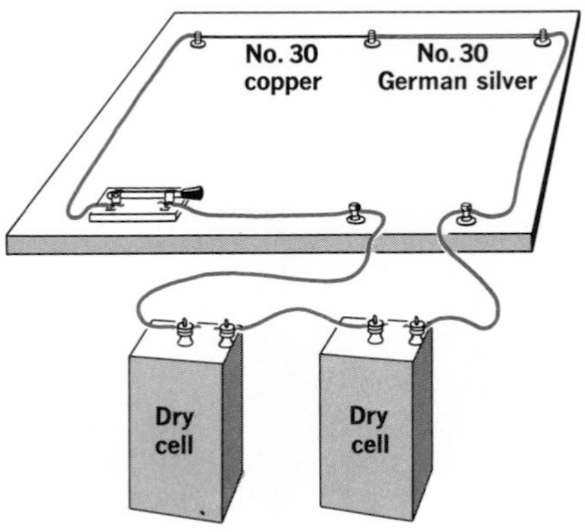

Figure 18-3. A circuit for studying the heating effect of an electric current.

Now suppose a single piece of No. 30 copper wire is connected in the circuit in place of the copper–German silver link. When the switch is closed, the copper wire may warm to a red heat and even melt. Because the resistance of this copper wire is less than that of the copper–German silver link, the current in the circuit is higher. We may infer that *the quantity of electric energy converted to heat in the wire is greater when the current is increased.* We can reason also that *the longer the switch is closed* (assuming the wire does not melt), *the greater will be the quantity of electric energy converted to heat.*

From Ohm's law, the potential difference across the resistance of an external circuit is the product of the current in the circuit and the resistance of the circuit.

$$V = IR$$

Substituting this product for V in Equation 1, we have an expression for the energy transferred to the resistance of a circuit.

$$W = I^2Rt \qquad \text{(Equation 2)}$$

Electric power is dissipated in a resistance as heat.

18.3 Joule's Law James Prescott Joule, the English physicist who first studied the relationship between heat and work, also pioneered the quantitative study of the heating effect of an electric current. From the results of his experiments he formulated the generalization known as ***Joule's law:*** *The heat developed in a conductor is directly proportional to the resistance of the conductor, the square of the current, and the time the current is maintained.*

Since all of the work done by a current in a resistance appears as heat, the electric energy expended in the resistance is equal to the heat energy that appears. Therefore

$$Q = I^2Rt \qquad \text{(Equation 3)}$$

By rearranging the terms and doing a unit analysis, we can see how I^2Rt is expressed in joules:

$$\begin{aligned}(\text{A}\cdot\Omega)(\text{A}\cdot\text{s}) &= \text{V}\cdot\text{C}\\ &= (\text{J/C})\text{C}\\ &= \text{J}\end{aligned}$$

Figure 18-4. An electrically-heated annealing oven on the assembly line of a car-manufacturing plant dries and hardens the coats of paint that were applied earlier.

The heat lost to resistance (I^2Rt) becomes the heat gained by another system. The following example illustrates an application of heat transfer using Joule's law.

EXAMPLE The heating element in a coffee maker has a resistance of 25.0 Ω. When it is plugged into a 120.0-V circuit and turned on, how long will it take to heat 1.00 kg of water from room temperature, 20.0°C, to the boiling point, 100.0°C? Assume that no heat is lost to the surroundings.

Given	Unknown	Basic equations
$R = 25.0\ \Omega$	t	$Q_{Lost} = Q_{Gained}$
$V = 120.0\ V$		$Q = I^2Rt$
$m = 1.00\ kg$		$Q = mc_w\Delta T$
$\Delta T = 80.0\ C°$		$I = \frac{V}{R}$
$c_w = 4.19\ J/g \cdot C°$		

Solution

$$\text{Working equation: } \overbrace{\left(\frac{V}{R}\right)^2 Rt}^{\text{Heat lost}} = \overbrace{mc_w\Delta T}^{\text{Heat gained}}$$

$$t = \frac{mc_w\Delta TR}{V^2}$$

$$= \frac{(1.00 \times 10^3\ g)(4.19\ J/g{\cdot}C°)(80.0\ C°)(25.0\ \Omega)}{(120\ V)^2}$$

$$= 582\ s$$

PRACTICE PROBLEMS **1.** A hot plate has a resistance of 30.0 Ω. It develops 1.00×10^6 J of heat in 60.0 s of operation. How much current does the hot plate draw? *Ans.* 23.6 A

2. The heating element of an electric teakettle has a resistance of 28.8 Ω. It is plugged into a 120-V circuit. How many minutes does it take to heat 750 g of water from 4.5°C to 100.0°C if there is no heat lost to the surroundings? *Ans.* 10.0 min

---

Many of the electric appliances we use in our homes are basically heating devices; irons, toasters, grills, ranges, and fryers are heated by electricity. The heating element in an appliance is usually a length of nichrome wire that, when used with the prescribed voltage, will provide the proper resistance according to Joule's law.

From the experiment in Section 18.2, we see one result of excessive currents in conductors. Electric circuits are supplied to houses by pairs of insulated copper wires that are strung through the walls. The appliances are connected in parallel across these lines. As more devices are connected, the equivalent resistance across the source of the current drops, and the current in the supply line rises.

When more current is drawn in a circuit than the conductors can safely carry, the circuit is said to be *overloaded*. If the circuit is not protected by a *circuit breaker,* the heat produced may burn away the insulation from the overloaded conductors. The conductors may even melt or make contact with one another, forming a *short circuit* that may set fire to the house.

The National Electric Code establishes safety standards for the interior wiring of homes. Circuits provided with circuit breakers according to this standard are protected from overloads. For example, the code requires that a branch circuit using 14-gauge copper wires contain a circuit breaker rated at 15 amperes. If an excessive load is placed across this circuit and the current rises above 15 amperes, the breaker will trigger, opening the circuit and preventing a dangerous overload. *Fuses* and *ground fault interrupters* provide the same protection and are sometimes used for overload protection in place of circuit breakers.

The energy delivered to a circuit element is the product of power and time..

Figure 18-5. High voltage in electric power lines minimizes energy losses to the lines and the atmosphere. But the static in your car radio when you drive under a power line shows that there is still some energy leakage.

18.4 Power in Electric Circuits

Power is defined as the rate of doing work. The symbol for power is P and the unit of power is the watt (W). The watt is a *rate* of one joule of work per second. In an electric circuit, it is convenient to think of power as *the rate at which electric energy is delivered to the circuit.*

$$P = \frac{W}{t} \qquad \text{(Equation 4)}$$

Recalling Equation 1 of Section 18.1, $W = VIt$, let us solve this expression for W/t.

$$\frac{W}{t} = VI \qquad \text{(Equation 5)}$$

Combining Equations 4 and 5 yields a basic power equation for electric circuits.

$$P = VI \qquad \text{(Equation 6)}$$

If I is in amperes and V is in volts, then P is in watts:

$$\frac{\textbf{joule}}{\textbf{coulomb}} \times \frac{\textbf{coulomb}}{\textbf{second}} = \frac{\textbf{joule}}{\textbf{second}}$$

$$= \textbf{watt}$$

By Ohm's law, the potential difference across a load resistance in a circuit is equal to the product IR_L. Substituting for V in Equation 6,

$$P_L = I \times IR_L$$

or

$$P_L = I^2R_L \qquad \text{(Equation 7)}$$

Can you derive Equation 7 from Equations 2 and 4 in just three steps?

This shows us that *the power expended by a current in a resistance is proportional to the square of the current in the resistance.*

Similarly, the power dissipated in the internal resistance of a source of emf is

$$P_r = I^2r \qquad \text{(Equation 8)}$$

Power dissipated in the internal resistance of a source of emf is wasted as heat.

The total power consumed in a circuit must be the sum of the power dissipated within the source of emf and the power expended in the load of the external circuit.

$$P_T = I_T{}^2R_T = I_T{}^2(R_L + r)$$

Recall that

$$\mathcal{E} = I_T(R_L + r)$$

Therefore

$$P_T = I_T\mathcal{E} \qquad \text{(Equation 9)}$$

There are instances in which the current in a circuit is unknown or of no interest. Power can than be expressed in terms of voltage and resistance since, by Ohm's law,

$$I_T = \frac{\mathcal{E}}{R_T}$$

Substituting for I_T in Equation 9, we get

$$P_T = \frac{\mathcal{E}^2}{R_T} \qquad \text{(Equation 10)}$$

and for the external circuit,

$$P_L = \frac{V^2}{R_L} \qquad \text{(Equation 11)}$$

18.5 Maximum Transfer of Power There are some electric-circuit applications in which the transfer of maximum power to the load is important. One example is the circuit of the starter motor of an automobile engine. To operate the starter, a large amount of electric energy must be supplied by a battery during a short interval of time. Other, less familiar examples would be the transfer of audio-frequency power from the output stage of a radio to the speaker and the transfer of radio-frequency power from the transmission line to the antenna.

Consider the circuits of Figure 18-6 in order to determine the circumstances under which maximum power from a given source of emf is expended in the load. Observe that the same source of emf is used in all three circuits and loads greater than, less than, and equal to the internal resistance of the source are supplied. In each circuit I_T, P_L, P_r, and P_T can be determined by using equations derived in Section 18.4. For comparison these values are listed below.

	Circuit **A**	Circuit **B**	Circuit **C**
I_T =	2.0 A	5.0 A	3.0 A
P_L =	4.0 W	2.5 W	4.5 W
P_r =	2.0 W	12.5 W	4.5 W
P_T =	6.0 W	15.0 W	9.0 W

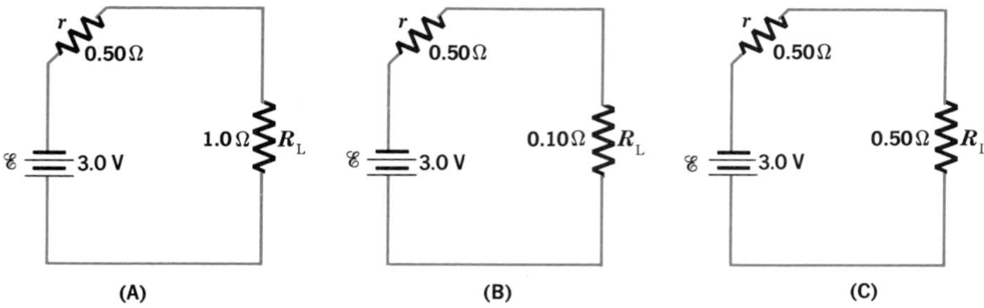

Figure 18-6. In which circuit is the most power expended in the load?

From such calculations it can be shown that the power expended in the load of circuit **C**, 4.5 W, is the maximum that can be transferred from this source of emf. *Maximum power is transferred to the load when the load resistance is equal to the internal resistance of the source of emf.* The source and the load are said to be *matched.* In circuit **B**, the load resistance is less than the internal resistance of the source and most of the power generated is dissipated as heat in the source itself. This power is wasted. In circuit **A**, the load resistance is greater than the internal resistance of the source and little power is lost in the internal resistance of the source. However, the power expended in the load is not as high as in circuit **C**.

18.6 Purchasing Electric Energy An operating circuit that requires 100 watts of power converts 100 joules of energy per second. Obviously, the longer this circuit continues to operate, the larger will be the total quantity of energy transformed. Since

$$\textbf{power} = \frac{\textbf{energy}}{\textbf{time}}$$

then

$$\textbf{energy} = \textbf{power} \times \textbf{time}$$

In an electric circuit, the current, potential difference, and resistance are measured in amperes, volts, and ohms respectively. If time is measured in seconds, then power is expressed in watts. Energy can thus be expressed in *watt-seconds* (W·s). The watt-second is inconveniently small for electric energy sold on a commercial scale. For example, a 100-watt lamp operated for 1 hour transforms 3.6×10^5 watt-seconds of energy. Dividing by 3.6×10^3 seconds/hour, this quantity is converted to 1×10^2 watt-hour (W·h). Dividing again by 10^3 watts/kilowatt, the energy expression is 1×10^{-1} kilowatt-hour (kW·h), or simply 0.1 kilowatt-hour. The kilowatt-hour is the practical electric energy unit in common use. Observe the magnitude of this combined conversion constant.

Your electric utility statement is based on the number of kilowatt-hours of energy used and the cost per kilowatt-hour.

$$(\mathbf{kW \cdot h}) \times \left(\frac{\mathbf{10^3\ W}}{\mathbf{kW}}\right) \times \left(\frac{\mathbf{3.6 \times 10^3\ s}}{\mathbf{h}}\right) = \mathbf{3.6 \times 10^6\ W \cdot s}$$

The domestic cost of electric energy may be in the range of 6¢ to 12¢ per kilowatt-hour, depending on location and primary energy source. The cost of electric energy is calculated in the example that follows.

EXAMPLE An electric furnace is connected across a 117-V circuit. The furnace dissipates 3.50 kW of power in the form of heat. Calculate the resistance of the furnace.

Given	Unknown	Basic equations
$V = 117$ V	R	$P = IV$
$P = 3.50 \times 10^3$ W		$V = IR$

Solution

Working equation: $R = \dfrac{V^2}{P}$

$$= \frac{(117\ \text{V})^2}{(3.50 \times 10^3\ \text{W})}$$

$$= 3.91\ \Omega$$

EXAMPLE How much heat is developed by the furnace in the previous example over a period of 3.00 hours?

Given	Unknown	Basic equations
$P = 3.50$ kW $t = 3.00$ h	Q	$Q = I^2Rt$ $P = I^2R$

Solution

Working equation: $Q = Pt$

$$= (3.50\ \text{kW})(10^3\ \text{W/kW})(3.00\ \text{h})(3600\ \text{s/h})$$

$$= 3.78 \times 10^7\ \text{J}$$

PRACTICE PROBLEM The heating element in a waffle iron dissipates 1.25 kW of power in the form of heat. How much heat does the iron produce in 1.00 minute when it is plugged into a 120.0-V circuit?

Ans. 7.5×10^4 J

QUESTIONS: GROUP A

1. How does the work done within the source of emf of a circuit differ from that done across a load?
2. What are the four types of energy that result from electric work being done in a circuit?
3. Upon what three factors does the heat developed in a conductor depend?
4. What energy changes are involved in lighting an incandescent bulb?
5. (a) What is power? (b) What is the SI unit for power? (c) What quantity is measured in kilowatt-hours?
6. How does the power expended by an unknown current in a resistor depend on (a) voltage drop (b) resistance?
7. Why do wires heat up when an electric current passes through them?
8. Which wire will become hotter in each of the following cases? (a) Equal lengths of nichrome and copper wire of the same cross-sectional area con-

nected in series (b) Two identical pieces of silver wire, one connected in series to a dry cell, the other connected in series across two dry cells.

9. State Joule's law (a) qualitatively and (b) quantitatively.

GROUP B

10. Why is power transmitted to your home along high-voltage rather than high-current lines?
11. (a) Use Joule's law to show that resistance increases as temperature increases. (b) How does kinetic theory help to explain this?
12. A 75-W light bulb is screwed into one of a pair of identical lamps, while a 100-W bulb is placed in the other. The switches are turned on. (a) Which bulb has more resistance? (b) Which has more current through it? (c) Which bulb is brighter?
13. House wiring is in parallel. Adding more appliances decreases the total resistance and increases the current. What prevents the wires from overheating and causing a fire?
14. When a small lamp is connected to a battery, the filament becomes hot enough to give off radiant energy, while the wires do not. What does this tell you about their relative resistances?
15. Why is the filament of an incandescent lamp made of a thin piece of nichrome wire?

PROBLEMS: GROUP A

1. A hair dryer draws a 9.1-A current. (a) How much charge flows through it in 3.5 min? (b) How many electrons does this represent?
2. (a) If the hair dryer in Problem 1 is connected across a 110-V line, how much work is being done? (b) What is the dryer's power rating?
3. A 75.0-W light bulb is connected across a 110-V household line. Calculate (a) the current flowing through the bulb (b) the resistance of the bulb filament.
4. (a) If the bulb in Problem 3 remains lit for 1.75 h, how much work will be done? (b) If you are charged 8.5 cents per kW·h, how much will it cost?
5. A 750.0-W microwave oven requires 3.50 min to pop a bag of popcorn. (a) How many joules of electric energy does this represent? (b) If 50.0% of this energy could be converted to heat energy, how many kilocalories would this represent?
6. You are watching a football game on a color television set with a power rating of 325 W. If the game lasts 3.10 h, (a) how much electric work is done? (b) At a cost of 6.8 cents per kW·h, how much will it cost you to watch the game?
7. The heating element of an electric iron has a resistance of 24 Ω and draws a current of 5.0 A. How many kilocalories are developed if the iron is used for 45 min?
8. A wire for use in an electric heater has a resistance of 5.25 Ω/m. What length of this wire is needed to make a heating element for a toaster that will draw 8.70 A across a $12\bar{0}$-V line?

GROUP B

9. If electric energy costs 9.2¢ per kW·h, what is the cost of heating 4.6 kg of water from 25°C to the boiling point, assuming no energy is wasted?
10. A percolator with a heating coil of 20.0-Ω resistance is connected across a $12\bar{0}$-V line. (a) What quantity of heat is liberated per second? (b) If it contains $50\bar{0}$ g of water at 22.5°C, how much time is required to heat the water to boiling, assuming no loss of heat?

11. An electric iron has a mass of 1.50 kg. The heating element is a 2.00-m length of No. 24 gauge nichrome wire. What time is required for the iron to be heated from 20.0°C to $15\bar{0}$°C when connected across a 115-V line, assuming no loss of heat?

12. An electric hot plate draws 10.0 A on a $12\bar{0}$-V circuit. In 7.00 min the hot plate can heat $60\bar{0}$ g of water at 20.0°C to boiling and boil away 60.0 g of water. What is the efficiency of the hot plate?

13. A lamp operates continuously across a 117-V line dissipating $10\bar{0}$ W. How many electrons flow through the lamp per day?

Physics Activity

Find out how much you pay per kilowatt-hour of electricity at your house. This information may be given on your monthly statement, or you may have to call the utility office for the rate. (The cost per kW·h possibly decreases as the total amount of electricity you use per month increases.) Monitor your electric meter for several different hours of the day to see when the most electricity is consumed. Which electric devices account for the most usage?

ELECTROLYSIS

18.7 Electrolytic Cells The close relationship between chemical energy and electric energy has already been discussed. **Spontaneous reactions** that involve the transfer of electrons from one reactant to the other were described in Section 17.3. Such reactions can be sources of emf. In the electrochemical cell, chemical energy is converted to electric energy as the spontaneous electron-transfer reaction proceeds. The products of the reaction have less energy than did the reactants.

Similar reactions *that are not spontaneous* can be forced to occur if electric energy is supplied from an external source, as in charging a storage battery. In such forced reactions, the products have more chemical energy than did the reactants. This means electric energy from the external source of emf is transformed into chemical energy as the reaction proceeds. An arrangement in which a forced electron-transfer reaction occurs is known as an **electrolytic cell.**

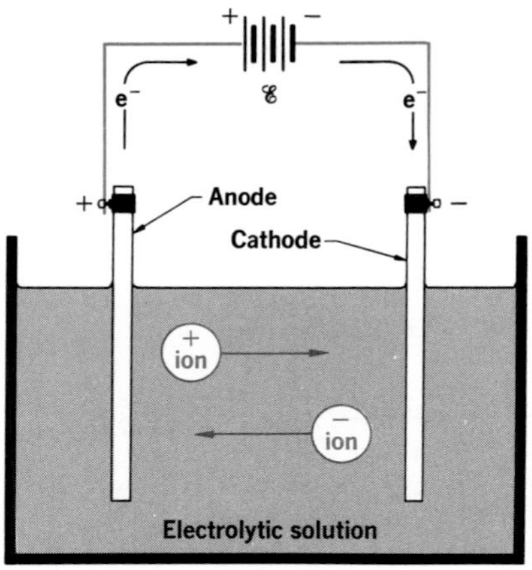

Figure 18-7. The essential components of an electrolytic cell.

The basic requirements for an electrolytic cell are shown in Figure 18-7. The conducting solution contains an **electrolyte** that furnishes positively and negatively charged ions. Two electrodes are immersed in the electrolytic solution. The negative terminal of a battery or other source of direct current is connected to one electrode to form the **cathode** of the cell; the positive terminal of the source is

The cathode is the electron-rich electrode.

connected to the other electrode to form the **anode** of the cell.

The anode is the electron-poor electrode.

When the circuit is closed, the cathode becomes negatively charged and the anode positively charged. Positive ions migrate to the cathode where they acquire electrons of high potential energy and are then discharged. Negatively charged ions migrate to the anode and are discharged by giving up electrons of low potential energy to the anode. Note that the electrode reactions occurring in electrochemical cells, presented in Sections 17.3 and 17.4, are just the reverse of those in an electrolytic cell.

The loss of electrons by the cathode and the acquisition of a like number of electrons by the anode constitute the conduction of an electric charge through the cell. *The conduction of an electric charge through a solution of an electrolyte or through a fused ionic compound, together with the resulting chemical changes, is called* ***electrolysis.***

The end result of electrolysis depends on the nature of the electrolyte, the kinds of electrodes, and to some extent the emf of the source. Electrolytic cells are useful in decomposing compounds, plating metals, and refining metals.

Chemical reactions in electrochemical cells are spontaneous; those in electrolytic cells are driven.

18.8 Electroplating Metals An electrolytic cell is set up with a metallic copper anode and positively charged copper(II) ions, Cu^{++}, in solution. When a small potential difference is placed across the cell, the Cu^{++} ions acquire electrons from the cathode and form copper atoms. The following chemical equation represents this reaction.

$$\mathbf{Cu^{++} + 2e^- \rightarrow Cu^0} \text{ (Cathode reaction)}$$

These copper atoms plate out on the cathode surface.

Simultaneously, atoms of the copper anode give up electrons to this positively charged anode and form Cu^{++} ions that go into the solution to replace those being plated out as copper atoms. The chemical reaction is

$$\mathbf{Cu^0 \rightarrow Cu^{++} + 2e^-} \text{ (Anode reaction)}$$

These Cu^{++} ions pass into the solution from the anode at the same rate as similar Cu^{++} ions leave the solution and form the plate on the cathode. To maintain the concentration of Cu^{++} ions in the electrolytic solution, Cu^{++} ions must continuously pass into the solution from the anode. The result is that the anode is used up in the plating process. A simplified diagram of a copper-plating cell is shown in Figure 18-8.

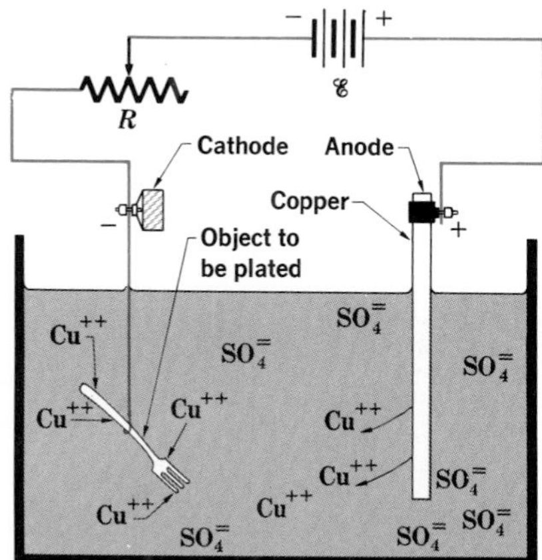

Figure 18-8. An electrolytic cell used for copper plating.

The reaction at the cathode makes an electrolytic cell useful for the electrolytic deposition, or **electroplating,** of one metal on the surface of another. The object to be plated is used as the cathode of the cell. The plating metal is used as the anode. The conducting solution contains a salt of the metal to be plated out onto the cathode.

Metallic copper of the highest purity is required for electric conductors. Even very small amounts of impurities in copper cause a marked increase in its electric resistance. These impurities can be removed almost completely from the copper by electrolysis. For this purpose an electroplating cell is used in which the anode is composed of impure copper. During electrolysis, refined copper is plated onto the cathode made of pure copper. This electrolytically refined copper metal deposited on the cathode is more than 99.99% pure.

Figure 18-9. Michael Faraday, the English scientist who formulated the laws of electrolysis and for whom the unit of electric capacitance is named. His lectures were popular events.

18.9 Faraday's Laws of Electrolysis Michael Faraday investigated the distribution of charge on isolated bodies as well as the nature of electric fields. He also performed experiments on the conduction of electric charge by solutions. Faraday discovered that a current that deposited a half-penny-weight of silver in ten minutes would deposit one penny-weight of silver in twenty minutes. This discovery suggests an important relationship between the quantity of electric charge that passes through an electrolytic cell and the quantity of a substance liberated by the chemical action.

Suppose we consider the quantitative significance of Faraday's discovery by imagining several cells connected in series and supplied by a source of emf. Each cell has a pair of inert platinum electrodes immersed in a water solution of an electrolyte. See Figure 18-10. Each cell has a different electrolyte: cell No. **1** has sulfuric acid (hydrogen sulfate), cell No. **2** has silver nitrate, cell No. **3** has copper(II) sulfate, and cell No. **4** has aluminum sulfate. Since this is a "thought" experiment we need not be concerned with the possibility of running out of ions in the cells.

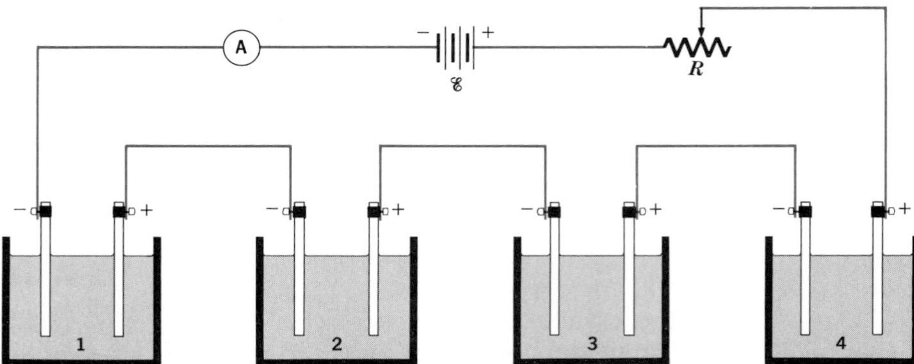

Figure 18-10. Electrolytic cells in series must have the same magnitude of current for the same length of time.

Since the cells are in series, a certain constant current in the circuit for a given time interval signifies that *the same quantity of charge has passed through each cell.*

Recall that there can be only one magnitude of current in a series circuit because there is only one conducting path.

Hydrogen, silver, copper, and aluminum are liberated at the respective cathodes. The quantity of each liberated substance must be proportional to the quantity of charge that passed through each cell: that is, to the product of the current and the time.

Faraday's first law: *The mass of an element deposited or liberated during electrolysis is proportional to the quantity of charge that passes.*

$$m \propto Q$$

where

$$Q = It$$

By varying either current or time interval, we merely vary the quantity of charge passing through each cell. By Faraday's first law, the mass of the element deposited or liberated varies accordingly. It follows that whatever the charge through the cells, the relationship among the quantities of the elements deposited remains *constant.* Such quantities of the elements are said to be *electrochemically equivalent.* The ***electrochemical equivalent,*** *z, of an element is the mass in grams of the element deposited or liberated by 1 coulomb of electric charge.* See Table 18-1. A more extensive table of electrochemical equivalents of common elements is given in Appendix B, Table 19.

Table 18-1
ELECTROCHEMICAL EQUIVALENTS

Element	*g-at wt* (g)	*Ionic charge (oxidation state)*	*z* (g/C)
aluminum	27.0	3	0.000 093 3
copper	63.5	2	0.000 329 1
hydrogen	1.01	1	0.000 010 4
silver	108	1	0.001 118 0

Suppose we allow the charge to flow through our cells until 1 g of hydrogen has been liberated in cell No. **1.** We would find 108 g of silver deposited in No. **2,** 31.8 g of copper in No. **3,** and 9 g of aluminum in No. **4.**

The hydrogen ion, H^+, and the silver ion, Ag^+, are each deficient by 1 electron and carry a net positive charge of 1 unit. The elements hydrogen and silver are said to have an ionic charge of 1, which means they lose 1 electron per

The number of electrons of an atom of an element participating in a chemical process is sometimes referred to as the valence of the element.

atom in a chemical reaction. Similarly the charge of the copper(II) ion, Cu^{++}, is 2, and that of the aluminum ion, Al^{+++}, is 3.

Using this information and the information previously obtained in the thought experiment, we can make two very interesting and important observations. *First, for a given quantity of charge, the quantity in grams of each element deposited or liberated is proportional to the ratio of its gram-atomic weight (g-at wt) to its ionic charge.* This ratio is known as the ***chemical equivalent*** of an element.

$$\textbf{chemical equivalent} = \frac{\textbf{g-at wt}}{\textbf{ionic charge}}$$

Thus the chemical equivalents of hydrogen, silver, copper, and aluminum are respectively 1, 108, 31.8, and 9 grams.

Second, the product of the current in amperes and the time in seconds during which the cells operated to deposit the chemical equivalents of these elements is found to be approximately 96 500 coulombs. This product is a constant and is called a *faraday.*

$$\textbf{1 faraday} = \textbf{96 500 coulombs}$$

*One **faraday,** 96 500 coulombs, is the quantity of electric charge required to deposit one chemical equivalent of an element.* This unit has many significant applications in electrolysis and electrochemistry.

Faraday's second law: *The mass of an element deposited or liberated during electrolysis is proportional to the chemical equivalent of the element.*

$$m \propto \textbf{chemical equivalent}$$

From our definition of the electrochemical equivalent of an element, we can see that

$$z = \frac{\textbf{chemical equivalent}}{\textbf{faraday}}$$

and has the dimensions g/C. The electrochemical equivalent z is a constant for a given element but is different for different elements.

We can combine Faraday's laws of electrolysis into the following equation:

$$m = zIt$$

where m is the mass in grams of an element deposited or liberated, z is the electrochemical equivalent in g/C of the element, I is the current in amperes, and t is the time in seconds.

Faraday's laws give us a precise method of measuring the quantity of electric charge flowing through a circuit. If we know the length of time the charge flows, we can determine the average current magnitude as shown in Figure 18-11. A platinum dish is connected with the negative terminal of a source of about 2 volts emf, and then the dish is partly filled with a solution of silver nitrate. A platinum spiral rod is connected to the positive terminal of the source of current and dipped into the silver solution.

One coulomb of charge (one ampere-second) flowing through such a silver solution will deposit on the walls of the dish 0.001 118 g of silver. If the average current in the circuit is one ampere for one hour, 3600 coulombs will deposit 4.025 g of silver. From the mass of the silver deposited, the number of coulombs of electric charge can be determined with good precision.

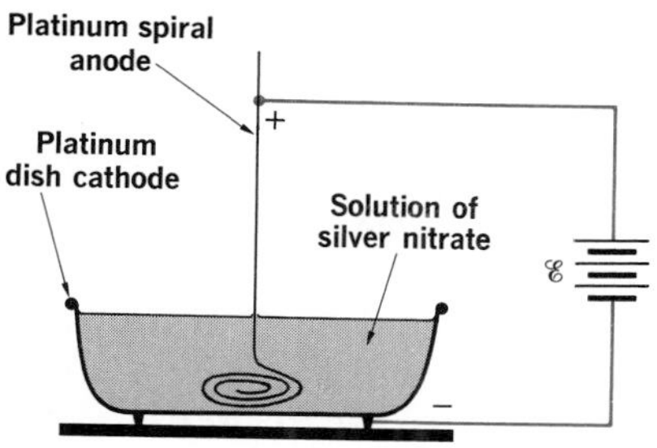

Figure 18-11. A silver coulombmeter.

Questions: Group A

1. How can electron-transfer reactions that are not spontaneous be forced to occur?
2. What is an electrolytic cell?
3. What may be the effect of the conduction of electric charge through a fused ionic compound?
4. What are the basic requirements for a silver-plating cell?
5. Why is electrolytically refined copper used to make electric conductors?
6. State Faraday's laws of electrolysis.
7. What part of a silver coulombmeter is used as the cathode?

Group B

8. Distinguish between an electrochemical cell and an electrolytic cell.
9. Compare the cathode and anode reactions in an electrolytic cell with the corresponding electrode reactions in an electrochemical cell.
10. The g-at wt of aluminum is 27, and the g-at wt of gold is 197; both ions have a charge of 3. Which has the higher electrochemical equivalent?
11. Show that the mass of an element deposited according to Faraday's law can be expressed in grams.
12. How could you define the ampere in relation to Faraday's laws and the silver coulombmeter?

Problems: Group A

1. How much silver can be deposited in 8 min by making use of a constant current of 0.500 A?
2. An electroplating cell connected across a source of direct current for 15.0 min deposits 0.750 g of copper. What is the average current in the cell?
3. How many hours are required for an electrolytic cell to liberate 0.160 g of oxygen using a constant current of 0.300 A?
4. A water-decomposition cell is maintained in continuous operation with a constant current of 25.0 A. How much hydrogen is recovered per day?
5. How many hours would it take to plate 25.0 g of nickel onto an automobile bumper if the current in the plating bath is 3.40 A?

GROUP B

6. An electrolysis-of-water cell is operated until 1 faraday of electricity has been passed. How many atoms of hydrogen are liberated?
7. (a) What magnitude of electric charge must pass through an electrolysis-of-water cell to liberate one g-at wt of hydrogen? (b) How much oxygen will be liberated during the same time?
8. How long will it take for a current of 4.50 A to produce 1.00 L of hydrogen by the electrolysis of water? Assume STP conditions.
9. Two coulombmeters, one of silver and the other of an unknown metal that forms ions having a +3 charge, are connected in series. A constant current of 2.00 A is maintained in the circuit for 2.00 h and 2.73 g of the unknown metal is deposited. How much silver is deposited?
10. (a) What is the g-at wt of the unknown metal that was deposited in the above problem? (b) Identify the unknown metal by using Table 22 of Appendix B on page 678.

SUMMARY

The electric energy expended in a resistance is converted to heat. The relationship among the heat developed in a circuit element, the resistance, and the time the current is maintained is expressed in terms of Joule's law.

Electric power dissipated in the resistance of a circuit is proportional to the square of the circuit current. Power dissipated in the internal resistance of the source of emf is wasted. When it is important to deliver maximum power to the load, the load resistance and that of the source of emf must be matched.

Electrolytic cells are driven by an external source of emf. Electric energy is then transformed to chemical energy through electron-transfer reactions at the cell electrodes. The products have higher chemical energy than the reactants. Electrolytic cells are useful in decomposing compounds, plating metals, and refining metals.

Faraday's laws of electrolysis provide an insight into the quantitative significance of energy transformations in chemical reactions. These laws relate the mass of an element deposited during electrolysis to the quantity of charge that passes through the cell and to the chemical equivalent of the element. The silver coulombmeter provides a precise method of measuring the quantity of charge in an electric circuit.

VOCABULARY

chemical equivalent
coulombmeter
electrochemical cell
electrochemical equivalent
electrolysis
electrolytic cell
electroplating
faraday
Faraday's first law
Faraday's second law
Joule's law
spontaneous reaction

19 Magnetic Effects

magnetism (MAG-ne-tizm) n.: the property associated with charged objects in motion, which give rise to a field of force.

MAGNETISM

OBJECTIVES

- Discuss the domain theory of magnetism.
- Describe the nature of the magnetic force.
- Describe a magnetic monopole.
- Discuss the techniques for mapping magnetic fields.
- Study magnetic induction.
- Describe the earth's magnetic field.
- Define the link between moving charges and magnetic fields of force.
- Study the magnetic field produced by the current in a straight wire and a solenoid.
- Describe electric meters and their use in d-c circuits.

19.1 Magnetic Materials Physicists believe that all magnetic phenomena result from forces between electric charges in motion. Vast quantities of electric energy are now generated as a consequence of relative motion between electric conductors and magnetic fields. Electric energy is transformed into mechanical energy by relative motion between electric currents and magnetic fields. The function of many electric measuring instruments depends on the relationship between electricity and magnetism.

The basic theory of electric generators and motors is presented in Chapter 20. Electric measuring instruments are discussed later in this chapter. Before undertaking the study of magnetic effects of electric currents, we shall examine the magnetic properties of substances and learn of the nature of magnetism and magnetic fields.

Deposits of a magnetic iron ore were discovered many centuries ago by the Greeks in a section of Turkey. The region was then known as Magnesia and the ore was called *magnetite.* Deposits of magnetite are found in the Adirondack Mountains of New York and in other regions of the world. Pieces of magnetite are known as *natural magnets.* A suspended piece of magnetite aligns itself with the magnetic field of the earth. These natural magnets, known as lodestones (leading stones), were first used as magnetic compasses during the twelfth century.

A few materials, notably iron and steel, are strongly attracted by magnets; cobalt and nickel are attracted to a lesser degree. These substances are said to have *ferromagnetic* properties. Special alloys such as *permalloy* and *alnico* have extraordinary ferromagnetic properties. Physicists have shown much interest in the structure of materials possessing the property of **ferromagnetism.**

The Latin word for iron is "ferrum"; thus the name "ferromagnetic."

Today very strong magnets are made from ferromagnetic substances. Alnico magnets may support a weight of over 1000 times that of the magnets themselves. Ferromagnetic substances are commonly referred to simply as "magnetic substances."

Alnico (Al Ni Co) consists mainly of aluminum, nickel, and cobalt plus iron.

Materials are commonly classified as magnetic or nonmagnetic. Those that do not demonstrate the strong ferromagnetism of the Iron Family of metals are said to be "nonmagnetic." However, if these materials are placed in the field of a very strong magnet, some are observed to be slightly repelled by the magnet while others are very slightly attracted.

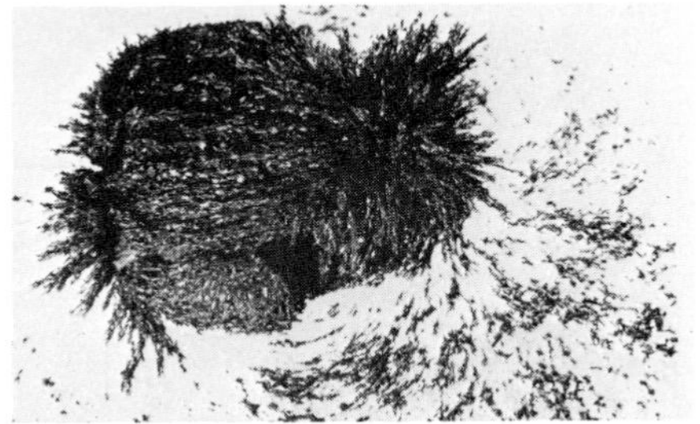

Figure 19-1. Iron filings attracted to a lodestone give evidence of the field of force surrounding the natural magnet.

Zinc, bismuth, sodium chloride, gold, and mercury are a few of the substances that are feebly repelled; they are *diamagnetic.* The property of **diamagnetism** is an important concept in the modern theory of magnetism, as we shall see in Section 19.2.

Wood, aluminum, platinum, oxygen, and copper(II) sulfate are examples of substances that are very slightly attracted by a strong magnet. Such materials are *paramagnetic,* and this magnetic behavior is called **paramagnetism.**

19.2 The Domain Theory of Magnetism William Gilbert's report on his experiments with natural magnets, published in 1600, probably represents the first scientific study of magnetism. In the years that followed, discoveries by Coulomb, Oersted, and Ampère added to our knowledge of the behavior of magnets and the nature of magnetic forces. Physicists believe, however, that it is only within this century that they have begun to understand the true nature of magnetism. The present view is that the magnetic properties of matter are electric in origin and result from the movements of electrons within the atoms of substances. Since the electron is an electrically charged particle, this theory suggests that *magnetism is a property of a charge in motion.* If so, we can account for the energy associated with magnetic forces by using known laws of physics.

Two kinds of electron motion are important in this modern concept of magnetism. *First, an electron revolving about the nucleus of an atom imparts a magnetic property to the atom structure.* See Figure 19-2. When the atoms of a substance are subjected to the magnetic force of a strong magnet, the

The orbiting electron produces a magnetic field because, in this motion, it constitutes an electric current.

force affects this magnetic property, opposing the motion of the electrons. The atoms are thus repelled by the magnet. This is diamagnetism. If the electron's only motion were its movement about the nucleus, all substances would be diamagnetic. Diamagnetic repulsion is quite feeble in its action on the total mass of a substance.

The second kind of motion is that of the electron spinning on its own axis. Each spinning electron acts as a tiny permanent magnet. Opposite spins are designated as + and − spins; electrons spinning in opposite directions tend to form **pairs** and so neutralize their magnetic character. See Figure 19-3. The magnetic character of an atom as a whole may be weak because of the mutual interaction between the electron spins.

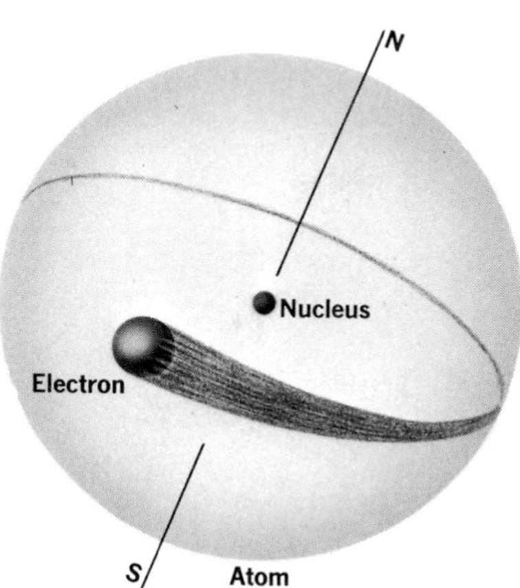

Figure 19-2. Revolving electrons impart a magnetic property to the atom.

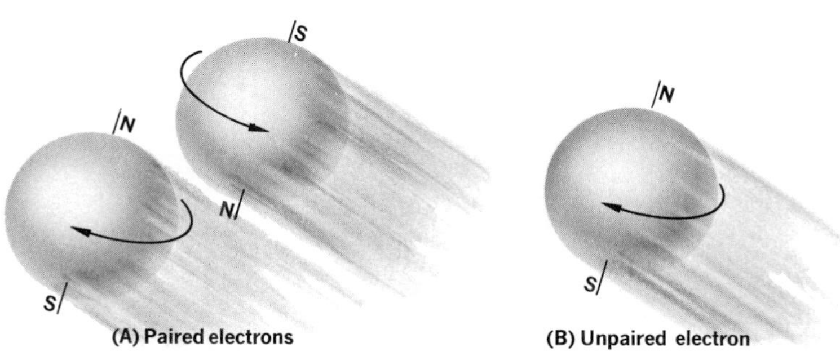

Figure 19-3. Ferromagnetism in matter from the spin of electrons.

Magnetic properties are associated with both kinds of electron motion. The atoms of some substances may possess permanent magnet characteristics because of an imbalance between orbits and spins. These atoms act like tiny magnets, called *dipoles,* and are attracted by strong magnets. Substances in which this attractive effect exceeds the diamagnetism common to all atoms show the property of paramagnetism.

In the atoms of ferromagnetic substances there are unpaired electrons whose spins are oriented in the same way. The common metals iron, cobalt, and nickel and the rare earth elements gadolinium and dysprosium show strong ferromagnetic properties. Some alloys of these and other elements, as well as certain metallic oxides called ferrites, also exhibit strong ferromagnetic properties.

The inner quantum levels, or shells, of the atom structures of most elements contain only paired electrons. The highest quantum level, or outer shell, of each of the noble gases (except helium) consists of a stable octet of electrons made up of four electron pairs. The atoms of other elements achieve this stable configuration by forming chemical bonds. Only in certain transition elements that have

Each iron atom has four unpaired inner-shell electrons.

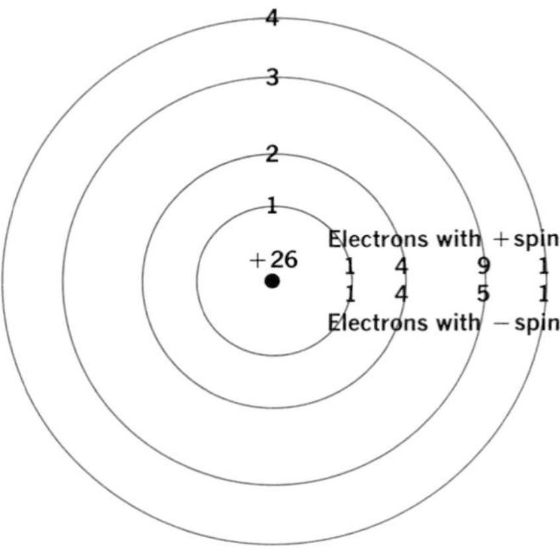

Figure 19-4. The iron atom has strong ferromagnetic properties.

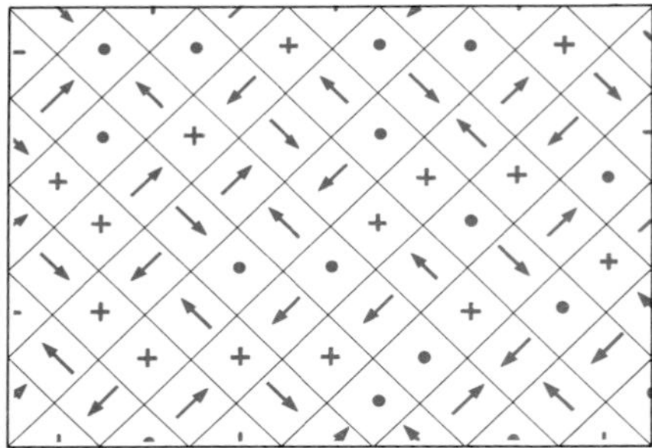

Figure 19-5. The domains of an unmagnetized ferromagnetic substance are polarized along the crystal axes. Dots and plus signs represent arrows going out of and into the page, respectively.

Table 19-1
CURIE POINTS OF FERROMAGNETIC ELEMENTS

Element	*Curie point*
iron	770°C
cobalt	1131°C
nickel	358°C
gadolinium	16°C

incomplete inner shells do unpaired electrons result in ferromagnetic properties. The electron configuration of the iron atom, Figure 19-4, shows four unpaired electrons in the third principal quantum level. The similarly oriented spins of these electrons, enhanced by the influence of nearby atoms in the metallic crystal, account for iron's strong ferromagnetism.

From the preceding discussion, it would seem that every piece of iron should behave as a magnet. However, such is not the case. Atoms are grouped in microscopic magnetic regions called **domains.** The atoms in each domain are magnetically polarized parallel to a crystal axis. In a polycrystalline specimen, ordinarily these axes (and the domains) are oriented in all possible directions. The domains effectively cancel one another and the net magnetism is essentially zero. In Figure 19-5 the polarity of each domain in an unmagnetized material is represented by an arrow.

When a ferromagnetic material is placed in an external magnetic field, two effects occur. The domains more favorably oriented in this magnetic field may increase in size at the expense of less favorably oriented adjacent domains. Other domains may rotate in order to become more favorably oriented with respect to the external field. The material becomes magnetized. If the domain boundaries remain extended to some degree even after the external magnetizing force is removed, the material is said to be "permanently" magnetized. When the direction of magnetization of a magnetic domain is rotated by an external magnetic field, it must be understood that the *material* of the domain does not change its position in the specimen. It is only its *direction* of magnetization that changes.

When the temperature of a ferromagnetic material is raised above a certain critical value, the domain regions disappear and the material becomes paramagnetic. This temperature is known as the **Curie point.** It is usually lower than the melting point of the substance. The Curie points for some ferromagnetic substances are given in Table 19-1.

When a single crystal of iron is sprinkled with colloidal particles of iron oxide, the microscopic domains become visible. Using this technique, physicists are able to photograph magnetic domains and observe the effects of external magnetic fields on them. Typical photomicrographs of magnetic domains are shown in Figure 19-6.

A recent magnet technology that makes use of a group of ferromagnetic substances known as *ferrites* yields strong hard magnets with unique properties. Ferrites are iron oxides combined with oxides of other metals such as manga-

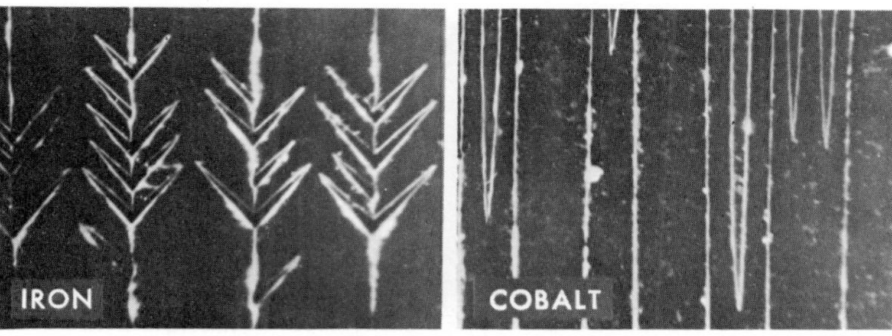

Figure 19-6. Photomicrographs of magnetic domains.

nese, cobalt, nickel, copper, or magnesium. The combined oxides are powdered, formed into the desired shape under pressure, and fired. The ferrites have very high electric resistance, a property that is extremely important in some applications of ferromagnetic materials. The original lodestone, commonly called magnetic iron oxide, is a material of this type. Chemically it is a combination of iron(II) oxide, FeO, and iron(III) oxide, Fe_2O_3. Its formula is considered to be $Fe(FeO_2)_2$.

19.3 Force Between Magnet Poles The fact that iron filings cling mainly to the ends of a bar magnet indicates that the magnetic force acts on the filings primarily in these regions, or *poles;* it does not mean that the middle region of the magnet is unmagnetized. The pole that points toward the north when the magnet is free to swing about a vertical axis is commonly called the *north-seeking pole,* or **N** pole. The opposite pole, which points toward the south, is called the *south-seeking pole,* or **S** pole.

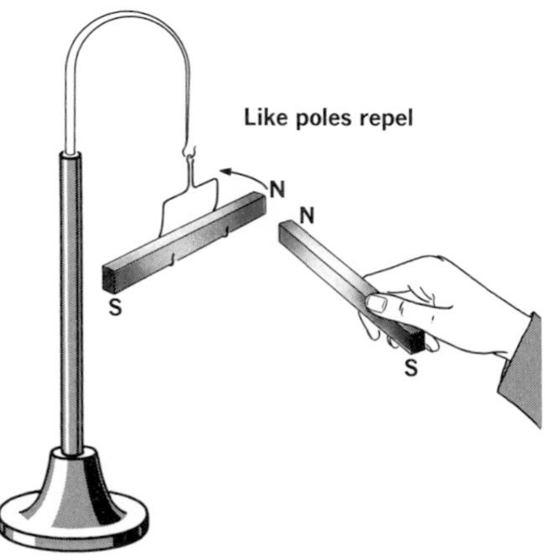

Figure 19-7. Like poles repel. Unlike poles attract.

Suppose a bar magnet is suspended as shown in Figure 19-7. When the **N** pole of a second magnet is brought near the **N** pole of the suspended magnet, the two repel each other. A similar action is observed with the two **S** poles. When the **S** pole of one magnet is placed near the **N** pole of the other magnet, they attract each other. Such experiments show that *like poles repel and unlike poles attract.*

Magnets usually have two well-defined poles—one **N** and one **S.** Sometimes long bar magnets acquire more than two poles, and an iron ring may have no poles at all when magnetized. Physicists have long speculated about the existence of single-pole magnetic particles called *monopoles.* Known magnetic poles, however, always come in pairs called *dipoles.* The most elementary magnet has an **S** pole and an **N** pole. If cut in half, each half is found to be dipolar. A magnet has an **S** pole for every **N** pole. An isolated **N** pole of unit strength is sometimes *assumed* in "thought" experiments. ***A unit pole*** *may be thought of as one that repels an exactly similar pole, placed one centimeter away, with a force of* 10^{-5} *N.*

Experimental evidence of the possible existence of magnetic monopoles has been reported but not verified. Physicists believe that proof of the existence of monopoles could help verify some of the basic concepts of physics.

The quantitative expression for Coulomb's law of magnetism is $F = k\frac{M_1M_2}{d^2}$*. Compare this equation with those in Sections 3.11 and 16.8.*

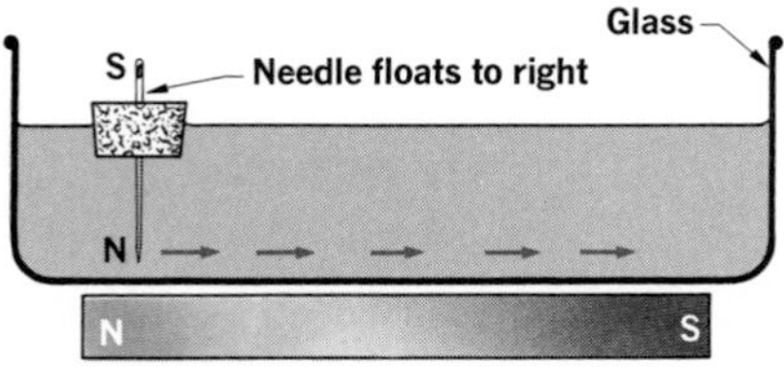

Figure 19-8. The path followed by the floating magnet in this experiment is approximately that of an independent N pole.

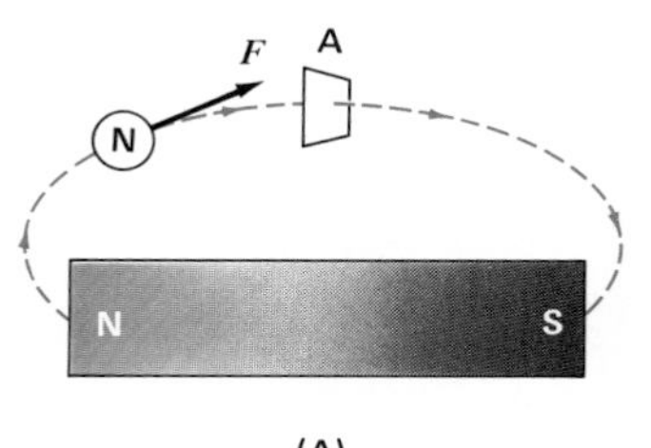

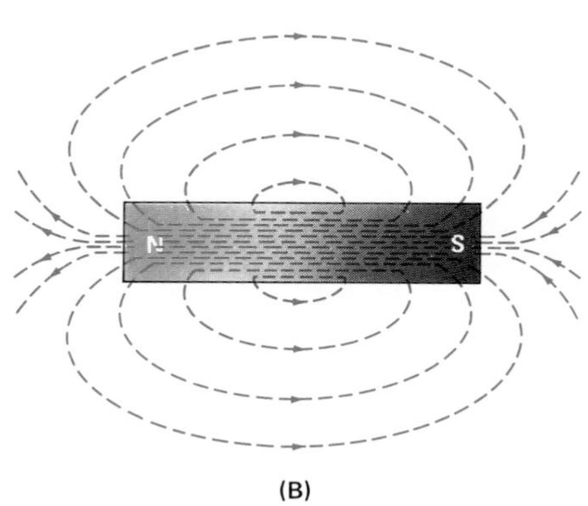

Figure 19-9. (A) The path taken by an independent N pole in a magnetic field suggests a line of flux. (B) Magnetic flux about a bar magnet.

The first quantitative study of the force between two magnetic poles is generally credited to Coulomb. He found this magnetic force governed by the same inverse-square relationship that applies to gravitational force and electrostatic force. ***Coulomb's law of magnetism is:*** *The force between two magnetic poles is directly proportional to the product of the strengths of the poles and inversely proportional to the square of the distance between them.* The force is one of repulsion or attraction, depending on whether the magnetic poles are alike or different.

19.4 Magnetic Fields of Force

In Section 16.9 we described the electric field of force near an electrically charged object. Electric forces are not the only forces that act on charged particles. Sometimes we observe the effect of a force that is both *perpendicular* and proportional to the velocity of a *moving* charge. This force identifies a *magnetic field.* A dipole magnet in such a region of space experiences a torque. We speak of a magnetic field in the space around a bar magnet in the same way we speak of an electric field around a charged rod. Furthermore, *we can represent a magnetic field by lines of flux,* just as we represented an electric field by lines of force.

The behavior of our imaginary independent **N** pole in a magnetic field can be approximated by using a magnetized darning needle as illustrated in Figure 19-8. The needle is supported by cork so that it floats with the **N** pole extended below the surface of the water. The **S** pole is far enough removed to have negligible influence on the movement of the needle. A bar magnet placed under the glass dish with its **N** pole near the needle causes the floating magnet to move along a path that approximates the path an isolated **N** pole would follow.

The path of an independent **N** pole in a magnetic field suggests a *line of flux. A* ***line of flux*** *is a line so drawn that a tangent to it at any point indicates the direction of the magnetic field.* Flux lines are assumed to emerge from a magnet at the **N** pole and to enter the magnet at the **S** pole. Every flux line is a closed path running from **S** pole to **N** pole within the magnet. See Figure 19-9.

The lines of flux perpendicular to a specified area in the magnetic field are collectively called the *magnetic flux,* for which the Greek letter Φ (phi) is used. The unit of magnetic flux is the *weber* (Wb).

The ***magnetic flux density, B,*** *is the number of flux lines per unit area that permeates the magnetic field.* The flux density B is a vector quantity; the direction of B at any point in the magnetic field is the direction of the field at that point.

$$B = \Phi/A$$

Flux density is expressed in *webers per square meter* (Wb/m^2). The flux density determines the *magnetizing force* at any point in the magnetic field. The weber per meter^2 is also called the *tesla* (T).

$$\textbf{1 weber/meter}^2 = \textbf{1 tesla}$$

The measurement of these quantities is in Section 19.9.

Flux lines drawn to indicate how tiny magnets would behave when placed at various points in a magnetic field provide a means of *mapping* the field. A line drawn tangent to a flux line at any point indicates the direction a very small magnet would assume if placed there. An arrowhead can be added to the tangent line to indicate the direction in which the **N** pole of the tiny magnet would point, thus giving the direction of the magnetic field, and the B vector, at that point.

Imaginary lines of magnetic flux are useful for mapping magnetic fields.

Using a suitable scale of flux lines per unit area perpendicular to the field, the flux density, B, at any point can be illustrated. Selection of a number of lines to represent a unit of magnetic flux is arbitrary. Usually, one flux line per square meter represents a flux density of 1 Wb/m^2. In this sense, one line of flux is a weber.

The magnetic field near a single bar magnet is suggested by the pattern formed by iron filings sprinkled on a glass plate laid over the magnet. A photograph of this field pattern is shown in Figure 19-10. Using a similar technique, the magnetic fields near the *unlike* poles and near the *like* poles of two bar magnets are illustrated in Figure 19-11. Observe that the magnetic force acting on the two unlike poles is one of attraction and that acting on the two like poles is one of repulsion. Figure 19-12 similarly illustrates an end-on view of the magnetic field between the poles of a horseshoe magnet.

19.5 Magnetic Permeability In Section 19.4 we described the effect of a magnetic field of force on iron filings and on a magnetized needle as experienced through glass

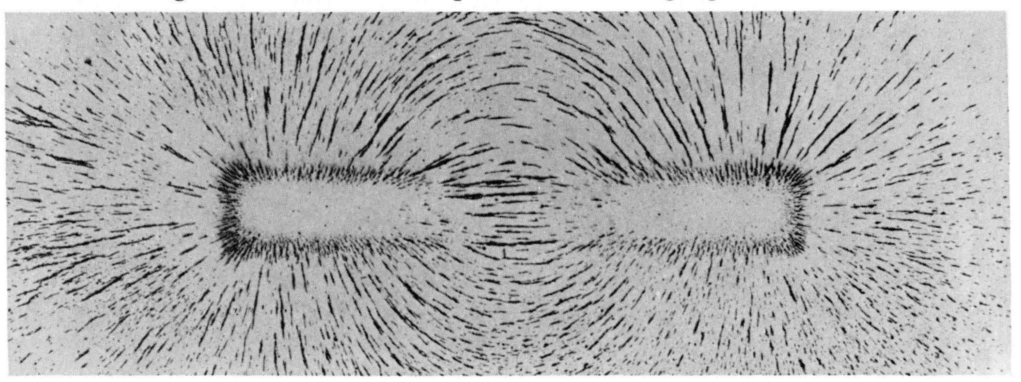

Figure 19-10. Iron filings near a single bar magnet.

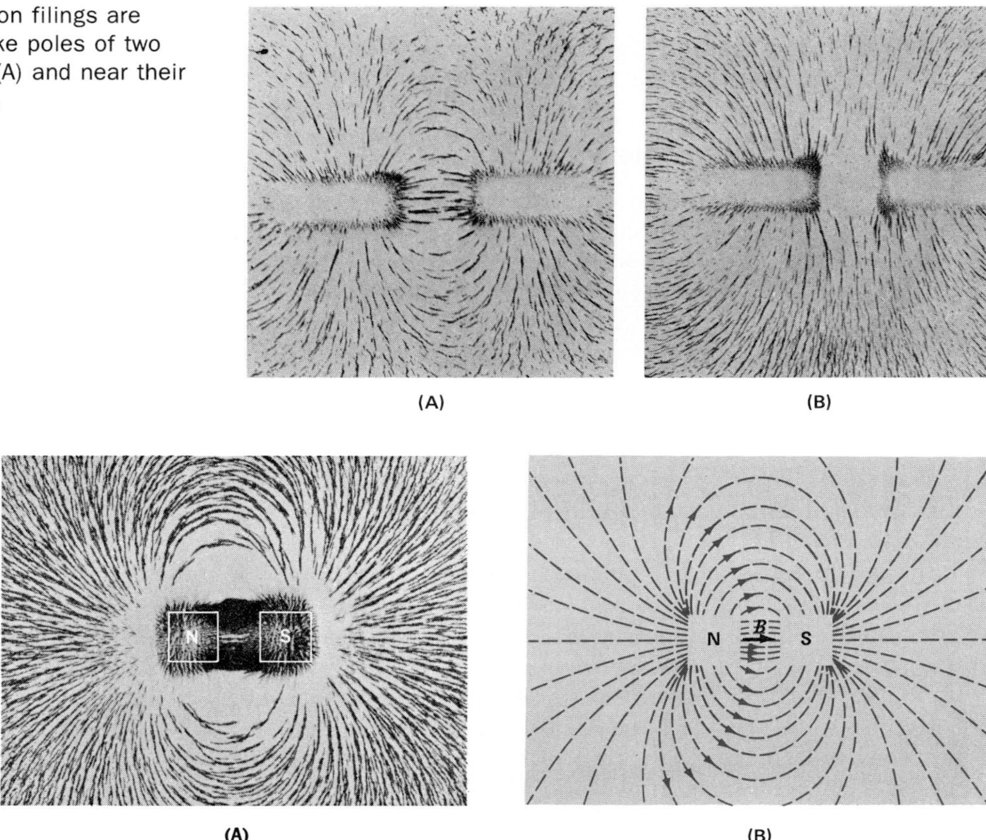

Figure 19-11. Iron filings are shown near unlike poles of two bar magnets in (A) and near their like poles in (B).

Figure 19-12. (A) Iron filings near the poles of a horseshoe magnet, end view. (B) An idealized drawing of (A) showing lines of flux.

In a practical sense, all materials except those that are ferromagnetic can be considered magnetically inert. In a magnetic field, they behave just like air.

and water. Nonmagnetic materials in general are transparent to magnetic flux; that is, their effect on the lines of flux is not appreciably different than that of air. *The property of a material by which it changes the flux density in a magnetic field from the value in air is called its* ***permeability,*** μ. Permeability is a ratio of flux densities and is without dimension. The permeability of empty space is taken as unity and that of air as very nearly the same. The permeabilities of diamagnetic substances are slightly less than unity; permeabilities of paramagnetic substances are slightly greater than unity. Permeabilities of ferromagnetic materials are many times that of air.

If a sheet of iron covers a magnet, there is little magnetic field above the sheet. The flux enters the iron and follows a path within the iron itself. Similarly, an iron ring placed between the poles of a magnet provides a better path than air for the magnetic flux. This effect is illustrated in Figure 19-13. The flux density in iron is greater than it is in air; therefore, iron is said to have a high permeability. The permeabilities of other ferromagnetic substances are also very high.

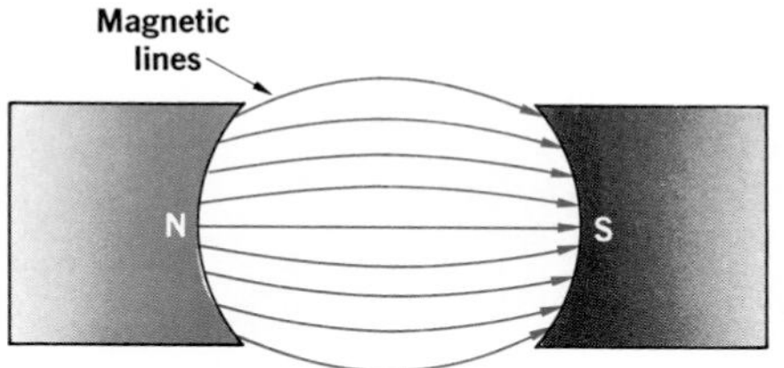

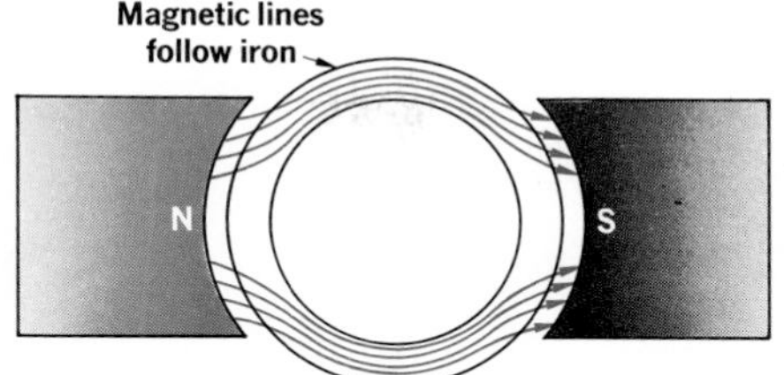

Figure 19-13. At left, magnetic flux crosses the air gap between the poles of a magnet. At right, magnetic flux follows the soft iron ring, which is more permeable than air.

Suppose a bar of soft iron lies in a magnetic field, as in Figure 19-14. Because of the high permeability of the iron, the field is distorted and the magnetic flux passes through the iron in preference to the air. Under these circumstances the soft iron bar becomes a magnet with end **A** as the **S** pole and end **B** as the **N** pole. Such a bar is said to be magnetized by *induction*. *Magnetism produced in a ferromagnetic substance by the influence of a magnetic field is called* ***induced magnetism.***

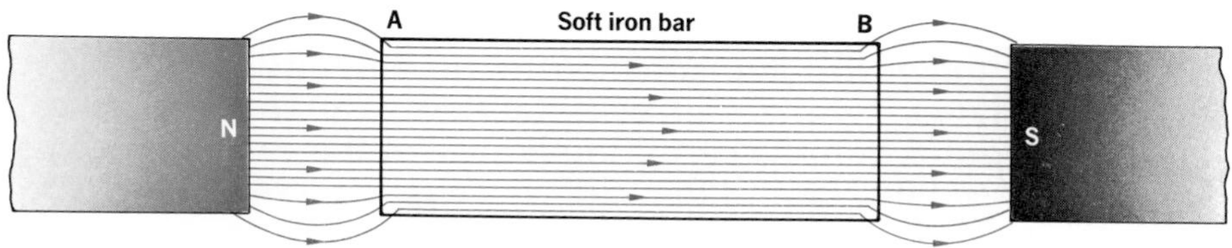

Figure 19-14. An iron bar magnetized by induction.

If the magnetic field is removed by withdrawing the two bar magnets, most of the induced magnetism will be lost. Magnets produced by induction are known as *temporary magnets*. A piece of hardened steel is not so strongly magnetized by induction but retains a greater *residual* magnetism when removed from the induction field.

There is no significant difference in the process if the iron bar in Figure 19-14 is brought into contact with one of the magnet poles. The magnetization process is somewhat more efficient due to the reduction of the air gap. See Figure 19-15.

Figure 19-15. The nail becomes a magnet by induction. Are the tacks also magnets?

19.6 Terrestrial Magnetism Suppose the earth contained a great bar magnet. See Figure 19-16. It would produce a magnetic field similar to its actual field. Over most of the earth's populated surface the north-seeking pole of a compass points northward. Although it is the south pole of our fictitious magnet that attracts the **N** pole of the compass, the pole region is conventionally called the *north magnetic pole* because it is located in the northern hemisphere. Similarly, the pole region in the southern hemisphere is called the *south magnetic pole*.

The earth's magnetic axis does not coincide with its polar (geographic) axis, but is inclined to the polar axis at a small angle. The north magnetic pole, at latitude 73°N and longitude 100°W, is about 2000 km (1200 miles) south

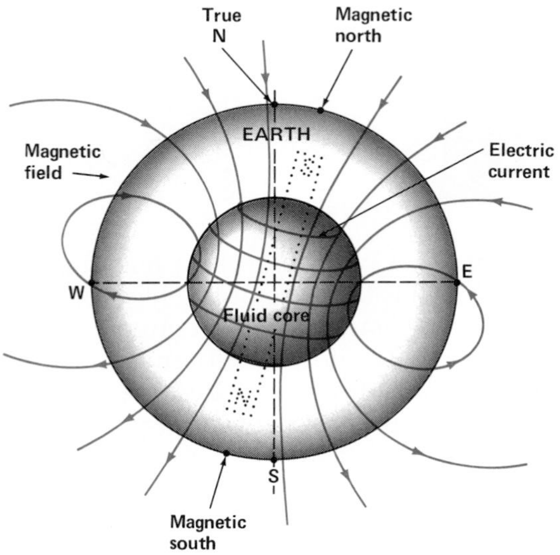

Figure 19-16. The magnetic field of the earth may be produced by electric currents within its fluid core. The field is oriented as though the earth contained a large magnet passing through its center, with the magnetic axis slightly inclined with respect to the axis of the earth.

of the north geographic pole. The south magnetic pole is located in Antarctica near the Ross Sea. Thus from most locations on the earth, the **N** pole of the compass needle does not point to the true geographic north. At any surface location the angle between magnetic north and the true north is called the **declination,** or *variation.* In the region of Los Angeles the compass variation is about 15°E. That is to say a compass needle points about 15° to the east of true north. In the region of Boston the variation is about 15°W. Cincinnati, Ohio, is located very near the line of zero declination. Here the compass needle points to the true north and the variation is 0°.

A compass needle mounted on a horizontal axis and provided with a means of measuring the angle the needle makes with the horizontal plane is called a **dipping *needle.*** At certain places on the earth's surface, about midway between the magnetic poles, the angle of dip is zero and the needle is horizontal. A line drawn through a succession of such points identifies the *magnetic equator.* The angle of dip is 90° at the magnetic pole. The dip, or deviation between the equilibrium position of a dipping needle and the horizontal, is known as the *magnetic inclination.*

In 1600 the English physicist William Gilbert (1540?–1603) published his scientific treatise *De magnete,* which deals with the magnetism of the earth. This is one of the earliest publications on the experimental treatment of a scientific topic. Gilbert inferred that the earth behaved as a large magnet because the interior consisted of permanently magnetic material. Today scientists believe the core of the earth is too hot to be a permanent magnet and is fluid rather than solid.

The German physicist Karl Friedrich Gauss (1777–1855) showed that the magnetic field of the earth must originate inside the earth. In 1939 the American theoretical physicist Walter M. Elsasser suggested that the earth's magnetic field results from electric currents generated by the flow of matter in the earth's fluid core. See Figure 19-16. Today physicists believe that the magnetic field is due primarily to electric currents within the earth, but they have not yet established the origin of these currents.

Electric current loops inside the earth are responsible for its magnetic field.

19.7 The Magnetosphere Because space vehicles now travel to the outer limits of the earth's atmosphere and beyond, there is a growing interest in a region of the outer atmosphere known as the **magnetosphere.** Located beyond 200 km, the magnetosphere is the region in which the motion of charged particles is governed primarily by the magnetic field of the earth. At lower altitudes, where the density of the atmosphere is much greater, the motion of charged particles is controlled largely by collisions.

The magnetosphere on the side facing the sun extends beyond the earth's surface approximately 57 000 km, or about 10 earth radii. On the side away from the sun, the magnetosphere probably extends outward for hundreds of earth radii. See Figure 19-17. The elongated shape results from the influence of the onrushing **solar wind,** or *solar plasma.* The solar wind, consisting mainly of protons and electrons emitted by the sun, compresses the magnetosphere on the side nearest the sun.

In 1958 regions of intense radiation were discovered within the magnetosphere by a team of physicists headed

Figure 19-17. The magnetosphere of the earth. The overall radiation regions are shown in color. The inner and outer Van Allen belts of intense radiation are the dark regions ranging outward to approximately 4 earth radii.

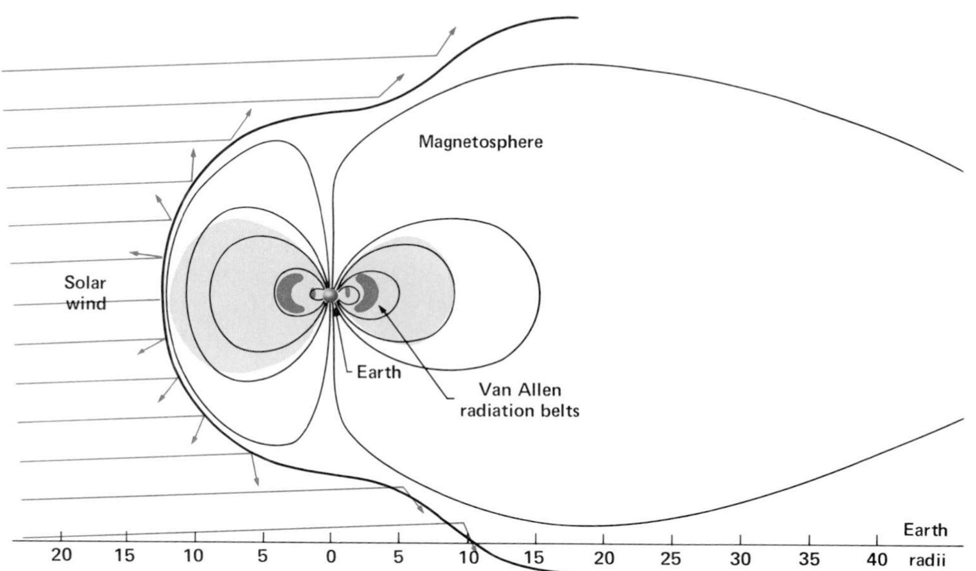

by Dr. J. A. Van Allen (b. 1914). These regions, now known as the Van Allen radiation belts, contain energetic protons and electrons trapped by the earth's magnetic field. Those trapped in the inner belts probably originate in the earth's atmosphere; those trapped in the outer belts probably have their origin in the sun. When these intense radiation belts were first discovered, scientists were concerned about the serious threat they appeared to present to space travel. Today, astronauts journeying into outer space are able to pass quickly through these regions with adequate protection from the Van Allen radiation.

Auroral displays over the polar regions are related to the escape of energetic particles from the radiation belts.

QUESTIONS: GROUP A

1. (a) What are natural magnets called? (b) Is this term appropriate? Explain.
2. Describe the two electron motions that affect an object's magnetic properties.
3. What is a magnetic domain?
4. Why are iron atoms so strongly affected by magnetic fields?
5. When a magnetized steel needle is heated strongly in a bunsen burner flame, it becomes demagnetized. Explain why.
6. What indicates that a piece of iron is magnetic—its attraction to or repulsion from another piece of iron?
7. Describe the magnetosphere.
8. (a) What is the minimum number of poles for a magnet? (b) Can a magnet have three poles? Explain.
9. (a) When you break a magnet in half, how many poles does each piece have? (b) How small would the pieces have to be for this not to be true?
10. What is the difference between the angles of declination and inclination?

GROUP B

11. What is the difference between a paramagnetic and a ferromagnetic material?
12. What do we mean when we say a piece of paper is magnetically transparent?
13. Why does a very strong magnet attract both poles of a weak magnet?
14. (a) How does solar wind affect the shape of the earth's magnetic field? (b) Name two other effects of charged atmospheric particles on the earth's magnetic field.
15. A strong magnet in a junkyard can lift a car; what does this tell you about the relative strength of the magnetic and gravitational forces on the car?
16. What happens on a subatomic level when a magnet attracts a steel needle?
17. If a small magnet is repeatedly dropped, it becomes demagnetized. Explain what is happening subatomically.
18. Compare and contrast the effects of electrostatic and magnetic forces.
19. Sir William Gilbert believed the earth contained an iron core that was a huge permanent magnet. Give some evidence to refute this theory.
20. What happens when you pass a magnet across a computer floppy disk?

ELECTROMAGNETISM

19.8 The Link Between an Electric Current and Magnetism It can be easily demonstrated that electrostatic charges and stationary magnets have no effect on one another. However, in 1820 Hans Christian Oersted (*er*-stet) (1777–1851), a Danish physicist and professor of physics at the University of Copenhagen, observed that a small compass needle is deflected when brought near a conductor carrying an electric current. This was the first evidence of a long-suspected link between electricity and magnetism. Oersted discovered that forces exist between a magnet and electric charges in motion. His famous experiment is so significant that a brief description of it is in order.

Figure 19-18. Hans Christian Oersted studied medicine before becoming professor of physics at the University of Copenhagen in 1806. Several years before he performed his famous experiment, he predicted that a link between electricity and magnetism would be found.

A dry cell, compass, switch, and conducting wire are arranged as shown in Figure 19-19(A). With the switch open, a straight section of the conductor is supported *above* the compass in the vertical plane of the compass needle. In Figure 19-19(B) the dry-cell connection is such that the electron flow will be from north to south. When the switch is closed, the **N** pole of the compass is deflected toward the west. When the dry-cell connections are reversed so electron flow is from south to north, the **N** pole of the compass is deflected to the east, as in Figure 19-19(C). It is evident that *a magnetic field exists in the region near the conductor when the circuit is closed.* Furthermore, *the direction of the field is dependent on the direction of the current in the conductor.*

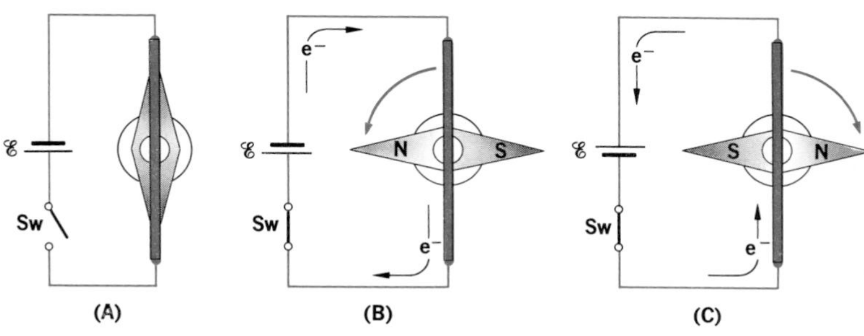

Figure 19-19. The Oersted experiment as viewed from above. In each diagram the compass needle is located below the conductor.

If the experiment is repeated with the conductor placed *below* the compass needle, the compass deflection is opposite to that in the first experiment. This suggests, but does not prove, that the magnetic field encircles the conductor.

19.9 Magnetic Field and a Charge in Motion Shortly after Oersted's discovery, the French physicist Ampère determined the shape of the magnetic field about a conductor carrying a current. He had discovered that forces exist between two parallel conductors in an electric circuit. If the two currents are in the same direction, the force is one of attraction; the force is one of repulsion if the currents are in opposite directions. See Figure 19-21.

Figure 19-20. André Ampère, the French physicist for whom the unit of electric current is named, did fundamental work in electromagnetism.

In a quantitative sense, two long, straight, parallel conductors of length l separated by a distance d and carrying currents I_1 and I_2 will each experience a force F of magnitude

$$F = \frac{2k\, l\, I_1 I_2}{d}$$

The constant k is exactly 10^{-7} N/A^2. If I_1 and I_2 are expressed in amperes and l and d in meters, the force F is given in newtons.

$$\left(\frac{\mathrm{N}}{\mathrm{A}^2}\right) \times \left(\frac{\mathrm{m} \cdot \mathrm{A} \cdot \mathrm{A}}{\mathrm{m}}\right) = \mathrm{N}$$

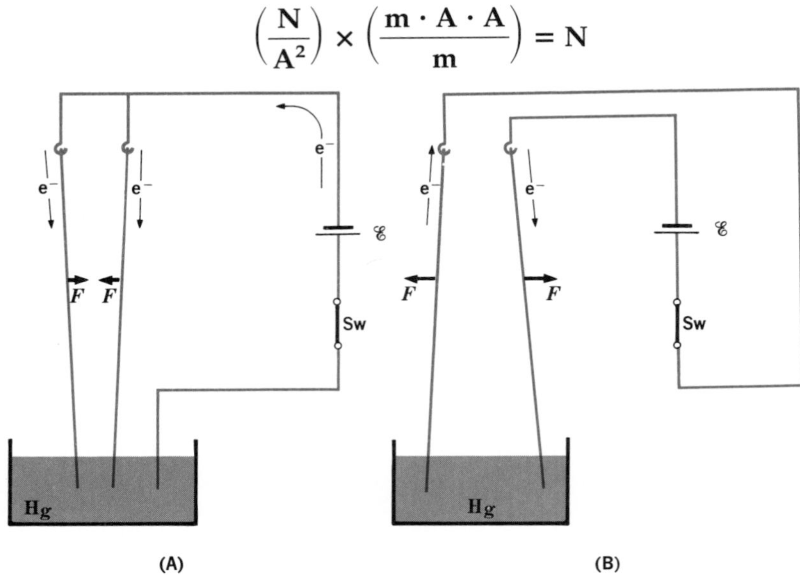

Figure 19-21. Forces between parallel currents (A) in the same direction and (B) in opposite directions.

Because these attractive and repulsive forces between current-carrying conductors are directly proportional to the currents in the conductors, they provide a precise method of defining the unit of current, the ampere. In this sense, the ***ampere*** may be defined as *the current in each of two long parallel conductors spaced one meter apart that causes a magnetic force of* 2×10^{-7} *newton per meter length of conductor.*

Following this scheme, the ***coulomb*** as a quantity of charge (an ampere-second) may be defined as *the quantity of electric charge that passes a given point on a conductor in one second when the conductor carries a constant current of one ampere.*

Ampère investigated the magnetic fields about conductors to find an explanation of the magnetic forces. Suppose a heavy copper wire passes vertically through the center of a horizontal sheet of stiff cardboard. When the ends of the vertical conductor are connected to a dry cell, iron filings sprinkled over the surface of the cardboard form a pattern of concentric circles around the conductor. See Figure 19-22. If a small compass is placed at successive points around a circle of filings, the needle always comes to rest

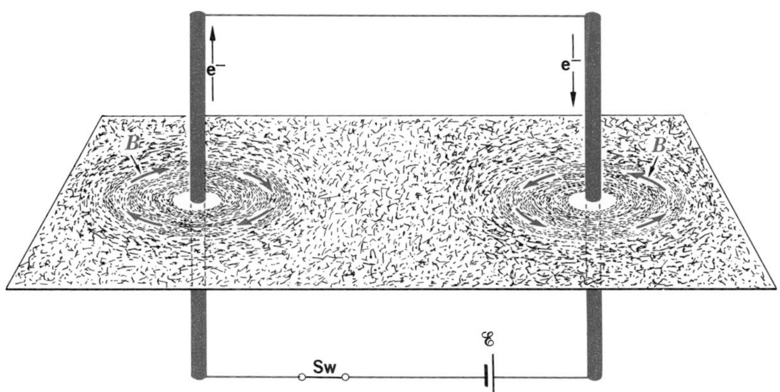

Figure 19-22. The magnetic field encircling a current in a straight conductor.

tangent to the circle and with the same tangential orientation of its **N** pole.

If the direction of current in the vertical conductor is reversed, the compass needle again becomes aligned tangent to the circle of filings, but with its **N**-pole orientation reversed. From these observations we conclude that *a magnetic field encircles an electric charge in motion.* The lines of flux are closed concentric circles in a plane perpendicular to the conductor with the axis of the conductor as their center. The direction of the magnetic field is everywhere tangent to the flux and is dependent on the direction of the current.

Magnetic phenomena are interpreted in terms of the forces associated with electric charges in motion.

Ampère devised a rule, known today as **Ampère's rule,** for determining the direction of the magnetic field around a current in a straight conductor when the direction of the electron flow is known.

Ampère's rule for a straight conductor: Grasp the conductor in the left hand with the thumb extended in the direction of the electron flow. The fingers then will circle the conductor in the direction of the magnetic flux. See Figure 19-23.

The left-hand rule for a straight conductor indicates the direction of the magnetic flux surrounding the conductor.

The flux density, B, also called the *magnetic induction,* at any point in the magnetic field of a long straight conductor carrying a current, I, *is directly proportional to the current in the conductor and inversely proportional to the radial distance, r, of the point from the conductor.*

$$B = 2k\frac{I}{r}$$

The constant k again is 10^{-7} N/A^2. When I is given in amperes and r is in meters, B is expressed in newtons per ampere meter, which is equivalent to webers per square meter.

In Section 19.4 flux density is defined in terms of the lines of flux per unit area that permeate the magnetic field.

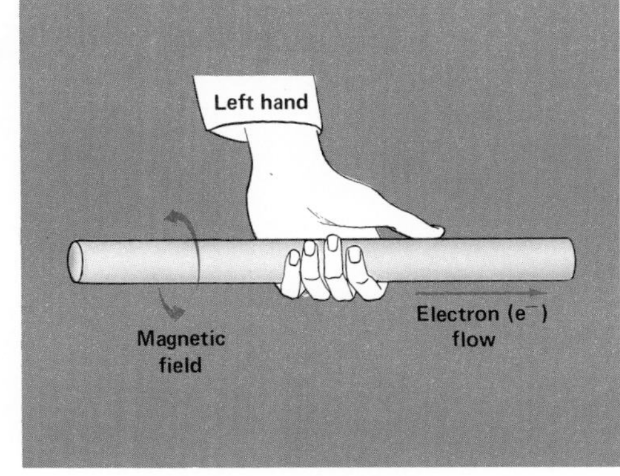

Figure 19-23. Ampère's rule for a straight conductor.

In this sense the expression for flux density is $B = \Phi/A$. Thus

$$\Phi = BA$$

When B is expressed in newtons per ampere meter, the unit for Φ, the weber, can be shown to be 1 newton-meter per ampere:

$$\text{Wb} = \left(\frac{\text{N}}{\text{A} \cdot \text{m}}\right) \times (\text{m}^2) = \frac{\text{N} \cdot \text{m}}{\text{A}}$$

Whether Φ is expressed in webers or newton meters per ampere and B is expressed in webers per square meter or newtons per ampere meter is a matter of convenience in each situation.

Observe that the definition of B given in Section 19.4 is based on the force exerted on an isolated unit pole. An isolated pole exists only in the fiction of a thought experiment; consequently, measurements based on this definition lack precision. The more practical definition given above involves quantities that can be measured precisely and is therefore generally preferred.

19.10 Magnetic Field and a Current Loop Keeping Ampère's rule in mind, let us consider a loop in a conductor carrying a current. The magnetic flux from all segments of the loop must pass through the inside of the loop in the same direction; that is, *the face of the loop must show polarity.* See Figure 19-24.

Figure 19-24. The magnetic field through a current loop.

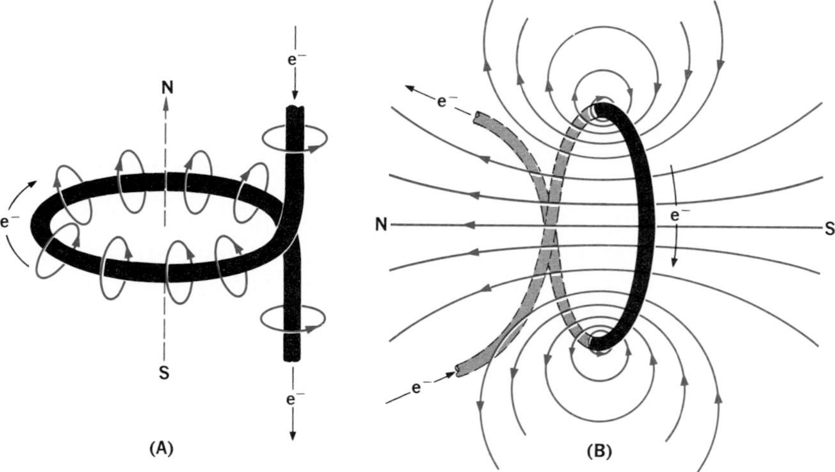

A magnetic tube is an imaginary tube bounded by magnetic lines. It always links the current that produces the field.

This loop magnet can be made stronger if the flux density can be increased. Because the magnetic field around a conductor varies with the current, the flux density can be increased by increasing the magnitude of the current in the conducting loop, by forming additional loops in the conductor, or by both.

A linear coil of such conducting loops takes the form of a helix and is called a *solenoid.* The cylindrical column of air inside the loops, extending the length of the coil, is called the *core.* When a current is in a solenoid, the core of each *turn* (loop) becomes a magnet; the core of the solenoid is a magnetic tube through which practically all the magnetic flux passes. See Figure 19-25.

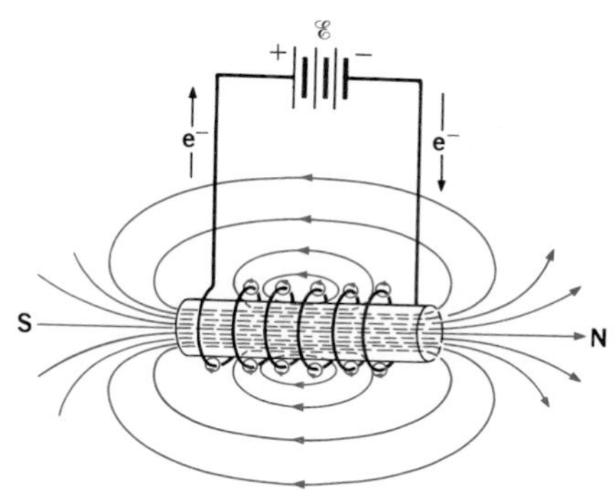

Figure 19-25. Magnetic field about a solenoid.

Because a solenoid conducting an electric current has the magnetic properties of a bar magnet, its polarity can be determined by means of a compass. However, the magnetic flux in the core of the solenoid is derived from the magnetic field of each turn of the conductor. Thus Ampère's rule is modified to adapt it to this special case of the solenoid.

Ampère's rule for a solenoid: Grasp the coil in the left hand with the fingers circling the coil in the direction of the electron flow. The extended thumb will point in the direction of the **N** pole of the core. See Figure 19-26.

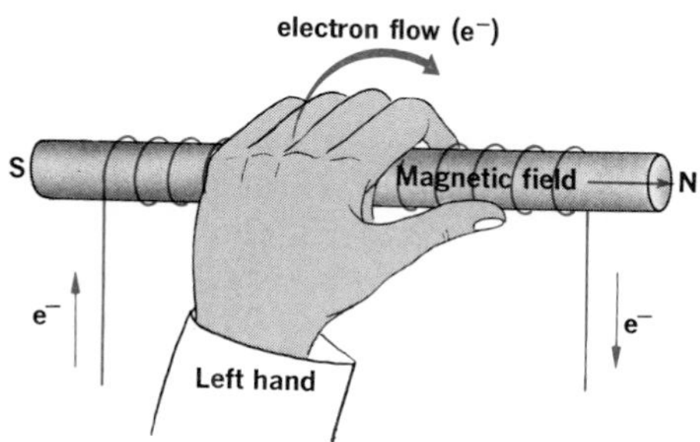

Figure 19-26. Ampère's rule for a solenoid.

19.11 The Electromagnet

A solenoid with a core of air, wood, or some other nonmagnetic material does not produce a very strong electromagnet because the permeability of all nonmagnetic substances is essentially equal to that of air—unity. Substitution of such materials for air does not appreciably change the flux density.

Soft iron, on the other hand, has a high permeability. If an iron rod is substituted for air as the core material, the flux density is greatly increased. Strong electromagnets therefore have ferromagnetic cores with high permeability. For a given core material, the strength of the electromagnet depends on the magnitude of the current and the number of turns. In other words, its strength is determined by *the number of ampere-turns.*

Figure 19-27. A superconducting electromagnet. In operation, the eight-foot cylinder is immersed in liquid helium (−232°C). At this temperature the niobium-titanium strips imbedded in the copper coils lose all electric resistance and the magnet produces a force field up to 5000 times greater than that of the earth.

19.12 The Galvanometer

Suppose we form a wire loop in a vertical plane, place a compass needle (free to rotate in a horizontal plane) in the center of it, and then introduce a current into the loop. The needle will be deflected. If we increase the number of turns sufficiently, even a feeble current will produce a deflection of the needle. Such a device, called a *galvanoscope,* may be used to detect the presence of an electric current or to determine its direction. A simple galvanoscope is shown in Figure 19-28.

A more versatile instrument for detecting feeble currents is the **galvanometer,** the essential parts of which are shown in Figure 19-29. A coil of wire wound on a soft iron core is pivoted on jeweled bearings between the poles of a

Figure 19-28. A simple galvanoscope.

Because a magnetic field exerts forces on moving charges, it exerts torques on current-carrying coils.

permanent horseshoe magnet. The coil becomes a magnet when current is in it. The instrument then has two magnets: a permanent horseshoe magnet in a fixed position and an electromagnet free to turn on its axis. Electric connections to the coil are made through two control springs (not shown), one above and one below the coil. These coiled springs also restrain the rotational motion of the coil so that the attached pointer returns to the zero scale position when no current is present in the coil. This zero position is often located at the midpoint of the scale, as in Figure 19-29.

When there is a current in the movable coil, its core is magnetized. The poles of the core are then attracted and repelled by the poles of the permanent magnet. A torque acts upon the coil and the coil rotates in an attempt to align its plane perpendicular to the line joining the poles of the permanent magnet. As the coil rotates, however, it does work against the two control springs. Its final position is reached when the torque acting on it is just neutralized by the reaction of the springs. Since the permanent field flux is constant, the torque on the coil is proportional to the current in it. We may assume, for small movements of the coil, that the reaction of the springs is proportional to the deflection angle. When the coil reaches its equilibrium position, these two opposing torques are equal, and the deflection angle of the coil is therefore proportional to the current in it.

The scale of the galvanometer is marked at intervals on either side of the zero center. Readings are made on this scale by means of a small, lightweight pointer attached to the coil. For a coil current in one direction, the needle deflection is to the left. If the current direction is reversed, the needle is deflected to the right.

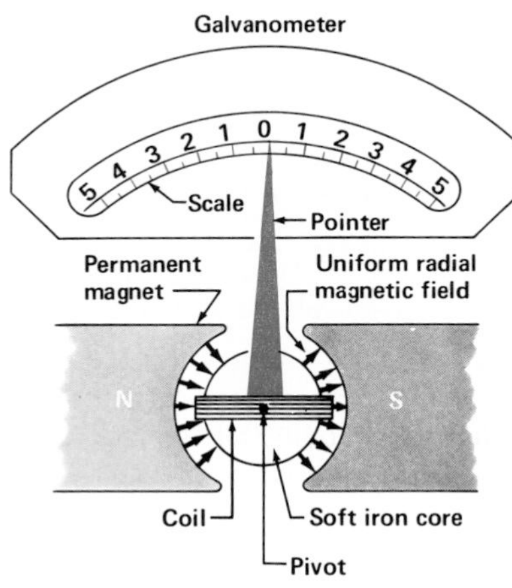

Figure 19-29. The basic components of a moving-coil galvanometer.

The galvanometer is a sensitive instrument for detecting feeble currents of the order of microamperes. For translation of a reading into absolute current values, the *current sensitivity* of the specific instrument must be known. Current sensitivity is usually expressed in *microamperes per scale division.*

The pointer deflection, d, of a galvanometer is proportional to the current, I_M, in the coil

$$I_M \propto d$$

or

$$I_M = kd$$

and

$$k = \frac{I_M}{d}$$

where k is the current sensitivity in microamperes per scale division.

The moving coil of the galvanometer has resistance. By Ohm's law, a potential difference appears across the resistance of the meter when a current is in the coil. We can express the *voltage sensitivity* of the instrument since it must be equal to the product of the meter resistance and the current per scale division.

Figure 19-30. Sensitive galvanometers use a mirror to indicate the position of a suspended coil by producing an image of a scale viewed through a telescope or by reflecting a beam of light onto a scale.

$$\textbf{Voltage sensitivity} = kR_M = \frac{I_M}{\text{div}} \times R_M$$

where R_M is the resistance of the meter movement. Voltage sensitivity is given in microvolts per scale division.

If provision is made to prevent excessive currents from entering the coil, the galvanometer can be adapted for service as either a d-c ammeter or d-c voltmeter.

The following examples illustrate calculations involving the galvanometer.

EXAMPLE A galvanometer has a current sensitivity of $2.50 \times 10^3\ \mu A$, and the scale goes from −50 to 0 to 50, with 0 at the very center. Calculate the current in amperes that is passing through the coil of the instrument when the reading is 18.5 on either side of zero.

Given	Unknown	Basic equation
$k = \dfrac{25\bar{0}0\ \mu A}{50.0\ \text{divisions}}$	I_M	$I_M = kd$
$d = 18.5$ divisions		

Solution

Working equation: $I_M = kd = \dfrac{(25\bar{0}0\ \mu A)(10^{-6}\ A/\mu A)(18.5\ \text{div})}{50.0\ \text{div}} = 9.25 \times 10^{-4}\ A$

EXAMPLE If the meter resistance in the previous example is 10.0 Ω, what is the voltage across the fully deflected meter?

Given	Unknown	Basic equations
$k = 50.0\ \mu A/\text{div}$ $d = 50.0$ div $R_M = 10.0\ \Omega$	V_M	$V_M = I_M R_M$ $I_M = kd$

Solution

Working equation: $V_M = kdR_M$

$$= (50.0\ \mu A/\text{div})(10^{-6}\ A/\mu A)(50.0\ \text{div})(10.0\ \Omega)$$

$$= 2.50 \times 10^{-2}\ V$$

PRACTICE PROBLEMS **1.** What magnitude of current will produce a full-scale deflection of a galvanometer with exactly 40 scale divisions and a sensitivity of 30.0 μA per scale division? *Ans.* 1.20×10^{-3} A

2. What is the voltage across the fully deflected galvanometer in Problem 1? (The meter resistance is 8.50 Ω.) *Ans.* 1.02×10^{-2} V

19.13 The d-c Voltmeter The potential difference across a galvanometer is quite small even when the needle is fully deflected. If a galvanometer is to be used to measure voltages of ordinary magnitudes, we must convert it to a high-resistance instrument. The essential parts of a d-c voltmeter are shown in Figure 19-31.

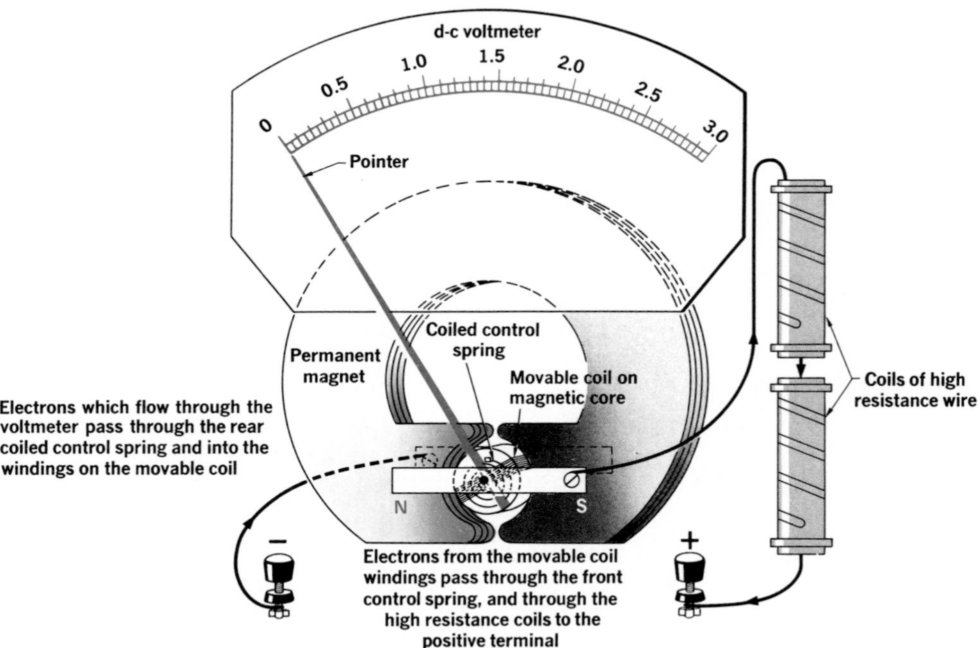

Figure 19-31. The construction of a d-c voltmeter showing the high resistance in series with the windings of the movable coil.

If a high resistance is added in series with the moving coil, most of the potential drop appears across this series resistor, or *multiplier*. Since *a voltmeter is connected in parallel* with the part of a circuit across which the potential difference is to be measured, a high resistance prevents an appreciable loading effect. By the proper choice of resistance, the meter can be calibrated to read any desired voltage.

Suppose we convert the galvanometer used in the examples in Section 19.12 to a voltmeter reading 15.0 volts on full-scale deflection. The current required for full deflection has been found to be 2.50×10^{-3} ampere, and the resistance of the meter coil is 10.0 ohms. We must determine the value of the resistor R_S to be placed in series with the moving coil. Figure 19-32 illustrates this problem.

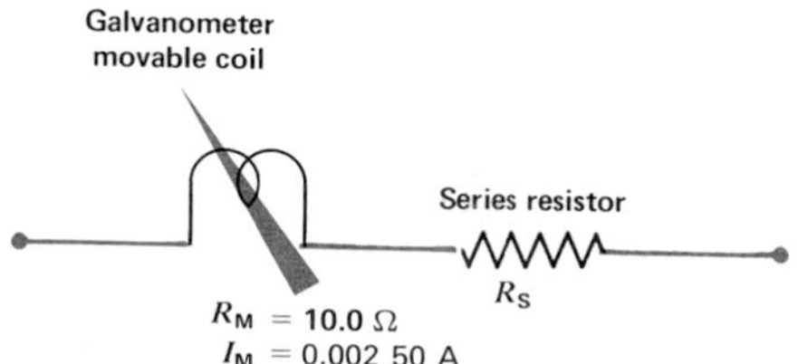

Figure 19-32. Converting a galvanometer to a voltmeter.

Since R_M and R_S (of Figure 19-32) are in series,

$$V = I_M R_M + I_M R_S$$

Then $$R_S = \frac{V}{I_M} - R_M$$

$$R_S = \frac{15.0\ \text{V}}{2.50 \times 10^{-3}\ \text{A}} - 10.0\ \Omega$$

$$R_S = 5990\ \Omega\text{, value of the series resistor}$$

Observe that the total resistance between the terminals of the meter is $60\bar{0}0$ ohms.

The voltmeter sensitivity is frequently expressed in terms of *ohms per volt*. When the ohms-per-volt sensitivity of a voltmeter is known, we can quickly estimate the loading effect it will have when placed across a known resistance component of a circuit. For example, at 400 ohms per volt, our meter, which reads from 0 to 15 volts, has $60\bar{0}0$ ohms between the terminals. If it were placed across resistances greater than about 600 ohms, the loading effect would result in serious meter errors.

19.14 The d-c Ammeter We could use the basic galvanometer in Section 19.12 as a microammeter by calibrating the graduated scale to read directly in microamperes. However, the meter would not be useful in circuits in which the current exceeded 2500 microamperes. Current in the resistance of a galvanometer coil produces I^2R heating, and an excessive current would burn out the meter.

To convert the galvanometer to read larger currents, an alternate (parallel) low-resistance path for current, called a *shunt*, must be provided across the terminals. By the proper choice of shunt resistance, we can calibrate the meter to read over the required range of current magnitudes. See Figure 19-33.

Suppose we wish to convert the same galvanometer to an ammeter reading 10.0 amperes full scale. As before, the current required for the full deflection of the moving coil is 2.50×10^{-3} ampere and the resistance is 10.0 Ω. We

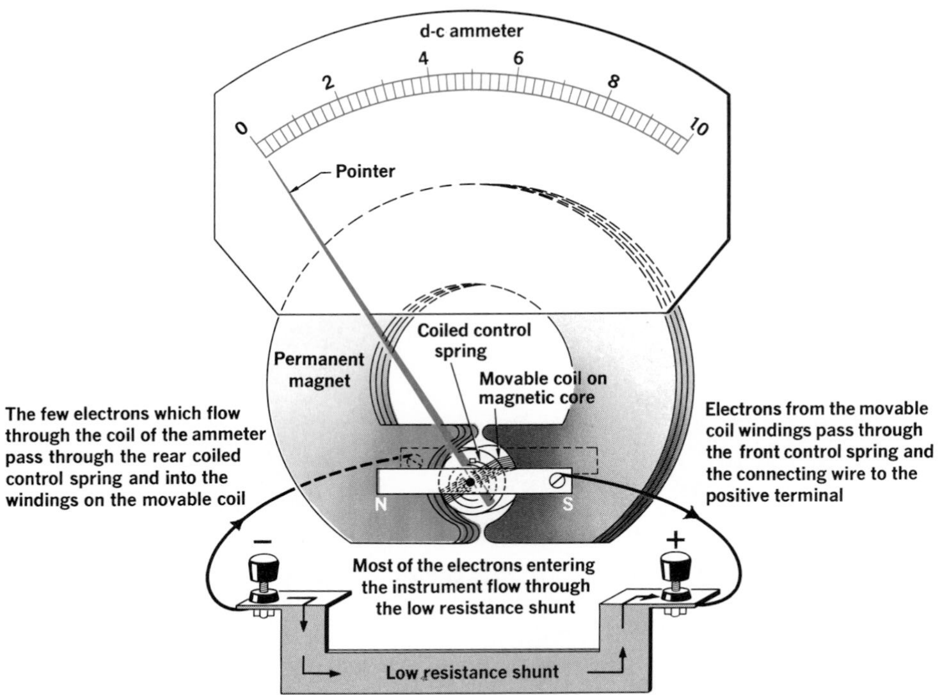

Figure 19-33. The construction of a d-c ammeter showing the low resistance in parallel with the windings of the movable coil.

must determine the resistance of the shunt to be used across the coil. Figure 19-34 applies.

Since R_M and R_S are in parallel,

$$I_M R_M = I_S R_S$$

But
$$I_S = I_T - I_M$$

Then
$$R_S = \frac{I_M R_M}{I_T - I_M}$$

$$R_S = \frac{2.50 \times 10^{-3} \text{ A} \times 10.0 \ \Omega}{10.0 \text{ A} - 0.002\ 50 \text{ A}}$$

$$R_S = 0.002\ 50\ \Omega, \text{ the value for the shunt resistor}$$

The total resistance of the ammeter is the equivalent value for 10.0 ohms and 0.002 50 ohm in parallel; it will be less than 0.002 50 ohm. This exercise demonstrates clearly why *an ammeter must be connected in series* in a circuit, and why it does not materially alter the magnitude of current in the circuit.

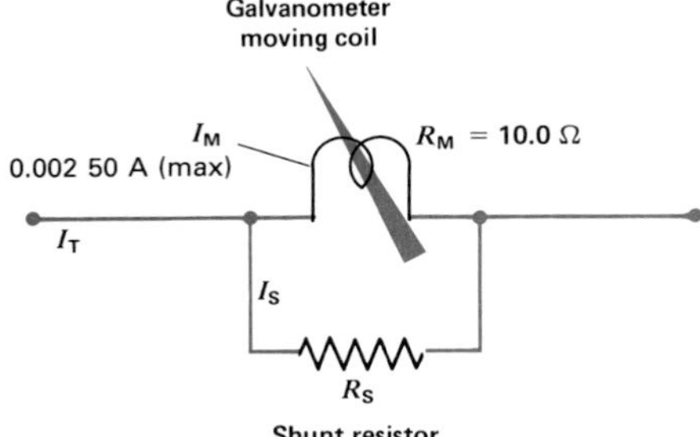

Figure 19-34. Converting a galvanometer to an ammeter.

19.15 The Ohmmeter An ohmmeter provides a convenient means of measuring the resistance of a circuit component. A basic ohmmeter circuit is shown in Figure 19-35. Its accuracy limitation is approximately the same as that of the voltmeter-ammeter method of measuring resistance; actually, it is a modified version of this method. The ohmmeter must be used only on a completely de-energized circuit.

The ohmmeter circuit of Figure 19-35 shows a milliammeter requiring 1 mA for full-scale deflection. With an emf of 4.5 volts, by Ohm's law, 4500 ohms of resistance will provide 1 mA of current when terminals **A-B** are short-circuited. A fixed resistor, R_2, of 4000 ohms and a rheostat, R_1, of 0-1000 ohms are provided.

To use this ohmmeter, **A** and **B** are short-circuited and R_1 is adjusted to give full deflection. If the emf is 4.5 V, R_1 will be set at 500 ohms. The pointer position at full-scale deflection is now marked as zero ohm (0 Ω). The rest position of the pointer is the open-circuit position with infinite resistance between **A** and **B**. This position is marked as ∞ Ω. Other resistance calibrations may be made from Ohm's law applications. For example, 4500 ohms between **A** and **B** will mean a total of 9000 ohms in the circuit and 0.5 mA of current. This mid-scale position of the pointer can be marked 4500 Ω. The meter, when recalibrated, will read the resistance between terminals **A-B**.

Each time the ohmmeter is used, it is first shorted across **A-B** and R_1 is adjusted to "zero" the meter. This operation calibrates the meter and accommodates any decrease in the terminal voltage of the battery with age. The resistance R_1 allows the ohmmeter to be used until $\mathscr{E}$ drops below 4.0 volts.

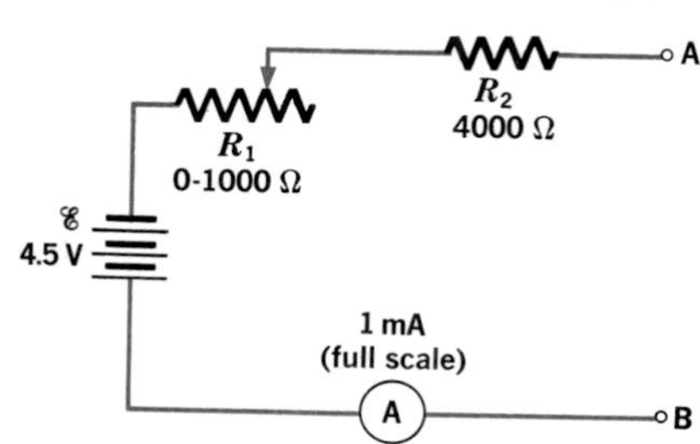

Figure 19-35. A basic ohmmeter circuit.

QUESTIONS: GROUP A

1. What did Oersted discover?
2. A conductor carrying a current is arranged so that electrons flow in one segment from north to south. If a compass is held over this segment of the wire, in what direction is the needle deflected?
3. Describe a simple experiment to show the nature of the magnetic field about a straight conductor carrying a current.
4. Suppose an electron flow in a conductor passing perpendicularly through this page is represented by a dot inside a small circle when the direction of flow is up out of the page. What is the direction of the magnetic flux about this current?
5. Upon what factors does the strength of an electromagnet depend?
6. What prevents the movable coil of a galvanometer from aligning its magnetic field parallel to that of the permanent magnet each time a current is in the coil?
7. (a) Why is it necessary that an ammeter be a low-resistance instrument? (b) Why must a voltmeter be a high-resistance instrument?

GROUP B

8. A solenoid with ends marked A and B is suspended by a thread so that the core can rotate in the horizontal plane. A current is maintained in the coil such that the electron flow is clockwise when viewed from end A

toward end B. How will the coil align itself in the earth's magnetic field?

9. A stream of electrons is projected horizontally to the right. A straight conductor carrying a current is supported parallel to the electron stream and above it. (a) What is the effect on the electron stream if the direction of the current in the conductor is from left to right, (b) if the current is reversed?
10. If the conductor in Question 9 is replaced by a magnet with a downward magnetic field, what is the effect on the electron stream?
11. Why might the potential difference indicated by a voltmeter placed across a circuit load be different from the potential difference with the meter removed?
12. Suppose the resistance of a high-resistance load is to be determined using the voltmeter-ammeter method. Considering the design characteristics of ammeters and voltmeters, how would you arrange the meters in the circuit to reduce the error to a minimum? Draw your circuit diagram and justify your arrangement.
13. Assume that the resistance of a low-resistance load is to be determined using the voltmeter-ammeter method. How would you arrange the meters in this circuit to reduce the error to a minimum? Draw your circuit diagram and justify your arrangement.

PROBLEMS: GROUP A

1. Two parallel conductors 2.0 m long and 1.0 m apart and carrying equal currents experience a total force of 1.6×10^{-6} N. What magnitude of current is in each conductor?
2. An ammeter that has a resistance of 0.01 ohm is connected in a circuit and indicates a current of 10 amperes. A shunt having a resistance of 0.001 ohm is then connected across the meter terminals. What is the new reading on the meter? Assume the introduction of the shunt does not affect the total circuit current.
3. A galvanometer has a zero-center scale with 20.0 divisions on each side of zero. The pointer deflects 15.0 scale divisions when a current of 375 μA is in the movable coil. (a) What is the current sensitivity of the meter? (b) What current will produce a full-scale deflection?

GROUP B

4. A galvanometer has a resistance of 50.0 Ω and requires 75.0 mA to produce a full-scale deflection. What resistance must be connected in series with the galvanometer in order to use it as a voltmeter for measuring a maximum of 300.0 V?
5. A galvanometer movement has a resistance of 2.5 ohms and when fully deflected has a potential difference of $\underline{5}0$ millivolts across it. What shunting resistance is required to enable the instrument to be used as an ammeter reading 7.5 amperes full scale?
6. A repulsive force of 9.6×10^{-4} N is experienced by each of two parallel conductors 5.0 m long when a current of 3.2 A is in each conductor. By what distance are they separated?

PHYSICS ACTIVITY

Try to measure the magnetic declination (if any) where you live. On a clear night, go outside with a compass and compare the direction of the compass needle with the direction of the North Star. If possible, conduct the experiment in an open space to avoid interference from local magnetic fields.

SUMMARY

Magnetite, a magnetic iron ore, is a natural magnet. Metals of the Iron Family and special metallic alloys and oxides are strongly attracted by magnets; they have ferromagnetic properties. Very strong magnets are made from ferromagnetic substances. Materials that are not ferromagnetic are commonly said to be nonmagnetic. Ferromagnetic materials have high permeabilities. Nonmagnetic materials in general are transparent to magnetic flux. These materials may be very feebly diamagnetic or paramagnetic. Magnetism is explained by the domain theory.

Coulomb's law for magnetism is a quantitative expression for the force acting between two magnetic poles. A magnetic field, and its influence on a fictitious **N** pole, shows similarities to an electric field and its influence on a positive test charge.

A charge in motion is surrounded by a magnetic field. The core of a coil carrying an electric current becomes a magnet. Strong electromagnets are produced by winding a conducting coil around a ferromagnetic core. The strength of an electromagnet depends on the number of turns of coil and the magnitude of the current in the coil.

The galvanometer is the basic meter for d-c measurements. The galvanometer can be calibrated as a voltmeter by placing a high resistance in series with the galvanometer coil. It can be calibrated as an ammeter by placing a very low resistance shunt across the galvanometer coil. An ohmmeter requires a source of emf, an adjustable resistance, and a sensitive ammeter. It is essentially a voltmeter-ammeter method of measuring the resistance of a circuit component.

VOCABULARY

ammeter
Ampère's rule
Coulomb's law of magnetism
Curie point
declination
diamagnetism
dipping needle
domain
electron pair
ferromagnetism
flux density
galvanometer
induced magnetism
line of flux
magnetic induction
magnetosphere
ohmmeter
paramagnetism
permeability
solar wind
solenoid
voltmeter

20

Electromagnetic Induction

induction (in-DUK-shun) n.: the process by which an electric or magnetic field is created in another object, usually a conductor.

OBJECTIVES

- Study the distinction between an applied emf and an induced emf.
- Discuss Faraday's laws.
- Describe the factors that affect an induced emf.
- Describe the generator principle.
- Study instantaneous current and voltage.
- Describe the motor principle.
- Study mutual inductance and self-inductance.
- Describe the transformer.

INDUCED CURRENTS

20.1 Discovery of Induced Current In Section 19.8 we discussed Oersted's discovery of the link between magnetism and electricity. Soon after Oersted's work, other scientists attempted to find out whether an electric current could be produced by the action of a magnetic field. In 1831 Michael Faraday discovered that *an emf is set up in a closed electric circuit located in a magnetic field whenever the total magnetic flux linking the circuit is changing.* The American physicist Joseph Henry (1797–1878) made a similar discovery at about the same time. This phenomenon is called **electromagnetic induction.** The emf is called an **induced emf,** and the resulting current in the closed conducting loop is called an **induced current.**

The production and distribution of ample electric energy are essential functions of a modern technological economy. The invention of electric generators and transformers has made it possible to provide these essential services. The discoveries of Faraday and Henry represent the initial step in the development of the broad knowledge base in electromagnetic induction that makes such inventions possible.

20.2 Faraday's Induction Experiments We shall examine some of Faraday's experiments in order to understand their significance. Suppose a sensitive galvanometer is connected in a closed conducting loop as shown in Figure 20-1. A segment of the conductor is poised in the field flux of a strong magnet. As the conductor in Figure 20-1(A) is moved down between the poles of the magnet, there is a momentary deflection of the galvanometer needle, indicating an induced current. The needle shows no deflection when the conductor is stationary in the magnetic flux. This observation suggests that *the induced current is related to the motion of the conductor in the magnetic flux.*

As the conductor is raised between the poles of the magnet, as in Figure 20-1(B), there is another momentary deflection of the galvanometer, but in the opposite direction. This suggests that *the direction of the induced current in the conductor is related to the direction of motion of the conductor in the magnetic field.* The emf induced in the conductor is of opposite polarity to that in the first experiment.

Magnetic lines of flux always link the current loop that sets up the magnetic field. Review Section 19.9.

Faraday found that he could induce an emf in a conductor either by moving the conductor through a stationary field or by moving the magnetic field near a stationary conductor. He observed that the direction of the induced current in the conducting loop is reversed with a change in either the direction of motion or the direction of the magnetic field.

Supporting the conducting loop of Figure 20-1 in a fixed position and lifting the magnet result in a deflection similar to that of Figure 20-1(A). When the magnet is lowered, the galvanometer needle is momentarily deflected as in

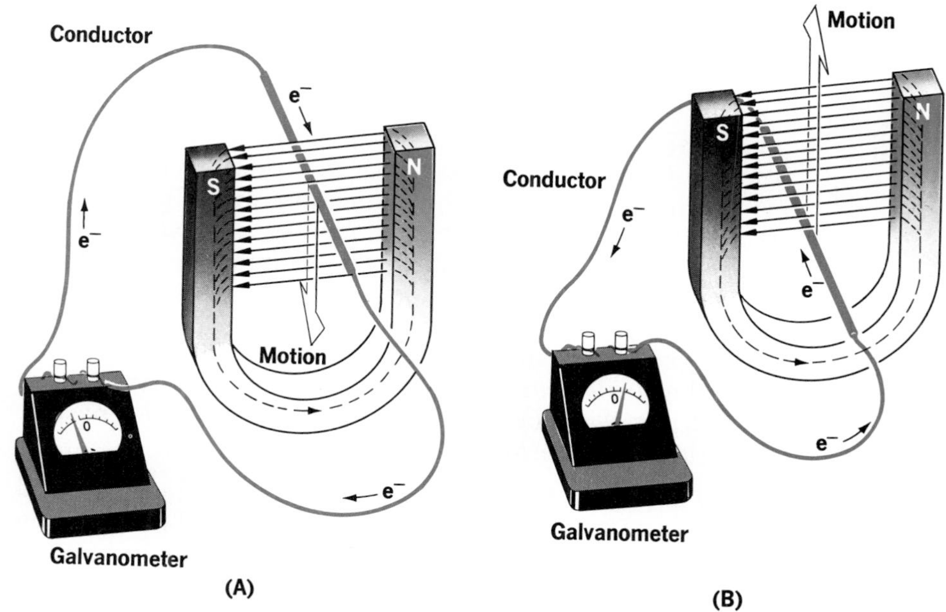

Figure 20-1. A current is induced in a closed conducting loop when the magnetic flux linked through the circuit is changing.

Figure 20-1(B). The relative motion between the conductor and the magnetic flux is the same whether the conductor is raised through the stationary field or the field is lowered past the stationary conductor.

The conducting segment of Figure 20-2 is part of a conducting loop that includes a galvanometer as shown in Figure 20-1.

Observe that the relative motion of the conductor in each of these experiments is *perpendicular* to the magnetic flux. If the conductor in the magnetic field is now moved *parallel* to the lines of flux, the galvanometer shows no deflection. A conductor that moves perpendicular to the magnetic flux can be construed to "cut through" lines of flux and experience *a change in flux linkage*. Conversely, a conductor that moves parallel to the magnetic field does not "cut through" lines of flux and does not experience a change in flux linkage. See Figure 20-2. These observations suggest that *electromagnetic induction results from those relative motions between conductors and magnetic fields which are accompanied by changes in magnetic flux linkage.*

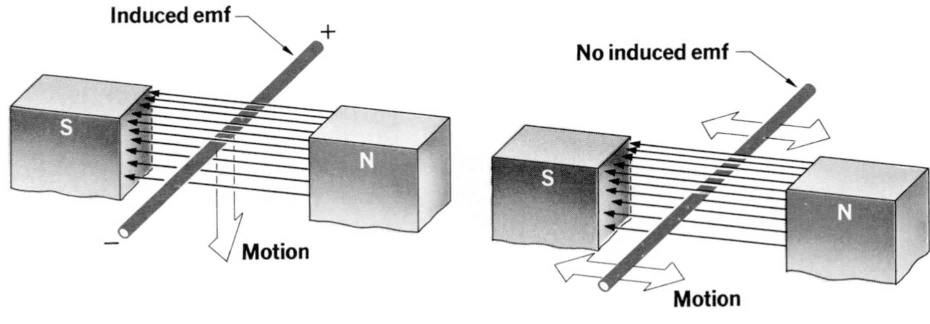

Figure 20-2. An emf is induced in a conductor when there is a change of flux linked by the conductor.

Suppose the conductor is looped so that several turns are poised in the magnetic field, as in Figure 20-3. When the coil is moved down between the poles of the magnet as before, there is a greater deflection on the galvanometer. By increasing the rate of motion of the coil across the magnetic flux or by substituting a stronger magnetic field, greater deflections are produced. In each of these cases the effect is to increase the number of flux lines "cut" by turns of the conductor in a given length of time. We can then state that *the magnitude of the induced emf, or of the induced current in a closed loop, is related to the rate at which the flux linked by the conductor changes.*

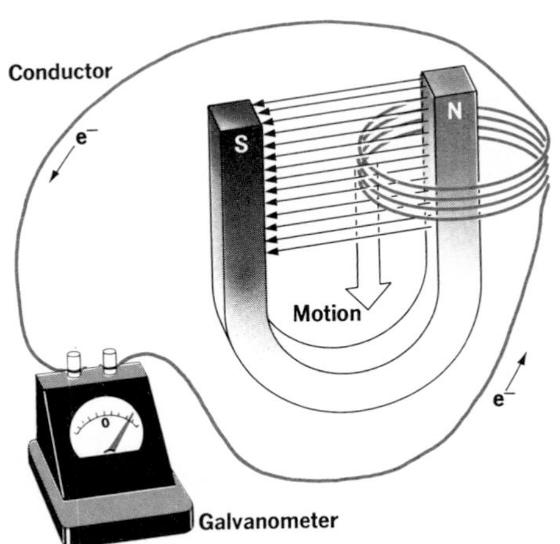

Figure 20-3. A greater change in flux linkage occurs when several turns of a conductor cut through the magnetic flux.

20.3 Factors Affecting Induced emf Faraday found that an emf is induced in a conductor whenever any change occurs in the magnetic flux (lines of induction) linking the conductor. Very precise experiments have shown that *the emf induced in each turn of a coiled conductor is proportional to the time rate of change of magnetic flux linking each turn of the coil.*

The total magnetic flux linking the coil is designated by the Greek letter Φ (phi). If $\Delta\Phi$ represents the *change* in magnetic flux linking the coil during the time interval Δt, the emf $\mathscr{E}$ induced in a single turn of the coil can be expressed as

$$\mathscr{E} \propto -\frac{\Delta\Phi}{\Delta t}$$

By introducing a proportionality constant k, the value of which depends on the system of units used, the expression can be written as

$$\mathscr{E} = -\mathbf{k}\frac{\Delta\Phi}{\Delta t}$$

Where $\mathscr{E}$ is measured in volts and Φ in webers, the numerical value of k is unity and the equation becomes

$$\mathscr{E} = -\frac{\Delta\Phi}{\Delta t}$$

Thus, a change in the magnetic flux linking a coil occurring at the rate of 1 weber per second induces an emf of 1 volt in a single turn of the coil.

A coil consists of turns of wire that are, in effect, connected in series. Therefore, the emf induced across the coil is simply the sum of the emfs induced in the individual turns. A coil of **N** turns has **N** times the emf of the separate turns. This relationship can be represented by the following equation

$$\mathscr{E} = -\mathrm{N}\frac{\Delta\Phi}{\Delta t}$$

The negative sign merely indicates the relative polarity of the induced voltage. It expresses the fact that *the induced emf is of such polarity as to oppose the change that induced it,* a basic energy conservation principle discussed in detail in Section 20.6.

To illustrate this concept, suppose a coil of 150 turns linking the flux of a magnetic field uniformly is moved perpendicular to the flux and a change in flux linkage of

3.0×10^{-5} weber occurs in 0.010 second. The induced emf is then

$$\mathscr{E} = -N\frac{\Delta\Phi}{\Delta t}$$

$$\mathscr{E} = -150 \times \frac{3.0 \times 10^{-5}\ \text{Wb}}{1.0 \times 10^{-2}\ \text{s}}$$

$$\mathscr{E} = -0.45\ \text{V}$$

$$\left(\frac{\text{Wb}}{\text{s}} = \frac{\text{N}\cdot\text{m/A}}{\text{s}} = \frac{\text{J}}{\text{C}} = \text{V}\right)$$

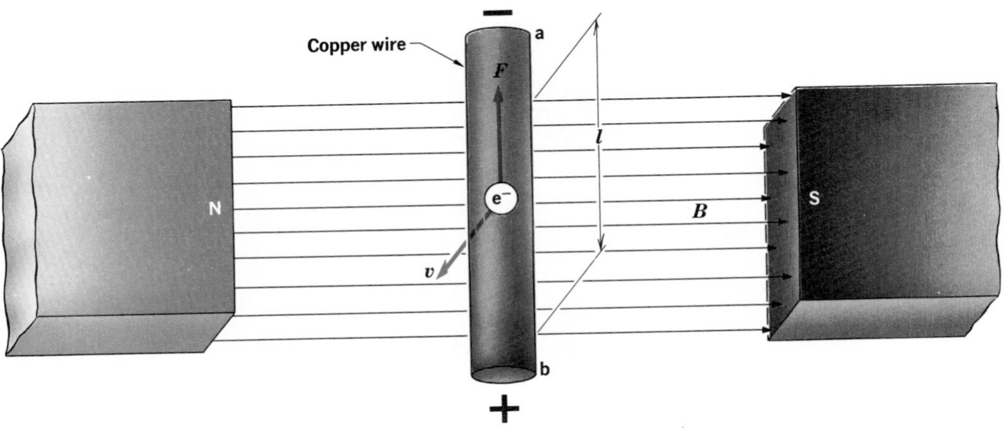

Figure 20-4. A force acts on a moving charge in a magnetic field.

20.4 The Cause of an Induced emf A length of conductor moving in a magnetic field has an emf induced across it that is proportional to the rate of change of flux linkage. However, an induced current persists *only* if the conductor is a part of a closed circuit. In order to understand the cause of an induced emf, we shall make use of several facts that have already been established.

A length of copper wire poised in a magnetic field, as shown in Figure 20-4, contains many free electrons, and these moving charges constitute an electric current. In Section 19.12 we recognized that a force acts on the movable coil of a galvanometer in a magnetic field when a current is in the coil. This force, in effect, acts on the moving charges themselves. We shall call this force a **magnetic force** to distinguish it from the force exerted on free electrons by the electrostatic field of a stationary charge.

Suppose the copper wire of Figure 20-4, at right angles to the uniform magnetic field, is pushed downward (into the page) with a velocity v through the magnetic field of flux density B. As a consequence of the motion of the wire, the free electrons of the copper conductor may be considered to move perpendicular to the flux with the speed v. A magnetic force, F, acts on them in a direction perpendicular to both B and v. In response to the magnetic force, these electrons move toward end **a** and away from end **b.** Because the two ends of the conductor are not connected in a circuit that would provide a closed path for the induced electron flow, end **a** acquires a growing negative charge while a residual positive charge builds up on end **b.** Thus a difference of potential is established across the conductor with **a** the negative end and **b** the positive end. If either the motion of the wire or the magnetic field is reversed, the direction of F in Figure 20-4 is reversed and an emf is induced so that end **a** is positive and end **b** is negative.

The accumulations of charges at the ends of the conductor establish an electric field that increasingly opposes the movement of electrons through the conductor. The force of this electric field acting on the electrons from **a** toward **b** soon balances the magnetic force arising from the motion of the conductor, and the flow of electrons ceases.

A changing magnetic field induces an electric field (Faraday's law).

The movement of electrons through the open conductor of Figure 20-4 comprises a "transient" induced current of extremely short duration.

The equilibrium potential difference across the open conductor is numerically equal to the induced emf and depends on the length, l, of wire linking the magnetic flux; the flux density, B, of the field; and the speed, v, with which the conductor is moved through the field.

$$\mathscr{E} = Blv$$

When B is in newtons/ampere meter (or webers/meter^2), l in meters, and v in meters/second, $\mathscr{E}$ is given in volts.

Flux density, B, is also called magnetic induction.

Assume that the length of the conductor linking the magnetic field in Figure 20-4 is 0.075 m and the flux density is 0.040 N/A · m. If the conductor is moved down through the flux with a velocity of 1.5 m/s, the induced emf is

$$\mathscr{E} = Blv$$
$$\mathscr{E} = 0.040\ \text{N/A}\cdot\text{m} \times 0.075\ \text{m} \times 1.5\ \text{m/s}$$
$$\mathscr{E} = 0.004\,5\ \text{V or } 4.5\ \text{mV}$$

In Figure 20-4 a magnetic force equal but opposite to F will act on the positively charged protons in the copper nuclei. Since these are in the bound parts of the copper atoms, they will not move in response to this force. But, positively charged ions in a liquid or gas would move.

The vectors B, v, and F are all mutually perpendicular as shown in Figure 20-5. If a charge Q moves with a velocity v

To compare the response of a moving electric charge to an electric force and to a magnetic force, see Figures 16-11 and 20-5.

Verify that if B is expressed in N/A · m, Φ *can be expressed in* N · m/A.

through a magnetic field of flux density B, the force, F, acting on the charge becomes

$$F = QvB$$

The force F is in newtons, Q is in coulombs, v is in meters/second, and B is in newtons/ampere-meter (or webers/meter^2).

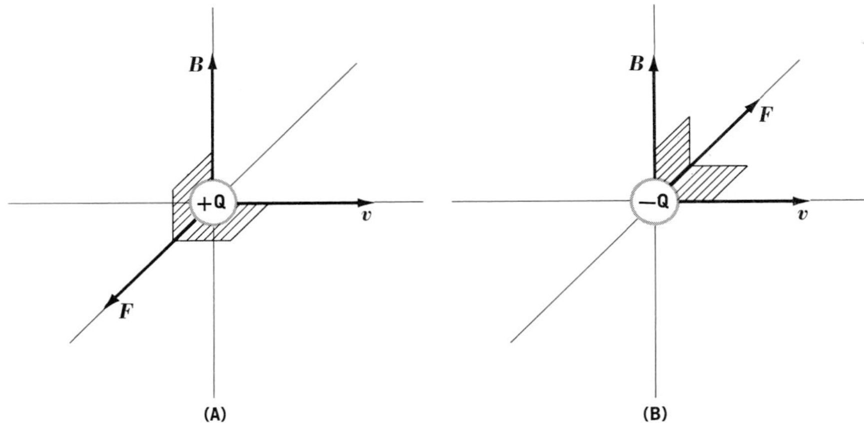

Figure 20-5. The magnetic force acting on a charge moving in a magnetic field. (A) The charge is positive and the force is directed out of the page. (B) The charge is negative and the force is directed into the page.

20.5 The Direction of Induced Current Now consider that a length of conductor is used outside the magnetic flux to connect ends **a** and **b** of the copper wire in Figure 20-4, thus providing a closed-loop path for electron flow. We shall call the length of wire linked with magnetic flux the *internal circuit* and the rest of the conducting path the *external circuit.*

When the straight conductor is pushed downward, as in Figure 20-4, electrons move from the negative end **a** through the external circuit to the positive end **b** and through the internal circuit from end **b** to end **a.** Over the external path from **a** to **b,** electrons transform the potential energy acquired in the internal circuit into kinetic energy that they expend in the external circuit. In the internal path, the electric force acting on the electrons from **a** to **b** is reduced below its equilibrium value because of the partial depletion of accumulated charges at **a** and **b.** A net magnetic force pumps electrons from **b** to **a,** maintaining a potential difference across the internal circuit. The value of this potential difference is less than the open-circuit potential difference which, you will recall, was numerically equal to the induced emf.

20.6 Lenz's Law The downward motion of the copper wire in Figure 20-4 is maintained by exerting a force on it in the direction of v. This force may be thought of as the action that generates the induced electron current in the closed circuit.

The relationship between the direction of an induced current and the action inducing it was recognized in 1834 by the German physicist Heinrich Lenz (1804–1865). His discovery, now referred to as *Lenz's law,* is true of all induced currents. Because Lenz's law refers to induced currents, it applies only to closed circuits. ***Lenz's law*** states that *an induced current is in such a direction that it opposes the change that induced it.* Thus, the magnetic effect of the induced current in Figure 20-4 must be such that it opposes the downward force (the pushing action) applied to the conductor segment in the external field of the magnet.

Faraday discovered how to determine the direction of an induced current also, but he did not express it as clearly as Lenz.

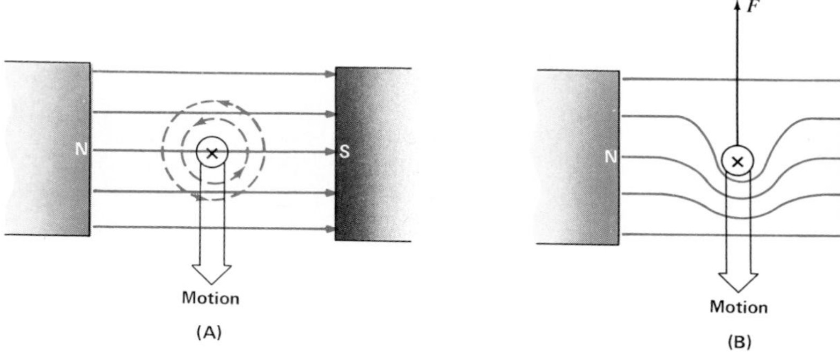

Figure 20-6. Two interacting magnetic fields. The separate magnetic lines of the fields of the magnet and of the induced electron flow (into the page) are shown in (A). The lines of the resultant field are shown in (B).

We can readily visualize what this effect must be. A cross-sectional view of a conductor poised in a magnetic field is shown in Figure 20-6(A). The conductor corresponds to end **b** of the wire in Figure 20-4. It is part of a closed circuit that cannot be seen in this cross-sectional diagram. As the wire is pushed downward through the magnetic field, the induced electron current is directed into the page as indicated by the × symbol (tail of the arrow). By Ampère's rule for a straight conductor, the magnetic field of this induced current encircles the current in the counterclockwise sense. Similar diagrams in Figure 20-7 represent the effects of the wire being pulled up through the magnetic field. Here the induced electron current is directed out of the page as indicated by the · symbol (head of the arrow). The composite, or resultant, fields are shown in Figures 20-6(B) and 20-7(B). They suggest that the agent moving the conductor always experiences an opposing force.

See Section 19.9 for Ampère's rule.

The fact that an induced current always opposes the motion that induces it illustrates the conservation-of-energy principle. Work must be done to induce a current

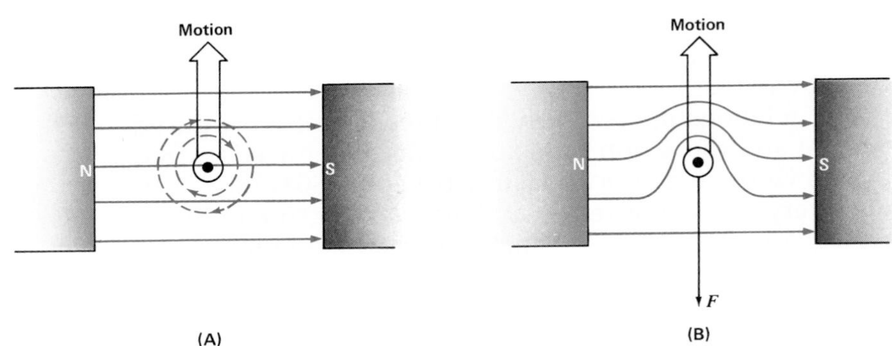

Figure 20-7. Two interacting magnetic fields. The separate magnetic lines of the fields of the magnet and of the induced electron flow (out of the page) are shown in (A). The lines of the resultant field are shown in (B).

in a closed circuit. The energy expended comes from outside the system and is a result of work done by the external force required to keep the conductor segment moving. The induced current can produce heat or do mechanical or chemical work in the external circuit as electrons of high potential energy fall through a difference of potential.

QUESTIONS: GROUP A

1. What is an induced current?
2. What contribution did each of the following scientists make to our knowledge about electromagnetic induction? (a) Joseph Henry (b) Michael Faraday (c) Hans Christian Oersted
3. What two conditions must be satisfied to produce an electric current with a magnetic field?
4. Referring to Ohm's law, upon what factor besides induced emf does induced current depend?
5. What three factors determine the value of the induced emf?
6. State Lenz's law.

GROUP B

7. Thrusting the north pole of a bar magnet into a 600-turn coil of copper wire causes a galvanometer attached to the coil to deflect to the right. What would happen if you: (a) Pull the magnet out of the coil? (b) Let the magnet sit at rest in the coil? (c) Slide the magnet left and right? (d) Turn the magnet around and thrust the south end of the magnet into the coil?
8. A student is turning the handle of a small generator attached to a lamp socket containing a 15-W bulb. The lamp barely glows. (a) What should she do to make it glow more brightly? (b) Why does this work?
9. What physical quantities are measured in (a) webers, (b) webers per meter^2, (c) webers per second, (d) joules per coulomb?
10. Explain how Lenz's law illustrates conservation of energy.
11. Demonstrate that the product Blv is properly expressed in volts.
12. Demonstrate that the product QvB is properly expressed in newtons.
13. A permanent magnet is moved away from a stationary coil as shown in the

accompanying diagram. (a) What is the direction of the induced current in the coil? (Indicate direction of electron flow in the straight section of conductor below the coil.) (b) What magnetic polarity is produced across the coil by the induced current? (c) Justify your answers to (a) and (b).

Stationary coil

Motion

N S

Permanent magnet

a b

PROBLEMS: GROUP A

1. A coil of 325 turns moving perpendicular to the flux in a uniform magnetic field experiences a change in flux linkage of 1.15×10^{-5} weber in 0.001 00 s. What is the induced emf?
2. How many turns are required to produce an induced emf of 0.25 volt for a coil that experiences a change in flux linkage at the rate of 5.0×10^{-3} Wb/s?
3. A straight conductor $1\bar{0}$ cm long is moved through a magnetic field perpendicular to the flux at a velocity of 75 cm/s. If the flux density is 0.025 weber/m^2, what emf is induced in the conductor?
4. A coil of 75 turns and an area of 4.0 cm^2 is removed from the gap between the poles of a magnet having a uniform flux density of 1.5 Wb/m^2 in 0.025 s. What voltage is induced across the coil?
5. A rod 15 cm long is perpendicular to a magnetic field of 4.5×10^{-1} N/A · m and is moved at right angles to the flux at the rate of $3\bar{0}$ cm/s. Find the emf induced in the rod.

GENERATORS AND MOTORS

20.7 The Generator Principle An emf is induced in a conductor whenever the conductor experiences a change in flux linkage. When the conductor is part of a closed circuit, an induced current can be detected in the circuit. By Lenz's law, work must be done to induce a current in a conducting circuit. Accordingly, this is a practical source of electric energy.

Moving a conductor up and down in a magnetic field is not a convenient method of inducing a current. A more practical way is to shape the conductor into a loop, the ends of which are connected to the external circuit by means of **slip rings,** and rotate it in the magnetic field. See Figure 20-8.

Such an arrangement is a basic *generator.* The loop across which an emf is induced is called the *armature.* The ends of the loop are connected to slip rings that rotate as the armature is turned. A graphite *brush* rides on each slip ring, connecting the armature to the external circuit. *An* ***electric generator*** *converts mechanical energy into electric energy.* The essential components of a generator are a *field magnet, an armature,* and *slip rings and brushes.*

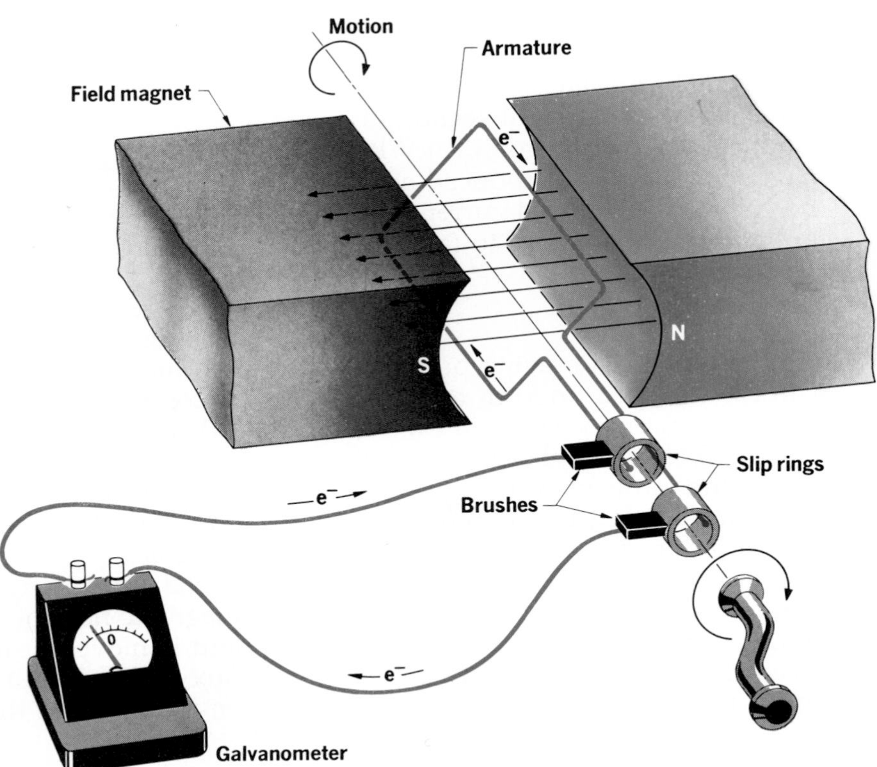

Figure 20-8. The essential components of an electric generator.

The induced emf across the armature and the induced current in the closed circuit result from relative motion between the armature and the magnetic flux that effects a change in the flux linkage. Thus either the armature or the magnetic field may be rotated. In some commercial generators the field magnet is rotated and the armature is the stationary element.

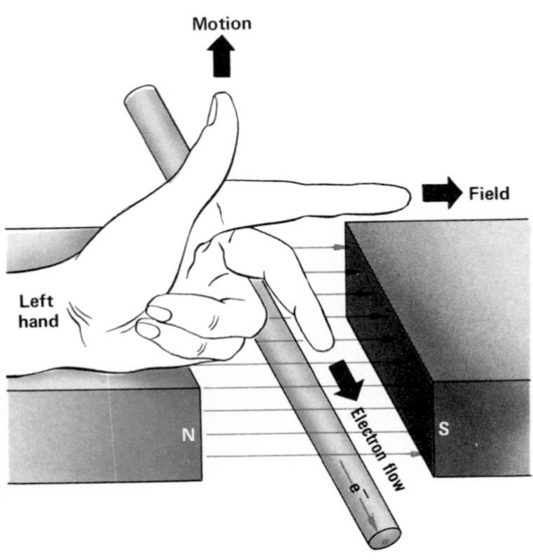

Figure 20-9. The left-hand generator rule.

According to Lenz's law, an induced current will appear in such a direction that the magnetic force on the electrons comprising the induced current opposes the motion producing it. The direction of induced current in the armature loop of a generator can be easily determined by use of a *left-hand rule* known as the **generator rule.** This rule takes into account Lenz's law and the fact that an electric current in a metal conductor consists of a flow of electrons. See Figure 20-9.

The generator rule: Extend the thumb, forefinger, and the middle finger of the left hand at right angles to each other. Let the forefinger point in the direction of the magnetic flux and the thumb in the direction the conductor is moving; the middle finger points in the direction of the induced electron flow.

20.8 The Basic a-c Generator The two sides of the conducting loop in Figure 20-8 move through the magnetic flux in opposite directions when the armature is rotated. By applying the generator rule to each side of the loop, the direction of the electron flow is shown to be toward one slip ring and away from the other. Thus a single-direction loop is established in the closed circuit. As the direction of each side of the loop changes with respect to the flux, the direction of the electron flow is reversed. As the armature rotates through a complete cycle, there are two such reversals in direction of the electron flow.

In Figure 20-10 one side of a conducting loop rotating in a magnetic field is shown cross-sectionally in color while the other side is shown in black. In (A) the colored side of the loop is shown moving down and cutting through the flux. By the generating rule, the electron flow is directed into the page. (The white × represents the tail of the arrow.) The motion of the black side of the loop induces a flow directed out of the page. (The white dot represents the head of the arrow.)

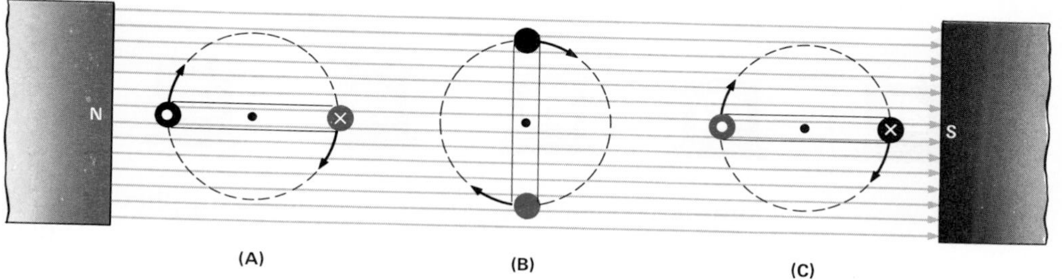

Figure 20-10. An emf is induced only when there is a change in the flux linking a conductor.

An emf is induced in a conductor as a result of a change in the flux linking the conductor. In (B) both sides of the loop are moving parallel to the flux and there is no change in linkage. Therefore, no emf is induced across the loop and there is no electron flow in the closed circuit.

In (C) the black side of the loop is moving down and cutting through the flux so that the electron flow is directed into the page. The colored side is moving up through the flux, and thus the electron flow is directed out of the page. The direction of the flow in the circuit is the reverse of that in (A). A quarter cycle later, the emf again drops to zero and there is no flow in the circuit. The emf induced across the conducting loop reaches a maximum value when the sides of the loop are moving perpendicular to the magnetic flux. See Section 20.2.

We can see from Figure 20-10 that the magnitude of the induced emf across the conducting loop must vary from zero through a maximum and back to zero during a half cycle of rotation. The emf must then vary in magnitude in a similar manner, from zero through the same magnitude maximum and back to zero, during the second half cycle. However, polarity across the loop is opposite. The emf across the loop thus *alternates* in polarity.

Similarly, the flow in a circuit connected to the rotating armature by way of the slip rings alternates in direction: electrons flowing in one direction during one half cycle and in the opposite direction during the other half cycle. *A current that has one direction during part of a generating cycle and the opposite direction during the remainder of the cycle is called an* ***alternating current,*** *ac.* A generator that produces an alternating current must, of course, produce an alternating emf. See Figure 20-11.

Figure 20-11. One cycle of operation of an a-c generator.

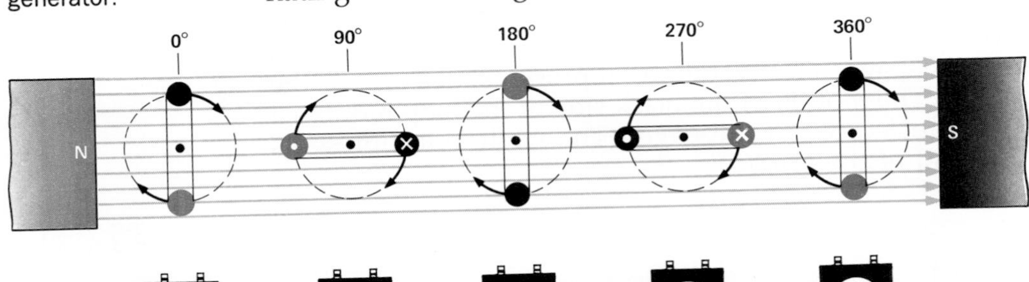

Commercial electric power is usually supplied by the generation of alternating currents and voltages. Such power is referred to as *a-c power*. The expressions *ac* and *dc* are often used to distinguish between alternating-current and direct-current properties: a-c voltage and d-c current are two examples.

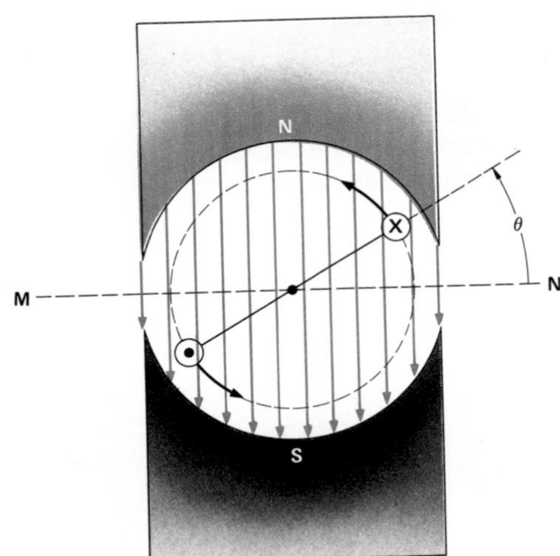

Figure 20-12. The instantaneous value of an induced voltage varies with the sine of the displacement angle of the loop in the magnetic field.

20.9 Instantaneous Current and Voltage The open-circuit voltage across a battery has a constant magnitude characteristic of the chemical makeup of the battery. The voltage across an armature rotating in a magnetic field, however, has no constant magnitude. It varies from zero through a maximum in one direction and back to zero during one half cycle. It then rises to a maximum in the opposite direction and falls back to zero during the other half cycle of armature rotation. At successive instants of time, different magnitudes of induced voltage exist across the rotating armature. *The magnitude of a varying voltage at any instant of time is called the* ***instantaneous voltage,*** *e.*

The maximum voltage, $\mathscr{E}_{max}$, is obtained when the conductor is moving perpendicular to the magnetic flux because the rate of change of flux linking the conductor is maximum during this time. *If the armature is rotating at a constant rate in a magnetic field of uniform flux density, the magnitude of the induced voltage varies sinusoidally (as a sine wave) with respect to time.*

In Figure 20-12 a single loop is rotating in a uniform magnetic field. When the plane of the loop is perpendicular to the flux (**MN** of Figure 20-12), the conductors are moving parallel to the flux lines; the displacement angle of the loop is said to be zero. We shall refer to this angle between the plane of the loop and the perpendicular to the magnetic flux as θ (theta). When $\theta = 0°$ and 180°, $e = 0$ V. When $\theta = 90°$, $e = \mathscr{E}_{max}$; and when $\theta = 270°$, $e = -\mathscr{E}_{max}$. These relationships are apparent from Figure 20-11. In general, the instantaneous voltage, e, varies with the sine of the displacement angle of the loop.

$$e = \mathscr{E}_{max} \sin \theta$$

The current in the external circuit of a simple generator consisting of pure resistance will vary in a similar way; the *maximum current*, I_{max}, occurs when the induced voltage is maximum. From Ohm's law

$$I_{max} = \frac{\mathscr{E}_{max}}{R}$$

The *instantaneous current*, i, is accordingly

$$i = \frac{e}{R}$$

but

$$e = \mathscr{E}_{max} \sin \theta$$

so

$$i = \frac{\mathscr{E}_{max}}{R} \sin \theta$$

and

$$i = I_{max} \sin \theta$$

20.10 Practical a-c Generators The simple generator consists of a coil rotating in the magnetic field of a permanent magnet. Any small generator employing a permanent magnet is commonly called a **magneto.** Magnetos are often used in the ignition systems of gasoline engines for lawn mowers, motorbikes, and boats.

The generator output is increased in a practical generator by increasing the number of turns on the armature or increasing the field strength. The field magnets of large generators are strong electromagnets; in a-c generators they are supplied with direct current from an auxiliary d-c generator called an **exciter.** See Figure 20-13.

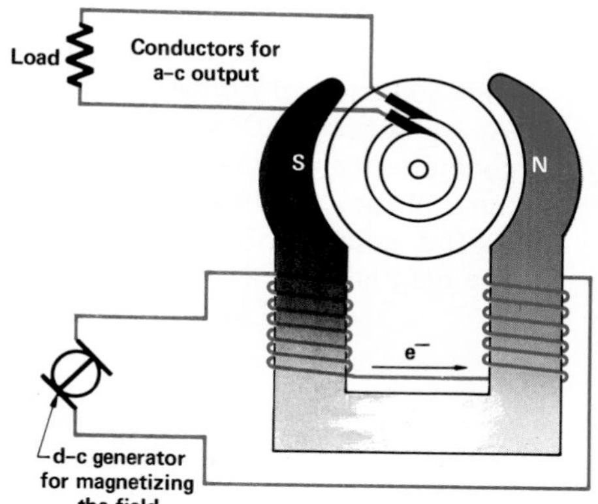

Figure 20-13. The field of a large alternating-current generator is produced by an electromagnet.

The performance of large a-c generators is generally more satisfactory if the armature is stationary and the field rotates inside the armature. Such stationary armatures are

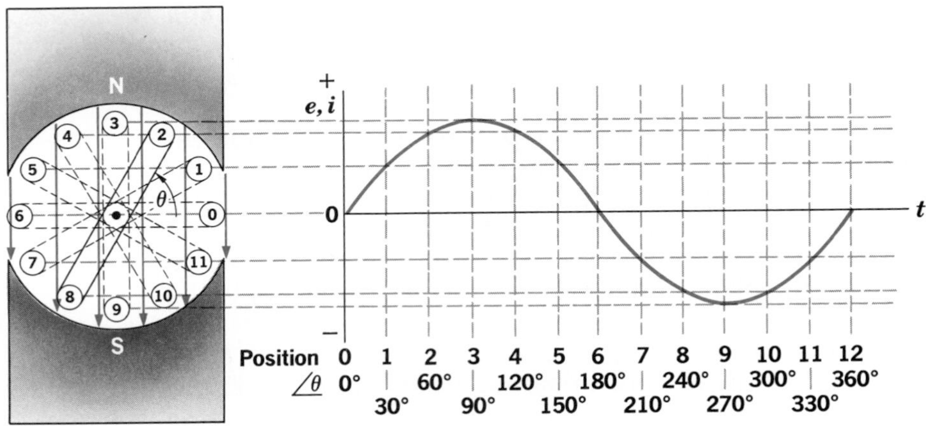

Figure 20-14. A sine curve representing current or voltage generated by a single-loop armature rotating at a constant rate in a uniform magnetic field. Positions 0 through 12 on the graph correspond to positions of the rotating armature in the magnetic field, as shown to the left.

referred to as **stators** and the rotating field magnets as **rotors.** Circuit current is taken from the stator at the high generated voltage without the use of slip rings and brushes. The exciter voltage, which is much lower than the armature voltage, is applied to the rotor through slip rings and brushes.

In a simple two-pole generator one cycle of operation produces one cycle or two alternations of induced emf, as shown in Figure 20-14. If the armature (or the field) rotates at the rate of 60 cycles per second, the frequency, f, of the generated voltage sine wave is 60 hertz and the period, T, is $\frac{1}{60}$ second.

The "frequency" of a d-c voltage or current is considered to be zero.

The ***frequency*** *of an alternating current or voltage is the number of cycles of current or voltage per second.* If the generator has a 4-pole field magnet, 2 cycles of emf are generated during 1 revolution of the armature or field. Such a generator turning at 30 rps would generate a 60-Hz voltage. In general

$$f = \textbf{No. of pairs of poles} \times \textbf{revolution rate}$$

Practically all commercial power is generated by **three-phase** ***generators*** having three armature coils spaced symmetrically and producing emfs spaced 120° apart. The coils are usually connected so that the currents are carried by three conductors.

Figure 20-15. The principle of the three-phase alternator. Armature coils schematically diagrammed in (A) are spaced 120° apart and generate peak voltage output in three phases as shown in (B). In (C) the output is shown as a voltage diagram.

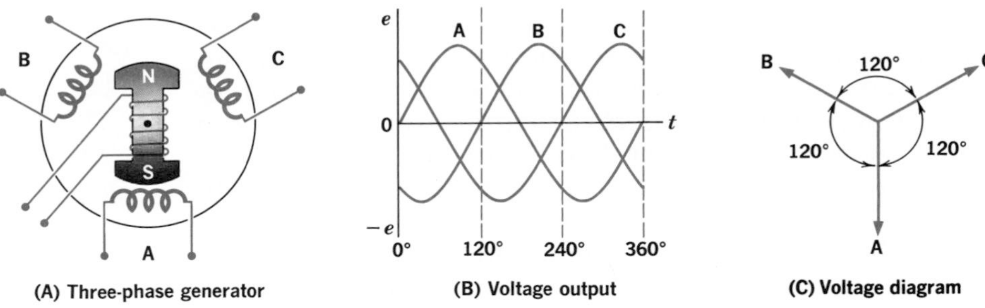

It is evident from Figure 20-15(B) that three-phase power is smoother than the single-phase power of Figure 20-14. Electric power is transmitted by a three-phase circuit but it is commonly supplied to the consumer by a single-phase circuit. A modern center for the control of production and distribution of electric energy is shown in Figure 20-16.

Figure 20-16. The control room of an electric power generating station. Computers monitor power production and transmission.

20.11 The d-c Generator The output of an a-c generator is not suitable for circuits that require a direct current. However, the a-c generator can be made to supply a *unidirectional* current to the external circuit by connecting the armature loops to a *commutator* instead of slip rings. *A* ***commutator*** *is a split ring, each segment of which is connected to an end of a corresponding armature loop.*

The current and voltage generated in the armature are alternating, as we would expect. By means of the commutator, the connections to the external circuit are reversed at the same instant that the direction of the induced emf reverses in the loop. See Figure 20-17. The alternating current in the armature appears as a *pulsating* direct current in the external circuit, and a pulsating d-c voltage appears across the load. A graph of the instantaneous values of the pulsating current from a generator with a two-segment commutator plotted as a function of time is shown in Figure 20-18. A graph of the voltage across a resistance load would have a similar form. Compare this pulsating d-c output to the a-c output of the generator in Figure 20-14.

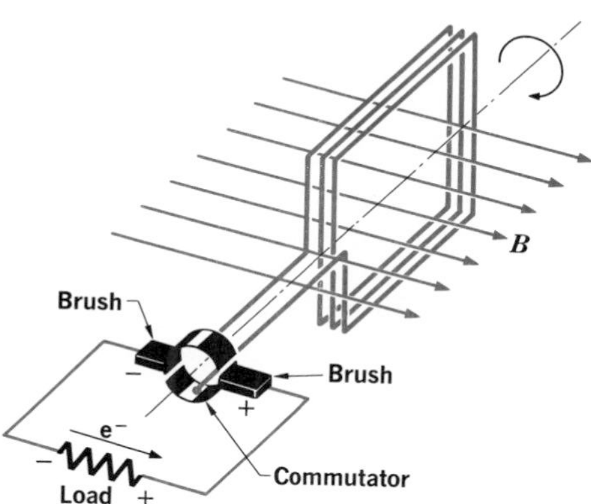

Figure 20-17. A split-ring commutator of two segments.

To secure from a d-c generator a more constant voltage and one having an instantaneous value that approaches the average value of the emf induced in the entire armature, many coils are wound on the armature. Each coil is connected to a different pair of commutator segments. The two brushes of each commutator segment are positioned so that they are in contact with successive pairs of commutator segments at the time when the induced emf in their respective coils is in the $\mathscr{E}_{max}$ region. See Figure 20-19.

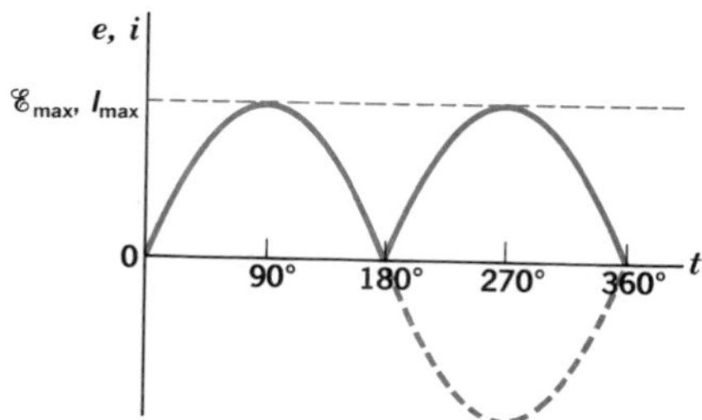

Figure 20-18. The variation of current or voltage with time in the external circuit of a simple generator with a two-segment commutator.

20.12 Field Excitation Most d-c generators use part of the induced power to energize their field magnets and are said to be *self-exciting*. The field magnets may be connected in series with the armature loops so all of the generator current passes through the coil windings. In the **series-wound** ***generator*** an increase in the load increases the magnetic field and hence the induced emf. See Figure 20-20.

The field magnets may also be connected in parallel with the armature so only a portion of the generated current is used to excite the field. In this **shunt-wound** ***generator***, Figure 20-21, an increase in load results in a decrease in the field and hence a decrease in the induced emf.

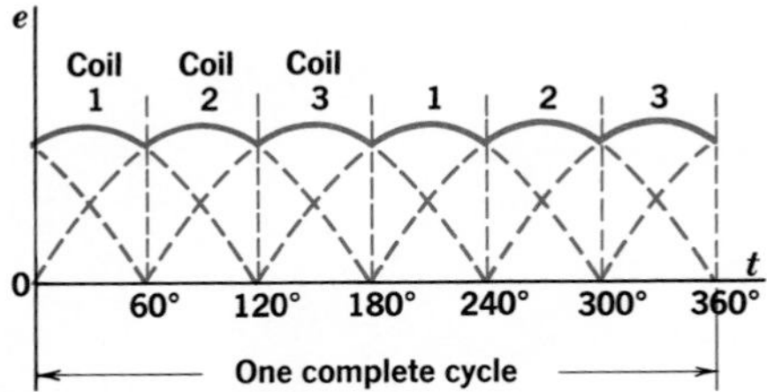

Figure 20-19. The output of a d-c generator having three armature coils and a six-segment commutator is fairly constant.

By using a combination of both series and shunt windings to excite the field magnets, the potential difference across the external circuit of a d-c generator may be maintained fairly constant; an increase in load causes an increase in current in the series windings and a decrease in current in the parallel windings. With the proper number of turns of each type of winding, a constant flux density can be maintained under varying loads. A **compound-wound generator** is shown in Figure 20-22.

20.13 Ohm's Law and Generator Circuits We can think of the armature turns as the source of emf in the d-c generator circuit. If the field magnet were separately excited, this emf would appear across the armature terminals on open-circuit operation since there would be no induced armature current.

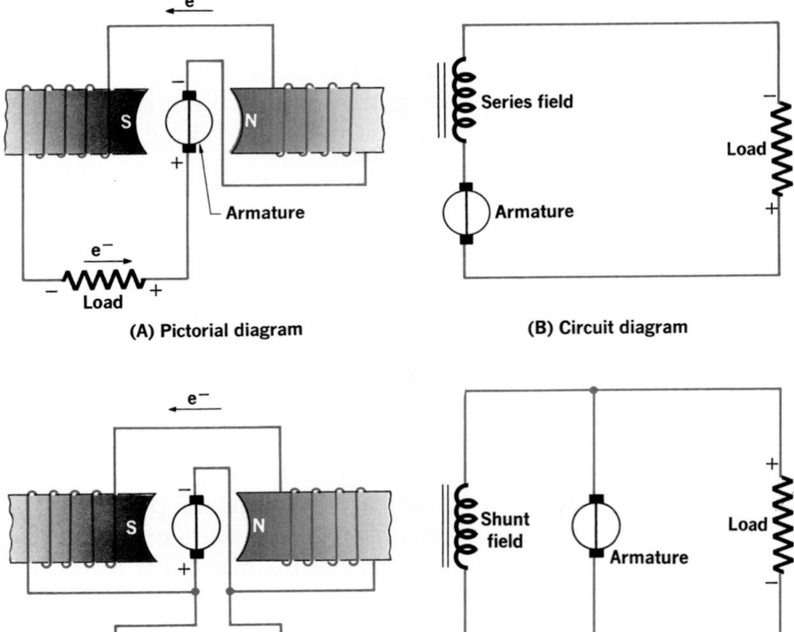

Figure 20-20. A series-wound d-c generator.

Figure 20-21. A shunt-wound d-c generator.

However, in a self-excited generator the armature circuit is completed through the field windings. The resistance of the armature turns, r_A, is in this current loop, and a situation analogous to that of a battery with internal resistance furnishing current to an external circuit results. A potential drop, $I_A r_A$, of opposite polarity to the induced

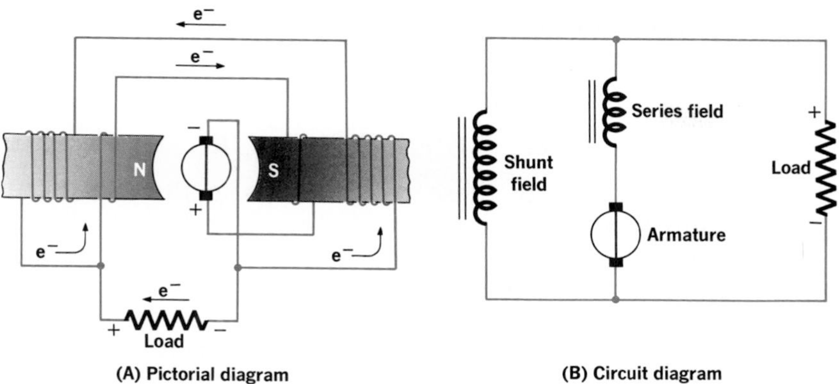

Figure 20-22. A compound-wound d-c generator.

emf, must appear across the armature. The armature potential difference, V, is

$$V = \mathscr{E} - I_A r_A$$

Figure 20-23 is a resistance circuit of a series-wound generator. The resistance of the field windings, R_F, is in series with the internal resistance of the armature and the load resistance so that the armature current is present in all three resistances.

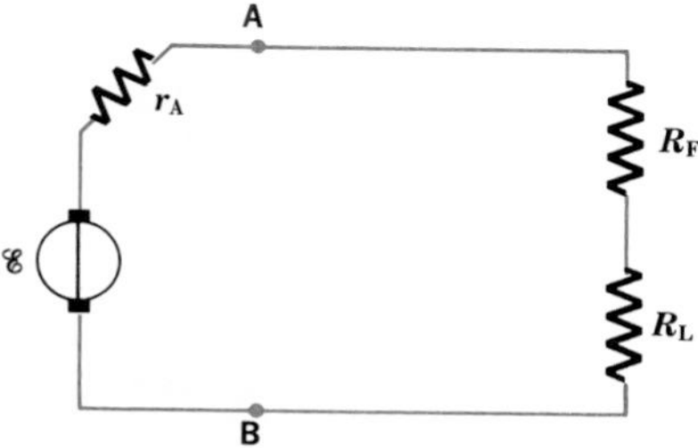

Figure 20-23. The resistance circuit of a series-wound d-c generator.

In Figure 20-21, the resistance of the shunt windings is in parallel with the load. The resistance circuit of a shunt-wound generator is diagrammed in Figure 20-24, with the resistance of the shunt windings shown as R_F. The total current of the circuit, I_A, must be in r_A, producing an $I_A r_A$ drop across the armature. The potential difference, V, which has been shown to be $\mathscr{E} - I_A r_A$, then appears across the network consisting of R_F and R_L in parallel. The following Ohm's law relationships hold:

$$I_F = \frac{V}{R_F}$$

$$I_L = \frac{V}{R_L}$$

and

$$I_A = I_F + I_L$$

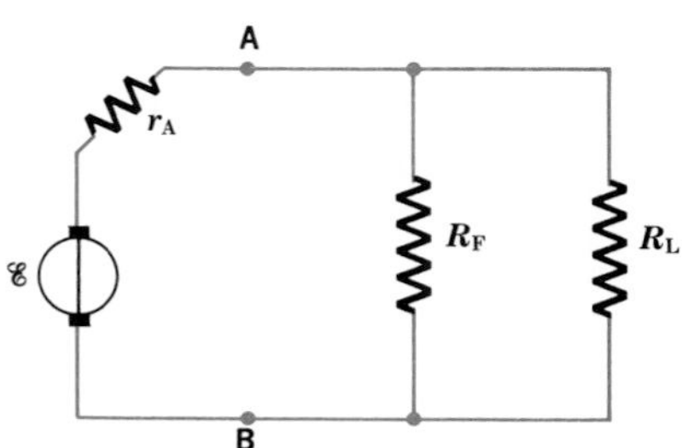

Figure 20-24. The resistance circuit of a shunt-wound d-c generator.

The total electric power in the generator circuit, P_T, derived from the mechanical energy source that turns the armature, is the product of the armature current, I_A, and the induced emf, $\mathscr{E}$.

$$P_T = \mathscr{E} I_A$$

Some of this power is dissipated as heat in the armature

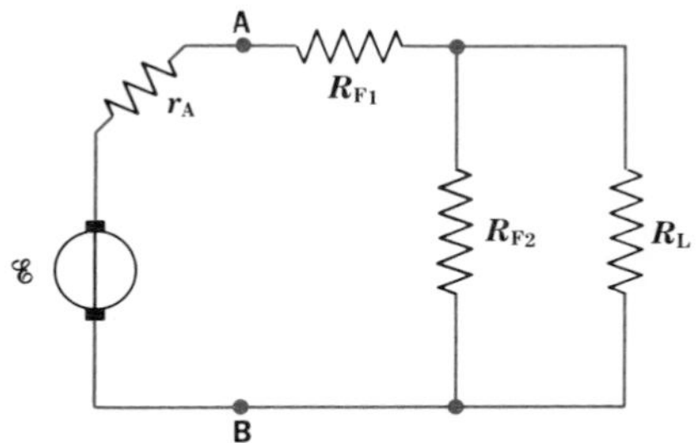

Figure 20-25. The resistance circuit of a compound-wound d-c generator.

resistance, and some is dissipated in the field windings; the remaining power is delivered to the load.

$$P_T = P_A + P_F + P_L$$

or

$$\mathscr{E}I_A = I_A^2 r_A + I_F^2 R_F + I_L^2 R_L$$

The application of these relationships to the d-c shunt-wound generator is illustrated in the example that follows.

Figure 20-25 is a resistance circuit of a compound-wound generator. Observe the similarity between this circuit and the resistance network shown in Figure 17-21, Section 17.10. How would you analyze this resistance circuit of the compound-wound d-c generator?

EXAMPLE A shunt-wound d-c generator has an armature resistance of 0.45 Ω. The shunt has a resistance of 65.2 Ω. The generator delivers 1250 W to a load with a resistance of 39.0 Ω. Calculate (a) the voltage across the load, (b) the current in the armature, and (c) the emf of the generator.

Given	Unknown	Basic equations
$r_A = 0.45\ \Omega$ $R_F = 65.2\ \Omega$ $R_L = 39.0\ \Omega$ $P_L = 1250\ \text{W}$	V_L I_A	$P_L = I^2R_L$ $V = \mathscr{E} - I_A r_A$ $V = IR$

Solution

(a) *Working equation:* $P_L = \dfrac{V_L^2}{R}$

$$V_L = \sqrt{(P_L)(R_L)} = \sqrt{(1250\ \text{W})(39.0\ \Omega)}$$
$$= 221\ \text{V}$$

(b) The armature current, I_A, is the sum of the shunt current, I_F, and the current in the load, I_L. Using Ohm's law, we get the

Working equation: $I_A = \dfrac{V_F}{R_F} + \dfrac{V_L}{R_L}$

$$= \frac{221\ \text{V}}{65.2\ \Omega} + \frac{221\ \text{V}}{39.0\ \Omega} = 9.06\ \text{A}$$

(c) *Working equation:* $\mathscr{E} = V + I_A r_A$

$$= 221\ \text{V} + (9.06\ \text{A})(0.45\ \Omega) = 2.3 \times 10^2\ \text{V}$$

PRACTICE PROBLEM A shunt-wound d-c generator has an armature resistance of 0.35 Ω and a field-winding resistance of 75.0 Ω. The generator delivers 1.44 kW to a load resistance of 22.5 Ω. (a) Draw the circuit diagram. (b) What is the potential difference across the load? (c) What is the magnitude of the armature current? (d) What is the emf of the generator? *Ans.* (a) See Figure 20-24; (b) 180 V; (c) 10.4 A; (d) 184 V

20.14 The Motor Effect We have learned that a current is induced in a conducting loop when the conductor is moving in a magnetic field so that the magnetic flux linking the loop is changing. This is the generator principle. By Lenz's law, we have seen that work must be done against a magnetic force as an induced current is generated in the conducting loop. Recall that an induced current always produces a magnetic force that opposes the force causing the motion by which the current is induced.

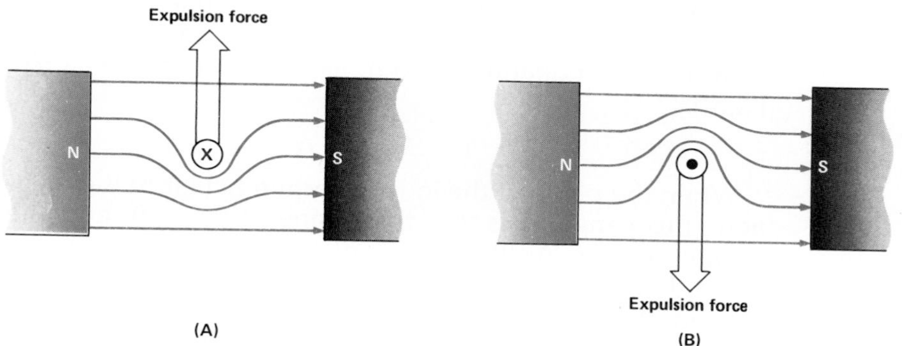

Figure 20-26. A visual representation of the motor effect.

Instead of employing a mechanical effort to move a conductor poised in the magnetic field, suppose a current is *supplied* to the conductor from an external source. The magnetic field is distorted, as discussed in Section 20.6, and the resulting magnetic force tends to expel the conductor from the magnetic field. This action is known as the **motor effect,** and it is illustrated in Figure 20-26.

If a current is supplied to an armature loop poised in a uniform magnetic field, as in Figure 20-27, the field around each conductor is distorted. A force acts on each side of the loop proportional to the flux density and the current in the armature loop. These two forces, equal in magnitude but opposite in direction, constitute a **couple** and produce a torque, T, that causes the armature loop to rotate about its axis. The magnitude of this torque is equal to the product of the force and the *perpendicular* distance between the two forces.

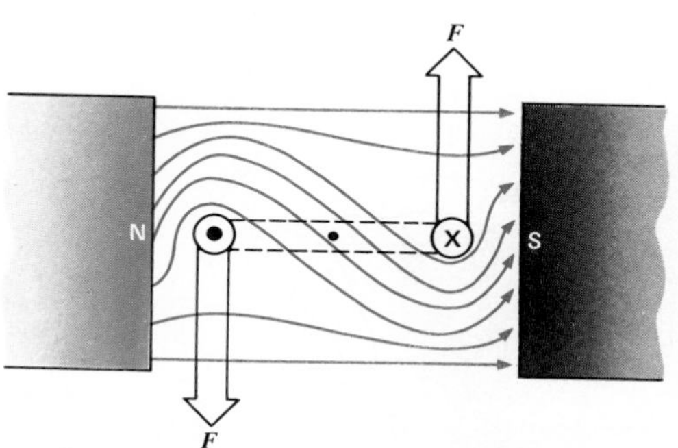

Figure 20-27. The resultant magnetic field of a current loop in the external field of a magnet gives a visual suggestion of the forces acting on the loop.

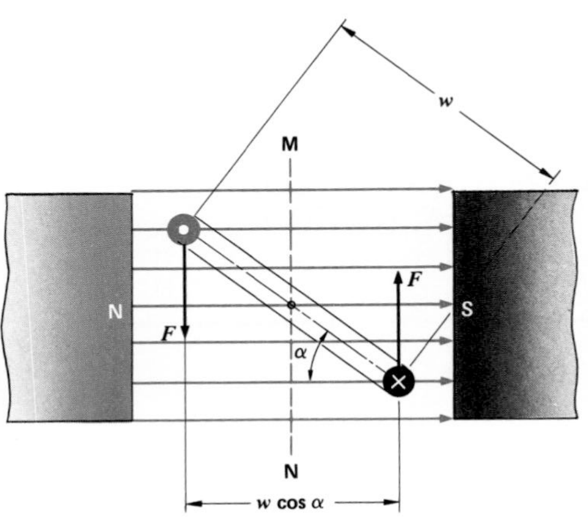

Figure 20-28. The torque on a current loop in a magnetic field is proportional to the perpendicular distance between the forces acting on the conductors.

It is evident from Figure 20-27 that the perpendicular distance between the torque-producing forces is maximum when the conductors are moving perpendicular to the magnetic flux. This distance is equal to the width of the armature loop. When the loop is in any other position with respect to the flux lines, the perpendicular distance between the forces is less than the width of the loop, and consequently the resulting torque must be less than the maximum value. See Figure 20-28.

We shall refer to the angle between the plane of the loop and the magnetic flux as angle α (alpha). When the angle is zero, the plane of the loop is parallel to the flux lines and the torque is maximum.

$$T_{max} = Fw$$

where F is the magnetic force acting on either conductor and w is the width of the conducting loop.

As the armature turns, the angle α approaches 90° and the torque diminishes since the perpendicular distance between the couple approaches zero. In general, the perpendicular distance between the forces acting on the two conductors is equal to $w \cos \alpha$, as shown in Figure 20-28. Hence

$$T = Fw \cos \alpha$$

When the plane of the loop is perpendicular to the magnetic flux, angle α is 90°, the cosine of 90° = 0, and the torque is zero. As the inertia of the conductor carries it beyond this point, a torque develops that reverses the motion of the conductor and returns it to the zero-torque position. In order to prevent this action, the direction of the current in the armature loop must be reversed at the proper instant. To reverse the current when the neutral position is reached, the conducting loop terminates in a commutator.

An electric motor performs the reverse function of a generator. *Electric energy is converted to mechanical energy* using the same electromagnetic principles employed in the generator. We can determine the direction of the motion of the conductor on a motor armature by use of a *right-hand rule* known as the **motor rule.** It is illustrated in Figure 20-29.

The motor rule: Extend the thumb, forefinger, and middle finger of the right hand at right angles to each other. Let the forefinger point in the direction of the magnetic flux and the middle finger in the direction of the electron flow; the thumb points in the direction of the motion of the conductor.

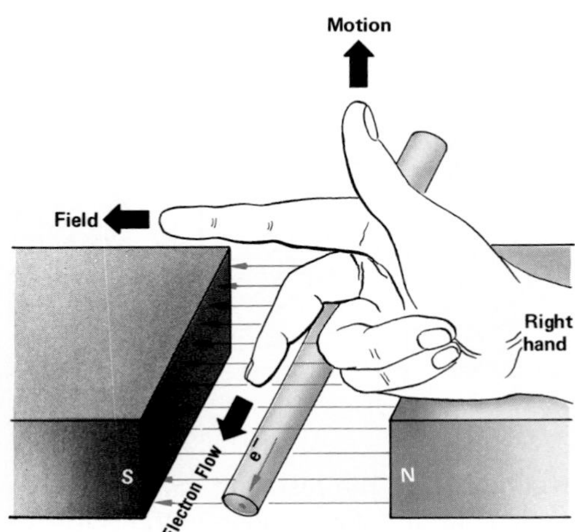

Figure 20-29. The right-hand motor rule.

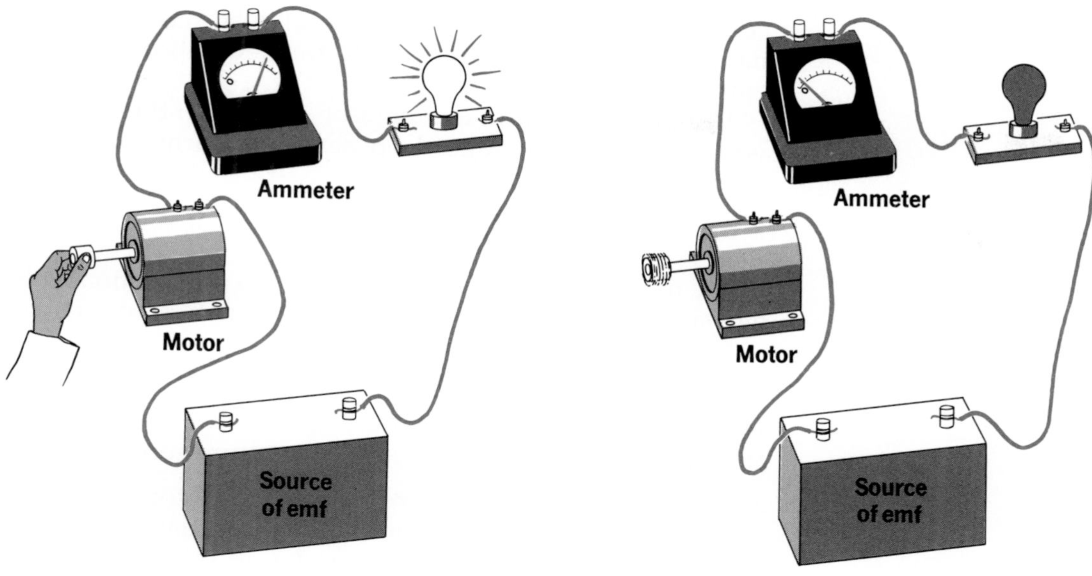

Figure 20-30. Demonstrating the back emf of a motor.

20.15 Back emf The simple d-c motor does not differ essentially from a generator; it has a field magnet, an armature, and a commutator ring. In fact, *the operating motor also acts as a generator.* As the conducting loop of the armature rotates in a magnetic field, an emf is induced across the armature turn. The magnitude of this emf depends on the speed of rotation of the armature.

Suppose that an incandescent lamp and an ammeter are connected in series with the armature of a small battery-driven motor, as illustrated in Figure 20-30. If the motor armature is held so that it cannot rotate as the circuit is closed, the lamp glows and the circuit current is indicated on the meter. Releasing the armature allows the motor to gain speed, and the lamp dims; the ammeter indicates a smaller current.

According to Lenz's law, the induced emf must oppose the motion inducing it. The emf induced by the generator action of a motor consequently opposes the voltage applied to the armature. *Such an induced emf is called the* ***back emf*** *of the motor.* The difference between the applied voltage and the back emf determines the current in the motor circuit.

A motor running at full speed under no load generates a back emf nearly equal to the applied voltage; thus a small current is required in the circuit. The more slowly the armature turns, the smaller is the back emf and consequently the larger is the voltage difference and the circuit current. A motor starting under a full load has a large initial current that decreases due to the generation of a back emf as the motor gains speed.

The induced emf in a generator is equal to the terminal voltage *plus* the voltage drop across the armature resistance.

$$\mathscr{E} = V + I_A r_A$$

In the motor, the induced emf is equal to the terminal voltage *minus* the voltage drop across the armature resistance.

$$\mathscr{E} = V - I_A r_A$$

Hence, the back emf in a motor must always be less than the voltage impressed across the armature terminals.

20.16 Practical d-c Motors A simple motor with a single armature coil would be impractical for many purposes because it has neutral positions and a pulsating torque. In practical motors a large number of coils is used in the armature; in fact there is little difference in the construction of motor and generator armatures. Multiple-pole field coils can be used to aid in the production of a uniform torque. The amount of torque produced in any given motor is proportional to the armature current and to the flux density.

Depending on the method used to excite the field magnets, practical d-c motors are of three general types. These are *series, shunt,* and *compound* motors. The excitation methods are similar to those of the d-c generators discussed in Section 20.12.

20.17 Practical a-c Motors Nearly all commercial distributors of electric power supply alternating-current power. Except for certain specialized applications, a-c motors are far more common than d-c motors. Much of the theory and technology of a-c motors is complex. We will briefly summarize the general characteristics of three common types of a-c motors: *the universal motor, the induction motor,* and *the synchronous motor.*

1. *The universal motor.* Any small d-c series motor can be operated from an a-c source. When this is done, the currents in the field windings and armature reverse direction simultaneously, maintaining torque in the same direction throughout the operating cycle. Heat losses in the field windings are extensive, however, unless certain design changes are incorporated into the motor. With laminated pole pieces and special field windings, small series motors

operate satisfactorily on either a-c or d-c power. They are known as *universal* motors.

2. *The induction motor.* Induction motors are the most widely used a-c motors because they are rugged, simple to build, and well adapted to constant speed requirements. They have two essential parts—a stator of field coils and a rotor. The rotor is usually built of copper bars laid in slotted, laminated iron cores. The ends of the copper bars are shorted by a copper ring to form a cylindrical cage; this common type is known as a *squirrel-cage* rotor.

Figure 20-31. A cut-away view of an induction motor.

By using three pairs of poles and a three-phase current, the magnetic field of the stator is made to rotate electrically and currents are induced in the rotor. In accordance with Lenz's law, the rotor will then turn so as to follow the rotating field. Since induction requires relative motion between the conductor and the field, the rotor must *slip* or lag behind the field in order for a torque to be developed. An increase in the load causes a greater slip, a greater induced current, and consequently a greater torque.

For its operation, an induction motor depends on a rotating magnetic field. Thus a single-phase induction motor is not self-starting because the magnetic field of its stator merely reverses periodically and does not rotate electrically. Single-phase induction motors can be made self-starting by a *split-phase winding*, a *capacitor*, a *shading coil*, or a *repulsion winding*. The name of the motor usually indicates the auxiliary method used for starting it.

3. *The synchronous motor.* We can illustrate the principle of synchronism by placing a thoroughly magnetized compass needle in a rotating magnetic field. The magnetized needle aligns itself with the magnetic field and rotates in synchronization with the rotating field.

The **synchronous motor** is a constant speed motor, running in synchronism with the a-c generator that supplies the stator current. However, it is not self-starting; the rotor must be brought up to synchronous speed by auxiliary means.

Electric clocks are operated by small single-phase synchronous motors that start automatically as a form of induction motor. Once synchronism is attained, they run as synchronous motors. Electric power companies maintain very accurate control of the 60-Hz frequency of commercial power.

Large industrial synchronous motors use an electromagnetic rotor supplied with d-c power, and a stator supplied with three-phase a-c power. The synchronous motors are used where the speed requirements are very exacting.

QUESTIONS: GROUP A

1. What are the essential components of an electric generator?
2. State the rule that helps us determine the direction of the induced current in the armature loops of a generator.
3. Distinguish between a direct current and an alternating current.
4. Under what circumstances does a simple generator produce a sine-wave variation of induced voltage?
5. What does the term θ (theta) represent in the expression $e = \mathscr{E}_{max} \sin \theta$?
6. (a)What is a magneto? (b) For what is it used?
7. What is the function of an *exciter* in the generation of a-c power?
8. (a) What is meant by the frequency of an alternating current? (b) Under what circumstances will the frequency of a generated current be the same as the rps of the armature?
9. How can a generator be made to supply a direct current to its external circuit?
10. In what way is the output of a d-c generator different from the d-c output of a battery?
11. (a) What methods are used to energize the field magnets of d-c generators? (b) Draw a circuit diagram of each method.
12. What are the three power-consuming parts of a d-c generator circuit?
13. What is meant by the term α (alpha) in the torque expression $T = Fw \cos \alpha$?
14. State the rule that helps us determine the direction of motion of the armature loops of a motor.
15. (a) What is meant by back emf? (b) How is it induced in an electric motor?
16. What two quantities influence the amount of torque produced in a motor?
17. What are the three common types of a-c motors?

GROUP B

18. What is the advantage of having a 4- or a 6-pole field magnet in a generator producing a 60-Hz output?
19. How does an increase in the load on a series-wound d-c generator affect the induced emf? Explain.
20. How does an increase in the load on a shunt-wound d-c generator affect the induced emf? Explain.
21. How is torque produced on the armature loops of a motor?
22. Why is it true that an operating motor is also a generator?
23. The torque in a series motor increases as the load increases. Explain.
24. Explain why a single-phase induction motor is not self-starting.
25. Why are synchronous motors used in electric clocks?
26. Two conducting loops, identical except that one is silver and the other aluminum, are rotated in a magnetic field. In which case is the larger torque required to turn the loop?

PROBLEMS: GROUP A

1. A series-wound d-c generator turning at its rated speed develops an emf of 28 V. The current in the external circuit is 16 A and the armature resistance is 0.25 Ω. What is the potential drop across the external circuit?
2. A d-c generator delivers $15\bar{0}$ A at $22\bar{0}$ V when operating at normal speed connected to a resistance load. The total losses are $32\bar{0}0$ W. Determine the efficiency of the generator.
3. A shunt-wound generator has an armature resistance of 0.15 Ω and a shunt winding of 75.0-Ω resistance. It delivers 18.0 kW at $24\bar{0}$ V to a load. (a) Draw the circuit diagram. (b) What is the generator emf? (c) What is the total power delivered by the armature?
4. A magnetic force of 3.5 N acts on one conductor of a conducting loop in a

magnetic field. A second force of equal magnitude but opposite direction acts on the opposite conductor of the loop. The conducting loop is 15 cm wide. Find the torque acting on the conducting loop (a) when the plane of the loop is parallel to the magnetic flux, (b) after the loop has rotated through 30°.

5. The maximum torque that acts on the armature loop of a motor is 10 m · N. At what positions of the loop with respect to the magnetic flux will the torque be 5 m · N?
6. A shunt-wound motor connected across a 117-V line generates a back emf of 112 V when the armature current is $1\bar{0}$ A. What is the armature resistance?

INDUCTANCE

20.18 Mutual Inductance An emf is induced across a conductor in a magnetic field when there is a change in the flux linking the conductor. In our study of the generator we observed that an induced emf appears across the armature loop whether the conductors move across a stationary field or the magnetic flux moves across stationary conductors. *In either action there is relative motion between conductors and magnetic flux.*

This relative motion can be produced in another way. By connecting a battery to a solenoid through a contact key, an electromagnet is produced that has a magnetic field similar to that of a bar magnet. When the key is open, there is no magnetic field. As the key is closed, the magnetic field builds up from zero to some steady value determined by the number of ampere-turns. The magnetic flux spreads out and permeates the region about the coil. *An expanding magnetic field is a field in motion.* When the key in the solenoid circuit is opened, the magnetic flux collapses to zero. *A collapsing magnetic field is also a field in motion.* Its motion, however, is in the opposite sense to that of the expanding field.

Suppose the solenoid is inserted into a second coil whose terminals are connected to a galvanometer as in Figure 20-32. The coil connected to a current source is the *primary coil;* its circuit is the *primary circuit.* The coil connected to the galvanometer (the load) is the *secondary coil;* its circuit is the *secondary circuit.* At the instant the contact key is closed, a deflection is observed on the galvanometer. There is no deflection, however, while the key remains closed. When the key is opened a galvanometer deflection again occurs, but in the opposite direction. An emf is induced across the secondary turns whenever the

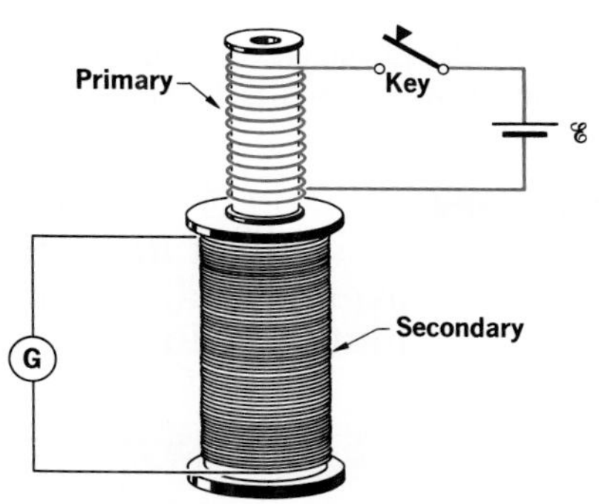

Figure 20-32. Varying the current in the primary induces an emf in the secondary.

flux linking the secondary is increasing or decreasing.

The relative motion between conductors and flux is in one direction when the field expands and in the opposite direction when the field collapses. Thus the emf induced across the secondary as the key is closed is of opposite polarity to that induced as the key is opened. The more rapid this relative motion, the greater the magnitude of the induced emf; the greater the number of turns in the secondary, the greater the magnitude of the induced emf. A soft-iron core placed in the primary greatly increases the flux density and the induced emf.

Two circuits arranged so that a change in magnitude of current in one causes an emf to be induced in the other show *mutual inductance. The* ***mutual inductance,*** *M, of two circuits is the ratio of the induced emf in one circuit to the rate of change of current in the other circuit.*

The unit of mutual inductance is called the *henry*, H, after the American physicist Joseph Henry. *The mutual inductance of two circuits is one henry when one volt of emf is induced in the secondary as the current in the primary changes at the rate of one ampere per second.*

Figure 20-33. Joseph Henry, the American physicist for whom the unit of inductance is named.

$$\boldsymbol{M = \frac{-\mathscr{E}_S}{\Delta I_P/\Delta t}}$$

Here, M is the mutual inductance of the two circuits in henrys, $\mathscr{E}_S$ is the average induced emf across the secondary in volts, and $\Delta I_P/\Delta t$ is the time rate of change of current in the primary in amperes per second. The negative sign indicates that the induced voltage opposes the change in current according to Lenz's law. From this equation the emf induced in the secondary becomes

$$\mathscr{E}_S = -M\frac{\Delta I_P}{\Delta t}$$

See the following example.

EXAMPLE When the switch is closed in the circuit of the primary coil of a pair of induction coils, the current reaches 9.50 $\underline{A}$ in 0.032 0 s. The average voltage induced in the secondary coil is 45$\bar{0}$ V. Calculate the mutual inductance of the two coils.

Given	Unknown	Basic equation
$\Delta I_P = 9.50$ A $\Delta t = 0.\underline{0}32\ 0$ s $\mathscr{E}_S = 45\bar{0}$ V	M	$\mathscr{E} = -M\frac{\Delta I_P}{\Delta t}$

Solution

$$\text{Working equation: } M = \frac{-\mathscr{E}_S \,\Delta t}{I_P}$$

$$= \frac{-(45\bar{0}\text{ V})(0.0320\text{ s})}{9.50\text{ A}}$$

$$= -1.52\text{ H}$$

PRACTICE PROBLEMS 1. Two adjacent coils have a mutual inductance of 0.880 H. What is the average induced emf in the secondary if the current in the primary changes from 0.00 A to 18.0 A in 0.055 0 s?

Ans. −288 V

2. The current in the primary of two adjacent coils changes from 0.0 A to 14 A in 0.025 s and the average induced emf in the secondary is 320 V. What is the mutual inductance of the two circuits? *Ans.* 0.57 H

20.19 Self-Inductance Suppose a coil is formed by winding many turns of insulated copper wire on an iron core and connected in a circuit to a 6-volt battery, a neon lamp, and a switch. See Figure 20-34. Since the neon lamp requires about 85 V dc to conduct, it acts initially as an open switch in a 6-volt circuit.

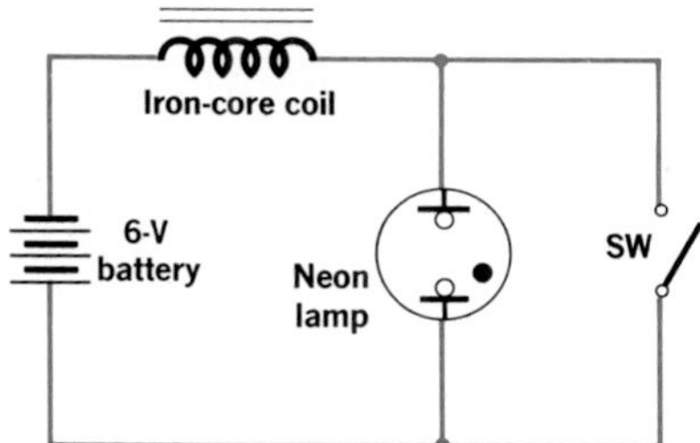

Figure 20-34. A circuit to demonstrate self-inductance.

When the switch is closed, a conducting path is completed through the coil; the fact that the lamp does not light is evidence that it is not in the conducting circuit. If now the switch is quickly opened, the neon lamp conducts for an instant, producing a flash of light. This means that the lamp is subjected to a potential difference considerably higher than that of the battery. What is the source of this higher voltage?

Any change in the magnitude of current in a conductor causes a change in the magnetic flux about the conductor. If the conductor is formed into a coil, a changing magnetic flux about one turn cuts across adjacent turns and induces a voltage across them. According to Lenz's law, the polarity of this induced voltage acts to oppose the motion of the flux inducing it. The sum of the induced voltages of all the turns constitutes a *counter* emf across the coil.

If a rise in current with its expanding magnetic flux is responsible for the counter emf, this rise in current will be opposed. Therefore, the counter emf is opposite in polarity to the applied voltage. When the switch in Figure 20-34 was closed, the rise in current from zero to the steady-state magnitude was opposed by the counter emf induced across the coil. Once the current reached a steady value, the magnetic field ceased to expand and the opposing voltage fell to zero.

If a fall in current with its collapsing flux is responsible for the induced voltage across the coil, the fall in current is opposed. In this case the induced emf has the same polarity as the applied voltage and tends to sustain the current in the circuit. A very rapid collapse of the magnetic field may induce a very high voltage. *It is the change in current, not the current itself, that is opposed by the induced emf.* It follows, then, that the greater the rate of change of current in a circuit containing a coil, the greater is the magnitude of the induced emf across the coil that opposes this change of current.

It is not the magnitude of current but the rate at which current is changing that relates to the magnitude of an induced emf.

The property of a coil that causes a counter emf to be induced across it by the change in current in it is known as *self-inductance,* or simply *inductance. The* ***self-inductance,*** *L, of a coil is the ratio of the induced emf across the coil to the rate of change of current in the coil.*

The unit of self-inductance is the *henry,* the same unit used for mutual inductance. Self-inductance is *one henry* if *one volt of emf* is induced across the coil when the current in the circuit changes at the rate of *one ampere per second.*

$$L = \frac{-\mathscr{E}}{\Delta I/\Delta t}$$

Here L is the inductance in henrys, $\mathscr{E}$ is the emf induced in volts, and $\Delta I/\Delta t$ is the rate of change of current in amperes per second. The negative sign merely shows that the induced voltage opposes the change of current. From this expression, the induced emf is expressed by

$$\mathscr{E} = -L\frac{\Delta I}{\Delta t}$$

The equation for inductance, L, is analogous to the defining equation for capacitance, C, as expressed in Section 16.14.

$$C = \frac{Q}{V}$$

The presence of a magnetic field in a coil conducting an electric current corresponds to the presence of an electric field in a charged capacitor.

Once a coil has been wound, its inductance is a constant property that depends on *the number of turns, the diameter of the coil, the length of coil,* and *the nature of the core.* Because of its property of inductance, a coil is commonly called an *inductor.*

Inductance in electricity is analogous to inertia in mechanics: the property of matter that opposes a *change* of velocity. If a mass is at rest, its inertia opposes a change that imparts a velocity; if the mass has a velocity, its inertia opposes a change that brings it to rest. The flywheel in mechanics illustrates the property of inertia. We know that energy is stored in a flywheel as its angular velocity is increased and is removed as its angular velocity is decreased.

In terms of mechanical quantities, impedance is analogous to mass, and current is analogous to velocity.

In an electric circuit inductance has no effect as long as the current is steady. An inductance does, however, oppose any change in the circuit current. An increase in current is opposed by the inductance; energy is stored in its magnetic field since work is done by the source against the counter emf induced. A decrease in current is opposed by the inductance; energy is removed from its field, tending to sustain the current. *We can think of inductance as imparting a flywheel effect in a circuit having a varying current.* This concept is very important in alternating-current circuit considerations.

20.20 Inductors in Series and Parallel The total inductance of a circuit consisting of inductors in series or parallel can be calculated in the same manner as total resistance. When inductors are connected in series, the total inductance, L_T, is equal to the sum of the individual inductances providing there is no mutual inductance between them.

$$L_T = L_1 + L_2 + L_3 + \text{etc.}$$

However, when two inductors are in series and arranged so that the magnetic flux of each links the turns of the other, the total inductance is

$$L_T = L_1 + L_2 \pm 2M$$

The ± sign is necessary in the general expression for the counter emf induced in one coil by the flux of the other may either aid or oppose the counter emf of self-induction. The two coils can be connected either in series "aiding" or series "opposing," depending on the manner in which their turns are wound.

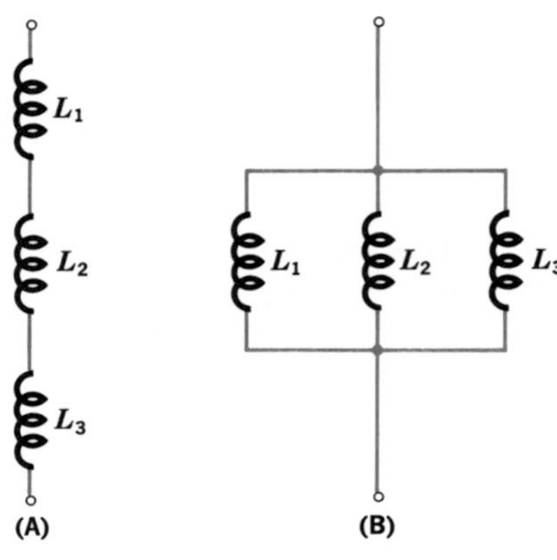

Figure 20-35. Inductors in series (A) and in parallel (B).

Inductors connected in parallel so that each is unaffected by the magnetic field of another provide a total inductance according to the following general expression:

$$\frac{1}{L_T} = \frac{1}{L_1} + \frac{1}{L_2} + \frac{1}{L_3} + \text{etc.}$$

See Figure 20-35 for schematic representation of these series and parallel connections for inductors.

20.21 The Transformer In principle the transformer consists of two coils, a *primary* and a *secondary,* electrically insulated from each other and wound on the same ferromagnetic core. Electric energy is transferred from the primary to the secondary by means of the magnetic flux in the core. A transformer is shown in its simplest forms in Figure 20-36.

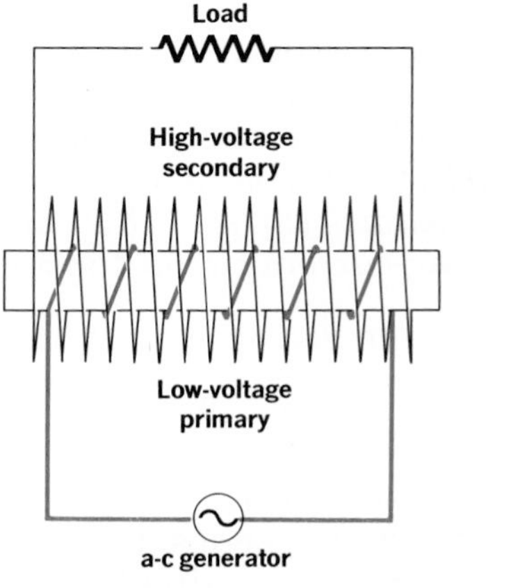

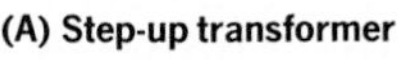

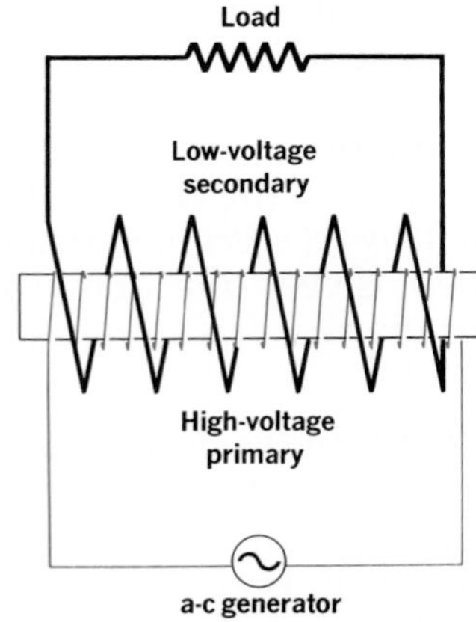

Figure 20-36. The transformer in its simplest forms.

Refer to Figure 20-14, in which the current sine wave produced by the a-c generator is given. The current is changing at its maximum rate as it passes through zero. Consequently, the emf induced across the secondary winding of a transformer is maximum as the primary current passes through zero. The polarity of the secondary emf reverses each time the primary current passes through a positive or negative maximum, since at these instants the current is changing at its minimum rate.

In a power transformer, a *closed* core is used to provide a continuous path for the magnetic flux, ensuring that practically all the primary flux links secondary turns. Since the same flux links both primary and secondary turns, the same emf per turn is induced in each and the ratio of secondary to primary emf is equal to the ratio of secondary to primary turns. Neglecting losses, we may assume the terminal voltages to be equal to the corresponding emfs. Thus

$$\frac{V_S}{V_P} = \frac{N_S}{N_P}$$

where V_S and V_P are the secondary and primary terminal voltages and N_S and N_P are the number of turns in the secondary and primary windings respectively.

If there are 20 turns in the secondary winding for every turn of the primary, the transformer is said to have a *turns ratio* of 20.

$$\textbf{Turns ratio} = \frac{\mathbf{N_S}}{\mathbf{N_P}} = \frac{\mathbf{20}}{\mathbf{1}} = \mathbf{20}$$

If the primary voltage is 110 volts, the secondary terminal voltage is 2200 volts and the transformer is called a *step-up transformer*. If the connections are reversed and the coil with the larger number of turns is made the primary, the transformer becomes a *step-down transformer*. The same primary voltage now produces a voltage across the secondary of 5.5 volts.

Figure 20-37. An external view, low-voltage side of a power transformer.

It can be shown that when power is delivered to a load in the secondary circuit, the product of the secondary current and secondary turns is essentially equal to the product of the primary current and primary turns.

$$I_S N_S = I_P N_P$$

or

$$\frac{I_S}{I_P} = \frac{N_P}{N_S}$$

Thus when the voltage is stepped up, the current is stepped down; there is no power gain as a result of transformer action. Ideally primary and secondary power are equal, but actually there are power losses as in any machine.

The efficiencies of practical transformers are high and constant over a wide power range. Transformer efficiencies above 95% are common. Efficiency can be expressed as the ratio of the power dissipated in the secondary circuit to the power used in the primary.

$$\textbf{Efficiency} = \frac{P_S}{P_P} \times 100\%$$

20.22 Transformer Losses While transformer efficiencies are high, the transfer of energy from the primary circuit to the secondary circuit does not occur without some loss. We shall consider two types: *copper losses* and *eddy-current losses*. They represent wasted energy and appear as heat.

1. Copper losses. These losses result from the resistance of the copper wires in the primary and secondary turns. Copper losses are I^2R heat losses. They cannot be avoided.

2. Eddy-current losses. When a mass of conducting metal is moved in a magnetic field or is subjected to a changing magnetic flux, induced currents circulate in the mass. These closed loops of induced current circulating in planes perpendicular to the magnetic flux are known as **eddy currents.**

Eddy currents in motor and generator armatures and transformer cores produce heat due to the I^2R losses in the resistance of the iron. They are induced currents that do no useful work, and they waste energy by opposing the change that induces them according to Lenz's law. Eddy-current losses are reduced by *laminating* the armature frames and cores. Thin sheets of metal with insulated surfaces are used to build up the armatures and cores. The laminations are set in planes parallel to the magnetic flux so that the eddy-current loops are confined to the width of the individual laminations. The high resistance associated with the narrow width of the individual laminations effectively reduces the induced currents and thus the I^2R heating losses.

Recall that resistance varies inversely with cross-sectional area.

QUESTIONS: GROUP A

1. What is mutual inductance?
2. Differentiate between the primary coil and the secondary coil.
3. (a) To what quantity in mechanics is inductance analogous? (b) Why?
4. Describe the principal parts of a transformer.
5. What is it that a transformer "transforms"?
6. What is the difference between a "step-up" and "step-down" transformer?
7. (a) Does a "step-up" transformer step up power? (b) Energy? (c) What law governs these transformations?

GROUP B

8. A calculator runs on a 9-V battery but has an adapter so you can plug it into a wall outlet and run it off 110-V alternating current. What are the functions of this "adapter"?
9. The power produced at a generating station is of relatively low voltage (several thousand volts), as is the power needed in homes (110-V). (a) What would happen if this low-voltage, high-current power were sent out over transmission lines? (b) How is this avoided?
10. A physics demonstration consists of dropping a bar magnet down a copper pipe. If the pipe is of sufficient length, it takes an appreciably longer time for the magnet to fall than if it had been dropped from the same height outside the tube. Why?

PROBLEMS: GROUP A

1. A pair of adjacent coils has a mutual inductance of 1.06 H. Determine the average emf induced in the secondary circuit when the current in the primary circuit changes from 0.00 A to 9.50 A in 0.033 6 s.
2. When the primary circuit of a pair of adjacent coils is activated, the current surges to 12 A in 0.048 s and the emf induced in the secondary circuit is 270 V. Determine the mutual inductance of the pair of inductors.
3. A step-up transformer is used on a 120-V line to provide a potential difference of 2400 V. If the primary has 75 turns, how many turns must the secondary have (neglecting losses)?
4. An initial rise in current in a coil occurs at the rate of 7.5 A/s at the instant a potential difference of 16.5 V is applied across it. (a) What is the self-inductance of the coil? (b) At the same instant a potential difference of $5\bar{0}$ V is induced across an adjacent coil. Find the mutual inductance of the two coils.
5. A coil with an inductance of 0.42 H and a resistance of 25 Ω is connected across a 110-V d-c line. What is the rate of current rise (a) at the instant the switch is closed; (b) at the instant the current reaches 85% of its steady-state value?
6. A 5:1 step-down transformer with negligible losses is connected to a $12\bar{0}$-V a-c source. The secondary circuit has a resistance of 15.0 Ω. (a) What is the potential difference across the secondary? (b) What is the secondary current? (c) How much power is dissipated in the secondary resistance? (d) What is the primary current?
7. Assume that the transformer of Problem 6 is replaced by one having an efficiency of 92.5%. What is the primary current?
8. A transformer with a primary of $40\bar{0}$ turns is connected across a $12\bar{0}$-V a-c line. The secondary circuit has a potential difference of $30\bar{0}0$ V. The secondary current is 60.0 mA and the primary current is 1.85 A. (a) How many turns are in the secondary winding? (b) What is the transformer efficiency?
9. A $27\bar{0}$-Ω resistor, a 2.50-H coil, and a switch are connected in series across a 12.0-V battery. At a certain instant after the switch is closed the current is 20.0 mA. What is the potential difference (a) across the resistor; (b) across the inductor? (c) What is the rate of change of current at this instant?

SUMMARY

An emf is induced in a conductor when relative motion between the conductor and a magnetic field produces a change in the flux linkage. The greater the rate of relative motion, the greater is the magnitude of the induced emf. If the conductor is part of a closed circuit, an electron current is induced in the circuit. The direction of an induced current is always in accord with Lenz's law.

An electric generator converts mechanical energy into electric energy. The direction of induced electron current in the armature turns is determined by use of the left-hand generator rule. The generator induces a sinusoidal emf across the armature turns. The current in the armature circuit alternates. The frequency of the generated current is expressed in hertz; one hertz is equivalent to one cycle per second. An a-c generator may be modified for a pulsating d-c output. The d-c generator is self-excited.

Electric motors convert electric energy

into mechanical energy. The motor effect is the result of an electric current in a magnetic field; it is the reverse of the generator effect. The direction of motion of the armature turns is determined by the use of the right-hand motor rule. An electric motor produces a back emf that subtracts from the applied voltage. Practical d-c motors are of three types: series, shunt, and compound wound. Three common types of a-c motors are the universal motor, the induction motor, and the synchronous motor. Of these, the induction motor is most widely used.

If a change in current in one circuit induces an emf in a second circuit, the two have a property of mutual inductance. An emf is induced across a coil by a change of current in the coil; this property is known as self-inductance. The unit of inductance is the henry. The induced emf across a given inductance depends on the time rate of change of current in the inductance. The property of inductance is described as a kind of electric inertia.

Transformers are alternating-current devices. Primary and secondary windings have a common core. For a given primary voltage, the turns ratio determines the secondary voltage. The transformer may either step up or step down the a-c voltage of the primary circuit. Efficient and economical distribution of a-c power is possible through the use of the transformer principle.

VOCABULARY

alternating current
back emf
commutator
compound-wound generator
eddy currents
electromagnetic induction
excitor
force couple
generator rule
induced current
induced emf
instantaneous current
instantaneous voltage
Lenz's law
magnetic force
magneto
motor effect
motor rule
mutual inductance
rotor
self-inductance
series-wound generator
shunt-wound generator
slip ring
stator
synchronous motor
three-phase generator

21 Alternating-Current Circuits

alternating current (AL-tur-nay-ting KUR-ent) n.: a current that has one direction during part of a generating cycle and the opposite direction during the remainder of the cycle.

a-c MEASUREMENTS

OBJECTIVES

- Study the distinction between a-c and d-c power.
- Define the effective values of alternating current and voltage.
- Analyze the effects of inductance and capacitance in a-c circuits.
- Study impedance in a-c circuits containing inductive and capacitive reactance.
- Study the effect of circuit impedance on the phase relation between current and voltage.
- Discuss the importance of resonant circuits.

21.1 Power in a-c Circuits A single conducting loop rotating at a constant speed in a uniform magnetic field generates an alternating emf. The magnitude of the emf varies with the sine of the angle that the plane of the loop makes with the perpendicular to the magnetic flux. The instantaneous value of the emf is

$$e = \mathscr{E}_{max} \sin \theta$$

The instantaneous current in a pure resistance load comprising the external circuit of the generator is similarly expressed as

$$i = I_{max} \sin \theta$$

This current is a consequence of the alternating voltages impressed across the load, and its maxima and minima occur at the same instants as the voltage maxima and minima. The alternating current and voltage are said to be *in phase*. This in-phase relationship is characteristic of a resistance load across which an alternating voltage has been applied. Sine curves of in-phase alternating current and voltage are shown in Figure 21-2.

In a d-c circuit, electric power, P, is the product $\mathscr{E} \times I$. In an a-c circuit, the instantaneous power, p, is the product $e \times i$. These relationships are shown in Figure 21-3.

The voltage and current in Figure 21-3(B) are shown in phase as they are in a resistive load. The instantaneous

Figure 21-1. Thomas A. Edison and Charles P. Steinmetz in Steinmetz's laboratory in Schenectady, New York, in 1922. Steinmetz was a brilliant mathematician and electrical engineer. He developed the mathematical analysis for alternating-current circuits and provided the firm mathematical base of electrical engineering.

power curve varies between some positive maximum and zero and has an average value P. Observe that the instantaneous power curve, p, has a frequency twice that of e and i. The ordinates of this power curve are always positive since e and i are in phase. During the first half-cycle, both e and i are positive, so their product is positive. During the second half-cycle, both e and i are negative, but their product is positive.

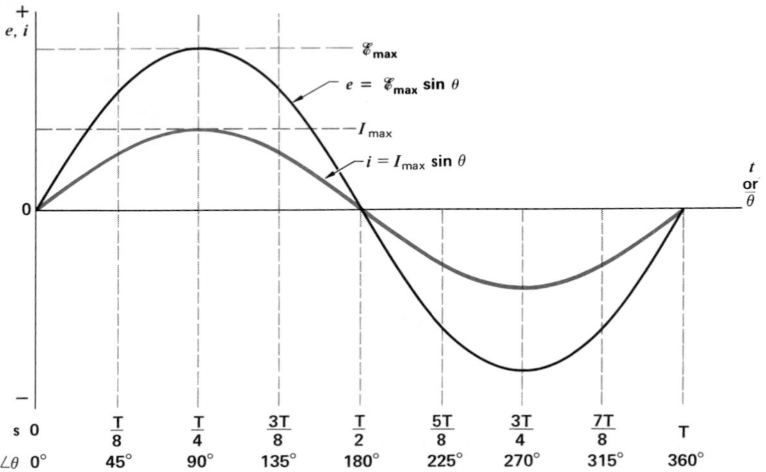

Figure 21-2. One cycle of alternating current and voltage in phase.

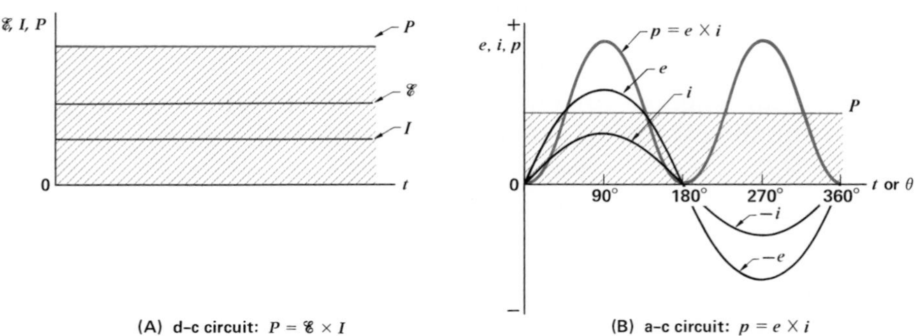

Figure 21-3. A comparison of d-c and a-c power.

21.2 Effective Values of Current and Voltage If a zero-centered d-c ammeter were placed in an a-c circuit, the pointer would tend to swing alternately in the positive and negative directions. At all but the very lowest frequencies, the inertia of the meter movement would cause the pointer to remain at zero. This, in effect, averages out the sine-wave variations. The same would be true of an a-c voltage measurement. *The average value of a sine curve of alternating current or voltage is zero*. This is true regardless of the magnitude of the maximum values attained.

The most useful value of an alternating current is based on its heating effect in an electric circuit and is commonly referred to as the *effective value*. *The* ***effective value*** *of an alternating current is the number of amperes that, in a given resistance, produces heat at the same average rate as that number of amperes of steady direct current.*

a-c meters indicate effective values of current and voltage, except in special applications.

Let us assume that a resistance element is immersed in water in a calorimeter and that a steady direct current of 1 ampere in the circuit raises the temperature of the water 25 C° in 10 minutes. An alternating current in the same resistance that would raise the temperature the same amount in the same time is said to have an effective value of 1 ampere. The symbol for the effective value of an alternating current is I, the same as that for steady direct current. When the magnitudes of alternating currents are given in amperes, they are understood to be effective values unless other values are clearly stated.

From Joule's law, $Q = I^2Rt$, we know that the heating effect of an electric current is proportional to I^2. The average rate of heat production by an alternating current is proportional to the mean (average) of the instantaneous values of current squared. From Figure 21-4 it is evident that the mean value of I^2 is $\frac{1}{2}I_{max}^2$ because the square of the instantaneous values of the alternating current of frequency f gives a curve of frequency $2f$ that varies between I_{max}^2 and zero. Thus

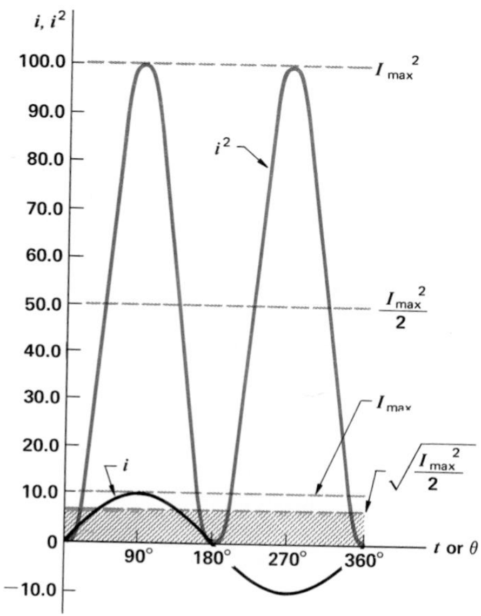

Figure 21-4. Current and current squared curves showing the effective value of an alternating current.

$$I^2 = \frac{I_{max}^2}{2}$$

and

$$I = \sqrt{\frac{I_{max}^2}{2}} = \frac{I_{max}}{\sqrt{2}}$$

$$I = 0.707\ I_{max}$$

Then

$$I_{max} = 1.414\ I$$

A sine curve of alternating current having a maximum value of 10.0 amperes is plotted in Figure 21-4. *The effective value is the square root of the mean of the instantaneous values squared* and is frequently called the *root-mean-square*, rms, value. In the example shown, the rms, or effective, value of current is

$$I = \sqrt{\frac{I_{max}^2}{2}} = \sqrt{\frac{100\ A^2}{2}} = \sqrt{50.0\ A^2} = 7.07\ A$$

The rms, or effective, value of an a-c emf is expressed similarly.

In special applications a-c meters can be calibrated to indicate maximum values.

$$\mathcal{E} = 0.707\ \mathcal{E}_{max}$$

Then $$\boldsymbol{\mathcal{E}_{max} = 1.414\ \mathcal{E}}$$

For a potential difference that does not include a source of emf, these expressions become

$$V = 0.707\ V_{max}$$

and $$\boldsymbol{V_{max} = 1.414\ V}$$

Thus a house-lighting circuit rated at 120 volts (effective) varies between +170 volts and −170 volts during each cycle. An appliance that draws 10.00 amperes (effective) when connected across this circuit has a current that varies between +14.14 amperes and −14.14 amperes. The meters used for measurements in a-c circuits are usually designed to indicate effective values rather than maximum values. Representative a-c meter movements are shown in Figure 21-5, (A, B, C, D).

If the load in an a-c circuit is a pure resistance, the current and voltage are in phase and the average power consumed is

$$P = I^2R = VI$$
$$\boldsymbol{P = 120\ V \times 10.00\ A}$$
$$\boldsymbol{P = 1200\ W}$$
$$\boldsymbol{(V \cdot A = (J/C) \times (C/s) = J/s = W)}$$

If, however, the current and voltage are not in phase, *the average power of the circuit is not a simple product of the*

Figure 21-5(A). Moving iron-vane meter. When a magnetizing current is in the actuating coil, a repelling force develops between the fixed and movable iron vanes. The moving vane exerts a force against the restoring spring, and the final pointer position is a measure of the current in the coil. The moving-vane mechanism is used in inexpensive a-c ammeters and voltmeters.

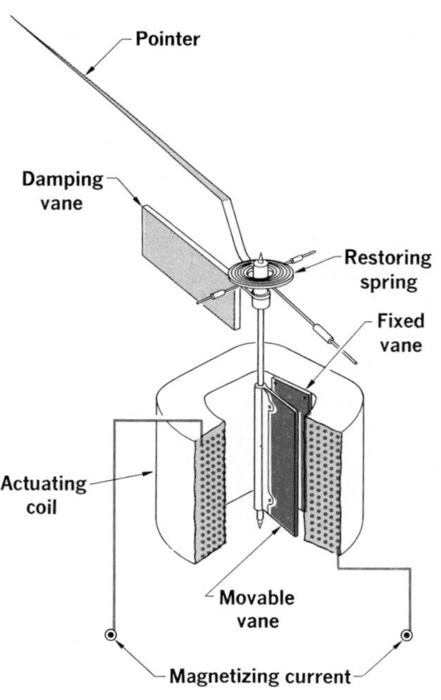

Figure 21-5(B). Hot-wire meter. When a current is carried in the meter circuit, the temperature of the platinum-alloy resistance wire **AB** rises because of I^2R heating. The resulting expansion reduces the tension on wire **CD** and allows the spring to pull thread **EF** to the left. The pulley rotates and moves the pointer slowly to the final deflection. The hot-wire mechanism is sensitive to temperature variations. A low-resistance shunt is added for ammeter functions. A high resistance is added in series with wire **AB** for voltmeter functions.

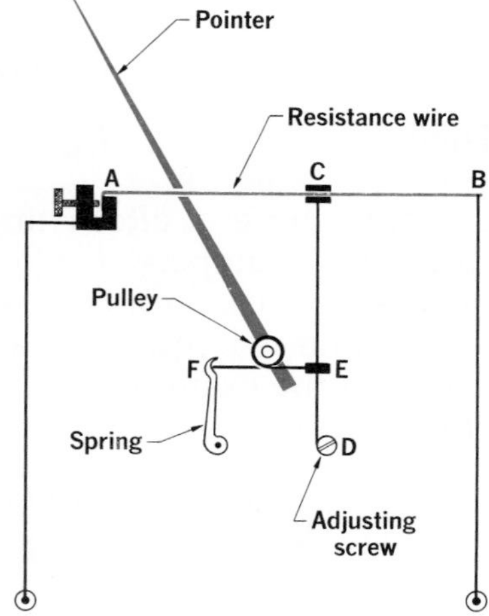

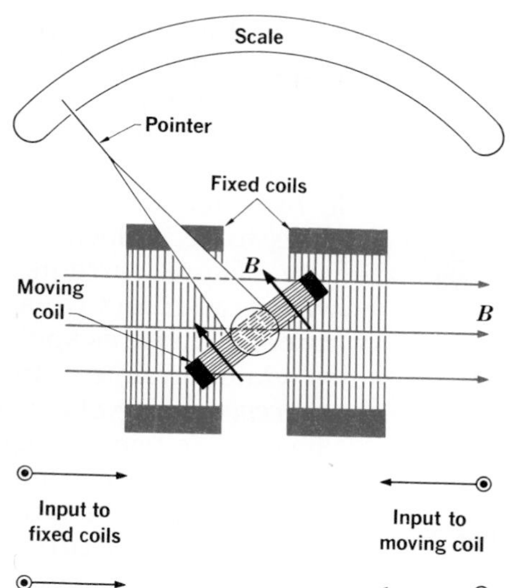

Figure 21-5(C). Electrodynamometer. The electrodynamometer is similar to the galvanometer movement of d-c meters except that it does not include a permanent magnet. The moving coil rotates in the magnetic field of a pair of fixed coils carrying magnetizing current. As a current in the fixed coils produces magnetic flux ***B***, a current in the moving coil produces a flux along its axis and the coil tends to align the two magnetic fields. The torque produces a pointer deflection related to the product of the two coil currents. The dynamometer mechanism is used in high-quality ammeters, voltmeters, and wattmeters.

effective values of voltage and current. The products I^2R and VI are no longer the same; I^2R gives the *actual* power dissipation and VI gives the *apparent* power. This is a basic difference between d-c and a-c circuits, and it stems from the fact that phase differences between current and voltage may be produced by capacitance and inductance in an a-c circuit. These effects have added a new dimension to alternating-current circuits known as *impedance*, which is considered in Section 21.6.

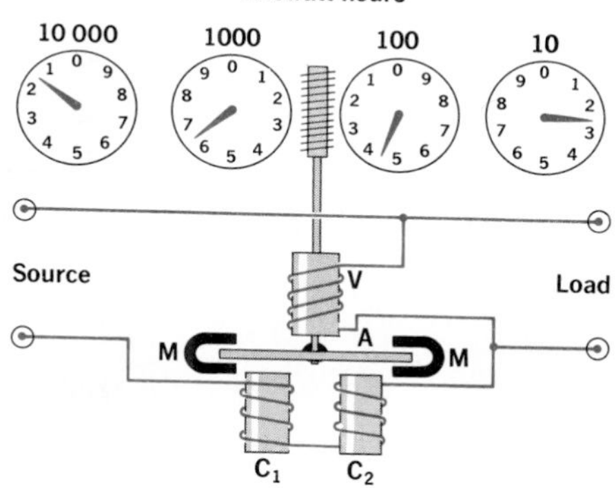

Figure 21-5(D). The induction watt-hour meter is used to record electric energy consumption. It is a small single-phase induction motor that turns at a rate proportional to the power being used. A coil **V** is connected in parallel with the circuit load and acts as a voltage winding. The coils $\mathbf{C_1}$ and $\mathbf{C_2}$ are connected in series with the load and act as current windings. An aluminum disk **A** turns as an induction motor armature to follow the sweeping flux set up by the combination of coils. Permanent magnets **M** induce eddy currents in the rotating disk that oppose the motion of the disk and produce the slippage required for operation.

21.3 Inductance in an a-c Circuit In Section 20.19 *inductance* was described as an inertia-like property that opposes any *change* in current in a coil and causes a counter emf proportional to the rate of change of current to be induced across the coil. Figure 21-6 illustrates the waveforms characteristic of a circuit consisting of pure inductance to which an alternating voltage is applied. The instant the current is passing through zero in a positive direction, it is changing at its maximum rate. Thus the counter emf induced across the coil at this instant must be at its negative maximum value. When the current reaches the positive maximum, the rate of change of current is zero and the induced emf is zero. Again 90° later, the current is changing in a negative direction at its maximum rate and the counter emf is at its positive maximum.

The counter emf follows behind the current inducing it by 90° and is opposite in polarity to the applied voltage, as shown in Figure 21-6. The current in the circuit must then

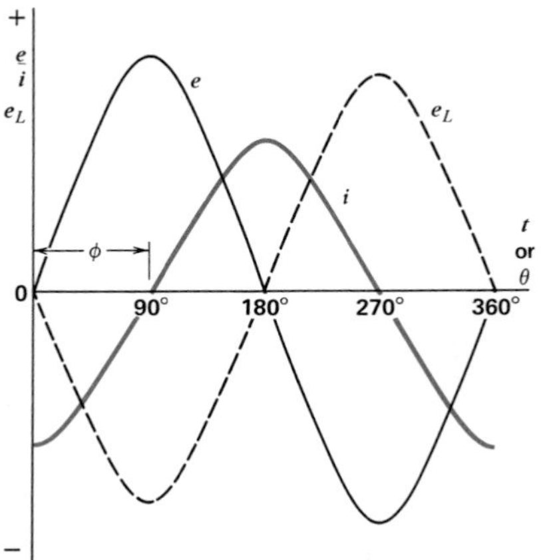

Figure 21-6. Current and voltages in an a-c circuit containing pure inductance.

lag behind the applied voltage by 90°. It is just as appropriate to consider that the voltage leads the current by 90°; the *difference in phase* is 90°.

Alternating currents and voltages do not have *direction* in the sense of vector quantities. They are not representable by directional components as are vector quantities such as force, velocity, momentum, and electric field intensity. These currents and voltages can be regarded, however, as rotors (rotating vectors), or *phasors*, rotating counterclockwise as spokes of a wheel and denoting quantities related in time or phase. *A **phasor** is a representation of the concepts of magnitude and direction in a reference plane.*

Phasors are used in Figure 21-7 to show the phase relation between current and voltage in a pure resistance (**A**) and in a pure inductance (**B**). Phasors usually represent effective values of current and voltage in *phase* diagrams such as those shown in Figure 21-7. These diagrams are analyzed and solved by geometric or vector methods.

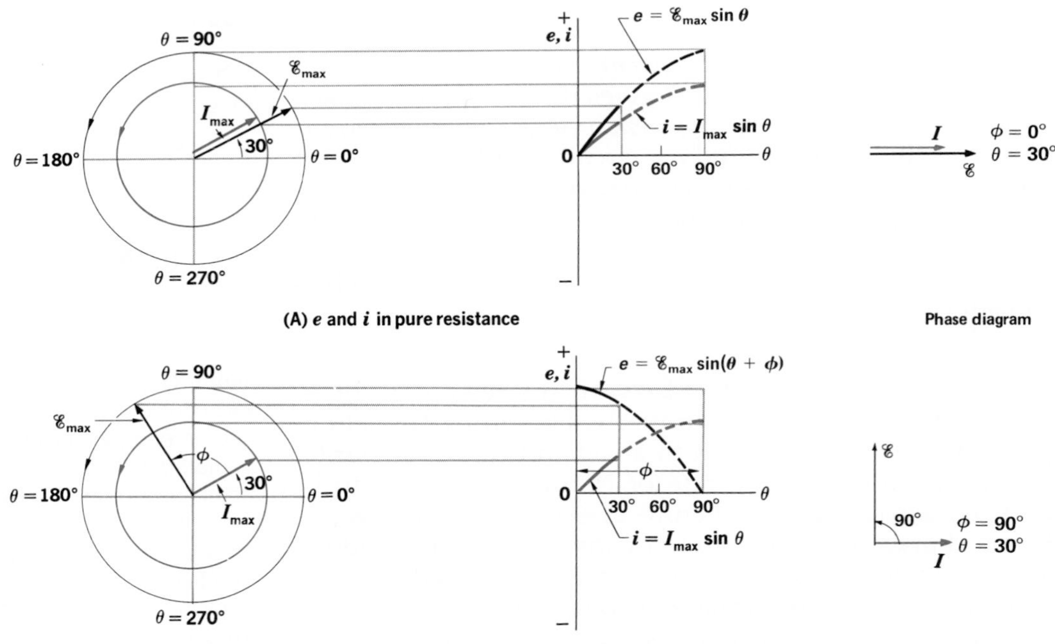

Figure 21-7. Phase relation of an a-c current and voltage in a pure resistance circuit (**A**) and in a pure inductance circuit (**B**).

The *phase angle* between voltage and current is commonly referred to as the angle ϕ (the lowercase Greek letter phi). This relation in an a-c circuit can be expressed as follows:

$$i = I_{max} \sin \theta$$
$$e = \mathscr{E}_{max} \sin (\theta + \phi)$$

The phase angle, ϕ, of a leading voltage is *positive*, and it indicates that the voltage is ahead of the current in phase. Suppose the effective value of voltage represented in Figure 21-7(B) is 10.0 volts. When the displacement angle, θ, is 30°, the instantaneous value e will be

$$e = \mathcal{E}_{max} \sin (\theta + \phi)$$
$$e = \mathcal{E}_{max} \sin (30° + 90°)$$
$$e = \mathcal{E}_{max} \sin 120°$$

But
$$\mathcal{E}_{max} = 1.414\ \mathcal{E} = 1.414 \times 10.0 \text{ V}$$
$$\mathcal{E}_{max} = 14.1 \text{ V}$$

Then
$$e = 14.1 \text{ V} \times \sin 120°$$
$$\sin 120° = \sin 60° = 0.866$$
$$e = 14.1 \text{ V} \times 0.866$$
$$e = 12.2 \text{ V}$$

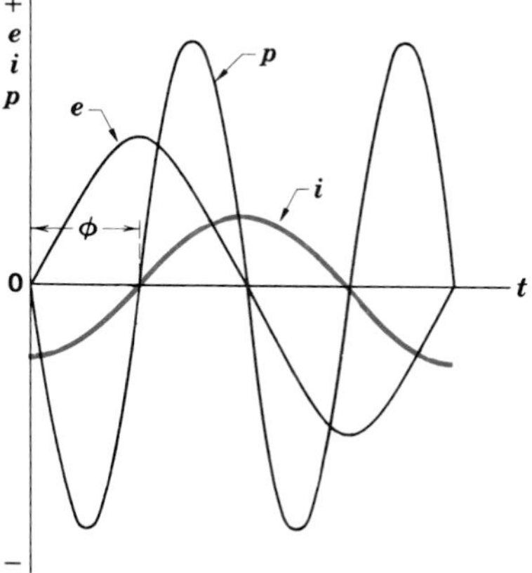

Figure 21-8. Power curve in a pure inductance.

Plotted as a function of time, the products of the instantaneous values of voltage and current in a pure inductance yield an instantaneous power curve as shown in Figure 21-8. Observe that the average power is zero. While the current is changing from zero to a positive maximum, energy is taken from the source and stored in the magnetic field of the *inductor*. These portions of the power curve are positive. When the current is changing from a positive maximum to zero, all the energy stored in the magnetic field is returned to the source. These portions of the power curve are negative. Therefore, the net energy removed from the source during a cycle by a pure inductance is zero; the product of the rms voltage and current can indicate only an *apparent power*.

21.4 Inductance and Resistance An inductor cannot be entirely without resistance since it is not possible to have pure inductance in a circuit. The ordinary resistance of the conductor is an inherent property of any coil. We can treat the resistance of an inductive circuit as a lumped value in series with a pure inductance. This arrangement is shown in Figure 21-9(A).

The same current must be present in all parts of a series circuit. The circuit current is thus used as a reference to show the phase relation between current and voltages in the resistance and inductance. The voltage across the resistance is in phase with the circuit current; this in-phase relation is shown in Figure 21-9(B). The voltage across the inductance leads the circuit current by 90°, as shown in Figure 21-9(C). The voltage across the series combination of R and L then leads the circuit current by some angle between 0° and 90°.

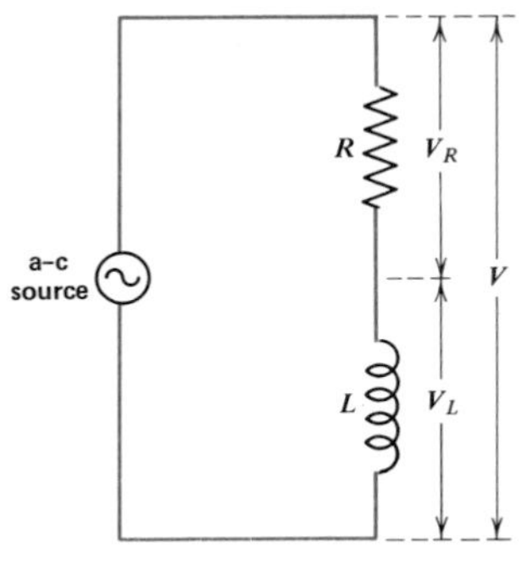

(A) Series circuit

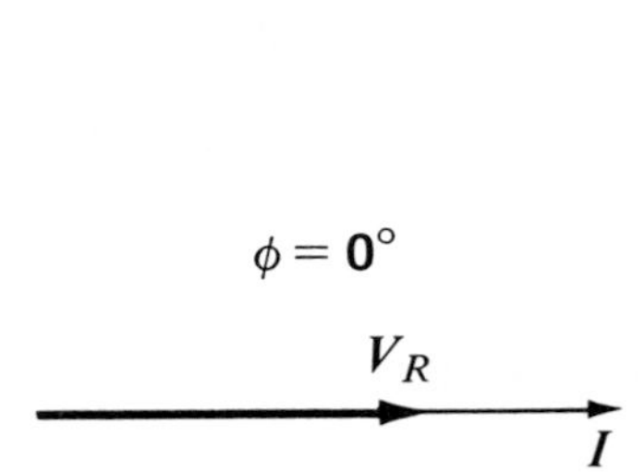

(B) Voltage and current in phase

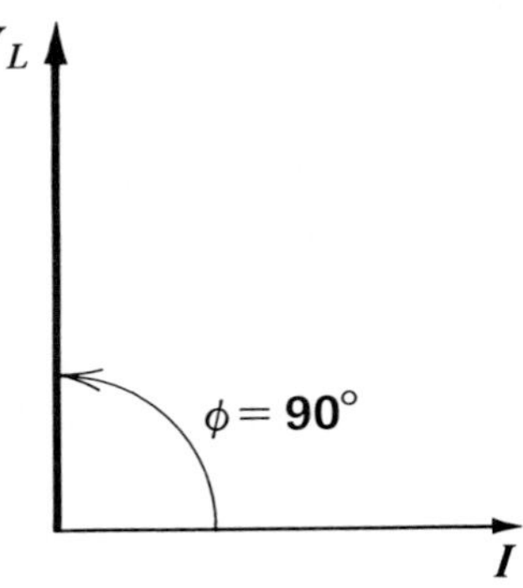

(C) Voltage and current 90° out of phase

Figure 21-9. An ***L-R*** circuit showing the phase relation between voltage and current in the resistance and the inductance.

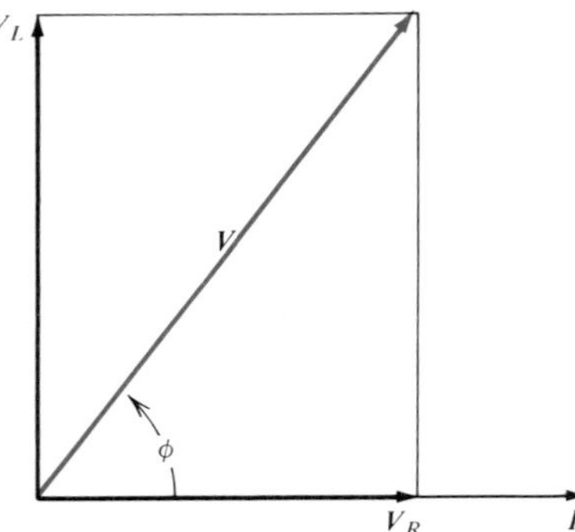

Figure 21-10. A phase diagram showing the addition of voltage phasors in a series circuit containing inductance and resistance.

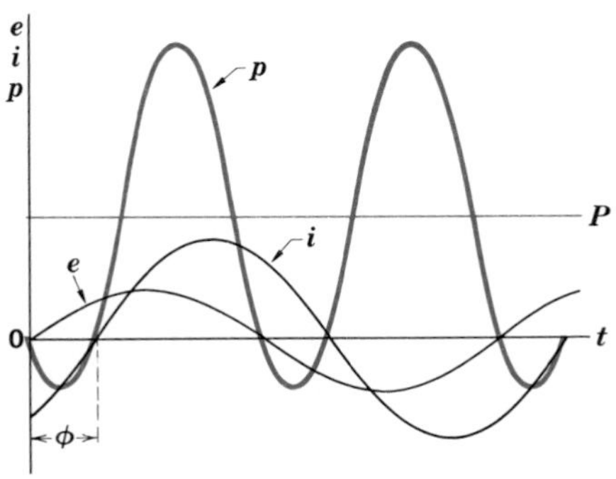

Figure 21-11. Power in an a-c circuit containing resistance and inductance.

With the circuit current of Figure 21-9 as a reference, it is apparent that V_R and V_L have a phase difference of 90°. These two voltages in series cannot be added algebraically but must be added vectorially to give the circuit voltage, V. This voltage V across the external circuit is the *resultant*, or vector sum, of voltages V_R and V_L shown in the phase diagram, Figure 21-10. The phase angle, ϕ, has a positive value less than 90°.

No power is dissipated in the inductance. The average power dissipated in the resistance is equal to the product of the effective values of the circuit current and the voltage across the resistance.

$$P_R = V_R I$$

From the voltage diagram shown in Figure 21-10 it is apparent that

$$V_R = V \cos \phi$$

Thus the actual power consumed in the external circuit is

$$P = VI \cos \phi$$

The total power consumed in the internal and external circuits is, of course,

$$P_T = \mathcal{E}I \cos \phi$$

When the phase angle, ϕ, is *zero* (a pure resistive load), then $\cos \phi = 1$ and the product $VI \cos \phi = VI$. When the phase angle, ϕ, is +90° (a pure inductive load), $\cos \phi = 0$ and the product $VI \cos \phi = 0$ (Figure 21-8). The cosine of ϕ is known as the *power factor*, pf, of an a-c circuit. See Figure 21-11.

$$\text{pf} = \cos \phi = \frac{V_R}{V}$$

Electric power companies strive to maintain a power factor near unity in their distribution lines. Why?

21.5 Inductive Reactance The potential drop across the resistance in the circuit of Figure 21-9 is represented by the phasor V_R in Figure 21-10. This potential difference is equal to the product IR, where R is the common resistance to current in the external circuit. As a consequence of its nonresistive opposition to the current in the circuit, the inductance has a potential difference across it. This opposition is *nonresistive* because no power is dissipated in the inductance and is called *reactance, X*. Reactance is expressed in *ohms*. When reactance is due to inductance, it is referred to as *inductive reactance, X_L*. ***Inductive reactance,*** *X_L, is the ratio of the effective value of inductive potential, V_L, to the effective value of current, I.*

$$X_L = \frac{V_L}{I}$$

Thus the voltage V_L of Figure 21-10 is equal to the product IX_L.

The inductive reactance of a coil is directly proportional to the inductance and to the frequency of the current in the circuit. This is true because the rate of change of a given current increases with frequency. The larger the inductance, the greater is the opposition to the change in current at a given frequency.

$$X_L = 2\pi fL$$

When f is in hertz and L is in henrys, X_L is expressed in ohms of inductive reactance; 2π is a proportionality constant.

21.6 Impedance The vector sum of IX_L and IR is equal to the voltage, V, applied to the external circuit, as in Figure 21-10. This potential difference is the product of the circuit current, I, and the combined effect of the inductive reactance, X_L, and the resistance, R, in the load. The joint effect of reactance and resistance in an a-c circuit is called *impedance, Z*. It is expressed in ohms. Thus V is equal to the product IZ. See Figure 21-12(A). If each voltage is divided by the current I, IX_L yields inductive reactance X_L, IR yields resistance R, and IZ yields impedance Z. This is illustrated in the *impedance diagram* that is shown in Figure 21-12(B).

The tangent of the phase angle ϕ is defined by the ratio of the side opposite ϕ to the side adjacent to ϕ. In the impedance diagram these are the sides X and R, respectively, of the right triangle. Therefore

$$\frac{X}{R} = \tan \phi$$

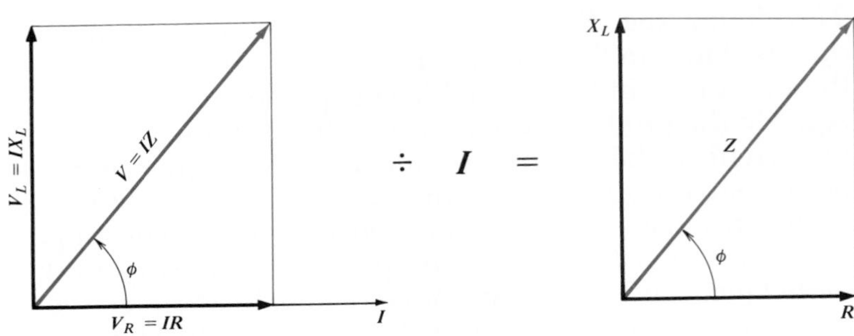

Figure 21-12. The derivation of an impedance diagram of an ***L-R*** series circuit.

If reactance is inductive, the phase angle is positive.

This equation gives X/R *as a function of* ϕ. It also enables us to determine ϕ *as a function of* X/R; that is, ϕ *is an angle whose tangent is* X/R. This expression, "ϕ is an angle whose tangent is X/R," gives the *inverse function* to the tangent in the equation. The inverse function to the tangent is called the **arc tangent.** The symbol for arc tangent is **arctan.** The inverse function to the tangent of the phase angle ϕ in the above equation is commonly written as

The term "arctan" is sometimes written as "tan^{-1}." They are equivalent notations for the arc tangent.

$$\phi = \text{arctan}\,\frac{X}{R}$$

The impedance, Z, has a magnitude equal to $\sqrt{R^2 + X^2}$ and a phase angle ϕ (direction angle with respect to R) whose tangent is X/R. It can be written in the polar form $Z\underline{|\phi}$ as

Relate $Z = \sqrt{R^2 + X^2}$ *to the Pythagorean theorem of geometry.*

$$Z = \sqrt{R^2 + X^2}\ \underline{|\text{arctan } X/R}$$

When the reactance is inductive, $X = X_L$ and ϕ is a positive angle whose tangent is X_L/R. In this polar expression, $\sqrt{R^2 + X^2}$ yields the *magnitude* of the impedance and the symbol $\underline{|\quad}$ is interpreted to mean *"at an angle of* — °*"*

We can now rewrite the Ohm's-law expression for d-c circuits in a form that applies to a-c circuits. For the entire circuit:

$$\mathscr{E} = IZ$$

For the external circuit:

$$V = IZ$$

EXAMPLE A coil has a resistance of 35 Ω and an inductance of 0.14 H. The coil is connected to a 60-Hz, 117-V power source. Calculate (a) the circuit current, (b) the phase angle, and (c) the power used in the external circuit. (See Figures 21-9 and 29-10.)

Given	Unknown	Basic equations
$R = 35\ \Omega$	I	$V = IZ$
$L = 0.14$ H	ϕ	$X_L = 2\pi fL$
$f = 60$ Hz	P	$Z = R^2 + X^2$
$V = 117$ V		$\phi = \arctan \dfrac{X}{R}$
		$P = VI \cos \phi$

Solution

(a) Substitute the value of X_L from the second into the third Basic equation; then substitute the value for Z into the first Basic equation. Solving for I gives the

$$\textit{Working equation: } I = \frac{V}{\sqrt{R^2 + (2\pi fL)^2}}$$

$$= \frac{117\text{ V}}{\sqrt{(35\ \Omega)^2 + [2(3.14)(6\bar{0}\text{ Hz})(0.14\text{ H})]^2}}$$

$$= 1.8\text{ A}$$

(b) $$\textit{Working equation: } \phi = \arctan \frac{2\pi fL}{R}$$

$$= \arctan \frac{2(3.14)(6\bar{0}\text{ Hz})(0.14\text{ H})}{35\ \Omega}$$

$$= \arctan 1.5$$

$$= 56^\circ$$

(c) $$\textit{Working equation: } P = VI \cos \phi$$

$$= (117\text{ V})(1.8\text{ A}) \cos 56^\circ$$

$$= 120\text{ W}$$

21.7 Capacitance in an a-c Circuit A capacitor in a d-c circuit charges to the applied voltage and effectively opens the circuit. The potential difference across a capacitor can change only as the charge on the capacitor changes. Since $Q = CV$, the charge Q, in coulombs, is directly proportional to the potential difference V, in volts, for a given capacitor. Thus we can refer to the charge in terms of the voltage across the capacitor.

Because $Q \propto V$, we describe the charge on a capacitor in terms of its voltage.

An uncharged capacitor offers no opposition to a charging current from a source of emf. However, as a charge builds up on the capacitor plates, a voltage develops across the capacitor that opposes the charging current. When the charge is such that the voltage across the capacitor is equal to the applied voltage, the charging current must be zero. Why? These two conditions are illustrated in Figure 21-13. We can conclude that *the current in a capacitor circuit is maximum when the voltage across the capacitor is zero and is zero when the capacitor voltage is maximum.*

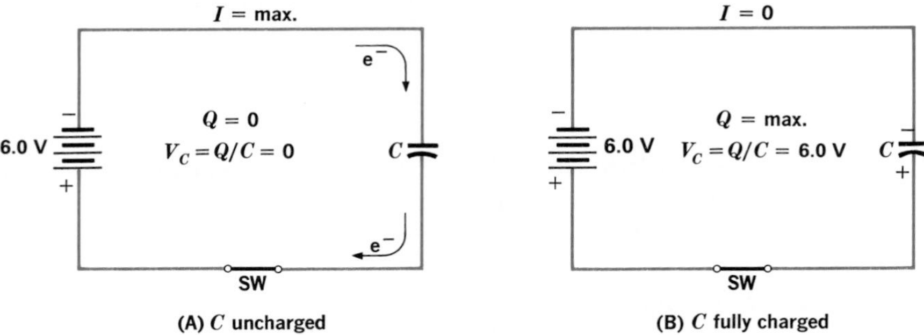

Figure 21-13. As a capacitor is charged, a voltage develops across it that opposes the charging current.

Let us examine the effect of pure capacitance in an a-c circuit. When the initial charging current is a positive maximum, the voltage across the capacitor plates is zero. As the capacitor charges, the potential difference builds up and the current decays; the charging current reaches zero when the capacitor voltage is maximum.

As the charging current reverses direction and increases toward a negative maximum, electrons flow from the negative plate of the capacitor through the circuit to the positive plate, thereby removing the charge. The capacitor then charges in the opposite sense, and the potential difference reaches a maximum with opposite polarity as the current returns to zero. Thus *the voltage across the capacitor lags the circuit current by 90°*. See Figure 21-14.

The phase angle, ϕ, in an a-c circuit with a pure capacitance load is $-90°$, and the circuit has a *lagging* voltage.

Figure 21-14. Pure capacitance in an a-c circuit.

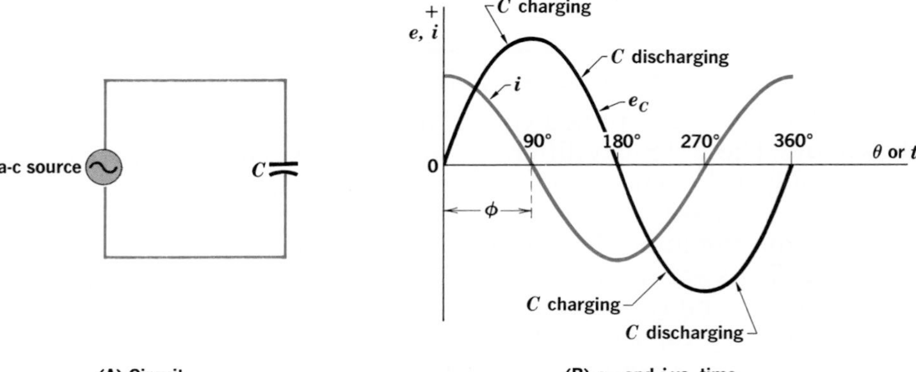

The phase relationship is expressed as

$$i = I_{max} \sin \theta \qquad \text{and} \qquad e = \mathscr{E}_{max} \sin (\theta + \phi)$$

These expressions are similar to those given for i and e in Section 21.3. However, the phase angle of a lagging voltage is *negative*. Hence the sign of the phase angle, ϕ, must be introduced properly.

Suppose the effective value of voltage across a capacitor is 10.0 volts. When the displacement angle, θ, is 30°, the instantaneous value of e is

$$e = \mathscr{E}_{max} \sin (\theta + \phi)$$
$$e = \mathscr{E}_{max} \sin [30° + (-90°)]$$
$$e = \mathscr{E}_{max} \sin (-60°)$$

Now $\mathscr{E}_{max} = 1.414\, \mathscr{E} = 1.414 \times 10.0 \text{ V}$

$$e = 14.1 \text{ V} \times \sin (-60°)$$

and $\sin (-60°) = -0.866$

Therefore $e = 14.1 \text{ V} \times (-0.866)$

$$e = -12.2 \text{ V}$$

Compare this result with the similar computation in Section 21.3.

Plotted as a function of time, the product of the instantaneous values of current and voltage in a circuit of pure capacitance results in the instantaneous power curve shown in Figure 21-15. Similar to a circuit with pure inductance, the average power is zero. As the current changes from each maximum to zero, energy is taken from the source of emf and stored in the electric field between the capacitor plates. All this energy is returned to the source as the current rises from zero to either maximum. Thus the product of the effective values of current and voltage indicates an *apparent power*. The phase angle of $-90°$ yields a power factor equal to zero [$\cos (-90°) = 0$], and so the *actual power* dissipated must be zero.

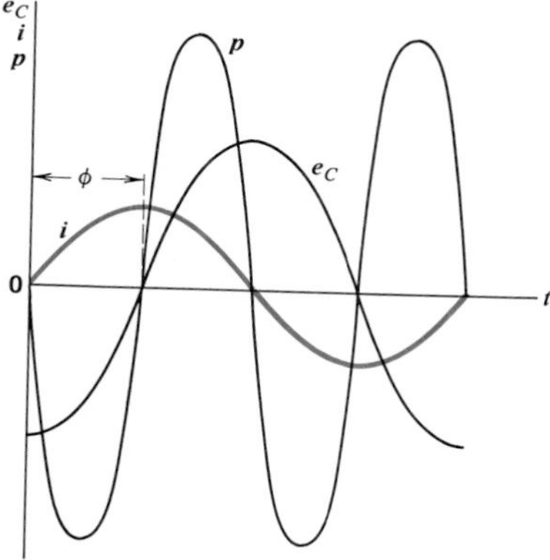

Figure 21-15. Power curve in a pure capacitance.

21.8 Capacitance and Resistance

In an a-c circuit the opposition to the current due to capacitance is *nonresistive* since no power is dissipated in the capacitor. This nonresistive opposition is called *capacitive reactance*. The symbol for capacitive reactance is X_C and the unit is the *ohm*.

If reactance is capacitive, the phase angle is negative.

The inherent resistance in any practical circuit containing capacitance can be represented in series with X_C. An *R-C* series circuit is shown in Figure 21-16. The potential difference, V, across the external circuit is the vector sum of V_C and V_R; the series current is used as the reference. There is, of course, no electron flow through the capacitor;

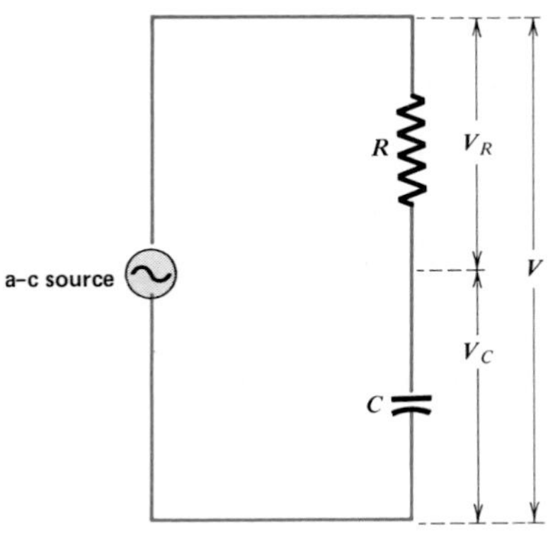

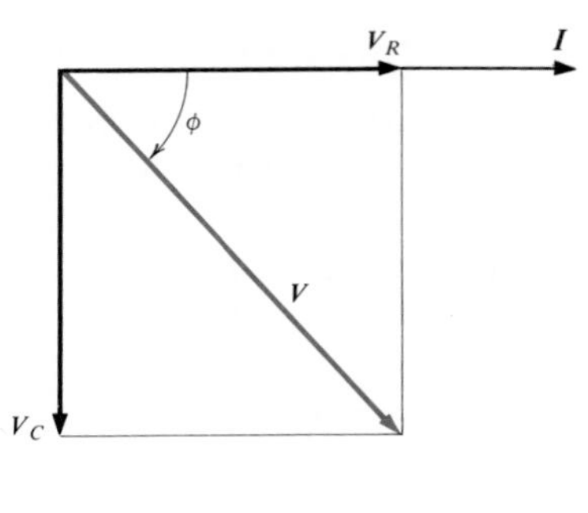

Figure 21-16. An ***R-C*** series circuit showing the phase relations of voltages V_R and V_C with the circuit current.

it is alternately charged, first in one sense and then in the other, as the charging current alternates. The phase angle, ϕ, is negative, which is characteristic of a lagging voltage.

The corresponding impedance diagram can be produced by dividing each voltage phasor by the circuit current. See Figure 21-17. Just as V_C is negative with respect to V_L in Figure 21-12(B), so X_C is plotted in the opposite direction to X_L.

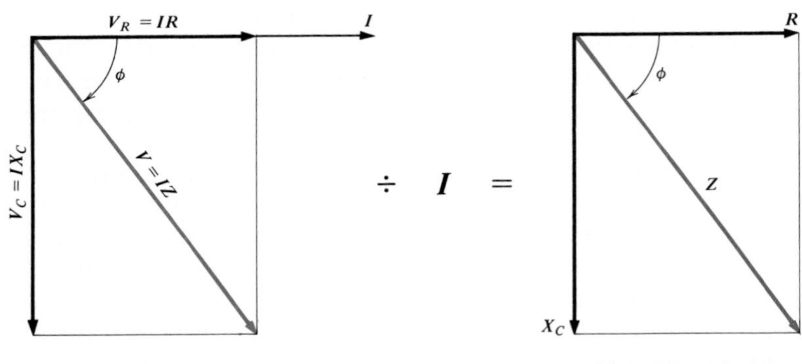

Figure 21-17. The derivation of an impedance diagram of an ***R-C*** series circuit.

Because $V_C = Q/C$, the higher the capacitance C, the lower is the potential difference V_C that develops across the capacitor for a given charge Q. *Thus the capacitive reactance to a circuit current varies inversely with the capacitance in the circuit.* As the frequency of a charging current is increased, a shorter time is available during each current cycle for the capacitor to charge and discharge. Smaller changes in voltage occur across the capacitor. *Hence the capacitive reactance to the circuit current varies inversely with the current frequency.*

$$X_C = \frac{1}{2\pi fC}$$

When f is in hertz and C is in farads, the capacitive reactance, X_C, is expressed in ohms.

The impedance of a circuit containing capacitance and resistance in series has a magnitude equal to $\sqrt{R^2 + X^2}$ and a phase angle, ϕ, whose tangent is $-X_C/R$. The impedance, in its general polar form $Z \underline{|\phi}$, is expressed as $\sqrt{R^2 + X^2} \underline{|\arctan X/R}$, where $X = -X_C$. See the following example.

EXAMPLE A 115-Ω resistor and a 30.0-μF capacitor are connected in series and then plugged into a 50.0-Hz, 117-V power source. Calculate (a) the magnitude and phase angle of the circuit impedance, (b) the circuit current, and (c) the voltage across the capacitor. (See Figures 21-16 and 21-17).

Given	Unknown	Basic equations
$R = 115\ \Omega$	Z	$Z = \sqrt{R^2 + X^2} \underline{|\arctan X/R}$
$C = 30.0 \times 10^{-6}$ F	ϕ	$X_C = \dfrac{1}{2\pi f C}$
$f = 50.0$ Hz	I	$V = IZ$
$V = 117$ V	V_C	$\phi = \arctan \dfrac{-X_C}{R}$

Solution

(a) *Working equation 1:*

$$Z = \sqrt{R^2 + \left(\frac{1}{2\pi fC}\right)^2}$$

$$= \sqrt{(115\ \Omega)^2 + \left(\frac{1}{2(3.14)(50.0\text{ Hz})(30.0 \times 10^{-6}\text{ F})}\right)^2}$$

$$= \sqrt{(115\ \Omega)^2 + (106\ \Omega)^2}$$

$$= 156\ \Omega$$

Working equation 2:

$$\phi = \arctan \frac{-X_C}{R}$$

$$= \arctan \frac{-106\ \Omega}{115\ \Omega}$$

$$= -42.7^\circ$$

(b) *Working equation:*

$$I = \frac{V}{Z} = \frac{117\text{ V}}{156\ \Omega} = 0.750\text{ A}$$

(c) *Working equation:* $V_C = IX_C$

$$= (0.750\ \text{A})(106\ \Omega)$$

$$= 79.5\ \text{V}$$

PRACTICE PROBLEM A 0.500-μF capacitor and a $30\bar{0}$-Ω resistor are connected in series across a $30\bar{0}$-V, 1.00-kHz source. (a) What is the load impedance? (b) Determine the magnitude of the circuit current. (c) What is the potential difference across the capacitor? (d) What is the potential difference across the resistor? (e) What power is expended in the resistor? (f) Demonstrate that the correct answer unit in part (e) is the watt. *Ans.* (a) 437 $\Omega \angle -46.7°$; (b) 0.686 A; (c) 218 V; (d) 206 V; (e) 141 W; (f) Students' proof

---

21.9 L, R, and C in Series Inductance produces a leading voltage and a positive phase angle in an a-c circuit. Capacitance produces a lagging voltage and a negative phase angle. Of course, the voltage across a resistance must be in phase with the current in the resistance.

If reactance is zero, the phase angle is zero.

Suppose values of L, R, and C are connected in series across an a-c source of emf as in Figure 21-18(A). The *total reactance* of the circuit is

$$X = X_L - X_C$$

since X_L and X_C are 180° apart.

The impedance has the general form $Z\angle\phi$.

$$Z = \sqrt{R^2 + X^2}\ \angle \arctan X/R$$

However, X can be either positive or negative (X^2 is always positive) since

$$X = X_L - X_C = 2\pi fL - \frac{1}{2\pi fC}$$

Therefore the magnitude of the impedance can be expressed as

$$Z = \sqrt{R^2 + (X_L - X_C)^2}$$

$$Z = \sqrt{R^2 + \left(2\pi fL - \frac{1}{2\pi fC}\right)^2}$$

Suppose the resistance R in Figure 21-18(A) is 16.0 Ω and the a-c source is a 60.0-Hz generator that applies a potential difference of $10\bar{0}$ V across the external circuit. At this frequency, X_L is determined to be 22.0 Ω and X_C to be 10.0 Ω. The impedance, Z, of the circuit is calculated from these known values of R, X_L, and X_C as illustrated in the following example.

The calculation of X_L is shown in Section 21.5.

The calculation of X_C is shown in Section 21.8.

$$Z = \sqrt{R^2 + X^2}\ \angle \arctan X/R$$

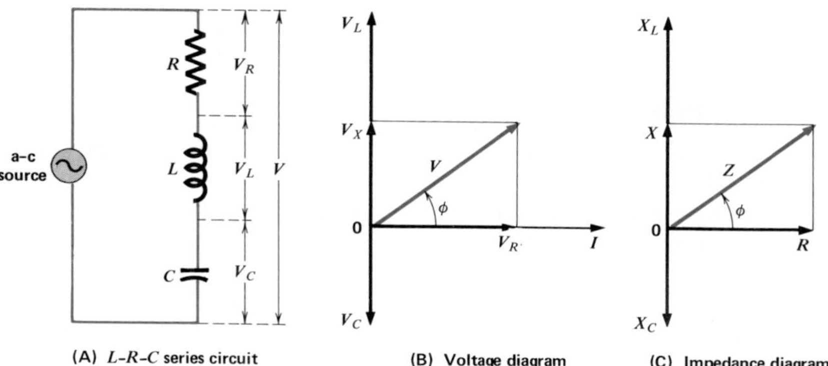

(A) *L-R-C* series circuit (B) Voltage diagram (C) Impedance diagram

Figure 21-18. An ***L-R-C*** series circuit with voltage and impedance diagrams.

But

$$X = X_L - X_C = 22.0\ \Omega - 10.0\ \Omega = 12.0\ \Omega$$
$$Z = \sqrt{(16.0\ \Omega)^2 + (12.0\ \Omega)^2}\ \underline{|\arctan 12.0\ \Omega/16.0\ \Omega}$$
$$Z = 20.0\ \Omega\ \underline{|37.0^\circ}$$

By calculating $Z\underline{|\phi}$ we found that the magnitude of the impedance is 20.0 Ω and its direction angle, ϕ, is 37.0° positive with respect to R. The impedance diagram is shown in Figure 21-18(C). This angle ϕ is positive because X_L is larger than X_C. Thus the voltage across the external circuit must *lead* the circuit current, I, by an angle of 37.0°. This voltage was given as $10\bar{0}$ V. Therefore the magnitude of the current is

Positive phase angle: voltage leads current.
Negative phase angle: voltage lags current.

$$I = \frac{V}{Z} = \frac{10\bar{0}\ \text{V}}{20.0\ \Omega} = 5.00\ \text{A}$$

Knowing the circuit current, we can determine the magnitudes of the voltages across L, R, and C.

$$V_L = IX_L = 11\bar{0}\ \text{V}$$
$$V_R = IR = 80.0\ \text{V}$$
$$V_C = IX_C = 50.0\ \text{V}$$

Observe that the vector sum of V_L, V_R, and V_C must be equal to V.

From the voltage diagram of Figure 21-18(B) it is evident that the magnitude of the voltage V across the external circuit is

$$V = \sqrt{V_R{}^2 + V_X{}^2}$$

where
$$V_X = V_L - V_C$$

Then
$$V = \sqrt{(80.0\ \text{V})^2 + (60.0\ \text{V})^2}$$
$$V = 10\bar{0}\ \text{V}$$

This result is in agreement with the value of the external circuit voltage V assumed at the beginning of this circuit-computation exercise. The direction angle of V must be 37.0° since this is the phase angle, ϕ, using I as the reference. This example of a series L-R-C circuit illustrates the fact that the magnitude of the voltage across a reactance in an a-c circuit can exceed the magnitude of the applied voltage across the entire external circuit.

Resistive and reactive components may also be connected in parallel in practical a-c circuits. The analysis of their parallel impedance networks is beyond the scope of this book.

QUESTIONS: GROUP A

1. What is the significant relationship between an alternating current and alternating voltage that are in phase?
2. In a-c circuits what type of load has a current in phase with the voltage across it?
3. What is meant by instantaneous power?
4. Upon what property of an alternating current is its effective value based?
5. Define the effective value of an alternating current.
6. What is the phase relation between (a) the current in a pure inductance and the applied voltage, (b) the current and the induced voltage, (c) the applied voltage and the induced voltage?
7. Express power factor (a) in terms of phase angle, (b) in terms of V and V_R. (c) What is the significance of the power factor pertaining to an a-c circuit?
8. Why is reactance described as being nonresistive?
9. (a) Define impedance. (b) Draw an impedance diagram for an inductor with reactance X_L and resistance R. (c) Draw an impedance diagram for a capacitor with reactance X_C and resistance R.
10. Describe the difference in the performance of a capacitor in a d-c circuit and in an a-c circuit.
11. What is the power factor of an a-c circuit containing (a) pure inductance, (b) pure resistance, (c) pure capacitance?
12. Express Ohm's law for an a-c circuit (a) that includes a source of emf, (b) that does not include a source of emf.

GROUP B

13. Explain why the instantaneous power curve for current in a resistive load varies between some positive maximum value and zero.
14. Demonstrate algebraically that the heating effect of an electric current is proportional to the effective value of current squared.
15. (a) Plot the current and current squared curves for an alternating current having a maximum value of 5.7 amperes. Use cross-section paper and a suitable scale for the coordinate axes to produce a graph similar to Figure 21-4. (b) Show that the rms value of the current is 4.0 amperes.
16. If an a-c circuit has a pure inductance load, the average power delivered to the load is zero. Does this mean that there is no energy transfer between the source and the load? Explain.

17. (a) Considering the expression for X_L in terms of frequency and inductance, draw a rough graph of X_L as a function of frequency for a given inductance. (b) What is the nature of the curve? (c) What is the significance of the fact that X_L is zero when f is zero, regardless of the value of L?

18. Considering the expression for X_C in terms of frequency and capacitance, what is the significance of the fact that when the frequency is zero, X_C is infinitely high for any capacitance?

19. (a) In an L-R-C series circuit operating at a certain frequency, what determines whether the load is inductive or capacitive? (b) Suggest a circumstance in which the load would be resistive.

20. What is the significance of the negative portions of the power curve shown in Figure 21-15?

PROBLEMS: GROUP A

Note: Compute angles to the nearest 0.5 degree.

1. A current in an a-c circuit measures 5.5 A. What is the maximum instantaneous magnitude of this current?

2. A capacitor has a voltage rating of $45\bar{0}$ V maximum. What is the highest rms voltage that can be impressed across it without danger of dielectric puncture?

3. The emf of an a-c source has an effective value of 122.0 V. Determine the instantaneous value when the displacement angle, θ, is 50.0°.

4. An alternating current in a 25.0-Ω resistance produces heat at the rate of $25\bar{0}$ W. What is the effective value (a) of current in the resistance, (b) of voltage across the resistance?

5. An inductor has an inductance of 2.20 H and a resistance of $22\bar{0}$ Ω. (a) What is the reactance if the a-c frequency is 25.0 Hz? (b) Draw the impedance diagram and determine graphically the magnitude and phase angle of the impedance.

6. A 2.00-μF capacitor is connected across a 60.0-Hz line and a current of 167 mA is indicated. (a) What is the reactance of the circuit? (b) What is the voltage across the line?

7. A capacitance of 2.65 μF is connected across a $12\bar{0}$-V line and is found to draw $12\bar{0}$ mA. What is the frequency of the source?

8. When a resistance of 4.0 Ω and an inductor of negligible resistance are connected in series across a 110-V, $6\bar{0}$-Hz line, the current is $2\bar{0}$ A. What is the inductance of the coil?

GROUP B

9. A 0.50-μF capacitor and a $3\bar{0}$-Ω resistor are connected in series across a source of emf whose frequency is 8.0×10^3 Hz. (a) Draw the circuit diagram. (b) Calculate the capacitive reactance. (c) Draw the impedance diagram and determine the magnitude and phase angle of the impedance.

10. The current in the circuit of Problem 9 is found to be $5\bar{0}$ mA. (a) What is the magnitude of the voltage across the series circuit? (b) What is the phase relation of this voltage to the circuit current? (c) What is the voltage across the capacitor? (d) What is the voltage across the resistor?

11. A 60.0-Hz circuit has a load consisting of resistance and inductance in series. A voltmeter, ammeter, and wattmeter, properly connected in the circuit, read respectively 117 V, 4.75 A, and $40\bar{0}$ W. (a) Determine the power factor. (b) What is the phase angle? (c) What is the resistance of the load? (d) What is the inductive reactance of the load? (e) Determine the voltage across the resistance. (f) Determine the voltage across the inductance. (g) Draw the voltage diagram with the circuit current as the reference phasor. What is the applied voltage as found graphically and in polar form?

12. A coil has a resistance of $9\bar{0}\ \Omega$ and an inductance of 0.019 H. (a) What is the impedance (complex) at a frequency of 1.0×10^3 Hz? (b) Determine the magnitude of the current when a potential difference of 6.0 V at this frequency is applied across it? (c) How much power is delivered to the coil?

13. A capacitance of $5\bar{0}\ \mu$F and a resistance of $6\bar{0}\ \Omega$ are connected in series across a 120-V, $6\bar{0}$-Hz line. (a) What is the magnitude of current in the circuit? (b) What power is dissipated? (c) What is the power factor? (d) Determine the voltage across the resistance. (e) Determine the voltage across the capacitance. (f) Draw the voltage diagram with the circuit current as the reference phasor. What is the applied voltage found graphically? Express it in polar form.

14. An inductance of 4.8 mH, a capacitance of 8.0 μF, and a resistance of $1\bar{0}\ \Omega$ are connected in series, and a 6.0-V, 1.0×10^3-Hz signal is applied across the combination. (a) What is the magnitude of the impedance of the series circuit? (b) What is the phase angle? (c) Determine the magnitude of the current in the circuit. (d) Find the potential drop across each component of the load, and draw the voltage diagram.

15. An inductor, a resistance, and a capacitor are connected in seriess across an a-c circuit. A voltmeter reads $9\bar{0}$ V when connected across the inductor, 16 V across the resistor, and 120 V across the capacitor. (a) What will the voltmeter read when placed across the series circuit? (b) Draw the voltage diagram, and graphically verify your answer to (a). (c) Compute the power factor.

RESONANCE

21.10 Inductive Reactance vs. Frequency An inductor in a d-c circuit has no effect on the steady, direct current except for that due to the ordinary resistance of the wire forming the coil. A steady, direct current can be thought of as an alternating current having a frequency of zero Hz. Since $X_L = 2\pi fL$, inductive reactance in the d-c circuit must be zero.

An alternating current changes from its positive maximum to its negative maximum in a time equal to one-half the period, T; if the frequency of this current is increased, the time rate of change of current is increased accordingly. See Figure 21-19. The inductive reactance presented to the current in an a-c circuit by an inductor increases as the frequency of this current is increased.

Consider the reactance of a certain inductance over a range of frequencies. A plot of the values of inductive reactance as ordinates against frequencies as abscissas produces the linear curve shown in Figure 21-20. Any value of

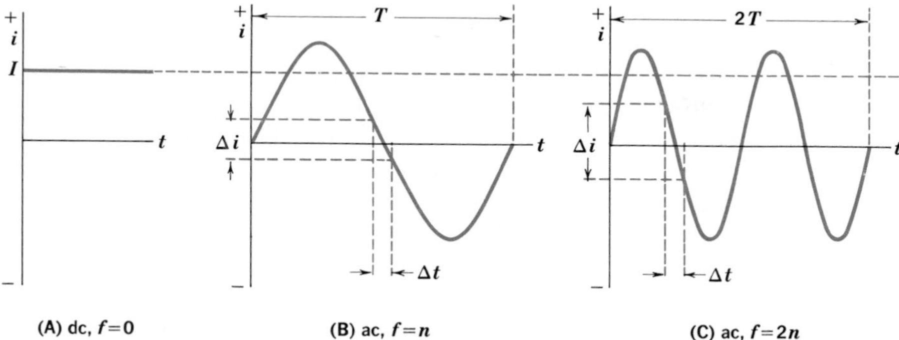

Figure 21-19. The time rate of change of current increases with frequency.

inductance yields a reactance curve that starts at the origin of the coordinate axes; the larger the value of inductance, the greater is the slope of the curve.

21.11 Capacitive Reactance vs. Frequency When a capacitor in a d-c circuit is charged to the applied voltage, it acts as an open switch and the current in the circuit is zero. While the capacitor remains in this fully charged state, the capacitive reactance in the circuit is infinitely high regardless of the value of C.

A capacitor in an a-c circuit has a very different effect on the alternating current and voltage. Electrons flow in one direction during one half-cycle, charging the capacitor in one sense. The electrons flow in the opposite direction during the other half-cycle, reversing the charge on the capacitor.

If the current has a very low frequency, the capacitive reactance is very large since $X_C = 1/(2\pi fC)$. This large reactance effectively limits the circuit current. At higher frequencies, less change occurs in the charge on the capacitor during each half-cycle and the reactive opposition to the circuit current is lower. At extremely high frequencies, the change in capacitor charge becomes negligible and the capacitive reactance in the circuit becomes a negligible factor in limiting the circuit current. The larger the capacitor in the circuit, the more rapidly the capacitive reactance decreases with increasing frequency. Observe that in the above equation for X_C, the magnitude of X_C varies inversely with frequency f, with capacitance C, and with their product fC.

In Figure 21-21 values of reactance for a given capacitor are plotted as a function of frequency. The slope of the curve is hyperbolic, which is typical of the inverse relation between capacitive reactance, X_C, and frequency, f. A larger capacitor will yield a reactance-frequency curve with a greater slope than that shown.

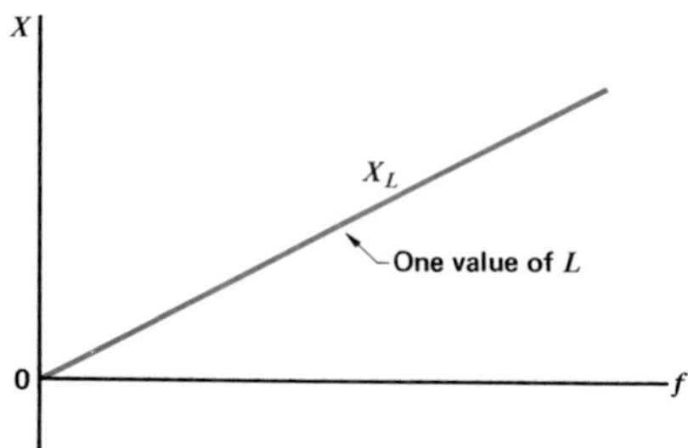

Figure 21-20. The reactance of a given inductor plotted as a function of frequency.

X_C for a given capacitor varies inversely with frequency.

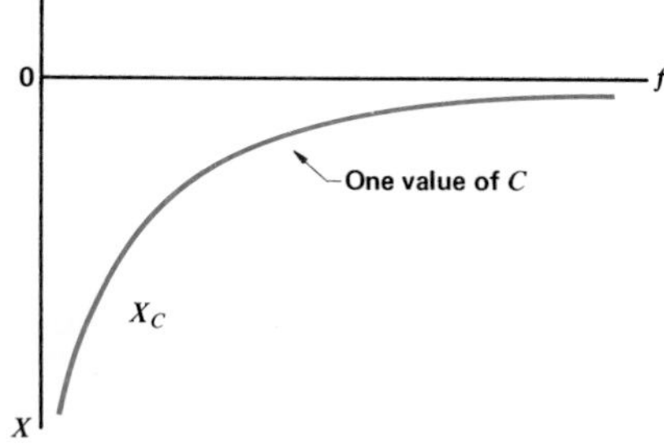

Figure 21-21. The reactance of a given capacitor plotted as a function of frequency. X_C is plotted as $-[1/(2\pi fC)]$ for comparison with X_L in Figure 21-20.

21.12 Series Resonance In Section 21.9 it was stated that the voltages across an inductance and a capacitance in a series circuit are of opposite polarity. If X_L is larger than X_C, the circuit is inductive and the voltage across the circuit leads the current. If X_C is the larger reactance, the circuit is capacitive and the circuit voltage lags behind the current. In either instance the impedance, Z, has a magnitude $\sqrt{R^2 + (X_L - X_C)^2}$ and a phase angle, ϕ, whose tangent is $(X_L - X_C)/R$.

The frequency of an a-c voltage applied to a series circuit with fixed values of L, R, and C determines the reactance of the circuit. If Figures 21-20 and 21-21 are combined, they will show that the reactance of the circuit is inductive over a range of high frequencies and capacitive over a range of low frequencies. At some intermediate frequency, the inductive and capacitive reactances are equal; the reactance of the circuit, $X_L - X_C$, equals zero. At this frequency the impedance, Z, is at its minimum value for the series circuit. It is equal to the inherent circuit resistance, R. As is true of all circuits with pure ohmic resistance, the voltage across the circuit and the current in the circuit are in phase. This special circuit condition is known as *series resonance*. The frequency at which resonance occurs is called the *resonant frequency*, f_R. ***Series resonance** is a condition in which the impedance of an L-R-C series circuit is equal to the circuit resistance, and the voltage across the circuit is in phase with the circuit current.*

Suppose an inductor of $\bar{1}\bar{0}$-millihenrys inductance and 25-ohms resistance is placed in series with an ammeter and a capacitor of 1.0-microfarad capacitance. A signal of $1\bar{0}$ volts rms from a variable frequency source, such as an audio signal generator, is applied. The circuit is shown in Figure 21-22. When the frequency of the applied signal is

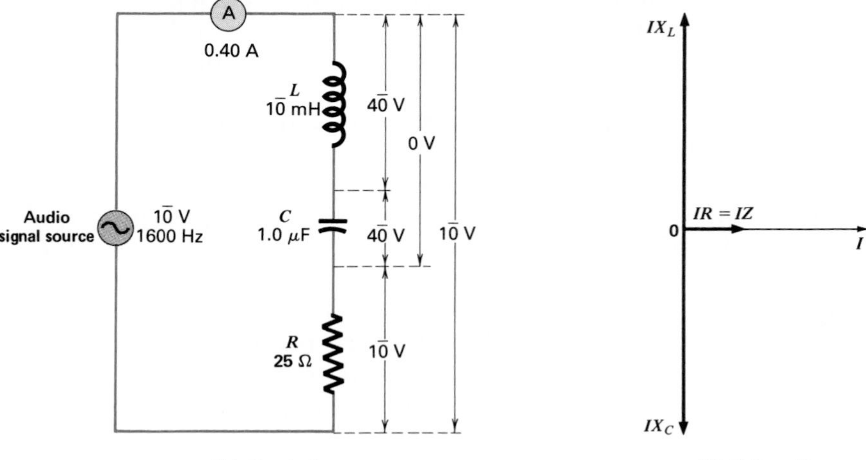

Figure 21-22. An ***L-R-C*** series circuit that has a resonant frequency of 1600 Hz.

varied through the audio range, a distinct rise in the current occurs as the frequency approaches 1600 Hz. The current magnitude quickly falls away as the frequency of the signal increases beyond this value. We can conclude that the series circuit is resonant at 1600 Hz.

At *resonance*, $X_L = X_C$ and the value of each is found as follows:

$$X_L = 2\pi f_R L$$
$$X_L = (2\pi \times 1.6 \times 10^3 \times 1.0 \times 10^{-2})\ \Omega$$
$$X_L = 1.0 \times 10^2\ \Omega$$
$$X_C = \frac{1}{2\pi f_R C}$$
$$X_C = \left(\frac{1}{2\pi \times 1.6 \times 10^3 \times 1.0 \times 10^{-6}}\right)\ \Omega$$
$$X_C = 1.0 \times 10^2\ \Omega$$

The impedance at resonance is

$$Z = \sqrt{R^2 + X^2}\ \underline{|\arctan X/R}$$

But
$$X = X_L - X_C = 0\ \Omega$$

and the phase angle
$$\phi = \arctan \frac{X}{R} = 0°$$

Thus
$$Z = R = 25\ \Omega$$

The impedance, Z, of the series-resonant circuit is the minimum value and is equal to R. The current, I, is the maximum value and is equal to V/R.

$$I_R = \frac{V}{R} = \frac{1\bar{0}\ \text{V}}{25\ \Omega} = 0.40\ \text{A}$$

$$V_L = I_R X_L = 0.40\ \text{A} \times 1.0 \times 10^2\ \Omega$$
$$V_L = 4\bar{0}\ \text{V}$$
$$V_C = I_R X_C = 0.40\ \text{A} \times 1.0 \times 10^2\ \Omega$$
$$V_C = 4\bar{0}\ \text{V}$$
$$I_R X_L - I_R X_C = 4\bar{0}\ \text{V} - 4\bar{0}\ \text{V} = 0\ \text{V}$$
$$V_R = I_R \times R = 0.40\ \text{A} \times 25\ \Omega = 1\bar{0}\ \text{V}$$

The phase angle, ϕ, at resonance is 0° and the power factor of the circuit, which is the cosine of ϕ, is 1. The series-resonant circuit has unity power factor. Thus maximum power, P, is dissipated in the circuit at resonance.

$pf = \cos\phi = \cos 0° = 1$

$$P = VI_R \cos\phi = VI_R$$

The frequency at which a series circuit resonates is determined by the combination of L and C used. For a particular combination, there is one resonant frequency, f_R. At series resonance

$$X_L = X_C$$

$$2\pi f_R L = \frac{1}{2\pi f_R C}$$

Solving for f_r,

$$f_R^2 = \frac{1}{4\pi^2 LC}$$

$$f_R = \frac{1}{2\pi\sqrt{LC}}$$

When L is in henrys and C is in farads, f_R is expressed in hertz.

At frequencies below the resonance point, the series circuit is capacitive, performing as one with an X_C and R in series. At the resonant frequency, f_R, the circuit performs as one containing pure resistance. Above the resonant frequency, it is inductive. These characteristics of the L-R-C series circuit are shown by the reactance curve X in Figure 21-23.

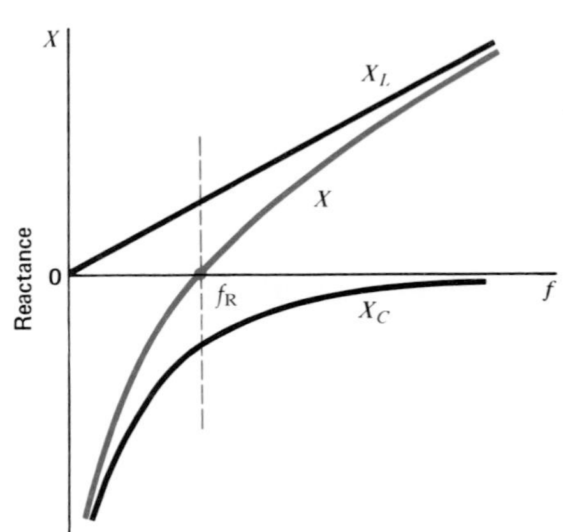

Figure 21-23. The reactance characteristic of an ***L-R-C*** series circuit on either side of the resonant frequency.

21.13 Selectivity in Series Resonance A resonant circuit responds to impressed voltages of different frequencies in a very selective manner. The circuit presents a high impedance to a signal voltage of a frequency above or below the resonant frequency of the circuit. The resulting current is correspondingly small. This same circuit presents a low impedance, consisting of the circuit resistance, to a signal voltage at the resonant frequency. The current is correspondingly large.

A tuned circuit thus *discriminates* among signal voltages of different frequencies. This property of signal discrimination is known as the circuit's *selectivity*. The lower the resistance of the series circuit, the higher is the resonant current and the more sharply selective is the circuit. The family of current-resonance curves shown in Figure 21-24 illustrates the influence of circuit resistance on selectivity. Curves are shown for three different values of circuit resistance.

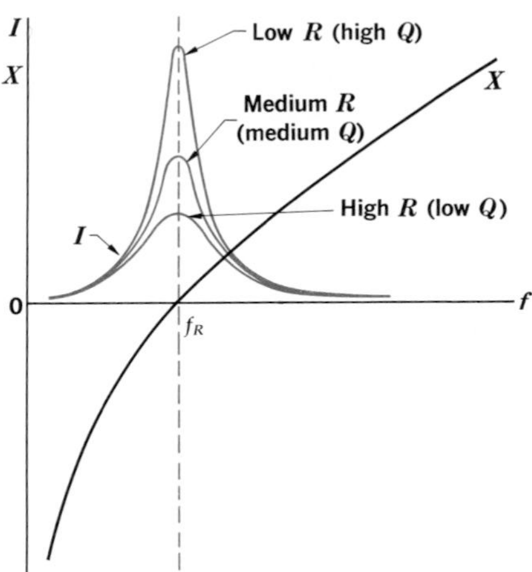

Figure 21-24. The magnitude of current in a series-resonant circuit plotted as a function of frequency.

The usual series-resonant circuit consists of an inductor and a capacitor. The only resistance is that inherent in the circuit, which is almost entirely the resistance of the coil. The characteristics of the series-resonant circuit depend primarily on the ratio of the inductive reactance to the circuit resistance, X_L/R. This ratio is called the Q, quality factor, of the circuit. It is essentially a design characteristic of the inductor. The higher the Q, the more sharply selective is the series-resonant circuit.

Resonant circuits can be tuned over a range of frequencies by making either the inductance or the capacitance variable. The position of a powdered-iron core in a coil can be varied to change the coil's inductance. A capacitor can be made variable by providing a means of changing its plate area or its plate separation. A variable capacitor of the type used in the tuning circuits of small radios is shown in Figure 21-25. Resonance is particularly important in communication circuits.

Figure 21-25. A two-gang variable capacitor of the type used in the tuning circuits of radio receivers. Each capacitor section has a set of stator plates and a set of rotor plates. The rotor plates are attached to the shaft. The capacitances are varied simultaneously by rotating the shaft and thus varying the amounts of effective plate areas of the two capacitor sections.

QUESTIONS: GROUP A

1. Select three different values of inductance, and determine the reactance of each at four different frequencies. Plot a curve of inductive reactance as a function of frequency for each using the same coordinate axes. (a) What is the shape of each curve? (b) Which curve has the greatest slope? (c) Why should each curve start at the origin of the coordinate axes?
2. Define series resonance.
3. What is the relation between the voltage across a series L-R-C circuit and the current in the circuit at a frequency below the resonant frequency? Explain.
4. What is the relation between the voltage across a series L-R-C circuit and the circuit current at a frequency above the resonant frequency? Explain.
5. What factor largely determines the sharpness of the rise of current near the resonance frequency in a series-resonant circuit?
6. What is the advantage of a resonant circuit having a high Q?
7. How is the Q of a resonant circuit affected by an increase in (a) resistance, (b) inductance, (c) frequency?

GROUP B

8. How can the resonant frequency of an L-R-C circuit be varied?
9. What would be the effect of having an inductor connected in series with a lamp in an a-c circuit? Explain.
10. A variable capacitor is connected in series in the circuit of Question 9 and when adjusted, the lamp glows normally. Explain.
11. A series L-R-C circuit is connected across an a-c signal source of variable frequency. Draw a curve to show the way in which the circuit current varies with the frequency of the source.

12. An a-c circuit has a lower power factor because of an inductive load. (a) What is the effect of this low power factor on the circuit current? (b) Suggest a way to raise the power factor closer to unity.

PROBLEMS: GROUP A

1. A resonant circuit has an inductance of 320 μH and a capacitance of $8\bar{0}$ pF. What is the resonant frequency?

2. Assume that the frequency of the signal voltage applied to the resonant circuit of Problem 1 is varied from a value of 25 kHz below the resonant frequency to 25 kHz above. (a) Obtain values of inductive reactance for every 5 kHz over this range of frequencies. Plot a curve showing inductive reactance as a function of frequency, using the frequency as the reference abscissa. (b) Obtain values of capacitive reactance for every 5 kHz, and plot a curve that shows the capacitive reactance as a function of frequency, using the same reference abscissa. (c) Determine the resonant frequency graphically from your curves by plotting the values for $(X_L - X_C)$.

3. The variable capacitor used to tune a broadcast receiver has a maximum capacitance of 350 pF. The lowest frequency to which we wish to tune the receiver is 550 kHz. What value of inductance should be used in the resonant circuit?

4. The minimum capacitance of the variable capacitor of Problem 3 is 15 pF. What is the highest resonant frequency that may be obtained from the circuit?

SUMMARY

The power expended in the load of an a-c circuit is determined by the average of the instantaneous values over the period of one cycle. Instantaneous power is the product of the instantaneous current and the instantaneous voltage. In a resistive load, the average power is the product of the effective values of current and voltage. In either a pure inductive load or a pure capacitive load, the average power is zero. Power in an a-c circuit is equal to the product of the effective value of the current, the effective value of the voltage, and the cosine of the phase angle.

An a-c circuit containing both inductance and resistance produces a current that lags behind the voltage. Such a load presents an impedance to the source. This impedance is the vector sum of the resistance and the inductive reactance in the circuit. The phase angle is the angle at which the current lags the voltage.

Capacitance and resistance in an a-c circuit produce an impedance equal to the vector sum of the capacitive reactance and resistance in the circuit. The phase angle in this case is due to a lagging voltage. When inductance and capacitance are both present, the phase angle is determined by the relative magnitudes of the two reactances. Inductive reactance is directly proportional to both frequency and inductance. Capacitive reactance is inversely proportional to both frequency and capacitance.

Series circuits containing inductance, capacitance, and resistance can be analyzed by the application of Ohm's law for a-c circuits.

At zero frequency the reactance of an inductor is zero, with a linear increase as the frequency increases. The reactance of a capacitor is infinitely large at zero frequency and decreases nonlinearly with increasing frequency. The resonant frequency of the circuit is the frequency at which the inductive and capacitive reactances of a circuit are equal.

At the resonant frequency the impedance of a series-resonant circuit is at a minimum and is equal to the circuit resistance. The resonant current is the maximum current possible in the series circuit and is in phase with the applied voltage. If the Q of the circuit is high, the resonant current is very large compared with the average circuit current over a range of frequencies. The current curve is sharply peaked. The selectivity of such a resonant circuit is high.

VOCABULARY

actual power
apparent power
arc tangent
capacitive reactance
effective value (of current or voltage)
impedance
impedance diagram
inductance
inductive reactance
inductor
phase angle
phase diagram
phasor
power factor
reactance
resonance
resonant frequency
root-mean-square
selectivity
series resonance

Magnetic Resonance Imaging (MRI)

Most people are familiar with X-ray devices that produce images of the bones. However, a new technology called magnetic resonance imaging, or MRI, has been developed that produces images of the tissues and organs. MRI devices produce these images with the combined use of magnetic fields, radio waves, and computers. A patient is placed inside the

hollow center of a huge box that houses a powerful magnet. The strong magnetic field produced by the magnet aligns the hydrogen atoms in the cells of the internal organs. Radio waves passed through the body cause the hydrogen atoms to realign. When the radio waves are turned off, the atoms generate low intensity radio waves. The waves are detected by a computer that generates an image of the body's anatomy.

MRI shows organs, detects chemical changes within the body, determines areas of cancer and other diseases, and distinguishes among different types of tissue.

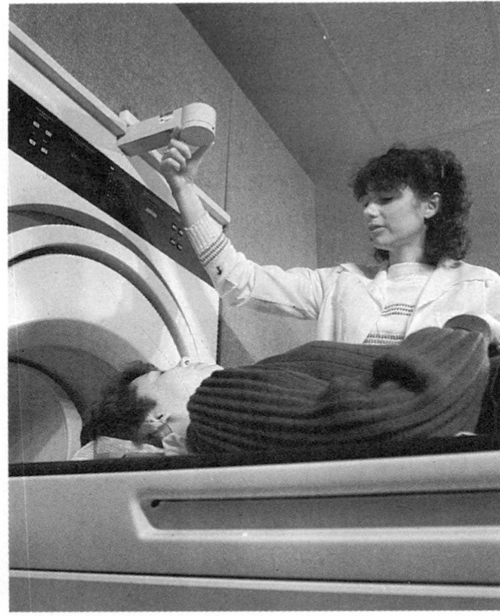

Adjustments being made by a technician prior to a patient's entry into an MRI scanner.

Medical diagnostic testing: technician at computer for MRI scan. Note scan of skull on screen. At the rear, a patient enters the scanner.

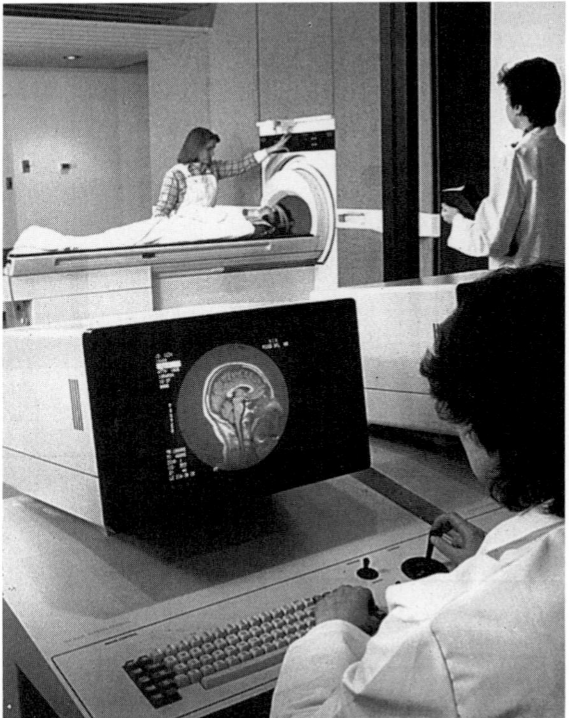

Medical Physicists

The development of theoretical concepts and methods used to diagnose and treat human diseases is the main concern of a medical physicist. Those medical physicists involved in research develop new medical instruments and technology such as MRI and X-ray imaging. They also pioneer new techniques to measure the radiation emitted by devices used in cancer therapy. Medical physicists working in a clinical environment are responsible for the performance specifications, testing, and quality of complex medical equipment and for the safety and accuracy of the methods of treatment used.

Medical physicists usually have a B.S. degree in physics and an M.S. or Ph.D. degree in physics or medical physics. Many have a subspecialty such as radiologic or nuclear physics. Most medical physicists are employed in hospitals or clinics. Some medical physicists hold teaching positions.

Biomedical Equipment Technicians

The complex medical equipment used in the health field must be maintained, repaired, installed, and operated. These are the responsibilities of biomedical equipment technicians. These technicians work with diagnostic devices, patient monitors, and thousands of other chemical, electronic, and mechanical medical devices. The lives of patients often depend on such medical equipment. Thus biomedical equipment technicians must rapidly determine the cause of a breakdown and make the necessary repairs. The technician must also test, disassemble, clean, and reassemble the equipment.

Biomedical equipment technicians need excellent mechanical ability and an associates degree in biomedical equipment technology. Course work in equipment technology, anatomy, physiology, and basic electronics is required. These technicians work in medical and research facilities and for medical equipment manufacturers. ■

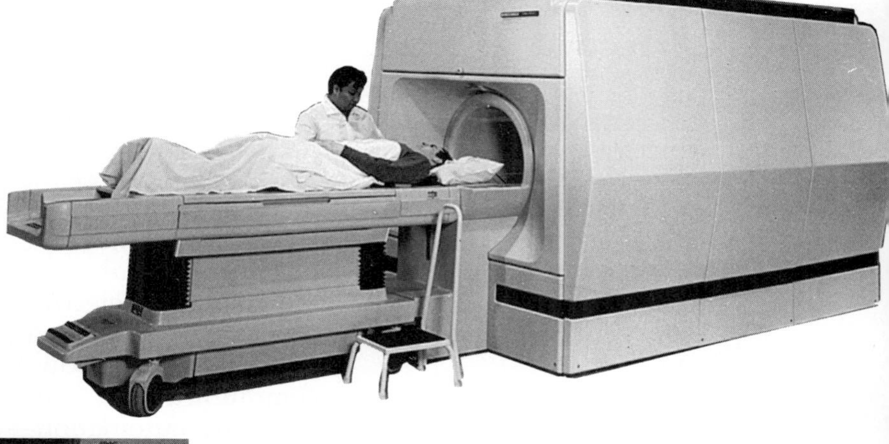

■ Patient being prepared to enter the MRI scanner.

■ Medical physicist monitoring a radiation-therapy device being used on a patient by closed-circuit TV.

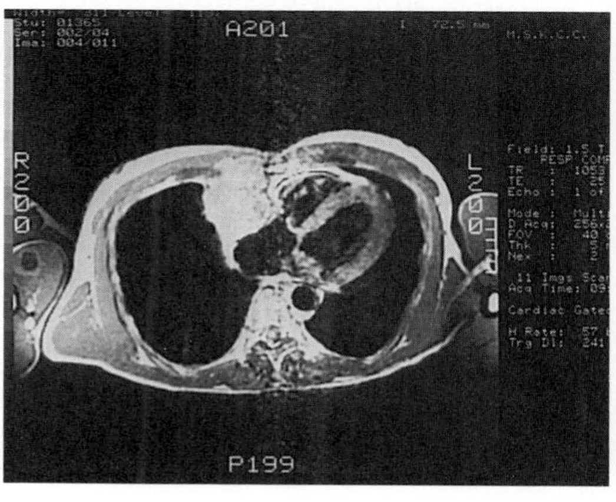

■ Computer generated image of a cross-sectional MRI scan of the chest cavity.

22 Electronic Devices

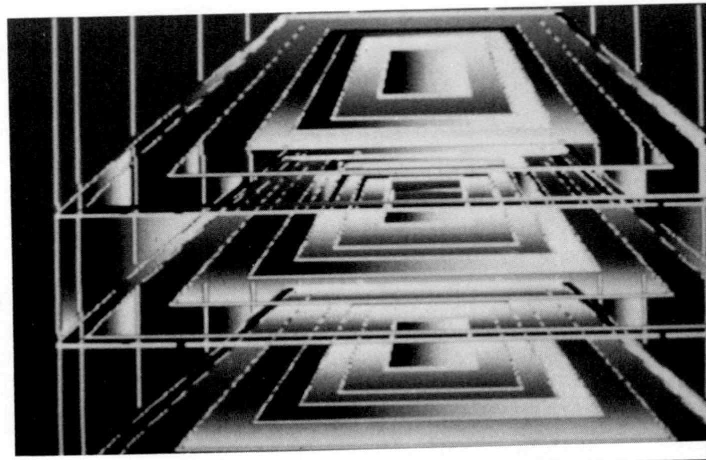

electronics (ee-lek-TRON-iks) n.: the branch of physics that is concerned with the emission, behavior, and effects of electrons in devices and with the utilization of these devices.

OBJECTIVES

- Study the role of the vacuum tube in the growth of electronics.
- Describe the thermionic emission of electrons.
- Describe the diode as a voltage rectifier and the triode as a voltage amplifier.
- Show how a cathode-ray tube works.
- Study the function of a photomultiplier tube.
- Describe the process of forming **P**- and **N**-type semiconductor materials.
- Analyze **P-N** junctions and their role in transistor circuits.

VACUUM TUBES

22.1 Electronics, a Branch of Physics The influence of *electronics* has become so commonplace that its role in our daily activities goes largely unnoticed. Many household appliances are electronically programmed to perform a sequence of timed operations at the push of a button. An increasing number of technical and industrial processes are controlled electronically by computers. Sophisticated electronic devices and circuits have an indispensable role in the advances now being achieved in medical science, communications, computer technology, space exploration, and automation—to name just a few. ***Electronics*** *is the branch of physics that is concerned with the emission, behavior, and effects of electrons in devices and with the utilization of these devices.*

The emergence of electronics as an important branch of physics began with the development of the *vacuum tube.* Vacuum tubes made possible such innovations as radio broadcasting, long-distance communications, sound motion pictures, public address systems, television, radar, electronic computers, and industrial automation.

Today, semiconductor devices have replaced vacuum tubes in many diverse applications of electronic circuits. However, a knowledge of vacuum tubes and basic vacuum-tube electronics is helpful in the study of semiconductor electronics.

22.2 Vacuum-Tube Applications

It may be useful to think of vacuum tubes as electron valves in electronic circuits. They have been designed to perform many different functions:

Vacuum tubes are also referred to as "electron tubes." In Great Britain they are called "valves."

1. As *rectifiers,* they convert alternating current to direct current.
2. As *mixers,* they combine separate signals to produce a different signal.
3. As *detectors,* they separate the useful component from a complex signal.
4. As *amplifiers,* they increase the strength of a signal.
5. As *oscillators,* they convert a direct current to an alternating current of a desired frequency.
6. As *wave-shapers,* they change a voltage waveform into a desired shape for a special use.

A vacuum tube usually contains a *cathode* as a source of electrons, an *anode,* commonly called a *plate,* that attracts electrons from the cathode, and one or more *grids* for controlling the flow of electrons between cathode and plate. These electrodes are enclosed in a highly evacuated, gas-tight envelope. Modified forms of the vacuum tube may contain only a cathode and a plate or two independent sets of electrodes in a single envelope. In some tubes, a small amount of a particular gas is introduced to obtain special operating characteristics.

Figure 22-1. Modern vacuum tubes are designed to perform a variety of highly specialized functions. Shown here is a magnetron oscillator, the radiation generator in a microwave oven. More powerful magnetrons serve as transmitter tubes in pulsed radar systems.

Common tubes are classified as *diodes, triodes, tetrodes,* and *pentodes,* according to the number of electrodes. High-vacuum tubes are known as *hard* tubes, and those containing a gas under very low pressure are known as *soft* tubes. Special tubes have special names. For example, the cathode-ray tube is a visual-indicating tube used in cathode-ray *oscilloscopes.* It serves also as the picture tube in television sets, the indicator tube in radar equipment, and the visual-display tube in computers.

22.3 Vacuum Tube Development

1. The Edison effect. While experimenting with the first electric lamp in 1883, Thomas Edison (1847–1931) was plagued by the frequent burning out of the carbon filament and the accompanying black deposit inside the bulb. In an effort to correct this difficulty Edison sealed a metal plate inside the lamp near the filament and connected it through a galvanometer to the filament battery. See Figure 22-2. He observed a deflection on the galvanometer when the plate was connected to the positive terminal of the battery. But no deflection occurred when the plate was connected to the negative terminal. Edison recorded these observations and continued his lamp experiments without attempting to explain the phenomenon.

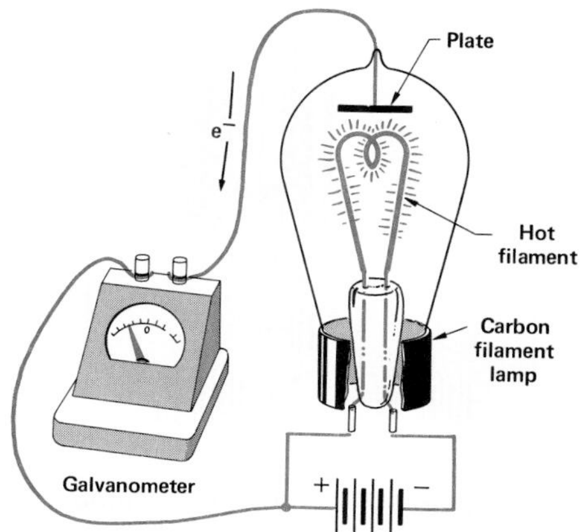

Figure 22-2. The Edison effect.

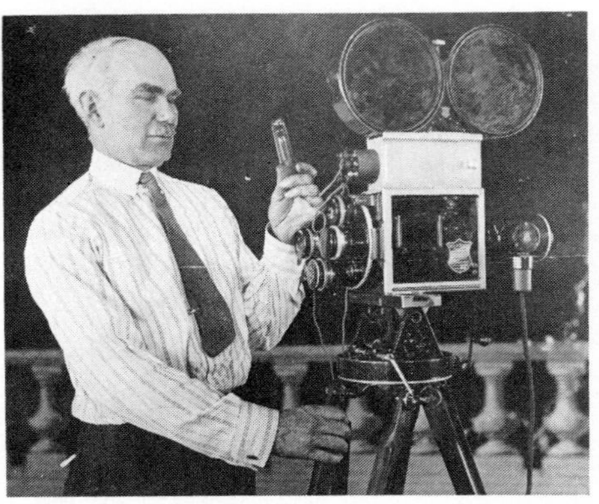

Figure 22-3. Lee De Forest was one of the pioneers in the development of wireless telegraphy in America. Among his many inventions was the triode, known as the audion, which made long-distance telephone and radio communications possible. De Forest also invented a system for recording sound on film which made sound motion pictures possible.

Several years later the English physicist Sir Joseph Thomson (1856–1940) discovered the electron and provided the explanation for the *Edison effect*. When the plate in Figure 22-2 was made positive, electrons escaping from the heated carbon filament were attracted through the evacuated space to the plate. The resulting electron flow in the galvanometer circuit produced the meter deflection that Edison had observed. When the plate was made negative, on the other hand, the escaping electrons were repelled from the plate and there was no electron flow in the galvanometer circuit.

2. *The Fleming valve.* Near the end of 1901 "wireless telegraphy," the forerunner of modern radio communications, was successfully used to transmit radio signals over long distances. In these early days of radio, crystals of semiconductor materials were used as receivers to *detect*, or recover, the message from the transmitted signals. These crystals were unsatisfactory in many ways. A better method of detection was clearly needed.

In 1904 the English physicist J. A. Fleming (1849–1945) patented the first *diode*, called the *Fleming valve*. He used the Edison effect to develop a crude detector for radio signals. Fleming's valve was so insensitive that it found little immediate application, and yet it was an important link in the evolution of the vacuum tube.

3. *The De Forest triode.* In 1906 the American inventor Lee De Forest (1873–1961) developed a vacuum tube that amplified feeble radio signals that neither the Fleming valve nor the crystal detector was sensitive enough to detect.

De Forest placed a third electrode, consisting of a *grid* of fine wire, between the filament and plate of the diode, as shown in Figure 22-4. He found that a small variation in voltage applied to the grid produced a large variation of current in the plate circuit of the tube. Thus the feeble radio signals from an antenna could be amplified by De Forest's *triode*. The amplified signals could then be successfully detected by the Fleming valve to provide the varying d-c signal necessary to operate headphones. The success of the De Forest triode led to the development of modern vacuum tubes.

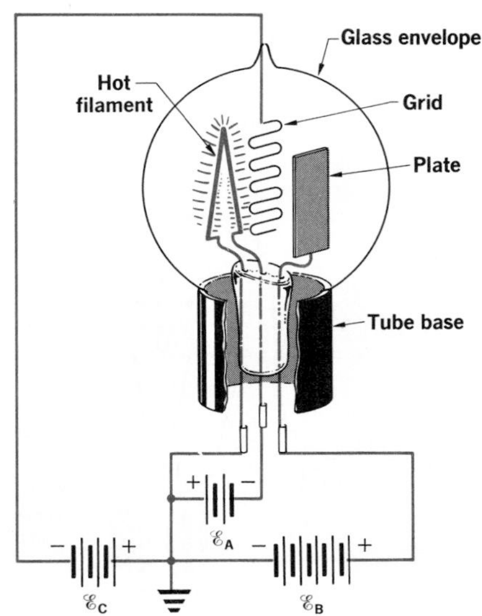

Figure 22-4. De Forest's three-element vacuum tube, the triode.

22.4 Thermionic Emission

Electric conduction involves a transfer of some kind of charged particles. Metallic substances conduct electricity because they have many free electrons. In a gas under low pressure, ionized gas molecules and electrons are responsible for conduction. The number of gas particles in a highly evacuated tube is so small, however, that conduction by gas ionization is virtually insignificant. Conduction in a high vacuum involves the introduction of charged particles into the electric field in the evacuated space between two electrodes. The charged particles released into the electric field are electrons. These electrons are released from a metallic surface by a process known as *thermionic emission.*

The free electrons of a metallic conductor are in continuous motion with speeds that increase with temperature. To escape from the surface of a conductor, an electron must do work to overcome the attractive forces at the surface of a material. At ordinary temperatures, the kinetic energies of electrons are not large enough to exceed the work function of the surface; the electrons cannot escape.

At a *high* temperature the average kinetic energy of the free electrons is large and a relatively large number will have enough energy to escape through the surface of the material. ***Thermionic emission*** *is the emission of electrons from a hot surface.* It is analogous to the evaporation of molecules from the surface of water.

The rate at which electrons escape from an emitting surface increases as the temperature of the emitter is raised. In the photoelectric effect, photons of light supply the energy required to eject electrons through the surface barrier of the photosensitive material. In thermionic emission, it is thermal energy that produces this same effect.

The high temperatures that are needed for satisfactory thermionic emission in vacuum tubes limit the number of suitable emitters to such substances as *tungsten, thoriated tungsten,* and certain *oxide-coated metals.* Tungsten and thoriated tungsten operate at very high temperatures and are connected directly to a source of filament current. These *directly heated cathodes* are known as *filament cathodes,* or simply *filaments.* They are used as electron emitters in large transmitting tubes.

Oxide-coated emitters consist of a metal such as nickel coated with a mixture of barium and strontium oxides, which in turn are covered with a surface layer of metallic barium and strontium. The electrons are emitted from this metallic surface layer. Such emitters can be heated to operating temperature either directly as a filament cathode or indirectly by radiation from an incandescent tungsten filament. *Indirectly heated cathodes* are commonly called *heater cathodes,* or simply *cathodes.* See Figure 22-6. Since practically all receiving tubes in use today have oxide-coated emitters, we shall concern ourselves principally with vacuum tubes of this type.

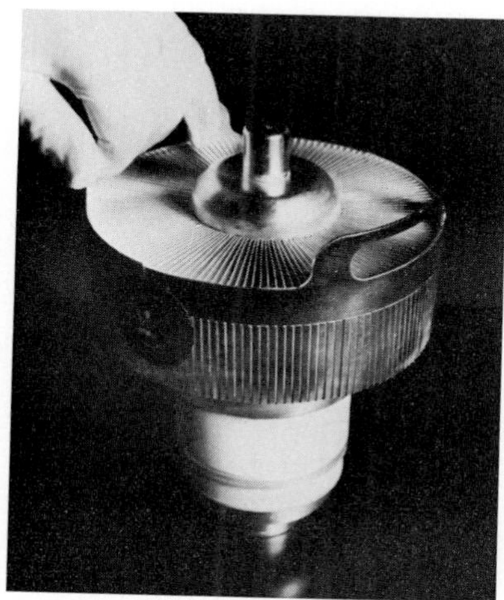

Figure 22-5. Vacuum tubes vary in their appearance because different designs are manufactured for different circuit requirements. This vacuum tube, designed for high-power applications, has radiation fins to dissipate heat.

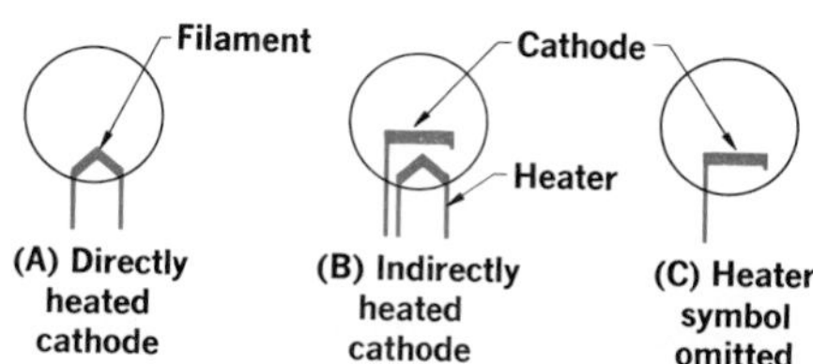

Figure 22-6. Circuit symbols for both directly and indirectly heated cathodes.

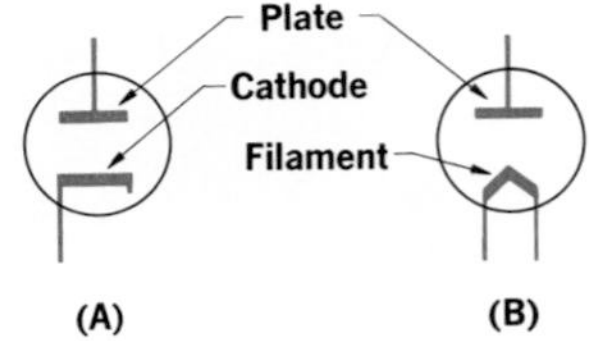

Figure 22-7. Circuit symbols for diodes with an indirectly heated cathode in **(A)** and a directly heated cathode in **(B)**.

22.5 Diode Characteristics The simplest vacuum tubes are diodes that contain two elements—a cathode and a plate. Diode circuit symbols are shown in Figure 22-7. If the plate is made positive with respect to the cathode, the diode conducts because electrons flow from the cathode to the plate inside the tube and from the plate through the plate circuit to the cathode outside the tube. If the plate is negative with respect to the cathode, the diode does not conduct and there is no plate current in the external circuit. Thus a diode acts as a one-way valve for electrons; it allows electrons to flow in one direction through the circuit, but not in the other.

As electrons are emitted from the hot cathode, the space about the cathode becomes negatively charged. This negatively charged space, called the *space charge,* opposes the escape of additional electrons from the cathode. The extent of electron emission depends on the temperature of the cathode.

When the plate charge is positive, electrons are accelerated toward the plate by the intervening electric field.

At low positive plate voltages, only those electrons near the plate are attracted to it, so a small plate current is produced. As the plate voltage is increased, greater numbers of electrons are attracted and the plate current is increased while the space charge is reduced.

In an ordinary circuit resistance, the proportionality of current to voltage is described by Ohm's law. The resistor is said to be a linear, or *ohmic,* element in the circuit. In the case of the diode, the plate current is not proportional to the plate voltage. The diode and other such devices are said to be nonlinear, or nonohmic, elements in the circuit.

A diode circuit from which we can determine the plate current, I_P, as a function of plate voltage, $\mathscr{E}_P$, for different filament temperatures is shown in Figure 22-8(A). The family of I_P vs. $\mathscr{E}_P$ curves is shown in Figure 22-8(B).

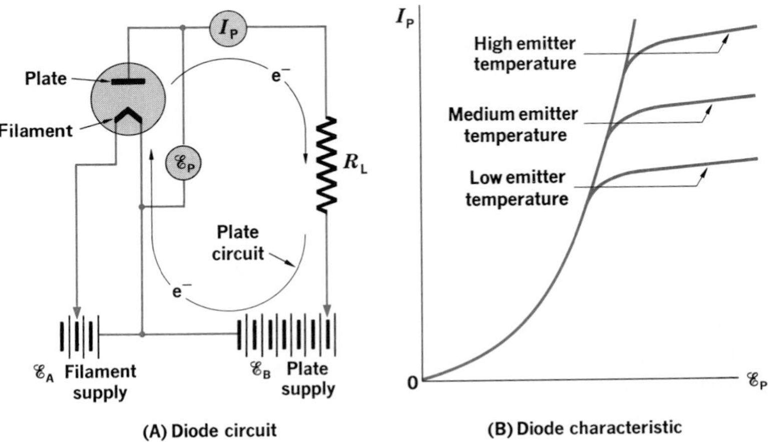

Figure 22-8. A diode circuit and characteristic curves showing the plate current as a function of plate voltage at different filament temperatures.

Diode tubes can be classified as *power* diodes or *signal* diodes. The difference is primarily one of size and power-handling capabilities. Power diodes are used as rectifiers in power-supply circuits and must be large enough to dissipate the heat generated during their operation. Signal diodes are small and handle signals of negligible power. They are used principally as detectors and wave-shapers. See Figure 22-9. They are frequently enclosed in the same envelope with triodes and pentodes.

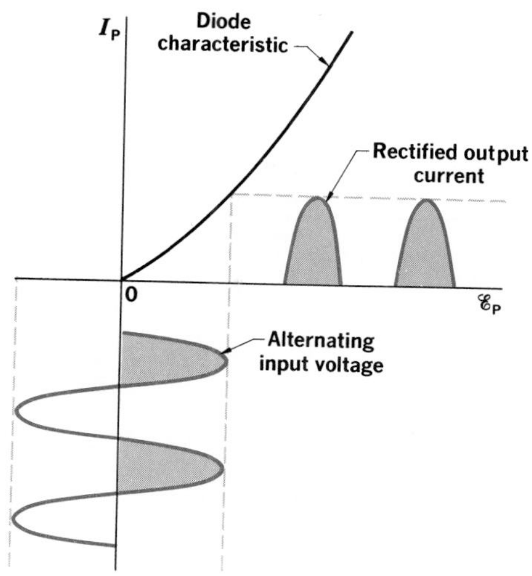

Figure 22-9. A diode characteristic used to show the rectifying action of a power diode.

22.6 The Triode Amplifier

De Forest perfected the first *triode* amplifier by inserting a third electrode between the cathode and plate of a diode. This electrode consists of an open spiral of fine wire that presents a negligible physical barrier to electrons flowing from cathode to plate. However, a small potential difference between this electrode and the cathode has an important controlling effect on the flow of electrons from cathode to plate, and thus on the current in the plate circuit. For this reason the third electrode is called a *control grid*. We may think of a triode as an adjustable electron valve. The electrode configuration of a triode and the triode circuit symbol are shown in Figure 22-10.

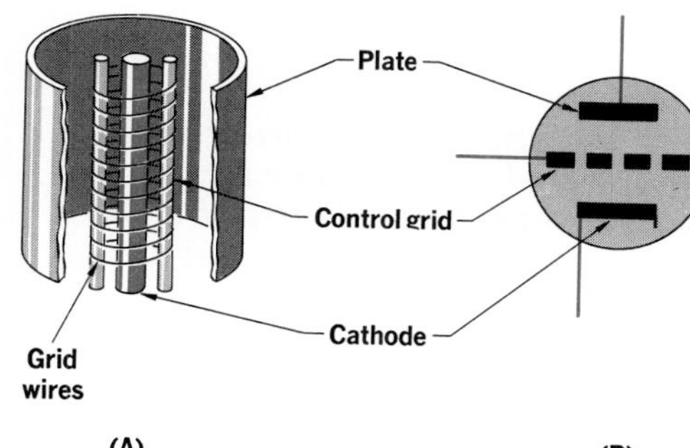

Figure 22-10. Sketch of the electrode configuration of a triode (A) and the triode circuit symbol (B).

If the control grid is made sufficiently negative with respect to the cathode, all electrons are repelled toward the cathode. In this condition the plate current is zero and the triode is said to be *cut off*. The grid-to-cathode voltage is called the *grid bias. The smallest negative grid voltage for a given plate voltage that causes the tube to cease to conduct is the* ***cutoff bias.*** This design characteristic of vacuum tubes containing control grids is important in the circuit applications of these tubes.

The effects of different grid-bias voltages on the plate current of a triode are shown in Figure 22-11. If a negative grid bias less than the cut-off value is used, some electrons pass through the grid and reach the plate to provide some magnitude of plate current. As the negative grid bias is

Figure 22-11. The effects of different control-grid bias voltages on the plate current of a triode.

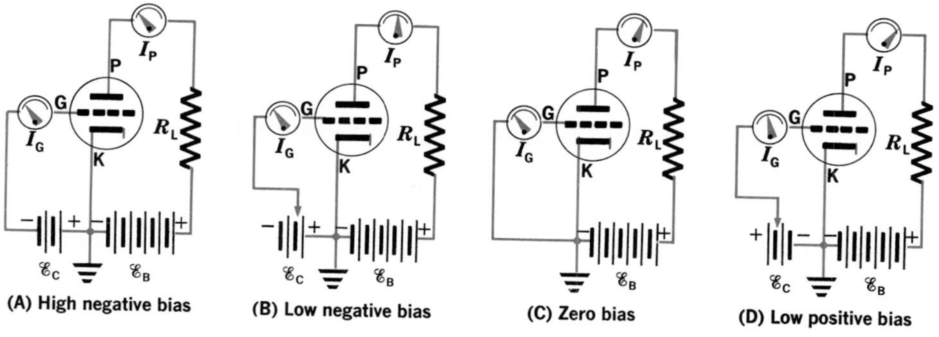

reduced toward zero, the plate current increases accordingly. At zero bias, the plate current is fairly large and depends primarily on the plate voltage. The performance of the triode is now essentially the same as that of a diode of similar construction.

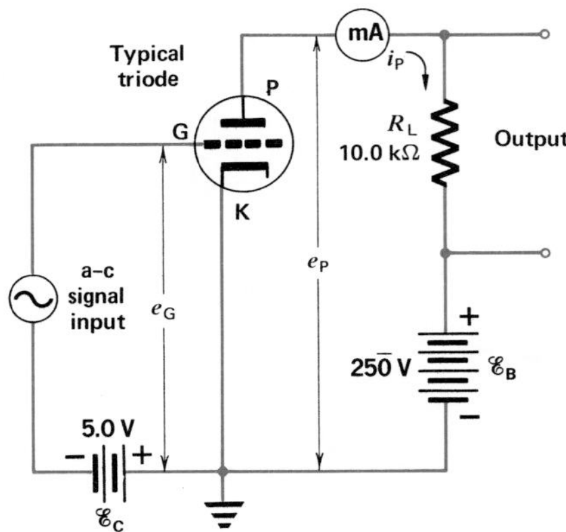

Figure 22-12. A basic triode amplifier circuit.

If, on the other hand, the grid is made positive with respect to the cathode, it no longer repels the electrons of the space charge but accelerates them in the direction of the plate. The plate current is increased by this action; however, *some electrons are attracted to the grid and produce an appreciable current in the grid circuit.* Grid current causes power dissipation in the grid circuit, which represents a waste of power. Thus the grid is normally maintained at some negative potential with respect to that of the cathode.

A basic triode amplifier circuit is shown in Figure 22-12. Small variations in grid potential cause larger variations in plate current than those caused by similar variations in plate potential. Thus voltage amplification across the circuit is possible. A small a-c voltage across the input circuit produces an amplified a-c voltage across the output circuit.

There is no d-c current in the grid circuit because the grid bias is negative with respect to the cathode. See Figure 22-11(B).

With batteries $\mathscr{E}_B$ and $\mathscr{E}_C$ in the circuit of Figure 22-12, a d-c current is in the plate circuit of the triode. Observe that the d-c plate circuit consists of the plate-supply battery $\mathscr{E}_B$, the cathode-to-plate resistance in the triode (an inherent value characteristic of the triode used), and the plate-load resistance R_L.

The sum of the potential drops around a series circuit must equal the applied emf.

The plate current I_P produces a drop in potential I_PR_L across the load resistance. Consequently, the d-c plate potential $\mathscr{E}_P$ is less than $\mathscr{E}_B$ by the magnitude of I_PR_L.

$$\mathscr{E}_P = \mathscr{E}_B - I_PR_L$$

In Figure 22-12, $\mathscr{E}_B$ is $25\bar{0}$ V, $\mathscr{E}_C$ is −5.0 V, and R_L is 10.0 kΩ. With no a-c signal on the grid, typical d-c operating conditions of the amplifier circuit can be established by assuming a d-c plate current I_P of 10.0 mA.

A d-c current of 10.0 mA in R_L produces a drop in potential across R_L of $10\bar{0}$ V.

$$I_PR_L = 10.0 \text{ mA} \times 10.0 \text{ k}\Omega = 10\bar{0} \text{ V}$$

The d-c plate potential $\mathscr{E}_P$ is therefore $15\bar{0}$ V.

$$\mathscr{E}_P = \mathscr{E}_B - I_PR_L = 25\bar{0} \text{ V} - 10\bar{0} \text{ V} = 15\bar{0} \text{ V}$$

These circuit values of I_P, $\mathscr{E}_P$, and I_PR_L represent the steady-state (d-c) operating conditions of the amplifier circuit. These steady-state values are shown in Figure 22-13.

Suppose we now apply an a-c signal of ±1.0 V across the grid circuit of Figure 22-12. This a-c voltage is imposed on the steady-state grid bias of −5.0 V. It gives an a-c component e_G that varies between −4.0 V and −6.0 V *at the frequency of the a-c signal.*

When the input signal is +1.0 V, e_G will be −4.0 V.

$$e_G = -5.0\ \text{V} + 1.0\ \text{V} = -4.0\ \text{V}$$

The plate current will *increase* because the grid bias is *less negative.* Assuming that the plate current increases to 15.0 mA, $i_P R_L$ will *increase* to $15\bar{0}$ V.

$$i_P R_L = 15.0\ \text{mA} \times 10.0\ \text{k}\Omega = 15\bar{0}\ \text{V}$$

Therefore, e_P will *decrease* to $10\bar{0}$ V.

$$e_P = 25\bar{0}\ \text{V} - 15\bar{0}\ \text{V} = 10\bar{0}\ \text{V}$$

Compare these results with the first half-period e_G, i_P, and e_P waveforms in Figure 22-13.

When the input signal is −1.0 V, e_G will be −6.0 V

$$e_G = -5.0\ \text{V} + (-1.0\ \text{V}) = -6.0\ \text{V}$$

and the plate current will *decrease* because the grid is *more negative.* If i_P decreases to 5.0 mA, $i_P R_L$ will *decrease* to $5\bar{0}$ V.

$$i_P R_L = 5.0\ \text{mA} \times 10.0\ \text{k}\Omega = 5\bar{0}\ \text{V}$$

As a result, e_P will *increase* to $20\bar{0}$ V.

$$e_P = 25\bar{0}\ \text{V} - 5\bar{0}\ \text{V} = 20\bar{0}\ \text{V}$$

Compare these results with the second half-period e_G, i_P, and e_P waveforms in Figure 22-13.

Figure 22-13 shows that a variation of ±1.0 V at the grid produces a variation of $\pm 5\bar{0}$ V at the plate; voltage amplification or "gain" across the amplifier circuit is $5\bar{0}$. Note that a change in grid voltage in a positive sense results in a change in plate voltage in a negative sense. The output voltage is opposite in *phase* to that of the input voltage. The a-c voltage gain across the amplifier circuit is accompanied by a voltage phase inversion.

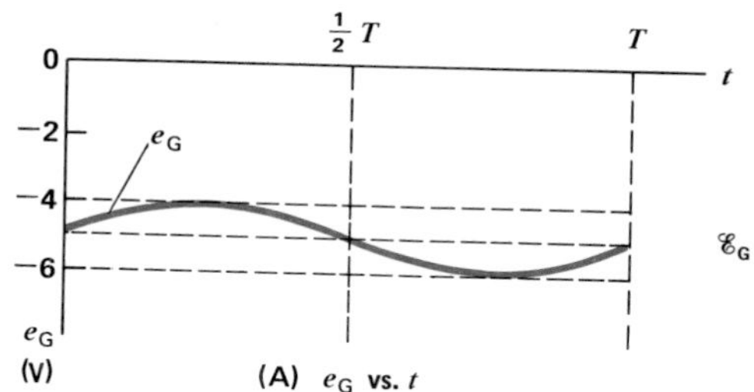

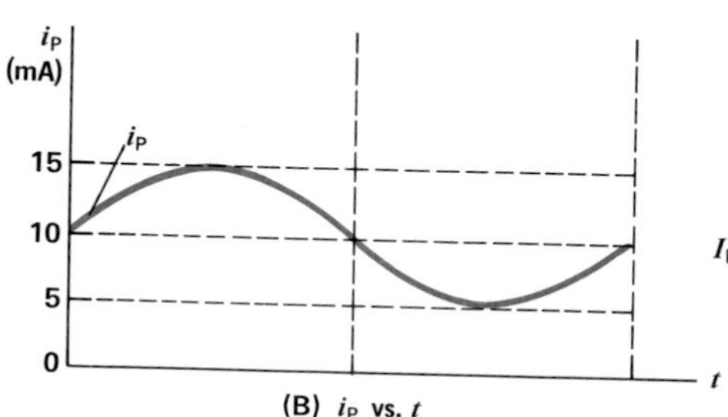

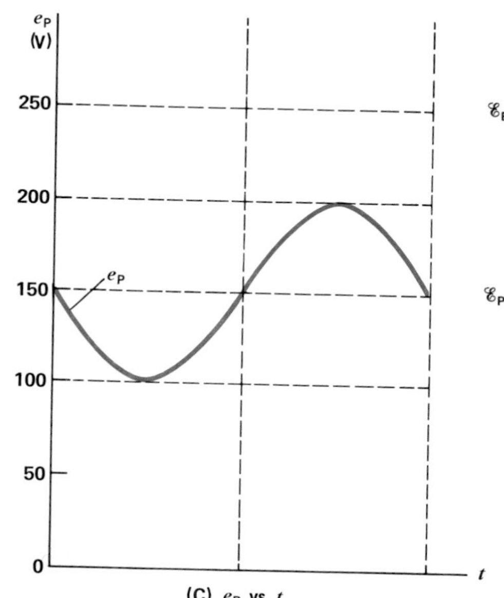

Figure 22-13. Grid voltage, plate current, and plate voltage variations of a vacuum triode amplifier showing phase inversion between e_G and e_P.

22.7 The Cathode-Ray Tube

The *cathode-ray tube* is a special kind of vacuum tube used in test equipment, radar, television receivers, and computers to present a visual display of information. Electrons are emitted from a cathode, accelerated to a high velocity, and brought to focus on a fluorescent screen. This screen is a translucent coating on the inside surface of the glass face of the tube. It *fluoresces,* emitting visible light at the point where the electron beam strikes.

When bombarded by electrons, fluorescent substances emit light of characteristic colors. All fluorescent substances have some *phosphorescence,* or afterglow, the persistence of which depends on the material and the energy of the bombarding electrons. Cathode-ray tubes that are used in test equipment such as the oscilloscope have short-persistence screens that emit predominantly green light. The picture tubes of television receivers have long-persistence screens.

The electron beam is moved over the fluorescent screen in response to the information signals being presented. This means that the information, in the form of electric signals, must be used to deflect the beam from its quiescent path within the tube.

An electron has negligible inertia since its mass is approximately 10^{-27} gram. Thus a beam of electrons can be deflected back and forth across the screen of a cathode-ray tube at very high speeds in response to weak signals. In a television receiver, for example, the electron beam sweeps across the face of the picture tube 15 750 times each second at a speed of the order of 24 000 km per hour.

There are two general methods of deflecting the electron beam in cathode-ray tubes: *electrostatic deflection* and *electromagnetic deflection.*

1. *Electrostatic deflection systems.* A cathode-ray tube employing electrostatic deflection is shown in Figure 22-14. The electrode arrangement that produces a focused beam of electrons is called the *electron gun.* It consists of an indirectly heated *cathode* capable of high emission, a *control grid* that regulates the intensity of the electron beam, a *focusing anode,* and an *accelerating anode.* The two anodes are maintained at high positive potentials with respect to the cathode. They produce an electric field that acts as an *electrostatic lens* to converge the electron paths at a point on the screen.

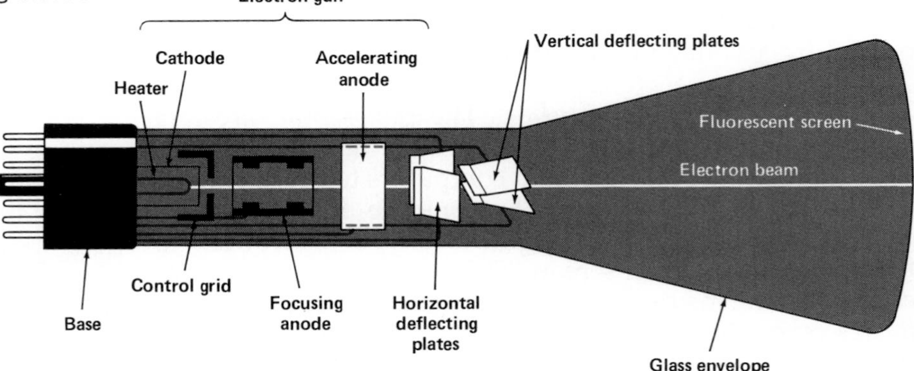

Figure 22-14. The construction of a cathode-ray tube using electrostatic deflection.

Deflection of the beam is accomplished by two pairs of *deflecting plates* set at right angles to each other. Signal voltages applied to these plates establish electrostatic fields through which the electrons move. The electron beam is subjected to deflecting forces in both the horizontal and vertical planes. Since electrons have negligible inertia, their response to these forces is practically instantaneous in the direction of the resultant force. Electrostatic deflection is generally used in the small cathode-ray tubes of test equipment.

2. *Electromagnetic deflection.* Large cathode-ray tubes have electromagnetic deflection systems. The beam can be focused either electrostatically, as shown in Figure 22-14, or magnetically by means of an external focusing coil. Beam control is accomplished with coils formed into a deflection yoke placed on the neck of the tube.

The most common color picture tubes have three electron guns, one for each primary-color phosphor. These color phosphors are arranged in clusters on a screen. The electron beams scan the screen to produce the primary colors of light on the face of the tube. The structure of a typical color television tube is shown in Figure 22-15. Cathode-ray tubes with more than three electron guns have been developed for use in highly specialized electronic equipment.

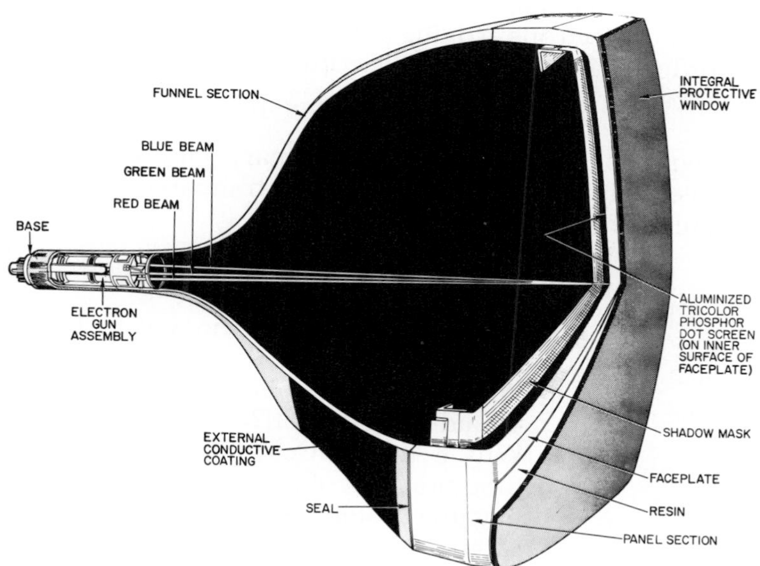

Figure 22-15. A cutaway view of a television color picture tube.

22.8 Photomultiplier Tube *Secondary emission,* the liberation of electrons from a metal surface by the impact of a high-speed electron, can be used in conjunction with

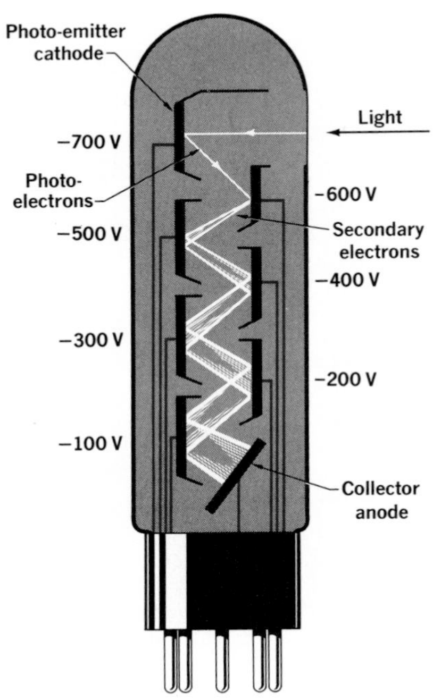

Figure 22-16. A photomultiplier tube with the six stages of secondary emission.

the photoelectric effect to detect light of very low intensity. This special form of photoelectric cell is known as a *photomultiplier tube.*

A diagram of a photomultiplier tube is shown in Figure 22-16. The number of photoelectrons emitted by the light-sensitive cathode in a given time is proportional to the intensity of the incident light. These photoelectrons are attracted to the first secondary-emission electrode, which is 100 volts positive with respect to the emitter cathode. On impact, additional electrons are released by secondary emission.

Successive secondary-emission electrodes, each more positive (less negative) than the one preceding it, contribute additional electrons as the electron stream progresses toward the collector anode. Thus a meager photoemission from a faint incident light can develop into an avalanche of electrons by the time it arrives at the collector plate. If each electron impact produces $\mathbf{n}$ secondary electrons, a photomultiplier tube with $\mathbf{x}$ secondary-emission electrodes will supply $\mathbf{n}^{\mathbf{x}}$ electrons to the collector plate for each photoelectron emitted.

Suppose a faint light incident on the photoemission cathode of the tube shown in Figure 22-16 will produce *five* photoelectrons per millisecond. Also suppose that each electron impact will produce *four* secondary electrons. Since the tube has *six* secondary-emission stages, the number of electrons arriving at the collector plate will be of the order of 2×10^4 per millisecond.

$$\mathbf{5n^x = 5 \times 4^6 = 2 \times 10^4}$$

This equation suggests the extraordinary gain that can be achieved with a photomultiplier tube compared to that of a regular photoelectric cell.

QUESTIONS: GROUP A

1. What device is primarily responsible for the development of electronics as a branch of engineering?
2. What discovery by Joseph Thomson explained the Edison effect?
3. How was De Forest's triode a significant improvement on previously developed diodes?
4. (a) Define thermionic emission. (b) What type of emitter is commonly used in receiving tubes?
5. Why is the grid of a triode called a control grid?
6. What is meant when a vacuum tube is said to be cut off?
7. Why is the control grid of a vacuum tube normally maintained at some negative potential with respect to the cathode?

GROUP B

8. Suggest two possible reasons why Edison did not offer an explanation

for the phenomenon known as the Edison effect.

9. Explain voltage inversion between the control grid and plate in a vacuum-tube circuit.
10. What is the electron gun of a cathode-ray tube?
11. What is the basic difference between the electrostatic and electromagnetic deflecting systems used in cathode-ray tubes?
12. (a) Explain why the first secondary-emission electrode of a photomultiplier tube is maintained at a positive (less negative) potential with respect to the photoemission electrode.
(b) Explain why each successive secondary-emission electrode is at a positive potential with respect to the preceding electrode.

TRANSISTORS

22.9 Crystal Diodes Metals are good *conductors* of electric currents. Many nonmetallic materials are classed as *insulators* because their resistivities are enormous compared with those of metals. For most practical purposes they are considered to be *nonconductors*. Some elements, including germanium, silicon, and selenium, and certain compounds, such as copper oxide, lead sulfide, and zinc oxide, are in a class of crystals called *semiconductors*. These substances have resistivities that fall between those of conductors and insulators.

Refer to the resistivity spectrum of Figure 17-23.

One of the devices from the early days of radio that is found in modern electronic circuits is the *crystal diode*. The rectifying property of silicon or galena (lead sulfide) crystals was used in early receivers before the development of the diode tube. During World War II, improved crystal diodes were developed for use in radar receivers. They are now used extensively in many types of circuits to perform a variety of specialized functions. They have the advantage of small size, and they require no heater or filament power.

A modern crystal diode can consist of a tiny wafer of silicon or germanium and a platinum *catwhisker*, all contained in a sealed capsule. See Figure 22-17. The crystal acts as the cathode of the diode and the catwhisker acts as the anode. The diode characteristic depends on the unique

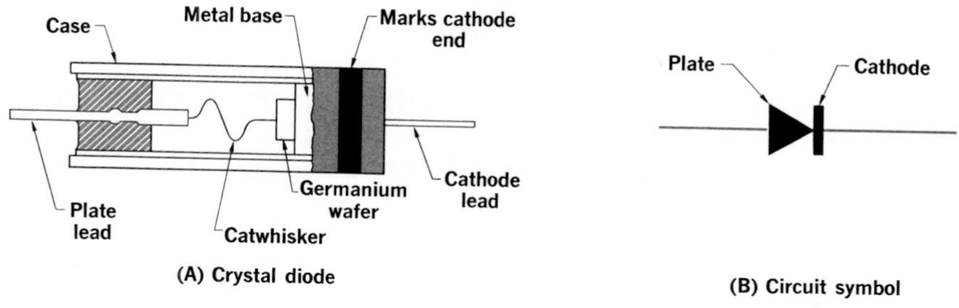

Figure 22-17. A crystal diode and its circuit symbol.

property of a semiconductor that permits a large flow of electrons in the *forward* direction with a small applied voltage, and only a very small flow of electrons in the *reverse* direction, even at a much larger applied voltage.

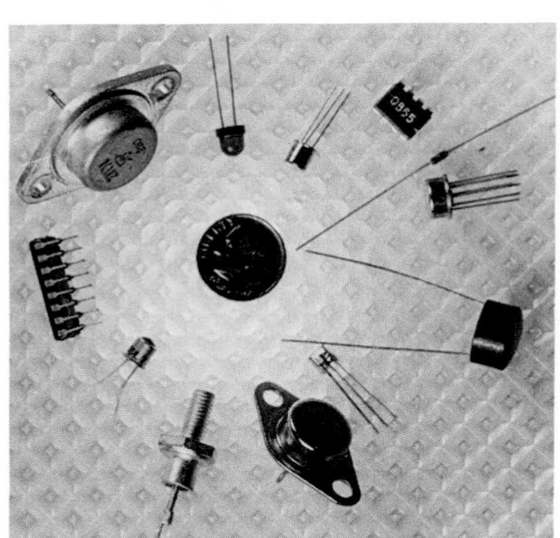

Figure 22-18. Semiconductor components are produced in a variety of circuit designs. The coin (dime) suggests component size.

22.10 Transistor Development After nearly a decade of basic research in the physics of semiconductors, physicists discovered how to produce a semiconductor device that could *amplify*. This device, known as a *transistor*, was developed by a team of physicists at the Bell Telephone Laboratories in 1948. Transistors are tiny, rugged structures that require very little power. The development of a family of transistors that can perform the functions of vacuum tubes has made possible the miniaturization of electronic circuits. See Figure 22-18.

The first transistors were known as *point-contact* transistors. These devices were replaced by *junction* transistors. Junction transistors can be further classified into several broad categories depending on the forming process and performance characteristics. Another addition to the semiconductor family is the *field-effect* transistor.

As the physicist's knowledge of semiconductor materials expands, new transistor configurations with improved frequency response and power-handling capabilities continue to appear. The key to the transistor effect is the introduction of specific impurities into silicon and germanium crystals to form **P**- and **N**-type semiconductor materials. The process is referred to as *doping*.

22.11 P- and N-Type Semiconductors Silicon and germanium are quite similar in their structure and chemical behavior. The atoms of both elements have *four* valence electrons bound in the same way in their respective crystals. In its transistor functions, germanium is the more versatile semiconductor. It has a diamond cubic crystal lattice in which each atom is bonded to four neighboring atoms through shared electrons, as illustrated in Figure 22-19. This arrangement binds the four valence electrons of each atom so that pure germanium would appear to be a nonconductor. *Germanium acquires the diode property of rectification and the transistor property of amplification through the presence of trace quantities of certain impurities in the crystal structure.* Two types of impurities are important: one is known as a *donor*, the other as an *acceptor*.

The controlled addition of donor or acceptor atoms to a semiconductor material during crystal growth is referred to as "doping" the semiconductor.

Antimony, arsenic, and phosphorus are typical donor elements. The atoms of each have *five* valence electrons. When minute traces of phosphorus are added to germanium, each phosphorus atom joins the crystal lattice by *donating* one electron to the crystal structure. Four of the

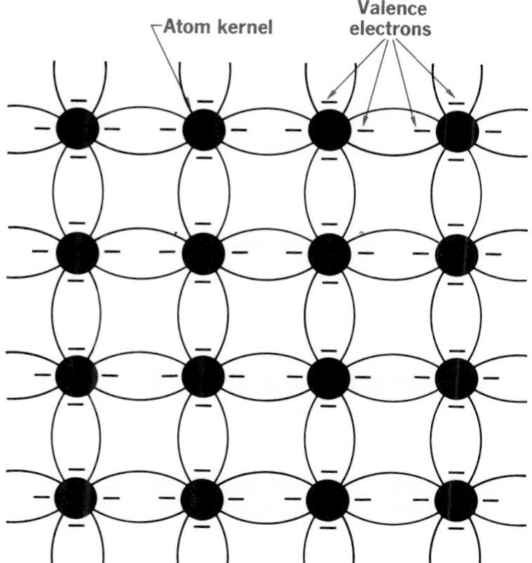

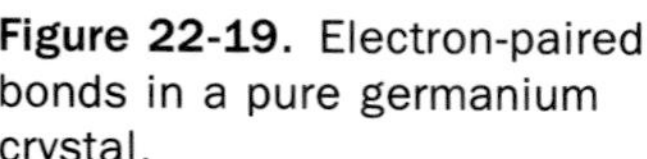
Figure 22-19. Electron-paired bonds in a pure germanium crystal.

five valence electrons are paired, but the fifth electron is relatively free to wander through the lattice like the free electrons of a metallic conductor. The detached electron leaves behind a phosphorus atom with a unit-positive charge bound into the crystal lattice. Germanium with this kind of crystal structure is known as *N-type*, or *electron-rich,* germanium because most of its conductivity results from *negative* charge carriers. ***N-type germanium*** *consists of germanium to which are added equal numbers of free electrons and bound positive charges so that the net charge is zero.* The added negative charge carriers (the free electrons) greatly increase conductivity over that of the pure semiconductor. An **N**-type semiconductor crystal is illustrated in Figure 22-20(A).

In **N**-*type semiconductors the charge carriers are negatively charged electrons.*

Figure 22-20. The **N**-type germanium crystal illustrated in **(A)** has donor atoms and free electrons in equal numbers. The **P**-type germanium crystal in **(B)** has acceptor atoms and free (positive) holes in equal numbers.

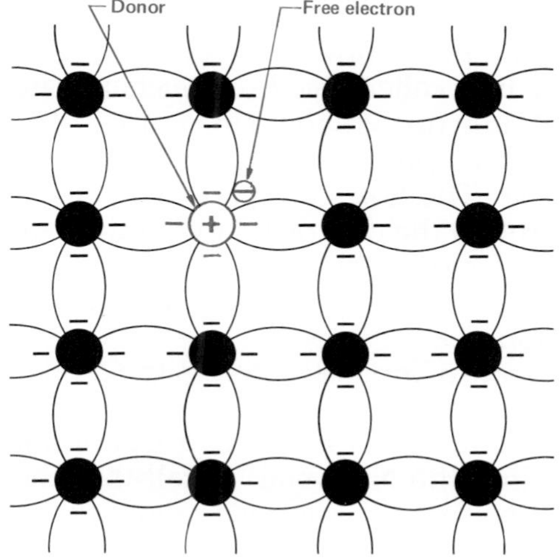

(A) N-type semiconductor

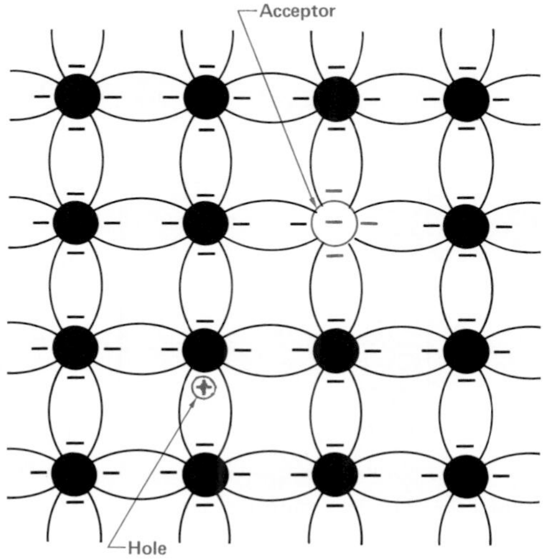

(B) P-type semiconductor

Atoms with *three* valence electrons, such as those of aluminum, boron, and gallium, will act as acceptors. When minute traces of aluminum are added to germanium, each aluminum atom joins the crystal lattice by *accepting* a single electron from a neighboring germanium atom. This leaves a *hole* in the electron-pair bond from which the electron is acquired.

We can think of this hole as the equivalent of a *positive charge* since it acts as a trap into which an electron can fall. As an electron fills the hole, it leaves another hole behind, into which another electron can fall. *In effect* then, the hole (positive charge) detaches itself and becomes free to move in the opposite sense to that of the electron, leaving behind the aluminum atom with a unit negative charge bound into the crystal lattice. Germanium with this crystal structure is called ***P**-type,* or *hole-rich,* germanium because most of the conductivity results from *positive* charge carriers. ***P-type germanium*** *consists of germanium to which are added equal numbers of free positive holes and bound negative charges so that the net charge is zero.* The added positive charge carriers (the holes) greatly increase conductivity over that of the pure semiconductor. A **P**-type semiconductor crystal is illustrated in Figure 22-20(B).

*In **P**-type semiconductors the charge carriers are, in effect, positively charged holes.*

22.12 The P-N Junction As a germanium crystal is formed, donor and acceptor atoms can be added in such a way that a definite boundary is created between **P**-type and **N**-type regions of the crystal. At this boundary some free electrons from the **N** region drift into the **P** region and a corresponding number of holes from the **P** region drift into the **N** region. An electric field, *E,* is established across the boundary layer as it becomes devoid of electron carriers on the **N** side and of hole carriers on the **P** side. In the resulting equilibrium state the boundary layer becomes a kind of electric barrier. It is called the ***P-N** junction.* The electric barrier of the **P-N** junction is shown in Figure 22-21.

Review electric fields in Sections 16.9 and 16.16.

*The **P-N** junction is the meeting zone of oppositely doped regions of a semiconductor crystal.*

The crystal as a whole has properties of a diode and is called a ***P-N** junction diode.* The hole-charge carriers of the **P** region cannot pass through the electric barrier of the **P-N** junction to reach the **N** region. The electron-charge carriers of the **N** region cannot pass through the electric barrier to reach the **P** region.

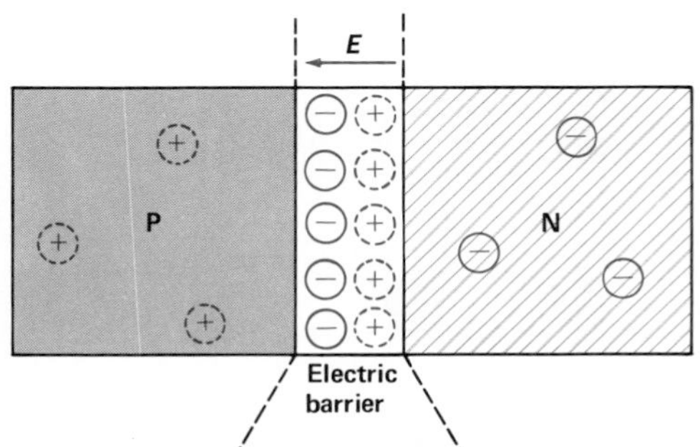

Figure 22-21. The electric barrier of the **P-N** junction.

A small potential difference impressed across the diode, as in Figure 22-22(A), enables the free electrons to cross the junction and pass into the **P** crystal. A corresponding number of holes cross into the **N** crystal. Recall that the apparent movement of holes is in the opposite sense to the actual movement of electrons that fall in the holes of the

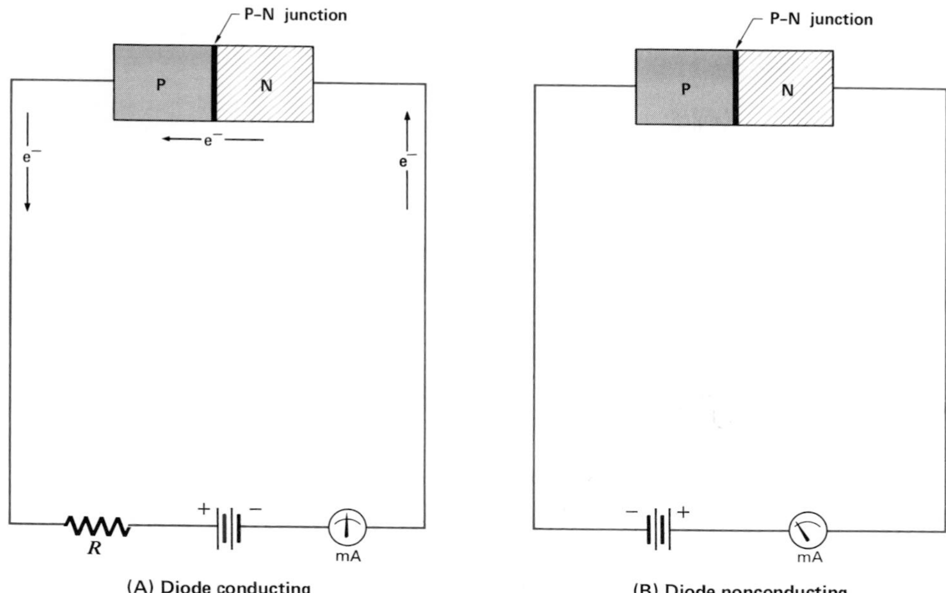

Figure 22-22. The **P-N** junction is the rectifying element of semiconductor crystals.

P-type crystal. Thus an electron current is established across the **P-N** junction and in the external circuit. The current-limiting resistance ***R*** is in the circuit to protect the conducting diode from excessive current.

If the battery connections are reversed, as in Figure 22-22(B), the free electrons in the **N** crystal and the free holes of the **P** crystal are attracted away from the **P-N** junction. The junction region is left without current carriers; consequently, there is no conduction across the junction and no current in the circuit.

The junction has the diode characteristic of permitting unidirectional electron flow with ease when a small voltage is applied in the proper sense. Thus the **P-N** junction is the rectifying element of semiconductor crystals.

22.13 Two Types of Junction Transistors Junction transistors consist of two **P-N** junctions back to back with a thin **P**-type or **N**-type region formed between them. If the center region is composed of **P**-type material, the transistor is designated **N-P-N.** The physical diagram of an **N-P-N** transistor and its circuit symbol are illustrated in Figure 22-23(A).

The **N**-type material shown at the left of the **P**-type center is the *emitter*. It is normally biased negatively with respect to the **P**-region, which is called the *base*. The **N**-type material on the right of the base is the *collector*. It is biased positively with respect to the base. The basic circuit of the **N-P-N** transistor is shown in Figure 22-24(A).

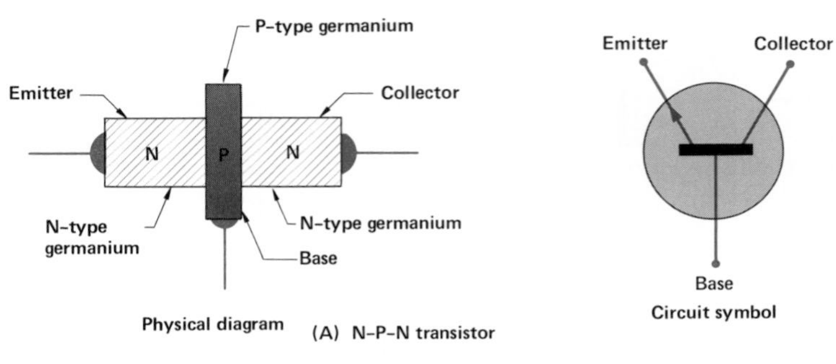

Figure 22-23. A comparison of **N-P-N** and **P-N-P** junction transistors.

Transistors in which the center region is composed of **N**-type semiconductor material are designated **P-N-P.** A diagram of a **P-N-P** transistor and its circuit symbol are shown in Figure 22-23(B).

The basic circuit of **P-N-P** transistors is given in Figure 22-24(B). Observe that it is the same as that for **N-P-N** transistors except that the battery connections are reversed.

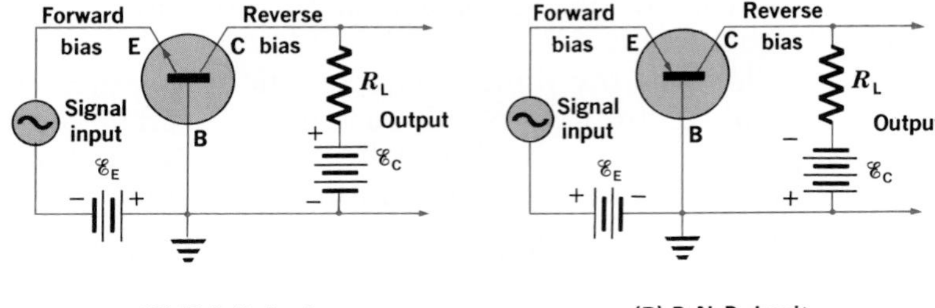

Figure 22-24. Basic **N-P-N** and **P-N-P** junction transistor circuits. Compare these circuits with that of the vacuum triode circuit of Figure 22-12.

In either **N-P-N** or **P-N-P** transistors, the emitter is biased to supply charges—electrons or holes—to be acquired by the collector through the two junctions of the base. An **N**-type emitter is biased negatively with respect to the base to supply electrons to its **N-P** junction and then to the **P**-type base. A **P**-type emitter is biased positively to withdraw electrons from its **P-N** junction and thus from the **N**-type base. This statement is equivalent to saying that the **P**-type emitter supplies holes to its **P-N** junction and then to the **N**-type base. Both emitter types are said to have *forward* bias. The emitter circuits of Figure 22-24(A) and Figure 22-24(B) have forward-bias voltages applied by batteries $\mathscr{E}_E$.

In the transistor circuit symbol, the arrowhead at the emitter is conventionally drawn to indicate the direction of the positive hole current. *Electron current is in the opposite direction.* This arrowhead in the transistor circuit symbol provides a way of showing which type of transistor is represented in a circuit diagram. In the **P-N-P** circuit symbol, note that the arrowhead at the emitter *points toward the base.* It points *away from the base* in the **N-P-N** circuit symbol.

An **N**-type collector is biased positively with respect to the base to attract electrons from its **N-P** junction and from the **P**-type base. A **P**-type collector is biased negatively to transfer electrons to its **P-N** junction and into the **N**-type base. Again, this is equivalent to saying it attracts holes from its **P-N** junction and thus from the **N**-type base. Both collector types are said to have *reverse* bias. The collector circuits of Figure 22-24(A) and of Figure 22-24(B) have reverse-bias voltages applied by batteries $\mathscr{E}_C$.

The forward bias on emitters and the reverse bias on collectors have the effect of producing an electron flow in the **N-P-N** transistor from emitter to base in the emitter circuit and from base to collector in the collector circuit. Forward emitter bias and reverse collector bias in the **P-N-P** transistor produce an electron flow from collector to base and from base to emitter.

The base section serves to isolate the emitter input circuit from the collector output circuit of a transistor. This isolation results from the barrier voltages established at the two junctions. The barrier polarities at these two junctions are reversed because the base has an **N-P** junction at one side and a **P-N** junction at the other.

Forward bias at the emitter junction overcomes the barrier voltage and enables the emitter to supply charge to the base. The reversed polarity of the barrier voltage at the collector junction, augmented by the reverse collector bias, aids the transfer of charge through the junction to the collector. Transistor performance is achieved when the collector current is controlled by the emitter current in the base.

22.14 *Transistor Amplification*

The transistor circuits shown in Figure 22-24 are *common-base* amplifiers. Small signals introduced into the emitter circuit are supplemented by energy from the collector battery and larger signals are transferred from the collector circuit. This effect is accomplished in spite of the fact that the collector current is slightly smaller than the emitter current.

Amplification in the common-base transistor amplifier is due to the ratio of output impedance to input impedance, which may be of the order of 1000:1. Low emitter impedance is obtained by the *forward* biasing of the emitter junction, that is, the junction between the emitter section and the base section. As stated in Section 22.12, the effect is to reduce the electric field across the emitter junction. High collector impedance is obtained by the *reverse* biasing of the junction between the collector section and the base section. This effect is to increase the electric field across the collector junction.

Only 1% to 5% of the emitter current passes through the base leg of the circuit. The remaining 99% to 95% passes through the collector circuit. This current can pass through the collector circuit because of the thinness of the base section and the consequent ease with which positive holes from the emitter diffuse across it to be collected.

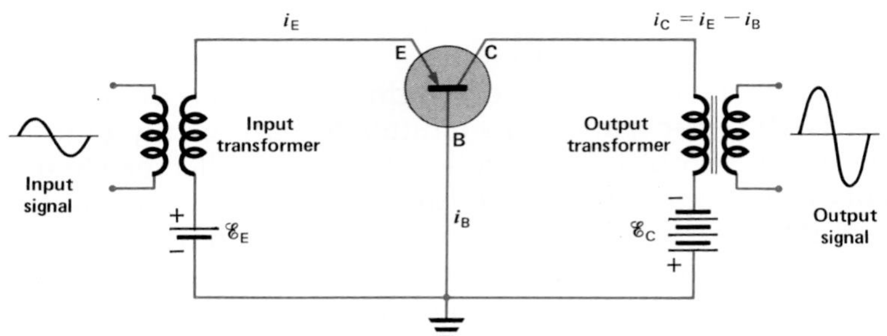

Figure 22-25. A **P-N-P** common-base amplifier.

A typical **P-N-P** common-base amplifier circuit is shown in Figure 22-25. A small signal current i_E in the emitter circuit results in a low input power $i_E^2 Z_E$ because the input impedance Z_E is low. The slightly smaller collector current ($i_C = i_E - i_B$) yields a higher output power $i_C^2 Z_C$ because of the high output impedance Z_C. The *power gain* of the circuit is the ratio of the output power to the input power.

$$P_{\text{Gain}} = \frac{P_{\text{Out}}}{P_{\text{In}}}$$

Suppose the input impedance of the amplifier is $5\bar{0}\ \Omega$, the a-c emitter current is 2.0 mA, the collector current is 1.9 mA, and the output impedance is $1\bar{0}$ kΩ.

The power input is

$$P_{\text{In}} = i_E^2 Z_E = (2.0 \text{ mA})^2 \times 5\bar{0}\ \Omega$$
$$P_{\text{In}} = 0.20 \text{ mW}$$

The power output is

$$P_{Out} = i_C^2 Z_C = (1.9 \text{ mA})^2 \times 1\bar{0} \text{ k}\Omega$$
$$P_{Out} = 36 \text{ mW}$$

The power gain is

$$P_{Gain} = \frac{P_{Out}}{P_{In}} = \frac{36 \text{ mW}}{0.20 \text{ mW}} = 180$$

A **P-N-P** transistor is used in Figure 22-25. An **N-P-N** transistor would serve just as well with the bias-voltage polarities reversed. In some circuits the two types of junction transistors are equivalent. In other circuits one type of junction transistor may have small advantages over the other.

The common-base transistor circuit that we have been considering is one of three general circuit configurations for junction transistors. Others are the common-emitter and the common-collector configurations. These configurations for both **N-P-N** and **P-N-P** transistors are shown in Figure 22-26. Observe that in all circuits the emitters have forward bias to provide low impedance and the collectors have reverse bias to provide high impedance.

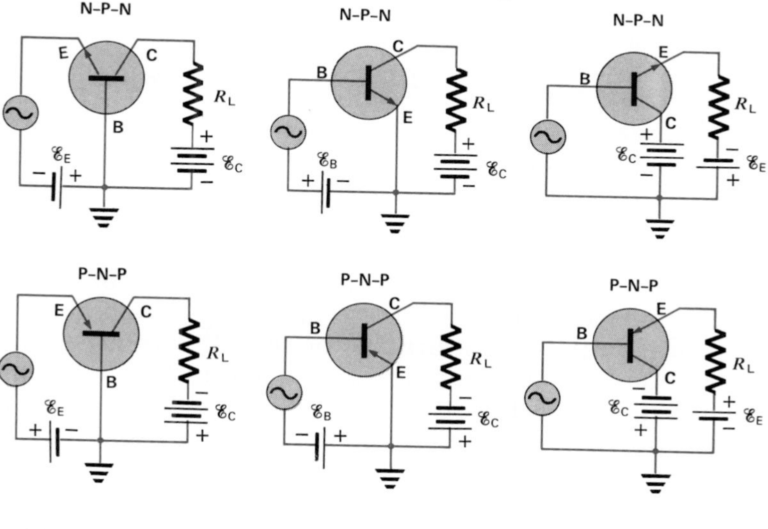

Figure 22-26. Circuit configurations for **N-P-N** and **P-N-P** junction transistors.

The amplifier output signal in Figure 22-25 is *in phase* with the input signal. The common-base amplifier *does not* produce phase inversion. However, the common-emitter amplifier *does* produce a phase shift, as does a vacuum-tube amplifier.

The amplifying property of a transistor is the result of current changes that occur in the different regions of the structure when proper voltages are applied. Consequently, the transistor is often referred to as a current amplifier. More precisely, it is a *current-operated device* in much the same way that a vacuum triode is a voltage-operated device. A transistor amplifier can produce a voltage gain, a current gain, or a power gain, depending on the circuit configuration.

When the operation of a transistor is compared with that of a vacuum triode, the emitter is equivalent to the cathode, the base to the control grid, and the collector to the plate. These relationships are illustrated in Figure 22-27. The emitter and base form the signal input circuit for a transistor amplifier in the same way that the grid and cathode form the input circuit of the vacuum-triode amplifier. The collector is the output section of the transistor just as the plate is the output of the vacuum triode.

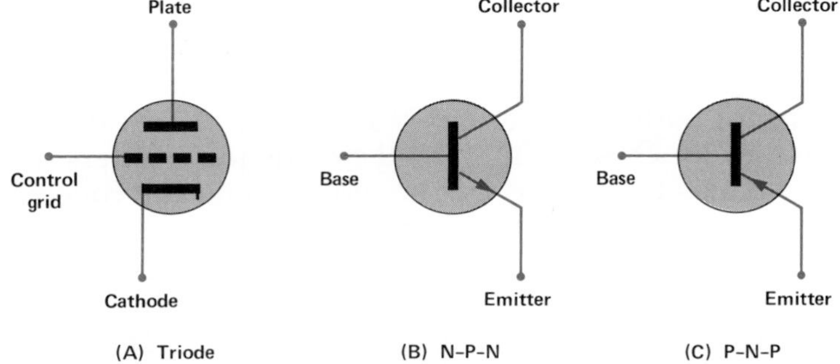

Figure 22-27. Transistor elements compared with those of a vacuum triode. (A) Triode circuit symbol. (B) **N-P-N** transistor circuit symbol. (C) **P-N-P** transistor circuit symbol.

22.15 Transistor Characteristics A small signal applied to the input circuit of a transistor can be amplified because the magnitude of collector current in its output circuit is controlled by the current in its input circuit. The performance of a transistor is indicated by its family of characteristic curves. These curves describe the behavior of the transistor in a specific circuit configuration.

Typical output characteristics for an **N-P-N** transistor in a common-base configuration are shown in Figure 22-28. Each curve shows how the collector current I_C varies with the output voltage V_{CB}, measured between the collector and base terminals, for a given value of emitter current I_E.

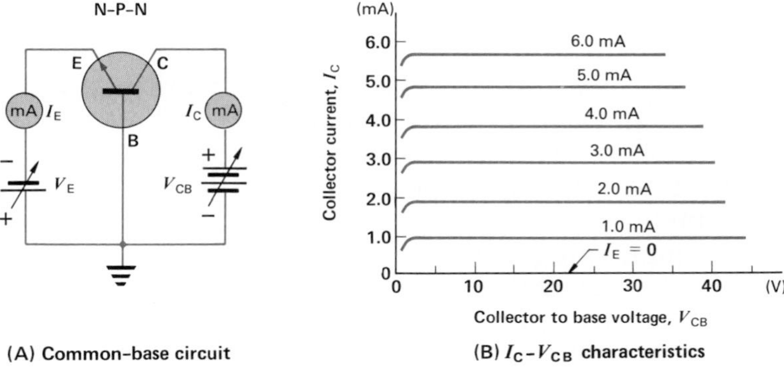

Figure 22-28. Typical common-base output characteristic curves for an **N-P-N** transistor. (A) Common-base test circuit. (B) $I_C - V_{CB}$ characteristics.

The family of curves is produced by using different values of emitter current.

These curves show that the collector current is independent of the collector voltage but is dependent on the magnitude of emitter current. Variations in I_E produce variations in I_C. At a constant value of V_{CB}, the ratio of a change in I_C to a change in I_E expresses the *current gain* of the common-base circuit. It is expressed as a positive number without dimension and is denoted by the Greek letter α (alpha).

$$\alpha = \frac{\Delta I_C}{\Delta I_E} \ (V_{CB} \text{ constant})$$

Referring to Figure 22-28(B), suppose the emitter current I_E is varied from 2.0 mA to 4.0 mA (a change of 2.0 mA) while the voltage V_{CB} is held constant at 15 V. Reading from the collector current axis, I_C will be found to vary from 1.9 mA to 3.8 mA (a change of 1.9 mA). The current gain can be determined from these data as follows:

$$\alpha = \frac{\Delta I_C}{\Delta I_E} = \frac{1.9 \text{ mA}}{2.0 \text{ mA}} = 0.95$$

The forward current gain, or alpha characteristic, of a common-base circuit is always slightly less than unity. Typical values of α range from 0.95 to 0.99. The closer α is to unity, the better the quality of the transistor.

Figure 22-29 shows a typical family of characteristic curves of transistor performance in a common-emitter configuration. Collector current I_C is plotted as a function of output voltage V_{CE}, measured between collector and emitter, for each value of base current I_B.

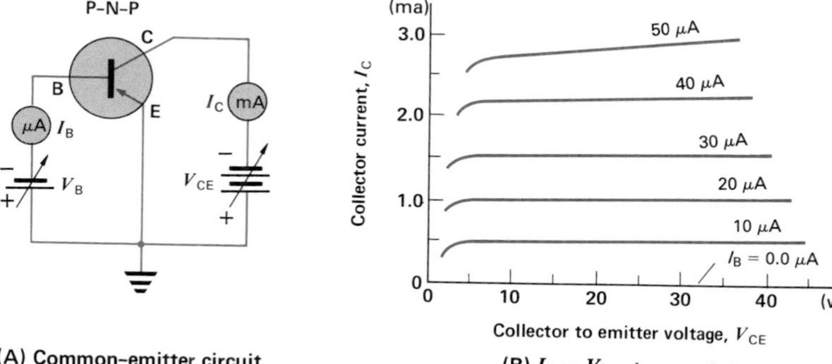

Figure 22-29. Typical common-emitter output characteristic curves for a **P-N-P** transistor. (A) Common-emitter test circuit. (B) $I_C - V_{CE}$ characteristics.

Base currents are very small compared to emitter and collector currents. In the common-emitter circuit, small changes in I_B produce large variations in I_C. At a constant value of V_{CE}, the ratio of a change in I_C to a change in I_B expresses the *current gain* for the common-emitter circuit.

This current-gain factor is denoted by the Greek letter β (beta).

$$\beta = \frac{\Delta I_C}{\Delta I_B} \text{ } (V_{CE} \text{ constant})$$

Using the typical curves of Figure 22-29(B), we will assume that the base current I_B is varied from $2\bar{0}$ μA to $4\bar{0}$ μA (a change of $2\bar{0}$ μA) while V_{CE} is held constant at $2\bar{0}$ V. Projecting to the collector current axis, we find that I_C will vary from 1.0 mA to 2.3 mA (a change of 1.3 mA). The current gain, or β characteristic, of this common-emitter circuit will be

$$\beta = \frac{\Delta I_C}{\Delta I_B} = \frac{1.3 \times 10^{-3}\text{ A}}{2.0 \times 10^{-5}\text{ A}} = 65$$

Some typical values of β range from 20 to 200. Either the common-base or the common-emitter transistor circuit will provide amplification of an input signal. With a low-impedance load, where high *current gain* is desired, the common-emitter circuit is preferred. With a high-impedance load, where large *power gain* is desired, the common-base circuit will serve as well and is even preferable.

The common-collector circuit is essentially the reverse of the common-emitter circuit. This configuration can be recognized in Figure 22-26. The common-collector circuit has a high input impedance and a low output impedance and a voltage gain of less than unity. Thus it has little value as an amplifier but can be used for impedance-matching between a high-impedance source and a low-impedance load.

22.16 The Photovoltaic Effect Energy-conversion processes are described in Section 17.3 as sources of electric currents. For example, heat is the source of electric current in thermoelectric action. Mechanical stress is the source of current in piezoelectric action. Chemical reactions are the source of current in fuel-cell action. These processes are characterized by a capability for *direct* conversion of thermal, mechanical, or chemical energy to electric energy.

Semiconductor materials can provide the direct conversion of radiant energy to electric energy under suitable conditions. The process is known as the *photovoltaic effect.* If a semiconductor **P-N** junction is designed to absorb incident light, a potential difference can be developed across the junction as a consequence of photovoltaic action. This potential difference can generate an electron current in a load resistance connected across the junction. Thus the

P-N junction can act as a source of emf while exposed to light radiations.

*The **P**- and **N**-type semiconductor sections of a photovoltaic cell can just as well be the reverse of the arrangement shown in Figure 22-30.*

The silicon *solar cell* is a photovoltaic converter of considerable interest. The components of a solar-cell photovoltaic circuit are shown schematically in Figure 22-30. The cell consists of a thin base layer of **P**-type silicon crystal over which a very thin (about 0.5 μm) **N**-type layer is formed. The **P-N** junction is a barrier layer with its electric field situated where the **P**- and **N**-type materials are joined. The external circuit of the cell consists of the resistance load R_L connected across the **P-N** junction at terminals in contact with the **P** and **N** surfaces.

*Review the formation of the **P-N** junction in Section 22.12.*

Figure 22-30. Photovoltaic action in a silicon solar cell.

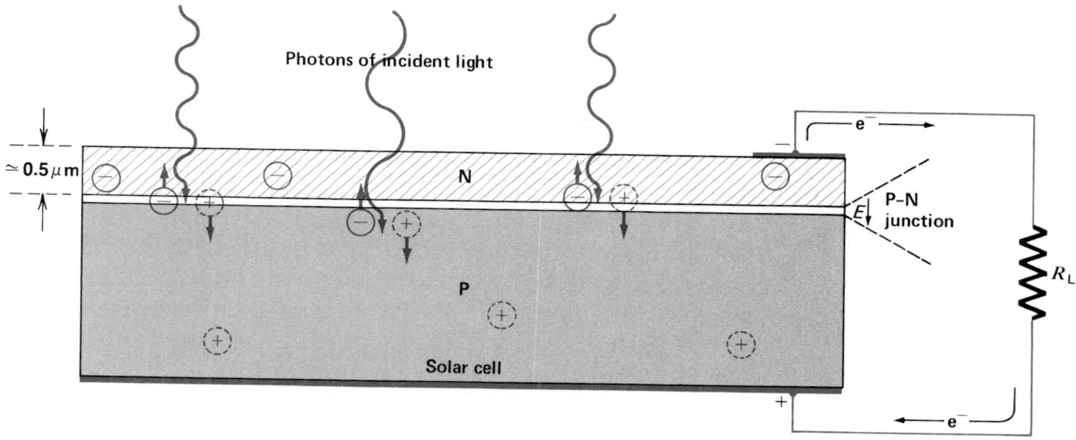

Photons of light radiations with energies corresponding to wavelengths shorter than approximately 10^{-6} m can dislodge electrons from silicon atoms in the crystal lattice of the solar cell. Reference to the radiation spectrum of Figure 12-8 shows that this minimum energy radiation ($\lambda = 10^{-6}$ m) lies in the near infrared region of the electromagnetic spectrum. Figure 12-8 also suggests that roughly half of the sun's radiation, including the visible and ultraviolet regions, has photon energies that can be converted to electric energy by the silicon solar cell.

$\lambda = 10^{-6}$ m $= 1$ μm *(1 micron)* $= 10^4$ Å *identifies radiation in the near infrared region of the solar spectrum.*

When light is incident on the exposed surface of the solar cell, most of the photons pass through the very thin **N** layer and are absorbed by atoms of the crystal lattice in the vicinity of the **P-N** junction. The energy *hf* of an absorbed photon is transferred to an electron, which is then ejected from the atom, leaving behind a positive hole. In this way, light photons generate *electron-hole pairs* in the region of the junction layer. The electric field of the **P-N** junction forces electrons that have not recombined into the **N** layer and a corresponding number of holes into the **P** layer. These electrons are excess negative charge carriers in the **N** layer and flow through the external circuit from the **N** terminal to the **P** terminal. They then combine with the excess holes in the **P** layer. Thus the illuminated solar cell acts as a source of emf and of electron current in its external circuit.

Batteries of solar cells provide electric energy for space vehicles. Earth-bound solar batteries serve many purposes; for example, they supply energy for telephone lines and recharge storage batteries that replace the solar batteries during hours of darkness. The silicon solar cell has an important role in the on-going search for practical and efficient ways to harness solar energy.

A photovoltaic cell of the type usually used in photographic exposure meters consists of a junction formed by a film of copper oxide deposited on metallic copper, or of a film of selenium deposited on metallic iron. Incident light photons pass through the selenium film to the iron. Electrons are moved from the iron into the selenium layer and an emf is generated across the junction. The iron becomes the positive electrode and the selenium the negative electrode of the photovoltaic cell. If a sensitive galvanometer is connected across these electrodes, the emf produces an electron current in the meter that is proportional to the illumination on the cell. By calibrating the galvanometer in light values, proper exposure conditions can be indicated directly by the meter.

The photoelectric effect is described in Section 12.6.

Photovoltaic action is a type of photoelectric effect in the sense that electron ejection from a parent atom is initiated by a light photon. The distinction between photoelectric action and photovoltaic action is that the former requires a vacuum environment and an external source of emf in the photoemission circuit. The latter does not require a vacuum environment, and it generates its own emf. Both types of cells are commonly referred to as *photocells.*

A photoelectric circuit is shown in Figure 17-5.

QUESTIONS: GROUP A

1. What is the property of a semiconductor crystal known as the diode characteristic?
2. Why is **N**-type germanium referred to as electron-rich germanium?
3. Why is **P**-type germanium referred to hole-rich germanium?
4. What is the **P-N** junction of a semiconductor?
5. How are the two types of junction transistors designated?
6. Draw the circuit symbol for an **N-P-N** transistor. Label the emitter, base, and collector.
7. Draw the circuit symbol for a **P-N-P** transistor. Label the emitter, base, and collector.
8. What is the significance of the arrowhead at the emitter in a transistor circuit symbol?

9. Which sections of a junction transistor are equivalent to the cathode, control grid, and plate of a vacuum triode?
10. What are the different circuit configurations used for transistor service?

GROUP B

11. Distinguish between a germanium diode and a germanium transistor.
12. Distinguish between a **P-N-P** transistor and an **N-P-N** transistor.
13. The emitter of an **N-P-N** transistor is biased negatively and the emitter of a **P-N-P** transistor is biased positively with respect to the base. In both cases they are described as forward bias. Explain.
14. What is the alpha characteristic of a transistor and to what transistor configuration does it apply?
15. Why must the current gain in a common-base transistor circuit always be less than unity?
16. Which transistor circuit configuration most closely resembles the basic vacuum triode circuit? Explain why.
17. To what does the beta characteristic of a transistor refer?
18. How can you account for amplification in a common-base transistor circuit when the collector current in the output circuit can never be as large as the emitter current in the input circuit?
19. Photovoltaic action is a type of photoelectric effect. How do the two actions differ?
20. Suggest a reason for the exposed layer of a solar cell being so thin that most of the incident photons will penetrate the layer and be absorbed by atoms in or very near the **P-N** junction layer.

SUMMARY

High-vacuum tubes are identified as diodes, triodes, tetrodes, and pentodes based on the number of electrodes present. Low-vacuum tubes contain a gas under reduced pressure. Diodes are used as detectors for radio signals and as rectifiers for converting alternating current to direct current. They are classed as signal or power diodes.

Electrons are emitted from the cathode of a vacuum tube by a process of thermionic emission. Electron emission is a function of emitter temperature. Emitters can be either directly or indirectly heated.

Voltage and power amplification are possible in triode circuits because a small signal on the grid circuit is capable of controlling a large current in the plate circuit. Conduction in a vacuum tube results from the thermionic emission of electrons from the cathode and the attraction of the electrons to the positively charged plate. The tube bias voltage between the control grid and the cathode establishes the operating condition of the tube. Bias voltages are produced by three methods. The a-c voltage gain across a vacuum-tube amplifier is accompanied by a voltage phase inversion.

P- and **N**-type semiconductors are formed by doping with trace quantities of acceptor and donor atoms respectively. **N**-type semiconductors are electron-rich materials. The free electrons are the negative charge carriers. **P**-type semiconductors are hole-rich materials. The holes act as positive charge carriers. The **P-N** junction acts as an electric barrier in the meeting zone of oppositely doped regions of a semiconductor crystal. The direction of the electric field across the junction is from the **N** side to the **P** side.

Junction transistors consist of two **P-N** junctions back to back with thin **P** or **N** regions in between. They are designated **N-P-N** and **P-N-P**. Transistor elements correspond functionally to the elements of a vacuum triode. In transistor circuits the emitter has a forward-bias voltage and the collector has a reverse-bias voltage. Three circuit configurations are used: common base, common emitter, and common collector.

The transistor is a current-operated device. The amplifying property is the result of current changes that occur in the different transistor regions when proper voltages are applied to the transistor elements. The performance of a transistor is indicated by its family of characteristic curves. The forward current gain of a common-base configuration is called its alpha characteristic. It is the ratio of the change in collector current to the change in emitter current for a constant collector-to-base voltage. The current gain of a common-emitter configuration is called its beta characteristic. It is expressed as the ratio of the change in collector current to the change in base current for a constant collector-to-emitter voltage.

The photovoltaic effect can provide the direct conversion of radiant energy to electric energy. Photovoltaic action is a type of photoelectric effect. The solar cell is a semiconductor diode in which the semiconductor crystal is doped in a special way. The construction and **P-N** junction provide the energy conversion effect when light strikes the cell's surface.

VOCABULARY

acceptor
amplifier
base
cathode-ray tube
collector
cutoff bias
diode
donor
doping
Edison effect
electronics
electron-rich semiconductor
electrostatic lens
emitter
forward bias
hole-rich semiconductor
N-P-N transistor
N-type semiconductor
photovoltaic effect
P-N junction
P-N-P transistor
P-type semiconductor
rectifier
reverse bias
secondary emission
solar cell
space charge
thermionic emission
transistor
triode

23 Atomic Structure

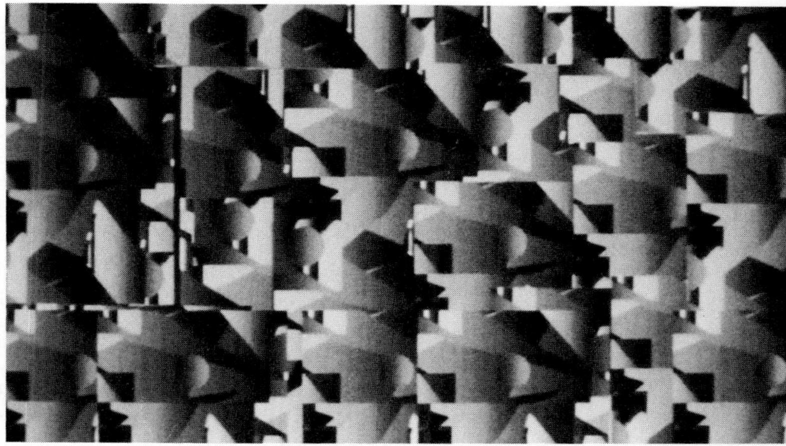

mass number n.: the sum of the number of protons and neutrons in the nucleus of an atom.

THE ELECTRON

OBJECTIVES

- Define the mass, size, and charge of the electron.
- Describe the wave-particle nature of the electron.
- Analyze the relative positions of the atomic nucleus and electron shells.
- Study the history behind the discovery of the proton and the neutron.
- Define the nature of isotopes.
- Discuss the relationship between nuclear binding force and nuclear mass defect.

23.1 Subatomic Particles Toward the end of the 19th century, evidence indicating that atoms were not the ultimate particles of matter began to appear. A simple way of demonstrating the existence of subatomic particles is shown in Figure 23-1. In each photo, the glass tube contains hydrogen gas under low pressure. The conductor leading to the center of the tube is connected to the negative terminal of a direct-current source of electricity. The conductor at the top of the right bulb in each picture is connected to the positive terminal. When a high voltage is impressed across the tubes, two beams shoot out in opposite directions from the hollow electrode in the center. The two beams have different colors.

Since the tubes contain only hydrogen atoms, the differently colored beams indicate that the hydrogen atom has at least two different parts. The fact that the beams flow in opposite directions shows that the two parts of the atom have opposite electric charges. One beam is negatively charged since it passes into the part of the tube where the positive electrode is located. The other beam is positively charged since it is attracted through the negative electrode at the center and into the left side of the tube.

The different colors in the tube also indicate that the electrons of the hydrogen atoms have been excited to different energy levels. Refer to Section 12.11.

The magnets in the pictures tell still more about the parts of the hydrogen atom. In the top picture, the magnet is held near the positive beam. The effect on the beam is

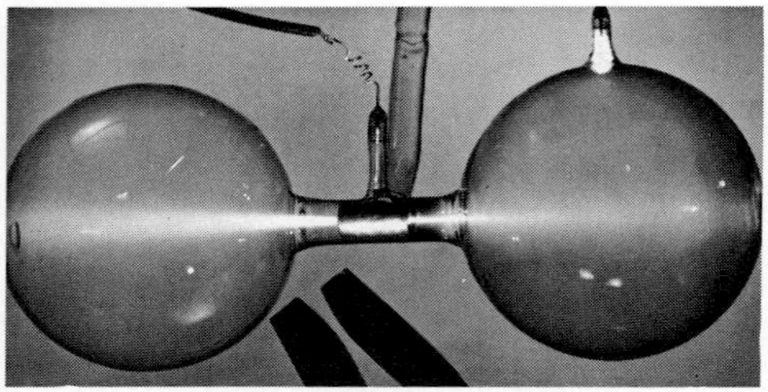

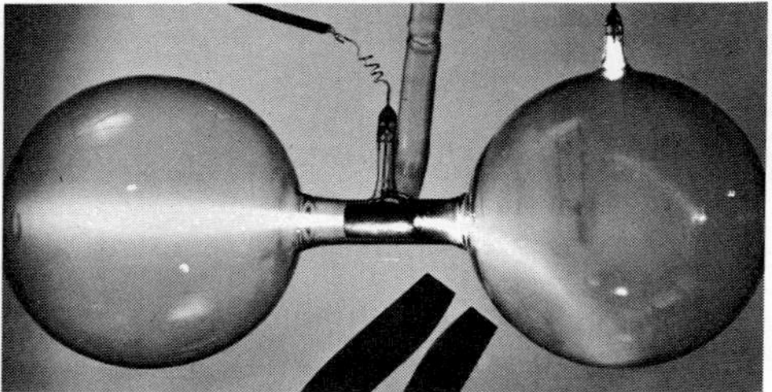

Figure 23-1. Evidence for subatomic particles. The dumbbell-shaped apparatus contains the gas, hydrogen. When d-c voltage is applied, hydrogen atoms separate into streams of oppositely charged particles. This shows that hydrogen atoms are composed of at least two different particles.

not very pronounced. In the bottom picture, the magnet is held near the negative beam. This time the bending effect is more noticeable. This suggests that the particles composing the two beams have differences in addition to that of opposite charge. Further studies show that the main reason for the difference in the way the beams react to magnetism is that the positive particles have much larger masses than do the negative particles.

With this and other experiments, scientists have developed a comprehensive theory of atomic structure. Various parts of this theory will be presented in this and the following chapters. It is important to remember, however, that all these discussions are based on current knowledge of atomic structure. Revisions are frequently necessary as new investigations and new interpretations are made. What sort of picture of the atom will finally emerge from this research no one can fully predict. It is certain that the atom is made up of a number of still smaller particles that are arranged in various complex ways. These subatomic particles interact with each other to make up various atoms.

23.2 Discovery of the Electron The first subatomic particle to be identified and studied by scientists was the electron. Actually, the discovery of this particle was a by-product of other investigations. Toward the end of the 19th century physicists were using tubes similar to the one just described to study the conduction of electricity through various gases. When such a tube was filled with air (at atmospheric pressure), electricity would not flow between the electrodes, even under high voltages. When some of the air was removed, however, electricity began to flow from one electrode to the other through the rarefied air. The air in the tube glowed with a purplish light. When still more air was pumped from the tube, the purplish light faded and the glass walls of the tube glowed with a greenish fluorescence.

Additional work with evacuated tubes showed that the greenish glow on the glass walls was produced by the particles emitted from the cathode, or negative electrode. Figure 23-2(A) shows how this was deduced. With the electrodes placed as shown in the figure, the walls of the tube glowed everywhere except where the cross inside the tube cast a shadow on the face of the tube. The shadow is in a direct line with the cross and the cathode. This means that the green glow is produced by something coming from the cathode in straight lines. These so-called *cathode rays* are invisible to the eye unless they produce a visible

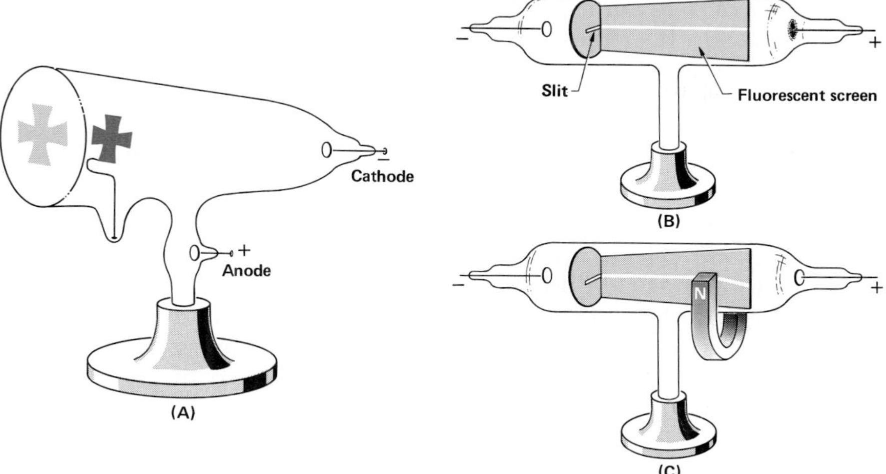

Figure 23-2. Properties of cathode rays. The shadow cast by the cross in (A) shows that cathode rays travel in straight lines. The coated screen in (B) shows the ability of cathode rays to produce fluorescence. In (C) the negative charge of cathode rays is identified by means of a magnet.

effect on some other substance, such as a glow on the glass walls of the tube. In addition, these cathode rays can be stopped by substances like the cross inside the tube.

In 1895, the French scientist Jean Perrin (1870–1942) discovered another property of cathode rays. He arranged an evacuated tube as shown in Figure 23-2(B). The slit in the tube forms a broad beam of cathode rays that strikes the fluorescent screen and causes it to glow. The screen is placed at an angle so that the beam is visible along the entire length of the screen. Ordinarily, the beam would trace a straight horizontal path on the screen. But when Perrin placed a magnet in the position shown in Figure 23-2(C), the beam was deflected downward. When he reversed the poles of the magnet, the beam was deflected upward. Since light beams are not deflected by magnets, Perrin concluded that cathode rays were really streams of negative electricity coming from the cathode.

The modern television picture tube makes use of many of the principles discovered many years ago by scientists working with cathode rays. The face of the picture tube glows when cathode rays strike it. In color television, different chemicals are used on the face of the tube so that multicolored glows are produced. The position of the picture on the tube is adjusted by magnets surrounding the neck of the tube inside the set.

Other scientists continued the investigation of cathode rays begun by Perrin. Among these were the English physicists Sir William Crookes (1832–1919) and Sir Joseph Thomson. Thomson was able to show that cathode-ray particles are much lighter than atoms and are present in all forms of matter. He arrived at this latter conclusion by observing that the nature of the particles did not change

The word "electron" is derived from a Greek word meaning "beaming sun."

when he changed the composition of the cathode or of the gas in his tubes. Thus Thomson was the first to discover a subatomic particle—the ***electron.***

23.3 Measuring the Mass of the Electron It is obviously impossible to use a laboratory balance to determine the mass of something as small as an electron. Even if you could pile as many as 10^{13} electrons on the balance, the total mass would still amount to only 10^{-17} kg. An indirect method must therefore be used. Thomson developed such a method in a famous series of experiments in 1897. The basic parts of this experiment are shown in Figure 23-3. The cathode rays (electrons) emanating from the electrode at the left are accelerated and focused by the positively charged disk that has a small hole in its center. The potential difference, V, between the cathode and the disk determines the kinetic energy of each electron since $eV = \frac{1}{2}mv^2$, where e is the electric charge on a single electron. A second disk further narrows the electron beam.

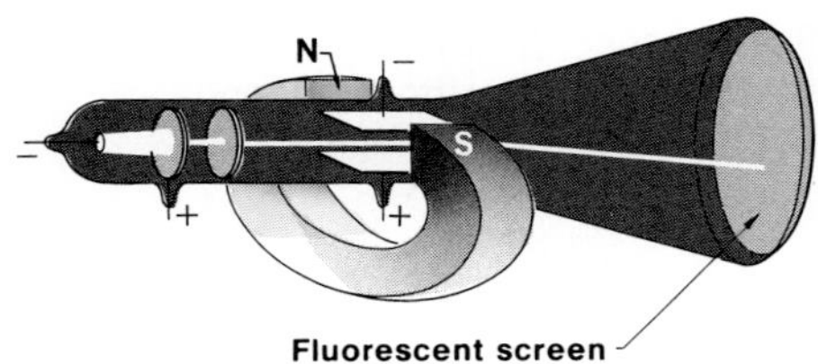

Figure 23-3. The Thomson experiment. By means of a cathode-ray tube, Thomson found that cathode radiation particles are subatomic in mass and are present in all forms of matter.

The beam then passes between two oppositely charged plates on its way to a fluorescent screen. These charged plates produce an electric field. Surrounding the neck of the tube is a magnetic field that is set at right angles to the electric field. In the arrangement of Figure 23-3, the electron beam is under the influence of a downward force due to the electric field and an upward force due to the magnetic field. Both forces depend on the charge of the electrons, but the force due to the magnetic field also depends on the speed of the electrons. By varying these forces until they exactly balance each other, the speed of the electrons, v, can be determined and used in $eV = \frac{1}{2}mv^2$. Two unknowns, e and m, still remain. Though neither e nor m can be determined individually by this method, their ratio e/m can be calculated from the measured values of the potential difference, the electric field, and the magnetic field. In other words, Thomson was able to calculate the quantity of electric charge per unit mass of electrons. This quantity is expressed in coulombs per kilogram of electrons. The presently accepted value for the ratio is $1.758\,819\,62 \times 10^{11}$ C/kg.

This measurement is a step in the right direction, but it still does not tell us the mass of an electron. If the charge on a single electron could be measured, then the charge-to-mass ratio could be used to calculate the mass. Similarly, if the mass could be determined, the ratio would tell us the charge. In 1912, the American physicist Robert A. Millikan succeeded in measuring the charge on the electron; the problem was solved. His experiment is described in the next section. First we shall take a look at a completely different way of measuring the electron's mass.

Millikan's work with photoelectric emissions is described in Section 12.7.

As mentioned in Chapter 1, Einstein's equation, $E = mc^2$, states the relationship between mass and energy. Consequently, if an electron could be converted into energy, Einstein's equation could be used to calculate the original mass of the electron from the resulting energy. Conversely, if a form of energy such as a beam of X rays could be converted into an electron, the mass could again be calculated from the known energy of the X rays. Figure 23-4 shows the results of such an experiment. As the diagram to the right of the photograph indicates, X rays had entered from the left and then changed into a pair of electrons, one with a positive charge, called a positron, and one with a negative charge. The bubble chamber in which the photograph was taken was in a magnetic field that caused the two particles to move in circular paths after their formation; the circles get smaller as the particles lose energy.

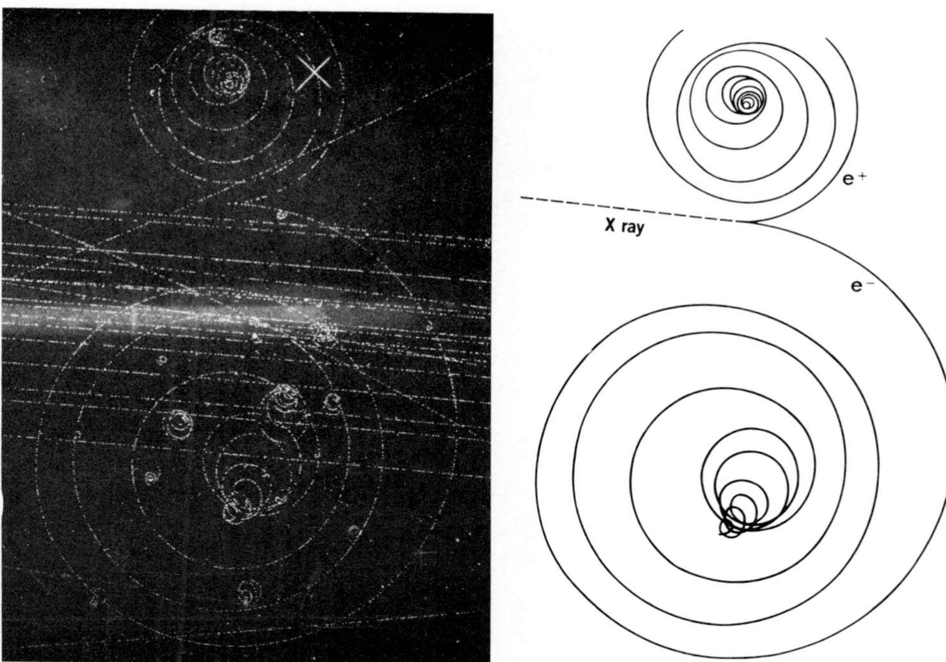

Figure 23-4. Pair production, as photographed in a particle detector. X rays entering from the left were converted into a positron-electron pair.

Repeated experiments have shown that to produce an electron pair, X rays must have energies greater than 1.64×10^{-13} joule. Since X rays with this energy produce two particles (a single electron is never produced), and since we assume that the two particles have equal masses, the mass of each particle is equivalent to half of this energy, or 8.2×10^{-14} joule. Solving Einstein's equation for m and substituting,

$$m = \frac{E}{c^2} = \frac{\mathbf{8.2 \times 10^{-14}\ J}}{\mathbf{(3.0 \times 10^8\ m/s)^2}}$$

$$m = \mathbf{9.1 \times 10^{-31}\ kg}$$

In Section 1.14 it was mentioned that the mass of an object varies with its velocity. The two particles in Figure 23-4 have very high velocities. But the mass computed above agrees with the accepted value of the *rest mass* to two significant figures. In terms of atomic mass units, the rest mass of the electron has been measured as 0.000 548 579 903 u or $\frac{1}{1837}$ of the mass of an atom of hydrogen. Obviously electrons do not account for a very large share of the mass of substances. Other subatomic particles must be responsible for most of the mass. These particles are identified later in this chapter.

23.4 The Electronic Charge The determination of the charge on the electron by Millikan is one of the classic experiments of physics. A diagram of the apparatus is shown in Figure 23-5. Charged capacitor plates were placed inside a metal box in which temperature and pressure conditions could be held constant. The top plate had a small hole in the center. The plates were connected to a source of high electric potential that could be varied at will. The plates were parallel and the distance between them was carefully measured. From the voltage and the plate spacing, the electric field, E_q, could be computed. The space between the plates was illuminated by a spotlight. Through a telemicroscope, a combination of two instruments used for viewing small objects at intermediate distances, this illuminated space could be viewed.

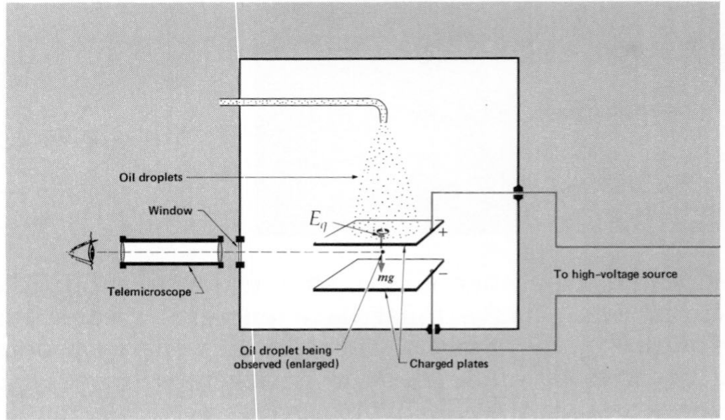

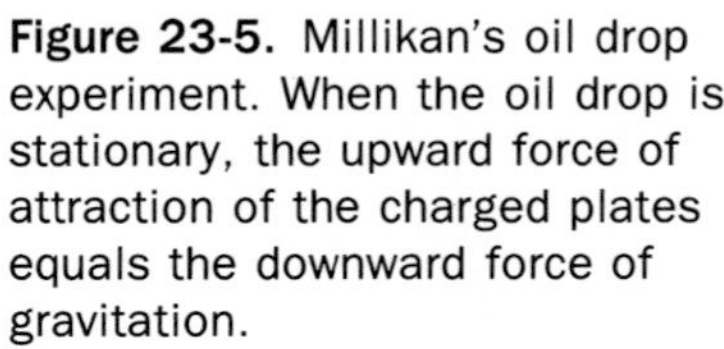
Figure 23-5. Millikan's oil drop experiment. When the oil drop is stationary, the upward force of attraction of the charged plates equals the downward force of gravitation.

A mist of light oil was sprayed into the region above the plates. Eventually a single oil drop fell through the opening in the top capacitor plate. The spray was then turned off so that this single drop could be studied as it fell between the plates. When the plates were uncharged, the drop fell through the field of view with a constant terminal velocity that was dependent on the drop's weight, mg. By means of crosshairs in the eyepiece of the telemicroscope, this velocity could be accurately measured. When the voltage was turned on, however, the behavior of the oil drop changed. The drop slowed its descent, stopped in midair, or even began to rise. These variations in behavior depended upon the weight of the drop, the amount of charge it acquired in the oil-spraying process, and the strength of the electric field between the capacitor plates.

Consider the case of a drop that was suspended motionless between the two plates. The force of gravity on the drop was exactly balanced by the force of the electric field between the two plates. In order for this to happen there had to be an electric charge on the oil drop upon which the electric field could exert a force. Hence the field strength between the capacitor plates needed to suspend the drop was a measure of the electric charge on the drop. This observation did not tell how many charged particles were on the drop, however, and so the value of the unit charge was still unknown.

Millikan studied the behavior of thousands of oil drops in the apparatus. By directing X rays at the drops, he was able to change the charge on a drop in midair as a result of the photoelectric effect. Finally he was able to arrive at two important conclusions:

1. Even though the equipment was capable of detecting much smaller values, the charges on the oil drops never fell below a certain minimum value.
2. All charges were integral multiples of the minimum charge.

Millikan correctly concluded that the minimum charge he had measured was the electronic charge—the charge on a single electron. This important fundamental unit of physics has the presently accepted value of $1.602\,177\,33 \times 10^{-19}$ C. The mass of the electron could now be computed by dividing the charge by the charge-to-mass ratio found by Thomson.

$$m = \frac{e}{e/m} = \frac{\mathbf{1.602\ 177\ 33 \times 10^{-19}\ C}}{\mathbf{1.758\ 819\ 62 \times 10^{11}\ C/kg}}$$

See Section 23.3 for the value of e/m.

$$m = \mathbf{9.109\ 3 \times 10^{-31}\ kg}$$

This result agrees with the mass of the electron as determined by the bubble chamber method described earlier. It is, however, much more precise.

The sizes of various atoms are given in Section 7.3.

23.5 Size of the Electron Since the electron is a subatomic particle, it must be smaller than the smallest atom, which is only about 1 Å in diameter. Suppose a narrow beam of electrons is directed through a thin metal foil and then allowed to strike a fluorescent screen. (The interatomic distances in the foil can be quite accurately measured by X rays.) If the electrons are very small, they can be expected to pass straight through the spaces between the foil atoms and to produce a bright narrow spot on the screen. Some of the electrons will come close enough to an atom in the foil to be partly deflected (scattered) by the atom, producing a fuzziness around the spot on the screen. A few may even come so close to hitting atoms head-on as to be strongly scattered, still further broadening the spot on the screen. Careful measurements of such scattering could be used to estimate the size of the electron. The actual result of such experiments, however, is a series of concentric rings in addition to a central spot. Figure 23-6 is a photograph of the image produced in such an experiment. The central spot is understandable, but where do the rings come from? In Chapter 15 we saw that a diffraction grating produces several discrete images of the incident light. The grating is a one-dimensional array of equally spaced lines. In a similar sense, the metal foil is a three-dimensional array of regularly spaced atoms. This should produce a set of discrete images that are symmetrical around a center spot. This is indeed the case when X rays, which are short wavelength photons, pass through a crystal. When X rays pass through a metal foil, the pattern will appear very much like Figure 23-6. In other words, the attempt to measure the size of the electron produced the unexpected evidence that the electron has wave properties!

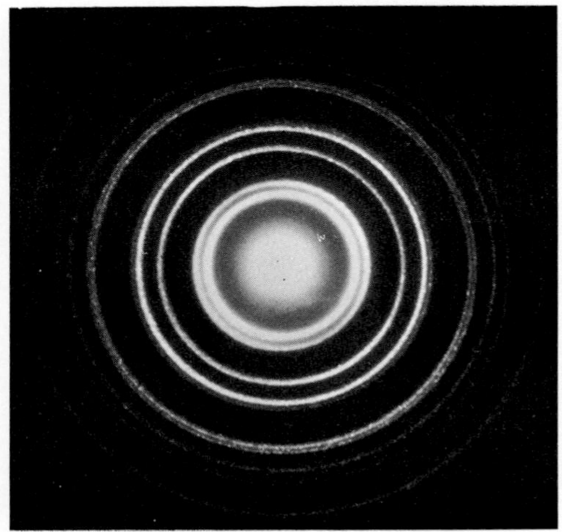

Figure 23-6. An electron diffraction pattern made with an electron microscope of a gold film about 40 Å thick. The pattern shows the rings characteristic of a crystalline substance.

Further experiments with electron patterns have shown that the patterns vary not only with the type of metal foil through which the electrons are fired but also with the momentum of the electron beam. If a description of the electron is given in terms of its wavelength, and if this wavelength is computed from the changing ring patterns, we must conclude that the size of the electron varies with its momentum.

This is a form of the wave-particle duality of matter that was mentioned in Section 1.14.

In the measurements of Thomson and Millikan the electron behaved as a charged particle in an electric or magnetic field. Now we see that it behaves as a wave. Does this mean that an electron is sometimes a particle and sometimes a wave? No. It means that *an electron sometimes exhibits particle characteristics and sometimes wave characteristics, depending upon the nature of the experiment and the measuring instruments used.* Thus an electron may at times be described as a wave and at times as a particle. Actually, it is neither a particle nor a wave—it is an electron.

Figure 23-7. A hydrogen atom has a nucleus consisting of one proton. One electron moves about this nucleus in the K shell. Electron paths are not as definite as these diagrams show.

23.6 Motions of the Electron Figures 23-7 and 23-8 are typical representations of hydrogen and carbon atoms. It is important to note that the electrons are not shown as either particles or waves; they are shown as rings of varying thickness around the center of each atom. Electrons are negatively charged. Hence a neutral atom must have a positive charge in its nucleus. If the electrons were not in motion, the force of attraction between these opposite charges would immediately produce a collapse of the atom. Our knowledge of atomic and nuclear sizes and of chemical and nuclear reactions precludes the possibility of the existence of collapsed atoms in ordinary matter. The force of attraction between the electron and the nucleus provides a centripetal force sufficient to hold the electron in its orbit.

The wave-particle duality of the electron makes it difficult to describe its path around the nucleus. Each electron has a definite amount of energy that determines the *orbital* in which it moves. An orbital is not an orbit or path in the sense that the orbit of a planet is. An orbital is a probable pattern of movement characteristic of the energy of the electron. Groups of orbitals in an atom are usually called *shells.* Shells are frequently designated by letters of the alphabet beginning with K and continuing sequentially to Q in the most complex elements. They also may be given numbers from 1 to 7. The number of orbitals in a shell is the square of the shell number (1st shell, 1 orbital; 2nd shell, 4 orbitals; 3rd shell, 9 orbitals, etc.). The maximum number of electrons that can occupy an orbital is 2; thus the maximum number of electrons that can occupy a shell is two times the number of orbitals in the shell.

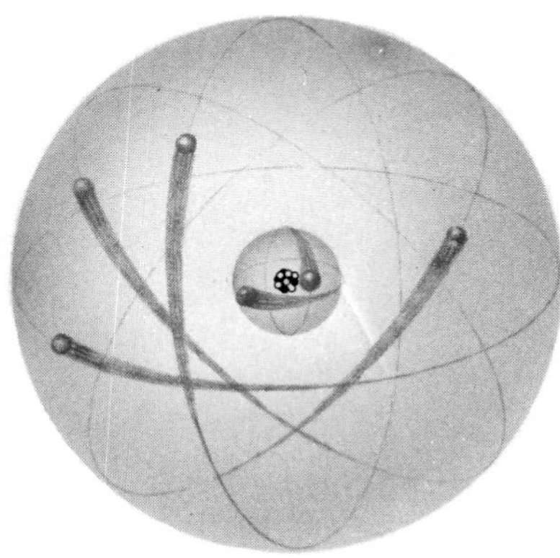

Figure 23-8. A carbon atom has a nucleus consisting of 6 protons and 6 neutrons. Two electrons are in the K shell and 4 electrons are in the L shell.

The aggregate of electrons is referred to as the *electron cloud.* The characteristics of the electron cloud about an atom give the atom its volume and prevent the interpenetration of one atom by another under ordinary conditions.

Experiments have also shown that, in addition to orbiting the nucleus, the electron spins on its own axis.

The evidence and consequence of electron spin were mentioned in Section 19.2.

QUESTIONS: GROUP A

1. Why were electron beams called *cathode rays?*
2. What evidence showed that these "rays" were really streams of charged particles?
3. Why is Thomson credited with discovering the electron?
4. Compare the two particles formed from an X ray in a bubble chamber.
5. (a) Define the term *orbital.* (b) How are orbitals related to electron shells?
6. How many (a) orbitals are in the 5th shell (b) electrons are in the 3rd shell?
7. What motions does an electron exhibit?

GROUP B

8. What evidence showed that electrons have wave and particle properties?
9. Why don't atoms collapse due to electron-proton attractions?

THE NUCLEUS

23.7 Discovery of the Atomic Nucleus In order to be electrically neutral, an atom must contain a positive charge of the same magnitude as the total negative charge of its electrons. But what is the exact location of the positive charge? At the time of the discovery of the electron, there were a number of theories about this. Some scientists suggested that the atom has a uniform structure in which the electrons are distributed evenly throughout a positively charged mass: something like the bits of fruit in a fruitcake. Others thought that the positive charge was concentrated at the center and that the electrons surrounded this positive core.

To test the validity of these theories, in 1911 the English physicist Ernest Rutherford conducted a series of experiments with particles emitted by radioactive materials. As will be explained more fully in Chapter 24, certain elements emit rays and particles that can be studied and identified as being subatomic in nature. One of these particles, called an *alpha particle,* was found to be the same as an atom of helium with its electrons missing. In other words, an alpha particle is a subatomic particle with a positive charge.

An alpha particle is a helium nucleus.

In his experiment, Rutherford used alpha particles that were emitted by a piece of radioactive polonium contained in a lead box, as shown in Figure 23-9. These alpha particles have a velocity of 1.60×10^7 m/s. This gave Rutherford subatomic "bullets" of known mass, charge, and velocity with which to explore the structure of other atoms.

Rutherford directed the alpha particles at thin metal foils of various kinds. First he used gold because it can be hammered into extremely thin sheets. The thickness of the foil was about 10^{-7} m, equivalent to only a few hundred

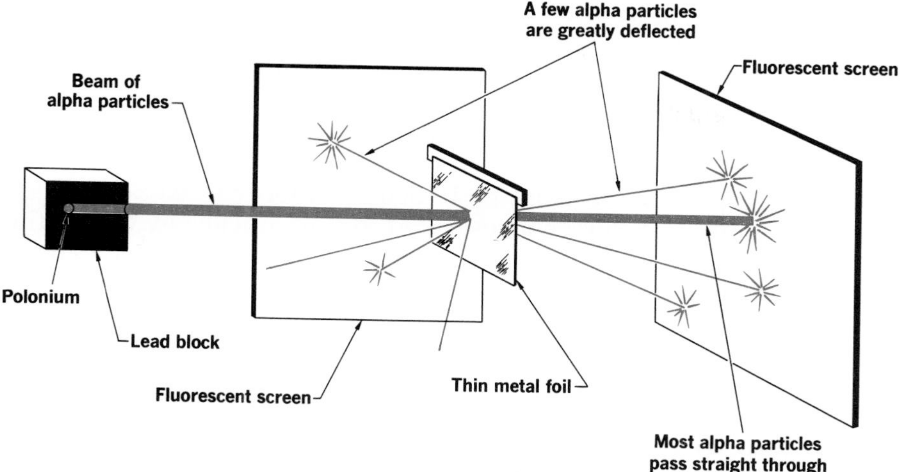

Figure 23-9. Rutherford's alpha scattering experiment. Fluorescent screens show whether the alpha particles from a sample of polonium are deflected when they are directed at a metal foil.

layers of gold atoms. Most of the alpha particles passed straight through the foil. This indicated that the metal foil was very porous to the alpha particles even though the atoms of the metal were tightly packed into many layers. Some of the alpha particles were deflected from their paths, however, at angles ranging from a very slight deflection to a direct rebound. The large deflections were a surprise. Could a massive alpha particle be deflected up to 180° by electrons of much smaller masses found inside the atom or by a positive charge spread out uniformly within the atom? In Rutherford's words: "It was about as credible as if you fired a 15-inch shell at a haystack and it came back and hit you."

The answer must be that the positive charge within the atom is not uniformly distributed as had been thought by some scientists. Instead, it is concentrated in a dense atomic *nucleus* that is able to deflect alpha particles the way a brick wall would deflect a tennis ball. Furthermore, the number of alpha particles rebounding compared with the number passing through the foil provided a clue to the size of the atomic nucleus compared with the size of the entire atom. Rutherford's students sat for hours in a darkened room to count and measure the deflections on a fluorescent screen. They found that about one out of every 8000 alpha particles was deflected by more than 90°.

From this and subsequent evidence, scientists have been able to determine the radius of the largest atomic nucleus at about 10^{-14} m, or 10^{-4} Å. This is less than 1/10 000th of the diameter of the smallest atom, and less than 1/1 000 000 000 000th of its volume. No wonder that alpha particles, which are themselves atomic nuclei, are able to pass through the atomic structure of metal foils with only an occasional collision! This situation is like

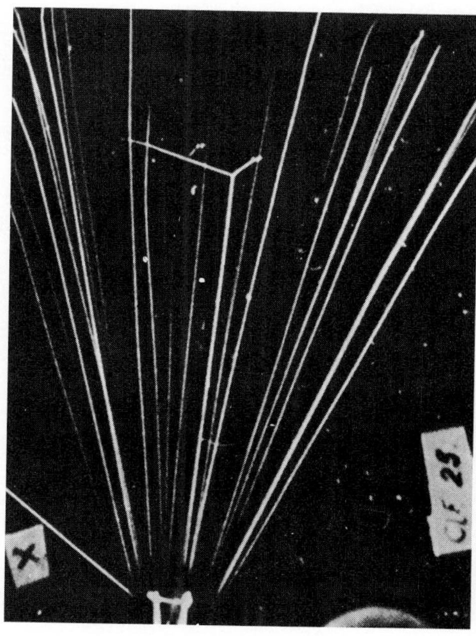

Figure 23-10. Alpha particle tracks in a sample of chlorine gas. Most of the particles are not scattered, but occasionally a particle is deflected through a large angle.

shooting at widely separated peas with an air rifle from the opposite side of a large football stadium.

Astronomers call a mass that consists only of atomic nuclei a "black hole" because not even light can escape from its concentrated gravitational field.

Yet this tiny nucleus accounts for more than 99.95% of the mass of the entire atom. Thus the density of the atomic nucleus is extremely high: about 10^{13} times the density of lead. One cubic centimeter of closely packed atomic nuclei would have a mass of 10 000 000 metric tons! Some stars consist of very concentrated atomic nuclei, in which the protons and electrons of the original atoms have been formed into neutrons by intense pressures.

23.8 The Proton The existence of positive charges in the atom was known even before the time of Rutherford's experiments. If a perforated disk is used as a cathode in a discharge tube, luminous rays are seen coming through the hole on the side opposite the anode, as in Figure 23-1. In 1895, Perrin tested this radiation with magnets and with electrically charged plates and showed that it consists of particles with a positive charge. In 1907, Thomson named them *positive rays.*

It was found that the properties of the positive rays, unlike those of cathode rays, depend on the nature of the gas in the tube. The charge-to-mass ratio of the particles composing positive rays indicated that they are charged particles formed from atoms of the gas in the tube. No positive particles with a charge-to-mass ratio similar to that of the electron were found. The lightest particle found in positive rays has the mass of a hydrogen atom and carries a charge equal in magnitude but opposite in sign to that carried by the electron. This positive particle was assumed to be a hydrogen atom from which one electron had been removed.

The measurement of the amount of positive charge in the nuclei of various elements was also the work of Rutherford and his colleagues. They developed equations stating that the scattering of alpha particles depends on three factors: the thickness of the metal foil, the velocity of the alpha particles, and the amount of positive charge in the atomic nuclei. Experiments soon showed that the first two of these relationships agreed remarkably with the results as predicted by the equations. So it could be assumed that the third relationship, scattering vs. relative nuclear charge, was also true. Tests were made with many different elements, and it was found that *the nuclear charge of the atoms of a given element is always the same and is characteristic of that element.*

The smallest nuclear charge is that of the hydrogen nucleus. In 1920, the name ***proton*** was given to the nuclear particle carrying this unit positive charge. The number of protons, and therefore the amount of positive charge (measured in units of electron charge), is called the ***atomic number.*** Atomic numbers range from 1 for hydrogen to over 100 for the most massive elements. The positive charge of a proton has the same magnitude as the negative charge of an electron. Consequently the atomic number of an element also gives the number of electrons surrounding the nucleus of neutral atoms.

The atomic number of an element is the number of protons in its nucleus.

The mass of the proton has been found to be $1.672\,623\,1 \times 10^{-27}$ kg, or 1.007 276 470 u on the atomic mass scale. This is $\frac{1836}{1837}$ of the mass of a hydrogen atom. Thus the mass of a hydrogen atom is contained almost entirely in its nucleus.

23.9 The Neutron

If an atom consisted entirely of electrons and protons and if the mass of the atom were almost entirely contained in its protons, then it should be possible to predict the atomic mass of an element from its atomic number. For example, since the atomic number of oxygen is 8, the atomic mass of oxygen should be eight times the atomic mass of hydrogen, which consists of a single proton and single electron. Similarly, the atomic mass of uranium ought to be 92. But this is not the case. The atomic mass of the most abundant form of oxygen is about 16 and that of the most common form of uranium is about 238. The same discrepancy shows up in all the elements except ordinary hydrogen. What explanation can be given to resolve this discrepancy?

One possibility is that the excess mass is due to protons whose positive charge is neutralized by electrons located within the nucleus. However, such an arrangement presents a number of complex problems that rule it out as a possible solution. For instance, the spin of such nuclear electrons would produce effects that disagree with observed nuclear behavior.

The answer was provided by the English physicist James Chadwick (1891–1974). In 1930 physicists found that when certain elements, such as beryllium or boron, are bombarded with alpha particles from the radioactive element polonium, a radiation with very high penetrating power is obtained. Two years later Chadwick discovered that this highly penetrating radiation is able to knock protons out of a block of paraffin, a compound containing carbon and hydrogen atoms. Figure 23-11 is a schematic diagram of Chadwick's experiment. The radiation that produces this proton emission is not a beam of X rays because it can be shown that X rays would require unusually high energies to expel protons from paraffin. Chadwick reasoned that the radiation is a stream of neutral particles.

In 1935 Chadwick received the Nobel Prize in Physics for his discovery of the neutron.

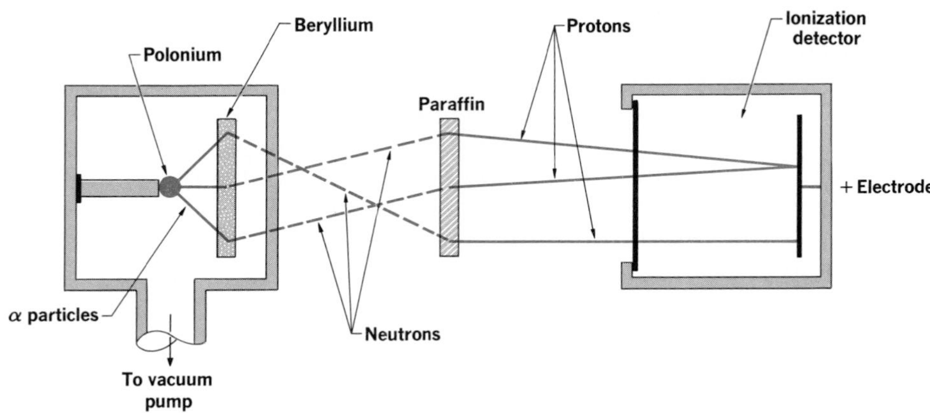

Figure 23-11. Chadwick's neutron experiment. Particles ejected from beryllium were in turn able to knock protons out of a sample of paraffin. The intermediate particles were found to have no charge but to have a mass similar to that of the proton. Chadwick called them neutrons.

Calculations showed that the particles have masses similar to that of the proton.

Additional experiments indicated that the same neutral particles can be released from a great variety of elements. This means that the neutral particles are regular components of atomic nuclei. Chadwick called these particles ***neutrons.*** They have a mass of $1.674\,928\,6 \times 10^{-27}$ kg (very slightly higher than that of the proton), or 1.008 664 904 *u* on the atomic mass scale. The two particles are also quite similar in size.

An interesting result of recent research with neutrons is shown in Figure 23-12. The bottom photo, which was made with the help of neutrons, shows more inner detail in the opaque toy locomotive than does the X-ray photo in the center. The neutrograph is made by passing high-intensity neutrons through the locomotive to a metallic converter, where they release the rays that expose the photographic film. Industrial and medical applications of neutron radiography are being developed.

Figure 23-12. Neutrons can be used to take pictures of opaque objects, revealing more inner detail than is obtainable with X rays.

23.10 Isotopes

Experiments show that the chemical properties of an element are related to its atomic number, Z. For hydrogen $Z = 1$, for carbon $Z = 6$, and for uranium $Z = 92$. The total number of protons and neutrons in the nucleus of an atom is equal to the *mass number, A.* Using the symbol N to represent the number of neutrons,

$$A = Z + N$$

Protons and neutrons are referred to collectively as *nucleons.* Thus in the formula above, A equals the number of nucleons.

All atoms of an element have the same chemical properties, but their masses may be different. The number of neutrons in the atomic nucleus of a given element may vary. Atoms of a given element that have different masses are called *isotopes.* ***Isotopes*** *of an element contain the same number of protons but a different number of neutrons. Each dif-*

ferent variety of atom as determined by the composition of its nucleus is called a ***nuclide.*** Nuclides having the same atomic number are isotopes. There are three isotopes of hydrogen: *protium* (ordinary hydrogen), 1 proton, no neutrons; *deuterium,* 1 proton, 1 neutron; and *tritium,* 1 proton, 2 neutrons. Isotopes of all chemical elements either occur naturally or have been produced artificially.

In general, nuclei with odd atomic numbers exist naturally in only one or two isotopic forms. Those with even atomic numbers exist in several isotopic forms; some elements have as many as eleven variations. Figure 23-13 is a graph of the nuclei of the naturally occurring light nuclides in which Z is plotted as the abscissa and N as the ordinate. Notice that Z and N are equal in many of the lightest nuclides, but that the neutrons outnumber the protons as the mass number increases. This fact is reflected in the properties of the elements, some of which will be discussed in the next chapter.

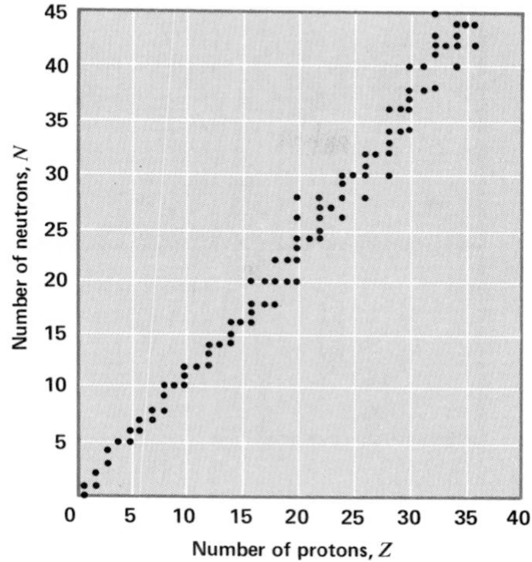

Figure 23-13. Nuclei of the stable isotopes of the light elements. Note that the lightest nuclides have equal numbers of neutrons and protons, but in the heavier nuclides there are more neutrons than protons.

Naturally occurring elements are mixtures of isotopes in quite definite proportions. Hydrogen consists of 99.985% protium, 0.015% deuterium, and less than 0.001% tritium. *The* ***gram-atomic weight*** *of an element is the mass in grams of one mole of naturally occurring atoms of an element.* Hydrogen has a gram-atomic weight of 1.007 97 g and carbon has a gram-atomic weight of 12.011 15 g. Similarly, the gram-atomic weights of all naturally occurring elements are mixed numbers.

The gram-atomic weight is the mass of a naturally occurring mole of an element.

23.11 The Mass Spectrograph The masses of ionized atoms can be measured with great precision in a device called a *mass spectrograph.* There are many different types of mass spectrographs, but in all types the operation depends on the slight difference in mass between isotopes. One type of spectrograph is shown in Figure 23-14.

A beam of ions enters from the ion gun and passes into a region of crossed electric and magnetic fields that allows only the ions with a specific velocity to pass through the slit **S.** The velocity of the ions is inversely proportional to their masses. (Ions with other velocities are deflected and blocked by the chamber wall around the slit.) As the ions enter the upper chamber they are influenced by another magnetic field perpendicular to the plane of the page. This magnetic field causes the ions to move in a circular path. If the ions have the same mass-charge ratio, they will all describe the same arc and will strike the photographic plate at the same place. However, if the beam entering the slit contains ions of various masses, the less massive ions will be deflected more than the more massive ones and will strike the photographic plate at different places. If the

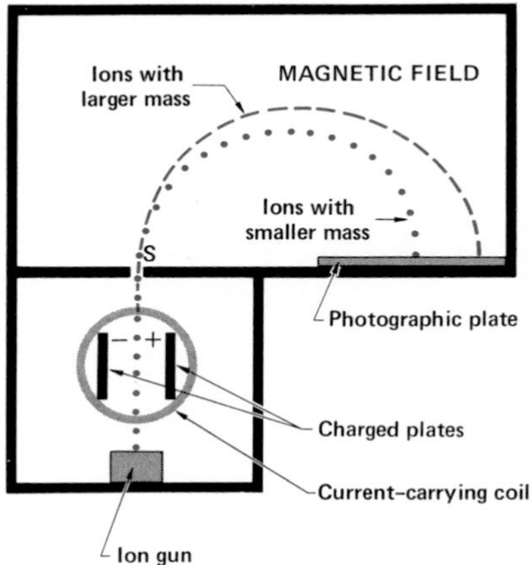

Figure 23-14. Diagram of a mass spectrograph. The magnetic field deflects the less massive ions more than the more massive ones.

photographic plate is replaced by collecting vessels, the mass spectrograph can be used to separate isotopes.

23.12 Nuclear Binding A very strong force holds protons and neutrons together in the nucleus. This force is not electrostatic because the similarly charged protons would then repel each other. The force cannot be gravitational either because calculations show that it would be much too weak. It is a completely unique force that acts only within the extremely small distances between nucleons. Experiments have shown that this *nuclear binding force* is effective to a distance of 2.0×10^{-15} m, or 2.0×10^{-5} Å. This is less than the diameter of a proton, which is 2.6×10^{-5} Å.

The magnitude of the nuclear binding force has been investigated in several ways. One way is to study the scattering pattern that results when nucleons are directed at each other, very much like the scattering experiments of Rutherford and Chadwick. Another way is to compare the mass of a nucleus with the sum of the masses of the uncombined nucleons that make up the nucleus. In each case, the magnitude of the nuclear force is derived from measurements of the energy involved.

The nuclear mass defect of a nucleus is the difference between its mass and the sum of the masses of its uncombined constituent particles.

Careful measurements have shown that *the mass of a nucleus is always less than the sum of the uncombined masses of its constituent particles. This difference in mass is called the* ***nuclear mass defect.*** The reason for the nuclear mass defect is found in Einstein's equation $E = mc^2$. When a nucleus is formed, energy is released. This is similar to the energy release that occurs when atoms combine to form a molecule. Einstein's equation states that a decrease in energy must be accompanied by a corresponding decrease in mass since the two quantities are directly proportional. This decrease in mass is the nuclear mass defect.

The nuclear mass defect can be used to compute the total nuclear *binding energy*. Binding energy is equivalent to the energy released during the formation of a nucleus and is the energy that must be applied to the nucleus in order to break it apart. Let us consider the formation of a helium nucleus (alpha particle) as an example. It contains 2 protons and 2 neutrons. Each proton has a mass of 1.007 276 u, and each neutron has a mass of 1.008 665 u. The mass of a helium nucleus is 4.001 509 u. The nuclear mass defect can now be computed as follows:

$$\begin{aligned} \textbf{2 protons} &= 2 \times 1.007\,276\ u = 2.014\,552\ u \\ \textbf{2 neutrons} &= 2 \times 1.008\,665\ u = \underline{2.017\,330\ u} \\ & \qquad\qquad\qquad\qquad\ \ 4.031\,882\ u \\ \textbf{helium nucleus} &= \underline{4.001\,509\ u} \\ \textbf{nuclear mass defect} &= 0.030\,373\ u \end{aligned}$$

In nuclear physics, nuclear binding energy is expressed in electron volts (eV). *An **electron volt** is the energy required to move an electron between two points that have a potential difference of one volt.* The mega electron volt (MeV) is also used. A single atomic mass unit is the equivalent of 931 MeV. Hence the binding energy of the helium nucleus is

One atomic mass unit has a nuclear binding energy equivalent to 931 mega electron volts.

$$\mathbf{0.030\ 373\ u \times 931\ MeV/u = 28.3\ MeV}$$

If the nuclear binding energy is divided by the number of nucleons, the binding energy per nucleon is obtained. For helium, this is 28.3 MeV/4 = 7.1 MeV/nucleon. The graph in Figure 23-15 shows the variation of binding energy per nucleon with the mass number. Note that elements of low or high mass number have lower binding energies per nucleon than those of intermediate mass number. The binding energy per nucleon is low for the lightest nuclei because the nucleons are all on the surface of the nucleus and are not held on all sides by other nucleons. This geometry reduces the binding energy per nucleon. As the number of nucleons increases, more and more of them are completely surrounded and the binding energy per nucleon increases. Beyond the maximum of 8.7 MeV/nucleon, the repelling force between the protons begins to cancel the binding force. Finally a point is reached at which the nucleus begins to fall apart.

The disintegration of a nucleus is a form of nuclear reaction. This is the topic of the next chapter.

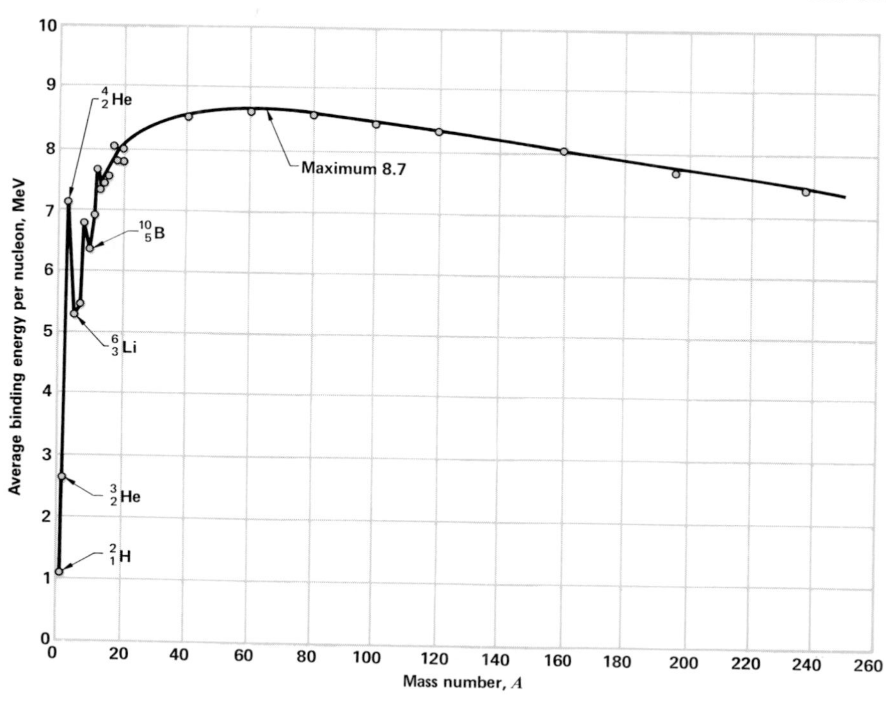

Figure 23-15. Graph showing the relationship between mass number and binding energy per nucleon.

QUESTIONS: GROUP A

1. (a) Who is credited with discovering the atomic nucleus? (b) What experimental evidence showed that positive charge could not be evenly spread out?
2. (a) What fraction of the volume of the atom is taken up by the nucleus? (b) What is the remainder of the atom's volume composed of? (c) How does the mass of the nucleus compare to the mass of the atom as a whole?
3. How is the atomic number related to the number of protons and electrons in an atom?
4. (a) What particles are called nucleons? (b) What is a nuclide?
5. (a) How do isotopes of an element differ? (b) How are they the same?
6. Why is the mass of a nucleus always less than the sum of the masses of its constituents?

GROUP B

7. (a) According to astrophysicists, what is a *black hole?* (b) Is this an appropriate term? Explain.
8. What is the equivalent of one electron volt in joules?
9. Is it possible to predict atomic mass from atomic number? Explain.
10. Would neutrons leave tracks in a bubble chamber? Why?
11. Why couldn't Rutherford's experimental apparatus be used to discover neutrons?
12. Why do ions with greater masses follow curves with greater radii in a mass spectrograph?
13. Why is binding energy low for nucleons with (a) low masses (b) the highest masses?
14. If atoms are mostly empty space, then why can't a physics student walk through a wall?
15. Copy and complete the following table on a separate sheet of paper.

Nuclide	*Z*	*A*	*N*	*Binding energy/nucleon (MeV/A)*
deuterium	1			
carbon-14	6			
oxygen-18	8			
sodium-23	11			
sulfur-32	16			
argon-40	18			
uranium-235	92			
uranium-238	92			

PROBLEMS: GROUP A

Note: Use Appendix B, Table 23, for information about nuclides.

1. If one atomic mass unit is converted to energy, how many joules of energy would be produced?
2. Determine *Z*, *N*, and *A* for krypton.
3. (a) How many nucleons does an atom of plutonium have? (b) How many electrons does it have?
4. (a) How many electron volts of energy would be produced if one neutron were annihilated? (b) How many joules of energy does this represent?
5. What particles, and how many of each, make up an atom of silver (atomic number 47, mass number 109)?
6. Calculate the binding energy per nucleon in MeV for carbon-12. This nuclide consists of 6 protons, 6 neutrons, and 6 electrons. Its atomic mass is 12.000 00.
7. Calculate the binding energy per nucleon in MeV for sulfur-32 that consists of 16 protons, 16 neutrons, and 16 electrons.

GROUP B

8. Calculate the gram-atomic weight for chlorine if naturally occurring chlorine consists of 75.4% Cl-35 and 24.6% Cl-37.

9. Naturally occurring magnesium consists of 78.6% Mg-24, 10.1% Mg-25, and 11.3% Mg-26. Calculate the gram-atomic weight.
10. An electron pair is produced from X rays with a total energy of 1.5 MeV. If the positron and electron each receive half of the extra energy, calculate what the resulting velocity of each particle would be if we could disregard relativity.

SUMMARY

The electron was the first subatomic particle to be identified by scientists. It was found to have a mass of 9.1×10^{-31} kg and a negative charge of 1.6×10^{-19} C. The electron exhibits both particle and wave characteristics. It orbits the nucleus of the atom in a path determined by its energy.

Almost all the mass of an atom is contained in its nucleus, which takes up only a very small percentage of the space occupied by an atom. The positive charge of the nucleus is due to protons. The magnitude of the charge of a proton is the same as that of an electron. The number of protons in a neutral atom is equal to the number of orbiting electrons. This number is known as the atomic number.

The atomic nucleus also contains uncharged particles called neutrons. The mass of a neutron is slightly greater than the mass of a proton. Protons and neutrons are collectively called nucleons, and their total number is the mass number of the atom. Isotopes are atoms with the same atomic number but different mass numbers. Each variety of an atom as determined by its nuclear composition is called a nuclide. The gram-atomic weight of an element is the average atomic mass of one mole of naturally occurring atoms of an element. The atomic weight scale is based on the carbon-12 isotope. Nuclides are separated by a mass spectrograph.

The mass of a nucleus is always less than the combined mass of its constituent particles by an amount called the nuclear mass defect. The nuclear mass defect is equivalent to the nuclear binding energy. Binding energies are expressed in mega electron volts. Elements with intermediate mass numbers have higher binding energies per nucleon than do elements with low or high mass numbers.

VOCABULARY

alpha particle
atomic number
cathode rays
electron
electron shell
electron volt
gram-atomic weight
isotope
neutron
nuclear binding force
nuclear mass defect
nucleon
nucleus
nuclide
orbital
proton
rest mass

24 Nuclear Reactions

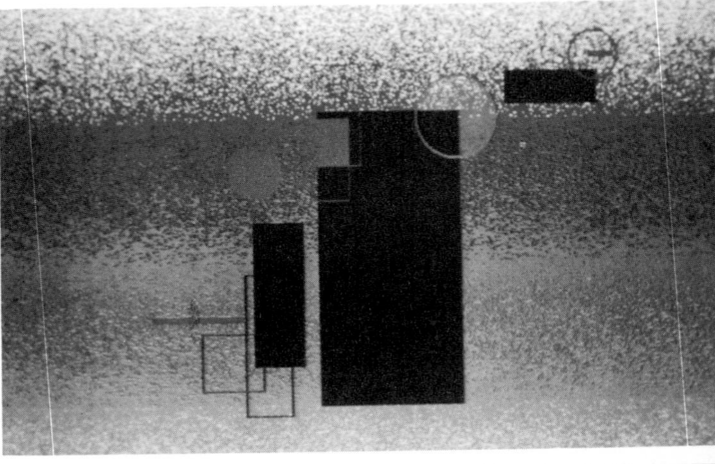

fusion (FYU-zhun) n.: a reaction in which light nuclei combine to form a nucleus with a larger mass number.

OBJECTIVES

- Study the effects of radioactive nuclides on various substances.
- Describe the nature of alpha, beta, and gamma emissions.
- Work with nuclear equations.
- Study the distinction between fission and fusion reactions.
- Describe the requirements for a nuclear chain reaction.
- Study the role of nuclear reactors and the uses of radioisotopes.

TYPES OF NUCLEAR REACTIONS

24.1 Discovery of Radioactivity At the time that Thomson was conducting his experiments with cathode rays, other scientists were gathering evidence that the atom could be subdivided. Some of this new evidence showed that certain atoms disintegrate by themselves. In 1896 Henri Becquerel (1852–1908) discovered this phenomenon while investigating the properties of *fluorescent* minerals. Fluorescent minerals glow after they have been exposed to strong light. Becquerel used photographic plates to record this fluorescence.

One of the minerals Becquerel worked with was a uranium compound. During a day when it was too cloudy to expose his mineral samples to direct sunlight, Becquerel stored some of the compound in the same drawer with the photographic plates. When he later developed these same plates for use in his experiments, he discovered that they were fogged. What could have produced this fogging? The plates were wrapped tightly before being used, so the fogging could not be due to stray light. Also, only the plates that were in the drawer with the uranium compound were fogged. Becquerel reasoned that the uranium compound must give off a type of radiation that could penetrate heavy paper and affect photographic film.

Elements that emit such radiation are called *radioactive* and possess the property of *radioactivity*. ***Radioactivity*** *is the spontaneous breakdown of an unstable atomic nucleus with the emission of particles and rays.* At the suggestion of Becquerel, Pierre (1859–1906) and Marie (1867–1934) Curie investigated uranium and its various ores. They found that all uranium ores were radioactive. However, one of them, pitchblende, was four times as radioactive as might be expected from the amount of uranium in it. Further studies showed that this extra radiation was actually due to the presence of two previously unknown elements, *polonium* and *radium*, both of which were intensely radioactive. The elements polonium and radium were discovered and named by the Curies.

Other radioactive elements have since been discovered or produced. All of the naturally occurring elements with atomic numbers greater than 83 are radioactive. Also, a few naturally radioactive isotopes of elements with atomic numbers smaller than 83 are known. Many artificial radioactive nuclides have been produced and put to use in different ways, as we shall see later in this chapter.

Figure 24-1. Henri Becquerel, the French physicist who discovered radioactivity.

In 1903, Becquerel and the Curies shared the Nobel Prize in Physics for their work in radioactivity.

24.2 Nature of Radioactivity All radioactive nuclides have certain common characteristics.

1. *Their radiations affect the emulsion on a photographic film.* Even though photographic film is wrapped in heavy black paper and kept in the dark, some of the radiations from radioactive nuclides penetrate the wrapping and affect the film. When the film is developed, a black spot can be seen where the invisible radiations struck the film. Some of the radiations penetrate wood, flesh, *thin* sheets of metal, and, as shown in Figure 24-3, even thick sheets of glass.

2. *Their radiations ionize the surrounding air molecules.* The radiations from radioactive nuclides knock out electrons from the atoms of the gas molecules in the air surrounding the radioactive material. This process leaves the gas molecules with a positive charge. An atom or a group of atoms having an electric charge is called an ion. The production of ions is termed ionization. An electroscope can detect ionized gas molecules in the air. As we noted in Section 7.23, an ionized gas is called plasma.

Figure 24-2. Marie and Pierre Curie discovered the radioactive elements radium and polonium.

3. *Their radiations make certain compounds fluoresce.* Radiations from radioactive nuclides produce bright flashes of light when they strike certain compounds. The combined effect of these flashes is a fluorescence, or glow, given off by the affected material. For example, radium compounds added to zinc sulfide cause the zinc sulfide to glow.

4. *Their radiations have special physiological effects.* Radiations from natural sources can destroy the germinating power of plant seeds, kill bacteria, and even injure and kill large animals. Burns from radioactive materials heal with great difficulty and may sometimes be fatal.

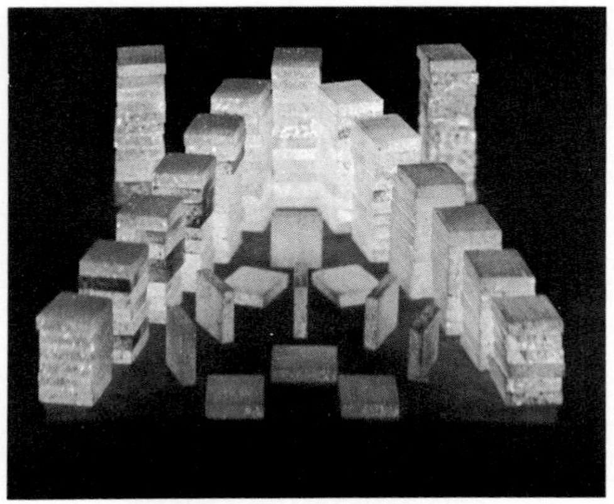

Figure 24-3. Self-portrait of a radioactive substance. The only source of illumination for this time exposure, which was taken through a 3-foot-thick heavy-glass window, was the natural radioactivity of the cesium-137 in the blocks of cesium chloride.

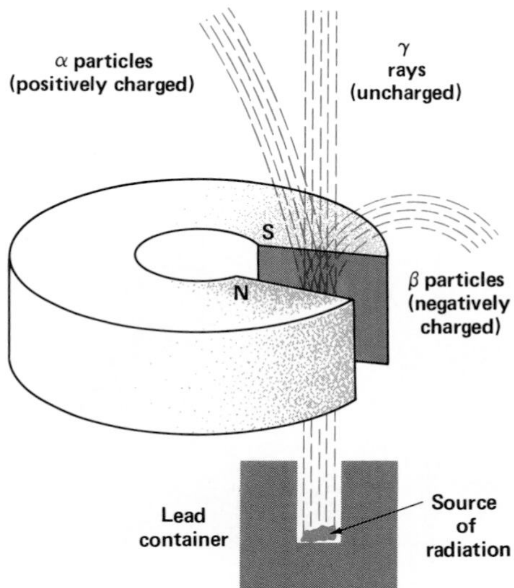

Figure 24-4. Types of radioactivity. Alpha and beta particles are deflected by a magnet, but gamma rays are not.

5. *They undergo radioactive decay.* The atoms of all radioactive elements continually decay into simpler atoms and simultaneously emit radiations. However, it is not possible to predict when a specific atom will decay. The rate of decay may be described by a rule that states how long it will take for half of the atoms in a given sample to decay. The accuracy of the rule is greater for larger samples. The time associated with radioactive decays of a certain type is called the *half-life* of the nuclide. ***Half-life*** *is the length of time during which, on the average, half of a given number of atoms of a radioactive nuclide decays.* For example, the half-life of radium-226 is 1620 years. This means that one-half of a given sample of radium-226 atoms can be expected to decay into simpler atoms in 1620 years.

24.3 Types of Natural Radioactivity Experiments show that the radiations from radioactive materials can be separated into three distinct types by means of a strong magnet. A diagram of this separation is shown in Figure 24-4.

1. *α (alpha) particles.* These are composed of two protons and two neutrons; hence they are helium nuclei. This conclusion can be proved by collecting large quantities of alpha particles in a partially evacuated tube and passing an electric discharge through the tube. The alpha particles acquire electrons from the residual air molecules in the tube. The spectrum of the resulting substance is identical with the spectrum of helium. Alpha particles have two positive electric charges, speeds about one-tenth the speed of light, and masses about four times that of the hydrogen atom. Hence they are deflected only slightly by the field of a magnet. Their penetrating power is not very great. They can be stopped by a thin piece of aluminum foil or by a thin sheet of paper.

2. *β (beta) particles.* These are electrons, just like cathode rays. They have single negative charges and may travel with nearly the speed of light. Their mass is only a small fraction of the mass of alpha particles. Hence, even though they have only half as much charge, they are deflected much more by a magnetic field and in the opposite direction. Beta particles are much more penetrating than alpha particles. The reason is that beta particles have less charge than alpha particles, and therefore they will lose less energy in passing through a substance.

3. *γ (gamma) rays.* These are high-energy photons. They are the same kind of radiation as visible light, but of much shorter wavelength and thus higher frequency. Gamma rays are produced by energy transitions in the nucleus

that do not change the composition of the nuclide. They are the most penetrating radiations given off by radioactive elements. They are not electrically charged and are not deflected by a magnetic field.

24.4 Nuclear Symbols and Equations Changes in matter can be classified into three types: *physical, chemical,* and *nuclear*. When a physical change occurs, the identity of the substance is not lost. When water freezes and forms ice, the composition of its molecules is not changed; if we heat the ice it changes back into water. When sugar dissolves in water it undergoes a physical change; if the water evaporates, the sugar remains behind. *In a* ***physical change*** *the composition of the substance is not changed.* The substance consists of the same molecules, atoms, or ions that it did before the change occurred.

When a chemical change occurs, new substances are formed from the atoms or ions of the original substances. These new substances have their own distinct chemical and physical properties. The rusting of iron and the souring of milk are examples of chemical changes. *In a* ***chemical change*** *the composition of the substance is changed, and new substances with new properties are produced.* The type and number of atoms or ions remain the same, but the atoms or ions are rearranged to form new substances. A chemical change does not alter the nuclei of the interacting atoms.

A nuclear change is similar to a chemical change in that new substances with new properties are formed. However, *in a* ***nuclear change*** *the new materials are formed by changes in the identities of the atoms themselves.* The radioactive decay of radium is a nuclear change, as are all forms of alpha, beta, and gamma emission.

Both chemical and nuclear changes are usually represented by equations. In equations, symbols or formulas are used to represent the particles that enter into a reaction and that are produced by it. You are probably familiar with the symbols of many of the chemical elements. In working with nuclear equations, you must know how to use the symbols for some of the subatomic particles and radiations. The most basic of these are given in Table 24-1.

Except for the gamma-ray symbol, all the symbols in the table have subscripts and superscripts. The subscript designates the electric charge of the particle. The superscript denotes the number of nucleons in the particle. These charge and nucleon designations are also used with the symbols of all other atomic nuclei presented in nuclear equations.

In a chemical equation, the same atoms appear on both

Table 24-1
SYMBOLS FOR SUBATOMIC PARTICLES

Symbol	*Particle*
${}^{1}_{0}n$	neutron
${}^{1}_{1}H$	proton (hydrogen nucleus)
${}^{0}_{-1}e$	electron (beta particle)
${}^{0}_{+1}e$	positron (positive electron)
${}^{4}_{2}He$	alpha particle (helium nucleus)
γ	gamma ray

Note: In the case of the proton and the alpha particle the atomic symbols for hydrogen and helium are meant to designate only the nuclei of these atoms.

The symbols of other subatomic particles are given in Chapter 25.

sides of the equation in equal numbers. This is true because the identity of the atoms is not changed. In a nuclear equation, however, the same atomic symbols do not necessarily appear on both sides. The factors that will remain constant in a nuclear equation are the total number of nucleons and the arithmetic sum of the electric charges on each side.

The following example will illustrate this.

$$^{14}_{7}N + ^{4}_{2}He \rightarrow ^{17}_{8}O + ^{1}_{1}H$$

In this equation, a nitrogen nucleus (charge, 7; nucleons, 14) reacts with an alpha particle (charge, 2; nucleons, 4) to form an oxygen nucleus (charge, 8; nucleons, 17) and a proton (charge, 1; nucleon, 1). There is a total of 9 electric charges and 18 nucleons on each side of the equation.

24.5 Radioactive Decay It was noted in Section 24.1 that the atoms of all naturally occurring elements with atomic numbers greater than 83 are unstable and decay spontaneously into lighter particles. This is also true of several naturally occurring nuclides and a great many artificially made nuclides with atomic numbers below 83. As a general rule, an atomic nucleus is stable only if the ratio of its neutrons to its protons is about 1 for the light elements and about $1\frac{1}{2}$ for the heavier elements. When there are more neutrons than this, a *nuclear transformation* will probably take place.

Figure 24-5 shows a series of nuclear transformations beginning with uranium-238 and ending with lead-206. Each step of the series can be represented by a nuclear equation in which nucleons and charge are conserved. For instance the equation for the first step in the chart may be written as

$$^{238}_{92}U \rightarrow ^{234}_{90}Th + ^{4}_{2}He$$

This reaction is called *alpha decay* because an alpha particle is emitted.

The next two steps in the series are examples of *beta decay*. In the first step, a neutron in the thorium nucleus ejects a beta particle and becomes a proton. This transformation increases the nuclear charge by one, and the nucleus becomes a protactinium nucleus.

$$^{234}_{90}Th \rightarrow ^{234}_{91}Pa + ^{0}_{-1}e$$

The chart also shows the half-life of each nuclide in the decay series. You will notice that there is a great variation in these values, from small fractions of a second to millions of years. In each case the exact length of the half-life

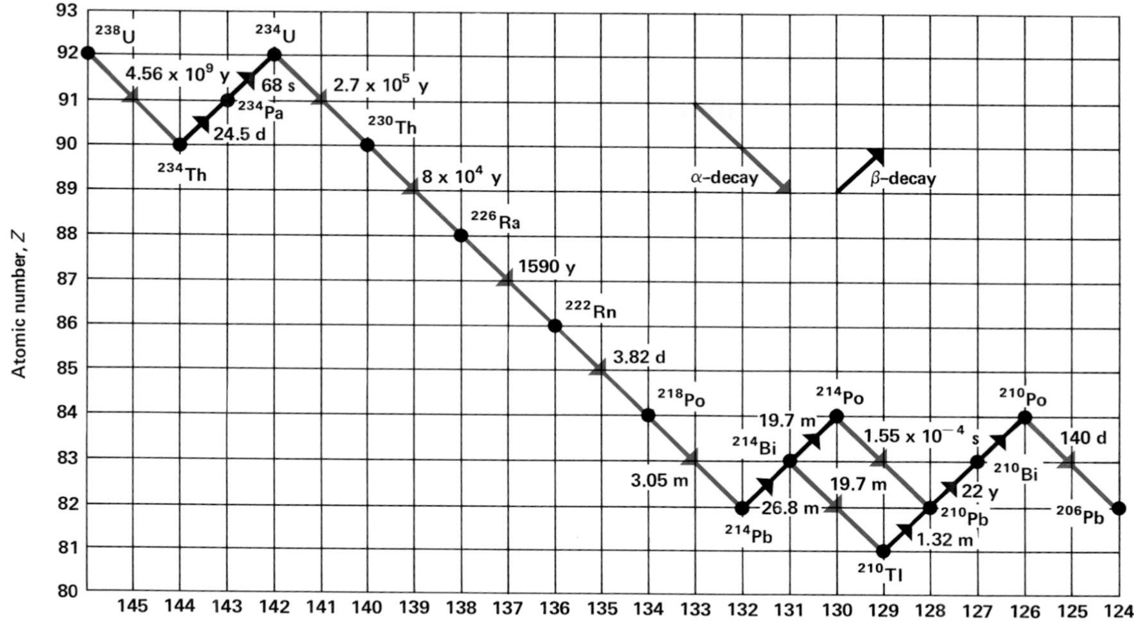

Figure 24-5. The uranium-238 disintegration series. Note that in a beta decay, the atomic number increases by one but the mass number remains constant.

is related to the energy change that accompanies the transformation. As a rule, the shorter the half-life of the nuclide, the greater is the kinetic energy of the alpha or beta particle it emits.

From the definition of half-life it is obvious that the radioactivity of a sample of material depends on the total number of nuclei present. The ratio between the number of nuclei decaying per unit time and the total number of original nuclei is known as the *decay constant*. If the half-life of a nuclide is known, its decay constant can be calculated by the following equation:

$$\lambda = \frac{0.693}{T_{1/2}}$$

In this equation λ (the Greek letter lambda) designates the decay constant in reciprocal seconds, 1/s or s^{-1}, if the half-life $T_{1/2}$ is given in seconds. For example, for $^{226}_{88}Ra$, $T_{1/2} =$ 1620 years and

$$\lambda = \frac{0.693}{1620 \text{ y} \times 365 \text{ d/y} \times 86\,400 \text{ s/d}}$$

$$\lambda = 1.36 \times 10^{-11} \text{ s}^{-1}$$

In 1.000 kg of radium there are 2.665×10^{24} nuclei. Consequently, the number of radium nuclei decaying per second is 2.665×10^{24} nuclei $\times\ 1.36 \times 10^{-11}\ s^{-1} = 3.62 \times 10^{13}$ nuclei/s. Observe that this equation does not indicate *when* a particular nucleus will decay. Instead it says that in a certain amount of time a definite proportion of nuclei will decay. A relationship of this kind represents the statistical behavior of a very large number of individual situations. It uses the mathematical methods of statistics or the theory of probability. The half-life probability law is shown in the form of a graph in Figure 24-6.

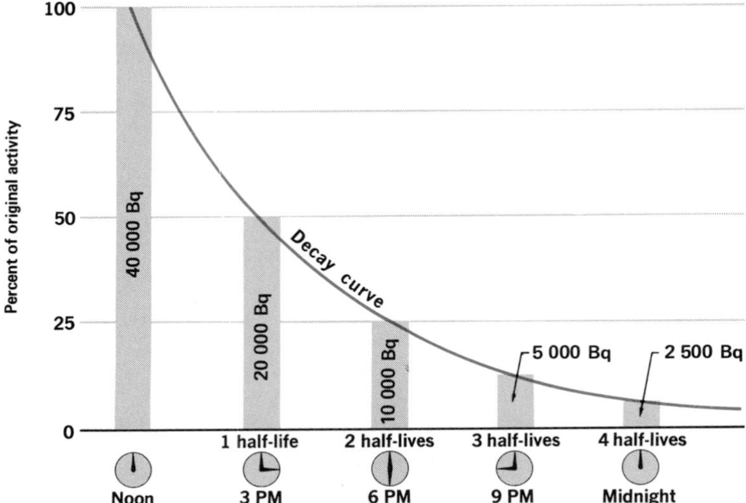

Figure 24-6. Activity curve of a radioactive nuclide with a half-life of three hours and an original activity of 40 000 Bq.

One radioactive disintegration per second is called a *becquerel* (Bq). A sample of radioactive material that emits 3.7×10^{10} Bq is said to have a strength of one *curie* (Ci). Since the curie is a very large unit, the millicurie (mCi) and microcurie (μCi) are frequently used. For example, a source with a strength of 4.00 μCi emits $(4.00 \times 10^{-6})(3.7 \times 10^{10})$, or 1.5×10^5 particles per second.

The *gray* (Gy) is the unit used to measure the amount of absorbed radiation. *One gray is equal to the absorption of one joule of radioactive energy per kilogram of absorbing matter.* Radioactive dosage is described in terms of the *sievert* (Sv). A sievert has the energy equivalent of a gray.

24.6 Nuclear Bombardment The study of radioactive decay led scientists to believe that transformations that would produce a different element could be achieved by adding protons to the nucleus. The first successful transformation of this kind was carried out by Rutherford in 1919. A diagram of his apparatus is shown in Figure 24-7. The container was filled with nitrogen gas. A radioactive source emitted alpha particles. A silver foil thick enough to absorb any alpha particles not absorbed by the nitrogen gas was used. A zinc sulfide screen recorded the scintillations of any particles with enough energy to pass through

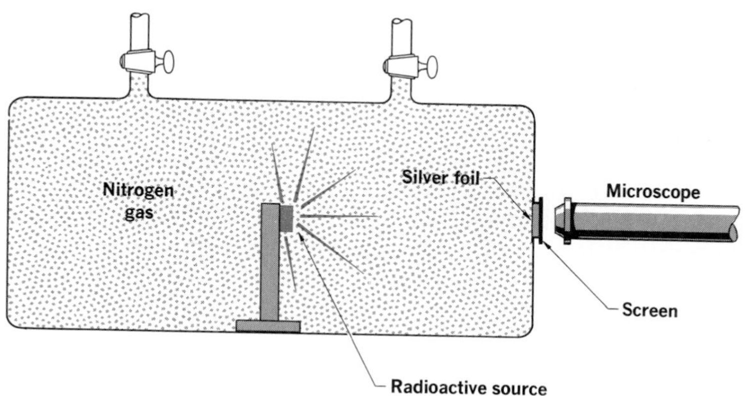

Figure 24-7. Rutherford's transformation experiment. Bombardments from the radioactive sample ejected protons from the nitrogen gas in the chamber. The protons produced scintillations on a fluorescent screen.

the silver foil. These scintillations could be observed through the microscope.

Rutherford observed scintillations on the screen when the chamber was filled with nitrogen and other light gases. However, there were no scintillations when he used oxygen and other heavy gasses such as carbon dioxide. He concluded that low density gases were changed by the alpha-particle bombardment with the accompanying emission of a high-energy particle such as a proton. The nuclear equation in Section 24.4 represents the transformation produced by the alpha bombardment of nitrogen. The detection of oxygen gas in the container after the alpha bombardment of pure nitrogen helps to verify the validity of the equation.

The alpha bombardment of nitrogen will result in the emission of protons. Other bombardments produce other subatomic particles. In 1932 J. D. Cockcroft and E. T. S. Walton bombarded lithium with high-speed protons.

$$^{7}_{3}\mathrm{Li} + {}^{1}_{1}\mathrm{H} \rightarrow {}^{4}_{2}\mathrm{He} + {}^{4}_{2}\mathrm{He} + \text{energy}$$

In this case all the products have lower mass numbers than the original target material. In addition there is the release of a considerable amount of binding energy. Actually, adjustments in binding energy are involved in all nuclear reactions. Cockcroft and Walton found that the measured energy in their experiments agreed closely with the theoretical value as derived from Einstein's equation, $E = mc^2$.

The method of computing binding energy is described in Section 23.12.

Another nuclear bombardment, also carried out in 1932, led to the discovery of the neutron. The nuclear equation for Chadwick's experiment is

Neutrons are described in Section 23.9.

$$^{9}_{4}\mathrm{Be} + {}^{4}_{2}\mathrm{He} \rightarrow {}^{12}_{6}\mathrm{C} + {}^{1}_{0}\mathrm{n} + \text{energy}$$

Many similar bombardments have since been found to produce neutrons. Since the neutron is not electrically charged, it penetrates atomic nuclei more easily than do charged particles like the proton or the alpha particle. Hence the neutron has become an important nuclear "bullet." In fact one of the goals in neutron bombardment is to get the neutrons to move *slowly* enough to produce a transformation. Fast neutrons may be slowed down by passage through materials called *moderators,* composed of elements of low atomic weight. Deuterium oxide or graphite are useful moderators. The equation for a typical neutron absorption is

$$^{10}_{5}\mathbf{B} + {}^{1}_{0}\mathbf{n} \rightarrow {}^{7}_{3}\mathbf{Li} + {}^{4}_{2}\mathbf{He} + \textbf{energy}$$

Bombardment reactions are usually designated by the symbols for the incident and emitted subatomic particles. For example, the preceding equation is called a (n, α) reaction. In bombardment reactions, the mass number of the target nucleus is increased or decreased by several units. Sometimes, however, a nucleus will split into two large segments when struck by a neutron moving at just the right speed.

24.7 Fission When a neutron is captured by $^{238}_{92}U$, the following reaction takes place:

$$^{238}_{92}\mathbf{U} + {}^{1}_{0}\mathbf{n} \rightarrow {}^{239}_{92}\mathbf{U}$$

This new isotope of uranium is unstable, and so it emits a beta particle.

$$^{239}_{92}\mathbf{U} \rightarrow {}^{239}_{93}\mathbf{Np} + {}^{0}_{-1}\mathbf{e}$$

$^{239}_{93}Np$ has a half-life of 2.3 days and decays by emitting another beta particle.

$$^{239}_{93}\mathbf{Np} \rightarrow {}^{239}_{94}\mathbf{Pu} + {}^{0}_{-1}\mathbf{e}$$

Neptunium and plutonium were the first two artificial *transuranium* elements that were discovered. *A transuranium element is one with an atomic number greater than 92, the atomic number of uranium.* At the time of this writing, seventeen artificial transuranium elements have been identified. All of them are radioactive, and some of them exist for only a small fraction of a second.

The fission of U-235 yields products with intermediate atomic numbers. Fission into two equal fragments is very unlikely, as Figure 24-8 shows.

Uranium-238 accounts for 99.3% of naturally occurring uranium. 0.7% is uranium-235. There are also traces of uranium-234. When $^{235}_{92}U$ absorbs a neutron, the reaction is quite different from that described above for $^{238}_{92}U$. Instead of emitting alpha or beta particles, the $^{235}_{92}U$ may split into segments of intermediate mass. *The splitting of a heavy nucleus into nuclei of intermediate mass is called* ***fission.*** During the process of fission, neutrons are emitted and a large

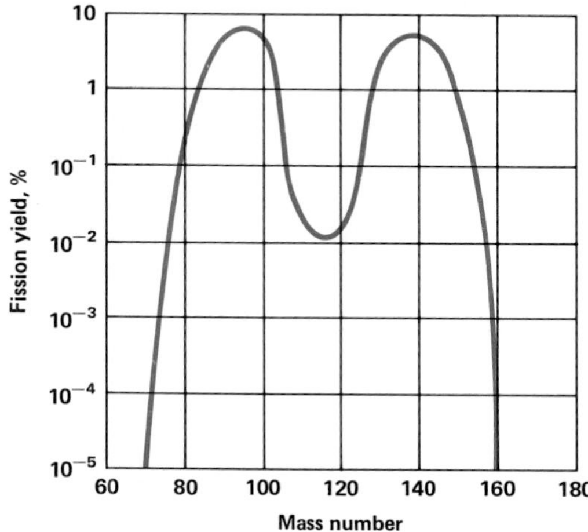

Figure 24-8. Graph of nuclides produced by the fission of the element uranium-235. The fission products vary in mass number from 72 to 161. The most probable products have mass numbers of 95 and 138.

amount of energy is released. One equation for the fission of ${}^{235}_{92}\text{U}$ is

$$\mathbf{{}^{235}_{92}U + {}^{1}_{0}n \rightarrow {}^{138}_{56}Ba + {}^{95}_{36}Kr + 3{}^{1}_{0}n + energy}$$

The energy involved in this reaction can be computed from the atomic masses of the particles as given in Appendix B, Table 23.

$$\begin{aligned}
{}^{235}_{92}\mathrm{U} &= 235.043\ 9\ \ u\\
{}^{1}_{0}\mathrm{n} &= \underline{\ \ 1.008\ 665\ u}\\
& \quad 236.052\ 6\ \ u\\
{}^{138}_{56}\mathrm{Ba} &= 137.905\ 0\ \ u\\
{}^{95}_{36}\mathrm{Kr} &= \ \ 94.9\ \ u\\
3 \times {}^{1}_{0}\mathrm{n} &= \underline{\ \ 3.025\ 995\ u}\\
& \quad 235.8\ \ u\\
\textbf{nuclear mass defect} &= 236.052\ 6\ u - 235.8\ u = 0.3\ u\\
\textbf{binding energy released} &= 0.3\ u \times 931\ \mathrm{MeV}/u = 300\ \mathrm{MeV}
\end{aligned}$$

The above calculation has been simplified by using the masses of the neutral atoms given in the table, rather than the masses of the nuclei by themselves. No error is introduced in this case because the masses of the electrons cancel. When a beta particle or a positron is emitted, however, the masses of the nuclei must be used in the calculation.

Note that the atomic mass of ${}^{95}_{36}\text{Kr}$ is not known with the same precision as the masses of the other particles in this reaction. This uncertainty is due to the fact that the atoms of the nuclide are so unstable that it is very difficult to measure their mass. Since the mass of krypton-95 is given in three significant figures, the mass defect and binding energy can be computed with a precision of only one significant figure.

Uranium-238 and uranium-235 can both be split with fast neutrons, while slow neutrons will split only the uranium-235 atoms. Hence fission in a mixture of uranium isotopes can be selectively controlled by regulating the speed of the neutrons with a moderator.

The first successful fission reaction was carried out in 1939. It was later discovered that the element plutonium also undergoes fission and produces more neutrons when bombarded with slow neutrons.

24.8 Fusion The graph of binding energy per nucleon in Section 23.12 suggests that a great deal of binding energy could be released by combining nuclei of the light elements into nuclei of medium mass. In fact, we could expect energies of about 7 MeV/nucleon as compared with about 1 MeV/nucleon for fission reactions. *A reaction in which light nuclei combine to form a nucleus with a larger mass number is called* ***fusion.*** Because fusion can take place only under extremely high temperature conditions, the process is also called a *thermonuclear reaction.*

Figure 24-9. Hans Bethe, the American scientist who proposed the thermonuclear fusion theory of energy production in the stars.

In 1967, Bethe won the Nobel Prize in Physics for his work on energy production in stars.

One of the first fusion reactions to be accomplished was the combination of two deuterons (nuclei of the isotope of hydrogen, deuterium).

$$^{2}_{1}\mathbf{H} + ^{2}_{1}\mathbf{H} \rightarrow ^{3}_{2}\mathbf{He} + ^{1}_{0}\mathbf{n} + \mathbf{3.3\ MeV}$$

The total energy released in this reaction is less than that of uranium fission, but the energy per nucleon is much greater. In other words, less material is required to produce the energy. In 1938 the American astrophysicist Hans Bethe (b. 1906) suggested that the tremendous and long-lasting rate of energy production of the sun and other stars is due to nuclear fusion. The sufficiently high temperatures and pressures of a star bring about the fusion of hydrogen nuclei to form nuclei of helium, with the accompanying release of energy. The net result of such a reaction is

$$\mathbf{4}^{1}_{1}\mathbf{H} \rightarrow ^{4}_{2}\mathbf{He} + \mathbf{2}\,^{0}_{+1}\mathbf{e} + \mathbf{25.7\ MeV}$$

The intermediate steps that take place in the thermonuclear reactions in the stars are not completely understood. Some scientists believe that the reaction is somewhat different in different parts of the interior of a star. But it is almost certain that several other atomic nuclei, especially those of oxygen and nitrogen, play an important role. Both oxygen and nitrogen have been identified in the sun and other stars.

Estimates of the sun's energy output indicate that about 6×10^{11} kg of hydrogen are converted into helium *every second,* with a mass defect of 4×10^{9} kg *per second*. It is comforting to know that the sun has a mass of 2×10^{30} kg, and so there is no immediate danger that its mass will disappear.

24.9 Cosmic Rays When a charged electroscope is exposed to the air, it will slowly discharge, even though no radioactive material is present. This indicates that there are always some ions in the atmosphere. Experiments have shown that most of this ionization is caused by high-energy particles coming into the atmosphere at great speeds from outer space. These particles are *cosmic rays.*

Cosmic rays have great penetrating power. Their intensity varies with latitude, indicating that they are charged and are affected by the earth's magnetic field. Scientists believe that cosmic rays consist largely of the nuclei of elements of low atomic weight, protons (hydrogen nuclei) being the most abundant type. Other nuclei, ranging up to those as heavy as iron and beyond, have also been detected in the upper atmosphere. A few of these particles from outer space reach the surface of the earth, but most of them collide with the gas particles in the upper atmosphere and produce showers of secondary particles. In effect, cosmic rays are furnishing the nuclear physicist with a constant supply of bombarding nuclei that would otherwise not be available for study.

The term "cosmic ray" was coined by the American physicist Robert Millikan.

Scientists believe that gigantic magnetic fields in space, in galaxies, and near certain stars may be influential in producing cosmic rays. Some cosmic radiation is also thought to originate from highly energetic explosions on certain stars, including the flares that erupt from our sun.

QUESTIONS: GROUP A

1. (a) What is radioactivity? (b) Who discovered it? (b) What were Marie and Pierre Curie's contributions to the study of radioactive materials?
2. What are the common properties of radioactive materials?
3. (a) Define *half-life.* (b) How long would it take before three-fourths of a sample of radium-226 decayed into simpler elements?
4. Which radioactive emissions (a) are the most penetrating (b) carry the highest charge (c) are deflected most by a magnetic field (d) are actually high-energy photons?
5. Why are alpha particles only slightly deflected by a magnetic field?
6. Why does the symbol for the gamma ray not have a subscript or superscript?
7. Categorize the following as physical, chemical, or nuclear changes: (a) a piece of zinc dissolving in sulfuric acid, (b) melting ice, (c) salt dissolv-

ing in water, (d) Chadwick's experiment, and (e) uranium fission.

8. How is a nuclear equation different from a chemical equation?
9. What conservation laws apply to a nuclear equation?
10. Differentiate between nuclear fission and nuclear fusion.
11. What is the difference between the quantity measured in curies and the quantity measured in grays?
12. Why is the neutron such an important "nuclear bullet"?
13. (a) What is a transuranium element? (b) Which were the first two discovered? (c) What characteristic do they all share?

GROUP B

14. (a) What are cosmic rays? (b) What property of cosmic rays indicates they are charged particles?
15. Why has it been more difficult to produce a thermonuclear (fusion) reactor than a fission reactor?
16. What happens to an atomic nucleus when it emits an alpha particle?
17. Is it possible for an atom to decay and yet move to a higher atomic number? Explain.
18. At the dentist, you wear a lead-coated shield to protect you from X rays. Would it also protect you from (a) alpha particles (b) beta particles (c) gamma rays?
19. How did the presence of oxygen gas indicate the validity of Rutherford's transformation experiment?
20. Categorize the following bombardment reactions according to incident and emergent particles: (a) the Cockcroft-Walton experiment, (b) Chadwick's experiment.
21. Look up the property of fluorescence. Describe how a radioactive source makes a material fluoresce.
22. If radiation can kill bacteria, suggest a way it could be used to improve food storage techniques.
23. What fundamental problems did the alchemists face when they tried to turn lead into gold?
24. Does radioactive decay of a material violate the principle of conservation of mass? Explain.
25. You have one atom of a radioactive nuclide with a half-life of two minutes. How long must you wait for it to decay? Explain.

PROBLEMS: GROUP A

Note: Consult Appendix B, Table 23, for the atomic masses of nuclides.

1. Complete the following nuclear equations:

   (a) $__ \rightarrow {}^{212}_{82}\mathrm{Pb} + {}^{4}_{2}\mathrm{He}$

   (b) ${}^{4}_{2}\mathrm{He} + __ \rightarrow {}^{12}_{6}\mathrm{C} + {}^{1}_{0}\mathrm{n}$

   (c) $__ \rightarrow {}^{36}_{16}\mathrm{S} + {}^{0}_{+1}\mathrm{e}$

   (d) ${}^{214}_{82}\mathrm{Pb} + 2__ \rightarrow {}^{214}_{84}\mathrm{Po}$

   (e) ${}^{235}_{92}\mathrm{U} + {}^{1}_{0}\mathrm{n} \rightarrow __ + {}^{95}_{36}\mathrm{Kr} + 3{}^{1}_{0}\mathrm{n}$

2. In the sun 4×10^9 kg of its mass is converted into energy every second. (a) How many joules of energy does this represent? (b) At this rate, in how many years will the 2×10^{30} kg of the sun's mass be turned completely into energy?
3. What is the decay constant of ${}^{222}_{86}\mathrm{Rn}$ that has a half-life of 3.82 days?
4. Determine the decay constant of ${}^{238}_{92}\mathrm{U}$ if its half-life is 4.51×10^9 years.

GROUP B

5. (a) What is the radioactivity in millicuries of $10\bar{0}$ g of pure ${}^{232}_{90}\mathrm{Th}$? Its half-life is 1.4×10^{10} years, and it decays to ${}^{228}_{88}\mathrm{Ra}$. (b) What is the rate of energy production?
6. (a) Calculate the radioactivity of 5.00×10^{-3} g of ${}^{222}_{86}\mathrm{Rn}$. (b) What is the rate of energy production?

7. In which of the following reactions is a greater proportion of the fuel elements converted into energy: ${}^{239}_{94}Pu + {}^{1}_{0}n \rightarrow {}^{137}_{52}Te + {}^{100}_{42}Mo + 3{}^{1}_{0}n$ + energy, or $2{}^{2}_{1}H \rightarrow {}^{4}_{2}He$ + energy?
8. The earth receives 3×10^{22} J of energy from the sun per day. If this energy is expressed as mass, what is the proportional increase in the earth's mass in one year? (The mass of the earth is presently 6×10^{24} kg.)
9. Compute the mass defect and binding energy per nucleon of ${}^{13}_{6}C$.

USES OF NUCLEAR ENERGY

24.10 Chain Reactions In a fission reaction the target nucleus may break up in a great many different ways. The reaction involving uranium-235 in Section 24.7, for example, is only one of the ways in which this nucleus may react under neutron bombardment. In most cases, however, two or three neutrons are emitted in the process. This is an example of a *chain reaction. In a* ***chain reaction*** *the material or energy that starts a reaction is also one of the products and can cause similar reactions.* Figure 24-10 shows the sequence of events in a typical chain reaction that involves uranium-235. The first controlled chain reaction was carried out at the University of Chicago in 1942 under the direction of the Italian physicist Enrico Fermi (1901–1954).

In 1938, Fermi received the Nobel Prize in Physics for his work in neutron bombardment.

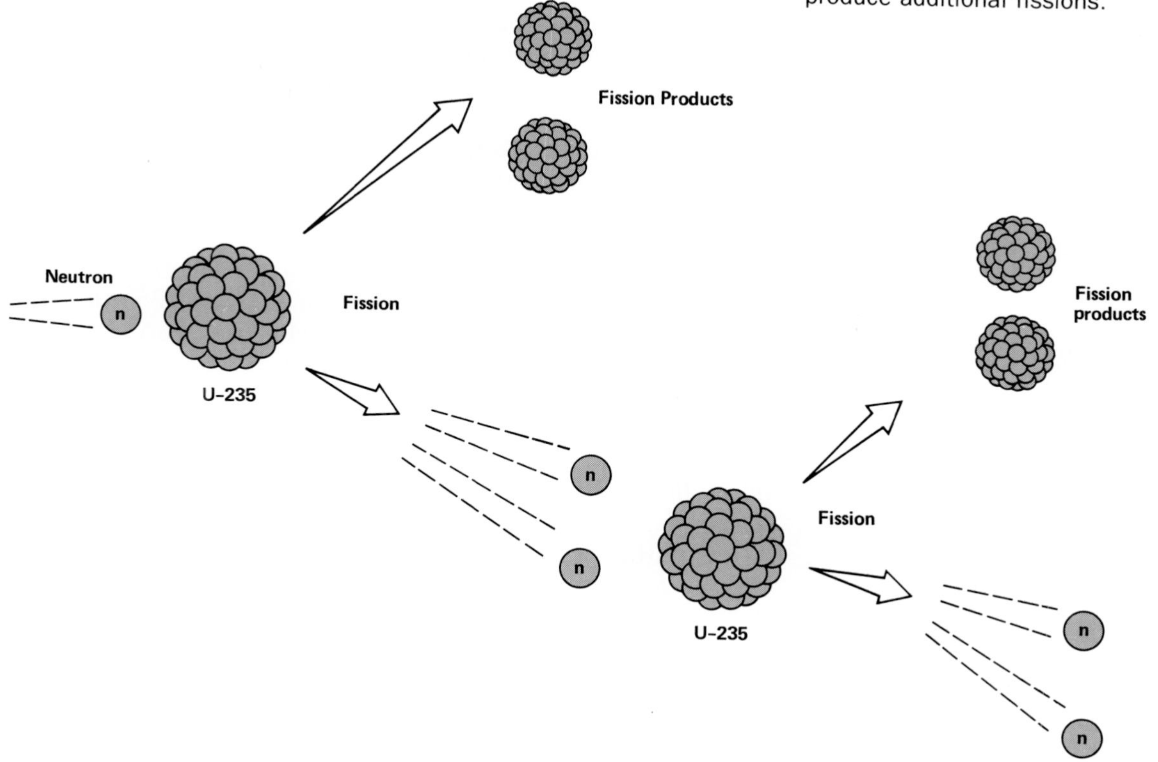

Figure 24-10. A nuclear chain reaction. When a neutron strikes a U-235 nucleus, it is absorbed. This makes the U-235 nucleus unstable, causing it to split into two lighter nuclei called fission products. Two or three other neutrons are also released. These neutrons can strike other U-235 nuclei to produce additional fissions.

If the material surrounding a fission reaction has the right dimensions and characteristics, the reaction is self-sustaining. *The amount of a particular fissionable material required to make a fission reaction self-sustaining is called the* ***critical mass.*** If the critical mass is exceeded and if the emitted neutrons are not absorbed by nonfissionable material, the reaction runs out of control and a nuclear explosion results. For example, suppose that two neutrons from each fission are able to produce further fission reactions and that 10^{-8} second elapses between successive reactions. In one millionth of a second, 100 successive reactions can take place. The total energy released is 200 MeV/nucleus $\times$ 2^{100} reactions, or 4×10^{19} J. This is an enormous amount of energy.

Actually, the chain reaction continues until all the fissionable nuclei have split or until the neutrons no longer strike the fissionable material. This is what happens in an uncontrolled chain reaction such as the explosion of a nuclear warhead. Two subcritical masses are placed a short distance from each other inside the warhead. To detonate it, the two masses are brought together quickly by means of explosive charges. The resulting mass exceeds the critical mass and an uncontrolled chain reaction takes place.

A thermonuclear bomb, sometimes called a hydrogen bomb or H-bomb, produces energy by a fusion reaction. The formation of alpha particles and tremendous amounts of energy from a compound of lithium and deuterium, ${}^{6}_{3}\mathrm{Li}{}^{2}_{1}\mathrm{H}$, is one possible reaction in a hydrogen bomb. This reaction is started by subjecting the compound to extremely high temperature and pressure and using a fission reaction as a detonator.

$$\mathbf{{}^{6}_{3}Li{}^{2}_{1}H \rightarrow 2{}^{4}_{2}He + 22.4\ MeV}$$

This reaction results in 22.4 MeV for 8 nucleons, or 2.80 MeV/nucleon, as compared with 186 MeV for 236 nucleons in uranium fission, or only 0.788 MeV/nucleon. In addition, a thermonuclear bomb is theoretically unlimited in possible size.

24.11 Nuclear Reactors

A ***nuclear reactor*** *is a device in which the controlled fission of certain substances is used to produce new substances and energy.* One of the earliest reactors using natural uranium as a fuel was built in 1943 at Oak Ridge, Tennessee. It had a lattice-type construction with blocks of graphite forming the framework and acting as a moderator. Rods of uranium were placed between the blocks of graphite. Control rods of neutron-absorbing boron steel were inserted in the lattice to regulate the number of free neutrons. The reactor contained a critical mass of uranium.

Two types of reactions occur in this kind of reactor. Neutrons cause $^{235}_{92}U$ nuclei to undergo fission. Fast neutrons from this fission are slowed down by passage through the graphite. Some neutrons strike other $^{235}_{92}U$ nuclei and continue the chain reaction. Other neutrons strike $^{238}_{92}U$ nuclei and initiate the changes that produce plutonium. Because great quantities of heat energy are generated, the reactor is cooled continuously by air blown through tubes in the lattice. The rate of the nuclear reactions is controlled by insertion or removal of control rods.

The nuclear equations for these changes are given in Section 24.7.

*A **breeder reactor** is one in which a fissionable material is produced at a greater rate than the fuel is consumed.* The fuel is a mixture of $^{238}_{92}U$ and $^{239}_{94}Pu$. Neutrons from the fission of $^{239}_{94}Pu$ strike the $^{238}_{92}U$ nuclei and convert them to $^{239}_{92}U$. These nuclei decay to $^{239}_{94}Pu$ by double beta decay, as described in Section 24.7. Thus $^{239}_{94}Pu$ is produced at the same time that it is being consumed.

Figure 24-11 illustrates the events in a breeder reactor. Since uranium-238 is much more plentiful than uranium-235, a breeder reactor can extract much larger amounts of

Figure 24-11. Reactions in a breeder reactor. When a neutron strikes a Pu-239 nucleus, the nucleus splits into lighter fission products, releasing additional neutrons. Some of these neutrons strike other Pu-239 nuclei to continue the chain reaction. Other neutrons strike U-238 nuclei, where they are absorbed to produce Pu-239. This produces additional fuel for the reactor.

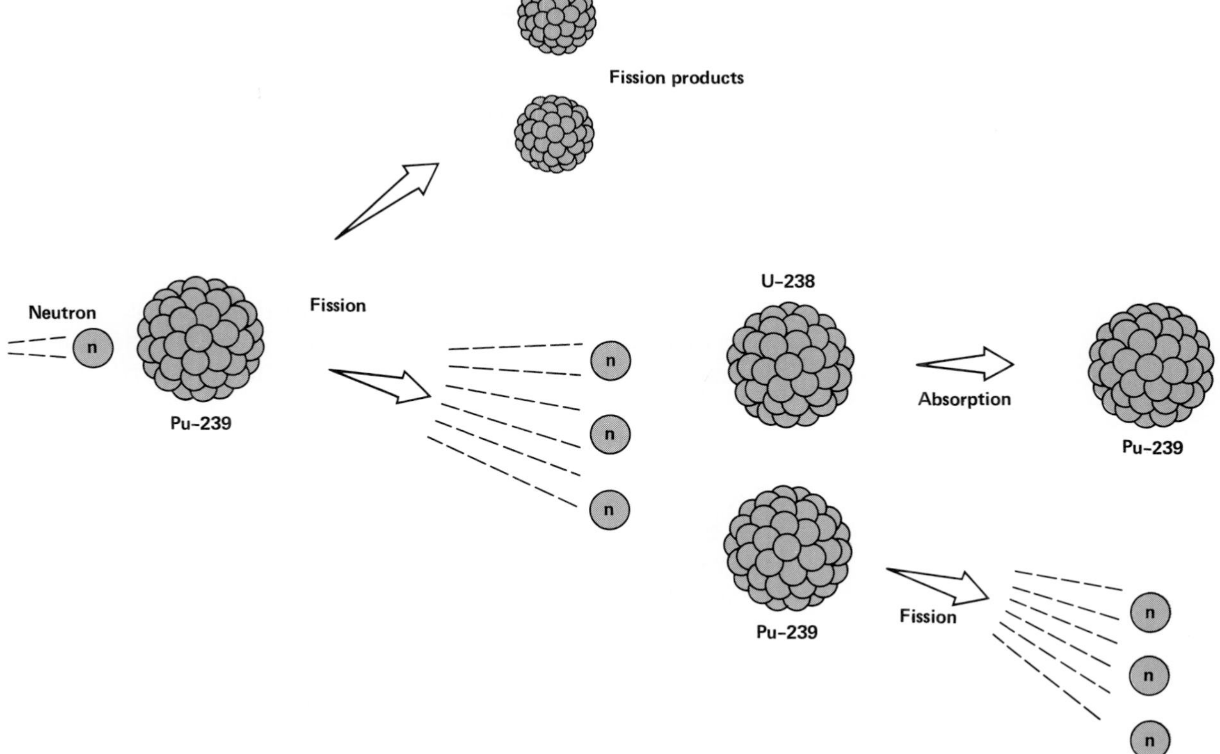

energy from natural uranium than can a reactor that utilizes the lighter isotope.

24.12 Nuclear Power A kilogram of uranium, when completely utilized in fission reactions, has about the same fuel values as 3 000 000 kg of coal or 12 000 000 kg of oil burned in the conventional way. Obviously, nuclear power is an important substitute for the world's dwindling supplies of fossil fuels. The challenge is to produce it safely and to dispose of radioactive wastes without damaging the environment.

At the present, breeder reactors are not used for power production in the U.S.

As Figure 24-12 shows, the basic difference between a nuclear power plant and a conventional one is in the way in which the steam that turns the turbines is produced. In an ordinary plant, the steam comes from a boiler fired with coal, oil, or gas. In a nuclear plant, the steam is generated by the heat released during the fission process in the reactor.

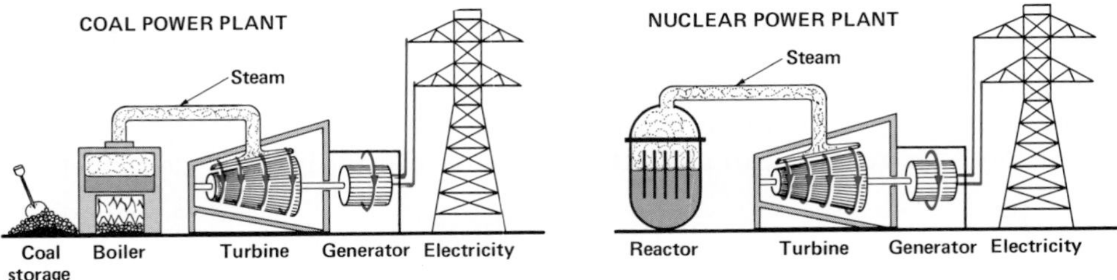

Figure 24-12. A nuclear power plant differs from other power plants only in the way in which steam is generated for the turbines.

The reactor in a nuclear power plant must be able to sustain a chain reaction. Several factors determine whether a chain reaction will continue. The relationship among these factors is described by the following equation:

$$k = \frac{P}{A + L}$$

In this equation k is known as the *multiplication factor*, P is the rate of production of neutrons in the reactor, A is the rate of fission-producing absorption, and L is the rate of leakage. When $k = 1$, the reactor is said to be *critical* and the reaction is self-sustaining.

The power production of a nuclear reactor depends largely on the rate of neutron production. Neutron production, in turn, reduces the amount of fissionable material in the reactor. Hence, to keep the reactor critical it is necessary to reduce the rate of absorption, A, or to increase the production of neutrons, P. The former is accomplished by gradually withdrawing the control rods, while the latter requires the addition of fissionable material.

Natural uranium contains only about 0.7% of the fissionable isotope U-235. Consequently, when natural uranium is used as a reactor fuel, the rate of neutron leakage,

L, is so high that the reactor can never become critical. L can be reduced, however, by using a moderator to slow down the neutrons so that they are more likely to strike U-235 nuclei. Another way to make the reactor critical is to use fuel that has been processed to increase the concentration of U-235.

A typical "neutron history" in a nuclear reactor is given in Table 24-2. It shows the relative number of neutrons involved in other events for every 100 neutrons that produce new fissions. Thus of a total of 256 neutrons, only the two that are absorbed by the control rods are available for increasing the power of the reactor. When the control rods are withdrawn, of course, these two neutrons will raise the multiplication factor of the reactor.

Table 24-2
NEUTRONS IN A REACTOR

Number of neutrons	*History*
100	produce new fissions
90	captured by U-238
20	captured by U-235
30	absorbed by moderator
5	absorbed by housing
9	escape from reactor
2	absorbed by control rods
256	

The first nonexplosive application of nuclear power was in the American submarine *Nautilus,* which was launched in 1954. Two years later, the first large-scale nuclear power plant was placed in operation in Shippingport, Pennsylvania. It had an output of 60 000 kilowatts of electric power. Table 24-3 lists the nuclear power plants in operation or under construction in various countries at the end of 1986. A modern nuclear plant has about 50% more power output than a typical non-nuclear plant.

Figure 24-13. Browns Ferry Nuclear Plant in northern Alabama.

The problem of releasing fusion energy in controlled amounts is being studied in a number of laboratories around the world. At Princeton University's Plasma Physics Laboratory, fusion reactions are produced in so-called *tokamaks* (a name derived from the Russian acronym for "toroidal magnetic chamber"). See Figure 24-14. In a tokamak, plasma is contained in an evacuated, doughnut-shaped stainless steel vessel. Magnetic fields, generated by massive coils located around the outside of the chamber, prevent the high-temperature plasma from striking the vessel walls. The objective is to hold the ions within the vacuum vessel long enough, pack them together densely enough, and heat them to a high enough temperature so that net fusion energy can be produced.

The performance of a fusion reactor is measured as particle density times confinement time (in $\text{cm}^{-3}\cdot\text{s}$) and is called the *Lawson parameter*. A higher temperature means a higher Lawson parameter. The "break even" point for practical power production is $3 \times 10^{14}\ \text{cm}^{-3}\cdot\text{s}$. In 1986, a plasma temperature of 200 million °C was produced in the Princeton tokamak. This is the highest temperature ever attained in a laboratory and is ten times hotter than the temperature that is thought to prevail at the center of the sun. The Lawson parameter achieved in the experiment was $10^{13}\ \text{cm}^{-3}\cdot\text{s}$.

Figure 24-14. With this magnetic-confinement fusion reactor, temperatures much higher than at the sun's core have been attained at Princeton University.

Table 24-3
NUCLEAR POWER PLANTS

Country	*Plants in operation*	*Plants under construction*	*Power output* (MW)	*Proportion of total power* (%)
Argentina	2	1	935	11
Belgium	7	0	5 500	61
Brazil	1	1	626	1
Bulgaria	5	2	2 585	36
Canada	18	4	12 185	16
China	0	2	—	—
Cuba	0	2	—	—
Czechoslovakia	8	8	3 264	28
Finland	4	0	2 310	35
France	55	9	52 588	75
Germany	30	6	24 818	29
Hungary	4	0	1 645	50
India	7	7	1 374	2
Italy	2	0	1 120	4
Japan	39	12	29 300	28
Korea, South	9	2	7 220	50
Mexico	1	1	654	<1
Netherlands	2	0	508	5
Pakistan	1	0	125	<1
Poland	0	2	—	—
Rumania	0	5	—	—
South Africa	2	0	1 842	7
Spain	10	0	7 544	38
Sweden	12	0	9 817	45
Switzerland	5	0	2 952	42
Taiwan	6	0	4 928	44
United Kingdom	39	1	11 242	22
United States	110	4	98 331	19
USSR	46	26	34 230	12
Yugoslavia	1	0	632	6

Another approach to controlled nuclear fusion is to direct lasers or beams of heavy ions at targets consisting of hydrogen isotopes. The resulting collisions implode the target to the densities and temperatures necessary for the fusion of the hydrogen nuclei. A device of this kind is called an inertial-confinement-fusion reactor.

24.13 Radioisotopes *A* ***radioisotope*** *is a radioactive isotope of an element.* Some radioisotopes are the products of natural radioactive decay, as in the case of the uranium decay series in Section 24.5. Since the advent of the nuclear reactor, however, a far greater number of radioisotopes has been prepared artificially. The reactor produces

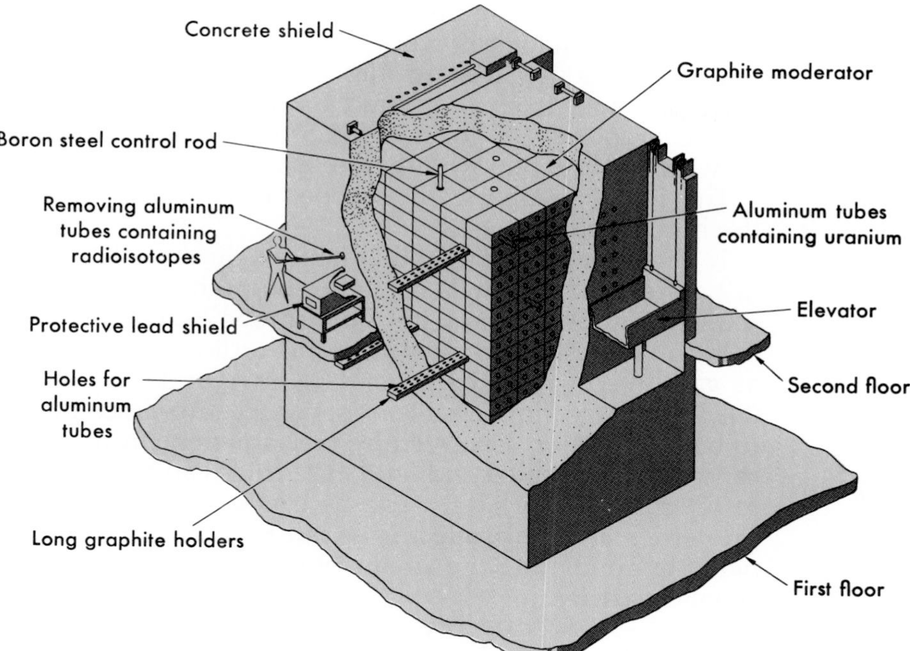

Figure 24-15. Cutaway view of a uranium reactor used for the preparation of radioisotopes. Note the thick concrete shield that is used to protect the technicians from radiation.

radioisotopes during the fission process. In addition, radioisotopes can be prepared by inserting elements into the reactor while it is operating. Radioisotopes of all the elements have been prepared in this way.

One way in which an element becomes radioactive in a reactor is by absorbing a neutron. For example, ${}^{13}_{6}C$ will turn into ${}^{14}_{6}C$, which is a radioisotope. In other cases, elements will absorb a neutron and immediately emit a proton. This process lowers the atomic number of the element but does not change its mass number. For example, natural sulfur may change into radiophosphorus.

$$^{32}_{16}\mathbf{S} + {}^{1}_{0}\mathbf{n} \rightarrow {}^{32}_{15}\mathbf{P} + {}^{1}_{1}\mathbf{H}$$

A large reactor can release neutrons at the rate of 10^{14} neutrons/cm^2·s. When the reactor is used to produce radioisotopes, it must be run at a rate that is somewhat higher than the rate needed for a chain reaction by itself. The extra neutrons are then available for radioisotope production. Special control and shielding arrangements make it possible to insert and remove substances without shutting down the reactor.

It is often desirable to incorporate radioisotopes into specific compounds before they are used. This can be done by chemically combining the radioisotope with other elements. The *radiocompound* that results is said to be tagged, or *labeled.* Complex compounds that cannot be tagged by direct combination can sometimes be labeled through the biological processes of plants and animals. Radioisotopes are injected into or fed to an animal and then removed from the blood or tissue in the form of radiocompounds. Using similar procedures, the process by which plants photosynthesize chemicals can be traced.

The many uses of radioisotopes can be divided into three categories:

1. Effects of radioisotope radiations on materials
2. Effects of materials on radioisotope radiations
3. Tracing materials with radioisotope radiations

Radiotherapy is an important tool in the fight against cancer.

In the first of these categories, the radioisotope is used as a source of radiation, just as X rays and radium are used. The target material consists of the substance that is to be changed or destroyed by the radiation. In this way cancer can be treated and food and drugs can be sterilized to retard spoilage. Plastic can be irradiated to change its properties and static electricity can be reduced or eliminated by irradiating the air surrounding the source of the static electricity.

In the second category of radioisotope utilization, the effect of the material on the radiation yields information about the material. Radiation is beamed at a target and the amount that is transmitted or reflected is recorded on photographic film or with some other detection device. Radiations from radioisotopes that are given to a patient can be used to trace or diagnose physiological processes. This method can also be used to measure the thickness of a moving sheet of metal or other material, to analyze the internal structure of a metal casting, to measure the level of a liquid inside a closed container, and even to sort out materials on a moving conveyor.

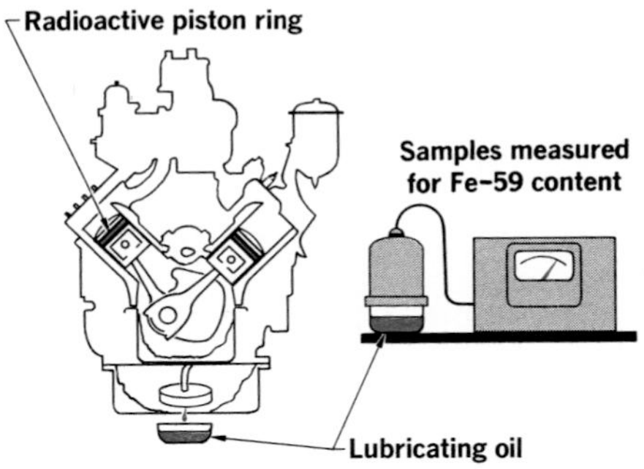

Figure 24-16. A tracer experiment. Radioactive iron (Fe-59) in the piston rings is transferred to the oil through friction while the motor is running. As little as 0.000 3 gram of iron in the oil can be detected by the counter at the right.

In the third category of radioisotope utilization, a radioisotope acts as a *tracer* that makes it possible to follow the course of a chemical or biological process. The target and the radioisotope are intimately mixed or combined. The radioisotope then serves as the label that indicates the location of the successive compounds or materials with which the radioisotope becomes associated. Detection instruments or photographic film again serve as measuring devices. The material that is labeled and traced may be food that is consumed by a human being, feed that is converted into milk inside the body of a cow, water that is running out of an undiscovered underground leak, atoms that are rearranged to form a complex molecule, wax that is worn away from the hood of a polished car, or any one of a great many other research materials or industrial products. Figure 24-16 shows how tracers are used to study the performance of piston rings in an automobile engine.

An interesting use of a natural radioisotope is the process known as *carbon-14 dating,* which was developed by the American chemist W. F. Libby (1908–1980) in 1952. In the atmosphere, neutrons produced by cosmic-ray bombardment strike $^{14}_{7}N$ atoms and convert them into the radioisotope $^{14}_{6}C$. In the process of photosynthesis, living plants absorb $^{14}_{6}C$ (as carbon dioxide) along with the more common and nonradioactive $^{12}_{6}C$. When the plant dies, it no longer replaces the carbon atoms in its cells with carbon atoms from the atmosphere and the intake of $^{14}_{6}C$ stops. The amount of radioactivity in the plant gradually decreases due to the fact that the half-life of $^{14}_{6}C$ is 5730 years. Measurements show that a gram of carbon from a living plant has an activity of 16 disintegrations/min. Consequently, a sample showing an activity of 8 disintegrations/g · min is assumed to be 5730 years old. A sample with 4 disintegrations/g · min is 11 460 years old, etc. See Figure 24-17. The precision of carbon-14 dating falls off rapidly after several half-life periods, although objects up to about 50 000 years old can be dated in this way.

Other nuclear dating methods are used to learn the age of rocks on the earth and the moon. Moon rocks as old as 4.5 billion years have been dated. It is believed that the solar system itself is not much older than this.

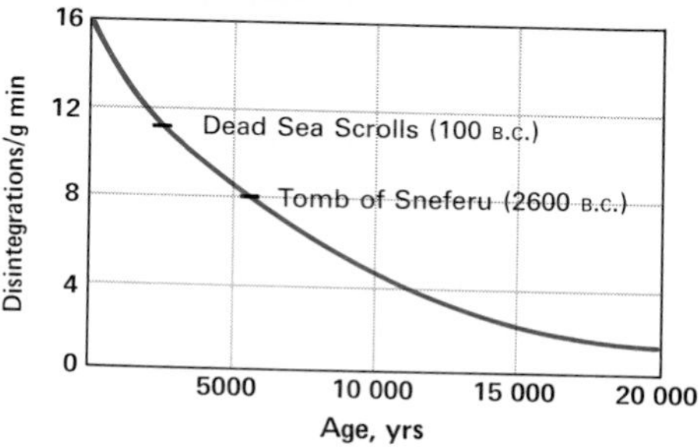

Figure 24-17. Carbon-14 dating. The age of organic compounds can be determined by measuring the remaining amount of radioactive carbon. The Dead Sea Scrolls, a portion of which is pictured above, were dated in this manner.

QUESTIONS: GROUP A

1. (a) What conditions are necessary for a chain reaction? (b) What is meant by critical mass?
2. Describe the two types of reactions in a nuclear reactor.
3. (a) How does a nuclear reaction produce electric power? (b) What is meant by a critical reactor? (c) How can natural uranium be used in a critical reactor?
4. How does a breeder reactor differ from conventional reactors?
5. What are the similarities and differences between a nuclear power plant and a fossil-fuel plant?
6. What is the distinction between a magnetic-confinement and an inertial-confinement fusion reactor?
7. (a) What are radioisotopes? (b) How are they made?
8. How could radioisotopes be used to determine the effectiveness of a fertilizer?
9. Estimate the age of a carbon sample showing an activity of 3 disintegrations/g·min.

SUMMARY

Radioactivity is the spontaneous decay of atomic nuclei with the emission of particles and rays. It is a property of all elements with atomic numbers greater than 83. Radioactive elements have these characteristics in common: effect on photographic film, ionization of air, production of fluorescence in certain compounds, special physiological effects, and radioactive decay.

The radiations from radioactive elements consist of alpha particles (helium nuclei), beta particles (high-speed electrons), and gamma rays (high-energy photons). In a nuclear change, new materials are formed by changes in the identity of the atoms themselves.

There are four types of energy-producing nuclear reactions: radioactive decay, nuclear bombardment, fission, and fusion. The rate at which a radioactive nuclide decays is indicated by its half-life. Radioactivity is also described in terms of becquerels, curies, and grays. In nuclear bombardment, a nuclide breaks into a large and a small fragment. In fission, two large fragments are produced. In fusion, two small fragments combine to form a larger one. More energy is produced from equal masses of fuel by fusion reactions than by fission reactions.

A chain reaction yields as a product the material or energy that started the reaction. The amount of material required to sustain a fission reaction is called the critical mass. The isotopes U-235 and Pu-239 capture slow neutrons and undergo fission, thereby producing neutrons to continue the chain reaction.

Nuclear reactors produce heat energy for power plants, new fissionable materials, and radioisotopes. The effects of these radioisotopes on certain materials and the effects of certain materials on radioisotopes are used in research. Radioisotopes are also used to trace materials in chemical or biological processes. Carbon-14 dating is an example of a process that employs the use of a natural radioisotope to learn the age of an ancient object.

VOCABULARY

alpha particle
becquerel
beta particle
breeder reactor
chain reaction
chemical change
cosmic ray
critical mass
curie
fission
fusion
gamma ray
gray
nuclear change
nuclear reactor
physical change
radioactivity
radioisotope
sievert

25 High-Energy Physics

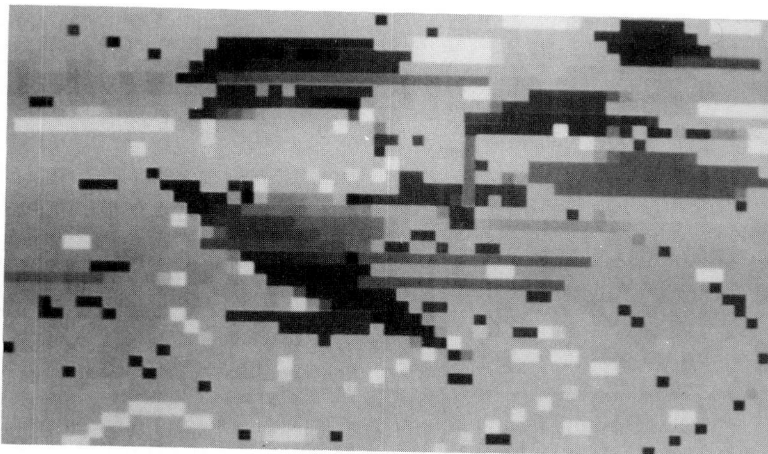

quantum mechanics (KWAN-tum me-KAN-iks) n.: the branch of physics that deals with the behavior of particles whose specific properties are given by quantum numbers.

QUANTUM MECHANICS

Objectives

- Define the uncertainty principle.
- Describe how quantum numbers are used in the description of electron motion.
- Study different types of particle accelerators and detectors.
- Classify the different types of fundamental particles.
- Study the four basic interactions between particles.
- Define the conservation laws of particle physics.
- Discuss recent developments in particle physics.

25.1 The Uncertainty Principle As seen in Section 12.9, the *quantum theory* states that light exhibits particle-like properties in addition to its wave-like properties. This means that light energy is transferred in discrete quantities, or quanta. Applying this idea to the atom, Bohr concluded that the electron is not free to assume any orbit whatever; rather, the size and shape of the electron orbit are governed by the quantum theory. An electron cannot move into a lower energy orbit except through the loss of energy exactly equal to the energy difference between the initial and the final orbit. An electron cannot move into a higher energy orbit unless it gains enough energy. An electron must stay in its orbit unless sufficient energy is gained or lost so that an abrupt orbital transition can take place. This process can occur spontaneously.

Bohr's model was remarkably successful in explaining the spectral lines of low-atomic-numbered elements. But the model ran into trouble when it was used to explain the behavior of helium atoms or atoms with large atomic numbers. Then, too, small but important differences were found between the predictions of the Bohr model regarding atomic spectra and the results of experiments based on the predictions. Attempts were made to account for these differences by suggesting that electron orbits are elliptical rather than circular and by using relativistic mass (a method that was successful in solving a similar problem

Figure 25-1. Werner Heisenberg, the German physicist, formulated the uncertainty principle, which provides a basis for quantum mechanics.

In 1932 Heisenberg received the Nobel Prize in Physics for his work in quantum mechanics.

with the orbit of the planet Mercury). These explanations resolved only some of the discrepancies.

There were other problems with the Bohr atom. The effect of the spin of the electron (Sections 19.2 and 23.6) on its orbital motion had not been taken into account by Bohr. Such spinning sets up electromagnetic forces that must be considered in computing the total energy of the orbiting electron.

The most serious blow to the Bohr model of the atom came in 1927 with the announcement of the *uncertainty principle* by the German physicist Werner Heisenberg (1901–1976). The ***uncertainty principle*** states that *it is impossible to specify simultaneously the exact position of an object, such as an electron, and its momentum.* Other pairs of physical quantities related to the motion of an object, such as time and energy, are also subject to the uncertainty principle. For bodies of ordinary mass, the inherent uncertainty in any such simultaneous determination is smaller than errors in measurement, and so the uncertainty effect is not observed. With electrons, however, the effect is pronounced. This principle means that it is possible to determine where an electron is at a given time, but it is not possible to determine its exact velocity at that same time. The reason is that the method used to determine the position of the electron, short-wavelength radiation, changes the electron's momentum. Likewise, to determine an electron's momentum, an instrument that changes the electron's direction of motion is used, and so the determination of the electron's position is prevented.

25.2 Quantum Numbers

To understand the description of the atom according to the quantum theory, it is necessary to be familiar with the idea of *quantum numbers.* As used in atomic physics, a ***quantum number*** *is a number that describes the allowable value of certain physical quantities.* For example, quantum numbers are used in special equations for computing the energy of a particle, such as the electron. *The branch of physics that deals with the behavior of particles whose specific properties are given by quantum numbers is called* ***quantum mechanics.***

The quantum-theory model of the atom is three-dimensional. The motion of the electron is not restricted to a single plane, as in the Bohr atom. Within the limits prescribed by its quantum numbers, the electron describes an orbit that completely encloses the nucleus.

The first quantum number for the electron describes the radius of its orbit. It is called the *principal quantum number* and is designated by the letter n. The principal quantum number may have the integral values 1, 2, 3, 4, etc., which correspond to the similarly numbered energy levels of the Bohr atom. For example, an electron with a principal quantum number 1 is in the first energy level, an electron with the number 2 is in the second level, and so on.

The equation for finding the energy of an electron from its principal quantum number is

$$E = -\frac{me^4}{8\epsilon_0{}^2h^2n^2}$$

where E is the energy of the electron, m is its mass, e is the electronic charge, ϵ_0 (the Greek letter epsilon) is a constant with the value 8.854×10^{-12} $C^2/N \cdot m^2$, h is Planck's constant, and n is the principal quantum number. The energy is expressed as a negative quantity to show that the electron is held by the nucleus.

Thus far the use of the principal quantum number yields exactly the same results for the orbits of electrons that Bohr obtained for his atomic model. But here the similarity ends. A second quantum number, called the *angular momentum quantum number* or *secondary quantum number*, describes the magnitude of the angular momentum of the orbiting electron. This quantum number is designated by the letter l. Unlike energy, which is a scalar quantity, angular momentum is a vector with both magnitude and direction. The equation for the magnitude of the angular momentum is

$$L = \sqrt{l(l+1)}\frac{h}{2\pi}$$

where L is the angular momentum, l is the angular momentum quantum number, and h is Planck's constant. l may have integral values ranging from zero to $n - 1$ and is therefore limited by the principal quantum number. In effect, a low value of l means that the electron's orbit is circular. The higher the value of l, the more elliptical the orbit is.

The magnitude of the angular momentum in a magnetic field is described by a third quantum number, the *magnetic quantum number*. It is designated as m_l and can have integral values ranging from $-l$ through zero to $+l$. The equation using m_l is

$$L_z = m_l\frac{h}{2\pi}$$

in which L_z is the component of the angular momentum that is parallel to the magnetic field.

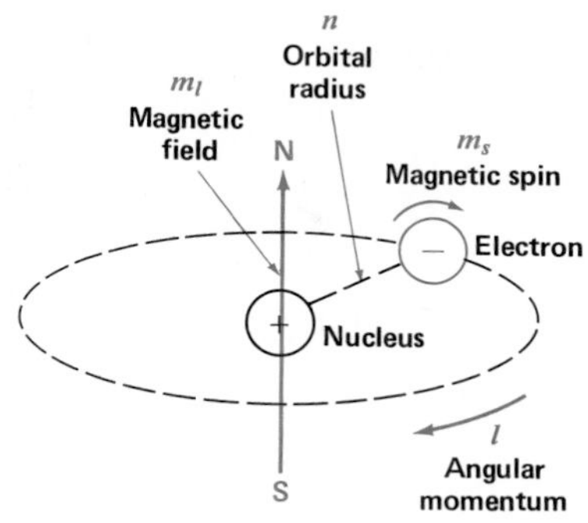

Figure 25-2. Quantum numbers describe the position and motion of an electron.

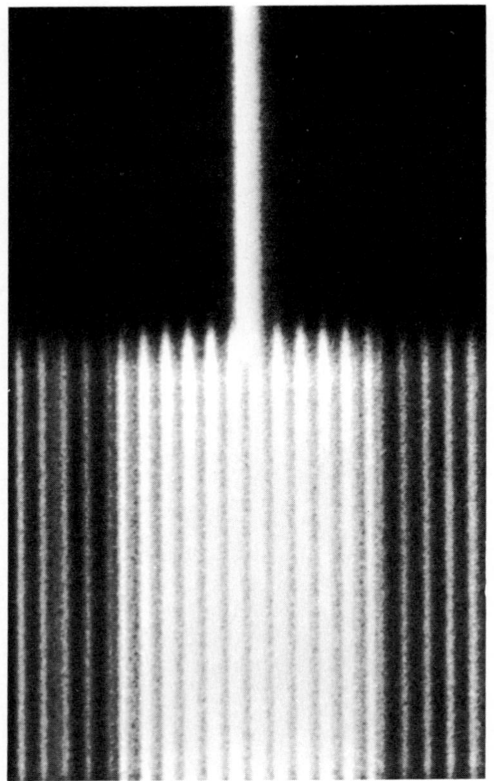

Figure 25-3. The Zeeman effect. The single line at the top is the 3779-Å line of the element curium. At the bottom is the spectrum of the same line in a strong magnetic field.

In 1945 Pauli received the Nobel Prize in Physics for his discovery of the exclusion principle.

Since m_l can have several values for all values of n except $n = 1$, the energy levels of an atom are split into two or more sublevels when the atom is subjected to a magnetic field. This effect is called the *Zeeman effect,* after the Dutch physicist Pieter Zeeman (1865–1943). The Zeeman effect can be detected in the spectra of excited atoms, as is shown in the photo of Figure 25-3.

In the 1920s it was discovered that many pairs of spectral lines could be described in terms of a model in which electrons spin on their axes. This electron spin model also harmonizes with the relativistic explanation of the behavior of matter. The spin of the electron is designated by the *spin quantum number, s,* which has the value $\frac{1}{2}$. The equation for the magnitude of spin angular momentum is

$$L_s = \sqrt{s(s + 1)}\frac{h}{2\pi}$$

The component of spin angular momentum that is parallel to a magnetic field is given by the equation

$$L_s = m_s\frac{h}{2\pi}$$

where m_s is the *spin magnetic quantum number* and can have the values $+\frac{1}{2}$ and $-\frac{1}{2}$. The + and − signs indicate that the electrons are spinning in opposite directions.

The assignment of four quantum numbers to each electron leads to the logical question of how the quantum numbers of any one electron compare with the numbers of other electrons. Does it ever happen that two electrons have the same set of numbers? The answer to this question is provided by the *exclusion principle* of the Austrian physicist Wolfgang Pauli (1900–1958). The ***exclusion principle*** states that *no two electrons in an atom can have the same set of quantum numbers.* The number of different energy levels that could be computed if there were no restrictions on the quantum numbers associated with each electron is far greater than those revealed by spectral evidence.

With the help of the four quantum numbers and the exclusion principle, it is possible to determine the electron configuration of any particular atom. Electron configurations for several elements are given in Table 25-1. In each case, the number of electrons in a particular sublevel is found by using the permissible values of each quantum number. For example, aluminum has 13 electrons. When $n = 1$, l equals 0, m_l equals 0, and m_s can have values of $+\frac{1}{2}$ and $-\frac{1}{2}$. In other words, two electrons can be listed in the column headed $n = 1$. When $n = 2$, l can be 0 or 1. When $l = 0$, m_l equals 0 and m_s can be $+\frac{1}{2}$ and $-\frac{1}{2}$, as before.

Table 25-1
ELECTRON CONFIGURATION OF THE LIGHT ELEMENTS

Atomic number	Symbol	Element	*Electron configuration of atom*							
			$n = 1$	$n = 2$		$n = 3$			$n = 4$	
			$l = 0$	$l = 0$,	$l = 1$	$l = 0$,	$l = 1$,	$l = 2$	$l = 0$,	$l = 1$
1	H	hydrogen	1							
2	He	helium	2							
3	Li	lithium	2	1						
4	Be	beryllium	2	2						
5	B	boron	2	2	1					
6	C	carbon	2	2	2					
7	N	nitrogen	2	2	3					
8	O	oxygen	2	2	4					
9	F	fluorine	2	2	5					
10	Ne	neon	2	2	6					
11	Na	sodium	2	2	6	1				
12	Mg	magnesium	2	2	6	2				
13	Al	aluminum	2	2	6	2	1			
14	Si	silicon	2	2	6	2	2			
15	P	phosphorus	2	2	6	2	3			
16	S	sulfur	2	2	6	2	4			
17	Cl	chlorine	2	2	6	2	5			
18	Ar	argon	2	2	6	2	6			
19	K	potassium	2	2	6	2	6		1	
20	Ca	calcium	2	2	6	2	6		2	
21	Sc	scandium	2	2	6	2	6	1	2	
22	Ti	titanium	2	2	6	2	6	2	2	
23	V	vanadium	2	2	6	2	6	3	2	
24	Cr	chromium	2	2	6	2	6	5	1	
25	Mn	manganese	2	2	6	2	6	5	2	
26	Fe	iron	2	2	6	2	6	6	2	
27	Co	cobalt	2	2	6	2	6	7	2	
28	Ni	nickel	2	2	6	2	6	8	2	
29	Cu	copper	2	2	6	2	6	10	1	
30	Zn	zinc	2	2	6	2	6	10	2	
31	Ga	gallium	2	2	6	2	6	10	2	1
32	Ge	germanium	2	2	6	2	6	10	2	2
33	As	arsenic	2	2	6	2	6	10	2	3
34	Se	selenium	2	2	6	2	6	10	2	4
35	Br	bromine	2	2	6	2	6	10	2	5
36	Kr	krypton	2	2	6	2	6	10	2	6

Figure 25-4. Wolfgang Pauli, the Austrian-born physicist, developed the exclusion principle, which states that no two electrons in an atom can have the same set of quantum numbers.

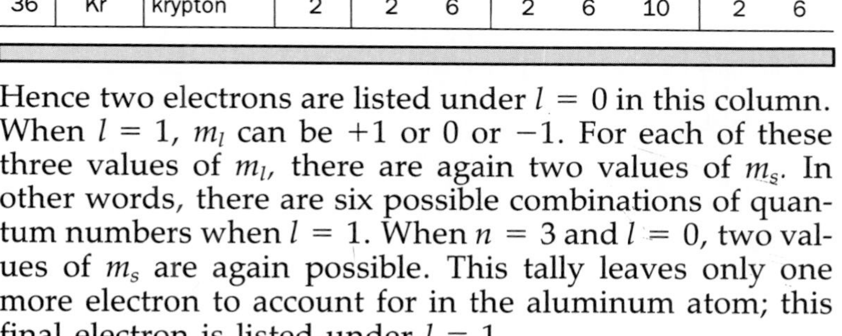

Hence two electrons are listed under $l = 0$ in this column. When $l = 1$, m_l can be $+1$ or 0 or -1. For each of these three values of m_l, there are again two values of m_s. In other words, there are six possible combinations of quantum numbers when $l = 1$. When $n = 3$ and $l = 0$, two values of m_s are again possible. This tally leaves only one more electron to account for in the aluminum atom; this final electron is listed under $l = 1$.

Notice that in the elements beyond argon, electrons may appear in the fourth energy level before the third level is complete. In such cases, the sublevel for which $n = 4$ has a lower energy than the remaining sublevels for which $n = 3$. For atoms with atomic numbers beyond 18, therefore, it is not simply a matter of successively filling in sublevels.

Quantum mechanics allows only a definite number of orbiting electrons for each value of the principal quantum number. For $n = 1$, it is 2 electrons, for $n = 2$, it is 8 electrons, and so on. When an atom has the maximum number of electrons for a particular electron shell (or subshell for heavy atoms), it is said to have a *complete shell.* In most cases, atoms are stable when they have complete shells.

The stability of an atom is described in terms of *ionization energy.* ***Ionization energy*** *is the energy required to remove an electron from an atom.* Obviously, the ionization energy must be equal to the energy of the electron as described by its quantum numbers, and it must also have the opposite sign. As n increases, the energy of the electron decreases. Consequently, the outermost electrons are usually the easiest to remove from an atom. Figure 25-5 shows the ionization energies of the outermost electrons of the elements. While there are variations in ionization energies within electron energy levels, you will notice that the ionization energies for completed levels (atomic numbers 2, 10, 18, 36, 54, and 86) decrease as the size of the atom increases.

25.3 Matter Waves A final word of explanation about the equations of the quantum theory is necessary at this

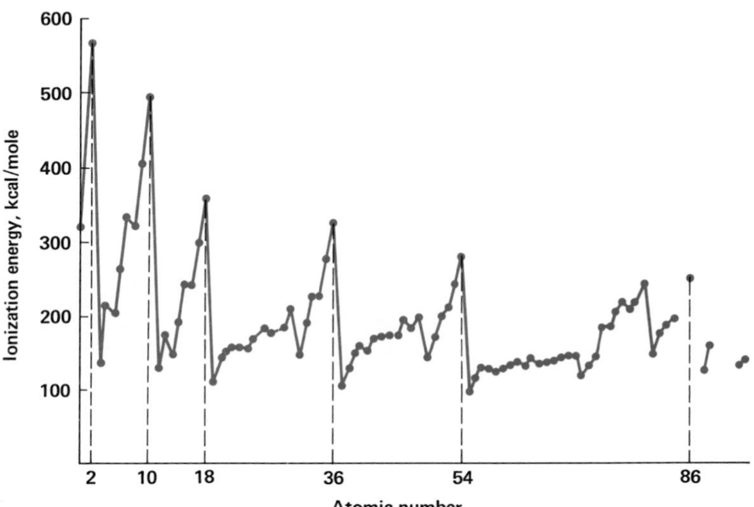

Figure 25-5. Ionization energies of the outermost electrons of the elements.

point. You will notice that Planck's constant, h, is used in each of the equations. In Section 12.11, we saw that h is associated with wave motion. The use of this same constant in describing the behavior of subatomic particles suggests that the particles have wave-like characteristics. This wave-particle duality was mentioned in the very first chapter and in the discussion of light. Now it is important to emphasize it once more. About 1924 the French physicist Louis de Broglie suggested that *all* matter possesses wave characteristics and that these *matter waves* become longer as the momentum of a particle of matter decreases.

In 1929 de Broglie received the Nobel Prize in Physics for his discovery of the wave nature of electrons.

In Section 12.11, the wavelength of a matter wave was given by the equation

$$\lambda = \frac{h}{mv}$$

where λ is the wavelength, h is Planck's constant, m is the mass of the particle, and v is its velocity. Since h has a very small value (6.63×10^{-34} J·s), it can be seen from the equation that m likewise must be extremely small before λ can become large enough to be detectable. This is why the wave properties of, let us say, a moving baseball are completely undetectable, while the wavelength of the electron is long enough to consider in the modern view of the atom.

It must not be assumed, however, that a de Broglie wave is like the physical disturbance of the water in a ripple tank. A de Broglie wave is not a motion, but rather a mathematical description of matter. Matter waves can exist in a vacuum in the same way that light waves can. They permeate space in the way a gravitational or magnetic field influences the area surrounding a mass or a magnet. In the final analysis, it is not possible to construct an analogy for matter waves. Furthermore, it should be recognized that de Broglie waves, like so many other descriptions in science, represent a particular model of the universe. As with all models in the history of science, this one can also be modified and improved as our understanding of the universe keeps growing.

Figure 25-6. Louis de Broglie, the French physicist who proposed the wave nature of matter.

QUESTIONS: GROUP A

1. (a) What is meant by the term *quanta*? (b) What physical phenomena that we have studied are said to be *quantized*?
2. What restrictions did the Bohr model place on the movement of an electron in an atom?
3. (a) State the uncertainty principle. (b) Who was the first to articulate this principle?
4. (a) What is a quantum number? (b) What branch of physics deals with

the behavior of particles that are assigned quantum numbers?

5. How is the principal quantum number related to Bohr's original model of the atom?
6. (a) Why is the energy of an electron given as a negative value? (b) What type of force does this negative energy represent?
7. (a) Which quantum number represents a vector quantity? (b) What does this number tell you about the electron?
8. (a) Describe the Zeeman effect. (b) With which quantum number is it associated?

GROUP B

9. What were some of the problems that arose with the Bohr atomic model?
10. Why is uncertainty not observed in objects with large masses?
11. Why is it not possible to know both the position *and* momentum of an electron?
12. How do the particles studied in quantum mechanics differ from those studied in classical (Newtonian) mechanics?
13. Can two electrons have the same set of quantum numbers? Explain.
14. Use de Broglie's equation to explain why the wave properties of an electron can be observed, while those of a speeding car cannot.

PROBLEMS: GROUP A

1. List the quantum numbers for each electron in the following atoms: (a) helium, (b) sodium, (c) argon, (d) magnesium.
2. (a) Compute the energy of an electron with a principal quantum number of 3. Use SI units. The mass and charge of the electron and the value of Planck's constant are given in Appendix B, Table 3. (b) Convert this energy to electron volts, and compare it with the ionization energy of the outer electron of the neutral sodium atom.

PARTICLE ACCELERATORS

25.4 Providing High-Energy Particles As we saw in Chapter 24, nuclear reactions can be caused by particles and radiations from naturally radioactive substances. But the subatomic research that can be conducted in this way is limited by the energy of the radioactive emissions. To provide nuclear "bullets" that have greater penetrating power, it is necessary to accelerate them. In the following sections, we shall describe several particle accelerators, among which are the largest scientific devices ever built.

25.5 Van de Graaff Generators One of the first particle accelerators was developed in 1931 by the American physicist R. J. Van de Graaff (1901–1967). As Figure 25-7 illustrates, a *Van de Graaff generator* projects electrons onto an insulated moving belt as a result of repulsion from a strongly negative electrode. The electrons ride on the belt to the other end of the generator, where they are picked off by a collector and transferred to a metal sphere. Since an electric charge always resides on the *outside* of a hollow conductor, the inside of the sphere is able to receive more charges. Thus a large negative charge is built up on the sphere.

Conversely, a large positive charge may be built up on the sphere by removing electrons from the belt by use of a strongly positive electrode. The large difference of electric potential thus obtained can be used to accelerate electrons or ions. Since the high charge on a Van de Graaff generator leaks off into the atmosphere if the air is humid, such generators are sometimes built under a large, pear-shaped shell that is filled with dry air under pressure to keep the charge on the sphere. Placing the generator in a vertical position helps to prevent the discharge of particles to the ground.

To provide maximum energy to particles, Van de Graaff generators are sometimes used in series with linear accelerators. The world's most powerful Van de Graaff installation is a pair of end-to-end generators at the Brookhaven National Laboratory in New York. The generators are well insulated to prevent grounding. This installation is capable of accelerating hydrogen ions to an energy of 30 MeV. Although this is not as high as the energies produced in other types of large accelerators, the Van de Graaff machines have certain other advantages. For example, the particle beam can be more precisely focused, energy variations are more easily controlled, and a wide range of particles can be employed.

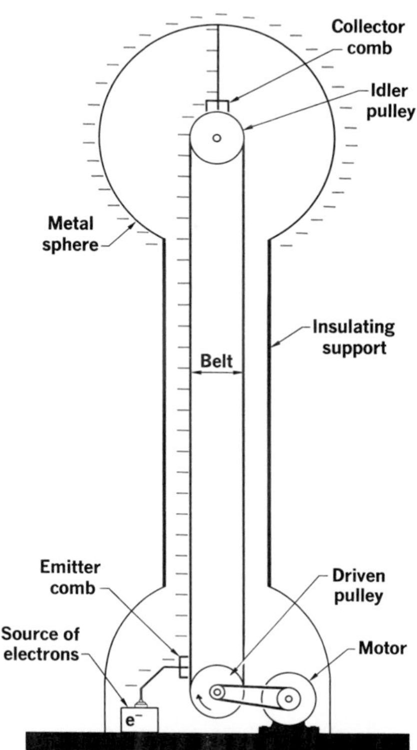

Figure 25-7. Schematic diagram of a Van de Graaff generator (top). Shown at the bottom are two such generators in tandem at Brookhaven National Laboratory. (The second generator is barely visible in the rear.)

25.6 Circular Accelerators Shortly after the development of the Van de Graaff generator, it was found that charged particles can be accelerated to very high velocities by driving them in a circular path with the aid of electromagnets. A variety of such accelerators have since been built.

1. *Cyclotrons.* In 1932 the American physicist E. O. Lawrence (1901–1958) developed a circular accelerator in which various kinds of charged particles could be used. He called it a *cyclotron.* It consists of a large cylindrical box placed between the poles of an electromagnet, as shown in Figure 25-8. The box is exhausted until a very high vacuum exists inside. Charged particles are produced by ionizing gases fed into the center of the box. Inside the box are two hollow, D-shaped electrodes called *dees* that are connected to a source of high-voltage electricity.

When the cyclotron is in operation, the electric charge on the dees is reversed very rapidly by an oscillator. The action of the field of the electromagnet causes the charged particles inside the cylindrical box to move in a semicircle in a fixed period of time. The potential applied to the dees

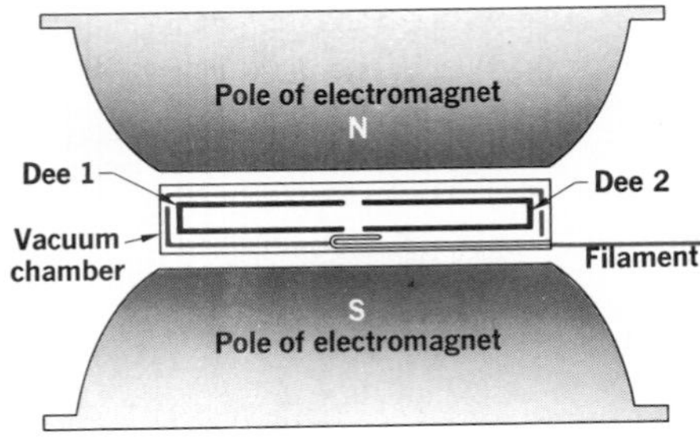

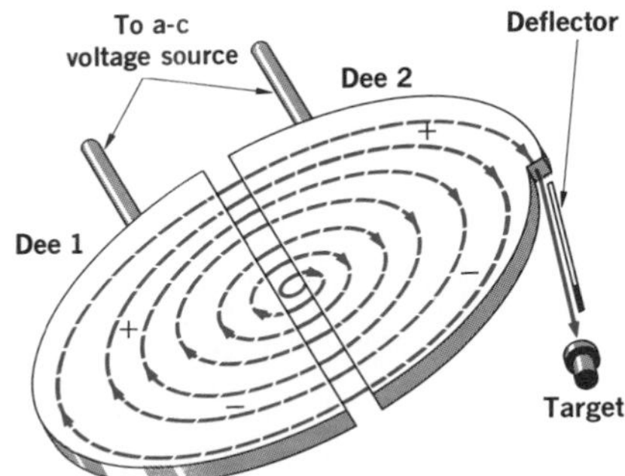

Figure 25-8. Schematic diagram of a cyclotron.

Figure 25-9. The inventor of the betatron, Donald Kerst, is shown with his original accelerator, a 2.5-MeV model built in 1940. In the background is a 340-MeV betatron used at the University of Illinois.

is adjusted to reverse itself in the same time period. Thus at each transit across the gap between the dees, the ions gain energy from the electric field. While the ions are inside the dees, their velocity is constant because there is no electric field on the inside. But at each transit, the ions move faster and approach the outside of the box. When they reach the outer rim of the box, they are deflected toward the target.

The energy of particles accelerated in a conventional cyclotron may reach 15 MeV. To obtain higher values, it is necessary to modify the accelerator to deal with the relativistic increase in mass of the particles as their velocity increases. This increased mass keeps the particles from gaining enough speed to reach the end of the dees by the time the oscillating voltage reverses. To overcome this difficulty, it is necessary to vary the oscillating frequency. Then as the mass of the particles increases and their acceleration decreases, the frequency of the oscillator can be reduced. The changes in voltage of the field are timed to coincide with the slower acceleration of the more massive particles. This type of accelerator is called a *synchrocyclotron.* It can boost the energy of particles to about 800 MeV.

2. *Betatrons.* As its name implies, the betatron is used to accelerate beta particles (electrons). The betatron uses the principle of the transformer; the primary is a huge electromagnet and the secondary is the stream of electrons that are being accelerated. The electrons are enclosed in a circular, evacuated tube called the *doughnut.* After the electrons have acquired maximum energy, they are directed toward a metal plate. The impact of the electrons on the metal plate produces high-energy X rays that are used for nuclear studies and for industrial and medical applications. In effect, the betatron is a large X-ray machine, producing energies up to about 300 MeV. See Figure 25-9. Beyond this range, electrons rapidly lose energy by electromagnetic radiation.

3. *Synchrotrons.* A newer type of particle accelerator is the *synchrotron.* Like the cyclotron, it operates on the principle of accelerating particles by making them move with increasing velocity in a circular path. However, the path of the particle is confined to a doughnut (as in the betatron) by varying both the oscillating voltage and the magnetic field.

The energies attainable in a synchrotron are theoretically unlimited. At present, the installation at the Fermi National Accelerator Laboratory (Fermilab) in Batavia, Illinois, is the United States' most powerful accelerator. It consists of four different kinds of accelerators in series. See Figure 25-10. The circular path of the synchrotron in the final stage is 6.6 km long! By using superconducting magnets in this synchrotron, it is possible to obtain proton energies of 1000 GeV (1 TeV) and velocities greater than 99.999% of the velocity of light. Some high-energy synchrotrons are also called *bevatrons* or *tevatrons*.

Figure 25-10. Sections of the particle accelerator at Fermilab. The voltage multiplier at the far left supplies an initial acceleration of protons to 750 keV. Second from left is a view of the linear accelerator that drives the protons up to 200 MeV. At the top right is part of the magnet ring of the booster accelerator that raises the energy to 8 GeV. The main accelerator tunnel, a section of which is shown at the bottom right, is 6.6 km long. Protons pass from the upper ring of 1000 conventional magnets to the lower ring of 1000 superconducting, helium-cooled magnets. They emerge with energies of nearly 1 TeV.

4. *Intersecting storage rings.* The circular accelerators mentioned thus far produce particles by the collision of beams of high-energy particles with stationary targets. The disadvantage of this arrangement is that most of the beam's energy is dissipated in pushing against the target material; only a small amount of the energy is available for nuclear reactions.

In the intersecting storage accelerator, particles are made to collide with each other head-on as they move in opposite directions. To increase the number of collisions, the two beams are first concentrated in *storage rings.* When the density of particles in the storage rings is sufficiently high, the beams are allowed to collide.

In an intersecting storage accelerator, the storage rings intersect in several places. See Figure 25-11. Superconducting magnets guide the beams inside the rings. Because of the efficiency of intersecting accelerators, they can provide much higher energies than are possible with the most powerful nonintersecting accelerators in operation today.

Since intersecting storage rings are best used with pro-

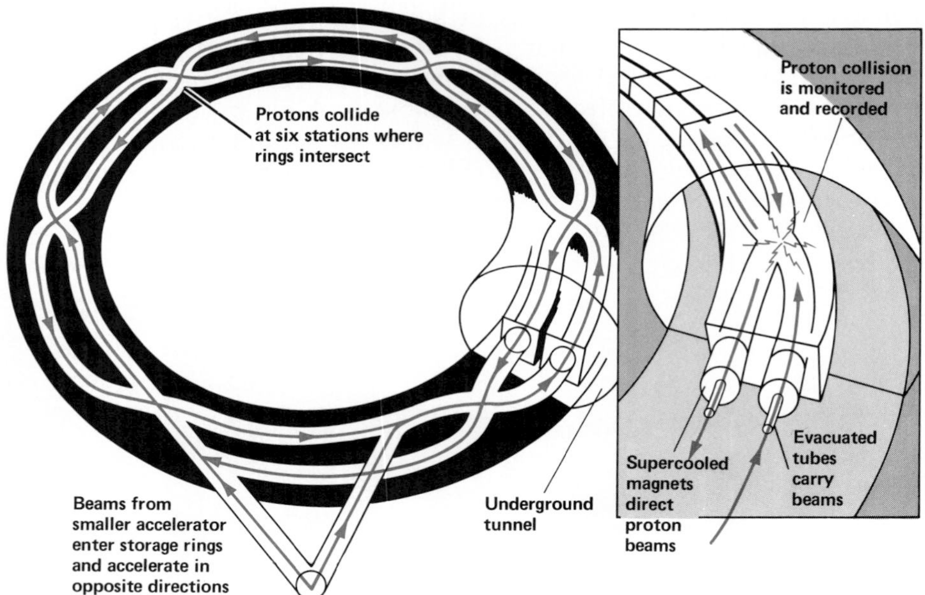

Figure 25-11. Schematic diagram of an intersecting storage accelerator. Almost all the energy of the proton beams is converted into particle production in the head-on collisions in this accelerator.

tons, however, and since the total number of collisions per second is not very large compared with that in other accelerators, intersecting storage accelerators cannot be used for all experiments.

Development of the world's largest accelerator is presently underway in Texas. Called the Superconducting Supercollider (SSC), this mammoth installation will be more than 80 km in circumference and will contain 10 000 superconducting magnets. When complete, the SSC is expected to accelerate protons in two opposing beams to energies of 40 TeV! With this powerful new tool, scientists hope to answer many of the questions that still exist about the nature of matter and energy.

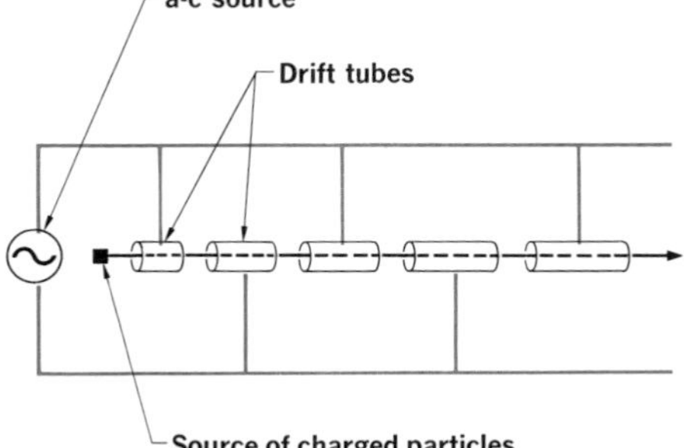

Figure 25-12. Schematic diagram of a linear accelerator. Charged particles accelerate as they are attracted by the alternately charged drift tubes.

25.7 Linear Accelerators

A *linear accelerator* is a device that moves charged particles to high velocities along a straight path, in contrast to the circular arrangements in cyclotrons and synchrotrons. Linear accelerators are used individually and in conjunction with other accelerators.

A simplified diagram of a linear accelerator is shown in Figure 25-12. A series of hollow *drift tubes* is mounted in a long, evacuated chamber. Charged particles are fed into one end of this arrangement and are accelerated by means of a high-frequency alternating voltage applied to the drift tubes. At any given instant, successive tubes have opposite charges. The alternating voltage is timed in such a way that the particle is repelled by the tube it is leaving and attracted by the tube it is approaching. In this way, the particle is accelerated every time it crosses the gap between two tubes. Because of this acceleration, the drift tubes are successively longer so that the particles will always be in a gap when the voltage is reversed.

A 3.3-km-long linear accelerator is located near Palo Alto, California. It is pictured in Figure 25-13. Inside a copper tube in a concrete tunnel buried 7.6 m underground, electrons are accelerated to 40 GeV. As mentioned previously, some circular accelerators are limited in their ability to accelerate electrons because of electromagnetic radiation losses. This difficulty is not encountered in linear accelerators. In addition, the beam of particles can be more easily controlled in a linear accelerator than in a circular one. Another advantage of a linear accelerator over a cyclotron is that the former can provide a higher frequency of impacts on the target material.

Figure 25-13. The world's largest linear accelerator. Electrons are injected at the far end of the 3.3-km-long structure (upper right). Particle production and analysis take place in the two buildings at the lower left.

QUESTIONS: GROUP A

1. Why are particle accelerators so important to nuclear physicists?
2. (a) Which type of particles can be accelerated with a Van de Graaff generator? (b) Which type cannot? Why?
3. What are some of the advantages of a Van de Graaff generator?
4. (a) How can particles be made to travel in circular paths? (b) With which type of particle will this system not work?
5. (a) What modification makes a cyclotron into a synchrocyclotron? (b) Why is this done?
6. In what fields is a betatron particularly useful?
7. How does a synchrotron accelerate protons to energies of 1000 GeV?
8. How does the use of superconducting magnets affect the maximum energy that can be produced in a particle accelerator?
9. What makes an intersecting storage accelerator more efficient than other types of circular accelerators?
10. (a) What is a drift tube? (b) How is it utilized in a linear accelerator? (c) Why are all the drift tubes in a linear accelerator not the same length?
11. What are some advantages of linear accelerators over circular accelerators?

DETECTION INSTRUMENTS

25.8 Types of Detectors Subatomic particles are far too small to be observed and measured directly. Most of these particles have extremely short half-lives. This, together with the high speed at which they usually travel, means that the time available to study them is very short. As is true of the construction of high-energy accelerators, the development of instruments capable of detecting such elusive events is one of the most impressive accomplishments of modern technology. Some particle detectors are able to photograph the effects of subatomic particles that spend as little as 10^{-11} second in the instruments!

In the following sections we will describe several low-energy detection instruments that have been available to physicists for many years. Then we shall examine several high-energy detectors, including the computerized electronic bubble chamber.

Figure 25-14. A gold leaf electroscope. Radioactive materials are placed in the hinged drawer at the bottom.

25.9 Low-Energy Detectors 1. *The electroscope.* One of the first instruments to be used to detect and measure radioactivity is the electroscope, shown in Figure 25-14. A negatively charged electroscope becomes discharged when ions in the air take electrons from the electroscope. Similarly, a positively charged electroscope becomes discharged as it takes electrons from ions in the air. Thus the rate at which an electroscope is discharged is a measure of the number of ions in the air near the electroscope.

Madame Curie used the electroscope to study the radioactivity of uranium ore. She found that some ores discharged the electroscope more rapidly than others. This finding suggested the presence of an undiscovered, highly radioactive element in some of the ores and led to the isolation of radium.

2. *The spinthariscope.* The spinthariscope (derived from the Greek word meaning "spark") is another instrument that was used in early studies of radioactivity. In this device, alpha and beta particles from radioactive materials or nuclear collisions strike a zinc sulfide screen. Where they strike the screen, they produce a momentary flash of light that can be observed through a magnifier. By counting the number of flashes in a given period of time, the intensity of the radiation can be determined.

3. *The scintillation counter.* As its name implies, the scintillation counter registers the flashes of light produced when subatomic particles hit a fluorescent screen. In this respect the scintillation counter is similar to both the spinthariscope and a television picture tube. However, the scintillation counter also provides information about the energy, and thus the identity of the particles it detects. This is done by having the screen of the counter coupled with a photomultiplier tube. (See Section 22.8.)

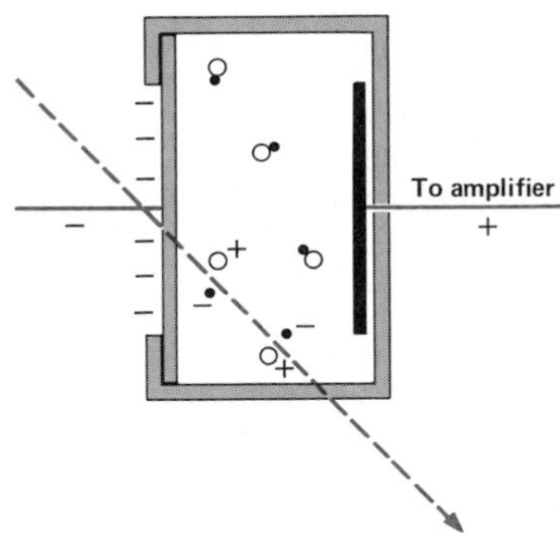

Figure 25-15. An ionization chamber. The passage of a subatomic particle (represented by the arrow) ionizes the gas in the chamber along its path. The negative ions are then attracted to the anode on the right side of the chamber.

4. *The ionization chamber.* The ionization chamber, shown in Figure 25-15, contains fixed electrodes. When the gas in the chamber becomes ionized because of the passage of a high-speed particle or ray, the electrons and ions drift to the electrodes. The electrons drift to the positive plate while the positively charged ions drift to the negative plate.

The charge produced by a single passing particle is much too small to activate the circuit in an ionization chamber, and so an amplifier is used in conjunction with the instrument. The resulting impulse can then be used to trigger a light or some other recording device.

5. *The Geiger tube.* The principle of the ionization chamber is also used in the Geiger tube, which is named after the German physicist Hans Geiger (1882–1945). A hollow cylinder acts as the negative electrode and a thin wire down the center is the positive electrode. The tube is filled with a gas at low pressure, and a potential difference of about 1000 volts is applied to the electrodes. This potential difference is slightly less than the voltage required to ionize the gas and to enable it to conduct an electric current between the electrodes.

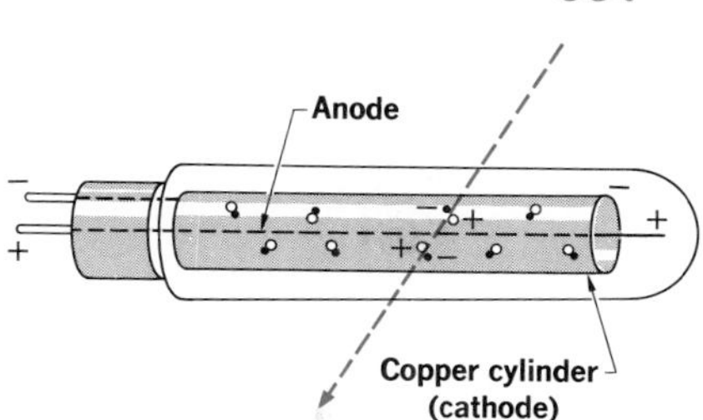

Figure 25-16. A Geiger tube. The ions produced by the subatomic particle (arrow) permit the passage of a flow of electrons between the cathode and the anode.

When a charged particle, such as a proton, enters the Geiger tube, it releases electrons from the atoms in its path and produces ions, as shown in Figure 25-16. These electrons are then attracted to the positively charged wire. As they move toward the wire, they collide with more gas atoms to produce additional free electrons and positive ions. Eventually, the surge of electrons toward the positive electrode is large enough to register as a detectable current flowing through the tube. Thus the Geiger tube acts as its own amplifier for the detection of nuclear events. The Geiger tube is generally used in conjunction with external circuits to activate flashing lights, a loudspeaker, or a counting device.

6. *The cloud chamber.* A device that makes the paths of ionizing particles clearly visible is the cloud chamber. It was invented by the British physicist Charles Wilson (1869–1959) in 1911. Its principle of operation is quite simple, and an operating cloud chamber can be assembled with a minimum of material and effort. Figure 25-17 shows a simple, homemade cloud chamber.

Figure 25-17. A simple cloud chamber. The radioactive source is placed in the eye of the needle that is suspended from the cork.

The chamber is a plastic or glass container. It is resting on a block of dry ice. A piece of dark cloth is saturated with alcohol and placed around the inside of the container near the top. A radioactive source is placed in the eye of a needle that is suspended from a cork placed on the lid of the container.

In the chamber, alcohol evaporates from the cloth and condenses again as it reaches the cold region at the bottom of the chamber. Just above the floor of the chamber there is a region where the alcohol vapor does not condense unless there are "seeds" around which drops of alcohol can form. (The situation is similar to a cloud that can be made to form rain by "seeding" it with a powdered chemical.) If there is no dust in the chamber, the only seeds available are ions produced by the passage of charged particles or rays. The resulting trail of alcohol droplets is then visible against the black bottom of the chamber and can be photographed.

The device we have just described is called a *diffusion cloud chamber*. Others, including the chamber developed by Wilson, rely on pressure changes instead of low temperatures to produce condensation. It is possible to use many other liquid-vapor arrangements to produce the tracks. In all cloud chambers, it is the ionizing effect of high-speed particles that makes a visible trail possible.

25.10 Photographic Film In addition to its use with detection instruments, photographic film can be used directly to record the passage of subatomic particles. When charged particles or gamma rays strike photographic film, they change the chemicals in the emulsion in much the same way that a ray of light changes them. When the film is developed, the passage of the particle or ray is recorded as a spot. If a stack of photographic plates is exposed to a nuclear event, the event is recorded as a series of streaks through the emulsions, as shown in Figure 25-18.

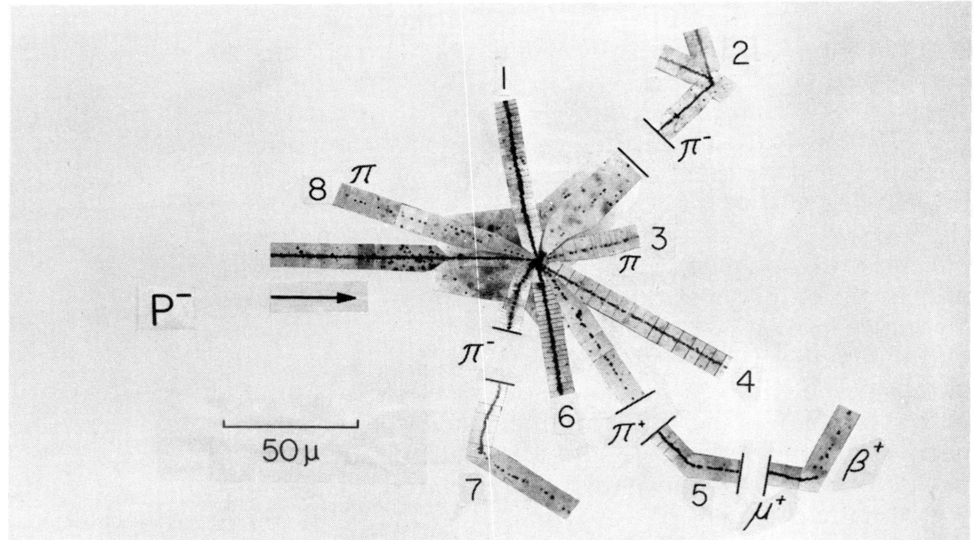

Figure 25-18. Impressions of subatomic particles on photographic film. An antiproton (p^-) entered from the left and produced a shower of eight other particles when it collided with a proton. This is the first photo of the reaction between a proton and its antiparticle. Other particles are identified by their Greek symbols.

25.11 Liquid Bubble Chambers In a cloud chamber, nuclear events are visible because of the condensation of a substance like alcohol. In a *bubble chamber*, exactly the opposite principle is used. A container is filled with a liquid, and the tracks of particles become visible as a series of bubbles when the liquid boils.

One substance used in bubble chambers is liquid hydrogen. As long as the temperature and pressure conditions are right, the hydrogen stays in liquid form. If the pressure is suddenly lowered, however, the hydrogen boils violently. If high-speed charged particles are allowed to pass through the hydrogen at this same instant, ions are formed. The hydrogen boils a few thousandths of a second sooner around these ions than it does in the rest of the chamber. Carefully timed photos, such as the one in Figure 25-19, are used to record the resulting bubble trails.

The advantage of the bubble chamber over the cloud chamber is that more nuclear collisions take place in the dense liquid of the bubble chamber. This difference makes it possible to study particles having speeds that are much higher than those that produce recordable events in a cloud chamber.

Figure 25-19. This photograph shows tracks of subatomic particles in a liquid hydrogen bubble chamber. Two charged particles arise from the collision shown at the upper left. One of these undergoes an additional collision at lower left center. The straight tracks were made by protons, some of which collided with electrons of the hydrogen atoms in the chamber. These electrons move in spiral paths under the influence of the strong magnetic field that is acting on the bubble chamber.

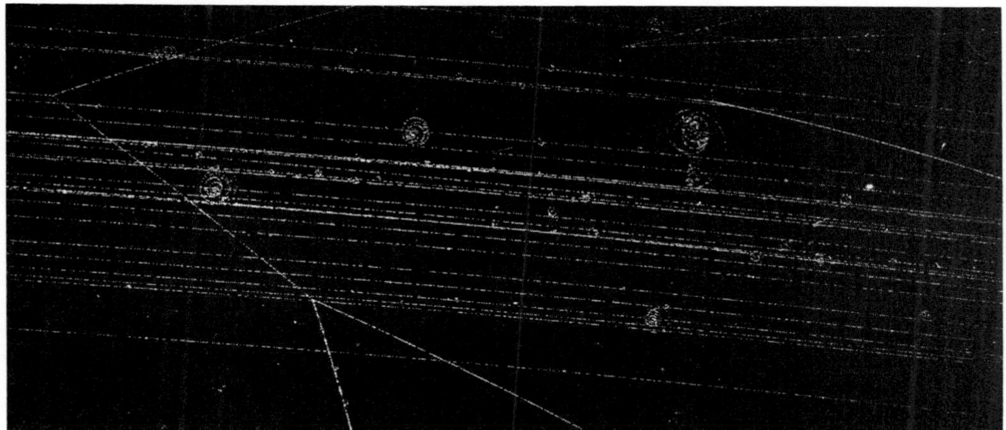

25.12 Electronic Bubble Chambers In 1962, the American physicist S. J. Lindenbaum (b. 1925) and his colleagues at Brookhaven National Laboratory developed a new technique that resulted in an electronic version of the bubble chamber. This detector, an exterior view of which is shown in Figure 25-20, is almost universally used today in high-energy physics experiments. The Lindenbaum detector consists of a series of picket-fence wire planes immersed in a gas such as argon. When a charged particle passes near one of the wires, it causes a signal that is fed into a computer. The computer then reconstructs and analyzes the events in the chamber in three dimensions. Such electronic chambers are more than 10 000 times faster than other particle detectors, and they have better resolution and identification capabilities.

Figure 25-20. Exterior of an electronic bubble chamber.

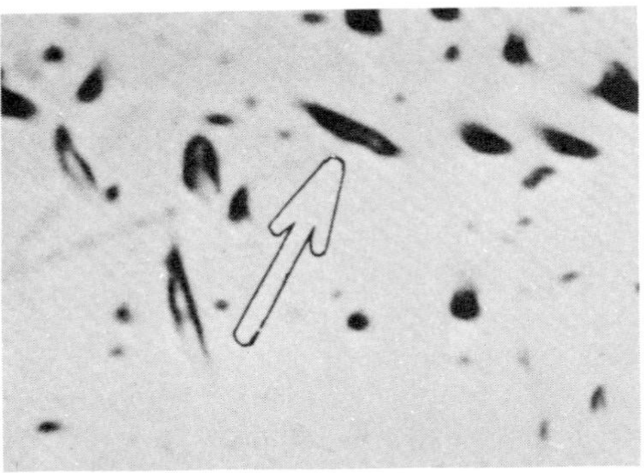

Figure 25-21. By etching fragments of lunar rocks collected by American astronauts, scientists have discovered that the rocks are marked with tracks left by cosmic rays. The arrow points to a typical cosmic-ray track.

25.13 Solid-State Detectors When certain nonconducting solids are exposed to charged nuclear particles, tracks that can be seen with an electron microscope are produced in the solids. The tracks are regions of distortion of the crystal lattices of the solids. Etching the solids with acid makes the tracks more permanent and more visible.

Cosmic rays and various radioactive decay products have left tracks in naturally occurring solids since the origin of the solar system. See Figure 25-21. Through the etching technique just described, scientists have examined these fossil tracks in rocks, in ancient artifacts, in meteorites, and in lava from the ocean floor. Important clues about the earth's history have been gathered from these studies. The same technique was used on lunar samples.

One of the first *solid-state detectors* made was a thin sheet of mica that was used by two British scientists in 1959 in connection with a fission experiment. Glass and certain plastics have also been used as detectors. Plastics are particularly sensitive to slow protons and alpha particles.

Each solid used as a solid-state detector has a threshold below which no tracks are produced. This property makes the detector particularly suited for the study of high-energy particles; the dense background of light-particle tracks that is present in other types of detectors is eliminated by the threshold property of the solid-state detector.

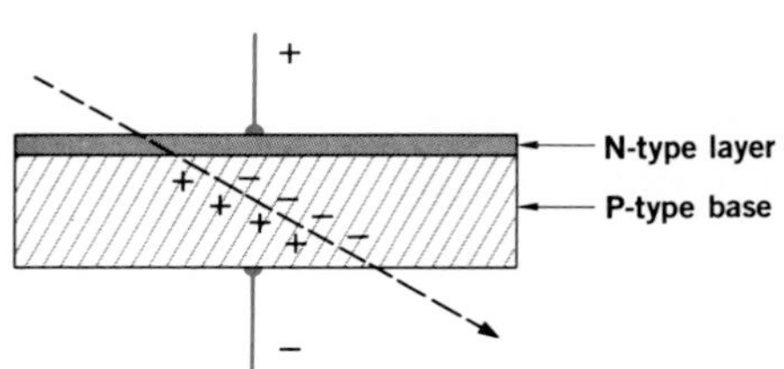

Figure 25-22. A junction detector. The passage of a charged particle (represented by the arrow) ionizes a region of the P-type base and produces an output signal.

Another type of solid-state detector, the *junction detector*, is based on the transistor principle. See Figure 25-22. An applied bias removes all free carriers from the detector. When charged particles enter the device, ionization is produced in the base and an output signal results. In essence, the junction detector is a solid-state ionization chamber. However, the junction detector is smaller in size, requires less power, has a faster response, is completely insensitive to magnetic fields, and produces an output signal that is directly proportional to the energy of the incident particle.

QUESTIONS: GROUP A

1. Why are special detectors needed for subatomic particles?
2. (a) Which detection instrument was used by Marie Curie? (b) How was it helpful to her in her work?
3. How does a spinthariscope indicate the intensity of radiation?
4. What advantage does a scintillation counter have over a spinthariscope?
5. How does photographic film react to charged particles or gamma rays?
6. How are the production of tracks in a cloud chamber similar to the formation of droplets in the atmosphere?
7. What is the purpose of establishing a potential difference across the ends of a Geiger tube?

SUBATOMIC REACTIONS

25.14 Classification of Subatomic Particles Through the use of the detection devices and particle accelerators just described, physicists have released or produced more than two hundred different subatomic particles. Almost all these particles are unstable, and many of them have extremely short half-lives. Nevertheless, these fleeting fragments of matter are essential pieces in the puzzle of nuclear structure. In the remainder of this chapter, we shall look at the picture of the atom that is emerging from these investigations.

More than one hundred years ago, scientists recognized relationships among the chemical elements that made it possible to arrange them in groups according to their properties. This arrangement is called the *periodic table of the chemical elements.* As more and more subatomic particles were discovered, it became evident that a similar grouping could be made of these "fundamental" particles. In 1957 such a table was proposed by the American physicist Murray Gell-Mann (b. 1929). Working independently, the Japanese scientist Kazuhiko Nishijima (b. 1926) also proposed such a table.

The first useful periodic table of the chemical elements was devised by the Russian chemist Dmitri Mendeleyev (1834–1907).

In 1969 Gell-Mann received the Nobel Prize in Physics for his work with subatomic particles.

Table 25-2 shows the main features of the table of fundamental particles. Although not all known particles in each category are listed, enough of them are shown so that the purpose and arrangement of the classification are evident.

The names and symbols of fundamental particles are taken from Greek words and letters. The names refer to the masses: *baryon* (*berry*-on) means heavy, *meson* (*mez*-on) means medium, and *lepton* (*lep*-ton) means small. Baryons and mesons are also called *hadrons* (*had*-rons); hadron comes from a Greek word meaning bulky. *Bosons* (*boz*-ons) are named after the Indian physicist Satyendra Bose (1894–1974). *Fermions* are named after the Italian physicist Enrico Fermi, who directed the research leading to the first controlled nuclear chain reaction. See Section 24.10.

Baryon numbers, lepton numbers, and hypercharge will be explained in Section 25.16.

The masses of the fundamental particles are given as the energy equivalents (in MeV) of the rest masses. Neutrinos, photons, and gravitons are given a mass of zero because they do not exist as stationary entities.

The main reasons for the construction of ever-larger particle accelerators are to study the interrelationships between nuclear particles and to explore the possibility that they are, in turn, composed of still simpler particles. Discoveries made with high-energy accelerators often require the revision of existing theories about the structure of mat-

Table 25-2
FUNDAMENTAL PARTICLES†

Category	*Name*	*Symbol*	*Antiparticle*	*Rest Mass* (MeV)	*Half-life* (s)
baryons (fermionic hadrons)	omega	Ω^-	Ω^+	1672	1.3×10^{-10}
	xi	Ξ^- Ξ^0	$\overline{\Xi^-}$ $\overline{\Xi^0}$	~1320 ~1310	1.2×10^{-10} 2.1×10^{-10}
	sigma	Σ^- Σ^0 Σ^+	$\overline{\Sigma^-}$ $\overline{\Sigma^0}$ $\overline{\Sigma^+}$	~1190 ~1190 ~1190	1.0×10^{-10} $\sim 10^{-20}$ 5.5×10^{-11}
	lambda	Λ^0	$\overline{\Lambda^0}$	1116	1.8×10^{-10}
	neutron	n	$\overline{n}$	940	700
	proton	p	$\overline{p}$	938	stable*
mesons (bosonic hadrons)	eta	η_c	$\overline{\eta_c}$	2980	$\sim 10^{-20}$
	kaon	K^0 K^+	$\overline{K^0}$ $\overline{K^+}$	498 494	8.5×10^{-7} 6.4×10^{-9} 3.5×10^{-8}
	pion	π^+ π^0	π^- π^0	140 135	1.0×10^{-8} 0.6×10^{-16}
fermions: quarks	up	u	$\overline{u}$	~4	stable
	down	d	$\overline{d}$	~7	stable
	charm	c	$\overline{c}$	~1500	stable
	strange	s	$\overline{s}$	150	stable
	top**	t	$\overline{t}$	>41 000	stable
	bottom	b	$\overline{b}$	~5000	stable
fermions: leptons	tau	τ	$\overline{\tau}$	1784	0.3×10^{-12}
	tau neutrino	ν_τ	$\overline{\nu_\tau}$	<40	stable
	muon	μ^-	μ^+	106	1.5×10^{-6}
	muon neutrino	ν_μ	$\overline{\nu_\mu}$	<0.3	stable
	electron	e^-	e^+	0.51	stable
	electron neutrino	ν_e	$\overline{\nu_e}$	$<2 \times 10^{-5}$	stable
bosons	photon	γ	γ	0	stable
	graviton**	g	g	0	stable
	intermediate bosons	W^+ Z^0	W^- Z^0	81 000 91 000	stable stable
	gluon	g	g	0	stable

*See Section 25.18. **Not yet observed. †Not all known particles are listed in this table.

ter. These studies of the building blocks of the universe are also called particle physics or "modern physics."

For many years, nuclear physicists have speculated that every particle in the universe has a mirror-image counterpart, or "antiparticle." Such antiparticles are listed in Table 25–2. A substance composed of antiparticles is called antimatter. The first evidence of the existence of antimatter was the pair production of electrons and antielectrons (positrons) as described in Section 23.3

When a particle and its antiparticle collide, all physical quantities associated with the particle or antiparticle (mass, charge, etc.) disappear, and pure energy is left. In 1981 scientists in Geneva, Switzerland, produced the first proton–antiproton collisions in a particle accelerator.

Astrophysicists think that large quantities of antimatter may exist in the universe in the form of stars or even entire galaxies (or "antistars" and "antigalaxies," to be more exact).

Matter and antimatter are converted into energy when they collide, in keeping with Einstein's equations, $E = mc^2$.

25.15 Particle Interactions As we saw in Section 4.1, a force produces a change in the state of motion of an object. In quantum mechanics, the broader concept of "interaction" replaces the Newtonian concept of force. *An* ***interaction*** *is any change in the amount or quantum numbers of particles that are near each other*. For example, an interaction occurs when an electron and a positron interact and change into two photons. An interaction is not merely another word for force. In an interaction, the identity of the particles colliding is changed as new particles are created and annihilated. The intermediate particles, or "carriers," of an interaction are called **bosons.**

1. *Strong interactions.* The nature of the strong interaction was discussed in Section 23.12. The strong interaction holds the particles of the nucleus together against the electromagnetic repulsion of the nucleus' similarly charged particles. The strong interaction is much stronger than charge. In other words, the attraction between two neutrons is virtually the same as that between two protons, even though the protons both have a positive charge.

Although the strong interaction is the strongest of all interactions, it has a very limited range. It is not effective beyond a distance of 2.0×10^{-5} Å. The carrier of the strong interaction is believed to be the meson; there is a constant exchange of mesons among the baryons of a stable nucleus. When the baryons are separated by the intervention of a high-energy particle from an accelerator, the carrier mesons are released.

The production and exchange of particles during an interaction is usually represented by a *Feynman diagram,*

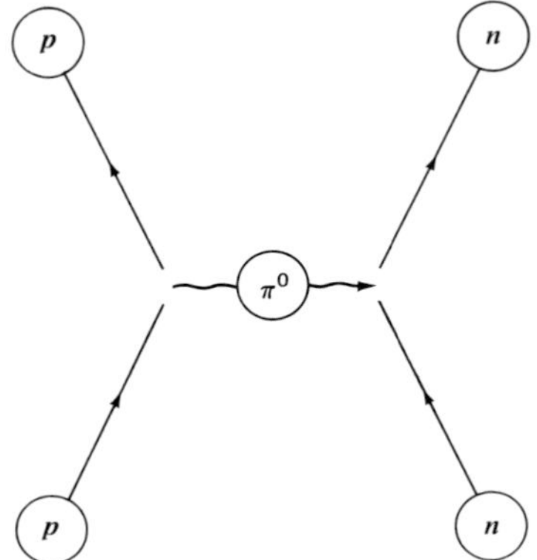

Figure 25-23. A Feynman diagram. In this interaction, a neutral pi meson is exchanged during the close approach of a proton and a neutron. The energy transferred by the carrier meson changes the momenta of the two baryons.

In 1949 Yukawa received the Nobel Prize in Physics for his prediction of mesons.

named after the American physicist Richard Feynman (1918–1988). In Figure 25-23, a Feynman diagram illustrates the reaction that takes place during the close approach of a proton and a neutron, the reaction in which a meson is emitted and absorbed. This carrier meson transfers the energy that is responsible for the change in momentum of the two baryons. The explanation of the strong "force" by meson exchange is the main reason the process is now called an interaction. In 1935 the Japanese physicist Hideki Yukawa (1907–1981) proposed the theory of the existence of energy-carrying mesons to explain the strong interaction. But mesons were not actually detected until 1947.

2. *Electromagnetic interactions.* These interactions differ from other interactions in several ways. (1) Electromagnetic interactions can either attract or repel the objects on which they are acting. The other three interactions are attractive only. (2) Electromagnetic interactions can act in directions other than along a straight line between objects. (3) The electromagnetic interactions depend on the relative velocities of particles. This is not true of other interactions. The carrier of the electromagnetic interaction is the photon.

Electromagnetism keeps electrons in orbit around the nucleus. The same interaction is also responsible for the bonding between atoms to form molecules and between molecules to form larger substances. Nevertheless, the electromagnetic interaction is less than 10^{-2} as strong as the strong interaction.

The elastic collisions discussed in Section 6.15 are examples of electromagnetic interactions between the orbital electrons of the atoms that make up the colliding objects.

3. *Weak interactions.* Radioactivity is an example of this interaction. The carriers responsible for weak interactions, which have a very short range and affect only leptons, are the three massive intermediate bosons, W^+, W^-, and Z^0. See Table 25-2. At very high energies, the intermediate bosons act in the same way as photons and the weak interaction becomes identical with the electromagnetic interaction.

4. *Gravitational interactions.* Gravitation was the first of the interactions to be studied, and yet it is the weakest of them all. Its ratio to the strong interaction is only 10^{-38} to 1. Gravitation has no known distance limitations, and so its effects are easily observed in the behavior of large objects on the earth and in the motions of celestial bodies. Within the atomic nucleus, the effect of gravitational interactions is negligible in comparison with the other interactions that are present.

The carrier for the gravitational interaction is called the *graviton*. Like the photon, its rest mass is thought to be zero. Because they are very weak, gravitational waves are extremely difficult to detect. In 1965, the American physicist Joseph Weber attempted to detect gravitational waves by suspending a 1000-kg cylindrical aluminum "antenna" in a vacuum chamber. He aimed his antenna at massive stars, hoping to pick up their gravitational radiation, but his results were inconclusive. In current research, to avoid the interference from the thermal energy of the aluminum, gravity wave antennas are cooled to within a few degrees of absolute zero. Efforts are also being made to detect the possible vibrational effects of gravitational waves on the earth, moon, and interplanetary spacecraft.

Figure 25-24. This 305-meter receiver near Arecibo, Puerto Rico, is the world's largest single radio telescope. It was used to explore the existence of gravitational waves in an experiment that extended over a 4-year period.

25.16 Subatomic Conservation Laws Repeatedly in your study of physics, the importance of the conservation principles has been emphasized. The conservation of mass-energy, the conservation of linear and angular momentum, the conservation of electric charge—these laws are basic to an understanding of all reactions above the atomic scale. Similar laws hold true in subatomic interactions. We have already seen in Section 24.4, for example, that both mass-energy and electric charge are conserved in nuclear reactions. As other subatomic particles were discovered, additional conservation laws were formulated.

1. *Conservation of baryons.* When a baryon decays or reacts with another particle, the number of baryons is the same on both sides of the equation. For example, if a neutron is left by itself, it will decay as follows:

$$\begin{array}{lcccccccc} & & \mathbf{n^0} & \rightarrow & \mathbf{p} & + & \mathbf{e^-} & + & \overline{\boldsymbol{\nu}} \\ \textbf{Baryon No.} & = & \mathbf{+1} & = & \mathbf{+1} & & \mathbf{+0} & & \mathbf{+0} \end{array}$$

The neutron and proton are both baryons, while the electron and antineutrino have baryon numbers of zero. Consequently, the baryon number of the equation is +1 on each side.

The conservation of baryons keeps the universe from deteriorating into mesons and leptons. If such a deterioration would take place, atoms as we know them would no longer exist and matter would become a rather homogeneous mixture of a few kinds of low-mass particles, such as electrons and neutrinos.

2. *Conservation of leptons.* The decay of the neutron illustrates another nuclear conservation principle. The lepton number of the neutron and proton is zero. The lepton number of the electron is +1. The lepton number of the

antineutrino is −1, since it is an antiparticle. Consequently, the arithmetic sum of the lepton numbers is the same on each side of the equation. This principle holds true for every interaction in which leptons are involved.

$$\begin{array}{rcccccc} & \mathbf{n^0} \rightarrow & \mathbf{p} & + & \mathbf{e^-} & + & \bar{\boldsymbol{\nu}} \\ \textbf{Lepton No.} = & \mathbf{0} = & \mathbf{0} & & \mathbf{+1} & & \mathbf{-1} \end{array}$$

Lepton conservation explains the necessity of pair production, an example of which is illustrated in Figure 23-4. When a lepton is created in an interaction, a particle with the opposite lepton number must also be created in order to conserve the lepton value of the reaction. Similarly, a lepton can be annihilated only by interacting with a particle of opposite lepton value.

The conservation of baryons and leptons does not prevent the change of a particle to another particle within the same category, however. In neutron decay, one kind of baryon (the neutron) changed into another kind (the proton). The decay of the antimuon is another example.

$$\begin{array}{rcccccc} & \boldsymbol{\mu^+} \rightarrow & \mathbf{e^+} & + & \boldsymbol{\nu} & - & \bar{\boldsymbol{\nu}} \\ \textbf{Lepton No.} = & \mathbf{-1} = & \mathbf{-1} & & \mathbf{+1} & & \mathbf{-1} \end{array}$$

3. *Conservation of hypercharge.* Bayrons and mesons have a property called *hypercharge.* Hypercharge can be either positive, negative, or zero. The hypercharge sign of a particle is the opposite of the hypercharge sign of its antiparticle. Hypercharge is conserved in strong and electromagnetic interactions, but it is not conserved in weak interactions.

Consider the following interaction involving hypercharge:

$$\begin{array}{rcccccc} & \boldsymbol{\pi^+} & + & \mathbf{p} \rightarrow & \mathbf{K^+} & + & \boldsymbol{\Lambda^0} \\ \textbf{Hypercharge} = & \mathbf{0} & & \mathbf{+1} = & \mathbf{+1} & & \mathbf{+0} \end{array}$$

The identity of the original particles has changed completely, but the sum of the hypercharges is +1 on each side of the reaction. (The sum of the baryon numbers is also +1 on each side.)

4. *Conservation of parity.* The distribution of matter in space is often described in terms of *parity. The* ***law of parity*** *states that for every process in nature there is a mirror-image process that is indistinguishable from the original process.* Another way to state this law is to say that there is no absolute and independent way of describing right and left, clockwise and counterclockwise, in the universe. Any experiment devised to establish one of these directions is also matched by an equally probable event in the opposite direction.

Parity experiments deal largely with the spin of nuclear particles. Spin is also responsible for the direction in which particles are emitted in subatomic interactions. The conservation of parity stipulates that particles will be emitted in opposite directions in equal numbers since nature shows no preference for spin direction. But in 1956 a series of experiments based on predictions by the Chinese physicists Tsung-Dao Lee (b. 1926) and Chen Ning Yang (b. 1922) showed that this was not the case. In the case of weak interactions involving beta decay, parity was not conserved!

Experiments were specifically designed to test the conservation of parity in other interactions. The question was whether parity by itself is conserved when particles are replaced by their antiparticles in subatomic interactions. In 1966 physicists at the Brookhaven National Laboratory in New York discovered a violation of parity with antiparticles as well. So perhaps there is an absolute sense of direction in the universe after all.

Figure 25-25. The Chinese physicists C. N. Yang and T. D. Lee shared the Nobel Prize in Physics in 1957 for their hypothesis that parity is not conserved in all nuclear reactions.

25.17 Quarks In 1963 Murray Gell-Mann and the American physicist George Zweig (b. 1937) proposed the theory that hadrons are made up of still smaller particles called *quarks,* each having a charge of $\pm\frac{1}{3}$ or $\pm\frac{2}{3}$. According to this theory, baryons are composed of three quarks each and mesons contain two quarks each. Three different quarks were postulated with the designations of "up," "down," and "strange." As shown in Table 25-2, for each of these three quarks there is also an antiquark.

The word "quark" does not have a specific meaning. It appears in the novel Finnegans Wake *by James Joyce.*

In 1974 two subnuclear particles were discovered that have unusually large masses and long half-lives. These so-called J/psi particles could not be explained with only three quarks, and so three more were added to the list. These were called "charm," "top," and "bottom." Also, in order to avoid possible violations of the exclusion principle in combining two and three quarks within a hadron, each quark is assigned one of three "colors": "red," "blue," and "green." (The "color" of a quark is the subnuclear counterpart of the quantum number of an electron.) In keeping with the terminology of the colors of visible light, the three "colors" of antiquarks are "cyan," "yellow," and "magenta."

Figure 25-26. The American physicists Richard Feynman (left) and Murray Gell-Mann. They won the Nobel Prize in Physics in 1965 and 1969, respectively, for their work with subatomic particles.

The carrier between quarks is the *gluon.* It is the exchange of "color" between quarks by gluons that provides the binding force between quarks within a hadron. Strangely enough, the gluon-grip between quarks is thought to increase, rather than decrease, with distance. That is probably why present particle accelerators have not been able to isolate individual quarks. According to one theory, there may even be as many as eight different gluons, each with a characteristic "color."

25.18 Unified Field Theories In 1974 it was found that at high energies, the boundaries between particle categories are not inviolate. At 300 GeV, collisions of electrons and positrons produced hadrons and vice versa. This suggests that the weak interaction is not a constant and that under certain conditions it may become stronger than the electromagnetic interaction.

In 1984 Carlo Rubbio and Simon van der Meer shared the Nobel Prize in Physics for their discovery of the W and Z particles.

One explanation of this phenomenon is that colliding leptons exchange a Z particle. This interaction is known as a *neutral weak current* (since there is no change in the charges of the colliding particles).

The discovery of neutral weak currents is an important step toward a *unified field theory,* which combines two or more of the interactions in the universe into a single concept, much as Maxwell's electromagnetic theory combined electricity and magnetism in the nineteenth century. A unified field theory proposed in 1967 combined the weak and electromagnetic interaction, and the neutral weak current helps to substantiate that concept.

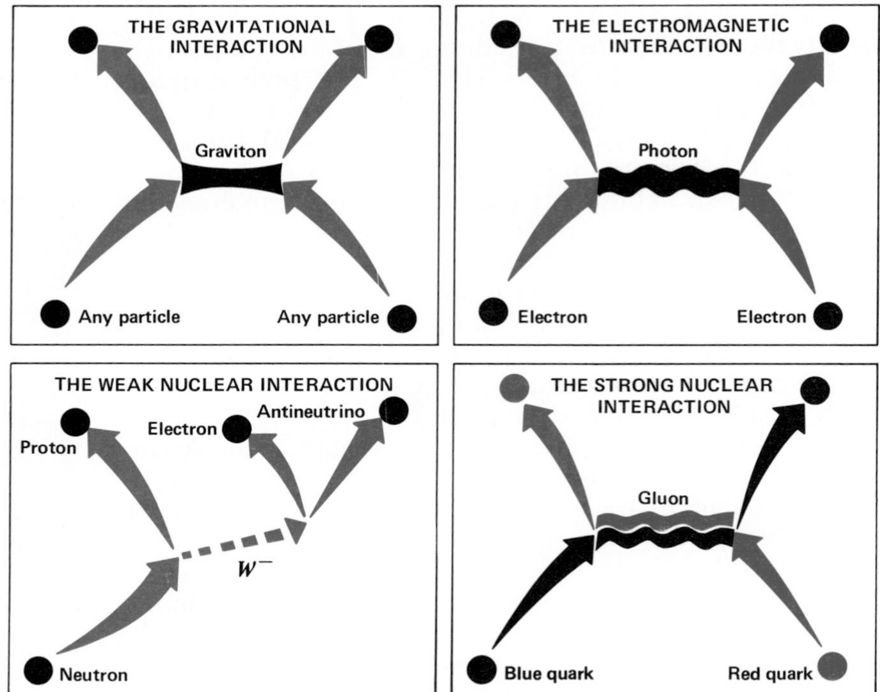

Figure 25-27. The four known interactions and their carriers. Unified field theories are attempts to combine two or more of these phenomena into a single concept.

Efforts are presently being made to include the strong interaction in a unified field theory (or *grand* unified field theory, as it has been called). Virtually every version of such a theory predicts the startling idea that the proton is unstable! The half-life of the proton, in this scheme, is more than 10^{30} years. Experiments to detect the decay of a proton are being carried out in several countries; and in 1982, physicists in Europe announced that such an event was recorded in a 134-ton iron detector inside a tunnel under the Alps. Even with this massive shielding, however, it is possible that the recorded nuclear event was produced by a cosmic ray and not by the spontaneous decay of a proton.

The acronym ***GUT*** *has been used in discussing grand unified theories.*

According to theory, an average of one proton will decay in every human being during a lifetime.

A successful grand unified field theory would combine the three nuclear interactions, but it would still not include gravitation. Until such a supertheory is found and successfully tested, no one really knows what the ultimate model of matter and its interactions will be, or even if such a model can ever be formulated. But one thing is certain: the search for new knowledge in high-energy physics requires keen and open minds and a close correlation between theory and technology.

Figure 25-28. The American physicists Sheldon Glashow and Steven Weinberg at a news conference after they won the 1979 Nobel Prize in Physics for their unified field theory of nuclear interactions. The third recipient was Abdus Salam of Pakistan.

QUESTIONS: GROUP A

1. You learned earlier that atoms are made up of protons, neutrons, and electrons. How many other subatomic particles have been discovered?

2. What properties do subatomic particles have in common?

3. (a) Who developed the table of fundamental particles? (b) What are the four major categories of fundamental particles?

4. Why are neutrinos thought to have no mass?

5. (a) What happens when a particle of matter meets its corresponding particle of antimatter? (b) What principle does this illustrate? (c) What equation is related to this principle?

6. What is the difference between a force and an interaction?

7. Which of the four types of interactions (a) is the weakest? (b) can attract or repel? (c) involves the property of strangeness? (d) is carried by the graviton? (e) acts over distances less than 2.0×10^{-15} m?

8. (a) Which particle is the carrier of the strong interaction? (b) Has this particle ever been detected? Explain.
9. What does a Feynman diagram illustrate?
10. (a) What conservation laws apply to subatomic particles? (b) Which of these laws is not applicable to all types of interactions?
11. What is a quark?
12. What charges do quarks carry?
13. What is the purpose of assigning each quark a color?
14. (a) What is a gluon? (b) What is its function?
15. Why have present accelerators not been able to isolate individual quarks?
16. Distinguish between a unified theory and a grand unified theory (GUT).
17. Why are proton decay experiments carried out underground?

SUMMARY

Modern atomic models take into account the uncertainty principle, which states that the exact position and momentum of a subatomic particle cannot be simultaneously specified. The allowable values of physical quantities in modern atomic theory are given in terms of quantum numbers. The first quantum number for the electron describes its orbital radius. The second quantum number gives its angular momentum. The third quantum number specifies its direction in terms of a magnetic field. The fourth quantum number designates the spin of the electron. The exclusion principle states that no two electrons in an atom may have the same set of quantum numbers.

The stability of an electron is described in terms of ionization energy, which is the energy required to remove an electron from its orbit. All the equations for quantum numbers include Planck's constant. This indicates that all matter has wave characteristics.

Particle accelerators fall into three main categories: Van de Graaff accelerators, circular accelerators, and linear accelerators. In Van de Graaff generators, energies of up to 20 MeV may be produced. Betatrons, cyclotrons, synchrotrons, and intersecting storage rings are types of circular accelerators. The world's largest accelerator is a synchrotron, in which proton energies of 1000 GeV are achieved after several preliminary stages of acceleration in other devices. Intersecting storage accelerators are more efficient than other types of circular accelerators. Linear accelerators avoid the radiation losses encountered in circular accelerators.

Subatomic particles can be detected and studied because they are able to ionize certain substances or to produce visible paths. Low-energy detectors include the electroscope, the spinthariscope, the scintillation counter, the ionization chamber, the Geiger tube, and the cloud chamber.

The bubble chamber detects high-energy particles by producing momentary trails of hydrogen or helium gas, which are then photographed. Electronic detectors use computers to analyze nuclear events in three dimensions. Solid-state detectors are more selective than bubble chambers. Some solid-state detectors work on the transistor principle.

The number of identified subatomic particles now exceeds two hundred. They can be classified into baryons, mesons, and leptons. Baryons and mesons are also called hadrons. Each particle has an antiparticle with opposite charge.

The four known particle interactions are strong, electromagnetic, weak, and gravitational. In strong interactions, the number of baryons and leptons remains constant. Hypercharge and parity are not conserved in weak interactions.

Hadrons are thought to consist of quarks. Six kinds of quarks, each having one of three "colors," have been postulated. Efforts are also being made to combine the four known interactions in the universe into a grand unified field theory.

VOCABULARY

antimatter
antiparticle
baryon
betatron
bevatron
boson
bubble chamber
cloud chamber
cyclotron
dees
drift tubes
electromagnetic interaction
electroscope
energy level
exclusion principle
Feynman diagram
Geiger tube
gravitational interaction
graviton
hypercharge
interaction
ionization chamber
ionization energy
junction detector
law of conservation of baryons
law of conservation of hypercharge
law of conservation of leptons
law of parity
lepton
linear accelerator
matter wave
meson
neutral weak current
orbital
quantum mechanics
quantum number
quantum theory
quark
solid-state detector
spinthariscope
storage ring
strangeness
strong interaction
synchrocyclotron
synchrotron
tevatron
uncertainty principle
unified field theory
Van de Graaff generator
W particle
weak interaction
Zeeman effect

INVESTIGATIONS
IN PHYSICS

The investigations that follow can be done individually or in groups.

They are designed with two purposes in mind. The first is to give you practice performing experiments in the laboratory. In each investigation, you will follow a set procedure, observe the phenomena produced, and analyze the data you collect to make inferences from the evidence seen in the experiment.

The second purpose is to provide an opportunity for further inquiry into the nature of the phenomena demonstrated by each investigation's procedure. You can establish your own hypotheses, design your own experiments, and draw your own conclusions from the results of your experiment. In this way, you will be working in the spirit of the scientist, using your curiosity and creativity to answer questions about how physical phenomena operate.

Process Skills in the Physics Laboratory

Fundamental to any scientist's work is the ability to identify and ask important questions. Once scientists have clearly identified what needs to be found out, they use specific methods to help them investigate the problem. In using the same scientific processes that professional scientists use, you will "rediscover" scientific facts in much the same way that scientists learned them when the facts were still unsolved problems. Listed below are the scientific process skills you will use in the investigations.

Observing is the use of one or more of the five senses to perceive objects or the course of events as they occur.

Measuring is the process of gathering quantitative data by making precise observations involving numbers.

Organizing is the process of placing data from observations and measurements in logical order.

Classifying is the process of grouping or ordering items according to the similarities and differences among the items by using established classes or by developing new organizational patterns.

Hypothesizing is the process of making statements that may answer questions about observations. A hypothesis is a possible, testable explanation for why something happens.

Predicting is the process of stating what the most likely outcome of an experiment will be. Predictions are based on hypotheses and specific data.

Experimenting is the process of designing and carrying out data-gathering procedures for the purpose of testing a hypothesis.

Analyzing is the process of determining how reliable the data are and whether they support or refute a hypothesis or prediction.

Inferring is the process of drawing conclusions about objects or events based on related facts or premises, but not on direct observation.

Modeling is the process of making visual or verbal constructs that help explain data.

Communicating is the process of sharing information with others in written or spoken form.

A Special Message on Safety

Safety should be FIRST in the minds of administrators, teachers, and students actively involved in a science program. The responsibility for safety and the enforcement of safety regulations and laws must be shared by everyone within the school community. With careful planning and instruction, a safe and healthful environment can be established within your science class.

You are encouraged to develop and maintain a safety program from the outset of this physics course. The information contained in this safety section will aid you in accomplishing this important task.

This information is not intended to be all-inclusive, for no publication can be prepared to list safe practices for science in every situation. Nor should the information be read as legal requirements; instead, the information contains suggestions and recommendations for establishing a safety base upon which to build. Proper planning, foresight, and care must be exercised continuously by everyone. By following the recommended safeguards and precautions described in these investigations, you will be practicing good safety skills.

Safety Regulations

To ensure that a safe and healthful environment is maintained when doing the investigations, read and follow the safety regulations listed below. You should check to see that your lab partner also understands and will abide by the safety regulations, for you will be working together closely in the laboratory.

1. NEVER work alone in the laboratory. Work in the laboratory only while under the supervision of your teacher and with your assigned class.
2. Prepare for each investigation by reading all instructions before you come to class. Follow all directions, and before you begin, review with your teacher the safety precautions needed to conduct the investigation safely. Use only materials and equipment authorized by your teacher.
3. Be alert and proceed with caution at all times in the laboratory. Take care not to bump other students, and remain in your lab station while performing an experiment. An unattended experiment can produce an accident.
4. Do not wear contact lenses in the laboratory when working with chemicals, because there is a distinct possibility that chemicals may get under the contact lenses and cause irreparable eye damage.
5. Wear proper safety equipment, such as safety goggles, apron, and gloves, when performing experiments that require precaution.
6. Your apparel should be appropriate for laboratory work. Long necklaces, bulky jewelry, and excessive and bulky clothing should not be worn in the laboratory. Cotton clothing is preferable to nylon, polyester, or wool.
7. Only this textbook and lab notebooks are permitted in the working areas. Other books, purses, and similar items should be placed in your desk or storage area.
8. No food, beverages, or cigarettes are permitted in any science laboratory.
9. NEVER taste chemicals. NEVER touch chemicals with your hands. NEVER smell chemicals unless specifically instructed to do so.
10. Exercise extreme caution when using a burner. Keep your head, hair, and clothing away from the flame, and turn off the burner when not in use. Gas burners should be lighted only with a sparker in accordance with your teacher's instructions. Check to see that all gas valves and hot plates are turned off before leaving the laboratory.
11. NEVER look directly into laser light or into its reflection off mirrors or polished objects. It is okay to look at its reflection off rough surfaces that diffuse the light, such as a wall, paper, or cloth.

12. You should know the proper fire drill procedures and the location of fire exits.
13. Work areas and apparatus should be kept clean and tidy. Always clean and wipe dry all apparatus, desks, tables, and laboratory work areas at the conclusion of each investigation.
14. Wash your hands thoroughly with soap at the conclusion of each investigation.
15. Know the locations and operations of all safety control equipment.
16. Recognize and heed all safety symbols and cautions incorporated into the procedures of the investigations.
17. Report all accidents to the teacher immediately, no matter how minor.

Safety Symbols

You should not be afraid to do experiments using equipment and chemicals, but you should respect them as potential hazards. To alert you to procedures in which added caution may be necessary, safety symbols will appear at the beginning of the procedure of an investigation. Before you begin, you and your teacher should review any safety rules and regulations needed to conduct the experiment safely.

This especially applies to student-designed inquiry investigations. Make sure your teacher approves all designs and reviews all safety concerns before you begin your investigation.

The figure below shows you the safety symbols that often appear in the investigations. A line is drawn from the person in the center to a safety symbol. The line connects the symbol with the area of the body most likely affected, should an accident occur. If you observe the safety precautions indicated by the symbol, accidents and injuries will be prevented.

SAFETY SYMBOLS

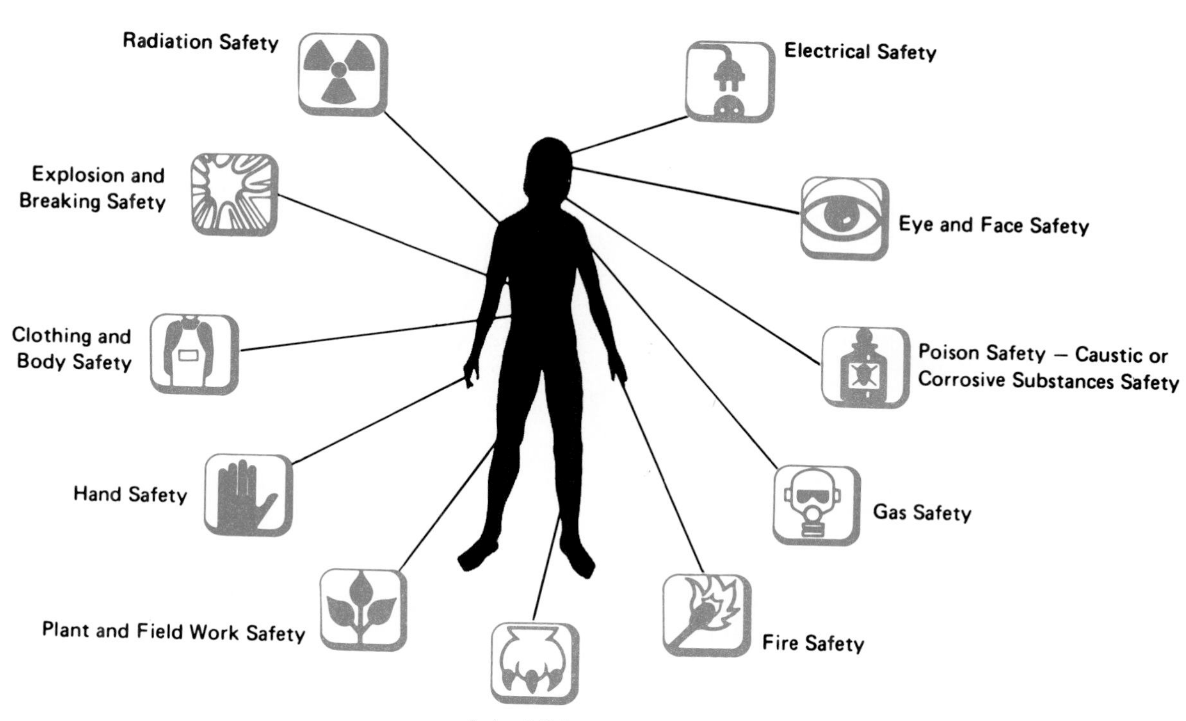

INVESTIGATION 1

What's So Scientific About Toys?

Objective

To investigate the need for a system of units, collection of data, and controls in experimental design

Materials

Meterstick, metric ruler, tape, string, stopwatch, graph paper, and a variety of toys (wind-up and battery-operated toys, tops, yo-yos, ball and paddle, etc.)

Preparation

Prepare a data table with columns for name of toy, kind of motion, direction of motion, form of energy, vertical and horizontal distance, time, and speed. Include rows for several trials with each toy.

Procedure

1. Observe and record the kind of energy input needed to start each of the toys.
2. Predict which toy will have the most kinetic energy and which will have the most gravitational potential energy.
3. Start the toys (one at a time) and observe the motion. Record the kinds of motion (straight-line, rotational, circular, etc.) observed for each toy.
4. Observe and record any changes in energy while the toys are moving.

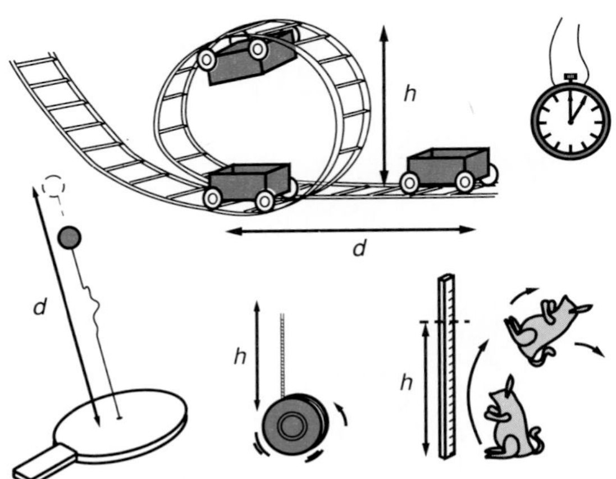

5. Separate the toys into several groups according to similarities in motion.
6. Select the toy from each group that you think can move the fastest. Set the toy in motion. Observe and record the direction(s) it travels and the number of changes in direction that occur.
7. Set it in motion again and measure the total distance it travels and the total time it moves. Do several trials.
8. Predict the direction the toy is going when it moves the fastest. Then measure distance and time intervals for all of the directions in which the toy moves. Do several trials for each direction.
9. Repeat steps 6–8 until you have measured distances and time intervals for all of the toys.

Analysis/Interpretation

1. Did any toy have more than one kind of motion at the same time? Explain.
2. Which distance and time intervals were the most difficult to measure? Why?
3. Were any of your measuring instruments or units unsatisfactory? Explain.
4. Why was it helpful to do several trials?
5. Calculate the average speed of each toy for its entire motion. Calculate the average speed of the toys during each interval of time.
6. Plot graphs of total distance *vs.* total time and of average speed *vs.* total time for each toy.
7. Make histograms of greatest height and greatest speed for each toy.

Conclusions

1. Which toy had the most kinetic energy? the most potential energy?
2. Do your experimental results support the law of conservation of energy? Explain.
3. How do histograms help you interpret data?

Inquiry and Experimentation

To design your own experiment, see page 681.

Extension

Prepare an oral report in which you try to prove to your classmates that you know which of the toys had the most kinetic energy.

INVESTIGATION 2

What Does a Graph of Motion Look Like?

Objective

To identify some techniques to use in the graphical analysis of motion

Materials

Polaroid® camera, black-and-white film, strobe light, battery-operated toy car, dynamics cart, pulley with table clamp, weights, metric ruler, white construction paper, string, tape or glue, graph paper, clock

Preparation

1. Prepare a data table with columns for position, time, time interval, velocity, and acceleration. Include enough rows to have one for each interval. Other tables will be needed for more trials.
2. Cut two triangles from the paper. Tape (or glue) one to the side of the toy car and one to the side of the cart.

Procedure

1. **(Constant velocity)** Cover the electric eye of the camera and place it 120 cm from the center of the path along which the car moves. Set the strobe light to provide 10 to 12 images, and take a photograph of the car as it moves at constant velocity across a table.
2. On the photograph, locate several consecutive images of the triangle that are equal distances apart. Measure the position of each image on the photo to the nearest 0.1 cm.
3. **(Accelerated Motion)** Set up the dynamics cart and pulley system as shown in the figure. With the camera 120 cm from the center of the path, release the weight and take a strobe photograph that has several images.
4. Measure and record the position of the images as you did in step 2.

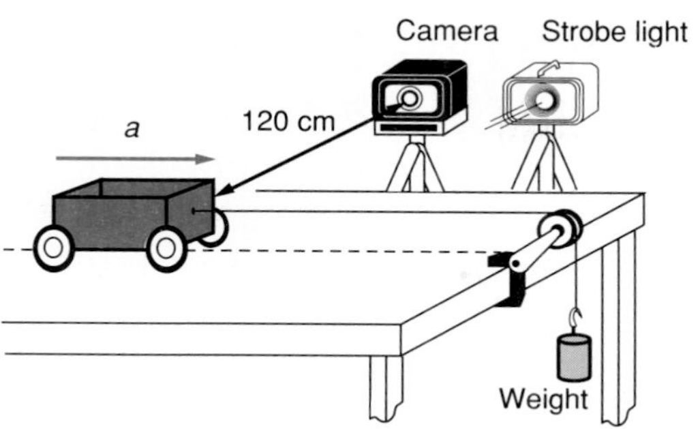

Analysis/Interpretation

1. Determine the period of the strobe.
2. For each photo, plot a graph of position (ordinate) *vs.* accumulated time (abscissa). Calculate the slope of the straight-line graph. Compare this slope to the average velocity of the corresponding motion.
3. Calculate change in position, average velocity, change in velocity, and average acceleration for each time interval.
4. Plot graphs of average velocity *vs.* accumulated time. Calculate the slopes of the graphs, and compare the slopes to the average accelerations of the corresponding motions.
5. Calculate the area between the velocity-time curves and the time axis for each graph.
6. Finally, plot a graph of position *vs.* time squared for the accelerating cart.

Conclusions

1. To what physical quantities do the following correspond? (a) the slope of a position *vs.* time graph, (b) the slope of a velocity *vs.* time graph, and (c) the area under a velocity *vs.* time curve
2. Do all of the graphs go through the origin? Should they? Explain.
3. Write equations for the motion of the car and cart in terms of position.
4. Why was a straight-line graph needed?

Inquiry and Experimentation

To design your own experiment, see page 681.

Extension

Make a chart using the data in the graph of Figure 2-12, page 32. Manipulate the data and plot a straight-line graph. Write a mathematical equation that expresses the relationship between the variables.

INVESTIGATION 3

Where Did Brand X Come From?

Objective

To investigate the relationship between gravitational acceleration and weight

Materials

Set of 20–30 weighted soft drink cans (supplied by instructor), 0–10.0 -N spring balance, stand to support spring balance, Table 3-3 (page 66), plastic bag

Preparation

Prepare a data table in which to record brand name, number, weight, and ratio of weight on the planet (or moon) to weight on the earth.

Procedure

1. Assume that each soft drink can in the set has a weight that is proportional to the weight it would have on the moon or on one of the planets in our solar system.
2. Separate the cans into groups according to the labels on the cans.
3. Count the cans in each group and write a different consecutive number on each of the cans in the same group.
4. Suspend the spring balance from the stand. Hang the bag on the hook of the balance. Place a can in the plastic bag, and measure its weight to the nearest 0.1 N. Weigh all of the cans.

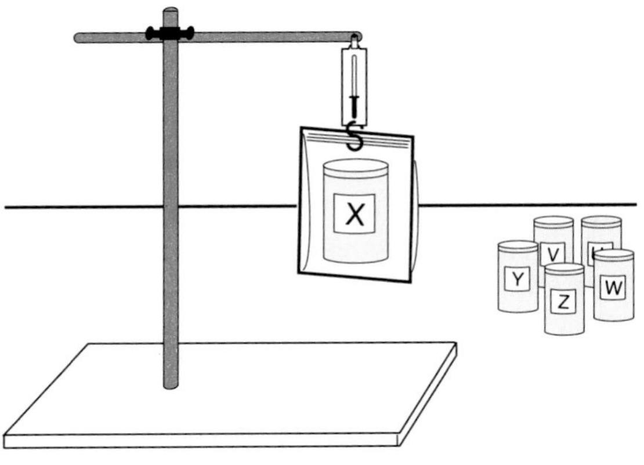

Analysis/Interpretation

1. Do all of the cans have the same weight? Do all of the cans with the same label have the same weight? What does this imply about the number of planets that may correspond to a given label?
2. Reorganize your data so that the cans are listed in order from greatest to least weight.
3. Which brand is the heaviest? Predict the planet or brand to which it should correspond. How will you test your prediction?
4. Predict which brand will correspond to that of the earth. Calculate the ratio between the weight of the heaviest brand and the weight of the brand you predicted corresponds to that of the earth.
5. If the value of the ratio is not what you expect it to be, select another brand that may correspond to the brand from the earth and calculate the ratio again.
6. Continue to calculate ratios between the weights until you are certain you know which brand has a relative weight corresponding to that of the earth.
7. Complete your data table by calculating the ratios between all of the weights and the weight of the brand from the earth.

Conclusions

1. Identify and record the planet to which each brand corresponds. Justify your conclusion.
2. Do any brands correspond to more than one of the planets in our solar system? Explain.
3. Do all of the brands correspond to one of the planets in our solar system? Explain.

Inquiry and Experimentation

To design your own experiment, see page 681.

Extension

The weight of an object varies with the distance the object is from the surface of a planet as well as with the mass and size of the planet. How would you design an experiment to investigate the relationship between weight and distance from a planet.

INVESTIGATION 4

How Does Standing on a Hill Help You Lose Weight?

Objective

To measure the weight of a student standing on an inclined plane

Materials

Two bathroom scales, board for inclined plane, blocks for support, meterstick

Preparation

Prepare a data table with columns for mass, weight, angle of inclination, force normal to the incline, and force parallel to the incline. Include rows for trials.

Procedure

1. **(Part I)** Place two bathroom scales side by side on the floor. Predict what the scales will read if you stand on them with one foot on each scale. Stand on the scales. Read and record both measurements.
2. Place one of the bathroom scales on a laboratory table and place the other scale on a second laboratory table. Pull the tables together, leaving just enough room to stand between them. Put one hand on each scale. Predict what the reading on each scale will be if you push hard enough to lift yourself off the floor. Ask your partner to read the measurements on the two scales.
3. **(Part II)** Put the board on the floor. Place two blocks under one end of the board to form an inclined plane with its lower end against a wall. The wall will keep the board from sliding, but you must still be careful that the ramp does not fall.
4. Measure the length and the height of the incline to the nearest cm.
5. Set the bathroom scale on the board near the lower end of the incline.
6. Predict what the reading on the scale will be when you stand on it. Then stand on it and record the scale reading.
7. Do two more trials, changing the angle of inclination each time.

Analysis/Interpretation

1. Calculate the angles of inclination for each trial in Part II.
2. Draw a vector diagram for each trial that shows the scale reading as the force normal to the incline and your weight as the force perpendicular to the floor.
3. Use the information in the diagrams to calculate your weight. Determine the average of the three values.
4. Explain the force parallel to the incline. Did you feel it?

Conclusions

1. Were the readings on the two bathroom scales in Part I the same? Did the two trials give the same reading? Explain.
2. Place a bathroom scale on a level floor. Stand on it and measure your weight. Does the result agree with the average of your computed values in Part I?
3. Explain why the three scale readings on the inclined plane did not agree.

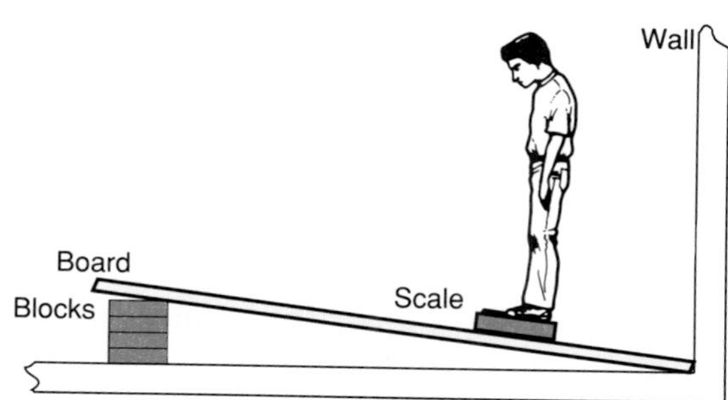

Inquiry and Experimentation

To design your own experiment, see page 681.

Extension

Design an experiment that will allow you to measure the component of your weight parallel to the inclined plane, or measure the force exerted by your hands on the scale as the tables are moved farther and farther apart.

INVESTIGATION 5

Why Does a Baseball Bat Sting Your Hands?

Objective

To locate the center of percussion of a baseball bat

Materials

Wooden baseball bat, fireplace matches with match heads removed, 3 wood blocks, metal laboratory pole, ¼-inch drill with bits, string or cord, weight for pendulum bob

Preparation

Prepare a data table with columns for period, center of oscillation, center of percussion, center of gravity, striking position, and other observations. Include rows for trials.

Procedure  

1. Drill a hole through the bat just large enough for the matchstick in the place it is normal for a batter to hold the bat. Label this point **O**.
2. Hold the bat as a batter would hold it. Have your partner strike the bat with the metal pole. Describe the sensation in your hands when the bat is struck.
3. Have your partner strike the bat in different places several times. Note the feeling in your hands each time.
4. Suspend the bat from point **O**. Make a simple pendulum and hang it parallel to the bat. The point of suspension of the pendulum should be at the same height as that of the bat.
5. Pull the bat to one side and allow it to swing. Adjust the length of the pendulum until its period matches that of the bat. See Figure 5-19, page 112.
6. Mark a point, **C**, on the side of the bat at the height of the center of the pendulum bob. Drill a hole through the bat at point **C**.
7. Suspend the bat like a pendulum from point **C** and determine its period of oscillation about point **C**.
8. Place the bat on the floor or table. Insert a fireplace match into the hole at point **O**. Rest the ends of the matchstick on two blocks and place a third block under the bat at point **C** (the center of percussion). Strike the bat at point **C** three or four times with the pole. Observe what happens to the matchstick.
9. Move the third block to three or four different positions under the bat. Each time the block is moved, strike the bat directly above the support block, and observe what happens to the matchstick.
10. Drill a hole about 10 cm below point **O**. Do another trial using the new hole instead of point **O**.

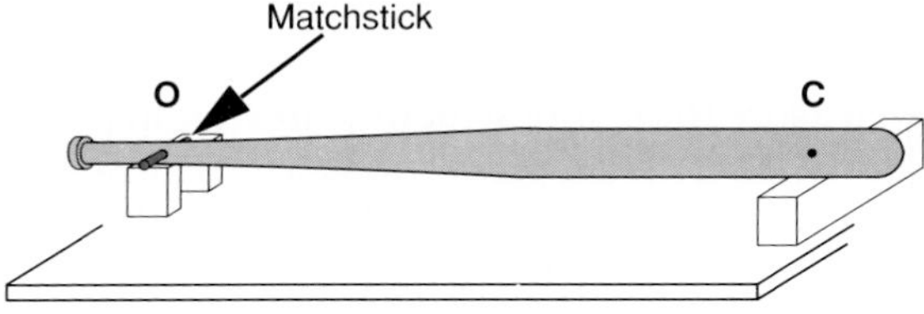

Analysis/Interpretation

1. Locate the center of gravity of the bat. Is the center of percussion (or oscillation) at the center of gravity?
2. How does the period of oscillation about point **O** compare to the period about point **C**?

Conclusions

1. Why is it desirable to hit a baseball at the center of percussion?
2. Does the position of your grip affect the center of percussion? Explain.
3. Does the center of percussion depend on the person swinging the bat? Explain.

Inquiry and Experimentation
To design your own experiment, see page 681.

Extension
Design an experiment to determine the center of percussion of a tennis racket, golf club, or hockey stick.

INVESTIGATION 6

How Do You Measure the Velocity of a Missile?

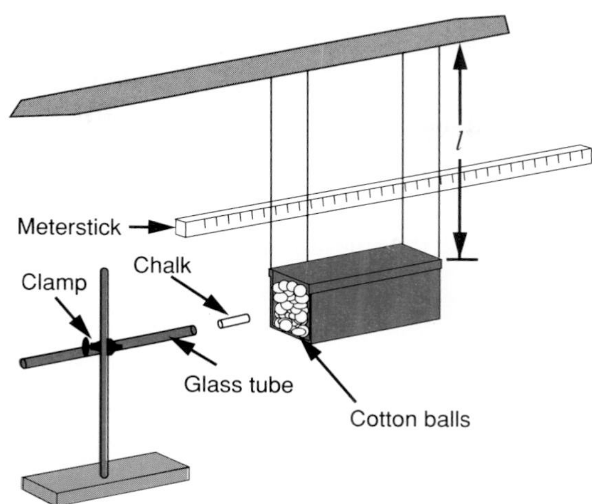

Objective

To measure the speed of a projectile

Materials

Piece of chalk about 3.0 cm long, fire-polished glass tube with large enough diameter to fit chalk loosely, stand with clamp to hold tube, box with one side removed, cotton balls, meterstick, laboratory balance, cord

Preparation

Prepare a data table in which to record mass of box and packing materials, mass of chalk, length of pendulum, horizontal and vertical distances the box moves, increase in potential energy, kinetic energy before and after collision, momentum of the box–chalk system, and speed of the chalk.

Procedure

1. Pack the box lightly with the cotton balls. Make a pendulum by suspending the box from a strong support by four parallel cords attached to the top of the box. Be sure the top of the box is horizontal.
2. Measure the height of the top of the box from the floor.
3. Clamp the glass tube to the stand and adjust the tube so that it is horizontal. Place the stand and tube about 10–15 cm from the box. Aim the tube at the open side of the box. See figure.
4. Put the piece of chalk inside the tube. Blow the chalk out of the tube into the box. With a little practice, you can make the chalk move very fast.
5. Measure the horizontal displacement of the box during its swing.

Analysis/Interpretation

1. Calculate the change in the height of the box from the bottom to the top of the swing. Use this value to calculate the gravitational potential energy gained by the box–chalk system.
2. Assume the energy of the box–chalk system was conserved, and calculate the kinetic energy and the velocity of the box–chalk system immediately after impact.
3. What other assumptions did you have to make in assuming the energy of the box was conserved?

Conclusions

1. Assuming the impact was completely inelastic, calculate the velocity of the chalk just before impact. What conservation law must you apply to determine the velocity?
2. Calculate the kinetic energy of the chalk before impact. Compare the kinetic energy of the chalk before impact to the kinetic energy of the pendulum after impact. Account for any discrepancy.
3. Would your experimental results be affected if the chalk bounced off the back of the box after the impact? Explain.

Inquiry and Experimentation

To design your own experiment, see page 681.

Extension

Design another experiment in which a different method is used to determine the velocity of a missile.

INVESTIGATION 7

Stress and Strain: What's the Difference?

Objective

To investigate the relationship between stress and strain

Materials

Two identical thin steel rods, thin wire, 2 ring stands, 2 right-angle clamps, slotted weights, slotted weight hanger, meterstick, cord, micrometer caliper, tape

Preparation

Prepare Data Tables 1 & 2 with columns for material, height of hanger, weight, distortion, and distortion per newton. Prepare Data Table 3 with columns for material, diameter, cross-sectional area, elongation, stress, and strain. Include rows for up to ten different weights.

Procedure

1. **(Part I)** Use right-angle clamps attached to support stands to support a thin steel rod at both ends. Adjust the rod so that it is horizontal.
2. Suspend a slotted weight hanger from the center of the rod with a strong cord.
3. Measure the distance between the bottom of the weight hanger and the tabletop to the nearest 0.1 cm.
4. Add enough weight to produce a measurable distortion (bending), and measure the height of the bottom of the weight hanger from the tabletop.
5. One by one, add 8–10 more weights and measure the amount of distortion that results from each addition.
6. **(Part II)** Use a stand and right-angle clamp to support the second thin steel rod from one end. Suspend a weight hanger from the other end.
7. Repeat steps 3–5.
8. **(Part III)** Suspend a 1.0-m length of thin wire from a strong support. Suspend a weight hanger from the lower end of the wire.
9. Measure the length and diameter of the wire.
10. Gradually add weights until the wire is permanently distorted, measuring the length and diameter of the wire as each weight is added.

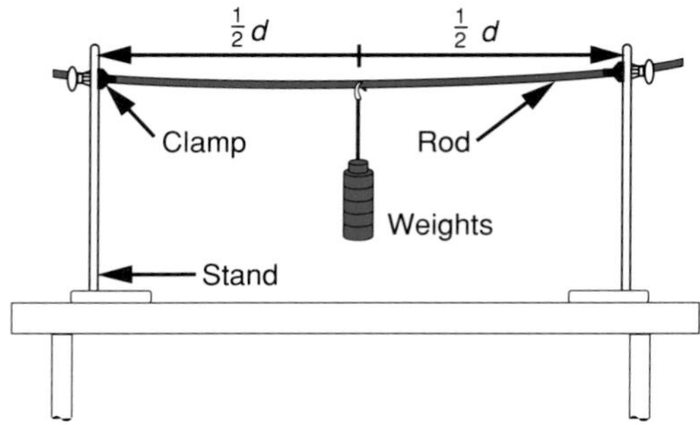

Analysis/Interpretation

1. Plot graphs of weight *vs.* distortion for Parts I and II of the experiment. What is the relationship between the added weights and the stress on the rod? between the amount of distortion and the strain?
2. Plot a graph of stress *vs.* strain for Part III.
3. Calculate the slope of each graph.
4. Use the data collected in Part III to calculate Young's modulus. Determine the percent error.

Conclusions

1. Judging from the results of your experiment, what is the relationship between stress and strain?
2. What forces tend to prevent strain in a substance when a force is applied?
3. Describe some real-life examples of stress *vs.* strain relationships.

Inquiry and Experimentation

To design your own experiment, see page 681.

Extension

Design an experiment to determine the tensile strength of a wire, rope, or thread.

INVESTIGATION 8

How Can You Boil Water with a Bag of Ice?

Objective

To investigate the effect of temperature on volume and pressure

Materials

Ring stand, laboratory burner, 500-mL round-bottom flask, thermometer, 1.0-L beaker, 30-cm glass tube, 2 one-hole stoppers, clamp to hold flask, food coloring, small bag of ice, boiling chips

Preparation

Carefully insert the glass tube and thermometer into the one-hole stoppers until they extend about 10 cm beyond the stoppers.

Procedure   

1. Fill the 1.0-L beaker with water. Add a few drops of food coloring to make the water more visible.
2. Insert the stopper and glass tube into the flask. Holding the flask with a clamp, heat the flask on all sides until the air inside is very hot. **CAUTION:** ***Excessive heating in one spot will crack the glass. Wear goggles.***
3. Remove the flask from the heat, quickly invert it, and place the end of the glass tube into the colored water in the beaker. Observe and record what happens.
4. Return the flask to an upright position and add 10–15 mL of water. Place the flask on the ring stand and heat it until the water boils and steam displaces all of the air in the flask.
5. Predict what will happen when the flask is again inverted with the glass tube and placed in the beaker of water. Try it. Observe and record what happens.
6. Remove the stopper and glass tube from the flask. Add water and insert the stopper and thermometer into the flask. Invert the flask and adjust the thermometer and the amount of water in the flask until the bulb of the thermometer is submerged about 1.0 cm below the surface of the water.
7. Set the flask upright and remove the stopper and thermometer from the flask. Add a few boiling chips. With the flask open, boil the water until all of the air has been displaced by steam.
8. Remove the flask from the heat and tightly cork it with the stopper and thermometer. Allow the flask to cool for a few minutes.
9. Invert the flask and place the bag of ice on top of it. Observe and record what happens to the water. Note the temperature of the water for several minutes.

Analysis/Interpretation

1. What caused the water to move up the glass tube into the flask?
2. Why did the water move up the tube more slowly when the heated flask was dry than it did when the flask was full of steam?
3. On what does the temperature at which a liquid boils depend?

Conclusions

1. Use the results of this experiment to explain how temperature affects the volume and pressure of a gas.
2. What do the results of this experiment indicate about the boiling temperature of water? Explain.

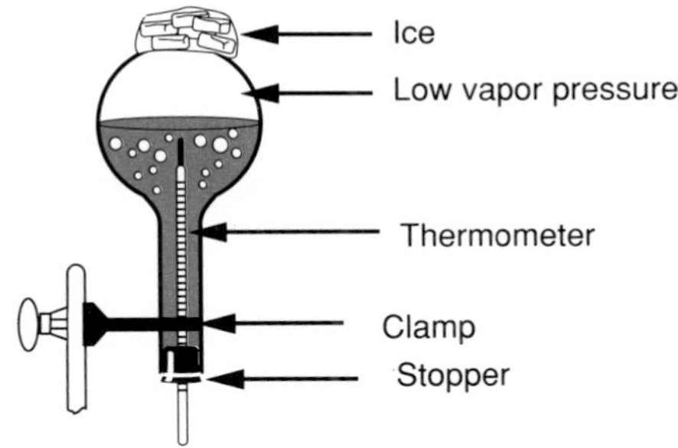

Inquiry and Experimentation

To design your own experiment, see page 681.

Extension

Air expands at a rate constant enough that it can be used in a thermometer. Make an air thermometer and use it to measure the temperature of several substances.

INVESTIGATION 9

How Simple Can a Heat Engine Get?

Objective

To investigate the similarities between heat engines and a toy drinking bird

Materials

Drinking bird, cup of water, ethanol, acetone or nail polish remover, clock with second hand

Preparation

Prepare a data table with columns for item tested, time interval for ten drinks, and rate of dipping.

Procedure   

1. Set the bird in front of a cup of water and try to make it dip for a drink. Wet the head of the bird and try to make it dip again. It may be necessary to adjust the fulcrum to make it operate smoothly. **CAUTION:** ***Handle the bird with care. Do not place it near an open flame.***
2. Observe the operation of the bird carefully. List as many physical properties as possible.
3. Once the bird starts dipping for a drink, measure the time for ten drinks. Do several trials.
4. Hold the bird's head under very cold water and get it to dip again. Measure and record the time for ten drinks.
5. Hold the bird's head under hot water. Time ten drinks and record.
6. Pat the head dry, and dunk its head into other liquids such as ethanol or acetone. The liquid in the cup should be the same as that used on the head. Repeat the time measurements.

Analysis/Interpretation

1. Calculate the rate of dipping for each trial and the average rate of dipping for all trials of each liquid.
2. What factors might affect the rate of dipping? Explain.
3. How many states of matter are present in the bird system?
4. What is the purpose of the fuzzy head? What is the purpose of the liquid on its head?
5. What causes the liquid to rise in the neck during a cycle?

Conclusions

1. Explain the operation of the bird in terms of temperature and vapor pressure.
2. What physical properties must a substance have to be used as a working fluid?
3. How is the drinking bird like a heat engine?

Inquiry and Experimentation

To design your own experiment, see page 681.

Extension

Design an experiment to determine the environment necessary for a continuous and maximum rate of dipping.

INVESTIGATION 10

How Does Sound Behave Like a Wave?

Objective

To investigate the nature of sound waves

Materials

Small speaker, variable oscillator, meterstick, stethoscope, plastic garbage bag, dry ice, box-sealing tape or duct tape

Preparation

1. Place about 200 g of dry ice in the garbage bag. **CAUTION:** ***Use gloves.*** Seal the bag with tape. Set the bag aside and allow it to expand as the dry ice vaporizes to form a CO_2 lens.
2. Prepare a table with columns for three frequencies and rows for position of maxima or minima, distance between maxima or minima, wavelength, and speed of sound.
3. Prepare a second table in which to record the distances of the speaker from the CO_2 lens and the distances of the maxima and minima from the lens.

Procedure 

1. **(Reflection)** Connect the speaker to the output of the oscillator. Adjust the oscillator to a low-volume frequency of about 1000 Hz.
2. Direct the speaker toward a smooth, hard surface 2.0 m away. Use the stethoscope to probe along a line between the speaker and the surface for changes in loudness. Make small adjustments to the distance until there are distinct positions of maxima and minima.
3. Measure the distance between adjacent maxima and between adjacent minima.
4. Repeat steps 1–3, using frequencies of about 500 Hz and 2000 Hz.
5. **(Refraction)** Suspend the CO_2-filled garbage bag from the ceiling or other support so that it forms a lens 0.50 m from the speaker.
6. Use the stethoscope to probe the region near the bag on the side away from the speaker. Locate a point where the sound appears loudest. Measure the distance from the bag to this point.
7. Move the CO_2 lens to a position 0.25 m from the speaker. Locate a point on the opposite side of the lens where the sound is loudest. Measure the distance from this point to the bag.
8. Remove the bag and probe the region where the loudest sound appeared.

Analysis/Interpretation

1. Use the distance between adjacent maxima or minima that lie between the speaker and the hard surface to determine the wavelength of the sound.
2. Does the wavelength depend on the frequency of the sound? Explain.
3. Use the data to calculate the speed of sound.
4. Does the distance of the lens from the speaker affect the position where the sound appears to focus?

Conclusions

1. Explain why the maxima and minima appear between the speaker and the hard surface.
2. Explain why the sound is loudest in one place on the side of the bag opposite the speaker.

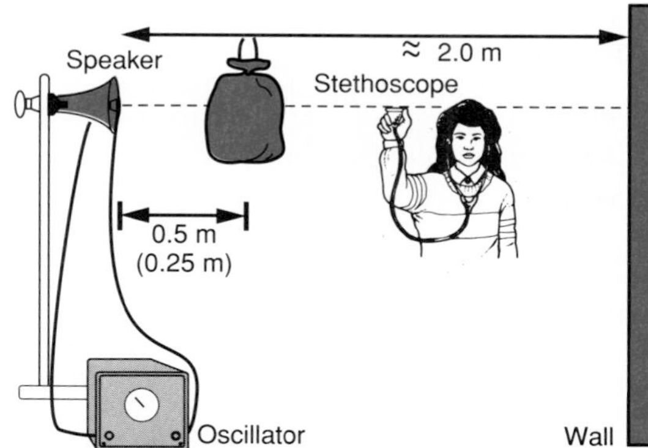

Inquiry and Experimentation

To design your own experiment, see page 681.

Extension

Design an experiment in which two speakers are used as sources of sound, and test for interference.

INVESTIGATION 11

How Can You Measure Sound with a Meterstick?

Objective

To measure the frequency of a tuning fork

Materials

Tuning fork (150 to 350 Hz), hooked mass (at least 500 g), 1.0-m length of paper tape, meterstick, felt-tip marking pen, masking tape, contact cement, thumbtacks, hole punch

Preparation

1. Prepare a data and calculations table in which to record initial and final position, distance interval, initial and final time, time interval, number of waves, and frequency. Include rows for trials.
2. Carefully remove the felt tip from the marking pen. Use one or two drops of contact cement to attach the felt tip to the side of one prong of the tuning fork. Save the ink for later use.
3. Reinforce one end of the paper tape with masking tape. Punch a hole in the end of the masking tape large enough for the hook of the 500-g mass.

Procedure

1. Stick two thumbtacks about 2.0 cm apart into one side of the laboratory table. Hold the paper tape vertically between the tacks with the masking tape at the lower end. The thumbtacks will guide the paper as the weight falls.
2. Suspend the 500-g mass from the paper tape so that the mass is about 5.0 cm below the tacks. Mark a point **X** on the tape about 2.0 cm above the 500-g mass.
3. Wet the felt tip with a few drops of ink from the pen. Strike the tuning fork and hold the felt tip lightly against the paper tape near point **X**.
4. Let go of the paper tape. As it falls, observe the waves made on the paper by the felt tip.
5. Locate an interval on the paper tape where the waves made by the felt tip are continuous. Label a wave crest near one end of the interval point **A** and a crest near the other end point **B**.
6. Count the number of waves between **A** and **B**.
7. Measure the distances between **A** and **B**, between **X** and **A**, and between **X** and **B**.

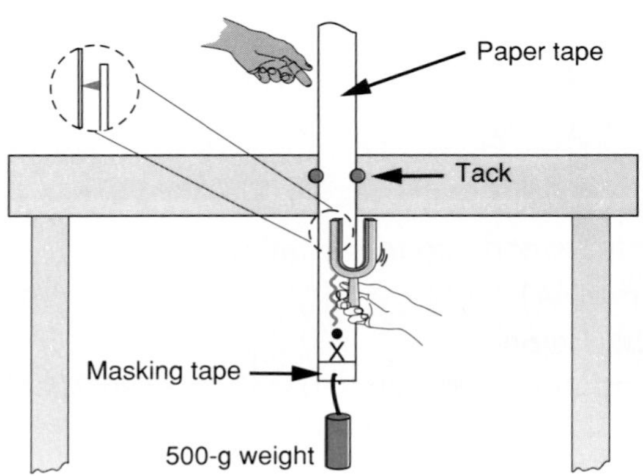

Analysis/Interpretation

1. Use the equation $\Delta d = \frac{1}{2}g\Delta t^2$ to calculate the time it took the tape to fall to the first crest (Δd = **XA**) and the time it took the tape to fall to the second crest (Δd = **XB**). Calculate the time it took to fall from **A** to **B**.
2. Calculate the frequency of the tuning fork.

Conclusions

1. How does the felt tip affect the frequency of the tuning fork? Explain.
2. How would the results be affected if a tuning fork of much smaller or much larger frequency were used?

Inquiry and Experimentation

To design your own experiment, see page 681.

Extension

Since a tuning fork can be used to measure small time intervals, design an experiment that uses a tuning fork to measure gravitational acceleration.

INVESTIGATION 12

Does Light Consist of Particles or Waves?

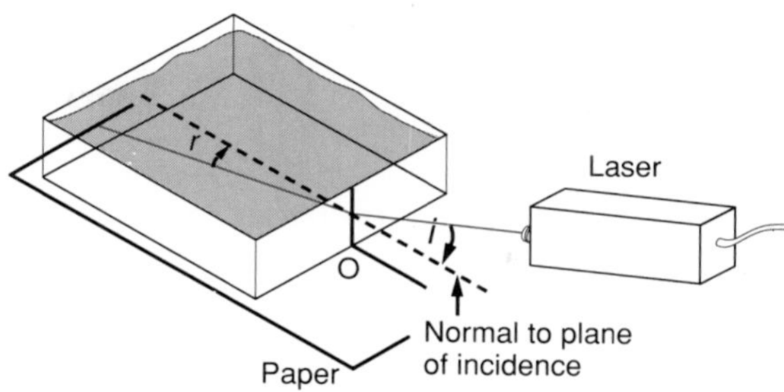

Objective

To compare the refraction of light to the refraction of particles

Materials

Steel ball, carbon paper, unlined paper, cardboard (30 × 35 cm), knife, book, laser, plastic box, protractor, transparent ruler, masking tape

Preparation

1. Design two data tables, one for particles and one for light. Provide columns for angles of incidence and refraction as well as the sines of the angles. Include rows for trials.
2. Score one side of the cardboard with a knife along a straight line about 5.0 cm from one edge. Fold the cardboard along the cut.
3. Place the cardboard on a book lying flat on a tabletop and allow the 5.0-cm strip to hang off the book's edge to create a ramp. See Figure 12-3, page 283. Place sheets of unlined paper topped with carbon paper (carbon-side down) along both edges of the ramp. Use masking tape to secure the paper.
4. Draw a straight line near the middle of another sheet of paper. Label a point on the line point **O**. Fill the plastic box with water and set the edge of the box on the line. Draw a line on the side of the box that is perpendicular to the paper at point **O**. See figure.

Procedure 

1. **(Part I)** Roll the steel ball at a constant speed across the cardboard. Let it roll down the ramp and across the carbon paper on the table.
2. Do four or five more trials, each with a different angle of incidence. Label the tracks after each trial.
3. Measure the angles between the path of the ball and the normal to the edge of the paper at the point where the ball left the top surface (incidence) and at the point where the ball entered the lower surface (refraction).
4. **(Part II)** Turn on the laser and aim the beam so that it enters the box above point **O**. **CAUTION:** ***Do not look into the laser.***
5. Measure the angle of incidence, i, (in air) and the angle of refraction, r, (in water).
6. Do four or five more trials, each with a different angle of incidence.

Analysis/Interpretation

1. Plot two graphs, one for particles and one for light, of the sines of the angles of incidence *vs.* the sines of the angles of refraction.
2. Calculate the slope of each graph.
3. What is the relationship between the angles of incidence and refraction for particles? for light?
4. If light behaves as a particle, how must its speed in water compare to its speed in air?

Conclusions

1. Can you assume that all particles refract the way the ball did?
2. Do the results of this experiment support the particle model of light? Explain.
3. Can the refraction of light be explained using a wave model of light? Explain.

Inquiry and Experimentation

To design your own experiment, see page 681.

Extension

Particles travel slower in dense liquids than in air. Design an experiment to investigate the refraction of a particle.

INVESTIGATION 13

How Many Images Can Two Mirrors Make?

Objective

To determine how the angle between two plane mirrors affects the number of images formed by the mirrors

Materials

Two plane mirrors about 5.0 × 15 cm, metric ruler, protractor, two wood blocks to hold mirrors, pencil, one-hole stopper, unlined paper

Preparation

1. Prepare a data table with columns for seven angles and rows for reciprocals of angles and number of images.
2. Draw a straight line on a sheet of unlined paper. Mark point **A** near the center of the line. Using point **A** as a vertex, draw angles of 120, 90, 75, 60, 45, and 30 degrees.

Procedure

1. Using the wood blocks for support, stand the mirrors vertically on the paper. Carefully place the reflecting surfaces (usually the back) of both mirrors on the line. The edges of the mirrors should touch, and the angle between them should be 180 degrees.
2. Insert a pencil into the hole in the stopper. Stand the pencil and stopper 5.0 cm in front of the mirrors. Count and record the number of images formed by the mirrors.
3. Move the mirrors so that the angle between the reflecting surfaces is 120 degrees and the mirrors touch at the vertex of the angle. Stand the pencil and stopper about midway between the mirrors. Count and record the number of images formed by the mirrors.
4. Do five more trials with angles of 90, 75, 60, 45, and 30 degrees between the mirrors.
5. Place the mirrors parallel to each other and about 25 cm apart. Predict the number of images that will be formed by the mirrors. Move your head along the lengths of the mirror fronts and observe the images.

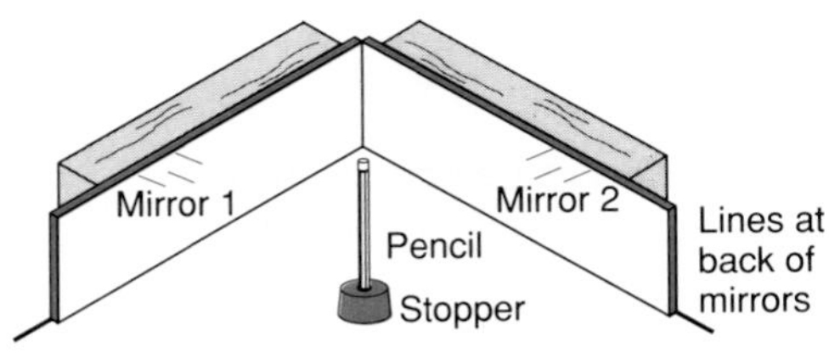

Analysis/Interpretation

1. Plot a graph of the number of images *vs.* the angles between the mirrors. Let the number of images be the ordinates and the angles be the abscissas.
2. Plot a second graph with the number of images as the ordinates and the reciprocals of the angles as the abscissas.
3. Calculate the slope of the straight-line graph.

Conclusions

1. Does the angle between the mirrors affect the number of images formed by the mirrors? Explain.
2. Write an equation that expresses the mathematical relationships between the number of images formed and the angles between the mirrors. (Hint: The slope-intercept equation for a straight line is $y = mx + b$.)
3. Calculate the number of images you would see if the angle between the mirrors were 36 degrees. Verify your answer by placing the mirrors at an angle of 36 degrees and counting the images that are formed.

Inquiry and Experimentation

To design your own experiment, see page 681.

Extension

A kaleidoscope can be made by joining rectangular mirrors along their edges and placing them in a tube with a peep hole at one end and a convex lens at the other end. Design an experiment to investigate the images formed by kaleidoscopes of various mirror arrangements.

INVESTIGATION 14

Just How White Is White?

Objective

To investigate basic methods of color mixing and the effect of the resulting combinations

Materials

Three flashlights or light ray boxes, color filters (2 red, 2 blue, 2 green, 1 yellow, 1 magenta, 1 cyan), white cardboard about 30 × 60 cm, wood blocks, support stands with clamps to hold flashlights, several brightly colored objects, masking tape

Preparation

1. Make a screen from a piece of white cardboard on which to project beams of light. Tape the screen to the blocks to hold it in a vertical position.
2. Cut a piece of the remaining cardboard into an irregularly shaped figure about 15 cm tall.
3. Prepare a data table in which to record incident colors and combined colors.

Procedure

1. **(Part I)** Cover the lens of each of the flashlights with one of the three primary color filters (red, blue, green). Clamp the flashlights to stands, and place them 10–15 cm from the screen.
2. Turn off the lights in the room. Shine each of the flashlights at a different spot on the screen. Observe the three spots of color.
3. Hold the six remaining color filters, one at a time, in front of the filters on each of the flashlights. Observe and record the colors that appear on the screen each time.
4. **(Part II)** Move the flashlight with the green filter so that its beam is shining on the same spot on the screen as the beam from the flashlight with the red filter. Next, shine the green light on the same spot as the blue light. Next, shine the red and blue lights at the same place on the screen. Each time, observe and record the colors that appear.
5. Finally, shine all three lights at the screen so that the spots partially, but not completely, overlap. Again, observe and record the colors on the screen.
6. **(Part III)** Turn off the flashlights and place the irregularly shaped object between the flashlights and the screen.
7. Predict the colors that will appear on the screen when the blue and green lights are turned on. Also, predict the colors that will appear when the red and green lights are used and when the red and blue lights are used. Test your predictions.
8. **(Part IV)** Shine the colored lights and combinations of the colored lights at some of the brightly colored objects. Observe and record what colors the objects appear to be.

Analysis/Interpretation

1. What is the purpose of a color filter?
2. What color is complementary to red? to blue? to green?
3. Are shadows necessarily gray or black? Explain.

Conclusions

1. Why are red, blue, and green considered to be primary colors?
2. How does the light incident on an opaque object affect the color that it appears to be?

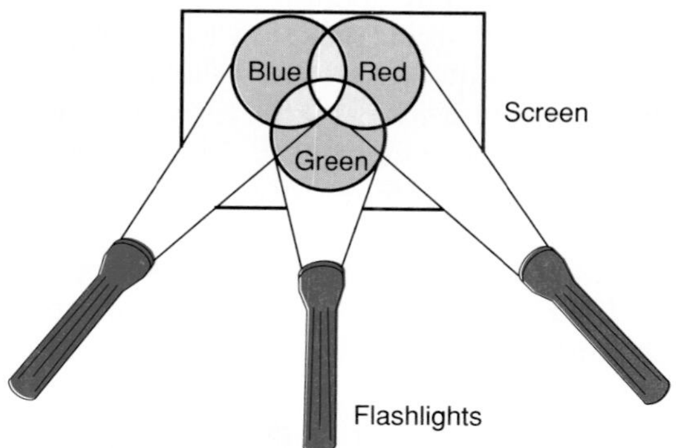

Inquiry and Experimentation

To design your own experiment, see page 681.

Extension

Design an experiment to investigate how colors are combined in pigments using paints, inks, or dyes.

INVESTIGATION 15

What Makes the Colors in Soap Bubbles?

Objective

To investigate color patterns due to thin-film interference

Materials

60–100-W incandescent light bulb in a socket with a cord, cardboard box, waxed paper, 15- × 15-cm color filters (red, blue, and green), shallow pan, soap-bubble solution, 2 plastic straws, cotton string, tape

Preparation 

1. Cut a hole in the center of the top of the box large enough to push the lamp-cord plug through it. Cut three or four more holes in the box top for ventilation. Cut a 5.5-cm square opening on one side of the box. Tape a piece of waxed paper over the square opening. Put the lamp into the box and push the cord through the center hole in the top. Tie a loose knot in the cord so that the lamp will hang vertically from the top of the box at the height of the side opening.
2. Construct a bubble maker by threading the string through the two straws to form a rectangle with straws at the top and bottom and 25 cm of string on each side. Tie two short pieces of string to opposite ends of one of the straws. Cut off any excess.
3. Prepare a data table with columns for color of filter, color of bands, and width of bands. Include rows for trials.

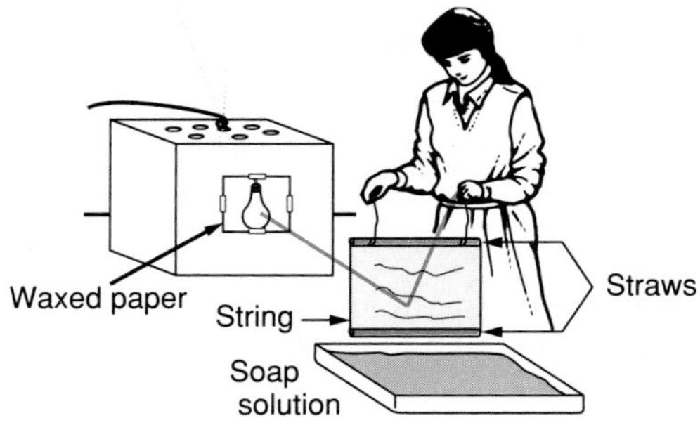

Procedure

1. Pour soap-bubble solution into the pan to a depth of 1.0 cm. Place the bubble maker in the soap solution so that all except the two short pieces of string are submerged. Turn on the lamp.
2. Tape the blue filter in front of the waxed-paper window.
3. Holding one short string in each hand, slowly lift the bubble maker out of the solution. A thin film should form between the straws.
4. Hold the bubble maker steady in front of the blue filter. Observe the reflected light from the soap film as it settles. Observe the color and widths of the bands that appear in the film.
5. Return the bubble maker to the pan, being careful to keep the short pieces of string dry.
6. Replace the blue filter with the red filter. Form another film and hold it in front of the red filter. Observe the color and width of the bands.
7. Repeat steps 5 and 6, using the green filter.
8. Remove the green filter. Form another film and hold it in front of the waxed-paper window. Observe the colors and widths of the bands that appear in the film with white light.

Analysis/Interpretation

1. Are all of the bands in the film the same width? Why or why not?
2. Compare the colors that appear in a rainbow with the colors that appear in the film when white light shines on it.

Conclusions

1. When a color filter is used, what causes the black bands? the color bands?
2. Explain the appearance of the magenta, cyan, and yellow bands when white light is used.

Inquiry and Experimentation

To design your own experiment, see page 681.

Extension

Design an experiment that will allow you to investigate thin-film interference in oil slicks.

INVESTIGATION 16

What Is the Shape of an Electric Field?

Objective

To map equipotential lines around various conductors and to use them to map an electric field

Materials

Ripple tank with metal sides, voltmeter, graph paper, 6-V power supply, banana plug, alligator clips, 18-gauge connecting wires, 14-gauge solid copper wire, soldering iron, solder, transparent tape

Preparation

Strip the insulation from the 14-gauge copper wire. Cut the wire into three lengths: two 30-cm lengths and one 15-cm length. Solder the ends of the 15-cm wire together to form a circular loop. Solder a length of connecting wire to the center of each 30-cm length, to the loop, to the banana plug, and to the clips.

Procedure 

1. Tape graph paper to the underside of a clean ripple tank. Pour water into the tank to a depth of 1.0 cm. Place other sheets of graph paper on the table near the ripple tank.
2. Connect the negative terminal of the 6-V power supply to the frame of the tank with a clip and connect the positive terminal to the loop of copper wire. Place the loop in the center of the ripple tank.
3. Connect the negative terminal of the voltmeter to the frame of the tank. Connect the positive terminal of the voltmeter to the banana plug.
4. Use the banana plug as a probe to explore the potentials in the ripple tank. First, locate all the points in the water where the potential is 2.0 V. Draw the 2.0-V potential line on one of the sheets of graph paper on the table, being careful that the position of the line corresponds to the line on the graph paper under the ripple tank.

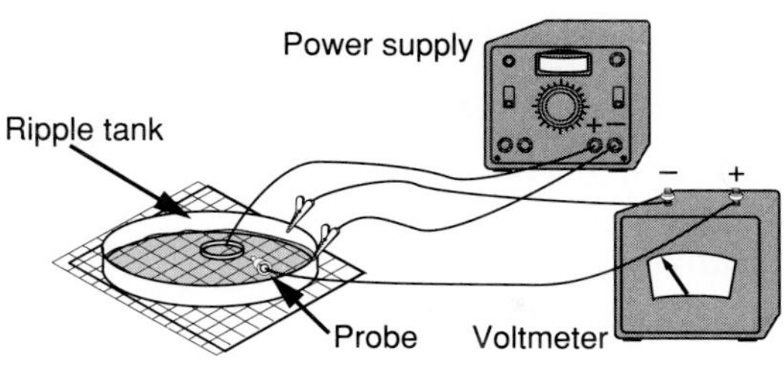

5. Use the probe to locate several other equipotential lines, testing inside as well as outside of the loop.
6. Bend the loop into different shapes and repeat steps 4 and 5.
7. Use a bare end of a piece of the 14-gauge wire as a point conductor and locate the equipotential lines around it.
8. Connect the positive terminal of the power supply to one of the 30-cm strips of 14-gauge wire and the negative terminal to the other 30-cm strip. Place the strips about 20.0 cm apart in the ripple tank and investigate the equipotential lines around them.

Analysis/Interpretation

1. What is the potential at the ripple tank frame? Why does it have this value?
2. Where is the highest potential found?
3. What does the field look like near a point conductor? where the conductor folds inward? where it folds outward? where it is straight?

Conclusions

1. What general rule can you suggest relating the shape of the field to the shape of the conductor?
2. Using your equipotential lines as a guide, draw the corresponding electric field lines for each conductor.

Inquiry and Experimentation
To design your own experiment, see page 681.

Extension
Design an experiment to investigate the electric field in a wire.

INVESTIGATION 17

What Makes for an Electric Relationship?

Objective

To relate potential difference and amount of resistance to the current in a resistor

Materials

Six 1.5-V dry cells or variable d-c power supply, d-c milliammeter, d-c voltmeter, six carbon resistors (½-W, 10% tolerance, varying from 47–330 Ω), switch, 18-gauge connecting wires, resistance color code

Preparation

1. Design a data table for Part I in which to record values for potential difference, current, and resistance as batteries are added to the circuit.
2. Design another data table in which to record potential difference, current, resistance, and the reciprocal of resistance as resistors are added to the circuit.

Procedure

1. **(Part I)** Connect a 1.5-V dry cell in series with a d-c milliammeter, a switch, and a resistor. Connect a d-c voltmeter in parallel with the resistor. *Be sure to observe the polarity markings on both meters.*
2. Close the switch. Measure the current in the circuit and the potential drop across the resistor.
3. Open the switch and connect a second 1.5-V dry cell in series with the first one.
4. Close the switch. Measure the current in the circuit and the potential drop across the resistor.

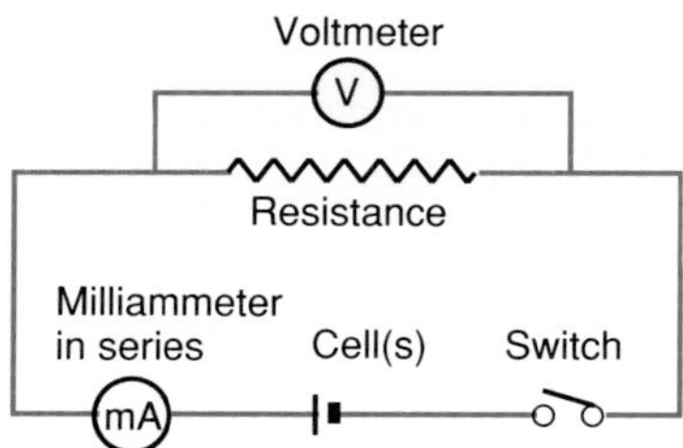

5. Do four more trials, each time with an additional dry cell in series with the others. *Keep the switch open except when taking readings.*
6. **(Part II)** Connect two 1.5-V dry cells in series with the milliammeter, one resistor, and the switch. Connect the voltmeter in parallel with the resistor.
7. Measure the current in the resistor and the potential difference across the resistor.
8. Do five more trials, each time with a different resistor in the circuit.

Analysis/Interpretation

1. Plot a graph of potential difference *vs.* current for Part I.
2. Use the resistance color code to determine the specified value of the resistance used in Part I, and compare the value to the slope of the graph.
3. Calculate the reciprocal of each resistance used in Part II.
4. Plot two graphs, using the data obtained in Part II. One graph should show current *vs.* resistance; the other should show current *vs.* the reciprocal of the resistance.
5. Calculate the slope of the graph of current *vs.* the reciprocal of the resistance.

Conclusions

1. To what physical quantities do the slopes of the graphs correspond?
2. What do the graphs indicate about the relationships between current and potential difference? between current and resistance?
3. What do the results of your experiment indicate about the relationship among potential difference, current, and resistance?

Inquiry and Experimentation

To design your own experiment, see page 681.

Extension

Design an experiment to show how resistors connected in various series and parallel combinations affect current, potential difference, and net effective resistance.

INVESTIGATION 18

How Is the Electrochemical Equivalent of Copper Determined?

Objective

To determine the factors that affect the mass of a metal deposited during electrolysis and to measure the electrochemical equivalent of copper

Materials

Battery jar, a copper strip (about 2.0 cm × 10 cm) for one electrode, a carbon rod for the other electrode, rheostat, switch, ammeter, d-c power supply, 2 alligator clips, set of weights, laboratory balance, fine sandpaper, 18-gauge connecting wires, copper sulfate solution, Bunsen burner

Preparation

Prepare a table in which to record data (initial and final mass of the carbon electrode, average current, initial and final time, potential difference) and calculations (mass of carbon deposited, elapsed time, electrochemical equivalent of copper, and percent error). Clean the copper strip with the sandpaper until it is bright and shiny.

Procedure

1. Determine the masses of the carbon rod and the copper strip to at least the nearest 0.01 g.
2. Rinse the battery jar and fill it with enough copper sulfate solution to nearly immerse the two electrodes.
3. Connect the circuit as shown in the figure. *Be careful to observe the polarity markings on the meter. Use the highest scale on the ammeter until you are certain it is safe to use a lower scale.* Place the electrodes about 4.0 cm apart in the copper sulfate solution. Note: The carbon rod must be the negative electrode (cathode), and the copper strip must be the positive electrode (anode).
4. Close the switch and record the exact time. Adjust the rheostat until the current is 0.30 A. Use the rheostat to keep the current constant.
5. Allow the current to flow in the circuit for about 25 minutes. Open the switch and record the time at which it was opened.
6. Remove both electrodes and rinse them in clean water. Dry them by holding them well above a Bunsen-burner flame.
7. Determine and record the final mass of the carbon and copper electrodes.
8. Pour the copper sulfate solution back into the original container and wash the battery jar.

Analysis/Interpretation

1. Calculate the elapsed time and the gain in mass of the carbon electrode.
2. Calculate the electrochemical equivalent of copper. Compute the percent error by comparing it with the value given in Table 18-1, page 461.

Conclusions

1. What happened to the mass of the anode during the experiment?
2. Explain what was taking place at the cathode when the copper ions came in contact with it.

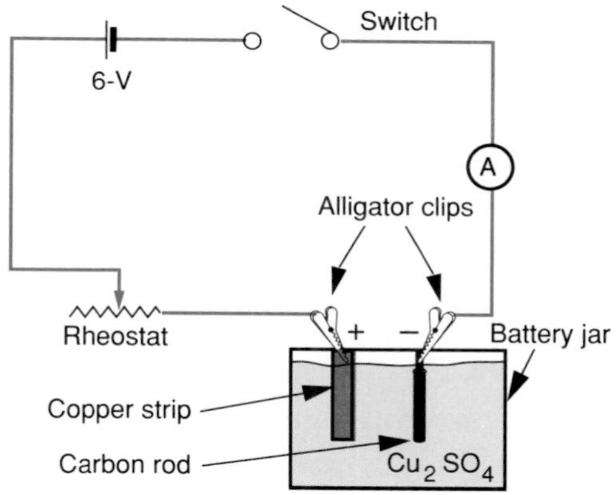

Inquiry and Experimentation

To design your own experiment, see page 681.

Extension

Design a similar experiment to determine the electrochemical equivalent of silver. Investigate the procedures used to plate a material with chromium or nickel. Why must they be copper-plated first?

INVESTIGATION 19

Just How Strong Is a Magnet?

Objective

To determine the force of repulsion between two magnetic poles

Materials

Two identical cylindrical magnets, smooth board to use as an inclined plane, support for the incline, laboratory balance, metric ruler, magnetic compass

Preparation

Prepare a data and calculations table in which to record length and height of the incline, angle of inclination, mass of the magnets, force parallel to the inclined plane, distances between like and unlike poles, reciprocals of the distances, force of attraction, force of repulsion, and net force.

Procedure

1. Support the board so that it forms an angle with the horizontal of about 30 degrees. Measure the length and height of the incline.
2. Measure the mass of the magnets.
3. Use the magnetic compass to determine the position of each magnetic pole as closely as possible. Measure the distance between the opposite poles of each magnet.
4. Tape one of the magnets to the board in such a position that it would roll down the incline if it were free.
5. Place the second magnet adjacent to the first one, but higher on the incline, and with like poles together.
6. Allow the second magnet to be repelled up the incline until it reaches equilibrium. After it comes to rest, measure the distances between the like poles and the unlike poles of the magnets.

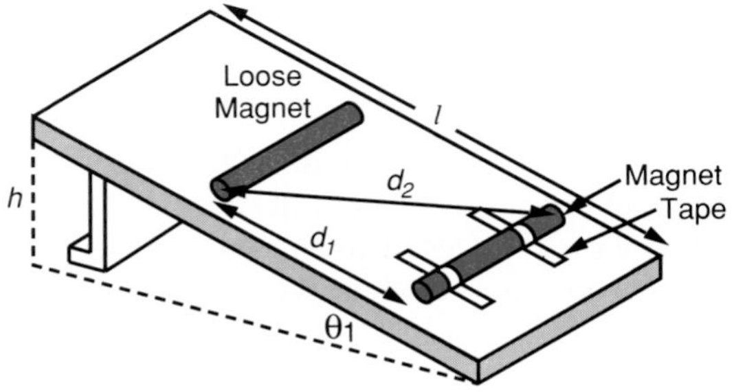

Analysis/Interpretation

1. Calculate the angle of inclination, θ_1, and the component of the weight of the upper magnet that is parallel to the incline, F_P.
2. Draw a vector diagram showing the force of attraction between unlike poles and the component of that force which pulls the magnet down the incline. Label the force of attraction F_A. Also show the force of repulsion between the like poles and label it F_R.
3. Write an equation that expresses the magnetic force pulling the upper magnet down the incline as a component of the force of attraction. Why does the factor 2 appear in the equation?
4. Write an equation for the force of repulsion between the magnets. Why is 2 a factor in the equation?
5. Write an equation that expresses the net magnetic force acting on the free magnet, F_M.

Conclusions

1. How do F_P and F_M compare to the total net force acting between the magnets?
2. Substitute the equation used to measure the force between two magnetic poles (Coulomb's law) into the equation for net force.
3. Assume the magnets have equal pole strength, and calculate the strength of the poles.

Inquiry and Experimentation

To design your own experiment, see page 681.

Extension

Design an experiment to verify that the force between the poles of a magnet is inversely proportional to the square of the distance between the poles.

INVESTIGATION 20

How Can Magnetism Generate Electricity?

Objective

To investigate the relationship between a magnetic field and an induced emf

Materials

Two bar magnets, 24-gauge magnet wire, galvanometer, dry cell, switch, resistor (10 000 Ω), 30-cm long soft iron rod for core, 30-cm long wood dowel, 18-gauge connecting wire

Preparation

Prepare two data tables, one with columns for magnet movements, magnitude of deflection, number of magnets, and number of turns, and the other with columns for position of switch, direction of needle deflection and electron flow, magnetic field in primary coil, and polarity of secondary coil.

Procedure

1. **(Part I)** Make three coils (25 turns, 50 turns, 100 turns) from the magnet wire. The coils must have a diameter large enough to insert a bar magnet without the magnet touching it.
2. Connect a galvanometer in series with the dry cell, switch, and resistor. *Observe the polarity markings on the meter.* Close the switch and observe the direction in which the galvanometer needle moves (left or right). Open the switch and reverse the connections to the dry cell. Close the switch and observe the deflection of the needle. Determine the direction of electron flow each time.
3. Draw diagrams of both circuits, showing the direction of electron flow and needle deflection.
4. Connect the leads from the 50-turn coil to the terminals of the galvanometer. Observe the deflection of the galvanometer needle when the **N** pole of a magnet is (a) pushed slowly into the coil, (b) held steady, and (c) slowly pulled out of the coil.
5. Observe the deflection of the needle when the magnet is pushed rapidly in and out of the coil.
6. Repeat steps 4–5, using two magnets with like poles together.
7. Predict what will happen to the needle deflection if the 25-turn and the 100-turn coils are used. Test your predictions.
8. **(Part II)** Insert one end of the soft iron core into the 25-turn coil. Connect the coil in series with the dry cell and the switch. Place the 100-turn coil onto the other end of the iron core and connect the leads to the terminals of the galvanometer.
9. Observe the deflection of the galvanometer needle (a) when the switch is open, (b) immediately after closing the switch, (c) when the switch is closed, and (d) immediately after opening the switch.
10. Replace the soft iron core with the wood dowel and repeat steps 8–9.

Analysis/Interpretation

1. What factors affect the deflection of the galvanometer needle? Explain.
2. How is an emf induced in a conductor?

Conclusions

Write a general statement explaining how turning a switch on and off can induce an emf.

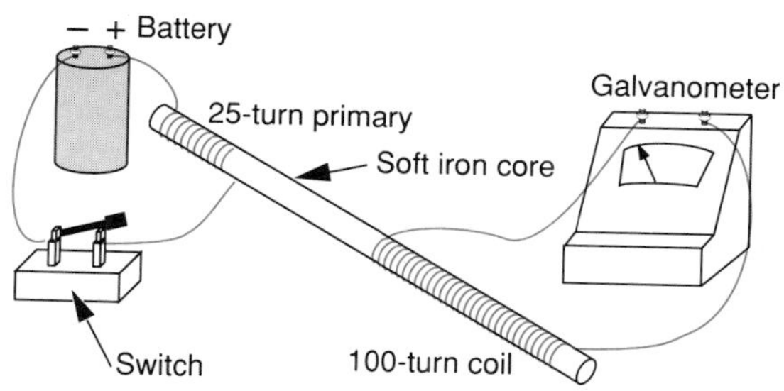

Inquiry and Experimentation

To design your own experiment, see page 681.

Extension

Design an experiment to investigate the effect that an induced current has on the magnetic phenomenon that caused it.

INVESTIGATION 21

What Does a Watt-Hour Meter Measure?

Objective

To investigate the operation of a watt-hour meter and to use it to determine the power consumption of various household appliances

Materials

Watt-hour meter, 0–120-V a-c voltmeter, 0–10-A a-c ammeter, 25-Ω rheostat, switch, extension cord with plug, connecting wires, noninductive loads (lightbulbs, irons, toasters, etc.)

Preparation

Prepare a data table with columns for appliances, potential difference, current, number of turns of watt-meter disk, time for disk turns, power (from watt-hour meter), power ($V \times I$), percent error, energy per month, and cost.

Procedure

1. Before plugging in the cord, make the connections as shown in the figure. Note there are four wires from the base of the watt-hour meter, two for the line, and two for the load. *Once your teacher has approved your connections, plug in the cord.*
   CAUTION: ***Do not touch connections once circuit is plugged in.***
2. Observe the appliances. Predict which one will consume the most power.
3. Turn the rheostat to maximum resistance, and connect an appliance in the plug at **P**. Adjust the range of the meters. Reduce the resistance of the rheostat to zero.
4. Close the switch and record the readings of the voltmeter and ammeter. Count the number of rotations of the disk for 3.0 minutes, and record the time and number of rotations. The dials will be too slow to detect any movement in only 3.0 minutes.
5. Repeat the procedure for as many different appliances as desired.

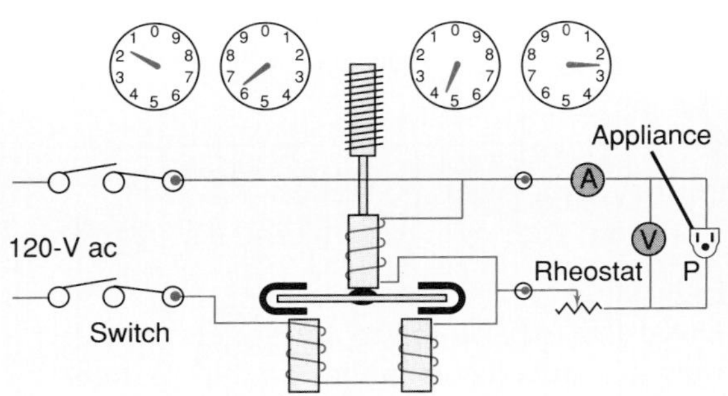

Analysis/Interpretation

1. Compute the power consumption for each appliance from the rate of rotation of the disk.
2. Calculate the power from the voltage and current readings. Then determine the percent error.
3. Assume each of the appliances will be used 1.0 hour per day for 30 days each month. Compute the energy in kilowatt-hours consumed per month.
4. Determine the local electric rate and calculate the cost of operating each of the appliances for one month.

Conclusions

1. Which type of electric appliance uses the most electricity?
2. If the power used by a consumer as indicated by the watt-hour meter is in error, what adjustment might be made to make the meter read correctly?
3. Sketch a diagram of the dials of the watt-hour meter at your home, and record the reading in kilowatt-hours.

Inquiry and Experimentation

To design your own experiment, see page 681.

Extension

Select two or three appliances at your home to test. Using the watt-hour meter at your home, time the rotating disk for the appliances, one at a time. Determine the power consumption for each.

INVESTIGATION 22

How Do Diodes and Transistors Direct Electron Traffic?

Objective

To investigate some of the functions of diodes and transistors

Materials

1-kΩ and 10-kΩ resistors, crystal diode, LED, 9-V d-c battery with snap connector, voltmeter, two 1.5-V batteries, **PNP** transistor, **NPN** transistor, 0–10 d-c microammeter, 0–100 d-c milliammeter, 50-kΩ variable resistor, 20-gauge solid copper connecting wire.

Preparation

Prepare two data tables, one with columns for device type, potential difference, and polarity, and another with columns for transistor circuit base current and collector current. Include rows for trials.

Procedure

1. **(Part I)** Connect the 1-kΩ resistor in series with the light-emitting diode (LED) and the 9-V d-c battery. Observe what happens.
2. Reverse the connections from the battery. Observe what happens.
3. Measure the potential difference across the LED and the resistor again.
4. Replace the LED with a crystal diode. Measure the potential difference across the diode and the resistor. Then reverse the connections of the crystal diode and measure the potential difference across the diode and the resistor again.
5. **(Part II)** Connect a circuit with the **PNP** transistor, 10-kΩ resistor, the 0–10 microammeter, and the 0–100 milliammeter, the 50-kΩ variable resistor, and the two 1.5-V batteries, as shown in the figure.
6. Turn the variable resistor all the way up for maximum resistance. Read and record the current.

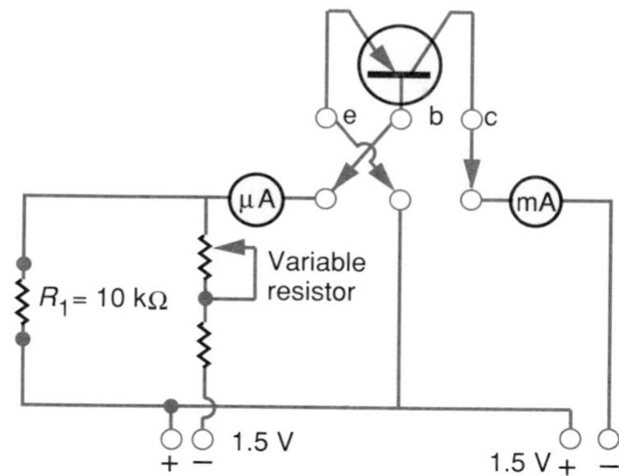

7. Adjust the variable resistor so as to obtain different collector currents. Record your observations.
8. Connect a circuit in a similar manner using the **NPN** transistor. *Be sure to reverse the polarity of the batteries. Repeat the experiment.*

Analysis/Interpretation

1. What do your measurements indicate about the LED and the crystal diode?
2. Plot a graph of collector current *vs.* base current, plotting the collector currents as ordinates and the base currents as abscissas. Calculate the slope of the graph.

Conclusions

1. What does your experiment indicate about a transistor? Explain.
2. What is the value of the direct-current gain? How did you determine the value?
3. Explain how a transistor is similar to a triode vacuum tube.

Inquiry and Experimentation

To design your own experiment, see page 681.

Extension

Design an experiment to investigate how an emitter amplifier circuit may be set into self-oscillation by transformer feedback.

INVESTIGATION 23

What Do Spectral Lines Say About Atoms?

Objective

To determine a relationship between the wavelengths of spectral lines of hydrogen and the energy of the hydrogen atom

Materials

Polaroid® camera and film, hydrogen spectrum tube, spectrum-tube power supply, diffraction grating, meterstick, metric ruler, tape

Preparation

Prepare a data table with columns for grating constant, distance from grating to spectrum tube, distance from spectrum tube to spectral line, angle, and wavelength. Include rows for spectral lines.

Procedure

1. Insert the hydrogen spectrum tube in the spectrum-tube power supply. Place the meterstick directly behind the spectrum tube.
2. Darken the room and turn on the power supply. Stand about 1.0 m from the spectrum tube so that the line between you and the tube is perpendicular to the meterstick. View the spectrum emitted by the hydrogen through the diffraction grating. The grating lines must be parallel to the spectrum tube. If you cannot see a series of discrete lines on both sides of the spectrum tube, rotate the grating 90 degrees.
3. Adjust the position of the meterstick so all of the spectral lines are seen just above it. Make marks on the meterstick at the positions of the first and the last spectral lines.
4. Turn off the power supply. Leave it off until you are ready to take a photograph.
5. Tape the diffraction grating in front of the lens. Remember to keep the grating lines parallel to the spectrum tube. Place the camera on a line perpendicular to the meterstick with the lens about 1.00 m from the spectrum tube.
6. Darken the room again and turn on the power supply. Take a photograph that shows both the spectral lines and the scale on the meterstick.

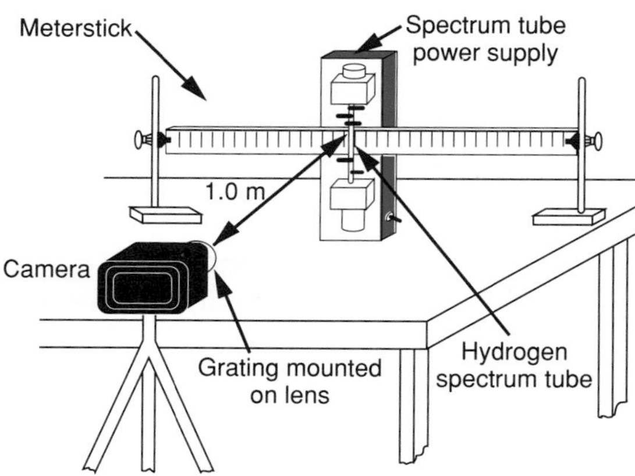

Analysis/Interpretation

1. There are more lines on the photograph than you were able to see by holding the grating to your eye. Explain.
2. Calculate the grating constant, the angle of diffraction, and the wavelength of the lines.
3. Using Planck's constant and the speed of light in a vacuum, calculate the energy of the photons of different wavelengths emitted when the hydrogen atoms changed from one energy level to another.

Conclusions

1. Assuming the final energy state is the same for all of the lines, draw an energy-level diagram showing the energy of the emitted photons from each excited state to the final state.
2. Calculate the energy a hydrogen atom loses when it emits the red line in its spectrum.

Inquiry and Experimentation

To design your own experiment, see page 681.

Extension

Do some research to learn how the energies of electrons are calculated. Then use the wavelengths of emitted photons and the electron energies to calculate Planck's constant.

INVESTIGATION 24

What Does Half-Life Mean?

Objective

To simulate radioactive decay and use the simulation to determine the half-life and decay constant

Materials

100 large washers, 100 small washers, 100 pennies, 50- × 50-cm sheet of unlined paper, box with 50- × 50-cm bottom, shaker cup, metric ruler

Preparation

1. Prepare a data table with columns for large washers removed, large washers remaining, small washers removed, small washers remaining, pennies in box. Include rows for number of tosses.
2. Draw grid lines on the unlined paper 2.5 cm apart. Place the paper into the box with the grid lines facing up.

Procedure

1. Shake the 100 large washers and toss them into the box. Remove all of the washers in which the intersection of two grid lines can be seen in the hole of the washer.
2. Count the large washers removed, set them to one side, and replace them with small washers. (These will represent nuclei that have experienced radioactive decay.)
3. Record the number of large washers remaining in the box as well as the number of small washers that were substituted. Total should be 100.
4. Shake the new combination of washers and toss them into the box. This time remove all of the large *and* small washers in which the intersection of two grid lines can be seen in the hole.
5. Count each type and set them aside. Replace the large washers that were removed with small washers, and replace the small washers that were removed with pennies. (These represent radioactive products that result from further decay.)

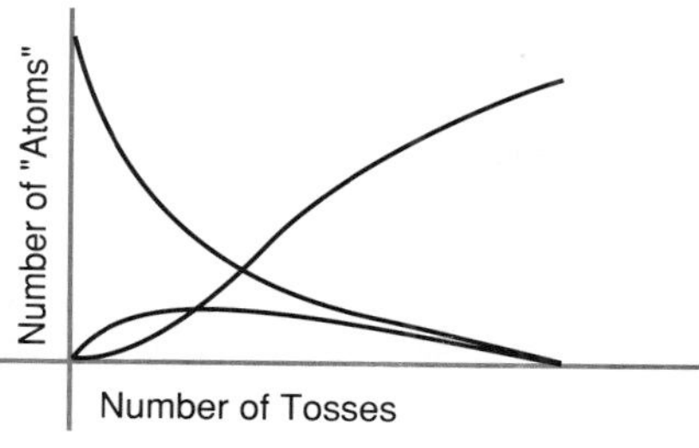

6. Record the number of large and small washers remaining in the box and the number of pennies substituted. Total should still be 100.
7. Shake the washers and pennies together, and toss them back into the box. Remove all of the large and small washers in which you can see intersecting grid lines through the holes. (The pennies are to be considered end products and will not be removed from the box.)
8. Count the washers, set them aside, and replace with the appropriate item.
9. Record the number of large and small washers and the number of pennies remaining in the box, as well as all substitutes. (Total = 100)
10. Continue steps 7–9 for twenty-five tosses.

Analysis/Interpretation

Plot graphs of number of large washers *vs.* the number of tosses, the number of small washers *vs.* the number of tosses, and the number of pennies *vs.* the number of tosses.

Conclusions

From the results of your experiment, calculate the half-life (in terms of number of tosses) and the decay constant for both the large and small washers.

Inquiry and Experimentation

To design your own experiment, see page 681.

Extension

Design an experiment that you can use to determine the half-life of a radioactive substance such as radon-220.

INVESTIGATION 25

How Random Is Random?

Objective

To learn to use data from random behavior to make predictions

Materials

Geiger tube, nuclear rate meter, support stand, beta source, meterstick, stopwatch

Preparation

Prepare a data table with columns for total counts, total number of intervals, and average counts per minute. Include rows for length of time interval. Also make a tally sheet with columns for each time interval and rows (from 1–15) in which to track number of counts.

Procedure 

1. Plug the Geiger tube into the nuclear rate meter and turn on the power. Counts will be registered by the meter. This is background radiation. Measure this count for three minutes.
2. Place the beta source 5.0 cm from the Geiger tube and measure the count for three minutes.
3. Move the source far enough from the Geiger tube that the rate drops to the background count. Measure the count for three minutes.
4. Move the source 10.0 cm closer to the Geiger tube than the position at which there is only background radiation.
5. Every 10.0 seconds, make a tick-mark on the tally sheet in the column next to the corresponding number of counts.
6. Continue to record counts at 10.0-s intervals for 10.0 minutes.
7. Repeat the procedure, but record the counts in 20.0-s intervals, and then, in 60.0-s intervals.

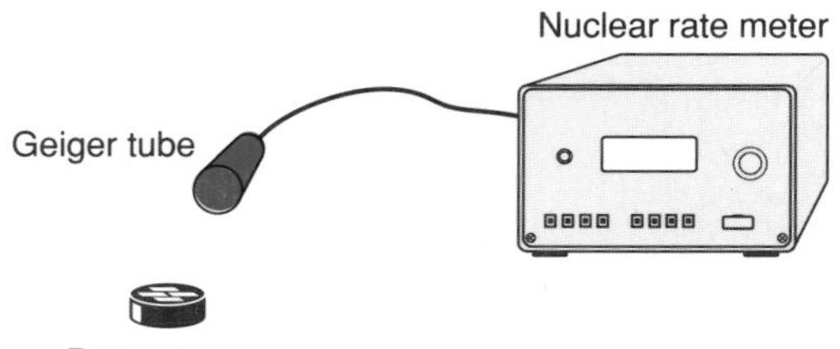

Analysis/Interpretation

1. What evidence do you observe of the random nature of radiation?
2. What part of the total number of disintegrations are you detecting in any time interval?
3. Calculate the total number of intervals and the total counts.
4. Calculate the average number of counts per minute for the 10.0-second intervals, the 20.0-second intervals, and the 60.0-second intervals.
5. Plot bar graphs of your results for the different time intervals. Plot the number of counts per interval as the abscissas and the number of intervals as the ordinates.

Conclusions

1. What is the most probable number of counts per 10.0-s interval? per 20.0-s interval? per 60.0-s interval?
2. How would the results of your experiment be affected if the counts per minute were based on the smallest number of counts in an interval? on the greatest number of counts? Explain.
3. Write a general statement that describes how predictions can be based on random events.

Inquiry and Experimentation

To design your own experiment, see page 681.

Extension

Design an experiment in which a different random event is used to predict the probability of the event occurring in a given time or situation.

INQUIRY AND EXPERIMENTATION

Once you are familiar with the basics of laboratory experimentation, you are ready to function as a research scientist.

Scientists endeavor to answer questions regarding the physical world around them by posing a hypothesis, or a statement of what they think might be happening, and then doing experiments to test their hypothesis. The experiments will either support (not prove) or refute their hypothesis. You, too, can be a scientist.

The following is a general method for conducting an inquiry investigation. You can use it to further your investigation into the phenomena you have observed in each of the chapter investigations. Once you have completed the investigation, you and your partners will want to design a similar experiment to test one of your own hypotheses regarding the same concept. Using the same materials and a similar procedure, test your hypothesis to see if it is valid. Your experiment should be designed to be safe, practical, and repeatable.

In your experimental design, you will establish variables that will indicate whether or not your experiment supports the hypothesis you established. There are two types of variables: independent (those that you manipulate to test your hypothesis) and dependent (those that change according to the effect of the independent variables). Your choice of variables is important to the outcome of your investigation.

Certain experiments will also require controls. Controls are parallel procedures in which you do not manipulate the independent variables and therefore can observe what happens if nothing changes in the setup.

Objective

To conduct an experiment to further inquire about the nature of phenomena observed in the chapter investigation you have just completed

Materials

See the appropriate chapter investigation for materials list and use the same in your own experiment design.

Preparation

1. Discuss the objective of the chapter investigation you have done with your partners.
2. Develop a hypothesis about the phenomenon you observed in your investigation.
3. Design an experiment to test your hypothesis that uses a procedure similar to the one described in the original investigation.
4. Make a table to record your data.

Procedural Guidelines

1. What are the independent and dependent variables?
2. How will you vary the independent variables?
3. How will you measure changes in the dependent variables?
4. What controls will you use as a comparison to your experiment?
5. Design a graph to show your results.
6. Proceed with your experiment after your design is approved by your teacher.

Analysis/Interpretation

1. Make calculations based on the data you collected.
2. Plot graphs or make charts to represent your measured and calculated data.
3. What (if any) is the relationship among your data?
4. What interpretation can you assign to your data?

Conclusions

1. Do your data support your hypothesis? Explain.
2. If your data offer support, what can you conclude about the phenomena you have observed?
3. If your data refute your hypothesis, what changes should you make in your approach?
4. Can you think of any sources of error in your experiment that may have affected the outcome?

Use of Electric Instruments

Introduction

Meters used for electric measurements are delicate instruments and must be handled with care. The greater the precision of the instrument, the more fragile it is and the more easily it is damaged. Ordinary commercial-grade meters are satisfactory for most laboratory experiments. Laboratory-grade meters may be required for experiments in which a high order of precision is needed.

Recall that the heating effect in an electric circuit increases as the *square* of the current. For this reason, excessive meter currents must be avoided so that the meter movement will not quickly burn out. If a circuit that contains a meter is closed and the meter pointer moves off scale, an excessive meter current is indicated. *The switch must be opened immediately.*

If a meter that is to be used in an electric circuit is a multirange instrument, a meter-protection procedure should be followed. First connect the highest meter range to the circuit. If a readout is not possible, open the circuit and connect the next highest range. Repeat this procedure until a meter readout is possible. Then connect the meter range that displays the readout in the middle region of the calibrated scale.

Meters used in a-c circuits are constructed differently from those used in d-c circuits. *These meters are not interchangeable between two kinds of circuits. Multiple-purpose meters, known as multimeters,* contain several instrument circuits in a single enclosure. Multimeters provide the means for measuring different electric quantities in a circuit. They may be designed to provide both a-c and d-c instrument capabilities.

Voltmeters

A meter designed to give readings in volts is used to measure the difference in potential between two points in an electric circuit. Thus a *voltmeter is always connected in parallel with the part of a circuit across which the potential difference is to be measured.* Suppose the potential difference across a 1.5-V dry cell is to be measured. A d-c voltmeter with a range of 0 to 3 volts is more suitable than one with a higher range. For a commercial lighting circuit, an a-c voltmeter with a range of 0 to 250 volts provides a midscale readout across the circuit. Assume you have a multirange voltmeter with ranges of 0 to 3 V, 0 to 15 V, 0 to 30 V, and 0 to 150 V, and you have no idea of the potential difference across the circuit in which the meter is to be used. To protect the meter, begin with the highest range in the circuit and then adjust to a range that provides an approximate midscale readout.

Because a voltmeter is connected in parallel with a circuit, it acts to *load* the circuit. The range of the meter together with its built-in resistance determines its sensitivity in *ohms-per-volt*. If the meter has a 0- to 3-volt range and is rated at 1000 ohms per volt, it has the same effect as a resistor of 3000 ohms when placed across a circuit. In practice, the loading effect of the circuit should be held as low as possible (the meter resistance should be high compared with the circuit resistance). The meter resistance should be at least 10 times the resistance of the circuit across which it is connected in order to avoid an excessive change in the circuit constants. The higher the ohms-per-volt rating of the meter, the lower will be its loading effect on a circuit.

Ammeters

A commercial-grade ammeter may have a movement that consists of a coil pivoting between the poles of a permanent magnet. Because the coil resistance is very low, a shunt is connected across the meter terminals to protect the coil from excessive current. The instrument may have different shunts to provide different ranges.

An ammeter is connected in series in a circuit. If the approximate current magnitude in a circuit is not known, precautions should be taken when connecting an ammeter in the circuit to assure that the current does not exceed the range of the instrument. A rheostat can be connected in series with the meter to reduce the current in the meter. This resistance can then be removed gradually as it is determined that the meter range is greater than the magnitude of current in the circuit.

Suppose an ammeter with different shunts is available for use in the circuit. One shunt provides the meter with a range of 0 to 5 A, and the other shunt gives a range of 0 to 25 A. Start with the 25-A range in the circuit. If the current readout is less than 5 A, change to the 5-A range to secure the advantage of a midscale readout. *Never connect an ammeter in parallel with a circuit component or another instrument.*

Galvanometers

A galvanometer is used for detecting feeble electric currents, for determining their directional sense, and for indicating their relative magnitudes. Zero-centered galvanometers are used in Wheatstone-bridge measurements and for testing induced currents. A galvanometer movement is not protected by either a high resistance in series or a low resistance shunt in parallel. Thus, it may be used safely *only with very small currents.* Your instructor may direct that a resistance coil or a shunt be used with the instrument.

Rheostats and Resistance Coils

A rheostat can be used in a circuit to introduce a variable resistance for the control of current, to control the voltage across the load, or to protect certain instruments. It may be calibrated to serve as a measuring instrument.

Although a rheostat, resistance coil, and resistance box (set of resistance coils) are only slightly affected by changes in temperature, excessive heating must be avoided. The maximum current or the power dissipation for such a resistance element is usually given. The terms "rheostat" and "resistance box" are sometimes used interchangeably. However, in this text, the term "resistance box" designates a calibrated measuring instrument, and the term "rheostat" designates any variable resistance inserted into a circuit to control current. Resistance boxes are designated for use with *small* currents, as in Wheatstone-bridge circuits, and should not be used as current-limiting resistances in general circuit applications where rheostats would be appropriate.

Mathematics Refresher

1. *Introduction.*

The topics selected for this Mathematics Refresher are those that sometimes trouble physics students. You may want to study this section before working on certain types of physics problems. The presentation here is not as detailed as that given in mathematics textbooks, but there is sufficient review to help you perform certain important mathematical operations.

The following references are given for your convenience in reviewing these topics:

Section 2.8 defines *significant figures* and gives rules for their notation.

Section 2.9 explains the *scientific notation* system.

Section 2.12 describes the *orderly procedure* you should use *in problem solving*, and the proper method of handling units in computations.

2. *Conversion of a fraction to a decimal.*

Divide the numerator by the denominator, expressing the answer in the required number of significant figures.

Example

Convert $\frac{45}{85}$ to a decimal.

Solution

Dividing 45 by 85, we find the equivalent decimal to be 0.53, rounded to two significant figures.

```
     0.529
85)45.000
   42 5
    2 50
    1 70
      800
      765
```

3. *Calculation of percentage.*

Uncertainty always exists in measurements of physical phenomena. Uncertainty exists because measuring instruments have limited precision. The instruments you use in laboratory experiments generally have less precision than those used by the physicists who obtained the values found in tables. *The precision of any measurement is limited to the measurement detail that the instrument is capable of providing.* In physics, an error is defined as the difference between experimentally obtained data and the accepted values for these data. In experimental work, the *relative error* (percentage error) is usually more meaningful than the *absolute error* (the actual difference between an observed value and its accepted value). Hence, as a physics student you must be able to calculate percentages and percentage errors, and convert fractions to percentages and percentages to decimals. Examples of these calculations are given in the following sections.

4. *Conversion of fractions to percentages.*

To convert a fraction to a percentage, divide the numerator by the denominator and multiply the quotient by 100%.

Example

Express $\frac{11}{13}$ as a percentage.

Solution

$\frac{11}{13} \times 100\% = 85\%$, rounded to two significant figures.

5. *Conversion of percentages to decimals.*

To convert a percentage to a decimal, move the decimal point two places to the left, and remove the percent sign.

Example

Convert 62.5% to a decimal.

Solution

If we express 62.5% as a fraction, it becomes 62.5/100. If we actually perform the indicated division, we obtain 0.625 as the decimal equivalent. Observe that the only changes have been to move the decimal point two places to the left, and remove the percent sign.

6. *Calculation of relative error.*

$$\text{Relative Error} = \frac{\text{Absolute Error}}{\text{Accepted Value}} \times 100\%$$

where the *absolute error* is the difference between the *observed value* and the *accepted value*.

Example

In a laboratory experiment carried out at 20.0° C, a student found the speed of sound in air to be 329.8 m/s. The accepted value at this temperature is 343.5 m/s. What was the relative error?

Solution

$$\text{Absolute Error} = 343.5 \text{ m/s} - 329.8 \text{ m/s}$$
$$= 13.7 \text{ m/s}$$

$$\text{Relative Error} = \frac{13.7 \text{ m/s}}{343.5 \text{ m/s}} \times 100\% = 3.99\%$$

7. *Proportions.*

Many physics problems involving temperature, pressure, and volume relationships of gases may be solved by using proportions.

Example

Solve the proportion $\frac{V}{225 \text{ mL}} = \frac{273°\text{C}}{298°\text{C}}$ for V.

Solution

Multiply both sides of the equation by 225 mL, the denominator of V, which yields

$V = \frac{225 \text{ mL} \times 273°\text{C}}{298°\text{C}}$. Solving, $V = 206$ mL.

8. *Fractional Equations.*

The equations for certain problems involving lenses, mirrors, and electric resistances produce fractional equations where the unknown is in the denominator. To solve such equations, clear them of fractions by multiplying each term by the lowest common denominator. Then isolate the unknown and complete the solution.

Example

Solve $\frac{1}{d_o} + \frac{1}{d_i} = \frac{1}{f}$ for f.

Solution

The lowest common denominator of d_o, d_i, and f is d_od_if. Multiplying the fractional equation by this product, we obtain

$$\frac{d_od_if}{d_o} + \frac{d_od_if}{d_i} = \frac{d_od_if}{f} \qquad \text{or} \qquad d_if + d_of = d_od_i$$

$$\text{Thus, } f(d_o + d_i) = d_od_i \qquad \text{and} \qquad f = \frac{d_od_i}{d_o + d_i}$$

9. *Equations.*

When an equation is used in solving a problem and the unknown quantity is not the one isolated, the equation should be solved algebraically to isolate the unknown quantity required. Then the values (with units) of the known quantitites can be substituted and the indicated operations performed.

Example

From the equation for potential energy, $E_P = mgh$, we are to calculate h.

Solution

Before substituting known values for E_P, m, and g, the unknown term h is isolated and expressed in terms of E_P, m, and g. This is accomplished by dividing both sides of the basic equation by mg.

$$\frac{E_P}{mg} = \frac{mgh}{mg} \qquad \text{or} \qquad h = \frac{E_P}{mg}$$

If E_P is given in joules, m in kilograms, and g in meters/second^2, the unit of h is

$$h = \frac{\text{J}}{\text{kg}\cdot\text{m/s}^2} = \frac{\text{kg}\cdot\text{m}^2/\text{s}^2}{\text{kg}\cdot\text{m/s}^2} = \text{m} \qquad \text{Thus } h \text{ is expressed in meters.}$$

10. *Laws of exponents.*

Operations involving exponents may be expressed in general fashion as $a^m \times a^n = a^{m+n}$; $a^m \div a^n = a^{m-n}$; $(a^m)^n = a^{mn}$; $\sqrt[n]{a^m} = a^{m/n}$. For example:

$x^2 \times x^3 = x^5$ $\quad 10^5 \div 10^{-3} = 10^8$ $\quad (t^2)^3 = t^6$ $\quad \sqrt[4]{4^2} = 4^{2/4}\ 4^{1/2} = 2$

11. *Quadratic equation.*

The two roots of the quadratic equation $ax^2 + bx + c = 0$ in which a does not equal zero are given by the quadratic equation

$$x = \frac{-b \pm \sqrt{b^2 - 4ac}}{2a}$$

12. *Triangles.*

In physics, triangles are used to solve problems about forces and velocities.

1. *30°—60°—90° right triangle.* It is useful to remember that the hypotenuse of such a triangle is twice as long as the side opposite the 30° angle. The length of the side opposite the 60° angle is $l\sqrt{3}$, where l is the length of the side opposite the 30° angle.

2. *45°—45°—90° right triangle.* The sides opposite the 45° angles are equal. The length of the hypotenuse is $l\sqrt{2}$, where l is the length of a side.

3. *Trigonometric functions.* Trigonometric functions are ratios of the lengths of sides of a right triangle and depend on the magnitude of one of its acute angles. Using right triangle **ABC**, below, we define the following trigonometric functions of ∠**A**:

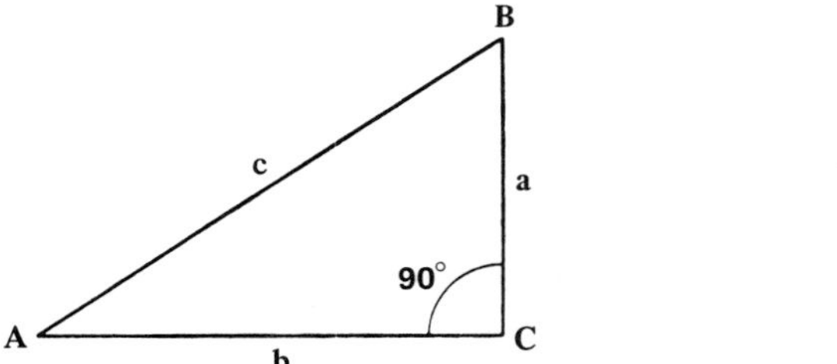

$$\text{sine } \angle \mathbf{A} = \frac{\mathbf{a}}{\mathbf{c}}$$

$$\text{cosine } \angle \mathbf{A} = \frac{\mathbf{b}}{\mathbf{c}}$$

$$\text{tangent } \angle \mathbf{A} = \frac{\mathbf{a}}{\mathbf{b}}$$

These functions are usually abbreviated as sin **A**, cos **A**, and tan **A**. Appendix B, Table 6, gives values of trigonometric functions.

Example

In a right triangle the length of one side is 25.3 cm and the length of the hypotenuse is 37.6 cm. Find the angle between these two sides.

Solution

Using the designations in the triangle shown above, **b** = 25.3 cm and **c** = 37.6 cm. You are to find ∠**A**. Hence you will use the trigonometric function cos **A** = **b**/**c**, or cos **A** = 25.3 cm/37.6 cm = 0.673. From Appendix B, Table 6, the cosine of 47.5° is 0.676 and the cosine of 48.0° is 0.669. The required angle is 3/7 × 0.5°, or 0.2°, greater than 47.5°. Thus ∠**A** = 47.7°.

Example

In a right triangle, one angle is 23.8° and the length of the adjacent side (not the hypotenuse) is 43.2 cm. Find the length of the other side.

Solution

Again using the designations in the above triangle, ∠**A** = 23.8° and **b** = 43.2 cm. The function involving these values and the unknown side, **a**, is tan **A** = **a**/**b**. Solving for **a**: **a** = **b** tan **A**. From Appendix B, Table 6, tan 23.8° = 0.441. Thus **a** = 43.2 cm × 0.441 = 19.1 cm.

4. *Sine law and cosine law.* For any triangle **ABC**, below, the sine law and the cosine law enable us to calculate the magnitudes of the remaining sides and angles if the magnitudes of one side and of any other two parts are given.

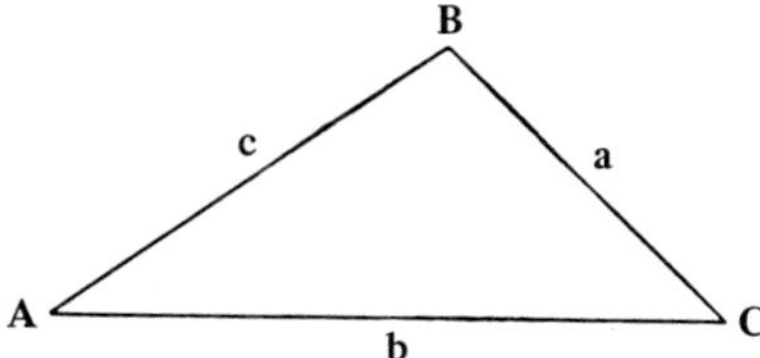

Sine law: $\frac{\mathbf{a}}{\sin \mathbf{A}} = \frac{\mathbf{b}}{\sin \mathbf{B}} = \frac{\mathbf{c}}{\sin \mathbf{C}}$

Cosine law: $\mathbf{a} = \sqrt{\mathbf{b}^2 + \mathbf{c}^2 - 2\mathbf{bc} \cos \mathbf{A}}$

$\mathbf{b} = \sqrt{\mathbf{c}^2 + \mathbf{a}^2 - 2\mathbf{ca} \cos \mathbf{B}}$

$\mathbf{c} = \sqrt{\mathbf{a}^2 + \mathbf{b}^2 - 2\mathbf{ab} \cos \mathbf{C}}$

13. *Circles.*

The circumference c of a circle of radius r or diameter d is

$$c = 2\pi r = \pi d$$

The area A of a circle of radius r or diameter d is

$$A = \pi r^2 = \pi d^2/4 = \tfrac{1}{4}\pi d^2$$

14. *Cylinders.*

The volume V of a right circular cylinder of height h and radius r or diameter d is

$$V = \pi r^2 h = \pi d^2 h/4 = \tfrac{1}{4}\pi d^2 h$$

15. *Spheres.*

The surface area A of a sphere of radius r or diameter d is

$$A = 4\pi r^2 = \pi d^2$$

The volume V of a sphere of radius r or diameter d is

$$V = \tfrac{4}{3}\pi r^3 = \tfrac{1}{6}\pi d^3$$

16. *Logarithms.*

The common logarithm of a number is the exponent or the power to which 10 must be raised in order to obtain the given number. A logarithm is composed of two parts: the *characteristic*, or integral part; and the *mantissa*, or decimal part. The characteristic of the logarithm *of any whole or mixed number* is one less than the number of digits to the left of its decimal point. The characteristic of the logarithm *of a decimal fraction* is always negative and is numerically one greater than the number of zeros immediately to the right of the decimal point. Mantissas are always positive and are read from tables such as Appendix B, Table 24. Proportional parts are used when numbers having four significant figures are involved. In determining the mantissa, the decimal point in the original number is ignored since its position is indicated by the characteristic.

Logarithms are exponents and follow the laws of exponents:

Logarithm of a product = sum of the logarithms of the factors

Logarithm of a quotient = logarithm of the dividend minus logarithm of the divisor

To find the number whose logarithm is given, determine the digits in the number from the table of mantissas. The characteristic indicates the position of the decimal point.

Example

Find the logarithm of 35.76.

Solution

There are two digits to the left of the decimal point. Therefore the characteristic of the logarithm of 35.76 is one less than two, or 1. To find the mantissa, ignore the decimal point and look up 3576 in Appendix B, Table 24. Read down the *n* column to 35. Follow this row across to the *7* column, where you will find 5527. This is the mantissa of 3570. Similarly, the mantissa for 3580 is 5539. To calculate the mantissa for 3576, take 6/10 of the difference between 5527 and 5539, or 7, and add it to 5527. Thus, the required mantissa is 5534, and the complete logarithm of 35.76 is 1.553 4.

Example

Find the logarithm of 0.469 2.

Solution

There are no zeros immediately to the right of the decimal point, hence the characteristic of the logarithm of 0.469 2 is −1. However, only the characteristic is negative; the mantissa is positive and is found from Appendix B, Table 24, to be 6714. The complete logarithm of 0.469 2 may be written as $\bar{1}.671\,4$, or 0.671 4 − 1, or 9.671 4 − 10.

Example

Find the logarithm of (1) 357 600 000 (2) 0.000 000 357 6

Solution

(1) $357\,600\,000 = 3.576 \times 10^8$ log = 8.553 4

(2) $0.000\,000\,357\,6 = 3.576 \times 10^{-7}$ log = $\bar{7}.553\,4$, or 0.553 4 − 7, or 3.553 4 − 10

Note: Trigonometric functions and logarithms can also be obtained with electronic calculators.

Appendix A—Equations

(The number indicates the section in which the equation is introduced.)

1.9 Mass Density

$$\text{mass density} = \frac{\text{mass}}{\text{volume}}$$

1.14 Mass-Energy Relationship

$$E = mc^2$$

E is energy; m is mass; c is speed of light.

2.6 Relative Error

$$E_r = \frac{E_a}{A} \times 100\%$$

E_r is relative error; E_a is absolute error; A is accepted value.

2.11 Vector addition—The resultant of two vectors acting at an angle of between 0° and 180° upon a given point is equal to the diagonal of a parallelogram of which the two vectors are sides.

3.2 Speed

$$\text{Average Speed} = \frac{\text{distance}}{\text{elapsed time}}$$

3.3 Velocity

$$v_{av} = \frac{d_f - d_i}{t_f - t_i} = \frac{\Delta d}{\Delta t}$$

v_{av} is average velocity; d_i is initial position at time t_i; d_f is final position at time t_f; Δd is displacement; Δt is elapsed time.

3.5 Acceleration

$$a_{av} = \frac{v_f - v_i}{t_f - t_i} = \frac{\Delta v}{\Delta t}$$

a_{av} is average accelaration; v_i is initial velocity at time t_i; v_f is final velocity at time t_f; Δv is change in velocity; Δt is elapsed time.

3.6 Uniformly Accelerated Motion

$$v_f = v_i + a\Delta t \quad \text{or} \quad v_f = v_i + g\Delta t$$

v_f is final velocity; v_i is initial velocity; a is acceleration or g is free-fall acceleration; Δt is elapsed time.

3.6 Accelerated Motion

$$\Delta d = v_i\Delta t + \tfrac{1}{2}a\Delta t^2 \quad \text{or} \quad \Delta d = v_i\Delta t + \tfrac{1}{2}g\Delta t^2$$

Δd is displacement; v_i is initial velocity; Δt is elapsed time; a is acceleration or g is free-fall acceleration.

3.6 Accelerated Motion

$$v_f = \sqrt{v_i^2 + 2a\Delta d} \quad \text{or} \quad v_f = \sqrt{v_i^2 + 2g\Delta d}$$

v_f is final velocity; v_i is initial velocity; a is acceleration or g is free-fall acceleration; Δd is displacement.

3.9 Newton's Second Law of Motion

$$F = ma$$

F is force; m is mass; a is acceleration.

3.9 Force and Acceleration on Bodies of Known Weight

$$F = \frac{F_w a}{g}$$

F is force; F_w is weight; a is acceleration; g is free-fall acceleration.

3.9 Weight and Mass

$$F_w = mg$$

F_w is weight; m is mass; g is free-fall acceleration.

3.11 Law of Universal Gravitation

$$F = \frac{G\, m_1 m_2}{d^2}$$

F is force of attraction; G is gravitational constant; m_1 and m_2 are masses of bodies; d is distance between their centers of mass.

3.12 Acceleration of Gravity

$$g = \frac{Gm_e}{d^2}$$

g is free-fall acceleration; G is gravitational constant; m_e is mass of earth; d is distance from center of earth.

4.7 Coefficient of Friction

$$\mu = \frac{F_f}{F_N}$$

μ is coefficient of friction; F_f is force of friction; F_N is force normal to surface.

5.2 Centripetal Force

$$F_c = \frac{mv^2}{r}$$

F_c is centripetal force; m is mass; v is velocity; r is radius of path.

5.3 Motion in Vertical Circle

$$v_{min} = \sqrt{rg}$$

v_{min} is critical velocity; r is radius of vertical circle; g is free fall acceleration.

5.6 Angular Velocity

$$\omega = \frac{\Delta\theta}{\Delta t}$$

ω is angular velocity; $\Delta\theta$ is angular displacement; Δt is elapsed time.

5.7 Angular Acceleration

$$\alpha = \frac{\Delta\omega}{\Delta t}$$

α is angular acceleration; $\Delta\omega$ is angular velocity; Δt is elapsed time.

5.7 Accelerated Rotary Motion

$$\omega_f = \omega_i + \alpha\Delta t$$
$$\Delta\theta = \omega_i\Delta t + \tfrac{1}{2}\alpha\Delta t^2$$
$$\omega_f = \sqrt{\omega_i^2 + 2\alpha\Delta\theta}$$

ω_f is final angular velocity; ω_i is initial angular velocity; α is angular acceleration; Δt is elapsed time; $\Delta\theta$ is angular displacement.

5.8 Rotational Inertia

$$T = I\alpha$$

T is torque; I is rotational inertia; α is angular acceleration.

5.8 Torque

$$T = Fr$$

T is torque; F is tangential force; r is distance from pivot point to point of application of force.

5.12 Pendulum

$$T = 2\pi\sqrt{\frac{l}{g}}$$

T is period; l is length; g is free-fall acceleration.

6.1 Work

$$W = F\Delta d$$

W is work; F is force; Δd is displacement.

6.3 Work (Rotary Motion)

$$W = T\Delta\theta$$

W is work; T is torque; $\Delta\theta$ is angular displacement.

6.4 Efficiency

$$\text{Efficiency} = \frac{W_{out}}{W_{in}}$$

W_{out} is work output; W_{in} is work input.

6.5 Power

$$P = \frac{F\Delta d}{\Delta t}$$

P is power; F is force; Δd is displacement; Δt is elapsed time.

6.6 Power (Rotary Motion)

$$P = T\omega$$

P is power; T is torque; ω is angular velocity.

6.7 Gravitational Potential Energy

$$E_P = mg\Delta h$$

E_P is gravitational potential energy; m is mass; g is free-fall acceleration; Δh is vertical distance.

6.8 Kinetic Energy

$$E_K = \tfrac{1}{2}mv^2$$

E_K is kinetic energy; m is mass; v is velocity.

6.9 Kinetic Energy (Rotary Motion)

$$E_K = \tfrac{1}{2}I\omega^2$$

E_K is kinetic energy; I is rotational inertia; ω is angular velocity.

6.10 Potential Energy (Elastic)

$$E_P = \tfrac{1}{2}k(\Delta d)^2$$

E_P is potential energy; k is elastic constant; Δd is displacement.

6.12 Impulse and Change of Momentum

$$F\Delta t = m\Delta v$$

F is force; Δt is elapsed time; the product $F\Delta t$ is impulse; m is mass; Δv is change of velocity; the product $m\Delta v$ is the change of momentum.

6.14 Elastic Collisions in One Dimension

$$mv_i + m'\,v'_i = mv_f + m'v'_f$$

m is mass of first body, v_i its initial velocity, v_f its final velocity; m' is mass of second body, v'_i its initial velocity, v'_f its final velocity.

6.15 Angular Impulse and Change of Angular Momentum

$$T\Delta t = I\omega_f - I\omega_i$$

T is torque; Δt is elapsed time; the product $T\Delta t$ is angular impulse; I is moment of inertia; ω_f is final angular velocity; ω_i is initial angular velocity; the product $I\omega$ is angular momentum.

7.11 Hooke's Law

$$Y = \frac{Fl}{\Delta lA}$$

Y is Young's modulus; F is distorting force; l is original length; Δl is change in length; A is cross-sectional area.

8.2 Kelvin Temperature

$$K = °C + 273°$$

K is Kelvin temperature; °C is Celsius temperature.

8.4 Linear Expansion

$$\Delta l = \alpha l\,\Delta T$$

Δl is change in length; α is coefficient of linear expansion; l is original length; ΔT is change in temperature.

8.5 Volume Expansion

$$\Delta V = \beta V\,\Delta T$$

ΔV is change in volume; β is coefficient of volume expansion; V is original volume; ΔT is change in temperature.

8.7 Charles' Law

$$\frac{V}{T_K} = \frac{V'}{T'_K}$$

V is original volume; T_K is original Kelvin temperature; V' is new volume; T'_K is new Kelvin temperature; provided pressure is constant.

8.8 Boyle's Law

$$pV = p'V'$$

p is original pressure; V is original volume; p' is new pressure; V' is new volume; provided temperature is constant.

8.8 Variation of Gas Density with Pressure

$$\frac{D}{D'} = \frac{p}{p'}$$

D is gas density at pressure p; D' is gas density at pressure p'.

8.9 Boyle's and Charles' Laws Combined

$$\frac{pV}{T_K} = \frac{p'V'}{T'_K}$$

p, V, and T_K are original pressure, volume, and Kelvin temperature, respectively; p', V', and T'_K are new pressure, volume, and Kelvin temperature, respectively.

8.10 Ideal Gas Equation

$$pV = nRT_K$$

p is pressure; V is volume; n is number of moles; R is universal gas constant; T_K is Kelvin temperature.

8.11 Heat Capacity

$$\text{heat capacity} = Q/\Delta T$$

Q is quantity of heat; ΔT is change in temperature.

8.12 Specific Heat

$$c = \frac{Q}{m\Delta T}$$

c is specific heat; Q is quantity of heat; m is mass of material; ΔT is change in temperature.

8.13 Heat Exchange

$$Q_{\text{Lost}} = Q_{\text{Gained}}$$

Q_{Lost} is heat lost; Q_{Gained} is heat gained.

9.6 Ideal Heat Engine Efficiency

$$\text{Efficiency} = \frac{T_1 - T_2}{T_1} \times 100\%$$

T_1 is input Kelvin temperature; T_2 is exhaust Kelvin temperature.

9.8 Change in Entropy

$$\Delta S = \Delta Q/T$$

ΔS is change in entropy; ΔQ is heat added or removed; T is Kelvin temperature.

10.6 Wave Equation

$$v = f\lambda$$

v is wave speed; f is frequency; λ is wavelength.

11.7 Intensity Level

$$\beta = 10 \log \frac{I}{I_0}$$

β is intensity level in db; I is intensity of sound; I_0 is intensity of threshold of hearing.

11.9 Doppler effect

1. *Source moving toward stationary listener.*

$$f_{LF} = f_S \frac{v}{v - v_S}$$

2. *Source moving away from stationary listener.*

$$f_{LB} = f_S \frac{v}{v + v_S}$$

3. *Listener moving toward stationary source.*

$$f_{LC} = f_S \frac{v + v_{LC}}{v}$$

4. *Listener moving away from stationary source.*

$$f_{LO} = f_S \frac{v - v_{LO}}{v}$$

f_{LF}, f_{LB}, f_{LC}, f_{LO} represent frequency of sound reaching the listener in the different listener-source conditions stated; f_s is the frequency of the source; v is the velocity of sound in the medium; v_s is the velocity of the source; v_{LC} is the closing velocity of the listener; v_{LO} is the opening velocity of the listener.

11.13 Laws of Strings

1. *Law of Lengths*

$$\frac{f}{f'} = \frac{l'}{l}$$

f and f' are frequencies corresponding to lengths l and l'.

2. *Law of Diameters*

$$\frac{f}{f'} = \frac{d'}{d}$$

f and f' are frequencies corresponding to diameters d and d'.

3. *Law of Tensions*

$$\frac{f}{f'} = \frac{\sqrt{F}}{\sqrt{F'}}$$

f and f' are frequencies corresponding to tensions F and F'.

4. *Law of Densities*

$$\frac{f}{f'} = \frac{\sqrt{D'}}{\sqrt{D}}$$

f and f' are frequencies corresponding to densities D and D'.

11.15 Resonance in Tubes

1. *Closed Tube*

$$\lambda = 4(l + 0.4d)$$

λ is wavelength; l is length of tube; d is diameter of tube.

2. *Open Tube*

$$\lambda = 2(l + 0.8d)$$

λ is wavelength; l is length of tube; d is diameter of tube.

11.16 Beats

$$f_{av} = \tfrac{1}{2}(f_L + f_H)$$

f_{av} is difference frequency; f_L is lower frequency; f_H is higher frequency.

12.9 Photon Energy

$$E = hf$$

E is energy; h is Planck's constant (6.63×10^{-34} joule-sec); f is frequency.

12.10 Photoelectric Equation

$$\tfrac{1}{2}mv^2_{max} = hf - w$$

$\tfrac{1}{2}mv^2_{max}$ is maximum kinetic energy of photoelectrons; h is Planck's constant; f is frequency of impinging radiation; w is work function of emitting material.

12.11 Photon Wavelength

$$\lambda = \frac{h}{mc}$$

λ is photon wavelength; h is Planck's constant; c is velocity of light; mc is photon momentum.

12.11 Wavelength of Particle Having a Velocity v

$$\lambda = \frac{h}{mv}$$

λ is particle wavelength; h is Planck's constant; m is mass of particle; v is velocity; (mv is particle momentum).

12.17 Illumination

1. *Uniformly Illuminated Surface*

$$E = \frac{F}{A}$$

E is illuminance; F is luminous flux; A is area uniformly illuminated.

2. *Surface Perpendicular to Luminous Flux*

$$E = \frac{I}{r^2}$$

E is illuminance; I is intensity of source; r is the distance from source to surface perpendicular to the beam.

3. *Surface at Any Angle to Luminous Flux*

$$E = \frac{I \cos \theta}{r^2}$$

E is illuminance; I is intensity of source; θ is the angle which the light beam makes with the normal to the illuminated surface; r is the distance from source to surface.

13.12 Focal Length of Mirrors and Lenses

$$\frac{1}{f} = \frac{1}{d_o} + \frac{1}{d_i}$$

f is focal length; d_o is object distance; d_i is image distance.

13.12 Images in Mirrors and Lenses

$$\frac{h_i}{h_o} = \frac{d_i}{d_o}$$

h_o is object size; h_i is image size; d_o is object distance; d_i is image distance.

14.3 Snell's Law

$$n = \frac{\sin i}{\sin r}$$

n is index of refraction; i is the angle of incidence; r is the angle of refraction.

14.12 Simple Magnifier

$$M = \frac{25 \text{ cm}}{f} \text{ (approx.)}$$

f is focal length of lens in cm.

14.13 Compound Microscope

$$M = \frac{25 \text{ cm} \times 1}{f_E \times f_o} \text{ (approx.)}$$

l is length of tube; f_E is focal length of eyepiece lens; f_o is focal length of objective lens in cm.

15.4 Diffraction Grating Equation

$$\lambda = \frac{d \sin \theta_n}{n}$$

λ is wavelength; d is grating constant; θ is diffraction angle; n is the order of image.

16.8 Coulomb's Law of Electrostatics

$$F = k\frac{Q_1 Q_2}{d^2}$$

F is force between two point charges; k is a proportionality constant; Q_1 and Q_2 are the two charges; d is distance separating the charges.

16.9 Electric Field Intensity

$$E = \frac{F}{q}$$

E is the electric field intensity; F is force acting on a charge; q is quantity of charge.

16.10 Potential Difference

$$V = \frac{W}{q}$$

V is potential difference; W is work done in moving a charge; q is quantity of charge moved.

16.14 Capacitance of a Capacitor

$$C = \frac{Q}{V}$$

C is capacitance of a capacitor; Q is charge on either plate; V is potential difference between the plates.

16.17 Capacitors in Parallel

$$C_T = C_1 + C_2 + C_3 + \text{etc.}$$

C_T is total capacitance; C_1, C_2, C_3, etc. are separate capacitances connected in parallel.

16.17 Capacitors in Series

$$\frac{1}{C_T} = \frac{1}{C_1} + \frac{1}{C_2} + \frac{1}{C_3} + \text{etc.}$$

C_T is total capacitance; C_1, C_2, C_3, etc. are separate capacitances connected in series.

17.1 Electric Current

$$I = \frac{Q}{t}$$

I is current; Q is quantity of charge; t is time.

17.6 Ohm's Law of Resistance

1. *Entire Circuit Including a Source of emf*

$$\mathscr{E} = IR \quad \text{or} \quad \mathscr{E} = IZ$$

$\mathscr{E}$ is emf of source; I is current in the circuit; R is resistance of the circuit; Z is impedance of a-c circuit.

2. *Any Part of a Circuit That Does Not Include a Source of emf*

$$V = IR \quad \text{or} \quad V = IZ$$

V is potential difference across a part of the circuit; I is the current in that part of the circuit; R is the resistance of that part of the circuit; Z is the impedance of that part of an a-c circuit.

17.8 Resistances in Series

$$R_T = R_1 + R_2 + R_3 + \text{etc.}$$

R_T is the total resistance; R_1, R_2, R_3, etc. are separate resistances connected in series.

17.9 Resistances in Parallel

$$\frac{1}{R_{EQ}} = \frac{1}{R_1} + \frac{1}{R_2} + \frac{1}{R_3} + \text{etc.}$$

R_{EQ} is equivalent resistance; R_1, R_2, R_3, etc. are separate resistances connected in parallel.

17.11 Law of Resistance

$$R = \rho\frac{l}{A}$$

R is resistance; ρ is resistivity, a proportionality constant; l is length; A is cross-sectional area.

18.3 Joule's Law

$$Q = I^2Rt$$

Q is quantity of heat energy; I is current; R is resistance; t is time.

18.4 Electric Power

1. *Total Power Consumed in a Circuit*

$$P_T = I_T{}^2R_T \quad \text{or} \quad P_T = \mathscr{E}I \cos \phi$$

P_T is total power; I_T is total current; R_T is total resistance; $\mathscr{E}$ is emf; ϕ is phase angle between current and voltage in an a-c circuit.

2. *Power Expended in the Circuit Load*

$$P_L = I^2R_L \quad \text{or} \quad P_L = I^2Z_L$$

P_L is load power; I is current; R_L is load resistance; Z_L is load impedance of an a-c circuit.

3. *Power Dissipated in a Source of emf*

$$P_r = I^2r$$

P_r is power dissipated in the internal resistance of source; I is current; r is internal resistance of source.

18.9 Faraday's Laws of Electrolysis

$$m = zIt$$

m is mass; z is the electrochemical equivalent; I is current; t is time.

19.4 Magnetic Flux Density

$$B = \Phi/A$$

B is magnetic flux density; Φ is magnetic flux; A is area.

20.3 Induced emf *(Coil in a Magnetic Field)*

$$\mathscr{E} = -N\frac{\Delta\Phi}{\Delta t}$$

$\mathscr{E}$ is induced emf; N is number of turns; $\Delta\Phi/\Delta t$ is the change in flux linkage in a given interval of time.

20.4 Induced emf *(Conductor in a Magnetic Field)*

$$\mathscr{E} = Blv$$

$\mathscr{E}$ is induced emf; B is flux density of the magnetic field; l is length of conductor; v is velocity of conductor across magnetic field.

20.9 Instantaneous Voltage

$$e = \mathscr{E}_{max} \sin \theta$$

e is instantaneous voltage; $\mathscr{E}_{max}$ is maximum voltage; θ is displacement angle, the angle between the plane of the conducting loop and the perpendicular to the magnetic flux.

20.9 Instantaneous Current

$$i = I_{max} \sin \theta$$

i is instantaneous current; I_{max} is maximum current; θ is displacement angle, the angle between the plane of the conducting loop and the perpendicular to the magnetic flux.

20.14 The Motor Effect

$$T = Fw \cos \alpha$$

T is torque; F is magnetic force acting on either conductor of loop; w is width of conducting loop; α is the angle between the plane of the loop and the magnetic flux.

20.18 Mutual Inductance

$$M = \frac{-\mathscr{E}_S}{\Delta I_P/\Delta t}$$

M is mutual inductance; $\mathscr{E}_S$ is average induced emf across the secondary; $\Delta I_P/\Delta t$ is time rate of change of current in the primary.

20.19 Self-inductance

$$L = \frac{-\mathscr{E}}{\Delta I/\Delta t}$$

L is self-inductance; $\mathscr{E}$ is the average emf induced across the coil; $\Delta I/\Delta t$ is time rate of change of current in the coil.

20.20 Inductors in Series

1. *No Mutual Inductance Between Inductors*

$$L_T = L_1 + L_2 + L_3 + \text{etc.}$$

L_T is total inductance; L_1, L_2, L_3, etc. are separate inductances connected in series.

2. *Mutual Inductance Between Two Inductors*

$$L_T = L_1 + L_2 \pm 2M$$

L_T is total inductance; L_1 and L_2 are inductances connected in series; M is mutual inductance between them.

20.20 Inductors in Parallel

$$\frac{1}{L_T} = \frac{1}{L_1} + \frac{1}{L_2} + \frac{1}{L_3} + \text{etc.}$$

L_T is total inductance; L_1, L_2, L_3, etc. are separate inductances connected in parallel so that each is unaffected by the magnetic field of the other.

20.21 Transformer

1. *Voltage Ratio*

$$\frac{V_S}{V_P} = \frac{N_S}{N_P}$$

V_S and V_P are secondary and primary terminal voltages; N_S/N_P is the turns ratio, secondary to primary.

2. *Current Ratio*

$$\frac{I_P}{I_S} = \frac{N_S}{N_P}$$

I_P and I_S are primary and secondary currents; N_S/N_P is the turns ratio, secondary to primary.

21.4 Power Factor

$$\text{pf} = \cos\phi$$

pf is power factor; ϕ is phase angle between voltage and current.

21.5 Inductive Reactance

$$X_L = 2\pi f L$$

X_L is inductive reactance; f is frequency; L is inductance.

21.8 Capacitive Reactance

$$X_C = \frac{1}{2\pi f C}$$

X_C is capacitive reactance; f is frequency; C is capacitance.

21.9 Impedance

$$Z = \sqrt{R^2 + X^2}\ \underline{|\arctan X/R}$$

Z is impedance; R is resistance; X is reactance $(X_L - X_C)$.

21.12 Resonance

$$f_R = \frac{1}{2\pi\sqrt{LC}}$$

f_R is resonant frequency; L is inductance; C is capacitance.

22.15 Transistor Characteristics

1. *Current gain, common-base circuit*

$$\alpha = \frac{\Delta I_C}{\Delta I_E}\ (V_{CB}\ \text{constant})$$

α is current gain; ΔI_C is change in collector current; ΔI_E is change in emitter current.

2. *Current gain, common-emitter circuit*

$$\beta = \frac{\Delta I_C}{\Delta I_B}\ (V_{CE}\ \text{constant})$$

β is current gain; ΔI_C is change in collector current; ΔI_B is change in base current.

23.10 Mass Number

$$A = Z + N$$

A is mass number of an atom; Z is atomic number; N is number of neutrons.

24.5 Decay Constant

$$\lambda = \frac{0.693}{T_{1/2}}$$

λ is decay constant; $T_{1/2}$ is half-life.

25.2 Energy of Electron

$$E = -\frac{me^4}{8\epsilon_0{}^2h^2n^2}$$

E is energy; m is mass; e is electronic charge; ϵ_0 is a constant; h is Planck's constant; n is principal quantum number.

Appendix B—Tables

Table 1 PREFIXES OF THE METRIC SYSTEM

Factor	Prefix	Symbol
10^{18}	exa	E
10^{15}	peta	P
10^{12}	tera	T
10^{9}	giga	G
10^{6}	mega	M
10^{3}	kilo	k
10^{2}	hecto	h
10	deka	da
10^{-1}	deci	d
10^{-2}	centi	c
10^{-3}	milli	m
10^{-6}	micro	μ
10^{-9}	nano	n
10^{-12}	pico	p
10^{-15}	femto	f
10^{-18}	atto	a

Table 2 GREEK ALPHABET

Greek letter	Greek name	English equivalent	Greek letter	Greek name	English equivalent
Α α	alpha	ä	Ν ν	nu	n
Β β	beta	b	Ξ ξ	xi	ks
Γ γ	gamma	g	Ο o	omicron	o
Δ δ	delta	d	Π π	pi	p
Ε ϵ	epsilon	e	Ρ ρ	rho	r
Ζ ζ	zeta	z	Σ σ	sigma	s
Η η	eta	ā	Τ τ	tau	t
Θ θ	theta	th	Υ υ	upsilon	ü, ōō
Ι ι	iota	ē	Φ ϕ	phi	f
Κ κ	kappa	k	Χ χ	chi	h
Λ λ	lambda	l	Ψ ψ	psi	ps
Μ μ	mu	m	Ω ω	omega	ō

Table 3 PHYSICAL CONSTANTS

Quantity	Symbol	Value
atmospheric pressure, normal	atm	$1.013\,25 \times 10^{5}$ Pa
atomic mass unit	u	$1.660\,540\,2 \times 10^{-27}$ kg
Avogadro number	N_A	$6.022\,136\,7 \times 10^{23}$/mole
charge to mass ratio for electron	e/m_e	$1.758\,819\,62 \times 10^{11}$ C/kg
electron rest mass	m_e	$9.109\,389\,7 \times 10^{-31}$ kg
		$5.485\,799\,03 \times 10^{-4}\ u$
electron volt	eV	$1.602\,177\,33 \times 10^{-19}$ J
electrostatic constant	k	8.987×10^{9} N·m²/C²
elementary charge	e	$1.602\,177\,33 \times 10^{-19}$ C
faraday	F	$9.648\,530\,9 \times 10^{4}$ C/mole
gas constant, universal	R	6.236×10^{4} mm·cm³/mole·K
		$8.205\,68 \times 10^{-2}$ L·atm/mole·K
		$8.314\,510 \times 10^{0}$ J/mole·K
gravitational acceleration, standard	g	$9.806\,65 \times 10^{0}$ m/s²
mechanical equivalent of heat	J	$4.186\,8 \times 10^{0}$ J/cal
molar volume of ideal gas at STP	V_m	$2.241\,410 \times 10^{1}$ L/mole
neutron rest mass	m_n	$1.674\,928\,6 \times 10^{-27}$ kg
		$1.008\,664\,904 \times 10^{0}\ u$
Planck's constant	h	$6.626\,075\,5 \times 10^{-34}$ J·s
proton rest mass	m_p	$1.672\,623\,1 \times 10^{-27}$ kg
		$1.007\,276\,470 \times 10^{0}\ u$
speed of light in a vacuum	c	$2.997\,924\,58 \times 10^{8}$ m/s
speed of sound in air at STP	v	$3.314\,5 \times 10^{2}$ m/s
temperature of the triple point of water	—	$2.731\,6 \times 10^{2}$ K (0.01°C)
universal gravitational constant	G	$6.672\,59 \times 10^{-11}$ N·m²/kg²

Table 4 SELECTED PHYSICAL QUANTITIES AND MEASUREMENT UNITS

Physical quantity	Quantity symbol	Measurement unit	Unit symbol	Unit dimensions
		Fundamental (base) units		
length	l	meter	m	m
mass	m	kilogram	kg	kg
time	t	second	s	s
electric charge*	Q	coulomb	C	C
temperature	T	kelvin	K	K
amount of substance	n	mole	mol	mol
luminous intensity	I	candela	cd	cd
		Derived units		
acceleration	a	meter per second per second	m/s^2	m/s^2
area	A	square meter	m^2	m^2
capacitance	C	farad	F	$C^2 \cdot s^2/kg \cdot m^2$
density	D	kilogram per cubic meter	kg/m^3	kg/m^3
electric current	I	ampere	A	C/s
electric field intensity	E	newton per coulomb	N/C	$kg \cdot m/C \cdot s^2$
electric resistance	R	ohm	Ω	$kg \cdot m^2/C^2 \cdot s$
emf	$\mathcal{E}$	volt	V	$kg \cdot m^2/C \cdot s^2$
energy	E	joule	J	$kg \cdot m^2/s^2$
force	F	newton	N	$kg \cdot m/s^2$
frequency	f	hertz	Hz	s^{-1}
heat	Q	joule	J	$kg \cdot m^2/s^2$
illuminance	E	lumen per square meter (lux)	lm/m^2 (lx)	cd/m^2
inductance	L	henry	H	$kg \cdot m^2/C^2$
luminous flux	F	lumen	lm	cd · sr
magnetic flux	Φ	weber	Wb	$kg \cdot m^2/C \cdot s$
magnetic flux density	B	weber per square meter (tesla)	Wb/m^2 (T)	$kg/C \cdot s$
potential difference	V	volt	V	$kg \cdot m^2/C \cdot s^2$
power	P	watt	W	$kg \cdot m^2/s^3$
pressure	p	newton per square meter (pascal)	N/m^2 (Pa)	$kg/m \cdot s^2$
velocity	v	meter per second	m/s	m/s
volume	V	cubic meter	m^3	m^3
work	W	joule	J	$kg \cdot m^2/s^2$

*The SI fundamental (base) electric unit is the ampere. For pedagogical reasons the coulomb of electric charge is introduced in this book as the fundamental unit and the ampere as a derived unit having the dimensions, coulomb per second.

Table 5 CONVERSION FACTORS

Length

$1\text{ m} = 10^{-3}\text{ km} = 10^{2}\text{ cm} = 10^{3}\text{ mm} = 10^{6}\ \mu\text{m} = 10^{9}\text{ nm} = 10^{10}\text{ Å}$
$1\ \mu = 1\ \mu\text{m} = 10^{-6}\text{ m} = 10^{-4}\text{ cm} = 10^{-3}\text{ mm} = 10^{3}\ \mu\text{m} = 10^{3}\text{ nm} = 10^{4}\text{ Å}$
$1\ \mu\text{m} = 1\text{ nm} = 10^{-9}\text{ m} = 10^{-7}\text{ cm} = 10^{-6}\text{ mm} = 10\text{ Å}$
$1\text{ Å} = 10^{-10}\text{ m} = 10^{-8}\text{ cm} = 10^{-7}\text{ mm} = 10^{-4}\ \mu\text{m} = 10^{-1}\text{ nm}$

Area

$1\text{ m}^2 = 10^{-6}\text{ km}^2 = 10^{4}\text{ cm}^2 = 10^{6}\text{ mm}^2$

Volume

$1\text{ m}^3 = 10^{-9}\text{ km}^3 = 10^{3}\text{ L} = 10^{6}\text{ cm}^3$
$1\text{ L} = 10^{3}\text{ mL} = 10^{3}\text{ cm}^3 = 10^{-3}\text{ m}^3$

Angular

$1° = 1.74 \times 10^{-2}\text{ radian} = 2.78 \times 10^{-3}\text{ revolution}$
$1\text{ radian} = 57.3° = 1.59 \times 10^{-1}\text{ revolution}$
$1\text{ revolution} = 360° = 6.28\text{ radians}$

Mass

$1\text{ kg} = 10^{3}\text{ g} = 10^{6}\text{ mg} = 6.02 \times 10^{26}\ u$
$1\text{ g} = 10^{-3}\text{ kg} = 10^{3}\text{ mg} = 6.02 \times 10^{23}\ u$
$1\ u = 1.66 \times 10^{-24}\text{ g} = 1.66 \times 10^{-21}\text{ mg} = 1.66 \times 10^{-27}\text{ kg}$

Time

$1\text{ h} = 60\text{ min} = 3.6 \times 10^{3}\text{s}$
$1\text{ min} = 60\text{ s} = 1.67 \times 10^{-2}\text{ h}$
$1\text{ s} = 1.67 \times 10^{-2}\text{ min} = 2.78 \times 10^{-4}\text{ h}$

Velocity

$1\text{ km/h} = 10^{3}\text{ m/h} = 16.7\text{ m/min} = 2.78 \times 10^{-1}\text{ m/s}$
$1\text{ m/min} = 10^{2}\text{ cm/min} = 1.67 \times 10^{-2}\text{ m/s} = 1.67\text{ cm/s}$
$1\text{ m/s} = 10^{-3}\text{ km/s} = 3.6\text{ km/h} = 10^{2}\text{ cm/s}$

Acceleration

$1\text{ cm/s}^2 = 10^{-2}\text{ m/s}^2 = 10^{-5}\text{ km/s}^2$
$1\text{ m/s}^2 = 10^{2}\text{ cm/s}^2 = 10^{-3}\text{ km/s}^2$
$1\text{ km/h/s} = 10^{3}\text{ m/h/s} = 2.78 \times 10^{-1}\text{ m/s}^2 = 2.78 \times 10^{1}\text{ cm/s}^2 = 2.78 \times 10^{2}\text{ mm/s}^2$

Pressure

$1\text{ atm} = 760.00\text{ mm Hg} = 1.013 \times 10^{5}\text{ N/m}^2 = 1.013 \times 10^{5}\text{ Pa} = 1.013 \times 10^{2}\text{ kPa}$
$1\text{ N/m}^2 = 10\text{ dynes/cm}^2 = 9.87 \times 10^{-6}\text{ atm} = 1\text{ Pa} = 10^{-3}\text{ kPa} = 7.50 \times 10^{-3}\text{ mm Hg}$

Energy

$1\text{ J} = 2.39 \times 10^{-1}\text{ cal} = 2.39 \times 10^{-4}\text{ kcal} = 2.78 \times 10^{-7}\text{ kW·h} = 6.25 \times 10^{18}\text{ eV}$
$1\text{ cal} = 10^{-3}\text{ kcal} = 4.19\text{ J} = 1.16 \times 10^{-6}\text{ kW·h}$
$1\text{ kcal} = 10^{3}\text{ cal} = 4.19 \times 10^{3}\text{ J} = 1.16 \times 10^{-3}\text{ kW·h}$
$1\text{ eV} = 10^{-6}\text{ MeV} = 1.60 \times 10^{-19}\text{ J}$
$1\text{ kW·h} = 10^{3}\text{ W·h} = 3.6 \times 10^{3}\text{ kW·s} = 3.6 \times 10^{6}\text{ W·s} = 8.6 \times 10^{5}\text{ cal}$
$1\text{ W·s} = 2.78 \times 10^{-4}\text{ W·h} = 2.78 \times 10^{-7}\text{ kW·h}$

Mass-Energy

$1\text{ J} = 1.11 \times 10^{-17}\text{ kg} = 1.11 \times 10^{-14}\text{ g} = 6.69 \times 10^{9}\ u$
$1\text{ eV} = 1.07 \times 10^{-9}\ u = 1.78 \times 10^{-33}\text{ g}$
$1\ u = 1.49 \times 10^{-10}\text{ J} = 931\text{ MeV} = 9.31 \times 10^{8}\text{ eV}$
$1\text{ kg} = 9.00 \times 10^{16}\text{ J}$

Table 6 NATURAL TRIGONOMETRIC FUNCTIONS

Angle (°)	Sine	Cosine	Tangent	Angle (°)	Sine	Cosine	Tangent
0.0	0.000	1.000	0.000				
0.5	0.009	1.000	0.009	**23.0**	0.391	0.921	0.424
1.0	0.017	1.000	0.017	**23.5**	0.399	0.917	0.435
1.5	0.026	1.000	0.026	**24.0**	0.407	0.914	0.445
2.0	0.035	0.999	0.035	**24.5**	0.415	0.910	0.456
2.5	0.044	0.999	0.044	**25.0**	0.423	0.906	0.466
3.0	0.052	0.999	0.052	**25.5**	0.431	0.903	0.477
3.5	0.061	0.998	0.061	**26.0**	0.438	0.899	0.488
4.0	0.070	0.998	0.070	**26.5**	0.446	0.895	0.499
4.5	0.078	0.997	0.079	**27.0**	0.454	0.891	0.510
5.0	0.087	0.996	0.087	**27.5**	0.462	0.887	0.521
5.5	0.096	0.995	0.096	**28.0**	0.470	0.883	0.532
6.0	0.104	0.995	0.105	**28.5**	0.477	0.879	0.543
6.5	0.113	0.994	0.114	**29.0**	0.485	0.875	0.554
7.0	0.122	0.992	0.123	**29.5**	0.492	0.870	0.566
7.5	0.131	0.991	0.132	**30.0**	0.500	0.866	0.577
8.0	0.139	0.990	0.141	**30.5**	0.508	0.862	0.589
8.5	0.148	0.989	0.149	**31.0**	0.515	0.857	0.601
9.0	0.156	0.988	0.158	**31.5**	0.522	0.853	0.613
9.5	0.165	0.986	0.167	**32.0**	0.530	0.848	0.625
10.0	0.174	0.985	0.176	**32.5**	0.537	0.843	0.637
10.5	0.182	0.983	0.185	**33.0**	0.545	0.839	0.649
11.0	0.191	0.982	0.194	**33.5**	0.552	0.834	0.662
11.5	0.199	0.980	0.204	**34.0**	0.559	0.829	0.674
12.0	0.208	0.978	0.213	**34.5**	0.566	0.824	0.687
12.5	0.216	0.976	0.222	**35.0**	0.574	0.819	0.700
13.0	0.225	0.974	0.231	**35.5**	0.581	0.814	0.713
13.5	0.233	0.972	0.240	**36.0**	0.588	0.809	0.726
14.0	0.242	0.970	0.249	**36.5**	0.595	0.804	0.740
14.5	0.250	0.968	0.259	**37.0**	0.602	0.799	0.754
15.0	0.259	0.966	0.268	**37.5**	0.609	0.793	0.767
15.5	0.267	0.964	0.277	**38.0**	0.616	0.788	0.781
16.0	0.276	0.961	0.287	**38.5**	0.622	0.783	0.795
16.5	0.284	0.959	0.296	**39.0**	0.629	0.777	0.810
17.0	0.292	0.956	0.306	**39.5**	0.636	0.772	0.824
17.5	0.301	0.954	0.315	**40.0**	0.643	0.766	0.839
18.0	0.309	0.951	0.325	**40.5**	0.649	0.760	0.854
18.5	0.317	0.948	0.335	**41.0**	0.656	0.755	0.869
19.0	0.326	0.946	0.344	**41.5**	0.663	0.749	0.885
19.5	0.334	0.943	0.354	**42.0**	0.669	0.743	0.900
20.0	0.342	0.940	0.364	**42.5**	0.676	0.737	0.916
20.5	0.350	0.937	0.374	**43.0**	0.682	0.731	0.932
21.0	0.358	0.934	0.384	**43.5**	0.688	0.725	0.949
21.5	0.366	0.930	0.394	**44.0**	0.695	0.719	0.966
22.0	0.375	0.927	0.404	**44.5**	0.701	0.713	0.983
22.5	0.383	0.924	0.414	**45.0**	0.707	0.707	1.000

Table 6 NATURAL TRIGONOMETRIC FUNCTIONS (cont'd)

Angle (°)	Sine	Cosine	Tangent	Angle (°)	Sine	Cosine	Tangent
45.5	0.713	0.701	1.018	**68.0**	0.927	0.375	2.475
46.0	0.719	0.695	1.036	**68.5**	0.930	0.366	2.539
46.5	0.725	0.688	1.054	**69.0**	0.934	0.358	2.605
47.0	0.731	0.682	1.072	**69.5**	0.937	0.350	2.675
47.5	0.737	0.676	1.091	**70.0**	0.940	0.342	2.747
48.0	0.743	0.669	1.111	**70.5**	0.943	0.334	2.824
48.5	0.749	0.663	1.130	**71.0**	0.946	0.326	2.904
49.0	0.755	0.656	1.150	**71.5**	0.948	0.317	2.983
49.5	0.760	0.649	1.171	**72.0**	0.951	0.309	3.078
50.0	0.766	0.643	1.192	**72.5**	0.954	0.301	3.172
50.5	0.772	0.636	1.213	**73.0**	0.956	0.292	3.271
51.0	0.777	0.629	1.235	**73.5**	0.959	0.284	3.376
51.5	0.783	0.622	1.257	**74.0**	0.961	0.276	3.487
52.0	0.788	0.616	1.280	**74.5**	0.964	0.267	3.606
52.5	0.793	0.609	1.303	**75.0**	0.966	0.259	3.732
53.0	0.799	0.602	1.327	**75.5**	0.968	0.250	3.867
53.5	0.804	0.595	1.351	**76.0**	0.970	0.242	4.011
54.0	0.809	0.588	1.376	**76.5**	0.972	0.233	4.165
54.5	0.814	0.581	1.402	**77.0**	0.974	0.225	4.331
55.0	0.819	0.574	1.428	**77.5**	0.976	0.216	4.511
55.5	0.824	0.566	1.455	**78.0**	0.978	0.208	4.705
56.0	0.829	0.559	1.483	**78.5**	0.980	0.199	4.915
56.5	0.834	0.552	1.511	**79.0**	0.982	0.191	5.145
57.0	0.839	0.545	1.540	**79.5**	0.983	0.182	5.396
57.5	0.843	0.537	1.570	**80.0**	0.985	0.174	5.671
58.0	0.848	0.530	1.600	**80.5**	0.986	0.165	5.976
58.5	0.853	0.522	1.632	**81.0**	0.988	0.156	6.314
59.0	0.857	0.515	1.664	**81.5**	0.989	0.148	6.691
59.5	0.862	0.508	1.698	**82.0**	0.990	0.139	7.115
60.0	0.866	0.500	1.732	**82.5**	0.991	0.131	7.596
60.5	0.870	0.492	1.767	**83.0**	0.992	0.122	8.144
61.0	0.875	0.485	1.804	**83.5**	0.994	0.113	8.777
61.5	0.879	0.477	1.842	**84.0**	0.994	0.104	9.514
62.0	0.883	0.470	1.881	**84.5**	0.995	0.096	10.38
62.5	0.887	0.462	1.921	**85.0**	0.996	0.087	11.43
63.0	0.891	0.454	1.963	**85.5**	0.997	0.078	12.71
63.5	0.895	0.446	2.006	**86.0**	0.998	0.070	14.30
64.0	0.899	0.438	2.050	**86.5**	0.998	0.061	16.35
64.5	0.903	0.431	2.097	**87.0**	0.999	0.052	19.08
65.0	0.906	0.423	2.145	**87.5**	0.999	0.044	22.90
65.5	0.910	0.415	2.194	**88.0**	0.999	0.035	28.64
66.0	0.914	0.407	2.246	**88.5**	1.000	0.026	38.19
66.5	0.917	0.399	2.300	**89.0**	1.000	0.017	57.29
67.0	0.921	0.391	2.356	**89.5**	1.000	0.009	114.6
67.5	0.924	0.383	2.414	**90.0**	1.000	0.000	. . .

Table 7 TENSILE STRENGTH OF METALS

Metal	Tensile strength (N/m^2)
aluminum wire	2.4×10^8
copper wire, hard drawn	4.8×10^8
iron wire, annealed	3.8×10^8
iron wire, hard drawn	6.9×10^8
lead, cast or drawn	2.1×10^8
platinum wire	3.5×10^8
silver wire	2.9×10^8
steel (minimum)	2.8×10^8
steel wire (maximum)	32×10^8

Table 8 YOUNG'S MODULUS

Metal	Elastic modulus (N/m^2)
aluminum, 99.3%, rolled	6.96×10^{10}
brass	9.02×10^{10}
copper, wire, hard drawn	11.6×10^{10}
gold, pure, hard drawn	7.85×10^{10}
iron, case	9.1×10^{10}
iron, wrought	19.3×10^{10}
lead, rolled	1.57×10^{10}
platinum, pure, drawn	16.7×10^{10}
silver, hard drawn	7.75×10^{10}
steel, 0.38% C, annealed	20.0×10^{10}
tungsten, drawn	35.5×10^{10}

Table 9 RELATIVE DENSITY AND VOLUME OF WATER

The mass of one cm^3 of water at 4°C is taken as unity.
The values given are numerically equal to the absolute density in g/mL.

T (°C)	D (g/cm^3)	V (cm^3/g)	T (°C)	D (g/cm^3)	V (cm^3/g)
−10	0.998 15	1.001 86	45	0.990 25	1.009 85
−5	0.999 30	1.000 70	50	0.998 07	1.012 07
0	0.999 87	1.000 13	55	0.985 73	1.014 48
+1	0.999 93	1.000 07	60	0.983 24	1.017 05
2	0.999 97	1.000 03	65	0.980 59	1.019 79
3	0.999 99	1.000 01	70	0.977 81	1.022 70
4	1.000 00	1.000 00	75	0.974 89	1.025 76
5	0.999 99	1.000 01	80	0.971 83	1.028 99
+10	0.999 73	1.000 27	85	0.968 65	1.032 37
15	0.999 13	1.000 87	90	0.965 34	1.035 90
20	0.998 23	1.001 77	95	0.961 92	1.039 59
25	0.997 07	1.002 94	100	0.958 38	1.043 43
30	0.995 67	1.004 35	110	0.951 0	1.051 5
35	0.994 06	1.005 98	120	0.943 4	1.060 1
40	0.992 24	1.007 82	150	0.917 3	1.090 2

Table 10 SURFACE TENSION OF VARIOUS LIQUIDS

Liquid	In contact with	Temp. (°C)	Surface tension (N/m)
carbon disulfide	vapor	20	3.233×10^{-2}
carbon tetrachloride	vapor	20	2.695×10^{-2}
ethyl alcohol	air	0	2.405×10^{-2}
ethyl alcohol	vapor	20	2.275×10^{-2}
water	air	20	7.275×10^{-2}

Table 11 MASS DENSITY

Gas	(T = 0°C; p = 1 atm) Formula	Density (g/L)	Liquid	(T = 0°C)	Density (g/cm³)
acetylene	C_2H_2	1.179 10	Alcohol, ethyl	C_2H_5OH	0.791
air, dry, CO_2 free		1.292 84	Carbon tetrachloride	CCl_4	1.60
ammonia	NH_3	0.771 26	Gasoline		0.66–0.69
argon	Ar	1.783 64	Mercury	Hg	13.6
chlorine	Cl_2	3.214	Water	H_2O	0.998
carbon dioxide	CO_2	1.976 9			
carbon monoxide	CO	1.250 04	**Solid**		**Density (g/cm³)**
ethane	C_2H_6	1.356 2			
helium	He	0.178 46			
hydrogen	H_2	0.089 88	Diamond		3.15–3.53
hydrogen chloride	HCl	1.639 2	Glass, flint		2.5–5.9
methane	CH_4	0.716 8	Gold	Au	19.32
neon	Ne	0.899 90	Ice	H_2O	0.917
nitrogen	N_2	1.250 36	Iron	Fe	7.87
oxygen	O_2	1.528 96	Lead	Pb	11.4
sulfur dioxide	SO_2	2.926 2	Rubber, hard		1.19
			Wood, maple		0.62–0.75

Table 12 EQUILIBRIUM VAPOR PRESSURE OF WATER

Temp. (°C)	Pressure (mm Hg)	Pressure (kPa)	Temp (°C)	Pressure (mm Hg)	Pressure (kPa)	Temp. (°C)	Pressure (mm Hg)	Pressure (kPa)
0	4.6	0.61	25	23.8	3.16	90	5[illegible]5.8	69.88
5	6.5	0.86	26	25.2	3.35	95	633.9	84.25
10	9.2	1.22	27	26.7	3.55	96	657.6	87.40
15	12.8	1.70	28	28.3	3.76	97	682.1	90.65
16	13.6	1.81	29	30.0	3.99	98	707.3	94.00
17	14.5	1.93	30	31.8	4.23	99	733.2	97.44
18	15.5	2.06	35	42.2	5.61	100	760.0	101.00
19	16.5	2.19	40	55.3	7.35	101	787.5	104.66
20	17.5	2.33	50	92.5	12.29	103	845.1	112.31
21	18.7	2.49	60	149.4	19.86	105	906.1	120.42
22	19.8	2.63	70	233.7	31.06	110	1074.6	142.81
23	21.1	2.80	80	355.1	47.19	120	1489.1	197.90
24	22.4	2.98	85	433.6	57.63	150	3570.5	474.52

Table 13 HEAT CONSTANTS

Material	Specific heat (J/g·C°)	Melting point (°C)	Boiling point (°C)	Heat of fusion (J/g)	Heat of vaporization (J/g)
alcohol, ethyl	2.43 (25°)	−115	78.5	104	855
aluminum	0.896 (20°)	660.2	2467	390	10 600
	0.909 (0-100°)				
	0.921 (20-100°)				
	0.942 (100°)				
ammonia, liquid	4.384 (−60°)	−77.7	−33.35	452.6	1370
liquid	4.710 (20°)				
gas	2.190 (20°)				
brass (40% Zn)	0.384	900			
copper	0.387	1083	2595	205	4819
glass, crown	0.674				
iron	0.450	1535	3000	33.0	6700
lead	0.128	327.5	1744	22.9	867
mercury	0.139	−38.87	356.58	11.8	295.64
platinum	0.133	1769	3827 ± 100	114	
silver	0.235	960.8	2212	109	2367
tin	0.227 (25°)	232	2270	60.3	2200
tungsten	0.135	3410 ± 10	5927	180	
water	4.19		100.00		2260
ice	2.22	0.00		334	
steam	2.01				
zinc	0.386	419.4	907	96.3	1800

Table 14 COEFFICIENT OF LINEAR EXPANSION

(Increase in length per unit length per Celsius degree)

Material	Coefficient (Δ1/1·C°)	Temperature (°C)	Material	Coefficient (Δ1/1·C°)	Temperature (°C)
aluminum	23.8×10^{-6}	20–100	invar (nickel steel)	0.9×10^{-6}	20
brass	19.30×10^{-6}	0–100	lead	29.40×10^{-6}	18–100
copper	16.8×10^{-6}	25–100	magnesium	26.08×10^{-6}	18–100
glass, tube	8.33×10^{-6}	0–100	platinum	8.99×10^{-6}	40
crown	8.97×10^{-6}	0–100	rubber, hard	80×10^{-6}	20–60
Pyrex	3.3×10^{-6}	20–300	quartz, fused	0.546×10^{-6}	0–800
gold	14.3×10^{-6}	16–100	silver	18.8×10^{-6}	20
ice	50.7×10^{-6}	−10–0	tin	26.92×10^{-6}	18–100
iron, soft	12.10×10^{-6}	40	zinc	26.28×10^{-6}	10–100
steel	10.5×10^{-6}	0–100			

The coefficient of cubical expansion may be taken as three times the linear coefficient.

Table 15 COEFFICIENT OF VOLUME EXPANSION

(Increase in volume per unit volume per C° at 20°C)

Liquid	Coefficient ($\Delta V/V \cdot C°$)
acetone	14.87×10^{-4}
alcohol, ethyl	11.2×10^{-4}
benzene	12.37×10^{-4}
carbon disulfide	12.18×10^{-4}
carbon tetrachloride	12.36×10^{-4}
chloroform	12.73×10^{-4}
ether	16.56×10^{-4}
gasoline	10.8×10^{-4}
glycerol	5.05×10^{-4}
mercury	1.82×10^{-4}
petroleum	9.55×10^{-4}
turpentine	9.73×10^{-4}
water	2.07×10^{-4}

Table 16 HEAT OF VAPORIZATION OF SATURATED STEAM

Temp.	Pressure		Heat of vaporization
(°C)	(mm Hg)	(kPa)	(J/g)
95	634.0	84.26	2269
96	657.7	87.41	2266
97	682.1	90.65	2263
98	707.3	94.00	2261
99	733.3	97.46	2258
100	760.0	101.0	2256
101	787.5	104.7	2253
102	815.9	108.4	2250
103	845.1	112.3	2248
104	875.1	116.3	2245
105	906.1	120.4	2243
106	937.9	124.6	$224\bar{0}$

Table 17 SPEED OF SOUND

Substance	Density (g/L)	Velocity (m/s)	$\Delta v/\Delta T$ (m/s·C°)
GASES (STP)			
air, dry	1.293	331.45	0.59
carbon dioxide	1.977	259	0.4
helium	0.178	965	0.8
hydrogen	0.089 9	1284	2.2
nitrogen	1.251	334	0.6
oxygen	1.429	316	0.56
LIQUIDS (25°C)	(g/cm^3)		
acetone	0.79	1174	
alcohol, ethyl	0.79	1207	
carbon tetrachloride	1.595	926	
glycerol	1.26	1904	
kerosene	0.81	1324	
water, distilled	0.998	1497	
water, sea	1.025	1531	
SOLIDS (thin rods)			
aluminum	2.7	5000	
brass	8.6	3480	
brick	1.8	3650	
copper	8.93	3810	
cork	0.25	500	
glass, crown	2.24	4540	
iron	7.85	5200	
lucite	1.18	1840	
maple (along grain)	0.69	4110	
pine (along grain)	0.43	3320	
steel	7.85	5200	

Table 18 INDEX OF REFRACTION

(λ = 589.3 nm; Temperature = 20°C except as noted)

Material	Refractive index
air, dry (STP)	1.000 29
alcohol, ethyl	1.360
benzene	1.501
calcite	1.658 3
	1.486 4
canada balsam	1.530
carbon dioxide (STP)	1.000 45
carbon disulfide	1.625
carbon tetrachloride	1.459
diamond	2.419 5
glass, crown	1.517 2
flint	1.627 0
glycerol	1.475
ice	1.310
lucite	1.50
quartz	1.544
	1.553
quartz, fused	1.458 45
sapphire (Al_2O_3)	1.768 6
	1.760 4
water, distilled	1.333
water vapor (STP)	1.000 25

Table 19 ELECTROCHEMICAL EQUIVALENTS

Element	g-at wt (g)	Ionic charge (oxidation number)	z (g/C)
aluminum	27.0	+3	0.000 093 3
cadmium	112	+2	0.000 580
calcium	40.1	+2	0.000 208
chlorine	35.5	−1	0.000 368
chromium	52.0	+6	0.000 089 8
chromium	52.0	+3	0.000 180
copper	63.5	+2	0.000 329
copper	63.5	+1	0.000 658
gold	197	+3	0.000 681
gold	197	+1	0.002 04
hydrogen	1.01	+1	0.000 010 4
lead	207	+4	0.000 536
lead	207	+2	0.001 07
magnesium	24.3	+2	0.000 126
nickel	58.7	+2	0.000 304
oxygen	16.0	−2	0.000 082 9
postassium	39.1	+1	0.000 405
silver	108	+1	0.001 118 0
sodium	23.0	+1	0.000 238
tin	119	+4	0.000 308
tin	119	+2	0.000 617
zinc	65.4	+2	0.000 339

Table 20 RESISTIVITY

(Temperature = 20°C)

Material	Resistivity (Ω·cm)	Melting point (°C)
advance	48×10^{-6}	1190
aluminum	2.824×10^{-6}	660
brass	7.00×10^{-6}	900
climax	87×10^{-6}	1250
constantan (Cu 60, Ni 40)	49×10^{-6}	1190
copper	1.724×10^{-6}	1083
german silver (Cu 55, Zn 25, Ni 20)	33×10^{-6}	1100
gold	2.44×10^{-6}	1063
iron	10×10^{-6}	1535
magnesium	4.6×10^{-6}	651
manganin (Cu 84, Mn 12, Ni 4)	44×10^{-6}	910
mercury	95.783×10^{-6}	−39
monel metal	42×10^{-6}	1300
nichrome	115×10^{-6}	1500
nickel	7.8×10^{-6}	1452
nickel silver (Cu 57, Ni 43)	49×10^{-6}	1190
platinum	10×10^{-6}	1769
silver	1.59×10^{-6}	961
tungsten	5.6×10^{-6}	3410

Table 21 **PROPERTIES OF COPPER WIRE**

American Wire Gauge (B&S)—for any metal (Temperature, 20°C)			for copper only	
Gauge number	**Diameter (mm)**	**Cross section (mm^2)**	**Resistance (Ω/km)**	**(m/Ω)**
0000	11.68	107.2	0.160 8	6219
000	10.40	85.03	0.202 8	4932
00	9.266	67.43	0.255 7	3911
0	8.252	53.48	0.322 4	3102
1	7.348	42.41	0.406 6	2460
2	6.544	33.63	0.502 7	1951
3	5.827	26.67	0.646 5	1547
4	5.189	21.15	0.815 2	1227
5	4.621	16.77	1.028	972.9
6	4.115	13.30	1.296	771.5
7	3.665	10.55	1.634	611.8
8	3.264	8.366	2.061	485.2
9	2.906	6.634	2.599	384.8
10	2.588	5.261	3.277	305.1
11	2.305	4.172	4.132	242.0
12	2.053	3.309	5.211	191.9
13	1.828	2.624	6.571	152.2
14	1.628	2.081	8.258	120.7
15	1.450	1.650	10.45	95.71
16	1.291	1.309	13.17	75.90
17	1.150	1.038	16.61	60.20
18	1.024	0.823 1	20.95	47.74
19	0.911 6	0.652 7	26.42	37.86
20	0.811 8	0.517 6	33.31	30.02
21	0.723 0	0.410 5	42.00	23.81
22	0.643 8	0.325 5	52.96	18.88
23	0.573 3	0.258 2	66.79	14.97
24	0.510 6	0.204 7	84.21	11.87
25	0.454 7	0.162 4	106.2	9.415
26	0.404 9	0.128 8	133.9	7.486
27	0.360 6	0.102 1	168.9	5.922
28	0.321 1	0.080 98	212.9	4.697
29	0.285 9	0.064 22	268.5	3.725
30	0.254 6	0.050 93	338.6	2.954
31	0.226 8	0.040 39	426.9	2.342
32	0.201 9	0.032 03	538.3	1.858
33	0.179 8	0.025 40	678.8	1.473
34	0.160 1	0.020 14	856.0	1.168
35	0.142 6	0.015 97	1079	0.926 5
36	0.127 0	0.012 67	1361	0.734 7
37	0.113 1	0.010 05	1716	0.582 7
38	0.100 7	0.007 967	2164	0.462 1
39	0.089 69	0.006 318	2729	0.366 4
40	0.079 87	0.005 010	3441	0.290 6

Table 22 THE CHEMICAL ELEMENTS

Name of element	Symbol	Atomic number	Atomic weight
actinium	Ac	89	227.027 8
aluminum	Al	13	26.981 54
americium	Am	95	[243]
antimony	Sb	51	121.75
argon	Ar	18	39.948
arsenic	As	33	74.921 6
astatine	At	85	[210]
barium	Ba	56	137.33
berkelium	Bk	97	[247]
beryllium	Be	4	9.012 18
bismuth	Bi	83	208.980 4
boron	B	5	10.81
bromine	Br	35	79.904
cadmium	Cd	48	112.41
calcium	Ca	20	40.08
californium	Cf	98	[251]
carbon	C	6	12.011
cerium	Ce	58	140.12
cesium	Cs	55	132.905 4
chlorine	Cl	17	35.453
chromium	Cr	24	51.996
cobalt	Co	27	58.933 2
copper	Cu	29	63.546
curium	Cm	96	[247]
dysprosium	Dy	66	162.50
einsteinium	Es	99	[254]
erbium	Er	68	167.26
europium	Eu	63	151.96
fermium	Fm	100	[257]
fluorine	F	9	18.998 403
francium	Fr	87	[223]
gadolinium	Gd	64	157.25
gallium	Ga	31	69.72
germanium	Ge	32	72.59
gold	Au	79	196.966 5
hafnium	Hf	72	178.49
hahnium	Ha	105	[263]
helium	He	2	4.002 60
holmium	Ho	67	164.930 4
hydrogen	H	1	1.007 9
indium	In	49	114.82
iodine	I	53	126.904 5
iridium	Ir	77	192.22
iron	Fe	26	55.847
krypton	Kr	36	83.80
lanthanum	La	57	138.900 5
lawrencium	Lr	103	[260]
lead	Pb	82	207.2
lithium	Li	3	6.941
lutetium	Lu	71	174.967
magnesium	Mg	12	24.305
manganese	Mn	25	54.938 0
mendelevium	Md	101	[258]
mercury	Hg	80	200.59
molybdenum	Mo	42	95.94
neodymium	Nd	60	144.24
neon	Ne	10	20.179
neptunium	Np	93	237.048 2
nickel	Ni	28	58.70
niobium	Nb	41	92.906 4
nitrogen	N	7	14.006 7
nobelium	No	102	[259]
osmium	Os	76	190.2
oxygen	O	8	15.999 4
palladium	Pd	46	106.4
phosphorus	P	15	30.973 76
platinum	Pt	78	195.09
plutonium	Pu	94	[244]
polonium	Po	84	[209]
potassium	K	19	30.098 3
praseodymium	Pr	59	140.907 7
promethium	Pm	61	[145]
protactinium	Pa	91	231.035 9
radium	Ra	88	226.025 4
radon	Rn	86	[222]
rhenium	Re	75	186.207
rhodium	Rh	45	102.905 5
rubidium	Rb	37	85.467 8
ruthenium	Ru	44	101.07
rutherfordium	Rf	104	[261]
samarium	Sm	62	150.4
scandium	Sc	21	44.955 9
selenium	Se	34	78.96
silicon	Si	14	28.0855
silver	Ag	47	107.868
sodium	Na	11	22.989 77
strontium	Sr	38	87.62
sulfur	S	16	32.06
tantalum	Ta	73	180.947 9
technetium	Tc	43	98.906 2
tellurium	Te	52	127.60
terbium	Tb	65	158.925 4
thallium	Tl	81	204.37
thorium	Th	90	232.038 1

Table 22 THE CHEMICAL ELEMENTS (cont'd)

Name of element	Sym-bol	Atomic number	Atomic weight	Name of element	Sym-bol	Atomic number	Atomic weight
thulium	Tm	69	168.934 2	vanadium	V	23	50.941 4
tin	Sn	50	118.69	xenon	Xe	54	131.30
titanium	Ti	22	47.90	ytterbium	Yb	70	173.04
tungsten	W	74	183.85	yttrium	Y	39	88.905 9
uranium	U	92	238.029	zinc	Zn	30	65.37
				zirconium	Zr	40	91.22

A value given in brackets denotes the mass number of the isotope of longest known half-life. The atomic weights of most of these elements are believed to have no error greater than ±1 of the last digit given.

Table 23 MASSES OF SOME NUCLIDES

Element	Symbol	Atomic mass* (*u*)**	Element	Symbol	Atomic mass* (*u*)**
hydrogen	$^{1}_{1}H$	1.007 825	sodium	$^{23}_{11}Na$	22.989 8
deuterium	$^{2}_{1}H$	2.014 0	magnesium	$^{24}_{12}Mg$	23.985 04
helium	$^{3}_{2}He$	3.016 03		$^{25}_{12}Mg$	24.985 84
	$^{4}_{2}He$	4.002 60		$^{26}_{12}Mg$	25.982 59
lithium	$^{6}_{3}Li$	6.015 12	sulfur	$^{32}_{16}S$	31.972 07
	$^{7}_{3}Li$	7.016 00	chlorine	$^{35}_{17}Cl$	34.968 85
beryllium	$^{6}_{4}Be$	6.019 7		$^{37}_{17}Cl$	36.965 90
	$^{8}_{4}Be$	8.005 3	potassium	$^{39}_{19}K$	38.963 71
	$^{9}_{4}Be$	9.012 18		$^{41}_{19}K$	40.961 84
boron	$^{10}_{5}B$	10.012 9	krypton	$^{95}_{36}Kr$	94.9
	$^{11}_{5}B$	11.009 31	molybdenum	$^{100}_{42}Mo$	99.907 6
carbon	$^{12}_{6}C$	12.000 00	silver	$^{107}_{47}Ag$	106.905 09
	$^{13}_{6}C$	13.003 35		$^{109}_{47}Ag$	108.904 7
nitrogen	$^{12}_{7}N$	12.018 8	tellurium	$^{137}_{52}Te$	136.91
	$^{14}_{7}N$	14.003 07	barium	$^{138}_{56}Ba$	137.905 0
	$^{15}_{7}N$	15.000 11	lead	$^{214}_{82}Pb$	213.998 2
oxygen	$^{16}_{8}O$	15.994 91	bismuth	$^{214}_{83}Bi$	213.997 2
	$^{17}_{8}O$	16.999 14	polonium	$^{218}_{84}Po$	218.008 9
	$^{18}_{8}O$	17.999 14	radon	$^{222}_{86}Rn$	222.017 5
fluorine	$^{19}_{9}F$	18.998 40	radium	$^{228}_{88}Ra$	228.030 3
neon	$^{20}_{10}Ne$	19.992 44	thorium	$^{232}_{90}Th$	232.038 2
	$^{22}_{10}Ne$	21.991 38	uranium	$^{235}_{92}U$	235.043 9
				$^{238}_{92}U$	238.050 8
			plutonium	$^{239}_{94}Pu$	239.052 2

*Atomic mass of neutral atom is given.
** 1 atomic mass unit (u) = $1.660\,540\,2 \times 10^{-27}$ kg.

Table 24 FOUR-PLACE LOGARITHMS

n	0	1	2	3	4	5	6	7	8	9
10	0000	0043	0086	0128	0170	0212	0253	0294	0334	0374
11	0414	0453	0492	0531	0569	0607	0645	0682	0719	0755
12	0792	0828	0864	0899	0934	0969	1004	1038	1072	1106
13	1139	1173	1206	1239	1271	1303	1335	1367	1399	1430
14	1461	1492	1523	1553	1584	1614	1644	1673	1703	1732
15	1761	1790	1818	1847	1875	1903	1931	1959	1987	2014
16	2041	2068	2095	2122	2148	2175	2201	2227	2253	2279
17	2304	2330	2355	2380	2405	2430	2455.	2480	2504	2529
18	2553	2577	2601	2625	2648	2672	2695	2718	2742	2765
19	2788	2810	2833	2856	2878	2900	2923	2945	2967	2989
20	3010	3032	3054	3075	3096	3118	3139	3160	3181	3201
21	3222	3243	3263	3284	3304	3324	3345	3365	3385	3404
22	3424	3444	3464	3483	3502	3522	3541	3560	3579	3598
23	3617	3636	3655	3674	3692	3711	3729	3747	3766	3784
24	3802	3820	3838	3856	3874	3892	3909	3927	3945	3962
25	3979	3997	4014	4031	4048	4065	4082	4099	4116	4133
26	4150	4166	4183	4200	4216	4232	4249	4265	4281	4298
27	4314	4330	4346	4362	4378	4393	4409	4425	4440	4456
28	4472	4487	4502	4518	4533	4548	4564	4579	4594	4609
29	4624	4639	4654	4669	4683	4698	4713	4728	4742	4757
30	4771	4786	4800	4814	4829	4843	4857	4871	4886	4900
31	4914	4928	4942	4955	4969	4983	4997	5011	5024	5038
32	5051	5065	5079	5092	5105	5119	5132	5145	5159	5172
33	5185	5198	5211	5224	5237	5250	5263	5276	5289	5302
34	5315	5328	5340	5353	5366	5378	5391	5403	5416	5428
35	5441	5453	5465	5478	5490	5502	5514	5527	5539	5551
36	5563	5575	5587	5599	5611	5623	5635	5647	5658	5670
37	5682	5694	5705	5717	5729	5740	5752	5763	5775	5786
38	5798	5809	5821	5832	5843	5855	5866	5877	5888	5899
39	5911	5922	5933	5944	5955	5966	5977	5988	5999	6010
40	6021	6031	6042	6053	6064	6075	6085	6096	6107	6117
41	6128	6138	6149	6160	6170	6180	6191	6201	6212	6222
42	6232	6243	6253	6263	6274	6284	6294	6304	6314	6325
43	6335	6345	6355	6365	6375	6385	6395	6405	6415	6425
44	6435	6444	6454	6464	6474	6484	6493	6503	6513	6522
45	6532	6542	6551	6561	6571	6580	6590	6599	6609	6618
46	6628	6637	6646	6656	6665	6675	6684	6693	6702	6712
47	6721	6730	6739	6749	6758	6767	6776	6785	6794	6803
48	6812	6821	6830	6839	6848	6857	6866	6875	6884	6893
49	6902	6911	6920	6928	6937	6946	6955	6964	6972	6981
50	6990	6998	7007	7016	7024	7033	7042	7050	7059	7067
51	7076	7084	7093	7101	7110	7118	7126	7135	7143	7152
52	7160	7168	7177	7185	7193	7202	7210	7218	7226	7235
53	7243	7251	7259	7267	7275	7284	7292	7300	7308	7316
54	7324	7332	7340	7348	7356	7364	7372	7380	7388	7396

Table 24 FOUR-PLACE LOGARITHMS (cont'd)

n	0	1	2	3	4	5	6	7	8	9
55	7404	7412	7419	7427	7435	7443	7451	7459	7466	7474
56	7482	7490	7497	7505	7513	7520	7528	7536	7543	7551
57	7559	7566	7574	7582	7589	7597	7604	7612	7619	7627
58	7634	7642	7649	7657	7664	7672	7679	7686	7694	7701
59	7709	7716	7723	7731	7738	7745	7752	7760	7767	7774
60	7782	7789	7796	7803	7810	7818	7825	7832	7839	7846
61	7853	7860	7868	7875	7882	7889	7896	7903	7910	7917
62	7924	7931	7938	7945	7952	7959	7966	7973	7980	7987
63	7993	8000	8007	8014	8021	8028	8035	8041	8048	8055
64	8062	8069	8075	8082	8089	8096	8102	8109	8116	8122
65	8129	8136	8142	8149	8156	8162	8169	8176	8182	8189
66	8195	8202	8209	8215	8222	8228	8235	8241	8248	8254
67	8261	8267	8274	8280	8287	8293	8299	8306	8312	8319
68	8325	8331	8338	8344	8351	8357	8363	8370	8376	8382
69	8388	8395	8401	8407	8414	8420	8426	8432	8439	8445
70	8451	8457	8463	8470	8476	8482	8488	8494	8500	8506
71	8513	8519	8525	8531	8537	8543	8549	8555	8561	8567
72	8573	8579	8585	8591	8597	8603	8609	8615	8621	8627
73	8633	8639	8645	8651	8657	8663	8669	8675	8681	8686
74	8692	8698	8704	8710	8716	8722	8727	8733	8739	8745
75	8751	8756	8762	8768	8774	8779	8785	8791	8797	8802
76	8808	8814	8820	8825	8831	8837	8842	8848	8854	8859
77	8865	8871	8876	8882	8887	8893	8899	8904	8910	8915
78	8921	8927	8932	8938	8943	8949	8954	8960	8965	8971
79	8976	8982	8987	8993	8998	9004	9009	9015	9020	9025
80	9031	9036	9042	9047	9053	9058	9063	9069	9074	9079
81	9085	9090	9096	9101	9106	9112	9117	9122	9128	9133
82	9138	9143	9149	9154	9159	9165	9170	9175	9180	9186
83	9191	9196	9201	9206	9212	9217	9222	9227	9232	9238
84	9243	9248	9253	9258	9263	9269	9274	9279	9284	9289
85	9294	9299	9304	9309	9315	9320	9325	9330	9335	9340
86	9345	9350	9355	9360	9365	9370	9375	9380	9385	9390
87	9395	9400	9405	9410	9415	9420	9425	9430	9435	9440
88	9445	9450	9455	9460	9465	9469	9474	9479	9484	9489
89	9494	9499	9504	9509	9513	9518	9523	9528	9533	9538
90	9542	9547	9552	9557	9562	9566	9571	9576	9581	9586
91	9590	9595	9600	9605	9609	9614	9619	9624	9628	9633
92	9638	9643	9647	9652	9657	9661	9666	9671	9675	9680
93	9685	9689	9694	9699	9703	9708	9713	9717	9722	9727
94	9731	9736	9741	9745	9750	9754	9759	9763	9768	9773
95	9777	9782	9786	9791	9795	9800	9805	9809	9814	9818
96	9823	9827	9832	9836	9841	9845	9850	9854	9859	9863
97	9868	9872	9877	9881	9886	9890	9894	9899	9903	9908
98	9912	9917	9921	9926	9930	9934	9939	9943	9948	9952
99	9956	9961	9965	9969	9974	9978	9983	9987	9991	9996

GLOSSARY

abscissa. The value corresponding to the horizontal distance of a point on a graph from the y axis. The x coordinate. (31)

absolute deviation. The difference between a single measured value and the average of several measurements made in the same way. (26)

absolute error. The actual difference between a measured value and its accepted value. (25)

absolute zero. The temperature of a body at which the kinetic energy of its molecules is at a minimum; 0 K or −273.16°C. (168)

absorption spectrum. A continuous spectrum interrupted by dark lines or bands that are characteristic of the medium through which the radiation has passed. (Plate IV)

acceleration. Time rate of change of velocity. (48)

acceptor. An element with three valence electrons per atom which when added to a semiconductor crystal provides electron "holes" in the lattice structure of the crystal. (566)

accuracy. Closeness of a measurement to the accepted value for a specific physical quantity; expressed in terms of error. (25)

adhesion. The force of attraction between unlike molecules. (146)

adiabatic process. A thermal process in which no heat is added to or removed from a system. (202)

alpha (α) particle. A helium-4 nucleus, especially when emitted from the nucleus of a radioactive atom. (590)(602)

alternating current. An electric current that has one direction during one part of a generating cycle and the opposite direction during the remainder of the cycle. (502)

ampere. The unit of electric current; one coulomb per second. (411)(478)

amplifier. A device consisting of one or more vacuum tubes (or transistors) and associated circuits, used to increase the strength of a signal. (555)

amplitude. The maximum displacement of a vibrating particle from its equilibrium position. (231)

angle of incidence. The angle between the incident ray and the normal drawn to the point of incidence. (235)

angle of reflection. The angle between the reflected ray and the normal drawn to the point of incidence. (235)

angle of refraction. The angle between the refracted ray and the normal drawn to the point of refraction. (334)

Ångstrom. A unit of linear measure equal to 10^{-10} m. (142)

angular acceleration. The time rate of change of angular velocity. (103)

angular impulse. The product of a torque and the time interval during which it acts. (136)

angular momentum. The product of the rotational inertia of a body and its angular velocity. (137)

angular velocity. The time rate of change of angular displacement. (102)

anode. (1) The positive electrode of an electric cell. (419) (2) The positive electrode or plate of an electronic tube. (555) (3) The electron-poor electrode. (419)

antimatter. A substance composed of antiparticles. (643)

antiparticle. A counterpart of a subatomic particle having opposite properties (except for equal mass). (643)

aperture. Any opening through which radiation may pass. The diameter of an opening that admits light to a lens or mirror. (320)

apparent power. The product of the effective values of alternating voltage and current. (531)

arc tangent. The inverse function to the tangent. Symbol: arctan or $\tan^{-1}$. Interpretation: "An angle whose tangent is ———." (534)

atom. The smallest particle of an element that can exist either alone or in combination with other atoms of the same or other elements. (143)

atomic mass unit. One-twelfth of the mass of carbon-12, or $1.660\ 540\ 2 \times 10^{-27}$ kg. (143)

atomic number. The number of protons in the nucleus of an atom. (593)

atomic weight. The weighted average of the atomic masses of an element's isotopes based on their relative abundance. (595)

average velocity. Total displacement divided by elapsed time. (43)

back emf. An induced emf in the armature of a motor that opposes the applied voltage. (511)

baryon. A subatomic particle with a large rest mass, e.g., the proton. (641)

basic equation. An equation that relates the unknown quantity with known quantities in a problem. (35)

basic law of electrostatics. Similarly charged objects repel each other. Oppositely charged objects attract each other. (381)

beam. Several parallel rays of light considered collectively. (304)

beat. The interference effect resulting from the superposition of two waves of slightly different frequencies propagating in the same direction. The amplitude of the resultant wave varies with time. (276)

becquerel. The rate of radioactivity equal to one disintegration per second. (606)

Bernoulli's principle. The sum of the pressure and the kinetic energy per unit volume of a flowing fluid is a constant. (163)

beta (β) particle. An electron emitted from the nucleus of a radioactive atom. (602)

betatron. A device that accelerates electrons by means of the transformer principle. (632)

bevatron. A high-energy synchrotron. (633)

binding energy. Energy that must be applied to a nucleus to break it up. (596)

boson. A subatomic particle with zero charge and rest mass, e.g., the photon. (643)

Boyle's law. The volume of a dry gas varies inversely with the pressure exerted upon it, provided the temperature is constant. (177)

breeder reactor. A nuclear reactor in which a fissionable material is produced at a greater rate than the fuel is consumed. (615)

Brownian movement. The irregular and random movement of small particles suspended in a fluid, known to be a consequence of the thermal motion of fluid molecules. (151)

bubble chamber. Instrument used for making the paths of ionizing particles visible as a trail of tiny bubbles in a liquid. (638)

buoyant force. The upward force that any fluid exerts upon an object placed in it. (162)

calorie. The quantity of heat equal to 4.19 joules. (170)

calorimeter. A heat-measuring device consisting of nested metal cups separated by an air space. (185)

candela. The unit of luminous intensity of a light source. (307)

capacitance. The ratio of the charge on either plate of a capacitor to the potential difference between the plates. (400)

capacitive reactance. Reactance in an a-c circuit containing capacitance which causes a lagging voltage. (537)

capacitor. A combination of conducting plates separated by layers of a dielectric that is used to store an electric charge. (400)

capillarity. The elevation or depression of liquids in small-diameter tubes. (154)

cathode. (1) The negative electrode of an electric cell. (419) (2) The electron-emitting electrode of an electronic tube. (555) (3) The electron-rich electrode. (419)

cathode rays. Particles emanating from a cathode; electrons. (582)

Celsius scale. The temperature scale using the ice point as 0° and the steam point as 100°, with 100 equal divisions, or degrees, between; formerly the centigrade scale. (168)

center of curvature. The center of the sphere of which the mirror or lens surface forms a part. (320)

center of gravity. The point at which all of the weight of a body can be considered to be concentrated. (86)

centrifugal force. Force that tends to move the particles of a rotating object away from the center of rotation. (99)

centripetal acceleration. Acceleration directed toward the center of a circular path. (95)

centripetal force. The force that produces centripetal acceleration. (96)

chain reaction. A reaction in which the material or energy that starts the reaction is also one of the products and can cause similar reactions. (613)

Charles' law. The volume of a dry gas is directly proportional to its Kelvin temperature, providing the pressure is constant. (176)

chemical change. A change in which new substances with new properties are formed. (603)

chemical equivalent. The quantity of an element, expressed in grams, equal to the ratio of its atomic weight to its valence. (462)

chromatic aberration. The nonfocusing of light of different colors. (359)

circular motion. Motion of a body along a circular path. (94)

cloud chamber. A chamber in which charged subatomic particles appear as trails of liquid droplets. (637)

coefficient of area expansion. The change in area per unit area of a solid per degree change in temperature. (172)

coefficient of cubic expansion. The change in volume per unit volume of a solid or liquid per degree change in temperature. (173)

coefficient of linear expansion. The change in length per unit length of a solid per degree change in temperature. (171)

coefficient of sliding friction. The ratio of the force needed to overcome sliding friction to the normal force pressing the surfaces together. (81)

coherence. The property of two wave trains with identical wavelengths and a constant phase relationship. (289)

cohesion. The force of attraction between like molecules. (146)

color. The visual perception of light associated with its frequency or wavelength. (356)

commutator. A split ring in a d-c generator, each segment of which is connected to an end of a corresponding armature loop. (505)

complementary colors. Two colors that combine to form white light. (358)

complete vibration. Back-and-forth motion of an object describing simple harmonic motion. (110)

component. One of the several vectors that can be combined geometrically to yield a resultant vector. (33)

composition of forces. The combining of two or more component forces into a single resultant force. (70)

compression. The region of a longitudinal wave in which the distance separating the vibrating particles is less than their equilibrium distance. (226)

concave. Surface with center of curvature on the same side as the observer. (318)

concave lens. See *diverging lens.*

concave mirror. A mirror that converges parallel light rays incident on its surface. (321)

concurrent forces. Forces with lines of action that pass through the same point. (72)

condensation. The change of phase from a gas or vapor to a liquid. (160)

conductor. A material through which an electric charge is readily transferred. (384)

conservative forces. Forces for which the law of conservation of mechanical energy holds true; gravitational forces and elastic forces. (129)

continuous spectrum. A spectrum without dark lines or bands or in which there is an uninterrupted change from one color to another. (Plate IV)

converging lens. A lens that is thicker in the middle than it is at the edge and bends incident parallel rays toward a common point. (341)

convex. Surface with center of curvature on the opposite side from the observer. (318)

convex lens. See *converging lens.*

convex mirror. A mirror that diverges parallel light rays incident on its surface. (321)

cosmic rays. High-energy nuclear particles apparently originating from outer space. (611)

coulomb. The quantity of electricity equal to the charge on 6.25×10^{18} electrons. (388)(478)

Coulomb's law of electrostatics. The force between two point charges is directly proportional to the product of their magnitudes and inversely proportional to the square of the distance between them. (388)

Coulomb's law of magnetism. The force between two magnetic poles is directly proportional to the strengths of the poles and inversely proportional to the square of their distance apart. (470)

couple. Two forces of equal magnitude acting in opposite directions in the same plane, but not along the same line. (89)

crest. A region of upward displacement in a transverse wave. (225)

critical angle. That limiting angle of incidence in the optically denser medium that results in an angle of refraction of 90°. (339)

critical mass. The amount of a particular fissionable material required to make a fission reaction self-sustaining. (614)

critical point. The upper limit of the temperature-pressure curve of a substance. (195)

critical pressure. The pressure needed to liquefy a gas at its critical temperature. (195)

critical temperature. The temperature to which a gas must be cooled before it can be liquefied by pressure. (195)

critical velocity. Velocity below which an object moving in a vertical circle will not describe a circular path. (98)

curie. The quantity of any radioactive nuclide that has a disintegration rate of 3.7×10^{10} becquerels. (606)

current sensitivity. Current per unit scale division of an electric meter. (482)

cutoff bias. The smallest negative grid voltage, for a given plate voltage, that causes a vacuum tube to cease to conduct. (559)

cutoff frequency. A characteristic threshold frequency of incident light below which, for a given material, the photoelectric emission of electrons ceases. (291)

cutoff potential. A negative potential on the collector of a photoelectric cell that reduces the photoelectric current to zero. (290)

cycle. A series of changes produced in sequence that recur periodically. (207)

cyclotron. A device for accelerating charged atomic particles by means of D-shaped electrodes. (631)

damping. The reduction in amplitude of a wave due to the dissipation of wave energy. (232)

decay constant. The ratio between the number of nuclei decaying per second and the total number of nuclei. (605)

decibel. A unit of sound intensity level. The smallest change of sound intensity that the normal human ear can detect. (258)

declination. The angle between magnetic north and the true north from any surface location; also called *variation*. (474)

dees. The electrodes of a cyclotron. (631)

density. See *mass density*.

derived unit. A unit of measure that consists of combinations of fundamental units. (21)

dew point. The temperature at which a given amount of water vapor will exert equilibrium vapor pressure. (160)

diamagnetism. The property of a substance whereby it is feebly repelled by a strong magnet. (466)

dichroism. A property of certain crystalline substances in which one polarized component of incident light is absorbed and the other is transmitted. (374)

dielectric. An electric insulator. A nonconducting medium. (402)

dielectric constant. The ratio of the capacitance with a particular material separating the plates of a capacitor to the capacitance with a vacuum between the plates. (402)

diffraction. The spreading of a wave disturbance into a region behind an obstruction. (241)(367)

diffraction angle. The angle that a diffracted wavefront forms with the grating plane. (369)

diffraction grating. An optical surface, either transmitting or reflecting, with several thousand equally spaced and parallel grooves ruled in it. (367)

diffusion. (1) The penetration of one type of particle into a mass of a second type of particle. (146) (2) The scattering of light by irregular reflection. (316)

dimensional analysis. The performance of indicated mathematical operations in a problem with the physical quantities alone. (36)

diode. A two-terminal device that will conduct electric current more easily in one direction than in the other. (555)

direct current. An essentially constant-value current in which the movement of charge is in only one direction. (419)

direct proportion. The relation between two quantities whose graph is a straight line. (32)

dispersion. The process of separating polychromatic light into its component wavelengths. (355)

displacement. (1) A change of position in a particular direction. (40) (2) Distance of a vibrating particle from the midpoint of its vibration. (110)

dissipative forces. Forces for which the law of conservation of mechanical energy does not hold true; frictional forces. (129)

distillation. The evaporation of volatile materials from a liquid or solid mixture and their condensation in a separate vessel. (192)

diverging lens. A lens that is thicker at the edge than it is in the middle and bends incident parallel rays so that they appear to come from a common point. (358)

domain. A microscopic magnetic region composed of a group of atoms whose magnetic fields are aligned in a common direction. (468)

donor. A substance with five valence electrons per atom which when added to a semiconductor crystal provides free electrons in the lattice structure of the crystal. (566)

Doppler effect. The change observed in the frequency with which a wave from a given source reaches an observer when the source and the observer are in relative motion. (261)

double refraction. The separation of a beam of unpolarized light into two refracted plane-polarized beams by certain crystals such as quartz and calcite. (375)

drift tubes. Charged cylinders used to accelerate charged subatomic particles in a linear accelerator. (634)

ductility. The property of a metal that enables it to be drawn through a die to form a wire. (146)

eddy currents. Closed loops of induced current set up in a piece of metal when there is relative motion between the metal and a magnetic field. The eddy currents are in such direction that the resulting magnetic forces oppose the relative motion. (522)

Edison effect. The emission of electrons from a heated metal in a vacuum. (556)

effective value of current. The magnitude of an alternating current that in a given resistance produces heat at the same average rate as that magnitude of steady direct current. (527)

efficiency. The ratio of the useful work output of a machine to total work input. (119)

elastic collision. A collision in which objects rebound from each other without a loss of kinetic energy. (134)

elastic limit. The condition in which a substance is on the verge of becoming permanently deformed. (147)

elastic potential energy. The potential energy in a stretched or compressed elastic object. (128)

elasticity. The ability of an object to return to its original size or shape when the external forces producing distortion are removed. (147)

electric current. The rate of flow of charge past a given point in an electric circuit. (411)

electric field. The region in which a force acts on an electric charge brought into the region. (390)

electric field intensity. The force per unit positive charge at a given point in an electric field. (392)

electrification. The process of charging a body by adding or removing electrons. (380)

electrochemical cell. A cell in which chemical energy is converted to electric energy by a spontaneous electron-transfer reaction. (418)

electrochemical equivalent. The mass of an element, in grams, deposited by one coulomb of electric charge. (461)

electrode. A conducting element in an electric cell, electronic tube, or semiconductor device. (419)

electrolysis. The conduction of electricity through a solution of an electrolyte or through a fused ionic compound, together with the resulting chemical changes. (459)

electrolyte. A substance whose solution conducts an electric current. (412)

electrolytic cell. A cell in which electric energy is converted to chemical energy by means of an electron-transfer reaction. (458)

electromagnetic induction. The process by which an emf is set up in a conducting circuit by a changing magnetic flux linked by the circuit. (415)(490)

electromagnetic interaction. The interaction that keeps electrons in orbit and forms bonds between atoms and molecules. (644)

electromagnetic waves. Transverse waves having an electric component and a magnetic component, each being perpendicular to the other and both perpendicular to the direction of propagation. (286)

electromotive force. See *emf.*

electron. A negatively charged subatomic particle having a rest mass of $9.109\ 389\ 7 \times 10^{-31}$ kg and a charge of $1.602\ 177\ 33 \times 10^{-19}$ C. (584)

electron shell. A region about the nucleus of an atom in which electrons move and which is made up of electron orbitals. (589)

electron volt. The energy required to move an electron between two points that have a difference of potential of one volt. (597)

electronics. The branch of physics concerned with the emission, behavior, and effects of electrons. (554)

electroscope. A device used to observe the presence of an electrostatic charge. (380)

elementary colors. The six regions of color in the solar spectrum observed by the dispersion of sunlight: red, orange, yellow, green, blue, and violet. (358)

elongation strain. The ratio of the increased in length to the unstretched length. (148)

emf. The energy per unit charge supplied by a source of electric current. (419)

endothermic. Referring to a process that absorbs energy. (184)

energy. The capacity for doing work. (10)

energy level. One of a series of discrete energy values that characterize a physical system governed by quantum rules. (296)

entropy. (1) The internal energy of a system that cannot be converted to mechanical work. (209) (2) The property that describes the disorder of a system. (211)

equilibrant force. The force that produces equilibrium. (75)

equilibrium. The state of a body in which there is no change in its motion. (74)

equilibrium position. Midpoint of the path of an object describing simple harmonic motion. (110)

equilibrium vapor pressure. The pressure exerted by vapor molecules in equilibrium with a liquid. (160)

evaporation. The change of phase from a liquid to a gas or vapor. (159)

exclusion principle. No two electrons in an atom can have the same set of quantum numbers. (626)

exothermic. Referring to a process that liberates energy. (184)

external combustion engine. A heat engine in which the fuel burns outside the cylinder or turbine chamber. (213)

farad. The unit of capacitance; one coulomb per volt. (401)

faraday. The quantity of electricity (96 500 coulombs) required to deposit one chemical equivalent of an element. (462)

Faraday's first law. The mass of an element deposited during electrolysis is proportional to the quantity of charge that passes through the electrolytic cell. (461)

Faraday's second law. The mass of an element deposited during electrolysis is proportional to the chemical equivalent of that element. (462)

ferromagnetism. The property of a substance by which it is strongly attracted by a magnet. (466)

Feynman diagram. A diagram showing the production and exchange of particles during a subatomic interaction. (643)

first law of photoelectric emission. The rate of emission of photoelectrons is directly proportional to the intensity of the incident light. (289)

first law of thermodynamics. When heat is converted to another form of energy, or when another form of energy is converted to heat, there is no loss of energy. (201)

fission. The splitting of a heavy nucleus into nuclei of intermediate mass. (608)

Fleming valve. The first vacuum-tube diode. (556)

fluorescence. The emission of light during the absorption of radiation from another source. (561)(600)

f-number. The ratio of the focal length of a lens to the effective aperture. (347)

focal length. The distance between the principal focus of a lens or mirror and its optical center or vertex. (320)(345)

focal plane. The plane perpendicular to the principal axis of a converging lens or mirror and containing the principal focus. (321)(345)

focus. A point at which light rays meet or from which rays of light appear to diverge. (304)

force. (1) A physical quantity that can affect the motion of an object. (56) (2) A measure of the momentum gained per second by an accelerating body. (131)

force of gravitation. See *gravitational force.*

forced vibration. Vibration that is due to the application of a periodic force, and not to the natural vibrations of the system. (272)

forward bias. Voltage applied to a semiconductor **P-N** junction that increases the electron current across the junction. (571)

fractional distillation. The process of separating the components of a liquid mixture by means of differences in their boiling points. (192)

frame of reference. Any system for specifying the precise location of objects in space. (99)

freezing point. The temperature at which a liquid changes to a solid. (154)

frequency. Number of vibrations, oscillations, or cycles per unit time. (229)

friction. A force that resists the relative motion of objects that are in contact with each other. (79)

fuel cell. An electrochemical cell in which the chemical energy of continuously supplied fuel is converted into electric energy. (419)

fundamental. The lowest frequency produced by a musical tone source. That harmonic component of a wave which has the lowest frequency. (266)

fundamental unit. Any one of seven basic units of measure. (20)

fusion. (1) The change of phase from a solid to a liquid; melting. (154) (2) A reaction in which light nuclei combine, forming a nucleus with a larger mass number. (608)

galvanometer. An instrument used to measure minute electric currents. (481)

gamma (γ) ray. High-energy photon emitted from the nucleus of a radioactive atom. (602)

Geiger tube. Ion-sensitive instrument used for the detection of subatomic particles. (637)

gram-atomic weight. See *atomic weight.*

gravitational field. Region of space in which each point is associated with a value of gravitational acceleration. (67)

gravitational force. The mutual force of attraction between particles of matter. (64)

gravitational interaction. The interaction between particles of matter that has no known distance limitations, but is the weakest interaction of all. (644)

gravitational potential energy. Potential energy acquired by an object when it is moved against gravity. (125)

graviton. The carrier for the gravitational interaction. (645)

gravity. The force of gravitation on an object on or near the surface of a celestial body. (66)

grid. An element of an electronic tube.

An electrode used to control the flow of electrons from the cathode to the plate. (555)

grid bias. The grid-to-cathode voltage. (559)

half-life. The length of time during which, on the average, half of a large number of radioactive nuclides decay. (602)

harmonics. The fundamental and the tones whose frequencies are whole-number multiples of the fundamental. (268)

heat. Thermal energy in the process of being added to, or removed from, a substance. (167)

heat capacity. The quantity of heat needed to raise the temperature of a body one degree. (183)

heat engine. Any device that converts heat energy into mechanical work. (207)

heat of fusion. The heat required per unit mass to change a substance from solid to liquid at its melting point. (189)

heat of vaporization. The heat required per unit mass to change a substance from liquid to vapor at its boiling point. (193)

heat pump. A device that absorbs heat from a cool environment and gives it off to a region of higher temperature. (219)

heat sink. A reservoir that absorbs heat without a significant increase in temperature. (207)

henry. The unit of inductance; one henry of inductance is present in a circuit when a change in the current of 1 ampere per second induces an emf of 1 volt. (516)

Hooke's law. Below the elastic limit, strain is directly proportional to stress. (149)

hyperbola. Graph of an inverse proportion. (32)

hypercharge. A property of some baryons and leptons that is conserved in strong and electromagnetic interactions but not in weak interactions. (646)

hypothesis. A plausible solution to a problem. (4)

ice point. The melting point of ice when in equilibrium with water saturated with air at standard atmospheric pressure. (168)

ideal gas. A theoretical gas consisting of infinitely small molecules that exert no forces on each other; also called perfect gas. (180)

illuminance. The luminous flux per unit area of a surface. (308)

impedance. (1) The ratio of applied wave-producing force to resulting displacement velocity of a wave-transmitting medium. (238) (2) The ratio of sound pressure to volume displacement at a given surface in a sound-transmitting medium. (267) (3) The ratio of the effective voltage to the effective current in an a-c circuit. (533)

impulse. The product of a force and the time interval during which it acts. (131)

index of refraction. The ratio of the speed of light in a vacuum to its speed in a given matter medium. (334)

induced magnetism. Magnetism produced in a ferromagnetic substance by the influence of a magnetic field. Magnetization. (473)

inductance. The property of an electric circuit by which a varying current induces a back emf in that circuit or a neighboring circuit. (518)

induction. The process of charging one body by bringing it into the electric field of another charged body. (385)

inductive reactance. Reactance in an a-c circuit containing inductance, which causes a lagging current. (533)

inelastic collision. A collision in which the colliding objects stick together after impact. (133)

inertia. The property of matter that opposes any change in its state of motion. (7)

inertial frame of reference. A nonaccelerating frame of reference in which Newton's first law holds true. (99)

infrared light. Electromagnetic waves longer than those of visible light and shorter than radio waves. (288)

infrasonic range. Vibrations in matter below 20 Hz. (253)

instantaneous current. The magnitude of a varying current at any instant of time. (503)

instantaneous velocity. Short displacement divided by elapsed time. Slope of the line that is tangent to a velocity graph at a given point. (44)

instantaneous voltage. The magnitude of a varying voltage at any instant of time. (502)

insulator. A material through which an electric charge is not readily transferred. (348)

intensity level. The logarithm of the ratio of the intensity of a sound to the intensity of the threshold of hearing. (256)

interaction. Any change in the amount or quantum numbers of particles that are near each other. (643)

interface. A surface that forms the boundary between two phases or systems. (240)

interference. (1) The superposing of one wave on another. (244) (2) The mutual effect of two beams of light, resulting in a reduction of energy in certain areas and an increase of energy in others. (362)

internal combustion engine. A heat engine in which the fuel burns inside the cylinder or turbine chamber. (213)

internal energy. Total potential and kinetic energy of the molecules and atomic particles of a substance. (200)

intersecting storage ring. An accelerator in which particles collide as they move in opposite directions. (633)

inverse photoelectric effect. The emission of photons of radiation from a material when bombarded with high-speed electrons. (300)

inverse proportion. The relation between two quantities whose product is a constant and whose graph is a hyperbola. (32)

ion. An atom or a group of atoms having an electric charge. (398)

ionization chamber. A device used to detect the passage of charged rays or particles by their ionizing effect on a gas. (636)

ionization energy. The energy required to remove an electron from an atom. (628)

irregular reflection. Scattering. Reflection in many different directions from an irregular surface. (316)

isothermal process. A thermal process that takes place at constant temperature. (204)

isotopes. Atoms whose nuclei contain the same number of protons but different numbers of neutrons. (594)

joule. The unit of work; the product of a force of one newton acting through a distance of one meter. (115)

Joule's law. The heat developed in a conductor is directly proportional to the resistance of the conductor, the square of the current, and the time the current is maintained. (450)

junction detector. A solid-state device based on the transistor principle that is used to detect the passage of charged particles. (640)

Kelvin scale. The scale of temperature having a single fixed point, the temperature of the triple point of water, which is assigned the value 273.16 K. (168)

kilogram. A unit of mass in the metric system; one of the seven fundamental units. (22)

kilowatt hour. A unit of electric energy equal to 3.6×10^6 W · s. (455)

kinetic energy. Energy possessed by an object because of its motion. (11)

kinetic theory. The molecules of matter are continuously in motion and the collisions between molecules are perfectly elastic. (144)

Kirchhoff's first law. The algebraic sum of the currents at any circuit junction is equal to zero. (430)

Kirchhoff's second law. The algebraic sum of all changes in potential occurring around any loop in a circuit equals zero. (429)

laser. An acronym for **l**ight **a**mplification by **s**timulated **e**mission of **r**adiation. (298)

law. A statement that describes a natural phenomenon; a principle. (3)

law of conservation of baryons. When a baryon decays or reacts with another particle, the number of baryons is the same on both sides of the equation. (645)

law of conservation of energy. The total quantity of energy in a closed system is constant. (12)

law of conservation of hypercharge. Hypercharge is conserved in strong and electromagnetic interactions, but not in weak interactions. (646)

law of conservation of leptons. In a reaction involving leptons, the arithmetic sum of the lepton numbers is the same on each side of the equation. (645)

law of conservation of mechanical energy. The sum of the potential and kinetic energies of an ideal energy system is constant. (129)

law of conservation of momentum. When no net external forces are acting on an object, the total vector momentum of the object remains constant. (132)

law of entropy. A natural process always takes place in such a direction as to increase the entropy of the universe. (210)

law of heat exchange. In any heat transfer system, the heat lost by hot materials equals the heat gained by cold materials. (184)

law of parity. For every process in nature there is a mirror-image process which is indistinguishable from the original process. (646)

Lenz's law. An induced current is in such a direction that its magnetic property opposes the change by which the current is induced. (497)

lepton. A subatomic particle with a small rest mass, e.g., the electron. (641)

line of flux. A line so drawn that a tangent to it at any point indicates the direction of the magnetic field. (470)

line of force. A line so drawn that a tangent to it at any point indicates the direction of the electric field. (391)

line spectrum. A spectrum consisting of monochromatic slit images having wavelengths characteristic of the atoms present in the source. (296) (Plate IV)

linear accelerator. A device for accelerating particles in a straight line through many stages of small potential difference. (634)

liter. A special name for the cubic decimeter. Symbol: L. (22)

longitudinal wave. A wave in which the vibrations are parallel to the direction of propagation of the wave. (226)

loop. A midpoint of a vibrating segment of a standing wave. (246)

loudness. The sensation that depends principally on the intensity of sound waves reaching the ear. (257)

lumen. The unit of luminous flux; the luminous flux on a unit surface all points of which are at unit distance from a point source of one candela. (307)

luminous. Visible because of the light emitted by its oscillating particles. (303)

luminous flux. The part of the total energy radiated per unit of time from a luminous source that is capable of producing the sensation of sight. (307)

machine. A device that multiplies force at the expense of distance or that multiplies distance at the expense of force. (119)

magnetic field. A region in which a magnetic force can be detected. (470)

magnetic flux. Lines of flux through a region of a magnetic field, considered collectively. (470)

magnetic flux density. The magnetic flux through a unit area normal to the magnetic field; also called magnetic induction. (470)

magnetic force. A force associated with motion of electric charges. (494)

magnetosphere. A region of the upper atmosphere in which the motion of charged particles is governed primarily by the magnetic field of the earth. (475)

magnification. The ratio of the image height to the object height. (350)

malleability. The property of a metal that enables it to be hammered or rolled into sheets. (147)

mass. A measure of the amount of material; a fundamental physical quantity. (7)

mass density. Mass per unit volume of a substance. (9)

mass number. (1) The sum of the number of protons and neutrons in the nucleus of an atom. (594) (2) The integer nearest to the atomic mass. (143)

mass spectrograph. Instrument used to determine the mass of ionized particles. (595)

matter. Anything that has mass and exhibits inertia. (9)

matter wave. A property of matter that is directly proportional to Planck's constant and inversely proportional to mass and velocity. (629)

mechanical equivalent of heat. The conversion factor that relates heat units to work units; 4.19 J/cal. (200)

mechanical wave. A wave that originates in the displacement of a portion of an elastic medium from its normal position, causing it to oscillate about an equilibrium position. (224)

melting point. The temperature at which a solid changes to a liquid. (154)

meniscus. The crescent-shaped surface at the edge of a liquid column. (152)

meson. A subatomic particle with a rest mass intermediate between that of a lepton and a baryon; the carrier of the strong interaction. (641)

meter. A unit of length in the metric system equal to the distance that light travels in a vacuum in 1 299 792 458th of a second. One of the seven fundamental units of measure. (22)

metric system. A system of measurement that is based on decimal multiples and subdivisions. (20)

moderator. A material that slows down neutrons. (608)

mole. Amount of substance containing the Avogadro number of particles such as atoms, molecules, ions, electrons, etc. (179)

molecule. The smallest chemical species of a substance that is capable of stable independent existence. (142)

momentum. The product of the mass and velocity of a moving body. (131)

monochromatic light. Light composed of a single color. (355)

mutual inductance. The ratio of the induced emf in one circuit to the rate of change of current in the coil of another circuit. (515)

neutral weak current. A subatomic reaction in which leptons collide without change in the charges of the colliding particles. (648)

neutron. A neutral subatomic particle having a mass of $1.674\,928\,6 \times 10^{-27}$ kg. (594)

newton. The unit of force; a derived unit having the dimensions kg · m/s^2. The force required to accelerate a one-kilogram mass at a rate of one meter per second each second. (23)

Newton's first law of motion. A body at rest or in uniform motion in a straight line will remain at rest or in the same uniform motion unless acted upon by an external force; also called the law of inertia. (57)

Newton's law of universal gravitation. The force of attraction between any two particles of matter in the universe is directly proportional to the product of their masses and inversely proportional to the square of the distance between their centers of mass. (64)

Newton's second law of motion. The acceleration of a body is directly proportional to the net force exerted on the body, is inversely proportional to the mass of the body, and has the same direction as the net force; also called the law of acceleration. (58)

Newton's third law of motion. If one body exerts a force on a second body, then the second body exerts a force equal in magnitude and opposite in direction on the first body; also called the law of interaction. (61)

node. A point of no disturbance in a standing wave. (246)

noise. Sound produced by irregular vibrations in matter which is unpleasant to the listener. (260)

noninertial frame of reference. An accelerating frame of reference in which Newton's first law of motion does not hold true. (99)

normal. A line drawn perpendicular to a line or surface. (76)

N-type germanium. "Electron-rich" germanium consisting of equal numbers of free electrons and bound positive charges so that the net charge is zero. (567)

nuclear binding force. The force that acts within the small distances between nucleons. (596)

nuclear change. A change in the identity of atomic nuclei. (603)

nuclear mass defect. The arithmetic difference between the mass of a nucleus and the larger sum of its uncombined constituent particles. (596)

nuclear reactor. A device in which the controlled fission of certain substances is used to produce new substances and energy. (614)

nucleon. A proton or neutron in the nucleus of an atom. (594)

nucleus. The positively charged dense central part of an atom. (591)

nuclide. An atom of a particular mass and of a particular element. (595)

octave. The interval between a given musical tone and one with double or half the frequency. (260)

ohm. The unit of electric resistance; one volt per ampere. (413)

Ohm's law. The ratio of the emf applied to a closed circuit to the current in the circuit is a constant. (424)

optical center. The point in a thin lens through which the secondary axes pass. (346)

optical density. A property of a transparent material that is a measure of the speed of light through it. (332)

orbital. The probability pattern of position of an electron about the nucleus of an atom. (589)

order of magnitude. A numerical approximation to the nearest power of ten. (29)

ordinate. The value corresponding to the vertical distance of a point on a graph from the x axis. The y coordinate. (31)

oscilloscope. A cathode-ray tube with associated electronic circuits that enable external voltages to deflect the electron beam of the cathode-ray tube simultaneously along both horizontal and vertical axes. (555)

parallel circuit. An electric circuit in which two or more components are connected across two common points in the circuit so as to provide separate conducting paths for the current. (429)

parallelogram method. The graphic method of finding the resultant of two vectors that do not act along a straight line. (34)

paramagnetism. The property of a substance by which it is feebly attracted by a strong magnet. (446)

pendulum. A body suspended so that it can swing back and forth about an axis. (111)

penumbra. The partially illuminated part of a shadow. (304)

period. (1) The time for one complete cycle, vibration, revolution, or oscillation. (8) (2) The time required for a single wavelength to pass a given point. (230)

periodic motion. Motion repeated in each of a succession of equal time intervals. (109)

permeability. The property of a material by which it changes the flux density in a magnetic field from its value in air. (471)

phase. (1) A condition of matter. (143) (2) In any periodic phenomenon, a number that describes a specific stage within each oscillation. (228) (3) The angular relationship between current and voltage in an a-c circuit. (525) (4) The number of separate voltage waves in a commercial a-c supply. (504)

phase angle. The angle between the voltage and current vectors. (530)

phasor. A representation of the concepts of magnitude and direction in a reference plane; a rotating vector. (530)

photoelastic. Pertaining to certain materials that become double refracting when strained. (376)

photoelectric effect. The emission of electrons by a substance when illuminated by electromagnetic radiation of sufficiently short wavelength. (289)

photoelectrons. Electrons emitted from a light-sensitive material when it is illuminated with light of sufficiently short wavelength. (289)

photometer. An instrument used for comparing the intensity of a light source with that of a standard source. (311)

photometry. The quantitative measurement of visible radiation from light sources. (306)

photon. A quantum of light energy; the carrier of the electromagnetic interaction. (293)

photovoltaic effect. The generation of a potential difference across a **P-N** junction as a consequence of the absorption of incident light of appropriate frequency. (576)

physical change. A change in which the composition and identifying properties of a substance remain unchanged. (603)

physical quantity. A measurable aspect of the universe, such as length. (20)

physics. The science that deals with the relationships between matter and energy. (1)

piezoelectric effect. The property of certain natural and synthetic crystals to develop a potential difference between opposite surfaces when subjected to a mechanical stress, and conversely. (418)

pitch. The identification of a certain sound with a definite tone; depends on the frequency which the ear receives. (260)

pivot point. The point from which the lengths of all torque arms are measured. (86)

Planck's constant. A fundamental constant in nature that determines what values are allowed for physical quantities in quantum mechanics; $h = 6.63 \times$

10^{-34} J · s. (293)

plasma. A gas that is capable of conducting an electric current. (161)

plate. The anode of an electronic tube. (555)

P-N junction. The boundary between **P-** and **N-**type materials in a semiconductor crystal. (568)

polarized light. Light radiations in which the vibrations of all light waves present are confined to planes parallel to each other. (373)

polarization angle. A particular angle of incidence at which polarization of reflected light is complete. (375)

polychromatic light. Light composed of several colors. (355)

positive rays. Rays coming through holes in a cathode on the side opposite the anode in a discharge tube. Positively charged ions. (592)

potential difference. The work done per unit charge as a charge is moved between two points in an electric field. (394)

potential energy. Energy that is the result of the position of an object. (11)

potential gradient. The change in potential per unit distance. (395)

power. The time rate of doing work. (121)

power factor. The cosine of the phase angle between current and voltage in an a-c circuit. (532)

precession. The motion that results from the application of a torque that tends to displace the axis of rotation of a rotating object. (106)

precision. The agreement between the numerical values of two or more measurements made in the same way and expressed in terms of deviation; the reproducibility of measured data. (26)

pressure. Force per unit area. (178)

primary. A transformer winding that carries current and normally induces a current in one or more secondary windings. (520)

primary cell. An electrochemical cell which must be replaced after a given amount of energy has been supplied to the external circuit. (418)

primary colors. Colors in terms of which all other colors may be described or from which all other colors may be evolved by mixtures. (358)

primary pigments. The complements of the primary colors. (359)

principal axis. (1) A line drawn through the center of curvature and the vertex of a curved mirror. (320) (2) A line drawn through the center of curvature and the optical center of a lens. (343)

principal focus. A point at which rays parallel to the principal axis converge or from which they diverge after reflection or refraction. (320)(343)

principle. See *law.*

property. A measurable aspect of matter, e.g., mass and inertia. (6)

proton. A positively charged subatomic particle having a mass of $1.672\ 623\ 1 \times 10^{-27}$ kg and a charge equal and opposite to that of the electron. (592)

P-type germanium. "Hole-rich" germanium consisting of equal numbers of free positive holes and bound negative charges so that the net charge is zero. (568)

pulse. A single nonrepeated disturbance. (225)

quality. The property of sound waves that depends on the number of harmonics and their prominence. (269)

quantum. An elemental unit of energy; a photon of energy *hf*. (293)

quantum mechanics. The branch of physics that deals with the behavior of particles whose specific properties are given by quantum numbers. (624)

quantum number. One of a set of notations used to characterize a discrete value that a quantized variable is allowed to assume. (624)

quantum theory. A unifying theory

based on the concept of the subdivision of radiant energy into discrete quanta (photons) and applied to the studies of structure at the atomic and molecular levels. (294)(623)

quark. One of a group of subatomic particles of which mesons and baryons are composed. (647)

radian. A unit of angular measurement. The angle that, when placed with its vertex at the center of a circle, subtends on the circumference an arc equal in length to the radius of the circle. Approximately 57.3°. (102)

radio waves. Also called Hertzian waves. Electromagnetic radiations produced by rapid reverses of current in a conductor. (287)

radioactivity. The spontaneous breakdown of an atomic nucleus with the emission of particles and rays. (600)

radioisotope. An isotope of an element that is radioactive. (618)

rarefaction. The region of a longitudinal wave in which the vibrating particles are farther apart than their equilibrium distance. (226)

ray. A single line of light from a luminous point. A line showing the direction of propagation of light. (304)

reactance. The nonresistive opposition to current in an a-c circuit. (533)

real image. An image formed by actual rays of light. (318)(343)

rectifier. A device for a changing alternating current to direct current. (555)

rectilinear propagation. Traveling in a straight line. (234)

reflectance. The ratio of the light reflected from a surface to the light falling on the surface. (316)

reflection. The return of a wave from the boundary of a medium. (235)

refraction. The bending of a wave disturbance as it passes obliquely from one medium into another in which the disturbance has a different velocity. (240)

regelation. The melting of a substance under pressure and the refreezing after the pressure is released. (157)

regular reflection. Reflection from a polished surface in which scattering effects are negligible. (316)

relative deviation. Percentage average deviation of a set of measurements. (26)

relative error. Percentage absolute error of a set of measurements. (25)

relative humidity. The ratio of the water vapor pressure in the atmosphere to the equilibrium vapor pressure at a given temperature. (160)

relativistic mass. The mass of an object in motion with respect to the observer. (14)

residual magnetism. Magnetism retained in a magnet after the magnetizing field has been removed. (473)

resistance. The ratio of the potential difference across a conductor to the magnitude of current in it. (413)

resistivity. A proportionality constant that relates the length and cross-sectional area of a given electric conductor to its resistance, at a given temperature. (437)

resolution of forces. The resolving of a single force into component forces acting in given directions on the same point. (76)

resonance. (1) The inducing of vibrations of a natural rate by a vibrating source having the same frequency. (272) (2) The condition in an a-c circuit in which the inductive reactance and capacitive reactance are equal. (547)

rest mass. The mass of an object not in motion. (14)

resultant. A vector representing the sum of several vector components. (33)

resultant force. The single force that has the same effect as two or more forces applied simultaneously at the same point. (72)

reverse bias. Voltage applied to a semiconductor **P-N** junction that reduces the electron current across the junction. (571)

rheostat. A variable resistance. (439)

root-mean-square (rms) current. The effective value of an alternating current; the square root of the mean of the instantaneous values squared. (527)

rotary motion. Motion of a body about an internal axis. (101)

rotational equilibrium. The state of a body in which the sum of all the clockwise torques in a given plane equals the sum of all the counterclockwise torques about a pivot point. (87)

rotational inertia. The property of a rotating object that resists changes in its angular velocity. (104)

scalar quantity. A quantity that is completely specified by a magnitude. (33)

scientific notation. A positive number expressed in the form of $M \times 10^n$ in which M is a number between 1 and 10 and n is an integral power of 10. (29)

scintillation counter. A device that counts the impacts of charged subatomic particles on a fluorescent screen by means of a photomultiplier tube. (636)

second. A unit of time; equivalent to 9 192 631 770 vibrations of cesium-133. One of the seven fundamental units of measure. (23)

second law of photoelectric emission. The kinetic energy of photoelectrons is independent of the intensity of the incident light. (291)

second law of thermodynamics. It is not possible for an engine to transfer heat from one body to another at a higher temperature unless work is done on the engine. (208)

secondary. A transformer output winding in which the current is due to inductive coupling with another winding called the primary. (520)

secondary axis. Any line other than the principal axis drawn through the center of curvature of a mirror or the optical center of a lens. (320)(345)

secondary emission. Emission of electrons as a result of the bombardment of an electrode by high-velocity electrons. (563)

selectivity. The property of a tuned circuit that discriminates between signal voltages of different frequencies. (548)

self-inductance. The ratio of the induced emf across a coil to the rate of change of current in the coil. (518)

series circuit. An electric circuit in which the components are arranged to provide a single conducting path for current. (427)

series resonance. A condition in which the impedance of a series circuit containing resistance, inductance, and capacitance is equal to the resistance of the circuit and the voltage across the circuit is in phase with the current. (546)

shear strain. The ratio of the amount of deformation of the side of a body to the length of the side. (148)

significant figures. Those digits in an observed quantity (measurement) that are known with certainty plus the first digit that is uncertain. (26)

simple harmonic motion. Motion in which the acceleration is proportional to the displacement from an equilibrium position and is directed toward that position. (110)

solar spectrum. The band of colors produced when sunlight is dispersed by a prism. (335)

solenoid. A long helically wound coil of insulated wire. (481)

solid-state detector. A device used to detect the passage of charged subatomic particles by their crystal-distorting or ionizing effects on a nonconducting or semiconducting solid. (640)

sonometer. A device, consisting of two or more wires or strings stretched over a sounding board, used for testing the frequency of strings and for showing how they vibrate. (267)

sound. The series of disturbances in matter to which the human ear is sensitive. Also similar disturbances in matter above and below the normal range of human hearing. (253)

sound intensity. The rate at which sound energy flows through a unit area. (256)

space charge. The negative charge in the space between the cathode and plate of a vacuum tube. (558)

specific gravity. The ratio of the mass density of a substance to that of water. (9)

specific heat. The heat capacity of a material per unit mass. (183)

speed. Time rate of motion. (41)

spherical aberration. The failure of parallel rays to meet at a single point on a spherical surface after reflection or refraction. (322) (346)

spinthariscope. A device used to detect subatomic particles by the light flashes they produce on a zinc sulfide screen. (636)

standard pressure. The pressure exerted by 1 atm or 1.01×10^5 Pa. (178)

standard temperature. 0°C; 273 K. (176)

standing wave. The resultant of two wave trains of the same wavelength, frequency, and amplitude traveling in opposite directions through the same medium. (247)

static electricity. Electricity at rest. (380)

steam point. The boiling point of water at standard atmospheric pressure. (168)

steradian. The ratio of the intercepted surface area of a sphere to the square of the radius. A unit of solid angle. (307)

storage cell. An electrochemical cell in which the reacting materials are regenerated by the use of a reverse current from an external source. (418)

strain. The relative amount of distortion produced in a body under stress. (148)

stress. The distorting force per unit area. (148)

strong interaction. The interaction that holds the particles of the nucleus together and is independent of charge. (643)

sublimation. The change of a solid to a gaseous phase without passing through the liquid phase. (159)

superconductivity. The condition of zero resistivity below the transition temperature of a substance. (438)

supercooling. The process of cooling a substance below its normal phase-change point without a change of phase. (191)

superposition. Combining the displacements of two or more waves vectorially to produce a resultant displacement. (243)

surface tension. The tendency of a liquid surface to contract; the measure of this tendency in newtons per meter. (153)

sympathetic vibration. See *resonance (1)*.

synchrotron. A particle accelerator in which the oscillating frequency varies. (632)

technology. The application of science to human needs and goals. (2)

temperature. The physical property that determines the direction in which heat will flow between substances. (167)

temporary magnet. A magnet produced by induction. (473)

tensile strength. The force required to break a rod or wire of unit cross-sectional area. (146)

theory. A plausible explanation of an observed event, supported experimentally and confirmed by experiments designed to test predictions based upon the explanation. (3)

thermal energy. The total potential and kinetic energy associated with the random motions of the particles of a material. (166)

thermionic emission. The liberation of electrons from the surface of a hot body. (557)

thermocouple. An electric circuit composed of two dissimilar metals whose junctions are maintained at different temperatures. (416)

thermodynamics. Study of quantitative relationships between heat and other forms of energy. (200)

thermoelectric effect. The production of an electron current in a closed circuit consisting of two dissimilar metals as a result of the emf developed when the two junctions are maintained at different temperatures. (416)

thermonuclear reaction. Nuclear fusion. (608)

third law of photoelectric emission. Within the region of effective frequencies, the maximum kinetic energy of photoelectrons varies directly with the difference between the frequency of the incident light and the cutoff frequency. (292)

thought experiment. An idealized experiment that cannot be performed under actual conditions. (57)

threshold of hearing. The intensity of the faintest sound audible to the average human ear, 10^{-16} W/cm^2 at 10^3 Hz. (257)

threshold of pain. For audible frequencies of sound, an intensity level above which pain results in the average human ear. (259)

tolerance. The property of a measuring device that determines its limits. (28)

torque. Product of a force and the effective length of its torque arm. (86)

torque arm. The perpendicular distance between the line of action of the torque producing force and the axis of rotation. (86)

total reflection. The reflection of light at the boundary of two transparent media when the angle of incidence exceeds the critical angle. (339)

transformer. A device for changing an alternating voltage from one potential to another. (520)

transistor. A semiconductor device used as a substitute for vacuum tubes in electronic applications. (438)(566)

transition temperature. A specific temperature at which the resistivity of some materials drops suddenly to zero. (438)

translational equilibrium. The state of a body in which there are no unbalanced forces acting on it. (75)

transuranium elements. Elements with atomic number greater than 92. (608)

transverse wave. A wave in which the vibrations are at right angles to the direction of propagation of the wave. (226)

triode. Vacuum tube consisting of a plate, grid, and cathode. (555)

triple point. The single condition of temperature and pressure at which the solid, liquid, and vapor phases of a substance can coexist in stable equilibrium. (168)

trough. A region of downward displacement in a transverse wave. (225)

ultrasonic range. Vibrations in matter above 20 000 Hz. (253)

ultraviolet light. Electromagnetic radiations of shorter wavelength than visible light but longer than X rays. (288)

umbra. The part of a shadow from which all light rays are excluded. (304)

uncertainty principle. It is impossible to specify simultaneously both the position of an object and its momentum. (624)

unified field theory. The principle that all forces in the universe are part of a single concept. (648)

unit analysis. The performance of indicated mathematical operations in a problem with the measurement units alone. (36)

unit magnetic pole. One that repels an exactly similar pole placed one centimeter away with a force of 10^{-5} N. (469)

universal gas constant. Constant of proportionality, R, in the ideal gas equation. (179)

universal gravitational constant. Constant of proportionality in Newton's law of universal gravitation; $G = 6.67 \times 10^{-11}\ N \cdot m^2/kg^2$. (64)

Van de Graaff generator. A particle accelerator that transfers a charge from an electron source to an insulated sphere by means of a moving belt composed of an insulating material. (630)

vapor. The gaseous phase of substance that exists as a liquid or solid under normal conditions. (144)

vaporization. The change in phase from a solid or a liquid to a gas. (159)

variation. See *declination*.

vector quantity. A quantity that is completely specified by a magnitude and a direction. (33)

velocity. Speed in a particular direction. (43)

vertex. The center of a curved mirror. (320)

virtual image. An image that only appears to be formed by rays of light. (318)(344)

viscosity. The ratio of shear stress to the rate of change of shear strain in a liquid or gas. (152)

volt. The unit of potential difference. The potential difference between two points in an electric field such that one joule of work moves a charge of one coulomb between these points. (394)

voltage sensitivity. Voltage per unit scale division of an electric instrument. (483)

volume strain. The ratio of the decrease in volume to the volume before stress is applied. (148)

$W^{\pm}$ particle. The carrier for weak subatomic interactions. (644)

watt. The unit of power; one joule per second. (121)

wavelength. In a periodic wave, the distance between consecutive points of corresponding phase. (230)

weak interaction. The interaction that produces pairs of particles that have unusually long half-lives. (644)

weber. The unit of magnetic flux. (470)

weight. The measure of the gravitational force acting on a substance. (66)

Wheatstone bridge. Instrument used to measure electric resistance. (440)

work. The product of a displacement and the force in the direction of the displacement. (115)

work function. The minimum energy required to remove an electron from the surface of a material and send it into field-free space. (289)

working equation. The equation derived from a basic equation as an expression of the unknown quantity in a problem directly in terms of the known quantities stated in the problem. The working equation is the mathematical expression of the solution to the problem. (35)

X rays. Invisible electromagnetic radiations of great penetrating power. (288)

Young's modulus. The ratio of stress to strain in a solid. (149)

Zeeman effect. The splitting of atomic energy levels into two or more sublevels by means of a magnetic field as seen in spectral lines under these conditions. (626)

INDEX

PHOTO CREDITS

Abbreviations used: (t) top; (c) center; (b) bottom; (l) left; (r) right.

Chapter 1: Page 2, Charles M. Falco/Photo Researchers, Inc.; 3(t), 3(b), General Electric Company; 4, Argonne National Laboratory; 7(b), Alex Mulligan; 8(l), 8(r), HRW Photos by Russell Dian; 12, John G. Zimmerman; 13, Deutsches Museum, Munich; 14, Courtesy of the Archives, California Institute of Technology.

Photo Essay LASERS: Page 18(l), Newport Corporation; 18(t), Anthony Howarth/Woodfin Camp & Associates; 18(r), Courtesy, General Motors Laboratories; 19(l), 19(c), Chuck O'Rear/Woodfin Camp & Associates; 19(r), Courtesy, 3M.

Chapter 2: Page 20(c), Tom McHugh/Photo Researchers, Inc.; 22, National Bureau of Standards; 23, Bureau International des Poids et Mesures; 24, National Museum of American History/Smithsonian Institution; 25, Alex Mulligan; 27, Cenco; 29, Palomar Observatory Photograph.

Chapter 3: Page 41, Project Physics, HRW; 54, Hatton/Woodfin Camp & Associates; 56, The Bettmann Archive/Painting-Kneller; 58(t), 58(b), Loomis Dean, Life Magazine, Time Inc.; 66, Jet Propulsion Laboratory, Pasadena, CA.

Chapter 4: Page 71, FMC Corporation Construction Equipment; 82, Austrian Press and Information Service.

Chapter 5: Page 94, PSSC Physics, D.C. Heath & Company; 99, Department of the Navy.

Chapter 6: Page 128, CSC Scientific; 131, Dr. Harold Edgerton, MIT; 132, NASA; 133, Ealing Corporation; 137(l), 137(r), Focus On Sports; 139, PSSC Physics, D.C. Heath & Company.

Chapter 7: Page 142, Brown Brothers; 143, Jutta Mueller Schwab; 146, USX Corporation; 152, 153, Ken Chen; 155, Courtesy, Citicorp; 156, John King; 157, Focus On Sports; 161, NASA; 164, Vandystadt/ALLSPORTS, USA.

Photo Essay SOUND: Page 250(l), New England Digital Corporation; 250(b), Todd Kaplin/Starfile; 250(r), HRW Photo by Lance Shriner, Courtesy, Acoustics Systems, Austin, TX; 251(l) Joseph Lynch/Medical Images, Inc.; 251(b), Ebet Roberts, 251(r), Kurzweil Music Systems.

Chapter 8: Page 168, National Bureau of Standards; 173, Alex Mulligan.

Chapter 9: Page 201, The Bettmann Archive; 208, AIP, Niels Bohr Library; 211, G. Hunter/National Film Board of Canada; 213, Westinghouse; 215, Detroit Diesel Allison; 218, Courtesy, NASA/HRW Photo By Russell Dian.

Chapter 10: Page 223(l), Alexander Lowry/Photo Researchers, Inc.; 223(r) NASA; 231(t), 231(b), UPI/Wide World Photos; 233, Fundamental Photographs, 234(l), PSSC Physics, D.C. Heath & Company; 234(r) Education Development Center; 235, PSSC Physics, D.C. Heath & Company; 236, Education Development Center; 237, 240(t), 240(b), PSSC Physics, D.C. Heath & Company; 242(t), 242(c), 242(b), Fundamental Photographs; 246, Education Development Center.

Chapter 11: Page 253, College Physics, HRW; 255, Bell Telephone Laboratories; 264, Education Development Center; 267, Ken Chen; 268, 269, C.G. Conn Ltd.; 270, Ken Chen.

Chapter 12: Page 286, Master and Fellows of Trinity College, Cambridge; 287, 288, The Bettmann Archive; 293, AIP, Niels Bohr Library; 295, Danish Information Office; 298, Howard Sochurek/Woodfin Camp & Associates; 306, National Bureau of Standards; 311, Stansi Scientific Company.

Chapter 13: 323 General Electric Company; 326(t), Wide World Photos; 326(b), Courtesy, Roland Cormier.

Chapter 14: Page 333, General Electric Company; 334, Bausch & Lomb; 339, Alex Mulligan; 343, Bausch & Lomb; 344, HSS, Inc. Bedford, MA; 353, Bausch & Lomb.

Color Plates: V(A), Adrian Davies/Bruce Coleman, Inc.; VI(A), FPG International; VI(B), Bausch & Lomb

Intra-Science: Page 1(tl), S.L. Craig Jr./Bruce Coleman, Inc.; 1(tc), St John and Logan Journal of Crystal Growth/Science Photo Library; 1(tr), Four By Five; 1(cl), Steven F. Grohe/The Picture Cube; 1(cr), M. Keller/FPG International; 1(bl), Howard Sochurek/Medichrome; 1(br), Chris Morris/Black Star; 2(l), M. Keller/FPG International; 2(r)Stacy Pick/Stock Boston, Inc.; 3(c), Charles Lightdale/Photo Researchers, Inc.; 3(b), J.T. Miller/The Stock Market; 4(t), S.L. Craig Jr./Bruce Coleman, Inc.; 4(b), Four By Five; 5(t), Walter Bibikow/The Image Bank; 5(b), S.L. Craig Jr./Bruce Coleman, Inc.; 6(t),Martin Dohrn/Science Photo Library; 6(b), Four By Five; 7(t), Merlin Tuttle/Photo Researchers, Inc.; 7(b), T. Qing/FPG International; 8(l), St John and Logan Journal of Crystal Growth/Science Photo Library; 8(r), Jan Hinsch/Science Photo Library; 9(t), NASA; 9(b), Four By Five; 10(t), Steven F. Grohe/The Picture Cube; 10(b), CNRI/Science Photo Library; 11(l), Stephen Dalton/Animals Animals; 11(r), David Frazier; 12(t), Bettmann Newsphotos; 12(b), Chris Morris/Black Star; 13(l), G. Davis/Sygma Newsphotos; 13(r), Dan McCoy/Rainbow; 14, Donald Dietz/Stock Boston, Inc.; 15(t) Shostal Associates; 15(bl), Howard Sochurek/Medichrome; 15(br), Peter Menzel/Stock Boston; 16, NASA.

Chapter 15: Page 363, Foundations of Physics, HRW; 366(l), 366(r), American Optical Corporation; 367, Bausch & Lomb; 376, Photo By George Garard with permission of Henry Semat, as published in Fundamentals of Physics 4th ed., HRW; 377(l), Bruce Roberts/Photo Researchers, Inc.; 377(r), NASA; 378, Carl Zeiss, West Germany.

Chapter 16: Page 394, General Electric Company, 399, Adam Woolfit/Woodfin Camp & Associates; 400, Cardwell Condenser Corporation.

Chapter 17: Page 418, NASA; 424, Burndy Library, Norwalk, CT; 439, Japanese National Railways; 441, Cenco.

Photo Essay SUPERCONDUCTORS: Page 446(l), Argonne National Laboratory; 446(r), Henry Groskinsky/Peter Arnold, Inc.; 447(l), IBM; 447(b), Argonne National Laboratory, 447(r), Marlin Minks/Biomagnetic Technologies, Inc. San Diego, CA.

Chapter 18: Page 449, General Electric Company; 450, William Rivelli/The Image Bank; 452, David Frazier; 460, Courtesy, Director of the Royal Institution of Great Britain.

Chapter 19: Page 466, Walter Dawn; 469(l), 469(r), Bell Telephone Laboratories; 481, Brookhaven National Laboratory; 482, Sargent-Welch Scientific Company; 483, Leeds & Northrup.

Chapter 20: Page 505, HRW Photo by Lance Schriner, Courtesy of City of Austin Utilities, TX; 513, General Electric Company; 516, The New York Public Library; 521, HRW Photo by Lance Shriner, Courtesy of City of Austin Utilities, TX.

Chapter 21: Page 526, The Granger Collection; 549, Ken Chen.

Photo Essay MRI: Page 552(l), Jon Riley/Medichrome; 552(r), Photo Researchers, Inc.; 553(l), S.I.U./Peter Arnold, Inc.; 553(tr), Cleveland Clinic Foundation; 553(br), Memorial Sloan Kettering.

Chapter 22: Page 555, Amana; 556, The New York Public Library; 557, Westinghouse Electric Corporation; 556, Ken Chen.

Chapter 23: Page 582(t), 582(b), Fritz Goro/Life Magazine, Time Inc.; 585, Brookhaven National Laboratory; 586, Margot Bergmann/Polytechnic Institute of Brooklyn; 594, General Electric Company.

Chapter 24: Page 601(t), The Bettmann Archive; 601(b), French Embassy Press & Information Division; 602, National Laboratory, Oak Ridge, TN; 610, Cornell University; 617(t), Cletus Mitchell/Tennessee Valley Authority; 617(b), Princeton University Plasma Physics Laboratory; 621, Zionist Archive and Library.

Chapter 25: Page 624, AIP, Niels Bohr Library; 626, University of California, Lawrence Berkeley Laboratory; 627, The Granger Collection; 629, AIP, Niels Bohr Library, Bainbridge Collection; 631, Brookhaven National Laboratory; 632, University of Illinois; 633(l), 633(c), 633(tr), 633(br), Fermi National Accelerator Laboratory; 635, Roland Quintero; 636, Sargent-Welch Scientific Company; 637, Education Development Center; 638, Brookhaven National Laboratory; 639(t), University of California, Lawrence Berkeley Laboratory; 639(b), Brookhaven National Laboratory; 640, US Atomic Energy Commission; 645, Barrett Gallagher; 647(t), AIP, Niels Bohr Library Courtesy, Alan Richards; 647(b), Michael R. Dressler; 649, Courtesy, Harvard University News Office.